FIFTH EDITION

Using Multivariate Statistics

Barbara G. Tabachnick

California State University, Northridge

Linda S. Fidell

California State University, Northridge

PEARSON

Boston ▪ New York ▪ San Francisco
Mexico City ▪ Montreal ▪ Toronto ▪ London ▪ Madrid ▪ Munich ▪ Paris
Hong Kong ▪ Singapore ▪ Tokyo ▪ Cape Town ▪ Sydney

To Friendship

Executive Editor: *Susan Hartman*
Editorial Assistant: *Therese Felser*
Marketing Manager: *Karen Natale*
Senior Production Administrator: *Donna Simons*
Editorial Production Service: *Omegatype Typography, Inc.*
Composition Buyer: *Andrew Turso*
Manufacturing Buyer: *Andrew Turso*
Electronic Composition: *Omegatype Typography, Inc.*
Cover Administrator: *Joel Gendron*

For related titles and support materials, visit our online catalog at www.ablongman.com.

Between the time web site information is gathered and published, it is not unusual for some sites to have
closed. Also, the transcription of URLs can result in typographical errors. The publisher would appreciate
notification where these occur so that they may be corrected in subsequent editions.

ISBN 0-205-45938-2

Printed in the United States of America

10   9   8   RRD   10

C O N T E N T S

3   Review of Univariate and Bivariate Statistics   33

10 Logistic Regression 437

PREFACE

Obesity threatened, and we've had to consider putting the book on a diet. We've added only one chapter this time around, Multilevel Linear Modeling (Chapter 15), and some spiffy new techniques for dealing with missing data (in Chapter 4). Otherwise, we've mostly streamlined and said good-bye to some old friends. We've forsaken the Time-Series Analysis chapter in the text, but you'll be able to download it from the publisher's web site at www.ablongman.com/tabachnick5e. Another sadly forsaken old friend is SYSTAT. We still love the program, however, for its right-to-the-point analyses and terrific graphics, and are pleased that most of the graphics have been incorporated into SPSS. Although absent from demonstrations, features of SYSTAT, and any other programs we've cut, still appear in the last sections of Chapters 5 through 16, and in online Chapter 18, where programs are compared. We've changed the order of some chapters: canonical correlation seemed rather difficult to appear as early as it did, and survival analysis seemed to want to snuggle up to logistic regression. Actually, the order doesn't seem to matter much; perusal of syllabi on the Web convinces us that professors feel free to present chapters in any order they choose—and that's fine with us.

Multilevel linear modeling (MLM) seems to have taken the world by storm; how did we ever live without it? Real life is hierarchical—students come to us within classrooms, teachers work within different schools, patients share wards and nursing staff, and audiences attend different performances. We hardly ever get to break these groups apart for research purposes, so we have to deal with intact groups and all their shared experiences. MLM lets us do this without violating all of the statistical assumptions we learned to know and hate. Now that SAS and SPSS can deal with these models, we're ready to tackle the real world. Hence, a new chapter.

SAS and SPSS also now offer reasonable ways to impute missing data through multiple-imputation techniques and fully assess missing data patterns, respectively. We expanded Chapter 4 to demonstrate these enhancements. SPSS and SAS keep adding goodies, which we'll try to show off. As before, we adapt our syntax from Windows menus whenever possible, and all of our data sets are available on the book's web page (www.ablongman.com/tabachnick5e). We've also paid more attention to effect sizes and, especially, confidence intervals around effect sizes. Michael Smithson of the Australian National University has kindly given us permission to include some nifty SPSS and SAS syntax and data files in our web page downloads. Jim Steiger and Rachel Fouladi have graciously given us permission to include their DOS program that finds confidence intervals around R^2.

One thing we'll never change is our practical bent, focusing on the benefits and limitations of applications of a technique to a data set—when, why, and how to do it. The math is wonderful, and we suggest (but don't insist) that students follow along through section four of each chapter using readily available software for matrix manipulations or spreadsheets. But we still feel that understanding the math is not enough to insure appropriate analysis of data. And our readers assure us that they really are able to apply the techniques without a great deal of attention to the math of section four. Our small-sample examples remain silly; alas, our belly dancing days are over. As for our most recent reviewers, kindly provided by our publisher, we had the three bears checking out beds: too hard, too soft, and just right. So we've not changed the tone or level of difficulty.

Some extremely helpful advice was offered by Steve Osterlind of the University of Missouri–Columbia and Jeremy Jewel of Southern Illinois University–Edwardsville. We also heartily thank Lisa Harlow of the University of Rhode Island, who wrote an extensive, insightful review of the entire fourth edition of our book in *Structural Equation Modeling* in 2002. We again thank the reviewers of earlier editions of our book, but fears of breaking the backs of current students dissuade us from listing them all once more. You know who you are; we still care. Our thanks to the reviewers of this edition: Joseph Benz, University of Nebraska–Kearney; Stanley Cohen, West Virginia University; Michael Granaas, University of South Dakota; Marie Hammond, Tennessee State University at Nashville; Josephine Korchmaros, Southern Illinois University; and Scott Roesch, San Diego State University.

As always, the improvements are largely due to reviewers and those colleagues who have taken the time to email us with suggestions and corrections. Any remaining errors and lack of clarity are due to us alone. As always, we hope the book provides a few smiles as well as help in analyzing data.

Barbara G. Tabachnick
Linda S. Fidell

CHAPTER

1

Introduction

1.1 Multivariate Statistics: Why?

Multivariate statistics are increasingly popular techniques used for analyzing complicated data sets. They provide analysis when there are many independent variables (IVs) and/or many dependent variables (DVs), all correlated with one another to varying degrees. Because of the difficulty of addressing complicated research questions with univariate analyses and because of the availability of canned software for performing multivariate analyses, multivariate statistics have become widely used. Indeed, a standard univariate statistics course only begins to prepare a student to read research literature or a researcher to produce it.

But how much harder are the multivariate techniques? Compared with the multivariate methods, univariate statistical methods are so straightforward and neatly structured that it is hard to believe they once took so much effort to master. Yet many researchers apply and correctly interpret results of intricate analysis of variance before the grand structure is apparent to them. The same can be true of multivariate statistical methods. Although we are delighted if you gain insights into the full multivariate general linear model,[1] we have accomplished our goal if you feel comfortable selecting and setting up multivariate analyses and interpreting the computer output.

Multivariate methods are more complex than univariate by at least an order of magnitude. But for the most part, the greater complexity requires few conceptual leaps. Familiar concepts such as sampling distributions and homogeneity of variance simply become more elaborate.

Multivariate models have not gained popularity by accident—or even by sinister design. Their growing popularity parallels the greater complexity of contemporary research. In psychology, for example, we are less and less enamored of the simple, clean, laboratory study in which pliant, first-year college students each provides us with a single behavioral measure on cue.

1.1.1 The Domain of Multivariate Statistics: Numbers of IVs and DVs

Multivariate statistical methods are an extension of univariate and bivariate statistics. Multivariate statistics are the *complete* or general case, whereas univariate and bivariate statistics are special cases

---

[1]Chapter 17 attempts to foster such insights.

of the multivariate model. If your design has many variables, multivariate techniques often let you perform a single analysis instead of a series of univariate or bivariate analyses.

Variables are roughly dichotomized into two major types—independent and dependent. Independent variables (IVs) are the differing conditions (treatment vs. placebo) to which you expose your subjects, or characteristics (tall or short) that the subjects themselves bring into the research situation. IVs are usually considered predictor variables because they predict the DVs—the response or outcome variables. Note that IV and DV are defined within a research context; a DV in one research setting may be an IV in another.

Additional terms for IVs and DVs are predictor-criterion, stimulus-response, task-performance, or simply input-output. We use IV and DV throughout this book to identify variables that belong on one side of an equation or the other, without causal implication. That is, the terms are used for convenience rather than to indicate that one of the variables caused or determined the size of the other.

The term *univariate statistics* refers to analyses in which there is a single DV. There may be, however, more than one IV. For example, the amount of social behavior of graduate students (the DV) is studied as a function of course load (one IV) and type of training in social skills to which students are exposed (another IV). Analysis of variance is a commonly used univariate statistic.

Bivariate statistics frequently refers to analysis of two variables where neither is an experimental IV and the desire is simply to study the relationship between the variables (e.g., the relationship between income and amount of education). Bivariate statistics, of course, can be applied in an experimental setting, but usually they are not. Prototypical examples of bivariate statistics are the Pearson product-moment correlation coefficient and chi square analysis. (Chapter 3 reviews univariate and bivariate statistics.)

With multivariate statistics, you simultaneously analyze multiple dependent and multiple independent variables. This capability is important in both nonexperimental (correlational or survey) and experimental research.

1.1.2 Experimental and Nonexperimental Research

A critical distinction between experimental and nonexperimental research is whether the researcher manipulates the levels of the IVs. In an experiment, the researcher has control over the levels (or conditions) of at least one IV to which a subject is exposed by determining what the levels are, how they are implemented, and how and when cases are assigned and exposed to them. Further, the experimenter randomly assigns subjects to levels of the IV and controls all other influential factors by holding them constant, counterbalancing, or randomizing their influence. Scores on the DV are expected to be the same, within random variation, except for the influence of the IV (Campbell & Stanley, 1966). If there are systematic differences in the DV associated with levels of the IV, these differences are attributed to the IV.

For example, if groups of undergraduates are randomly assigned to the same material but different types of teaching techniques, and afterward some groups of undergraduates perform better than others, the difference in performance is said, with some degree of confidence, to be caused by the difference in teaching technique. In this type of research, the terms *independent* and *dependent* have obvious meaning: the value of the DV depends on the manipulated level of the IV. The IV is manipulated by the experimenter and the score on the DV depends on the level of the IV.

In nonexperimental (correlational or survey) research, the levels of the IV(s) are not manipulated by the researcher. The researcher can define the IV, but has no control over the assignment of subjects to levels of it. For example, groups of people may be categorized into geographic area of residence (Northeast, Midwest, etc.), but only the definition of the variable is under researcher control. Except for the military or prison, place of residence is rarely subject to manipulation by a researcher. Nevertheless, a naturally occurring difference like this is often considered an IV and is used to predict some other nonexperimental (dependent) variable such as income. In this type of research, the distinction between IVs and DVs is usually arbitrary and many researchers prefer to call IVs *predictors* and DVs *criterion variables.*

In nonexperimental research, it is very difficult to attribute causality to an IV. If there is a systematic difference in a DV associated with levels of an IV, the two variables are said (with some degree of confidence) to be related, but the cause of the relationship is unclear. For example, income as a DV might be related to geographic area, but no causal association is implied.

Nonexperimental research takes many forms, but a common example is the survey. Typically, many people are surveyed, and each respondent provides answers to many questions, producing a large number of variables. These variables are usually interrelated in highly complex ways, but univariate and bivariate statistics are not sensitive to this complexity. Bivariate correlations between all pairs of variables, for example, could not reveal that the 20 to 25 variables measured really represent only two or three "supervariables."

Or, if a research goal is to distinguish among subgroups in a sample (e.g., between Catholics and Protestants) on the basis of a variety of attitudinal variables, we could use several univariate *t* tests (or analyses of variance) to examine group differences on each variable separately. But if the variables are related, which is highly likely, the results of many *t* tests are misleading and statistically suspect.

With the use of multivariate statistical techniques, complex interrelationships among variables are revealed and assessed in statistical inference. Further, it is possible to keep the overall Type I error rate at, say, 5%, no matter how many variables are tested.

Although most multivariate techniques were developed for use in nonexperimental research, they are also useful in experimental research in which there may be multiple IVs and multiple DVs. With multiple IVs, the research is usually designed so that the IVs are independent of each other and a straightforward correction for numerous statistical tests is available (see Chapter 3). With multiple DVs, a problem of inflated error rate arises if each DV is tested separately. Further, at least some of the DVs are likely to be correlated with each other, so separate tests of each DV reanalyze some of the same variance. Therefore, multivariate tests are used.

Experimental research designs with multiple DVs were unusual at one time. Now, however, with attempts to make experimental designs more realistic, and with the availability of computer programs, experiments often have several DVs. It is dangerous to run an experiment with only one DV and risk missing the impact of the IV because the most sensitive DV is not measured. Multivariate statistics help the experimenter design more efficient and more realistic experiments by allowing measurement of multiple DVs without violation of acceptable levels of Type I error.

One of the few considerations not relevant to choice of *statistical* technique is whether the data are experimental or correlational. The statistical methods "work" whether the researcher manipulated the levels of the IV. But attribution of causality to results is crucially affected by the experimental-nonexperimental distinction.

1.1.3   Computers and Multivariate Statistics

One answer to the question "Why multivariate statistics?" is that the techniques are now accessible by computer. Only the most dedicated number cruncher would consider doing real-life-sized problems in multivariate statistics without a computer. Fortunately, excellent multivariate programs are available in a number of computer packages.

Two packages are demonstrated in this book. Examples are based on programs in SPSS (Statistical Package for the Social Sciences) and SAS.

If you have access to both packages, you are indeed fortunate. Programs within the packages do not completely overlap, and some problems are better handled through one package than the other. For example, doing several versions of the same basic analysis on the same set of data is particularly easy with SPSS whereas SAS has the most extensive capabilities for saving derived scores from data screening or from intermediate analyses.

Chapters 5 through 16 (the chapters that cover the specialized multivariate techniques) and Chapter 18 (available at www.ablongman.com/tabachnick5e) offer explanations and illustrations of a variety of programs[2] within each package and a comparison of the features of the programs. We hope that once you understand the techniques, you will be able to generalize to virtually any multivariate program.

Recent versions of the programs are implemented in Windows, with menus that implement most of the techniques illustrated in this book. All of the techniques may be implemented through syntax, and syntax itself is generated through menus. Then you may add or change syntax as desired for your analysis. For example, you may "paste" menu choices into a syntax window in SPSS, edit the resulting text, and then run the program. Also, syntax generated by SPSS menus is saved in the "journal" file (spss.jnl) which also may be accessed and copied into a syntax window. Syntax generated by SAS menus is recorded in a "log" file. The contents may then be copied to an interactive window, edited, and run. Do not overlook the help files in these programs. Indeed, SAS and SPSS now provide the entire set of user manuals on CD, often with more current information than is available in printed manuals.

Our demonstrations in this book are based on syntax generated through menus whenever feasible. We would love to show you the sequence of menu choices, but space does not permit. And, for the sake of parsimony, we have edited program output to illustrate the material that we feel is the most important for interpretation. We have also edited out some of the unnecessary (because it is default) syntax that is generated through menu choices.

With commercial computer packages, you need to know which version of the package you are using. Programs are continually being changed, and not all changes are immediately implemented at each facility. Therefore, many versions of the various programs are simultaneously in use at different institutions; even at one institution, more than one version of a package is sometimes available.

Program updates are often corrections of errors discovered in earlier versions. Occasionally, though, there are major revisions in one or more programs or a new program is added to the package. Sometimes defaults change with updates, so that output looks different although syntax is the same. Check to find out which version of each package you are using. Then be sure that the manual you are using is consistent with the version in use at your facility. Also check updates for error correction in previous releases that may be relevant to some of your previous runs.

Except where noted, this book reviews Windows versions of SPSS Version 13 and SAS Version 9.1. Information on availability and versions of software, macros, books, and the like changes almost daily. We recommend the Internet as a source of "keeping up."

---

[2]We have retained descriptions of features of SYSTAT in these sections despite the removal of detailed demonstrations of that program in this edition.

1.1.4 Garbage In, Roses Out?

The trick in multivariate statistics is not in computation; that is easily done as discussed above. The trick is to select reliable and valid measurements, choose the appropriate program, use it correctly, and know how to interpret the output. Output from commercial computer programs, with their beautifully formatted tables, graphs, and matrices, can make garbage look like roses. Throughout this book, we try to suggest clues that reveal when the true message in the output more closely resembles the fertilizer than the flowers.

Second, when you use multivariate statistics, you rarely get as close to the raw data as you do when you apply univariate statistics to a relatively few cases. Errors and anomalies in the data that would be obvious if the data were processed by hand are less easy to spot when processing is entirely by computer. But the computer packages have programs to graph and describe your data in the simplest univariate terms and to display bivariate relationships among your variables. As discussed in Chapter 4, these programs provide preliminary analyses that are absolutely necessary if the results of multivariate programs are to be believed.

There are also certain costs associated with the benefits of using multivariate procedures. Benefits of increased flexibility in research design, for instance, are sometimes paralleled by increased ambiguity in interpretation of results. In addition, multivariate results can be quite sensitive to which analytic strategy is chosen (cf. Section 1.2.4) and do not always provide better protection against statistical errors than their univariate counterparts. Add to this the fact that occasionally you still cannot get a firm statistical answer to your research questions, and you may wonder if the increase in complexity and difficulty is warranted.

Frankly, we think it is. Slippery as some of the concepts and procedures are, these statistics provide insights into relationships among variables that may more closely resemble the complexity of the "real" world. And sometimes you get at least partial answers to questions that could not be asked at all in the univariate framework. For a complete analysis, making sense of your data usually requires a judicious mix of multivariate and univariate statistics.

And the addition of multivariate statistical methods to your repertoire makes data analysis a lot more fun. If you liked univariate statistics, you will love multivariate statistics![3]

1.2 Some Useful Definitions

In order to describe multivariate statistics easily, it is useful to review some common terms in research design and basic statistics. Distinctions were made in preceding sections between IVs and DVs and between experimental and nonexperimental research. Additional terms that are encountered repeatedly in the book but not necessarily related to each other are described in this section.

1.2.1 Continuous, Discrete, and Dichotomous Data

In applying statistical techniques of any sort, it is important to consider the type of measurement and the nature of the correspondence between numbers and the events that they represent. The distinction made here is among continuous, discrete, and dichotomous variables; you may prefer to substitute the terms *interval* or *quantitative* for *continuous* and *nominal, categorical* or *qualitative* for *dichotomous* and *discrete*.

---

[3]Don't even think about it.

Continuous variables are measured on a scale that changes values smoothly rather than in steps. Continuous variables take on any value within the range of the scale, and the size of the number reflects the amount of the variable. Precision is limited by the measuring instrument, not by the nature of the scale itself. Some examples of continuous variables are time as measured on an old-fashioned analog clock face, annual income, age, temperature, distance, and grade point average (GPA).

Discrete variables take on a finite and usually small number of values, and there is no smooth transition from one value or category to the next. Examples include time as displayed by a digital clock, continents, categories of religious affiliation, and type of community (rural or urban).

Sometimes discrete variables are used in multivariate analyses as if continuous if there are numerous categories and the categories represent a quantitative attribute. For instance, a variable that represents age categories (where, say, 1 stands for 0 to 4 years, 2 stands for 5 to 9 years, 3 stands for 10 to 14 years, and so on up through the normal age span) can be used because there are a lot of categories and the numbers designate a quantitative attribute (increasing age). But the same numbers used to designate categories of religious affiliation are not in appropriate form for analysis with many of the techniques[4] because religions do not fall along a quantitative continuum.

Discrete variables composed of qualitatively different categories are sometimes analyzed after being changed into a number of dichotomous or two-level variables (e.g., Catholic vs. non-Catholic, Protestant vs. non-Protestant, Jewish vs. non-Jewish, and so on until the degrees of freedom are used). Recategorization of a discrete variable into a series of dichotomous ones is called *dummy variable coding*. The conversion of a discrete variable into a series of dichotomous ones is done to limit the relationship between the dichotomous variables and others to linear relationships. A discrete variable with more than two categories can have a relationship of any shape with another variable, and the relationship is changed arbitrarily if assignment of numbers to categories is changed. Dichotomous variables, however, with only two points, can have only linear relationships with other variables; they are, therefore, appropriately analyzed by methods using correlation in which only linear relationships are analyzed.

The distinction between continuous and discrete variables is not always clear. If you add enough digits to the digital clock, for instance, it becomes for all practical purposes a continuous measuring device, whereas time as measured by the analog device can also be read in discrete categories such as hours or half hours. In fact, any continuous measurement may be rendered discrete (or dichotomous) with some loss of information, by specifying cutoffs on the continuous scale.

The property of variables that is crucial to application of multivariate procedures is not the type of measurement so much as the shape of distribution, as discussed in Chapter 4 and elsewhere. Non-normally distributed continuous variables and dichotomous variables with very uneven splits between the categories present problems to several of the multivariate analyses. This issue and its resolution are discussed at some length in Chapter 4.

Another type of measurement that is used sometimes produces a rank order (ordinal) scale. This scale assigns a number to each subject to indicate the subject's position vis-à-vis other subjects along some dimension. For instance, ranks are assigned to contestants (first place, second place, third place, etc.) to provide an indication of who was best—but not by how much. A problem with ordinal measures is that their distributions are rectangular (one frequency per number) instead of normal, unless tied ranks are permitted and they pile up in the middle of the distribution.

---

[4]Some multivariate techniques (e.g., logistic regression, SEM) are appropriate for all types of variables.

In practice, we often treat variables as if they are continuous when the underlying scale is thought to be continuous but the measured scale actually is ordinal, the number of categories is large—say, seven or more—and the data meet other assumptions of the analysis. For instance, the number of correct items on an objective test is technically not continuous because fractional values are not possible, but it is thought to measure some underlying continuous variable such as course mastery. Another example of a variable with ambiguous measurement is one measured on a Likert-type scale in which consumers rate their attitudes toward a product as "strongly like," "moderately like," "mildly like," "neither like nor dislike," "mildly dislike," "moderately dislike," or "strongly dislike." As mentioned previously, even dichotomous variables may be treated as if continuous under some conditions. Thus, we often use the term "*continuous*" throughout the remainder of this book whether the measured scale itself is continuous or the variable is to be treated as if continuous. We use the term "*discrete*" for variables with a few categories, whether the categories differ in type or quantity.

1.2.2   Samples and Populations

Samples are measured in order to make generalizations about populations. Ideally, samples are selected, usually by some random process, so that they represent the population of interest. In real life, however, populations are frequently best defined in terms of samples, rather than vice versa; the population is the group from which you were able to randomly sample.

Sampling has somewhat different connotations in nonexperimental and experimental research. In nonexperimental research, you investigate relationships among variables in some predefined population. Typically, you take elaborate precautions to ensure that you have achieved a representative sample of that population; you define your population, then do your best to randomly sample from it.[5]

In experimental research, you attempt to create different populations by treating subgroups from an originally homogeneous group differently. The sampling objective here is to ensure that all subjects come from the same population before you treat them differently. Random sampling consists of randomly assigning subjects to treatment groups (levels of the IV) to ensure that, before differential treatment, all subsamples come from the same population. Statistical tests provide evidence as to whether, after treatment, all samples still come from the same population. Generalizations about treatment effectiveness are made to the type of subjects who participated in the experiment.

1.2.3   Descriptive and Inferential Statistics

Descriptive statistics describe samples of subjects in terms of variables or combinations of variables. Inferential statistical techniques test hypotheses about differences in populations on the basis of measurements made on samples of subjects. If reliable differences are found, descriptive statistics are then used to provide estimations of central tendency, and the like, in the population. Descriptive statistics used in this way are called *parameter estimates.*

Use of inferential and descriptive statistics is rarely an either-or proposition. We are usually interested in both describing and making inferences about a data set. We describe the data, find

---

[5]Strategies for random sampling are discussed in many sources, including Levy and Lemenshow (1999), Rea and Parker (1997), and de Vaus (2002).

reliable differences or relationships, and estimate population values for the reliable findings. However, there are more restrictions on inference than there are on description. Many assumptions of multivariate statistical methods are necessary only for inference. If simple description of the sample is the major goal, many assumptions are relaxed, as discussed in Chapters 5 through 16 and 18 (online).

1.2.4   Orthogonality: Standard and Sequential Analyses

Orthogonality is a perfect nonassociation between variables. If two variables are orthogonal, knowing the value of one variable gives no clue as to the value of the other; the correlation between them is zero.

Orthogonality is often desirable in statistical applications. For instance, factorial designs for experiments are orthogonal when two or more IVs are completely crossed with equal sample sizes in each combination of levels. Except for use of a common error term, tests of hypotheses about main effects and interactions are independent of each other; the outcome of each test gives no hint as to the outcome of the others. In orthogonal experimental designs with random assignment of subjects, manipulation of the levels of the IV, and good controls, changes in value of the DV can be unambiguously attributed to various main effects and interactions.

Similarly, in multivariate analyses, there are advantages if sets of IVs or DVs are orthogonal. If all pairs of IVs in a set are orthogonal, each IV adds, in a simple fashion, to prediction of the DV. Consider income as a DV with education and occupational prestige as IVs. If education and occupational prestige are orthogonal, and if 35% of the variability in income may be predicted from education and a different 45% is predicted from occupational prestige, then 80% of the variance in income is predicted from education and occupational prestige together.

Orthogonality can easily be illustrated in Venn diagrams, as shown in Figure 1.1. Venn diagrams represent shared variance (or correlation) as overlapping areas between two (or more) circles. The total variance for income is one circle. The section with horizontal stripes represents the part of income predictable from education, and the section with vertical stripes represents the part predictable from occupational prestige; the circle for education overlaps the circle for income 35% and the circle for occupational prestige overlaps 45%. Together, they account for 80% of the variability in income because education and occupational prestige are orthogonal and do not themselves overlap. There are similar advantages if a set of DVs is orthogonal. The overall effect of an IV can be partitioned into effects on each DV in an additive fashion.

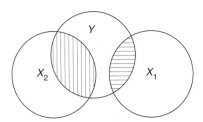

**FIGURE 1.1   Venn diagram for Y
(income), X_1 (education), and X_2
(occupational prestige).**

Usually, however, the variables are correlated with each other (nonorthogonal). IVs in nonexperimental designs are often correlated naturally; in experimental designs, IVs become correlated when unequal numbers of subjects are measured in different cells of the design. DVs are usually correlated because individual differences among subjects tend to be consistent over many attributes.

When variables are correlated, they have shared or overlapping variance. In the example of Figure 1.2, education and occupational prestige correlate with each other. Although the independent contribution made by education is still 35% and that by occupational prestige is 45%, their joint contribution to prediction of income is not 80% but rather something smaller due to the overlapping area shown by the arrow in Figure 1.2(a). A major decision for the multivariate analyst is how to handle the variance that is predictable from more than one variable. Many multivariate techniques have at least two strategies for handling it; some have more.

In standard analysis, the overlapping variance contributes to the size of summary statistics of the overall relationship but is not assigned to either variable. Overlapping variance is disregarded in assessing the contribution of each variable to the solution. Figure 1.2(a) is a Venn diagram of a standard analysis in which overlapping variance is shown as overlapping areas in circles; the unique contributions of X_1 and X_2 to prediction of Y are shown as horizontal and vertical areas, respectively, and the total relationship between Y and the combination of X_1 and X_2 is those two areas plus the area with the arrow. If X_1 is education and X_2 is occupational prestige, then in standard analysis, X_1 is "credited with" the area marked by the horizontal lines and X_2 by the area marked by vertical lines. Neither of the IVs is assigned the area designated with the arrow. When X_1 and X_2 substantially overlap each other, very little horizontal or vertical area may be left for either of them despite the fact that they are both related to Y. They have essentially knocked each other out of the solution.

Sequential analyses differ in that the researcher assigns priority for entry of variables into equations, and the first one to enter is assigned both unique variance and any overlapping variance it has with other variables. Lower-priority variables then are assigned on entry their unique and any remaining overlapping variance. Figure 1.2(b) shows a sequential analysis for the same case as Figure 1.2(a), where X_1 (education) is given priority over X_2 (occupational prestige). The total variance explained is the same as in Figure 1.2(a), but the relative contributions of X_1 and X_2 have changed;

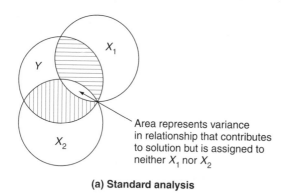

Area represents variance
in relationship that contributes
to solution but is assigned to
neither X_1 nor X_2

(a) Standard analysis

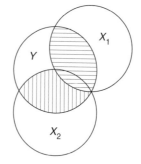

**(b) Sequential analysis in which
X_1 is given priority over X_2**

**FIGURE 1.2   Standard (a) and sequential (b) analyses of the relationship
between Y, X_1, and X_2. Horizontal shading depicts variance assigned to X_1.
Vertical shading depicts variance assigned to X_2.**

education now shows a stronger relationship with income than in the standard analysis, whereas the relation between occupational prestige and income remains the same.

The choice of strategy for dealing with overlapping variance is not trivial. If variables are correlated, the overall relationship remains the same but the apparent importance of variables to the solution changes depending on whether a standard or a sequential strategy is used. If the multivariate procedures have a reputation for unreliability, it is because solutions change, sometimes dramatically, when different strategies for entry of variables are chosen. However, the strategies also ask different questions of the data, and it is incumbent on the researcher to determine exactly which question to ask. We try to make the choices clear in the chapters that follow.

1.3   Linear Combinations of Variables

Multivariate analyses combine variables to do useful work such as predict scores or predict group membership. The combination that is formed depends on the relationships among the variables and the goals of analysis, but in most cases, the combination that is formed is a linear combination. A linear combination is one in which each variable is assigned a weight (e.g., W_1), and then products of weights and variable scores are summed to predict a score on a combined variable. In Equation 1.1, Y' (the predicted DV) is predicted by a linear combination of X_1 and X_2 (the IVs).

$$Y' = W_1X_1 + W_2X_2 \tag{1.1}$$

If, for example, Y' is predicted income, X_1 is education, and X_2 is occupational prestige, the best prediction of income is obtained by weighting education (X_1) by W_1 and occupational prestige (X_2) by W_2 before summing. No other values of W_1 and W_2 produce as good a prediction of income.

Notice that Equation 1.1 includes neither X_1 or X_2 raised to powers (exponents) nor a product of X_1 and X_2. This seems to severely restrict multivariate solutions until one realizes that X_1 could itself be a product of two different variables or a single variable raised to a power. For example, X_1 might be education squared. A multivariate solution does not produce exponents or cross-products of IVs to improve a solution, but the researcher can include Xs that are cross-products of IVs or are IVs raised to powers. Inclusion of variables raised to powers or cross-products of variables has both theoretical and practical implications for the solution. Berry (1993) provides a useful discussion of many of the issues.

The size of the W values (or some function of them) often reveals a great deal about the relationship between DV and IVs. If, for instance, the W value for some IV is zero, the IV is not needed in the best DV–IV relationship. Or if some IV has a large W value, then the IV tends to be important to the relationship. Although complications (to be explained later) prevent interpretation of the multivariate solution from the sizes of the W values alone, they are nonetheless important in most multivariate procedures.

The combination of variables can be considered a supervariable, not directly measured but worthy of interpretation. The supervariable may represent an underlying dimension that predicts something or optimizes some relationship. Therefore, the attempt to understand the meaning of the combination of IVs is worthwhile in many multivariate analyses.

In the search for the best weights to apply in combining variables, computers do not try out all possible sets of weights. Various algorithms have been developed to compute the weights. Most algorithms involve manipulation of a correlation matrix, a variance-covariance matrix, or a sum-of-squares and cross-products matrix. Section 1.6 describes these matrices in very simple terms and shows their development from a very small data set. Appendix A describes some terms and manipulations appropriate to matrices. In the fourth sections of Chapters 5 through 16 and 18 (online), a small hypothetical sample of data is analyzed by hand to show how the weights are derived for each analysis. Though this information is useful for a basic understanding of multivariate statistics, it is not necessary for applying multivariate techniques fruitfully to your research questions and may, sadly, be skipped by those who are math aversive.

1.4   Number and Nature of Variables to Include

Attention to the number of variables included in analysis is important. A general rule is to get the best solution with the fewest variables. As more and more variables are included, the solution usually improves, but only slightly. Sometimes the improvement does not compensate for the cost in degrees of freedom of including more variables, so the power of the analyses diminishes.

A second problem is *overfitting*. With overfitting, the solution is very good, so good in fact, that it is unlikely to generalize to a population. Overfitting occurs when too many variables are included in an analysis relative to the sample size. With smaller samples, very few variables can be analyzed. Generally, a researcher should include only a limited number of uncorrelated variables in each analysis,[6] fewer with smaller samples. We give guidelines for the number of variables that can be included relative to sample size in the third section of Chapters 5–16 and 18.

Additional considerations for inclusion of variables in a multivariate analysis include cost, availability, meaning, and theoretical relationships among the variables. Except in analysis of structure, one usually wants a small number of valid, cheaply obtained, easily available, uncorrelated variables that assess all the theoretically important dimensions of a research area. Another important consideration is reliability. How stable is the position of a given score in a distribution of scores when measured at different times or in different ways? Unreliable variables degrade an analysis whereas reliable ones enhance it. A few reliable variables give a more meaningful solution than a large number of less reliable variables. Indeed, if variables are sufficiently unreliable, the entire solution may reflect only measurement error. Further considerations for variable selection are mentioned as they apply to each analysis.

1.5   Statistical Power

A critical issue in designing any study is whether there is adequate power. Power, as you may recall, represents the probability that effects that actually exist have a chance of producing statistical significance in your eventual data analysis. For example, do you have a large enough sample size to show a significant relationship between GRE and GPA if the actual relationship is fairly large? What if the relationship is fairly small? Is your sample large enough to reveal significant effects of treatment on your DV(s)? Relationships among power and errors of inference are discussed in Chapter 3.

Issues of power are best considered in the planning state of a study when the researcher determines the required sample size. The researcher estimates the size of the anticipated effect (e.g., an expected mean difference), the variability expected in assessment of the effect, the desired alpha

---

[6]The exceptions are analysis of structure, such as factor analysis, in which numerous correlated variables are measured.

level (ordinarily 0.05), and the desired power (often .80). These four estimates are required to determine necessary sample size. Failure to consider power in the planning stage often results in failure to find a significant effect (and an unpublishable study). The interested reader may wish to consult Cohen (1965, 1988), Rossi (1990), or Sedlmeier and Gigerenzer (1989) for more detail.

There is a great deal of software available to help you estimate the power available with various sample sizes for various statistical techniques, and to help you determine necessary sample size given a desired level of power (e.g., an 80% probability of achieving a significant result if an effect exists) and expected sizes of relationships. One of these programs that estimates power for several techniques is PASS (NCSS, 2002). Many other programs are reviewed (and sometimes available as shareware) on the Internet. Issues of power relevant to each of the statistical techniques are discussed in chapters covering those techniques.

1.6 Data Appropriate for Multivariate Statistics

An appropriate data set for multivariate statistical methods consists of values on a number of variables for each of several subjects or cases. For continuous variables, the values are scores on variables. For example, if the continuous variable is the GRE (Graduate Record Examination), the values for the various subjects are scores such as 500, 650, 420, and so on. For discrete variables, values are number codes for group membership or treatment. For example, if there are three teaching techniques, students who receive one technique are arbitrarily assigned a "1," those receiving another technique are assigned a "2," and so on.

1.6.1 The Data Matrix

The data matrix is an organization of scores in which rows (lines) represent subjects and columns represent variables. An example of a data matrix with six subjects[7] and four variables is in Table 1.1. For example, X_1 might be type of teaching technique, X_2 score on the GRE, X_3 GPA, and X_4 gender, with women coded 1 and men coded 2.

Data are entered into a data file with long-term storage accessible by computer in order to apply computer techniques to them. Each subject starts with a new row (line). Information identifying the subject is typically entered first, followed by the value of each variable for that subject.

TABLE 1.1   A Data Matrix of Hypothetical Scores

Student	X_1	X_2	X_3	X_4
1	1	500	3.20	1
2	1	420	2.50	2
3	2	650	3.90	1
4	2	550	3.50	2
5	3	480	3.30	1
6	3	600	3.25	2

[7]Normally, of course, there are many more than six subjects.

Scores for each variable are entered in the same order for each subject. If there are more data for each subject than can be accommodated on a single line, the data are continued on additional lines, but all of the data for each subject are kept together. All of the computer package manuals provide information on setting up a data matrix.

In this example, there are values for every variable for each subject. This is not always the case with research in the real world. With large numbers of subjects and variables, scores frequently are missing on some variables for some subjects. For instance, respondents may refuse to answer some kinds of questions, or some students may be absent the day that one of the tests is given, and so forth. This creates missing values in the data matrix. To deal with missing values, first build a data file in which some symbol is used to indicate that a value on a variable is missing in data for a subject. The various programs have standard symbols, such as a dot (.), for this purpose. You can use other symbols, but it is often just as convenient to use one of the default symbols. Once the data set is available, consult Chapter 4 for various options to deal with this messy (but often unavoidable) problem.

1.6.2   The Correlation Matrix

Most readers are familiar with **R,** a correlation matrix. **R** is a square, symmetrical matrix. Each row (and each column) represents a different variable, and the value at the intersection of each row and column is the correlation between the two variables. For instance, the value at the intersection of the second row, third column, is the correlation between the second and the third variables. The same correlation also appears at the intersection of the third row, second column. Thus, correlation matrices are said to be symmetrical about the main diagonal, which means they are mirror images of themselves above and below the diagonal going from top left to bottom right. For this reason, it is common practice to show only the bottom half or the top half of an R matrix. The entries in the main diagonal are often omitted as well, since they are all ones—correlations of variables with themselves.[8]

Table 1.2 shows the correlation matrix for X_2, X_3, and X_4 of Table 1.1. The .85 is the correlation between X_2 and X_3, and it appears twice in the matrix (as do others). Other correlations are as indicated in the table.

Many programs allow the researcher a choice between analysis of a correlation matrix and analysis of a variance-covariance matrix. If the correlation matrix is analyzed, a unit-free result is produced. That is, the solution reflects the relationships among the variables but not in the metric in which they are measured. If the metric of the scores is somewhat arbitrary, analysis of **R** is appropriate.

TABLE 1.2   Correlation Matrix for Part of Hypothetical Data for Table 1.1

		X_2	X_3	X_4
	X_2	1.00	.85	−.13
R =	X_3	.85	1.00	−.46
	X_4	−.13	−.46	1.00

[8]Alternatively, other information such as standard deviations is inserted.

1.6.3 The Variance-Covariance Matrix

If, on the other hand, scores are measured along a meaningful scale, it is sometimes appropriate to analyze a variance-covariance matrix. A variance-covariance matrix, Σ, is also square and symmetrical, but the elements in the main diagonal are the variances of each variable, and the off-diagonal elements are covariances between pairs of different variables.

Variances, as you recall, are averaged squared deviations of each score from the mean of the scores. Because the deviations are averaged, the number of scores included in computation of a variance is not relevant, but the metric in which the scores are measured is. Scores measured in large numbers tend to have large numbers as variances; scores measured in small numbers tend to have small variances.

Covariances are averaged cross-products (the deviation between one variable and its mean times the deviation between a second variable and its mean). Covariances are similar to correlations except that they, like variances, retain information concerning the scales in which the variables are measured. The variance-covariance matrix for the continuous data of Table 1.1 appears in Table 1.3.

1.6.4 The Sum-of-Squares and Cross-Products Matrix

This matrix, **S**, is a precursor to the variance-covariance matrix in which deviations are not yet averaged. Thus, the size of the entries depends on the number of cases as well as on the metric in which the elements were measured. The sum-of-squares and cross-products matrix for X_2, X_3, and X_4 in Table 1.1 appears in Table 1.4.

The entry in the major diagonal of the **S** matrix is the sum of squared deviations of scores from the mean for that variable, hence, "sum of squares," or SS. That is, for each variable, the value in the major diagonal is

$$SS(X_i) = \sum_{i=1}^{N} (X_{ij} - \overline{X}_j)^2 \qquad (1.2)$$

where $i = 1, 2, \ldots, N$
$N =$ the number of subjects
$j =$ the variable identifier
$X_{ij} =$ the score on variable j by subject i
$\overline{X}_j =$ the mean of all scores on the jth variable

TABLE 1.3 Variance-Covariance Matrix for Part of Hypothetical Data of Table 1.1

		X_2	X_3	X_4
	X_2	7026.66	32.80	–6.00
$\Sigma =$	X_3	32.80	.21	–.12
	X_4	–6.00	–.12	.30

TABLE 1.4 Sum-of-Squares and Cross-Products Matrix for Part of Hypothetical Data of Table 1.1

		X_2	X_3	X_4
	X_2	35133.33	164.00	–30.00
$S =$	X_3	164.00	1.05	–0.59
	X_4	–30.00	–0.59	1.50

For example, for X_4, the mean is 1.5. The sum of squared deviations around the mean and the diagonal value for the variable is

$$\sum_{i=1}^{6} (X_{i4} - \overline{X}_4)^2 = (1 - 1.5)^2 + (2 - 1.5)^2 + (1 - 1.5)^2 + (2 - 1.5)^2 + (1 - 1.5)^2 + (2 - 1.5)^2$$

$$= 1.50$$

The off-diagonal elements of the sum-of-squares and cross-products matrix are the cross-products—the sum of products (SP)—of the variables. For each pair of variables, represented by row and column labels in Table 1.4, the entry is the sum of the product of the deviation of one variable around its mean times the deviation of the other variable around its mean.

$$SP(X_j X_k) = \sum_{i=1}^{N} (X_{ij} - \overline{X}_j)(X_{ik} - \overline{X}_k) \tag{1.3}$$

where j identifies the first variable, k identifies the second variable, and all other terms are as defined in Equation 1.1. (Note that if $j = k$, Equation 1.3 becomes identical to Equation 1.2.)

For example, the cross-product term for variables X_2 and X_3 is

$$\sum_{i=1}^{N} (X_{i2} - \overline{X}_2)(X_{i3} - \overline{X}_3) = (500 - 533.33)(3.20 - 3.275) + (420 - 533.33)(2.50 - 3.275)$$

$$+ \cdots + (600 - 533.33)(3.25 - 3.275) = 164.00$$

Most computations start with **S** and proceed to Σ or **R**. The progression from a sum-of-squares and cross-products matrix to a variance-covariance matrix is simple.

$$\Sigma = \frac{1}{N-1} S \tag{1.4}$$

The variance-covariance matrix is produced by dividing every element in the sum-of-squares and cross-products matrix by $N - 1$, where N is the number of cases.

The correlation matrix is derived from an **S** matrix by dividing each sum of squares by itself (to produce the 1s in the main diagonal of **R**) and each cross-product of the **S** matrix by the square root of the product of the sum of squared deviations around the mean for each of the variables in the pair. That is, each cross-product is divided by

$$\text{Denominator}(X_j X_k) = \sqrt{\sum (X_{ij} - X_j)^2 \sum (X_{ik} - X_k)^2} \tag{1.5}$$

where terms are defined as in Equation 1.3.

For some multivariate operations, it is not necessary to feed the data matrix to a computer program. Instead, an **S** or an **R** matrix is entered, with each row (representing a variable) starting a new line. Often, considerable computing time and expense are saved by entering one or the other of these matrices rather than raw data.

1.6.5   Residuals

Often a goal of analysis or test of its efficiency is its ability to reproduce the values of a DV or the correlation matrix of a set of variables. For example, we might want to predict scores on the GRE (X_2 of Table 1.1) from knowledge of GPA (X_3) and gender (X_4). After applying the proper statistical operations—a multiple regression in this case—a predicted GRE score for each student is computed by applying the proper weights for GPA and gender to the GPA and gender scores for each student. But because we already obtained GRE scores for our sample of students, we are able to compare the predicted with the obtained GRE score. The difference between the predicted and obtained values is known as the *residual* and is a measure of error of prediction.

In most analyses, the residuals for the entire sample sum to zero. That is, sometimes the prediction is too large and sometimes it is too small, but the average of all the errors is zero. The squared value of the residuals, however, provides a measure of how good the prediction is. When the predictions are close to the obtained values, the squared errors are small. The way that the residuals are distributed is of further interest in evaluating the degree to which the data meet the assumptions of multivariate analyses, as discussed in Chapter 4 and elsewhere.

1.7   Organization of the Book

Chapter 2 gives a guide to the multivariate techniques that are covered in this book and places them in context with the more familiar univariate and bivariate statistics where possible. Included in Chapter 2 is a flow chart that organizes statistical techniques on the basis of the major research questions asked. Chapter 3 provides a brief review of univariate and bivariate statistical techniques for those who are interested.

Chapter 4 deals with the assumptions and limitations of multivariate statistical methods. Assessment and violation of assumptions are discussed, along with alternatives for dealing with violations when they occur. Chapter 4 is meant to be referred to often, and the reader is guided back to it frequently in Chapters 5 through 16 and 18 (online).

Chapters 5 through 16 and 18 (online) cover specific multivariate techniques. They include descriptive, conceptual sections as well as a guided tour through a real-world data set for which the analysis is appropriate. The tour includes an example of a Results section describing the outcome of the statistical analysis appropriate for submission to a professional journal. Each technique chapter includes a comparison of computer programs. You may want to vary the order in which you cover these chapters.

Chapter 17 is an attempt to integrate univariate, bivariate, and multivariate statistics through the multivariate general linear model. The common elements underlying all the techniques are emphasized, rather than the differences among them. Chapter 17 is meant to pull together the material in the remainder of the book with a conceptual rather than pragmatic emphasis. Some may wish to consider this material earlier, for instance, immediately after Chapter 2.

CHAPTER

2

A Guide to Statistical Techniques

Using the Book

2.1 Research Questions and Associated Techniques

This chapter organizes the statistical techniques in this book by major research question. A decision tree at the end of this chapter leads you to an appropriate analysis for your data. On the basis of your major research question and a few characteristics of your data set, you determine which statistical technique(s) is appropriate. The first, most important criterion for choosing a technique is the major research question to be answered by the statistical analysis. Research questions are categorized here into degree of relationship among variables, significance of group differences, prediction of group membership, structure, and questions that focus on the time course of events. This chapter emphasizes differences in research questions answered by the different techniques described in nontechnical terms, whereas Chapter 17 provides an integrated overview of the techniques with some basic equations used in the multivariate general linear model.[1]

2.1.1 Degree of Relationship among Variables

If the major purpose of analysis is to assess the associations among two or more variables, some form of correlation/regression or chi square is appropriate. The choice among five different statistical techniques is made by determining the number of independent and dependent variables, the nature of the variables (continuous or discrete), and whether any of the IVs are best conceptualized as covariates.[2]

2.1.1.1 *Bivariate* **r**

Bivariate correlation and regression, as reviewed in Chapter 3, assess the degree of relationship between two continuous variables such as belly dancing skill and years of musical training. Bivariate correlation measures the association between two variables with no distinction necessary between IV and DV. Bivariate regression, on the other hand, predicts a score on one variable from knowledge of the score on another variable (e.g., predicts skill in belly dancing as measured by a single index such as knowledge of steps, from a single predictor such as years of musical training).

---

[1]You may find it helpful to read Chapter 17 now instead of waiting for the end.

[2]If the effects of some IVs are assessed after the effects of other IVs are statistically removed, the latter are called *covariates*.

The predicted variable is considered the DV, whereas the predictor is considered the IV. Bivariate correlation and regression are not multivariate techniques, but they are integrated into the general linear model in Chapter 17.

2.1.1.2   Multiple R

Multiple correlation assesses the degree to which one continuous variable (the DV) is related to a set of other (usually) continuous variables (the IVs) that have been combined to create a new, composite variable. Multiple correlation is a bivariate correlation between the original DV and the composite variable created from the IVs. For example, how large is the association between belly dancing skill and a number of IVs such as years of musical training, body flexibility, and age?

Multiple regression is used to predict the score on the DV from scores on several IVs. In the preceding example, belly dancing skill measured by knowledge of steps is the DV (as it is for bivariate regression), and we have added body flexibility and age to years of musical training as IVs. Other examples are prediction of success in an educational program from scores on a number of aptitude tests, prediction of the sizes of earthquakes from a variety of geological and electromagnetic variables, or stock market behavior from a variety of political and economic variables.

As for bivariate correlation and regression, multiple correlation emphasizes the degree of relationship between the DV and the IVs, whereas multiple regression emphasizes the prediction of the DV from the IVs. In multiple correlation and regression, the IVs may or may not be correlated with each other. With some ambiguity, the techniques also allow assessment of the relative contribution of each of the IVs toward predicting the DV, as discussed in Chapter 5.

2.1.1.3   Sequential R

In sequential (sometimes called hierarchical) multiple regression, IVs are given priorities by the researcher before their contributions to prediction of the DV are assessed. For example, the researcher might first assess the effects of age and flexibility on belly dancing skill before looking at the contribution that years of musical training makes to that skill. Differences among dancers in age and flexibility are statistically "removed" before assessment of the effects of years of musical training.

In the example of an educational program, success of outcome might first be predicted from variables such as age and IQ. Then scores on various aptitude tests are added to see if prediction of outcome is enhanced after adjustment for age and IQ.

In general, then, the effects of IVs that enter first are assessed and removed before the effects of IVs that enter later are assessed. For each IV in a sequential multiple regression, higher-priority IVs act as covariates for lower-priority IVs. The degree of relationship between the DV and the IVs is reassessed at each step of the sequence. That is, multiple correlation is re-computed as each new IV (or set of IVs) is added. Sequential multiple regression, then, is also useful for developing a reduced set of IVs (if that is desired) by determining when IVs no longer add to predictability. Sequential multiple regression is discussed in Chapter 5.

2.1.1.4   Canonical R

In canonical correlation, there are several continuous DVs as well as several continuous IVs, and the goal is to assess the relationship between the two sets of variables. For example, we might study the relationship between a number of indices of belly dancing skill (the DVs, such as knowledge of steps, ability to play finger cymbals, responsiveness to the music) and the IVs (flexibility, musical training, and age). Thus, canonical correlation adds DVs (e.g., further indices of belly dancing skill) to the single index of skill used in bivariate and multiple correlation, so that there are multiple DVs as well as multiple IVs in canonical correlation.

Or we might ask whether there is a relationship among achievements in arithmetic, reading, and spelling as measured in elementary school and a set of variables reflecting early childhood development (e.g., age at first speech, walking, toilet training). Such research questions are answered by canonical correlation, the subject of Chapter 12.

2.1.1.5 Multiway Frequency Analysis

A goal of multiway frequency analysis is to assess relationships among discrete variables where none is considered a DV. For example, you might be interested in the relationships among gender, occupational category, and preferred type of reading material. Or the research question might involve relationships among gender, categories of religious affiliation, and attitude toward abortion. Chapter 16 deals with multiway frequency analysis.

When one of the variables is considered a DV with the rest serving as IVs, multiway frequency analysis is called *logit analysis,* as described in Section 2.1.3.3.

2.1.1.6 Multilevel Modeling

In many research applications, cases are nested in (normally occurring) groups, which may, in turn, be nested in other groups. The quintessential example is students nested in classrooms which are, in turn, nested in schools. (Another common example involves repeated measures where, for example, scores are nested within students who are, in turn, nested in classrooms, and then nested in schools.) However, students in the same classroom are likely to have scores that correlate more highly than those of students in general. This creates problems with an analysis that pools all students into one very large group, ignoring classroom and school designations. Multilevel modeling (Chapter 15) is a somewhat complicated but increasingly popular strategy for analyzing data in these situations.

2.1.2 Significance of Group Differences

When subjects are randomly assigned to groups (treatments), the major research question usually is the extent to which statistically significant mean differences on DVs are associated with group membership. Once significant differences are found, the researcher often assesses the degree of relationship (effect size or strength of association) between IVs and DVs.

The choice among techniques hinges on the number of IVs and DVs and whether some variables are conceptualized as covariates. Further distinctions are made as to whether all DVs are measured on the same scale and how within-subjects IVs are to be treated.

2.1.2.1 One-Way ANOVA and t Test

These two statistics, reviewed in Chapter 3, are strictly univariate in nature and are adequately covered in most standard statistical texts.

2.1.2.2 One-Way ANCOVA

One-way analysis of covariance is designed to assess group differences on a single DV after the effects of one or more covariates are statistically removed. Covariates are chosen because of their known association with the DV; otherwise, there is no point to their use. For example, age and degree of reading disability are usually related to outcome of a program of educational therapy (the DV). If groups are formed by randomly assigning children to different types of educational therapies (the IV), it is useful to remove differences in age and degree of reading disability before examining the relationship between outcome and type of therapy. Prior differences among children in age and reading disability are used as covariates. The ANCOVA question is: Are there mean differences in outcome associated with type of educational therapy after adjusting for differences in age and degree of reading disability?

ANCOVA gives a more powerful look at the IV–DV relationship by minimizing error variance (cf. Chapter 3). The stronger the relationship between the DV and the covariate(s), the greater the power of ANCOVA over ANOVA. ANCOVA is discussed in Chapter 6.

ANCOVA is also used to adjust for differences among groups when groups are naturally occurring and random assignment to them is not possible. For example, one might ask if attitude toward abortion (the DV) varies as a function of religious affiliation. However, it is not possible to randomly assign people to religious affiliation. In this situation, there could easily be other systematic differences among groups, such as level of education, that are also related to attitude toward abortion. Apparent differences among religious groups might well be due to differences in education rather than differences in religious affiliation. To get a "purer" measure of the relationship between attitude and religious affiliation, attitude scores are first adjusted for educational differences, that is, education is used as a covariate. Chapter 6 also discusses this somewhat problematical use of ANCOVA.

When there are more than two groups, planned or post hoc comparisons are available in ANCOVA just as in ANOVA. With ANCOVA, selected and/or pooled group means are adjusted for differences on covariates before differences in means on the DV are assessed.

2.1.2.3 Factorial ANOVA

Factorial ANOVA, reviewed in Chapter 3, is the subject of numerous statistics texts (e.g., Brown, Michels, & Winer, 1991; Keppel & Wickens, 2004; Myers & Well, 2002; Tabachnick & Fidell, 2007) and is introduced in most elementary texts. Although there is only one DV in factorial ANOVA, its place within the general linear model is discussed in Chapter 17.

2.1.2.4 Factorial ANCOVA

Factorial ANCOVA differs from one-way ANCOVA only in that there is more than one IV. The desirability and use of covariates are the same. For instance, in the educational therapy example of Section 2.1.2.2, another interesting IV might be gender of the child. The effects of gender, type of educational therapy and their interaction on outcome are assessed after adjusting for age and prior degree of reading disability. The interaction of gender with type of therapy asks if boys and girls differ as to which type of educational therapy is more effective after adjustment for covariates.

2.1.2.5 Hotelling's T^2

Hotelling's T^2 is used when the IV has only two groups and there are several DVs. For example, there might be two DVs, such as score on an academic achievement test and attention span in the classroom, and two levels of type of educational therapy, emphasis on perceptual training versus emphasis on academic training. It is not legitimate to use separate *t* tests for each DV to look for differences between groups because that inflates Type I error due to unnecessary multiple significance tests with (likely) correlated DVs. Instead, Hotelling's T^2 is used to see if groups differ on the two DVs combined. The researcher asks if there are non-chance differences in the centroids (average on the combined DVs) for the two groups.

Hotelling's T^2 is a special case of multivariate analysis of variance, just as the *t* test is a special case of univariate analysis of variance, when the IV has only two groups. Multivariate analysis of variance is discussed in Chapter 7.

2.1.2.6 One-Way MANOVA

Multivariate analysis of variance evaluates differences among centroids (composite means) for a set of DVs when there are two or more levels of an IV (groups). MANOVA is useful for the educational therapy example in the preceding section with two groups and also when there are more than two groups (e.g., if a nontreatment control group is added).

With more than two groups, planned and post hoc comparisons are available. For example, if a main effect of treatment is found in MANOVA, it might be interesting to ask post hoc if there are differences in the centroids of the two groups given different types of educational therapies, ignoring the control group, and, possibly, if the centroid of the control group differs from the centroid of the two educational therapy groups combined.

Any number of DVs may be used; the procedure deals with correlations among them, and the entire analysis is accomplished within the preset level for Type I error. Once statistically significant differences are found, techniques are available to assess which DVs are influenced by which IV. For example, assignment to treatment group might affect the academic DV but not attention span.

MANOVA is also available when there are within-subject IVs. For example, children might be measured on both DVs three times: 3, 6, and 9 months after therapy begins. MANOVA is discussed in Chapter 7 and a special case of it (profile analysis, in which the within-subjects IV is treated multivariately) in Chapter 8. Profile analysis is an alternative to one-way between-subjects MANOVA when the DVs are all measured on the same scale. Discriminant analysis is an alternative to one-way between-subjects designs, as described in Section 2.1.3.1 and Chapter 9.

2.1.2.7 One-Way MANCOVA

In addition to dealing with multiple DVs, multivariate analysis of variance can be applied to problems when there are one or more covariates. In this case, MANOVA becomes multivariate analysis of covariance—MANCOVA. In the educational therapy example of Section 2.1.2.6, it might be worthwhile to adjust the DV scores for pretreatment differences in academic achievement and attention span. Here the covariates are pretests of the DVs, a classic use of covariance analysis.

After adjustment for pretreatment scores, differences in posttest scores (DVs) can be more clearly attributed to treatment (the two types of educational therapies plus control group that make up the IV).

In the one-way ANCOVA example of religious groups in Section 2.1.2.2, it might be interesting to test political liberalism versus conservatism and attitude toward ecology, as well as attitude toward abortion, to create three DVs. Here again, differences in attitudes might be associated with both differences in religion and differences in education (which, in turn, varies with religious affiliation). In the context of MANCOVA, education is the covariate, religious affiliation the IV, and attitudes the DVs. Differences in attitudes among groups with different religious affiliations are assessed after adjustment for differences in education.

If the IV has more than two levels, planned and post hoc comparisons are useful, with adjustment for covariates. MANCOVA (Chapter 7) is available for both the main analysis and comparisons.

2.1.2.8   *Factorial MANOVA*

Factorial MANOVA is the extension of MANOVA to designs with more than one IV and multiple DVs. For example, gender (a between-subjects IV) might be added to type of educational therapy (another between-subjects IV) with both academic achievement and attention span used as DVs. In this case, the analysis is a two-way between-subjects factorial MANOVA that provides tests of the main effects of gender and type of educational therapy and their interaction on the centroids of the DVs.

Duration of therapy (3, 6, and 9 months) might be added to the design as a within-subjects IV with type of educational therapy a between-subjects IV to examine the effects of duration, type of educational therapy, and their interaction on the DVs. In this case, the analysis is a factorial MANOVA with one between- and one within-subjects IV.

Comparisons can be made among margins or cells in the design, and the influence of various effects on combined or individual DVs can be assessed. For instance, the researcher might plan (or decide post hoc) to look for linear trends in scores associated with duration of therapy for each type of therapy separately (the cells) or across all types of therapies (the margins). The search for linear trend could be conducted among the combined DVs or separately for each DV with appropriate adjustments for Type I error rate.

Virtually any complex ANOVA design (cf. Chapter 3) with multiple DVs can be analyzed through MANOVA, given access to appropriate computer programs. Factorial MANOVA is covered in Chapter 7.

2.1.2.9   *Factorial MANCOVA*

It is sometimes desirable to incorporate one or more covariates into a factorial MANOVA design to produce factorial MANCOVA. For example, pretest scores on academic achievement and attention span could serve as covariates for the two-way between-subjects design with gender and type of educational therapy serving as IVs and posttest scores on academic achievement and attention span serving as DVs. The two-way between-subjects MANCOVA provides tests of gender, type of educational therapy, and their interaction on adjusted, combined centroids for the DVs.

Here again procedures are available for comparisons among groups or cells and for evaluating the influences of IVs and their interactions on the various DVs. Factorial MANCOVA is discussed in Chapter 7.

2.1.2.10 *Profile Analysis of Repeated Measures*

A special form of MANOVA is available when all of the DVs are measured on the same scale (or on scales with the same psychometric properties) and you want to know if groups differ on the scales. For example, you might use the subscales of the Profile of Mood States as DVs to assess whether mood profiles differ between a group of belly dancers and a group of ballet dancers.

There are two ways to conceptualize this design. The first is as a one-way between-subjects design in which the IV is the type of dancer and the DVs are the Mood States subscales; one-way MANOVA provides a test of the main effect of type of dancer on the combined DVs. The second way is as a profile study with one grouping variable (type of dancer) and the several subscales; profile analysis provides tests of the main effects of type of dancer and of subscales as well as their interaction (frequently the effect of greatest interest to the researcher).

If there is a grouping variable and a repeated measure such as trials in which the same DV is measured several times, there are three ways to conceptualize the design. The first is as a one-way between-subjects design with several DVs (the score on each trial); MANOVA provides a test of the main effect of the grouping variable. The second is as a two-way between- and within-subjects design; ANOVA provides tests of groups, trials, and their interaction, but with some very restrictive assumptions that are likely to be violated. Third is as a profile study in which profile analysis provides tests of the main effects of groups and trials and their interaction, but without the restrictive assumptions. This is sometimes called *the multivariate approach to repeated-measures ANOVA.*

Finally, you might have a between- and within-subjects design (groups and trials) in which several DVs are measured on each trial. For example, you might assess groups of belly and ballet dancers on the Mood States subscales at various points in their training. This application of profile analysis is frequently referred to as *doubly multivariate.* Chapter 8 deals with all these forms of profile analysis.

2.1.3 Prediction of Group Membership

In research where groups are identified, the emphasis is frequently on predicting group membership from a set of variables. Discriminant analysis, logit analysis, and logistic regression are designed to accomplish this prediction. Discriminant analysis tends to be used when all IVs are continuous and nicely distributed, logit analysis when IVs are all discrete, and logistic regression when IVs are a mix of continuous and discrete and/or poorly distributed.

2.1.3.1 *One-Way Discriminant Analysis*

In one-way discriminant analysis, the goal is to predict membership in groups (the DV) from a set of IVs. For example, the researcher might want to predict category of religious affiliation from attitude toward abortion, liberalism versus conservatism, and attitude toward ecological issues. The analysis tells us if group membership is predicted at a rate that is significantly better than chance. Or the researcher might try to discriminate belly dancers from ballet dancers from scores on Mood States subscales.

These are the same questions as those addressed by MANOVA, but turned around. Group membership serves as the IV in MANOVA and the DV in discriminant analysis. If groups differ significantly on a set of variables in MANOVA, the set of variables significantly predicts group membership in discriminant analysis. One-way between-subjects designs can be fruitfully analyzed through either procedure and are often best analyzed with a combination of both procedures.

As in MANOVA, there are techniques for assessing the contribution of various IVs to prediction of group membership. For example, the major source of discrimination among religious groups might be abortion attitude, with little predictability contributed by political and ecological attitudes.

In addition, discriminant analysis offers classification procedures to evaluate how well individual cases are classified into their appropriate groups on the basis of their scores on the IVs. One-way discriminant analysis is covered in Chapter 9.

2.1.3.2   Sequential One-Way Discriminant Analysis

Sometimes IVs are assigned priorities by the researcher, so their effectiveness as predictors of group membership is evaluated in the established order in sequential discriminant analysis. For example, when attitudinal variables are predictors of religious affiliation, variables might be prioritized according to their expected contribution to prediction, with abortion attitude given highest priority, political liberalism versus conservatism second priority, and ecological attitude lowest priority. Sequential discriminant analysis first assesses the degree to which religious affiliation is predicted from abortion attitude at a better-than-chance rate. Gain in prediction is then assessed with addition of political attitude, and then with addition of ecological attitude.

Sequential analysis provides two types of useful information. First, it is helpful in eliminating predictors that do not contribute more than predictors already in the analysis. For example, if political and ecological attitudes do not add appreciably to abortion attitude in predicting religious affiliation, they can be dropped from further analysis. Second, sequential discriminant analysis is a covariance analysis. At each step of the hierarchy, higher-priority predictors are covariates for lower-priority predictors. Thus, the analysis permits you to assess the contribution of a predictor with the influence of other predictors removed.

Sequential discriminant analysis is also useful for evaluating sets of predictors. For example, if a set of continuous demographic variables is given higher priority than an attitudinal set in prediction of group membership, one can see if attitudes significantly add to prediction after adjustment for demographic differences. Sequential discriminant analysis is discussed in Chapter 9. However, it is usually more efficient to answer such questions through sequential logistic regression, particularly when some of the predictor variables are continuous and others discrete (see Section 2.1.3.5).

2.1.3.3   Multiway Frequency Analysis (Logit)

The logit form of multiway frequency analysis may be used to predict group membership when all of the predictors are discrete. For example, you might want to predict whether someone is a belly dancer (the DV) from knowledge of gender, occupational category, and preferred type of reading material (science fiction, romance, history, statistics).

This technique allows evaluation of the odds that a case is in one group (e.g., belly dancer) based on membership in various categories of predictors (e.g., female professors who read science fiction). This form of multiway frequency analysis is discussed in Chapter 16.

2.1.3.4   Logistic Regression

Logistic regression allows prediction of group membership when predictors are continuous, discrete, or a combination of the two. Thus, it is an alternative to both discriminant analysis and logit analysis. For example, prediction of whether someone is a belly dancer may be based on gender, occupational category, preferred type of reading material, and age.

Logistic regression allows one to evaluate the odds (or probability) of membership in one of the groups (e.g., belly dancer) based on the combination of values of the predictor variables (e.g., 35-year-old female professors who read science fiction). Chapter 10 covers logistic regression analysis.

2.1.3.5   *Sequential Logistic Regression*

As in sequential discriminant analysis, sometimes predictors are assigned priorities and then assessed in terms of their contribution to prediction of group membership given their priority. For example, one can assess how well preferred type of reading material predicts whether someone is a belly dancer after adjusting for differences associated with age, gender, and occupational category. Sequential logistic regression is also covered in Chapter 10.

2.1.3.6   *Factorial Discriminant Analysis*

If groups are formed on the basis of more than one attribute, prediction of group membership from a set of IVs can be performed through factorial discriminant analysis. For example, respondents might be classified on the basis of both gender and religious affiliation. One could use attitudes toward abortion, politics, and ecology to predict gender (ignoring religion) or religion (ignoring gender), or both gender and religion. But this is the same problem as addressed by factorial MANOVA. For a number of reasons, programs designed for discriminant analysis do not readily extend to factorial arrangements of groups. Unless some special conditions are met (cf. Chapter 9), it is usually better to rephrase the research question so that factorial MANOVA can be used.

2.1.3.7   *Sequential Factorial Discriminant Analysis*

Difficulties inherent in factorial discriminant analysis extend to sequential arrangements of predictors. Usually, however, questions of interest can readily be rephrased in terms of factorial MANCOVA.

2.1.4   Structure

Another set of questions is concerned with the latent structure underlying a set of variables. Depending on whether the search for structure is empirical or theoretical, the choice is principal components, factor analysis, or structural equation modeling. Principal components is an empirical approach, whereas factor analysis and structural equation modeling tend to be theoretical approaches.

2.1.4.1   *Principal Components*

If scores on numerous variables are available from a group of subjects, the researcher might ask if and how the variables group together. Can the variables be combined into a smaller number of super-variables on which the subjects differ? For example, suppose people are asked to rate the effectiveness of numerous behaviors for coping with stress (e.g., "talking to a friend," "going to a movie," "jogging," "making lists of ways to solve the problem"). The numerous behaviors may be empirically related to just a few basic coping mechanisms, such as increasing or decreasing social contact, engaging in physical activity, and instrumental manipulation of stress producers.

Principal components analysis uses the correlations among the variables to develop a small set of components that empirically summarizes the correlations among the variables. It provides a description of the relationship rather than a theoretical analysis. This analysis is discussed in Chapter 13.

2.1.4.2   Factor Analysis

When there is a theory about underlying structure or when the researcher wants to understand underlying structure, factor analysis is often used. In this case, the researcher believes that responses to many different questions are driven by just a few underlying structures called *factors*. In the example of mechanisms for coping with stress, one might hypothesize ahead of time that there are two major factors: general approach to problems (escape vs. direct confrontation) and use of social supports (withdrawing from people vs. seeking them out).

Factor analysis is useful in developing and assessing theories. What is the structure of personality? Are there some basic dimensions of personality on which people differ? By collecting scores from many people on numerous variables that may reflect different aspects of personality, researchers address questions about underlying structure through factor analysis, as discussed in Chapter 13.

2.1.4.3   Structural Equation Modeling

Structural equation modeling combines factor analysis, canonical correlation, and multiple regression. Like factor analysis, some of the variables can be latent, whereas others are directly observed. Like canonical correlation, there can be many IVs and many DVs. And like multiple regression, the goal may be to study the relationships among many variables.

For example, one may want to predict birth outcome (the DVs) from several demographic, personality, and attitudinal measures (the IVs). The DVs are a mix of several observed variables such as birth weight, a latent assessment of mother's acceptance of the child based on several measured attitudes, and a latent assessment of infant responsiveness; the IVs are several demographic variables such as socioeconomic status, race, and income, several latent IVs based on personality measures, and prebirth attitudes toward parenting.

The technique evaluates whether the model provides a reasonable fit to the data and the contribution of each of the IVs to the DVs. Comparisons among alternative models are also possible, as well as evaluation of differences between groups. Chapter 14 covers structural equation modeling.

2.1.5   Time Course of Events

Two techniques focus on the time course of events. Survival/failure analysis asks how long it takes for something to happen. Time-series analysis looks at the change in a DV over the course of time.

2.1.5.1   Survival/Failure Analysis

Survival/failure analysis is a family of techniques dealing with the time it takes for something to happen: a cure, a failure, an employee leaving, a relapse, a death, and so on. For example, what is the life expectancy of someone diagnosed with breast cancer? Is the life expectancy longer with chemotherapy? Or, in the context of failure analysis, what is the expected time before a hard disk fails? Do DVDs last longer than CDs?

Two major varieties of survival/failure analysis are life tables, which describe the course of survival of one or more groups of cases, for example, DVDs and CDs; and determination of whether survival time is influenced by some variables in a set. The latter technique encompasses a set of regression techniques in which the DV is survival time. Chapter 11 covers this analysis.

2.1.5.2   *Time-Series Analysis*

Time-series analysis is used when the DV is measured over a very large number of time periods—at least 50; time is the major IV. Time-series analysis is used to forecast future events (stock markets' indices, crime statistics, etc.) based on a long series of past events. Time-series analysis also is used to evaluate the effect of an intervention, such as implementation of a water-conservation program, by observing water usage for many periods before and after the intervention. Chapter 18 is available on the publisher's website (www.ablongman.com/tabachnick5e).

2.2   Some Further Comparisons

When assessing the degree of relationship among variables, bivariate *r* is appropriate when only two variables (one DV and one IV) are involved, while multiple *R* is appropriate when there are several variables on the IV side (one DV and several IVs). The multivariate analysis adjusts for correlations that are likely present among the IVs. Canonical correlation is available to study the relationship between several DVs and several IVs, adjusting for correlations among all of them. These techniques are usually applied to continuous (and dichotomous) variables. When all variables are discrete, multiway frequency analysis (vastly expanded chi square) is the choice.

Numerous analytic strategies are available to study mean differences among groups, depending on whether there is a single DV or multiple DVs, and whether there are covariates. The familiar ANOVA (and ANCOVA) is used with a single DV while MANOVA (and MANCOVA) is used when there are multiple DVs. Essentially, MANOVA uses weights to combine multiple DVs into a new DV and then performs ANOVA.

A third important issue when studying mean differences among groups is whether there are repeated measures (the familiar within-subjects ANOVA). You may recall the restrictive and often-violated assumption of sphericity with this type of ANOVA. The two multivariate extensions of repeated-measures ANOVA (profile analysis of repeated measures and doubly multivariate profile analysis) circumvent this assumption by combining the DVs; MANOVA combines different DVs while profile analysis combines the same DV measured repeatedly. Another variation of profile analysis (called here profile analysis of repeated measures) is a multivariate extension of the familiar "mixed" (between-within-subjects) ANOVA. None of the multivariate extensions is quite as powerful as its univariate "parent."

The DV in both discriminant analysis and logistic regression is a discrete variable. In discriminant analysis, the IVs are usually continuous variables. A complication arises with discriminant analysis when the DV has more than two groups because there can be as many ways to distinguish the groups from each other as are there are degrees of freedom for the DV. For example, if there are three levels of the DV, there are two degrees of freedom and therefore two potential ways to combine the IVs to separate the levels of the DV. The first combination might, for instance, separate members of the first group from the second and third groups (but not them from each other); the second combination might, then, separate members of group two from group three. Those of you familiar with comparisons in ANOVA probably recognize this as a familiar process for working with more than two groups; the difference is that in ANOVA *you* create the comparison coefficients used in the analysis while in discriminant analysis, the analysis tells *you* how the groups are best discriminated from each other (if they are).

Logistic regression analyzes a discrete DV, too, but the IVs are often a mix of continuous and discrete variables. For that reason, the goal is to predict the probability that a case will fall into various levels of the DV rather than group membership, per se. In this way, the analysis closely resembles the familiar chi-square analysis. In logistic regression, as in all multivariate techniques, the IVs are combined, but in an exponent rather than directly. That makes the analyses conceptually more difficult, but well worth the effort, especially in the medical/biological sciences where risk ratios, a product of logistic regression, are routinely discussed.

There are several procedures for examining structure (that become increasingly "speculative"). Two very closely aligned techniques are principal components and factor analyses. These techniques are interesting because there is no DV (or, for that matter, IVs). Instead, there is just a bunch of variables, with the goal of analysis to discover which of them "go" together. The idea is that some latent, underlying structure (e.g., several different factors representing components of personality) is driving similar responses to correlated sets of questions. The trick for the researcher is to divine the "meaning" of the factors that are developed during analysis. Principal components provides an empirical solution while factor analysis provides a more theoretical solution.

Structural equation modeling combines multiple regression with factor analysis. There is a DV in this technique, but the IVs can be both discrete and continuous, both latent and observed. That is, the researcher tries to predict the values on an observed DV (continuous or discrete) using both observed variables (continuous and discrete) and latent ones (factors derived from many observed variables during the analysis). Structural equation modeling is undergoing rapid development at present, with expansion to MANOVA-like analyses, sophisticated procedures for handling missing data, and the like.

Multilevel modeling assesses the significance of variables where the cases are nested into different levels (e.g., students nested in classes nested in schools; patients nested in wards nested in hospitals). There is a DV at the lowest (student) level, but some IVs pertain to students, some to classes, and some to schools. The analysis takes into account the (likely) higher correlations among scores of students nested in the same class and of classes nested in the same school. Relationships (regressions) developed at one level (e.g., predicting student scores on the SAT from parental educational level) become the DVs for the next level, and so on.

Finally, we present two techniques for analyzing the time course of events, survival analysis and time-series analysis. One underlying IV for both of these is time; there may be other IVs as well. In survival analysis, the goal is often to determine whether a treated group survives longer than an untreated group given the current standard of care. (In manufacturing, it is called failure analyses, and the goal, for instance, is to see if a part manufactured from a new alloy fails later than the part manufactured from the current alloy.) One advantage of this technique, at least in medicine, is its ability to analyze data for cases that have disappeared for one reason or another (moved away, gone to another clinic for treatment, died of another cause) before the end of the study; these are called censored cases.

Time-series analysis tracks the pattern of the DV over multiple measurements (at least 50) and may or may not have an IV. If there is an IV, the goal is to determine if the pattern seen in the DV over time is the same for the group in one level of the IV as for the group in the other level. The IV can be naturally occurring or manipulated.

Generally, statistics are like tools—you pick the wrench you need to do the job.

2.3   A Decision Tree

A decision tree starting with major research questions appears in Table 2.1. For each question, choice among techniques depends on number of IVs and DVs (sometimes an arbitrary distinction) and whether some variables are usefully viewed as covariates. The table also briefly describes analytic goals associated with some techniques.

TABLE 2.1   Choosing among Statistical Techniques

Major Research Question	Number (Kind) of Dependent Variables	Number (Kind) of Independent Variables	Covariates	Analytic Strategy	Goal of Analysis

Degree of relationship among variables

- One (continuous)
  - One (continuous) ——————— Bivariate r ——— Create a linear combination of IVs to optimally predict DV.
  - Multiple (continuous)
    - None — Multiple R
    - Some — Sequential multiple R

- Multiple (continuous) — Multiple (continuous) ——————— Canonical R — Maximally correlate a linear combination of DVs with a linear combination of IVs.

- One (may be repeated) — Multiple (continuous and discrete; cases and IVs are nested) ——— Multilevel modeling — Create linear combinations of DVs and IVs at one level to serve as DVs at another level.

- None — Multiple (discrete) ——— Multiway frequency analysis — Create a log-linear combination of IVs to optimally predict category frequencies.

Significance of group differences

- One (continuous)
  - One (discrete)
    - None — One-way ANOVA or t test
    - Some — One-way ANCOVA
  - Multiple (discrete)
    - None — Factorial ANOVA
    - Some — Factorial ANCOVA

  — Determine reliability of mean group differences.

- Multiple (continuous)
  - One (discrete)
    - None — One-way MANOVA or Hotelling's T^2
    - Some — One-way MANCOVA
  - Multiple (discrete)
    - None — Factorial MANOVA
    - Some — Factorial MANCOVA

  — Create a linear combination of DVs to maximize mean group differences.

- One (continuous) — Multiple (one discrete within S) — Profile analysis of repeated measures

- Multiple (continuous/commensurate) — One (discrete) — Profile analysis

- Multiple (continuous) — Multiple (one discrete within S) — Doubly-multivariate profile analysis

  — Create linear combinations of DVs to maximize mean group differences and differences between levels of within-subjects IVs.

(continued)

TABLE 2.1 Continued

Major Research Question	Number (Kind) of Dependent Variables	Number (Kind) of Independent Variables	Covariates	Analytic Strategy	Goal of Analysis
Prediction of group membership	One (discrete)	Multiple (continuous)	None	One-way discriminant function	Create a linear combination of IVs to maximize group differences.
			Some	Sequential one-way discriminant function	
		Multiple (discrete)		Multiway frequency analysis (logit)	Create a log-linear combination of IVs to optimally predict DV.
		Multiple (continuous and/or discrete)	None	Logistic regression	Create a linear combination of the log of the odds of being in one group.
			Some	Sequential logistic regression	
	Multiple (discrete)	Multiple (continuous)	None	Factorial discriminant function	Create a linear combination of IVs to maximize group differences (DVs).
			Some	Sequential factorial discriminant function	
Structure	Multiple (continuous observed)	Multiple (latent)		Factor analysis (theoretical)	Create linear combinations of observed variables to represent latent variables.
	Multiple (latent)	Multiple (continuous observed)		Principal components (empirical)	
	Multiple (continuous observed and/or latent)	Multiple (continuous observed and/or latent)		Structural equation modeling	Create linear combinations of observed and latent IVs to predict linear combinations of observed and latent DVs.

TABLE 2.1   Continued

Major Research Question	Number (Kind) of Dependent Variables	Number (Kind) of Independent Variables	Covariates	Analytic Strategy	Goal of Analysis
Time course of events	One (time)	None —— None		Survival analysis (life tables)	Determine how long it takes for something to happen.
		One or more	None or some	Survival analysis (with predictors)	Create a linear combination of IVs and CVs to predict time to an event.
	One (continuous)	Time	None or some	Time-series analysis (forecasting)	Predict future course of DV on basis of past course of DV.
		One or more (including time)	None or some	Time-series analysis (intervention)	Determine whether course of DV changes with intervention.

The paths in Table 2.1 are only recommendations concerning an analytic strategy. Researchers frequently discover that they need two or more of these procedures or, even more frequently, a judicious mix of univariate and multivariate procedures to answer fully their research questions. We recommend a flexible approach to data analysis in which both univariate and multivariate procedures are used to clarify the results.

2.4   Technique Chapters

Chapters 5 through 16 and Chapter 18 (online), the basic technique chapters, follow a common format. First, the technique is described and the general purpose briefly discussed. Then the specific kinds of questions that can be answered through application of that technique are listed. Next, both the theoretical and practical limitations of the technique are discussed; this section lists assumptions particularly associated with the technique, describes methods for checking the assumptions for your data set, and gives suggestions for dealing with violations. Then a small hypothetical data set is used to illustrate the statistical development of the procedure. Most of the data sets are deliberately silly and too small to produce significant differences. It is recommended that students follow the matrix calculations using a matrix algebra program available in SPSS, SAS/IML, or a spreadsheet program such as Excel or Quattro. Simple analyses by both computer packages follow.

The next section describes the major types of the techniques, when appropriate. Then some of the most important issues to be considered when using the technique are covered, including special statistical tests, data snooping, and the like.

The next section shows a step-by-step application of the technique to actual data gathered, as described in Appendix B. Because the data sets are real, large, and fully analyzed, this section is often more difficult than the preceding sections. Assumptions are tested and violations dealt with, when necessary. Major hypotheses are evaluated, and follow-up analyses are performed as indicated. Then a Results section is developed, as might be appropriate for submission to a professional journal. The Results section is in APA format; we recommend close attention to the publication manual (APA, 2001) for advice about clarity, simplification of presentation, and the like. These Results sections provide a model for presentation to a fairly sophisticated audience. It is a good idea to discuss the analysis technique and its appropriateness early in the Results section when writing for an audience that is expected to be unfamiliar with the technique. When more than one major type of technique is available, there are additional complete examples using real data. Finally, a detailed comparison of features available in the SPSS, SAS, and SYSTAT programs is made.

In working with these technique chapters, it is suggested that the student/researcher apply the various analyses to some interesting large data set. Many data banks are readily accessible through computer installations.

Further, although we recommend methods of reporting multivariate results, it may be inappropriate to report them fully in all publications. Certainly, one would at least want to mention that univariate results were supported and guided by multivariate inference. But the details associated with a full disclosure of multivariate results at a colloquium, for instance, might require more attention than one could reasonably expect from an audience. Likewise, a full multivariate analysis may be more than some journals are willing to print.

2.5   Preliminary Check of the Data

Before applying any technique, or sometimes even before choosing a technique, you should determine the fit between your data and some very basic assumptions underlying most of the multivariate statistics. Though each technique has specific assumptions as well, most require consideration of material in Chapter 4.

3

Review of Univariate
and Bivariate Statistics

This chapter provides a brief review of univariate and bivariate statistics. Although it is probably too "dense" to be a good source from which to learn, it is hoped that it will serve as a useful reminder of material already mastered and will help in establishing a common vocabulary. Section 3.1 goes over the logic of the statistical hypothesis test, and Sections 3.2, 3.3, and 3.4 skim many topics in analysis of variance and are background for Chapters 6 to 9. Section 3.5 summarizes correlation and regression, which are background for Chapters 5, 12, 14, and 15, and Section 3.6 summarizes chi square which is background for Chapters 10, 14, and 16.

3.1 Hypothesis Testing

Statistics are used to make rational decisions under conditions of uncertainty. Inferences (decisions) are made about populations based on data from samples that contain incomplete information. Different samples taken from the same population probably differ from one another and from the population. Therefore, inferences regarding the population are always a little risky.

The traditional solution to this problem is statistical decision theory. Two hypothetical states of reality are set up, each represented by a probability distribution. Each distribution represents an alternative hypothesis about the true nature of events. Given sample results, a best guess is made as to which distribution the sample was taken from using formalized statistical rules to define "best."

3.1.1 One-Sample z Test as Prototype

Statistical decision theory is most easily illustrated through a one-sample z test, using the standard normal distribution as the model for two hypothetical states of reality. Suppose there is a sample of 25 IQ scores and a need to decide whether this sample of scores is a random sample of a "normal" population with $\mu = 100$ and $\sigma = 15$, or a random sample from a population with $\mu = 108$ and $\sigma = 15$.

First, note that hypotheses are tested about means, not individual scores. Therefore, the distributions representing hypothetical states of reality are distributions of means rather than distributions of individual scores. Distributions of means produce "sampling distributions of means" that differ systematically from distributions of individual scores; the mean of a population distribution, μ, is equal to the mean of a sampling distribution, μ, but the standard deviation of a population of individual scores, σ, is not equal to the standard deviation of a sampling distribution, σ_Y. Sampling distributions

$$\sigma_{\overline{Y}} = \frac{\sigma}{\sqrt{N}} \tag{3.1}$$

have smaller standard deviations than distributions of scores, and the decrease is related to N, the sample size. For the sample, then,

$$\sigma_{\overline{Y}} = \frac{15}{\sqrt{25}} = 3$$

The question being asked, then, is, "Does our mean, taken from a sample of size 25, come from a sampling distribution with $\mu_{\overline{Y}} = 100$ and $\sigma_{\overline{Y}} = 3$ or does it come from a sampling distribution with $\mu_{\overline{Y}} = 108$ and $\sigma_{\overline{Y}} = 3$?" Figure 3.1(a) shows the first sampling distribution, defined as the null hypothesis, H_0, that is, the sampling distribution of means calculated from all possible samples of size 25 taken from a population where $\mu = 100$ and $\sigma = 15$.

The sampling distribution for the null hypothesis has a special, fond place in statistical decision theory because it alone is used to define "best guess." A decision axis for retaining or rejecting H_0 cuts through the distribution so that the probability of rejecting H_0 by mistake is small. "Small" is defined probabilistically as α; an error in rejecting the null hypothesis is referred to as an α, or Type I, error. There is little choice in picking α. Tradition and journal editors decree that it is .05 or smaller, meaning that the null hypothesis is rejected no more than 5% of the time when it is true.

With a table of areas under the standard normal distribution (the table of z scores or standard normal deviates), the decision axis is placed so that the probability of obtaining a sample mean above that point is 5% or less. Looking up 5% in Table C.1, the z corresponding to a 5% cutoff is 1.645 (between 1.64 and 1.65). Notice that the z scale is one of two abscissas in Figure 3.1(a). If the decision axis is placed where $z = 1.645$, one can translate from the z scale to the $\overline{Y}$ scale to properly position the decision axis. The transformation equation is

$$\overline{Y} = \mu + z\sigma_{\overline{Y}} \tag{3.2}$$

Equation 3.2 is a rearrangement of terms from the z test for a single sample:[1]

$$z = \frac{\overline{Y} - \mu}{\sigma_{\overline{Y}}} \tag{3.3}$$

Applying Equation 3.2 to the example,

$$\overline{Y} = 100 + (1.645)(3) = 104.935$$

The null hypothesis that the mean IQ of the sampling distribution is 100 is rejected if the mean IQ of the sample is equal to or greater than 104.935; call it 105.

Frequently, this is as far as the model is taken—the null hypothesis is either retained or rejected. However, if the null hypothesis is rejected, it is rejected in favor of an alternative hypothe-

---

[1]The more usual procedure for testing a hypothesis about a single mean is to solve for z on the basis of the sample mean and standard deviation to see if the sample mean is sufficiently far away from the mean of the sampling distribution under the null hypothesis. If z is 1.645 or larger, the null hypothesis is rejected.

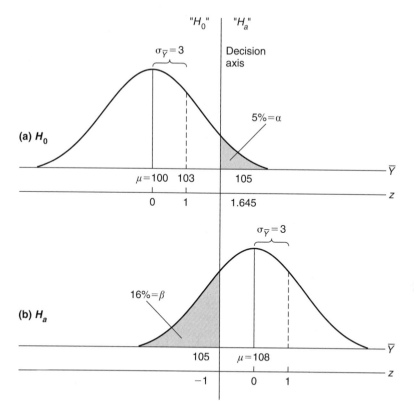

**FIGURE 3.1   Sampling distribution for means with $N = 25$
and $\sigma = 15$ under two hypotheses: (a) H_0: $\mu = 100$ and
(b) H_a: $\mu = 108$.**

sis, H_a. The alternative hypothesis is not always stated explicitly,[2] but when it is, one can evaluate the probability of retaining the null hypothesis when it should be rejected because H_a is true.

   This second type of error is called a β, or Type II, error. Because in the example the μ for H_a is 108, the sampling distribution of means for H_a can be graphed, as shown in Figure 3.1(b). The decision axis is placed with regard to H_0, so we need to find the probability associated with the place it crosses H_a. The first step is to find z corresponding to an IQ score of 105 in a distribution with $\mu_{\overline{Y}} = 108$ and $\sigma_{\overline{Y}} = 3$. Applying Equation 3.3, we find that

$$z = \frac{105 - 108}{3} = -1.00$$

   By looking up $z = -1.00$, about 16% of the time sample means are equal to or less than 105 when the population $\mu = 108$ and the alternative hypothesis is true. Therefore, $\beta = .16$.

---

[2]Often, the alternative hypothesis is simply that the sample is taken from a population that is not equal to the population represented by the null hypothesis. There is no attempt to specify "not equal to."

H_0 and H_a represent alternative realities, only one of which is true. When the researcher is forced to decide whether to retain or reject H_0, four things can happen. If the null hypothesis is true, a correct decision is made if the researcher retains H_0 and an error is made if the researcher rejects it. If the probability of making the wrong decision is α, the probability of making the right decision is $1 - \alpha$. If, on the other hand, H_a is true, the probability of making the right decision by rejecting H_0 is $1 - \beta$, and the probability of making the wrong decision is β. This information is summarized in a "confusion matrix" (aptly named, according to beginning statistics students) showing the probabilities of each of these four outcomes:

		Reality	
		H_0	H_a
Statistical	"H_0"	$1 - \alpha$	β
decision	"H_a"	α	$1 - \beta$
		1.00	1.00

For the example, the probabilities are

		Reality	
		H_0	H_a
Statistical	"H_0"	.95	.16
decision	"H_a"	.05	.84
		1.00	1.00

3.1.2 Power

The lower right-hand cell of the confusion matrix represents the most desirable outcome and the power of the research. Usually, the researcher believes that H_a is true and hopes that the sample data lead to rejection of H_0. Power is the probability of rejecting H_0 when H_a is true. In Figure 3.1(b), power is the portion of the H_a distribution that falls above the decision axis. Many of the choices in designing research are made with an eye toward increasing power because research with low statistical power usually is not worth the effort.

Figure 3.1 and Equations 3.1 and 3.2 suggest some ways to enhance power. One obvious way to increase power is to move the decision axis to the left. However, it cannot be moved far or Type I error rates reach an unacceptable level. Given the choice between .05 and .01 for α error, though, a decision in favor of .05 increases power. A second strategy is to move the curves farther apart by applying a stronger treatment. Other strategies involve decreasing the standard deviation of the sampling distributions either by decreasing variability in scores (e.g., exerting greater experimental control) or by increasing sample size, N.

This model for statistical decisions and these strategies for increasing power generalize to other sampling distributions and to tests of hypotheses other than a single sample mean against a hypothesized population mean.

There is occasionally the danger of *too much power*. The null hypothesis is probably never exactly true and any sample is likely to be slightly different from the population value. With a large enough sample, rejection of H_0 is virtually certain. For that reason, a "minimal meaningful difference" and acceptable effect size should guide the selection of sample size (Kirk, 1995). The sample size should be large enough to be likely to reveal a minimal meaningful difference. Rejection of the null hypothesis may be trivial if the sample is large enough to reveal any difference whatever. This issue is considered further in Section 3.4.

3.1.3 Extensions of the Model

The z test for the difference between a sample mean and a population mean readily extends to a z test of the difference between two sample means. A sampling distribution is generated for the difference between means under the null hypothesis that $\mu_1 = \mu_2$ and is used to position the decision axis. The power of an alternative hypothesis is calculated with reference to the decision axis, just as before.

When population variances are unknown, it is desirable to evaluate the probabilities using Student's t rather than z, even for large samples. Numerical examples of use of t to test differences between two means are available in most univariate statistics books and are not presented here. The logic of the process, however, is identical to that described in Section 3.1.1.

3.1.4 Controversy Surrounding Significance Testing

While the statistical significance test is pervasive in the social sciences, its use is not without controversy. The latest round of arguments against use of statistical significance testing began with an article by Carver in 1978, updated in 1993. In these articles, Carver argues that the significance test, used by itself, does not answer most research questions. These articles, and many others in a rather large literature, are summarized by McLean and Ernest (1998). The significance test, they assert, tells whether the result was likely obtained by chance, but does not convey information about the practical importance of the difference (effect size), the quality of the research design, the reliability and validity of the measures, the fidelity of the treatment, and whether the results are replicable. Thus, a significance test is properly only one among many criteria by which a finding is assessed.

Because of the controversy, the Task Force on Statistical Inference was convened by the American Psychological Association in 1996 and produced a final report in 1999 (Wilkinson et al., 1999). In it, the authors stress the importance of the factors listed above, along with the importance of data screening prior to analysis. Like those who oppose use of statistical significance testing, they urge reporting effect size, and particularly confidence intervals around effect size estimates. We take this recommendation to heart in the chapters that follow and try to provide guidance regarding how that is to be accomplished. Another approach (Cummings and Finch, 2005) involves plots of means with error bars as a way of accomplishing statistical inference by eye. They propose "7 *rules of eye* to guide the inferential use of figures with error bars" (p. 170).

3.2 Analysis of Variance

Analysis of variance is used to compare two or more means to see if there are any statistically significant differences among them. Distributions of scores for three hypothetical samples are provided in Figure 3.2. Analysis of variance evaluates the differences among means relative to the dispersion in the sampling distributions. The null hypothesis is that $\mu_1 = \mu_2 = \cdots = \mu_k$ as estimated from $\overline{Y}_1 = \overline{Y}_2 = \cdots = \overline{Y}_k$, with k equal to the number of means being compared.

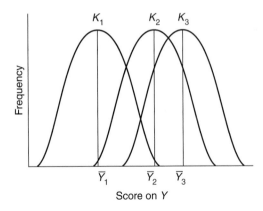

FIGURE 3.2 Idealized frequency distribution of three samples and their means.

Analysis of variance (ANOVA) is really a set of analytic procedures based on a comparison of two estimates of variance. One estimate comes from differences among scores within each group; this estimate is considered random or error variance. The second estimate comes from differences in group means and is considered a reflection of group differences or treatment effects plus error. If these two estimates of variance do not differ appreciably, one concludes that all of the group means come from the same sampling distribution of means, and that the slight differences among them are due to random error. If, on the other hand, the group means differ more than expected, it is concluded that they were drawn from different sampling distributions of means, and the null hypothesis that the means are the same is rejected.

Differences among variances are evaluated as ratios, where the variance associated with differences among sample means is in the numerator, and the variance associated with error is in the denominator. The ratio between these two variances forms an F distribution. F distributions change shape depending on degrees of freedom in both numerator and denominator of the F ratio. Thus, tables of critical F, for testing the null hypothesis, depend on two degree-of-freedom parameters (cf. Appendix C, Table C.3).

The many varieties of analysis of variance are conveniently summarized in terms of the partition of *sums of squares,* that is, sums of squared differences between scores and their means. A sum of squares (SS) is simply the numerator of a variance, S^2.

$$S^2 = \frac{\sum (Y - \overline{Y})^2}{N - 1} \tag{3.4}$$

$$SS = \sum (Y - \overline{Y})^2 \tag{3.5}$$

The square root of variance is standard deviation, S, the measure of variability that is in the metric of the original scores.

$$S = \sqrt{S^2} \tag{3.6}$$

3.2.1   One-Way Between-Subjects ANOVA

DV scores appropriate to one-way between-subjects ANOVA with equal n are presented in a table, with k columns representing groups (levels of the IV) and n scores within each group.[3] Table 3.1 shows how subjects are assigned to groups within this design.

Each column has a mean, $\overline{Y}_j$, where $j = 1, 2,..., k$ levels of treatment. Each score is designated Y_{ij}, where $i = 1, 2,..., n$ scores within each treatment. Each case provides a single score on the DV. The symbol GM represents the grand mean of all scores over all groups.

The difference between each score and the grand mean $(Y_{ij} - \text{GM})$ is considered the sum of two component differences, the difference between the score and its own group mean and the difference between that mean and the overall mean.

$$Y_{ij} - \text{GM} = (Y_{ij} - \overline{Y}_j) + (\overline{Y}_j - \text{GM}) \tag{3.7}$$

This result is achieved by first subtracting and then adding the group mean to the equation. Each term is then squared and summed separately to produce the sum of squares for error and the sum of squares for treatment, respectively. The basic partition holds because, conveniently, the cross-product terms produced by squaring and summing cancel each other out. Across all scores, the partition is

$$\sum_i \sum_j (Y_{ij} - \text{GM})^2 = \sum_i \sum_j (Y_{ij} - \overline{Y})^2 + n \sum_j (\overline{Y} - \text{GM})^2 \tag{3.8}$$

Each of these terms is a sum of squares (SS)—a sum of squared differences between scores (with means sometimes treated as scores) and their associated means. That is, each term is a special case of Equation 3.5.

The term on the left of the equation is the total sum of squared differences between scores and the grand mean, ignoring groups with which scores are associated, designated SS_{total}. The first term on the right is the sum of squared deviations between each score and its group mean. When summed over all groups, it becomes the sum of squares within groups, SS_{wg}. The last term is the sum of

TABLE 3.1   Assignment of Subjects in a One-Way Between-Subjects ANOVA

Treatment		
K_1	K_2	K_3
S_1	S_4	S_7
S_2	S_5	S_8
S_3	S_6	S_9

---

[3]Throughout the book, n is used for sample size within a single group or cell, and N is used for total sample size.

squared deviations between each group mean and the grand mean, the sum of squares between groups, SS_{bg}. Equation 3.8 is also symbolized as

$$SS_{total} = SS_{wg} + SS_{bg} \qquad (3.9)$$

Degrees of freedom in ANOVA partition the same way as sums of squares:

$$df_{total} = df_{wg} + df_{bg} \qquad (3.10)$$

Total degrees of freedom are the number of scores minus 1. The 1 df is lost when the grand mean is estimated. Therefore,

$$df_{total} = N - 1 \qquad (3.11)$$

Within-groups degrees of freedom are the number of scores minus k, lost when the means for each of the k groups are estimated. Therefore,

$$df_{wg} = N - k \qquad (3.12)$$

Between-groups degrees of freedom are k "scores" (each group mean treated as a score) minus 1, lost when the grand mean is estimated, so that

$$df_{bg} = k - 1 \qquad (3.13)$$

Verifying the equality proposed in Equation 3.10, we get

$$N - 1 = N - k + k - 1$$

As in the partition of sums of squares, the term associated with group means is subtracted out of the equation and then added back in.

Another common notation for the partition of Equation 3.7 is

$$SS_{total} = SS_K + SS_{S(K)} \qquad (3.14)$$

as shown in Table 3.2(a). In this notation, the total sum of squares is partitioned into a sum of squares due to the k groups, SS_K, and a sum of squares due to subjects within the groups, $SS_{S(K)}$. (Notice that the order of terms on the right side of the equation is the reverse of that in Equation 3.9.)

TABLE 3.2   Partition of Sums of Squares and Degrees of Freedom for Several ANOVA Designs

(a) One-way between-subjects ANOVA

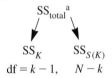

TABLE 3.2 Continued

(b) Factorial between-subjects ANOVA

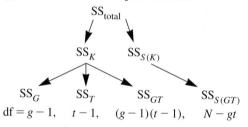

$$\text{df} = g - 1, \quad t - 1, \quad (g-1)(t-1), \quad N - gt$$

(c) One-way within-subjects ANOVA

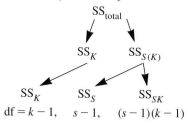

$$\text{df} = k - 1, \quad s - 1, \quad (s-1)(k-1)$$

(d) One-way matched-randomized ANOVA

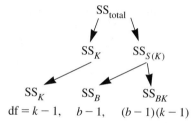

$$\text{df} = k - 1, \quad b - 1, \quad (b-1)(k-1)$$

(e) Factorial within-subjects ANOVA

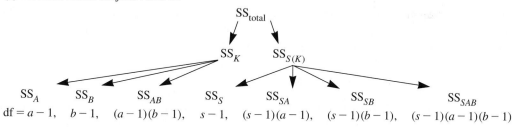

$$\text{df} = a - 1, \quad b - 1, \quad (a-1)(b-1), \quad s - 1, \quad (s-1)(a-1), \quad (s-1)(b-1), \quad (s-1)(a-1)(b-1)$$

(f) Mixed within-between-subjects ANOVA

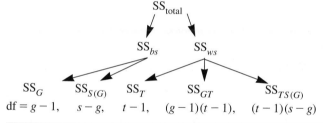

$$\text{df} = g - 1, \quad s - g, \quad t - 1, \quad (g-1)(t-1), \quad (t-1)(s-g)$$

[a]For all SS_{total}, df $= N - 1$.

   The division of a sum of squares by degrees of freedom produces variance, called mean square (MS), in ANOVA. Variance, then, is an "average" sum of squares. ANOVA produces three variances: one associated with total variability among scores, MS_{total}; one associated with variability within groups, MS_{wg} or $MS_{S(K)}$; and one associated with variability between groups, MS_{bg} or MS_K. MS_K and $MS_{S(K)}$ provide the variances for an F ratio to test the null hypothesis that $\mu_1 = \mu_2 = \cdots = \mu_k$.

$$F = \frac{MS_K}{MS_{S(K)}} \qquad df = (k - 1), N - k \qquad (3.15)$$

Once F is computed, it is tested against critical F obtained from a table, such as Table C.3, with numerator $df = k - 1$ and denominator $df = N - k$ at a desired alpha level. If obtained F exceeds critical F, the null hypothesis is rejected in favor of the hypothesis that there is a difference among the means in the k groups.

   Anything that increases obtained F increases power. Power is increased by decreasing the error variability or increasing the sample size in the denominator ($MS_{S(K)}$) or by increasing differences among means in the numerator (MS_K).

3.2.2   Factorial Between-Subjects ANOVA

If groups are formed along more than one dimension, differences among means are attributable to more than one source. Consider an example with six groups, three of women and three of men, in which the DV is scores on a final examination in statistics. One source of variation in means is due to gender, SS_G. If the three groups within each gender are exposed to three different methods of teaching statistics, a second source of differences among means is teaching method, SS_T. The final source of known differences among means is the interaction between gender and teaching methods, SS_{GT}. The interaction tests whether effectiveness of teaching methods varies with gender.

   Allocation of subjects in this design is shown in Table 3.3. Sums of squares and degrees of freedom are partitioned as in Table 3.2(b). Error is estimated by variation in scores within each of the six cells, $SS_{S(GT)}$. Three null hypotheses are tested using the F distribution.

   The first test asks if means for men and women are likely to have come from the same sampling distribution of means. Scores are averaged across teaching methods to eliminate that source of variability. Gender differences are tested in the F ratio:

Rejection of the null hypothesis supports an interpretation of differences between women and men

$$F = \frac{MS_G}{MS_{S(GT)}} \qquad df = (g - 1), N - gt \qquad (3.16)$$

in performance on the final exam.

   The second test asks if means from the three teaching methods are likely to have come from the same sampling distribution of means, averaged across women and men. This is tested as

$$F = \frac{MS_T}{MS_{S(GT)}} \qquad df = (t - 1), N - gt \qquad (3.17)$$

Rejection of the null hypothesis supports an interpretation of differences in effectiveness of the three teaching methods.

TABLE 3.3   Assignment of Subjects in a Factorial Between-Subjects Design

		Teaching Techniques		
		T_1	T_2	T_3
Gender	G_1	S_1 S_2	S_5 S_6	S_9 S_{10}
	G_2	S_3 S_4	S_7 S_8	S_{11} S_{12}

The third test asks if the cell means, the means for women and the means for men within each teaching method, are likely to have come from the same sampling distribution of *differences* between means.

$$F = \frac{MS_{GT}}{MS_{S(GT)}} \qquad df = (g-1)(t-1), N-gt \qquad (3.18)$$

Rejection of the null hypothesis supports an interpretation that men and women differ regarding the most effective teaching methods.

In each case, the estimate of normally occurring variability in test scores, error, is $MS_{S(GT)}$, or within-cell variance. In each case, critical F is read from a table with appropriate degrees of freedom and desired alpha, and if obtained F (Equation 3.16, 3.17, or 3.18) is greater than critical F, the null hypothesis is rejected. When there is an equal number of scores in each cell, the three tests are independent (except for use of a common error term): the test of each main effect (gender and teaching method) is not related to the test of the other main effect or the test of the interaction.

In a one-way between-subjects design (Section 3.2.1), $k-1$ degrees of freedom are used to test the null hypothesis of differences among groups. If k is equal to the number of cells in a two-way design, tests of G, T, and GT use up the $k-1$ degrees of freedom. With proper partitioning, then, of a two-way factorial design, you get three tests for the price of one.

With higher-order between-subjects factorial designs, variation due to differences among groups is partitioned into main effects for each IV, two-way interactions between each pair of IVs, three-way interactions among each trio of IVs, and so on. In any between-subjects factorial design, error sum of squares is the sum of squared differences within each cell of the design.

3.2.3   Within-Subjects ANOVA

In some designs, the means that are tested are derived from the same subjects measured on different occasions, as shown in Table 3.4, rather than from different groups of subjects.[4] In these designs,

---

[4]This design is also called repeated measures, one score per cell, randomized blocks, matched-randomized changeover, or crossover.

TABLE 3.4 Assignment of Subjects in a One-Way Within-Subjects Design

		Treatment	
	K_1	K_2	K_3
	S_1	S_1	S_1
Subjects S_1	S_2	S_2	S_2
S_2	S_3	S_3	S_3
S_3			

computation of sum of squares and mean square for the effect of the IV is the same as for the between-subject designs. However, the error term is further partitioned into individual differences due to subjects, SS_S, and interaction of individual differences with treatment, SS_{SK}. Because subjects are measured repeatedly, their effect as a source of variability in scores is estimated and subtracted from $SS_{S(K)}$, the error term in a corresponding between-subjects design. The interaction of individual differences with treatment, MS_{SK}, is used as the error term:

$$F = \frac{MS_K}{MS_{SK}} \qquad df = (k-1),\ (k-1)(s-1) \qquad (3.19)$$

The partition of sums of squares for a one-way within-subjects design with k levels is shown in Table 3.2(c), where s is the number of subjects.

MS_{SK} is used as the error term because once SS_S is subtracted, no variation is left within cells of the design—there is, in fact, only one score per cell. The interaction of individual differences with treatment is all that remains to serve as an estimate of error variance. If there are individual differences among subjects in scores, and if individuals react similarly to the IV, the interaction is a good estimate of error. Once individual differences are subtracted, the error term is usually smaller than the error term in a corresponding between-subjects design, so the within-subjects design is more sensitive than the between-subjects design.

But if there are no consistent individual differences in scores,[5] or if there is an interaction between subjects and treatment, the error term may be larger than that of a between-subjects design. The statistical test is then conservative, it is more difficult to reject the null hypothesis of no difference between means, and the power of the test is reduced. Because Type I error is unaffected, the statistical test is not in disrepute, but, in this case, a within-subjects design is a poor choice of research design.

A within-subjects analysis is also used with a matched-randomized blocks design, as shown in Table 3.5 and Table 3.2(d). Subjects are first matched on the basis of variables thought to be highly

---

[5]Notice that the degrees of freedom for error, $(k-1)(s-1)$, are fewer than in the between-subjects design. Unless the reduction in error variance due to subtraction of SS_S is substantial, the loss of degrees of freedom may overcome the gain due to smaller SS when MS_{SK} is computed.

related to the DV. Subjects are then divided into blocks, with as many subjects within each block as there are levels of the IV. Finally, members of each block are randomly assigned to levels of the IV. Although the subjects in each block are actually different people, they are treated statistically as if they were the same person. In the analysis, there is a test of the IV, and a test of blocks (the same as the test of subjects in the within-subjects design), with the interaction of blocks and treatment used as the error term. Because matching is used to produce consistency in performance within blocks and the effect of blocks is subtracted from the error term, this design should also be more sensitive than the between-subjects design. It will not be, however, if the matching fails.

Factorial within-subjects designs, as shown in Table 3.6, are an extension of the one-way within-subjects design. The partition of a two-way within-subjects design is in Table 3.2(e).

In the analysis, the error sum of squares is partitioned into a number of "subjects-by-effects" interactions just as the sum of squares for effects is partitioned into numerous sources. It is common (though not universal) to develop a separate error term for each F test; for instance, the test of the main effect of A is

$$F = \frac{MS_A}{MS_{SA}} \qquad df = (a-1), (a-1)(s-1) \tag{3.20}$$

For the main effect of B, the test is

$$F = \frac{MS_B}{MS_{SB}} \qquad df = (b-1), (b-1)(s-1) \tag{3.21}$$

and for the interaction,

$$F = \frac{MS_{AB}}{MS_{SAB}} \qquad df = (a-1)(b-1), (a-1)(b-1)(s-1) \tag{3.22}$$

For higher-order factorial designs, the partition into sources of variability grows prodigiously, with an error term developed for each main effect and interaction tested.

TABLE 3.5 Assignment of Subjects in a Matched-Randomized Design[a]

		Treatment		
		A_1	A_2	A_3
	B_1	S_1	S_2	S_3
Blocks	B_2	S_4	S_5	S_6
	B_3	S_7	S_8	S_9

[a]Where subjects in the same block have been matched on some relevant variable.

TABLE 3.6 Assignment of Subjects in a Factorial Within-Subjects Design

		Treatment A		
		A_1	A_2	A_3
	B_1	S_1	S_1	S_1
		S_2	S_2	S_2
Treatment B				
	B_2	S_1	S_1	S_1
		S_2	S_2	S_2

There is controversy in within-subject analyses concerning conservatism of the *F* tests and whether separate error terms should be used. In addition, if the repeated measurements are on single subjects, there are often carry-over effects that limit generalizability to situations where subjects are tested repeatedly. Finally, when there are more than two levels of the IV, the analysis has the assumption of sphericity. One component of sphericity—homogeneity of covariance—roughly speaking, is the assumption that subjects "line up" in scores the same for all pairs of levels of the IV. If some pairs of levels are close in time (e.g., trial 2 and trial 3) and other pairs are distant in time (e.g., trial 1 and trial 10), the assumption is often violated. Such violation is serious because Type I error rate is affected. Sphericity is discussed in greater detail in Chapters 7 and, especially, 8, and in Tabachnick and Fidell (2006) and Frane (1980).

For these reasons, within-subjects ANOVA is sometimes replaced by profile analysis, where repetitions of DVs are transformed into separate DVs (Chapter 8) and a multivariate statistical test is used.

3.2.4   Mixed Between-Within-Subjects ANOVA[6]

Often in factorial designs, one or more IVs are measured between subjects, whereas other IVs are measured within subjects.[7] The simplest example of the mixed between-within-subjects design involves one between-subjects and one within-subjects IV, as shown in Table 3.7.[8]

To show the partition, the total SS is divided into a source attributable to the between-subjects part of the design (Groups), and a source attributable to the within-subjects part (Trials), as shown in Table 3.2(f). Each source is then further partitioned into effects and error components: between-subjects into groups and subjects-within-groups error term; and within-subjects into trials, the group-by-trials interaction; and the trial-by-subjects-within-groups error term.

TABLE 3.7   Assignment of Subjects in a Between-Within-Subjects Design

		Trials		
		T_1	T_2	T_3
Groups	G_1	S_1 S_2	S_1 S_2	S_1 S_2
	G_2	S_3 S_4	S_3 S_4	S_3 S_4

[6]This design is also called a split-plot, repeated-measures, or randomized-block factorial design.

[7]Mixed designs can also have "blocks" rather than repeated measures on individual subjects as the within-subjects segment of the design.

[8]When the between-subjects variables are based on naturally occurring differences among subjects (e.g., age, sex), the design is said to be "blocked" on the subject variables. This is a different use of the term *blocks* from that of the preceding section. In a mixed design, both kinds of blocking can occur.

As more between-subjects IVs are added, between-subjects main effects and interactions expand the between-subjects part of the partition. For all the between-subjects effects, there is a single error term consisting of variance among subjects confined to each combination of the between-subjects IVs. As more within-subjects IVs are added, the within-subjects portion of the design expands. Sources of variability include main effects and interactions of within-subjects IVs and interactions of between- and within-subjects IVs. Separate error terms are developed for each source of variability in the within-subjects segment of the design.[9]

Problems associated with within-subjects designs (e.g., homogeneity of covariance) carry over to mixed designs, and profile analysis is sometimes used to circumvent some of these problems.

3.2.5   Design Complexity

Discussion of analysis of variance has so far been limited to factorial designs where there are equal numbers of scores in each cell, and levels of each IV are purposely chosen by the researcher. Several deviations from these straightforward designs are possible. A few of the more common types of design complexity are mentioned below, but the reader actually faced with use of these designs is referred to one of the more complete analysis of variance texts such as Brown et al. (1991), Keppel & Wickens (2004), Myers and Well (2002), and Tabachnick and Fidell (2007).

3.2.5.1   *Nesting*

In between-subjects designs, subjects are said to be nested within levels of the IV. That is, each subject is confined to only one level of each IV or combination of IVs. Nesting also occurs with IVs when levels of one IV are confined to only one level of another IV, rather than factorially crossing over the levels of the other IV.

Take the example where the IV is various levels of teaching methods. Children within the same classroom cannot be randomly assigned to different methods but whole classrooms can be so assigned. The design is one-way between-subjects where teaching methods is the IV and classrooms serve as subjects. For each classroom, the mean score for all children on the test is obtained, and the means serve as DVs in one-way ANOVA.

If the effect of classroom is also assessed, the design is nested or hierarchical, as shown in Table 3.8(a). Classrooms are randomly assigned to and nested in teaching methods, and children are nested in classrooms. The error term for the test of classroom is subjects within classrooms and teaching method, and the error term for the test of teaching method is classrooms within teaching technique. Nested models also are analyzed through multilevel modeling (Chapter 15).

3.2.5.2   *Latin-Square Designs*

The order of presentation of levels of an IV often produces differences in the DV. In within-subjects designs, subjects become practiced or fatigued or experiment wise as they experience more levels of the IV. In between-subjects designs, there are often time-of-day or experimenter effects that change

---

[9]"Subjects" are no longer available as a source of variance for analysis. Because subjects are confined to levels of the between-subjects variable(s), differences between subjects in each group are used to estimate error for testing variance associated with between-subjects variables.

TABLE 3.8   **Some Complex ANOVA Designs**

(a) Nested designs				(b) Latin-square designs[a]			

	Teaching Techniques					**Order**	

T_1	T_2	T_3			*1*	*2*	*3*
Classroom 1	Classroom 2	Classroom 3		S_1	A_2	A_1	A_3
Classroom 4	Classroom 5	Classroom 6	**Subjects**	S_2	A_1	A_3	A_2
Classroom 7	Classroom 8	Classroom 9		S_3	A_3	A_2	A_1

[a]Where the three levels of treatment *A* are experienced by different subjects in different orders, as indicated.

scores on the DV. To get an uncontaminated look at the effects of the IV, it is important to counter-balance the effects of increasing experience, time of day, and the like, so that they are independent of levels of the IV. If the within-subjects IV is something like trials, counterbalancing is not possible because the order of trials cannot be changed. But when the IV is something like background color of slide used to determine if background color affects memory for material on the slide, a Latin-square arrangement is often used to control order effects.

A Latin-square design is shown in Table 3.8(b). If A_1 is a yellow background, A_2 a blue background, and A_3 a red background, then subjects are presented the slides in the order specified by the Latin square. The first subject is presented the slide with the blue background, then yellow, then red. The second subject is presented with yellow, then red, then blue, and so on. The yellow slide (A_1) appears once in first position, once in second, and once in third, and so on for the other colors, so that order effects are distributed evenly across the levels of the IV.

The simple design of Table 3.8(b) produces a test of the IV (*A*), a test of subjects (if desired), and a test of order. The effect of order (like the effect of subjects) is subtracted out of the error term, leaving it smaller than it is when order effects are not analyzed. The error term itself is composed of fragments of interactions that are not available for analysis because effects are not fully crossed in the design. Thus, the design is more sensitive than a comparable between-subjects design when there are order effects and no interactions and less sensitive when there are no order effects but there are interactions. Consult Tabachnick and Fidell (2007) or one of the other ANOVA texts for greater detail on this fascinating topic.

3.2.5.3   *Unequal* n *and Nonorthogonality*

In a simple one-way between-subjects ANOVA, problems created by unequal group sizes are relatively minor. Computation is slightly more difficult, but that is no real disaster, especially if analysis is by computer. However, as group sizes become more discrepant, the assumption of homogeneity of variance is more important. If the group with the smaller *n* has the larger variance, the *F* test is too liberal, leading to increased Type I error rate and an inflated alpha level.

In factorial designs with more than one between-subjects IV, unequal sample sizes in each cell create difficulty in computation and ambiguity of results. With unequal *n,* a factorial design is nonorthogonal. Hypotheses about main effects and interactions are not independent, and sums of squares are not additive. The various sources of variability contain overlapping variance, and the same variance can be attributed to more than one source, as discussed in Chapter 1. If effects are tested without taking the overlap into account, the probability of a Type I error increases because systematic variance contributes to more than one test. A variety of strategies are available to deal with the problem, none of them completely satisfactory.

The simplest strategy is to randomly delete cases from cells with greater n until all cells are equal. If unequal n is due to random loss of a few subjects in an experimental design originally set up for equal n, deletion is often a good choice. An alternative strategy with random loss of subjects in an experimental design is an unweighted-means analysis, described in Chapter 6 and ANOVA textbooks such as Tabachnick and Fidell (2007). The unweighted-means approach has greater power than random deletion of cases and is the preferred approach as long as computational aids are available.

But in nonexperimental work, unequal n often results from the nature of the population. Differences in sample sizes reflect true differences in numbers of various types of subjects. To artificially equalize n is to distort the differences and lose generalizability. In these situations, decisions are made as to how tests of effects are to be adjusted for overlapping variance. Standard methods for adjusting tests of effects with unequal n are discussed in Chapter 6.

3.2.5.4 *Fixed and Random Effects*

In all the ANOVA designs discussed so far, levels of each IV are selected by the researchers on the basis of their interest in testing significance of the IV. This is the usual fixed-effects model. Sometimes, however, there is a desire to generalize to a population of levels of an IV. In order to generalize to the population of levels of the IVs, a number of levels are randomly selected from the population, just as subjects are randomly selected from the population of subjects when the desire is to generalize results to the population of subjects. Consider, for example, an experiment to study effects of word familiarity[10] on recall where the desire is to generalize results to all levels of word familiarity. A finite set of familiarity levels is randomly selected from the population of familiarity levels. Word familiarity is considered a random-effects IV.

The analysis is set up so that results generalize to levels other than those selected for the experiment—generalize to the population of levels from which the sample was selected. During analysis, alternative error terms for evaluating the statistical significance of random-effects IVs are used. Although computer programs are available for analysis of random-effects IVs, use of them is fairly rare. The interested reader is referred to one of the more sophisticated ANOVA texts, such as Tabachnick and Fidell (2007) or Brown et al. (1991), for a full discussion of the random-effects model.

3.2.6 Specific Comparisons

When an IV has more than one degree of freedom (more than two levels) or when there is an interaction between two or more IVs, the overall test of the effect is ambiguous. The overall test, with $k - 1$ degrees of freedom, is pooled over $k - 1$ single-degree-of-freedom subtests. If the overall test is significant, so usually are one or more of the subtests, but there is no way to tell which one(s). To find out which single-degree-of-freedom subtests are significant, comparisons are performed.

---

[10]Word familiarity is usually operationalized by frequency of usage of words in the English language.

In analysis, degrees of freedom are best thought of as a nonrenewable resource. They are analyzed once with conventional alpha levels, but further analyses require very stringent alpha levels. For this reason, the best strategy is to plan the analysis very carefully so that the most interesting comparisons are tested with conventional alpha levels. Unexpected findings or less interesting effects are tested later with stringent alpha levels. This is the strategy used by the researcher who has been working in an area for a while and knows precisely what to look for.

Regrettably, research is often more tentative; so the researcher "spends" the degrees of freedom on omnibus (routine) ANOVA testing main effects and interactions at conventional alpha levels and then snoops the single-degree-of-freedom comparisons of significant effects at stringent alpha levels. Snooping through data after results of ANOVA are known is called *conducting post hoc comparisons*.

We present here the most flexible method of conducting comparisons, with mention of other methods as they are appropriate. The procedure for conducting comparisons is the same for planned and post hoc comparisons up to the point where an obtained F is evaluated against a critical F.

3.2.6.1 Weighting Coefficients for Comparisons

Comparison of treatment means begins by assigning a weighting factor (w) to each of the cell or marginal means so the weights reflect your null hypotheses. Suppose you have a one-way design with k means and you want to make comparisons. For each comparison, a weight is assigned to each mean. Weights of zero are assigned to means (groups) that are left out of a comparison, although at least two of the means must have nonzero weights. Means that are contrasted with each other are assigned weights with opposite signs (positive or negative) with the constraint that the weights sum to zero, that is,

$$\sum_{j=1}^{k} w_j = 0$$

For example, consider an IV with four levels, producing $\overline{Y}_1$, $\overline{Y}_2$, $\overline{Y}_3$, and $\overline{Y}_4$. If you want to test the hypothesis that $\mu_1 - \mu_3 = 0$, weighting coefficients are 1, 0, −1, 0, producing $1\overline{Y}_1 + 0\overline{Y}_2 + (-1)\overline{Y}_3 + 0\overline{Y}_4$. $\overline{Y}_2$ and $\overline{Y}_4$ are left out while $\overline{Y}_1$ is compared with $\overline{Y}_3$. Or if you want to test the null hypothesis that $(\mu_1 + \mu_2)/2 - \mu_3 = 0$ (to compare the average of means from the first two groups with the mean of the third group leaving out the fourth group), weighting coefficients are $1/2$, $1/2$, −1, 0 (or any multiple of them, such as 1, 1, −2, 0), respectively. Or if you want to test the null hypothesis that $(\mu_1 + \mu_2)/2 - (\mu_3 + \mu_4)/2 = 0$ (to compare the average mean of the first two groups with the average mean of the last two groups), the weighting coefficients are $1/2, 1/2, -1/2, -1/2$ (or 1, 1, −1, −1).

The idea behind the test is that the sum of the weighted means is equal to zero when the null hypothesis is true. The more the sum diverges from zero, the greater the confidence with which the null hypothesis is rejected.

3.2.6.2 Orthogonality of Weighting Coefficients

In a design with an equal number of cases in each group, any pair of comparisons is orthogonal if the sum of the cross-products of the weights for the two comparisons is equal to zero. For example, in the following three comparisons,

	w_1	w_2	w_3
Comparison 1	1	−1	0
Comparison 2	1/2	1/2	−1
Comparison 3	1	0	−1

the sum of the cross-products of weights for comparison 1 and comparison 2 is

$$(1)(1/2) + (-1)(1/2) + (0)(-1) = 0$$

Therefore, the two comparisons are orthogonal.

Comparison 3, however, is orthogonal to neither of the first two comparisons. For instance, checking it against comparison 1,

$$(1)(1) + (-1)(0) + (0)(-1) = 1$$

In general, there are as many orthogonal comparisons as there are degrees of freedom. Because $k = 3$ in the example, df $= 2$. There are only two orthogonal comparisons when there are three levels of an IV, and only three orthogonal comparisons when there are four levels of an IV.

There are advantages to use of orthogonal comparisons, if they suit the needs of the research. First, there are only as many of them as there are degrees of freedom, so the temptation to "overspend" degrees of freedom is avoided. Second, orthogonal comparisons analyze nonoverlapping variance. If one of them is significant, it has no bearing on the significance of another of them. Last, because they are independent, if all $k - 1$ orthogonal comparisons are performed, the sum of the sum of squares for the comparisons is the same as the sum of squares for the IV in omnibus ANOVA. That is, the sum of squares for the effect has been completely broken down into the $k - 1$ orthogonal comparisons that comprise it.

3.2.6.3 Obtained F for Comparisons

Once the weighting coefficients are chosen, the following equation is used to obtain F for the comparison if sample sizes are equal in each group:

$$F = \frac{n_c (\sum w_j \overline{Y}_j)^2 / \sum w_j^2}{\text{MS}_{\text{error}}} \tag{3.23}$$

where n_c = the number of scores in each of the means to be compared,

$(\sum w_j \overline{Y}_j)^2$ = the squared sum of the weighted means,

$\sum w_j^2$ = the sum of the squared coefficients,

MS_{error} = the mean square for error in the ANOVA.

The numerator of Equation 3.23 is both the sum of squares and the mean square for the comparison because a comparison has only one degree of freedom.

For factorial designs, comparisons are done on either marginal or cell means, corresponding to comparisons on main effects and interactions, respectively. The number of scores per mean and the error term follow from the ANOVA design used. However, if comparisons are made on within-subjects effects, it is customary to develop a separate error term for each comparison, just as separate error terms are developed for omnibus tests of within-subjects IVs.

Chapter 6 has much more information on comparisons of both main effects and interactions, including syntax for performing them through some of the more popular computer programs.

Once you have obtained F for a comparison, whether by hand calculation or computer, the obtained F is compared with critical F to see if it is statistically reliable. If obtained F exceeds critical F, the null hypothesis for the comparison is rejected. But which critical F is used depends on whether the comparison is planned or performed post hoc.

3.2.6.4   Critical F for Planned Comparisons

If you are in the enviable position of having planned your comparisons prior to data collection, and if you have planned no more of them than you have degrees of freedom for effect, critical F is obtained from the tables just as in routine ANOVA. Each comparison is tested against critical F at routine alpha with one degree of freedom in the numerator and degrees of freedom associated with the MS_{error} in the denominator. If obtained F is larger than critical F, the null hypothesis represented by the weighting coefficients is rejected.

With planned comparisons, omnibus ANOVA is not performed;[11] the researcher moves straight to comparisons. Once the degrees of freedom are spent on the planned comparisons, however, it is perfectly acceptable to snoop the data at more stringent alpha levels (Section 3.2.6.5), including main effects and interactions from omnibus ANOVA if they are appropriate.

Sometimes, however, the researcher cannot resist the temptation to plan more comparisons than degrees of freedom for effect. When there are too many tests, even if comparisons are planned, the α level across all tests exceeds the α level for any one test and some adjustment of α for each test is needed. It is common practice to use a Bonferroni-type adjustment where slightly more stringent α levels are used with each test to keep α across all tests at reasonable levels. For instance, when 5 comparisons are planned, if each one of them is tested at $\alpha = .01$, the alpha across all tests is an acceptable .05 (roughly .01 times 5, the number of tests). However, if 5 comparisons are each tested at $\alpha = .05$, the alpha across all tests is approximately .25 (roughly .05 times 5)—unacceptable by most standards.

If you want to keep overall α at, say, .10, and you have 5 tests to perform, you can assign each of them $\alpha = .02$, or you can assign two of them $\alpha = .04$ with the other three evaluated at $\alpha = .01$, for an overall Type I error rate of roughly .11. The decision about how to apportion α through the tests is also made prior to data collection.

As an aside, it is important to realize that routine ANOVA designs with numerous main effects and interactions suffer from the same problem of inflated Type I error rate across all tests as planned comparisons where there are too many tests. Some adjustment of alpha for separate tests is needed in big ANOVA problems as well, if all effects are evaluated even if the tests are planned.

---

[11]You might perform routine ANOVA to compute the error term(s).

3.2.6.5   Critical F for Post Hoc Comparisons

If you are unable to plan your comparisons and choose to start with routine ANOVA instead, you want to follow up significant main effects (with more than two levels) and interactions with post hoc comparisons to find the treatments that are different from one another. Post hoc comparisons are needed to provide adjustment to α level because of two considerations. The first is that you have already spent your degrees of freedom, and your "cheap" α level, conducting routine ANOVA; therefore, you run into rapidly increasing overall error rates if you conduct additional analyses without adjustment. Second, you have already seen the means, and it is much easier to identify comparisons that are likely to be significant. These mean differences, however, may have come from chance fluctuations in the data unless you have some theoretical reason to believe they are real—and if you believe they are real, you should plan to test them ahead of time.

Many procedures for dealing with an inflated Type I error rate are available as described in standard ANOVA texts such as Tabachnick and Fidell (2007). The tests differ in the number and type of comparisons they permit and the amount of adjustment required of α. The tests that permit more numerous comparisons have correspondingly more stringent adjustments to critical F. For instance, the Dunnett test, which compares the mean from a single control group with each of the means of the other groups, in turn, has a less stringent correction than the Tukey test, which allows all pairwise comparisons of means. The name of this game is to choose the most liberal test that permits you to perform the comparisons of interest.

The test described here (Scheffé, 1953) is the most conservative and most flexible of the popular methods. Once critical F is computed with the Scheffé adjustment, there is no limit to the number and complexity of comparisons that can be performed. You can perform all pairwise comparisons and all combinations of treatment means pooled and contrasted with other treatment means, pooled or not, as desired. Some possibilities for pooling are illustrated in Section 3.2.6.1. Once you pay the "price" in conservatism for this flexibility, you might as well conduct all the comparisons that make sense, given your research design.

The Scheffé method for computing critical F for a comparison on marginal means is

$$F_s = (k - 1)F_c \tag{3.24}$$

where F_s is adjusted critical F, $(k - 1)$ is degrees of freedom for the effect, and F_c is tabled F with $k - 1$ degrees of freedom in the numerator and degrees of freedom associated with the error term in the denominator.

If obtained F is larger than critical F_s, the null hypothesis represented by the weighting coefficients for the comparison is rejected. (See Chapter 8 for a more extended discussion of the appropriate correction.)

3.3   Parameter Estimation

If a statistically significant difference among means is found, one is usually interested in reporting the likely population value for each mean. Since sample means are unbiased estimators of population means, the best guess about the size of a population mean (μ) is the mean of the sample randomly selected from that population. In most reports of research, therefore, sample mean values are reported along with statistical results.

Sample means are only estimations of population means. They are unbiased because they are systematically neither large nor small, but they are rarely precisely at the population value—and there is no way to know when they are. Thus, the error in estimation, the familiar confidence interval of introductory statistics, is often reported along with the estimated means. The size of the confidence interval depends on sample size, the estimation of population variability, and the degree of confidence one wishes to have in estimating μ. Alternatively, cell standard deviations or standard errors are presented along with sample means so that the reader can compute the confidence interval if it is desired.

3.4 Effect Size[12]

Although significance testing, comparisons, and parameter estimation help illuminate the nature of group differences, they do not assess the degree to which the IV(s) and DV are related. It is important to assess the degree of relationship to avoid publicizing trivial results as though they had practical utility. As discussed in Section 3.1.2, overly powerful research sometimes produces results that are statistically significant but realistically meaningless.

Effect size reflects the proportion of variance in the DV that is associated with levels of an IV. It assesses the amount of total variance in the DV that is predictable from knowledge of the levels of the IV. If the total variances of the DV and the IV are represented by circles as in a Venn diagram, effect size assesses the degree of overlap of the circles. Statistical significance testing assesses the *reliability* of the association between the IV and DV. Effect size measures *how much* association there is.

A rough estimate of effect size is available for any ANOVA through η^2 (eta squared).

$$\eta^2 = \frac{SS_{effect}}{SS_{total}} \tag{3.25}$$

When there are two levels of the IV, η^2 is the (squared) point biserial correlation between the continuous variable (the DV) and the dichotomous variable (the two levels of the IV).[13] After finding a significant main effect or interaction, η^2 shows the proportion of variance in the DV (SS_{total}) attributable to the effect (SS_{effect}). In a balanced, equal-n design, η^2s are additive; the sum of η^2 for all significant effects is the proportion of variation in the DV that is predictable from knowledge of the IVs.

This simple popular measure of effect size is flawed for two reasons. The first is that η^2 for a particular IV depends on the number and significance of other IVs in the design. η^2 for an IV tested

---

[12]This is also called *strength of association* or *treatment magnitude.*

[13]All effect size values are associated with the particular levels of the IV used in the research and do not generalize to other levels.

in a one-way design is likely to be larger than η^2 for the same IV in a two-way design where the other IV and the interaction increase the size of the total variance, especially if one or both of the additional effects are large. This is because the denominator of η^2 contains systematic variance for other effects in addition to error variance and systematic variance for the effect of interest.

Therefore, an alternative form of η^2, called partial η^2, is available in which the denominator contains only variance attributable to the effect of interest plus error.

$$\text{Partial } \eta^2 = \frac{\text{SS}_{\text{effect}}}{\text{SS}_{\text{effect}} + \text{SS}_{\text{error}}} \tag{3.26}$$

With this alternative, η^2s for all significant effects in the design do not sum to proportion of systematic variance in the DV. Indeed, the sum is sometimes greater than 1.00. It is imperative, therefore, to be clear in your report when this version of η^2 is used.

A second flaw is that η^2 describes proportion of systematic variance in a sample with no attempt to estimate proportion of systematic variance in the population. A statistic developed to estimate effect size between IV and DV in the population is $\hat{\omega}^2$ (omega squared).

$$\hat{\omega}^2 = \frac{\text{SS}_{\text{effect}} - (\text{df}_{\text{effect}})(\text{MS}_{\text{error}})}{\text{SS}_{\text{total}} + \text{MS}_{\text{error}}} \tag{3.27}$$

This is the additive form of $\hat{\omega}^2$, where the denominator represents total variance, not just variance due to effect plus error, and *is limited to between-subjects analysis of variance designs with equal sample sizes in all cells.* Forms of $\hat{\omega}^2$ are available for designs containing repeated measures (or randomized blocks), as described by Vaughn and Corballis (1969).

A separate measure of effect size is computed and reported for each main effect and interaction of interest in a design. Confidence intervals also may be developed around effect sizes using recent software (Smithson, 2003; Steiger and Fouladi, 1992). These are demonstrated in subsequent chapters.

Effect sizes described can range from 0 to 1 because they are proportions of variance. Another type of effect size is Cohen's *d,* basically a difference between standardized means (i.e., means divided by their common standard deviation). That measure becomes less convenient in multivariate designs in which comparisons are more complex than simply the difference between two means. Further, Cohen (1988) shows equations for converting *d* to η^2. Therefore, the measures described in this book are based on η^2.

A frequent question is "Do I have (or expect to find) a big effect?" The answer to this question depends on the research area and type of study. Simple experiments typically account for less variance than do nonexperimental studies (nature generally exhibits more control over people than we do in our roles as experimenters). Clinical/personality/social psychology and education tend to have smaller effects than found in sociology, economics, and perception/physiological psychology. Cohen (1988) has presented some guidelines for small ($\eta^2 = .01$), medium ($\eta^2 = .09$), and large ($\eta^2 = .25$) effects. These guidelines apply to experiments and social/clinical areas of psychology; larger values could be expected for nonexperimental research, sociology, and the more physiological aspects of psychology.

3.5 Bivariate Statistics: Correlation and Regression

Effect size as described in Section 3.4 is assessed between a continuous DV and discrete levels of an IV. Frequently, however, a researcher wants to measure the effect size between two continuous variables where the IV–DV distinction is blurred. For instance, the association between years of education and income is of interest even though neither is manipulated and inferences regarding causality are not possible. Correlation is the measure of the size and direction of the linear relationship between the two variables, and squared correlation is the measure of strength of association between them.

Correlation is used to measure the association between variables; regression is used to predict one variable from the other (or many others). However, the equations for correlation and bivariate regression are very similar, as indicated in what follows.

3.5.1 Correlation

The Pearson product-moment correlation coefficient, r, is easily the most frequently used measure of association and the basis of many multivariate calculations. The most interpretable equation for Pearson r is

$$r = \frac{\sum Z_X Z_Y}{N - 1} \tag{3.28}$$

where Pearson r is the average cross-product of standardized X and Y variable scores.

$$Z_Y = \frac{Y - \overline{Y}}{S} \quad \text{and} \quad Z_X = \frac{X - \overline{X}}{S}$$

and S is as defined in Equations 3.4 and 3.6.

Pearson r is independent of scale of measurement (because both X and Y scores are converted to standard scores) and independent of sample size (because of division by N). The value of r ranges between 1.00 and +1.00, where values close to .00 represent no linear relationship or predictability between the X and Y variables. An r value of $+1.00$ or -1.00 indicates perfect predictability of one score when the other is known. When correlation is perfect, scores for all subjects in the X distribution have the same relative positions as corresponding scores in the Y distribution.[14]

The raw score form of Equation 3.28 also sheds light on the meaning of r:

$$r = \frac{N \sum XY - (\sum X)(\sum Y)}{\sqrt{[N \sum X^2 - (\sum X)^2][N \sum Y^2 - (\sum Y)^2]}} \tag{3.29}$$

---

[14]When correlation is perfect, $Z_X = Z_Y$ for each pair, and the numerator of Equation 3.28 is, in effect, $\sum Z_X Z_X$. Because $\sum Z_X^2 = N - 1$, Equation 3.28 reduces to $(N - 1)/(N - 1)$, or 1.00.

Pearson r is the covariance between X and Y relative to (the square root of the product of) the X and Y variances. Only the numerators of variance and covariance equations appear in Equation 3.29 because the denominators cancel each other out.

3.5.2   Regression

Whereas correlation is used to measure the size and direction of the linear relationship between two variables, regression is used to predict a score on one variable from a score on the other. In bivariate (two-variable) regression (simple linear regression) where Y is predicted from X, a straight line between the two variables is found. The best-fitting straight line goes through the means of X and Y and minimizes the sum of the squared distances between the data points and the line.

To find the best-fitting straight line, an equation is solved of the form

$$Y' = A + BX \tag{3.30}$$

where Y' is the predicted score, A is the value of Y when X is 0.00, B is the slope of the line (change in Y divided by change in X), and X is the value from which Y is to be predicted.

The difference between the predicted and the observed values of Y at each value of X represents errors of prediction or residuals. The best-fitting straight line is the line that minimizes the squared errors of prediction.

To solve Equation 3.30, both B and A are found.

$$B = \frac{N \sum XY - (\sum X)(\sum Y)}{N \sum X^2 - (\sum X)^2} \tag{3.31}$$

The bivariate regression coefficient, B, is the ratio of the covariance of the variables and the variance of the one from which predictions are made.

Note the differences and similarities between Equation 3.29 (for correlation) and Equation 3.31 (for the regression coefficient). Both have the covariance between the variables as a numerator but differ in denominator. In correlation, the variances of both are used in the denominator. In regression, the variance of the predictor variable serves as the denominator; if Y is predicted from X, X variance is the denominator, whereas if X is predicted from Y, Y variance is the denominator. To complete the solution, the value of the intercept, A, is also calculated.

$$A = \overline{Y} - B\overline{X} \tag{3.32}$$

The intercept is the mean of the observed value of the predicted variable minus the product of the regression coefficient times the mean of the predictor variable.

Figure 3.3 illustrates many of the relationships among slopes, intercepts, predicted scores, and residuals.

The intercept for the small data set is 2.16; at a value of zero on the X axis, the regression line crosses the Y axis at 2.16. The slope is .60; when the value on the X axis increases by 1 unit, the value on the Y axis goes up .60 units. The equation for the predicted Y score (Equation 3.30 above) is

X	Y
3.00	2.00
5.00	4.00
2.00	5.00
6.00	4.00
8.00	7.00
4.00	5.00
9.00	8.00
3.00	4.00
5.00	6.00
6.00	7.00

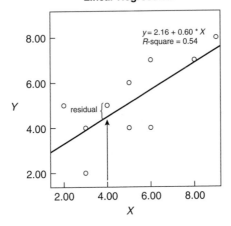

Linear Regression

FIGURE 3.3   Slopes, intercepts, predicted scores, and residuals in bivariate regression; figure created using SPSS graphs.

$Y' = 2.16 + .60X$. For an X score of 4, the predicted Y score is [2.16 + .60(4) =] 4.56. This is the value indicated by the up arrow in the figure. But 5 is the actual Y score associated with an X score of 4 in the data set. The residual, then, for the X score of 4 is the difference between the actual Y score and the predicted Y score ($5 - 4.56 = .44$). The bracket indicates the residual for this value in the data set. Linear regression minimizes the squared residuals across the whole data set.

3.6   Chi-Square Analysis

Analysis of variance examines the relationship between a discrete variable (the IV) and a continuous variable (the DV), correlation and regression examine the relationship between two continuous variables, and the chi-square (χ^2) test of independence is used to examine the relationship between two discrete variables. If, for instance, one wants to examine a potential relationship between region of the country (Northeast, Southeast, Midwest, South, and West) and approval versus disapproval of current political leadership, χ^2 is the appropriate analysis.

In χ^2 analysis, the null hypothesis generates expected frequencies against which observed frequencies are tested. If the observed frequencies are similar to the expected frequencies, then the value of χ^2 is small and the null hypothesis is retained; if they are sufficiently different, then the value of χ^2 is large and the null hypothesis is rejected. The relationship between the size of χ^2 and the difference in observed and expected frequencies can be seen readily from the computational equation for χ^2 (Equation 3.33), which follows.

$$\sum_{ij} (f_o - F_e)^2 / F_e \qquad\qquad (3.33)$$

where f_o represents observed frequencies, and F_e represents the expected frequencies in each cell. Summation is over all the cells in a two-way table.

Usually, the expected frequencies for a cell are generated from its row sum and its column sum.

$$\text{Cell } F_e = (\text{row sum})(\text{column sum})/N \qquad\qquad (3.34)$$

When this procedure is used to generate the expected frequencies, the null hypothesis tested is that the variable on the row (say, region of the country) is independent of the variable on the column (attitude toward current political leadership). If the fit to the observed frequencies is good (so that χ^2 is small), then one concludes that the two variables are independent; a poor fit leads to a large χ^2, rejection of the null hypothesis, and the conclusion that the two variables are related.

The techniques discussed in this chapter for making decisions about differences, estimating population means, assessing association between two variables, and predicting a score on one variable from a score on another are important to, and widely used in, the social and behavioral sciences. They form the basis for most undergraduate—and some graduate—statistics courses. It is hoped that this brief review reminds you of material already mastered, so that, with common background and language, we begin in earnest the study of multivariate statistics.

4

Cleaning Up Your Act

Screening Data Prior to Analysis

This chapter deals with a set of issues that are resolved after data are collected but before the main data analysis is run. Careful consideration of these issues is time consuming and sometimes tedious; it is common, for instance, to spend many days in careful examination of data prior to running the main analysis that, itself, takes about 5 minutes. But consideration and resolution of these issues before the main analysis are fundamental to an honest analysis of the data.

The first issues concern the accuracy with which data have been entered into the data file and consideration of factors that could produce distorted correlations. Next, missing data, the bane of (almost) every researcher, are assessed and dealt with. Next, many multivariate procedures are based on assumptions; the fit between your data set and the assumptions is assessed before the procedure is applied. Transformations of variables to bring them into compliance with requirements of analysis are considered. Outliers, cases that are extreme, create other headaches because solutions are unduly influenced and sometimes distorted by them. Finally, perfect or near-perfect correlations among variables can threaten a multivariate analysis.

This chapter deals with issues that are relevant to most analyses. However, the issues are not all applicable to all analyses all the time; for instance, multiway frequency analysis (Chapter 16) and logistic regression (Chapter 10), procedures that use log-linear techniques, have far fewer assumptions than the other techniques. Other analyses have additional assumptions that are not covered in this chapter. For these reasons, assumptions and limitations specific to each analysis are reviewed in the third section of the chapter describing the analysis.

There are differences in data screening for grouped and ungrouped data. If you are performing multiple regression, canonical correlation, factor analysis, or structural equation modeling, where subjects are not subdivided into groups, there is one procedure for screening data. If you are performing analysis of covariance, multivariate analysis of variance or covariance, profile analysis, discriminant analysis, or multilevel modeling where subjects are in groups, there is another procedure for screening data. Differences in these procedures are illustrated by example in Section 4.2. Other analyses (survival analysis and time-series analysis) sometimes have grouped data and often do not, so screening is adjusted accordingly.

You may find the material in this chapter difficult from time to time. Sometimes it is necessary to refer to material covered in subsequent chapters to explain some issue, material that is more understandable after those chapters are studied. Therefore, you may want to read this chapter now to get an overview of the tasks to be accomplished prior to the main data analysis and then read it again after mastering the remaining chapters.

4.1   Important Issues in Data Screening

4.1.1   Accuracy of Data File

The best way to ensure the accuracy of a data file is to proofread the original data against the computerized data file in the data window. In SAS, data are most easily viewed in the Interactive Data Analysis window. With a small data file, proofreading is highly recommended, but with a large data file, it may not be possible. In this case, screening for accuracy involves examination of descriptive statistics and graphic representations of the variables.

The first step with a large data set is to examine univariate descriptive statistics through one of the descriptive programs such as SPSS FREQUENCIES, or SAS MEANS or UNIVARIATE or Interactive Data Analysis. For continuous variables, are all the values within range? Are means and standard deviations plausible? If you have discrete variables (such as categories of religious affiliation), are there any out-of-range numbers? Have you accurately programmed your codes for missing values?

4.1.2   Honest Correlations

Most multivariate procedures analyze patterns of correlation (or covariance) among variables. It is important that the correlations, whether between two continuous variables or between a dichotomous and continuous variable, be as accurate as possible. Under some rather common research conditions, correlations are larger or smaller than they should be.

4.1.2.1   Inflated Correlation

When composite variables are constructed from several individual items by pooling responses to individual items, correlations are inflated if some items are reused. Scales on personality inventories, measures of socioeconomic status, health indices, and many other variables in social and behavioral sciences are often composites of several items. If composite variables are used and they contain, in part, the same items, correlations are inflated; do not overinterpret a high correlation between two measures composed, in part, of the same items. If there is enough overlap, consider using only one of the composite variables in the analysis.

4.1.2.2   Deflated Correlation

Sample correlations may be lower than population correlations when there is restricted range in sampling of cases or very uneven splits in the categories of dichotomous variables.[1] Problems with distributions that lead to lower correlations are discussed in Section 4.1.5.

A falsely small correlation between two continuous variables is obtained if the range of responses to one or both of the variables is restricted in the sample. Correlation is a measure of the extent to which scores on two variables go up together (positive correlation) or one goes up while the other goes down (negative correlation). If the range of scores on one of the variables is narrow because of restricted sampling, then it is effectively a constant and cannot correlate highly with

---

[1]A very small coefficient of determination (standard deviation/mean) was also associated with lower correlations when computers had less computational accuracy. However, computational accuracy is so high in modern statistical packages that the problem is unlikely to occur, unless, perhaps, to astronomers.

another variable. In a study of success in graduate school, for instance, quantitative ability could not emerge as highly correlated with other variables if all students had about the same high scores in quantitative skills.

When a correlation is too small because of restricted range in sampling, you can estimate its magnitude in a nonrestricted sample by using Equation 4.1 if you can estimate the standard deviation in the nonrestricted sample. The standard deviation in the nonrestricted sample is estimated from prior data or from knowledge of the population distribution.

$$\tilde{r}_{xy} = \frac{r_{t(xy)}[S_x/S_{t(x)}]}{\sqrt{1 + r_{t(xy)}^2[S_x^2/S_{t(x)}^2] - r_{t(xy)}^2}} \tag{4.1}$$

where $\tilde{r}_{xy}$ = adjusted correlation

$r_{t(xy)}$ = correlation between X and Y with the range of X truncated

S_x = unrestricted standard deviation of X

$S_{t(x)}$ = truncated standard deviation of X

Many programs allow analysis of a correlation matrix instead of raw data. The estimated correlation is inserted in place of the truncated correlation prior to analysis of the correlation matrix. (However, insertion of estimated correlations may create internal inconsistencies in the correlation matrix, as discussed in Section 4.1.3.3.)

The correlation between a continuous variable and a dichotomous variable, or between two dichotomous variables (unless they have the same peculiar splits), is also too low if most (say, over 90%) responses to the dichotomous variable fall into one category. Even if the continuous and dichotomous variables are strongly related in the population, the highest correlation that could be obtained is well below 1. Some recommend dividing the obtained (but deflated) correlation by the maximum it could achieve given the split between the categories and then using the resulting value in subsequent analyses. This procedure is attractive, but not without hazard, as discussed by Comrey and Lee (1992).

4.1.3 Missing Data

Missing data is one of the most pervasive problems in data analysis. The problem occurs when rats die, equipment malfunctions, respondents become recalcitrant, or somebody goofs. Its seriousness depends on the pattern of missing data, how much is missing, and why it is missing. Recent summaries of issues surrounding missing data are provided by Schafer and Graham (2002) and by Graham, Cumsille, and Elek-Fisk (2003).

The pattern of missing data is more important than the amount missing. Missing values scattered randomly through a data matrix pose less serious problems. Nonrandomly missing values, on the other hand, are serious no matter how few of them there are because they affect the generalizability of results. Suppose that in a questionnaire with both attitudinal and demographic questions several respondents refuse to answer questions about income. It is likely that refusal to answer questions about income is related to attitude. If respondents with missing data on income are deleted, the sample values on the attitude variables are distorted. Some method of estimating income is needed to retain the cases for analysis of attitude.

Missing data are characterized as MCAR (missing completely at random), MAR (missing at random, called ignorable nonresponse), and MNAR (missing not at random or nonignorable). The distribution of missing data is unpredictable in MCAR, the best of all possible worlds if data must be missing. The pattern of missing data is predictable from other variables in the data set when data are MAR. In MNAR, the missingness is related to the DV and, therefore, cannot be ignored.

If only a few data points, say, 5% or less, are missing in a random pattern from a large data set, the problems are less serious and almost any procedure for handling missing values yields similar results. If, however, a lot of data are missing from a small to moderately sized data set, the problems can be very serious. Unfortunately, there are as yet no firm guidelines for how much missing data can be tolerated for a sample of a given size.

Although the temptation to assume that data are missing randomly is nearly overwhelming, the safest thing to do is to test it. Use the information you have to test for patterns in missing data. For instance, construct a dummy variable with two groups, cases with missing and nonmissing values on income, and perform a test of mean differences in attitude between the groups. If there are no differences, decisions about how to handle missing data are not so critical (except, of course, for inferences about income). If there are significant differences and η^2 is substantial (cf. Equation 3.25), care is needed to preserve the cases with missing values for other analyses, as discussed in Section 4.1.3.2.

SPSS MVA (Missing Values Analysis) is specifically designed to highlight patterns of missing values as well as to replace them in the data set. Table 4.1 shows syntax and output for a data set with missing values on ATTHOUSE and INCOME. A TTEST is requested to see if missingness is related to any of other variables, with $\alpha = .05$ and tests done only for variables with at least 5 PERCENT of data missing. The EM syntax requests a table of correlations and a test of whether data are missing completely at random (MCHR).

The Univariate Statistics table shows that there is one missing value on ATTHOUSE and 26 missing values on INCOME. Separate Variance t Tests show no systematic relationship between missingness on INCOME and any of the other variables. ATTHOUSE is not tested because fewer than 5% of the cases have missing values. The Missing Patterns table shows that Case number 52, among others, is missing INCOME, indicated by an S in the table. Case number 253 is missing ATTHOUSE. The last table shows EM Correlations with missing values filled in using the EM method, to be discussed. Below the table is Little's MCAR test of whether the data are missing completely at random. A statistically nonsignificant result is desired: $p = .76$ indicates that the probability that the pattern of missing diverges from randomness is greater than .05, so that MCAR may be inferred.

MAR can be inferred if the MCAR test is statistically significant but missingness is predictable from variables (other than the DV) as indicated by the Separate Variance t Tests. MNAR is inferred if the *t* test shows that missingness is related to the DV.

The decision about how to handle missing data is important. At best, the decision is among several bad alternatives, several of which are discussed in the subsections that follow.

4.1.3.1 *Deleting Cases or Variables*

One procedure for handling missing values is simply to drop any cases with them. If only a few cases have missing data and they seem to be a random subsample of the whole sample, deletion is a good alternative. Deletion of cases with missing values is the default option for most programs in the SPSS and SAS packages.[2]

---

[2]Because this is the default option, numerous cases can be deleted without the researcher's knowledge. For this reason, it is important to check the number of cases in your analyses to make sure that all of the desired cases are used.

TABLE 4.1   SPSS MVA Syntax and Output for Missing Data

MVA
  timedrs attdrug atthouse income mstatus race emplmnt
  /TTEST PROB PERCENT=5
  /MPATTERN
  /EM.
MVA

Univariate Statistics

	N	Mean	Std. Deviation	Missing Count	Missing Percent	No. of Extremes[a,b] Low	No. of Extremes[a,b] High
TIMEDRS	465	7.90	10.948	0	.0	0	34
ATTDRUG	465	7.69	1.156	0	.0	0	0
ATTHOUSE	464	23.54	4.484	1	.2	4	0
INCOME	439	4.21	2.419	26	5.6	0	0
MSTATUS	465	1.78	.416	0	.0	.	.
RACE	465	1.09	.284	0	.0	.	.
EMPLMNT	465	.47	.500	0	.0	0	0

[a]Number of cases outside the range (Q1 − 1.5*IQR, Q3 + 1.5*IQR).

[b]indicates that the interquartile range (IQR) is zero.

Separate Variance t Tests[a]

		TIMEDRS	ATTDRUG	ATTHOUSE	INCOME	MSTATUS	RACE	EMPLMNT
INCOME	t	.2	−1.1	−.2	.	−1.0	−.4	−1.1
	df	32.2	29.6	28.6	.	29.0	27.3	28.0
	P(2-tail)	.846	.289	.851	.	.346	.662	.279
	# Present	439	439	438	439	439	439	439
	# Missing	26	26	26	0	26	26	26
	Mean(Present)	7.92	7.67	23.53	4.21	1.77	1.09	.46
	Mean(Missing)	7.62	7.88	23.69		1.85	1.12	.58

For each quantitative variable, pairs of groups are formed by indicator variables (present, missing).

[a]Indicator variables with less than 5% missing are not displayed.

TABLE 4.1   Continued

Missing Patterns (cases with missing values)

Case	# Missing	% Missing	TIMEDRS	ATTDRUG	MSTATUS	RACE	EMPLMNT	ATTHOUSE	INCOME
					Missing and Extreme Value Patterns[a]				
52	1	14.3			−	−			S
64	1	14.3	+		−	−			S
69	1	14.3			−	−			S
77	1	14.3			−	−			S
118	1	14.3			−	−			S
135	1	14.3			−	−			S
161	1	14.3			−	−			S
172	1	14.3			−	−			S
173	1	14.3			−	−			S
174	1	14.3			−	−			S
181	1	14.3			−	−			S
196	1	14.3			−	+			S
203	1	14.3	+		−	−			S
236	1	14.3			−	−			S
240	1	14.3			−	−			S
258	1	14.3			−	+			S
304	1	14.3			−	−			S
321	1	14.3			−	−			S
325	1	14.3			−	−			S
352	1	14.3			−	−			S
378	1	14.3			−	−			S
379	1	14.3			−	−			S
409	1	14.3	+		−	−			S
419	1	14.3			−	−			S
421	1	14.3			−	−			S
435	1	14.3			−	+		−	S
253	1	14.3			−	+		S	

− indicates an extreme low value, and + indicates an extreme high value. The range used is (Q1 − 1.5*IQR, Q3 + 1.5*IQR).

[a]Cases and variables are sorted on missing patterns.

(continued)

TABLE 4.1   Continued

EM Estimated Statistics

EM Correlations[a]

	timedrs	attdrug	atthouse	income	mstatus	race	emplmnt
timedrs	1						
attdrug	.104	1					
atthouse	.128	.023	1				
income	.050	−.005	.002	1			
mstatus	−.065	−.006	−.030	−.466	1		
race	−.035	.019	−.038	.105	−.035	1	
emplmnt	.059	.085	−.023	−.006	.234	−.081	1

[a]Little's MCAR test: Chi Square = 19.550, DF = 12, Sig. = 0.76.

If missing values are concentrated in a few variables and those variables are not critical to the analysis, or are highly correlated with other, complete variables, the variable(s) with missing values are profitably dropped.

But if missing values are scattered throughout cases and variables, deletion of cases can mean substantial loss of subjects. This is particularly serious when data are grouped in an experimental design because loss of even one case requires adjustment for unequal n (see Chapter 6). Further, the researcher who has expended considerable time and energy collecting data is not likely to be eager to toss some out. And as previously noted, if cases with missing values are not randomly distributed through the data, distortions of the sample occur if they are deleted.

4.1.3.2   Estimating Missing Data

A second option is to estimate (impute) missing values and then use the estimates during data analysis. There are several popular schemes for doing so: using prior knowledge; inserting mean values; using regression; expectation-maximization; and multiple imputation.

Prior knowledge is used when a researcher replaces a missing value with a value from a well-educated guess. If the researcher has been working in an area for a while, and if the sample is large and the number of missing values small, this is often a reasonable procedure. The researcher is often confident that the value would have been about at the median, or whatever, for a particular case. Alternatively, the researcher can downgrade a continuous variable to a dichotomous variable (e.g., "high" vs. "low") to predict with confidence into which category a case with a missing value falls. The discrete variable replaces the continuous variable in the analysis, but it has less information than the continuous variable. An option with longitudinal data is to apply the last observed value to fill in data missing at a later point in time. However, this requires the expectation that there are no changes over time.

TABLE 4.2   Missing Data Options Available in Some Computer Programs

		Program						
Strategy		SPSS MVA	SOLAS MDA	SPSS REGRESSION	NORM	SAS STANDARD	SAS MI and MIANALYZE	AMOS, EQS, and LISREL
Mean substi-tution	Grand mean	No	Group Means[a]	MEAN SUBSTITUTION	No	REPLACE	No	No
	Group mean	No	Group Means	No	No	No	No	No
Regression		Regression	No	No	No	No	No	No
Expectation Maximization (EM)		EM	No	No	Yes[c]	No	PROC MI with NIMPUTE =	Yes
Multiple imputation		No[b]	Multiple Imputation	No	Yes	No	Yes	No

[a]With omission of group identification.

[b]May be done by generating multiple files through the EM method and computing additional statistics.

[c]For preparation of data prior to multiple imputation; provides missing values for one random imputation.

Remaining options for imputing missing data are available through software. Table 4.2 shows programs that implement missing data procedures.

Mean substitution has been a popular way to estimate missing values, although it is less commonly used now that more desirable methods are feasible through computer programs. Means are calculated from available data and used to replace missing values prior to analysis. In the absence of all other information, the mean is the best guess about the value of a variable. Part of the attraction of this procedure is that it is conservative; the mean for the distribution as a whole does not change and the researcher is not required to guess at missing values. On the other hand, the variance of a variable is reduced because the mean is closer to itself than to the missing value it replaces, and the correlation the variable has with other variables is reduced because of the reduction in variance. The extent of loss in variance depends on the amount of missing data and on the actual values that are missing.

A compromise is to insert a group mean for the missing value. If, for instance, the case with a missing value is a Republican, the mean value for Republicans is computed and inserted in place of the missing value. This procedure is not as conservative as inserting overall mean values and not as liberal as using prior knowledge. However, the reduction in within-group variance can make differences among groups spuriously large.

Many programs have provisions for inserting mean values. SAS STANDARD allows a data set to be created with missing values replaced by the mean on the variable for complete cases. SOLAS MDA, a program devoted to missing data analysis, produces data sets in which group means are used to replace missing values. SPSS REGRESSION permits MEANSUBSTITUTION. And, of course, transformation instructions can be used with any program to replace any defined value of a variable (including a missing code) with the mean.

Regression (see Chapter 5) is a more sophisticated method for estimating missing values. Other variables are used as IVs to write a regression equation for the variable with missing data serving as DV. Cases with complete data generate the regression equation; the equation is then used to predict missing values for incomplete cases. Sometimes the predicted values from a first round of regression are inserted for missing values and then all the cases are used in a second regression. The predicted values for the variable with missing data from round two are used to develop a third equation, and so forth, until the predicted values from one step to the next are similar (they converge). The predictions from the last round are the ones used to replace missing values.

An advantage to regression is that it is more objective than the researcher's guess but not as blind as simply inserting the grand mean. One disadvantage to use of regression is that the scores fit together better than they should; because the missing value is predicted from other variables, it is likely to be more consistent with them than a real score is. A second disadvantage is reduced variance because the estimate is probably too close to the mean. A third disadvantage is the requirement that good IVs be available in the data set; if none of the other variables is a good predictor of the one with missing data, the estimate from regression is about the same as simply inserting the mean. Finally, estimates from regression are used only if the estimated value falls within the range of values for complete cases; out-of-range estimates are not acceptable. Using regression to estimate missing values is convenient in SPSS MVA. The program also permits adjustment of the imputed values so that overconsistency is lessened.

Expectation maximization (EM) methods are available for randomly missing data. EM forms a missing data correlation (or covariance) matrix by assuming the shape of a distribution (such as normal) for the partially missing data and basing inferences about missing values on the likelihood under that distribution. It is an iterative procedure with two steps—expectation and maximization—for each iteration. First, the E step finds the conditional expectation of the "missing" data, given the observed values and current estimate of the parameters, such as correlations. These expectations are then substituted for the missing data. Second, the M step performs maximum likelihood estimation as though the missing data had been filled in. Finally, after convergence is achieved, the EM variance-covariance matrix may be provided and/or the filled-in data saved in the data set.

However, as pointed out by Graham et al. (2003), analysis of an EM-imputed data set is biased because error is not added to the imputed data set. Thus, analyses based on this data set have inappropriate standard errors for testing hypotheses. The bias is greatest when the data set with imputed values filled in is analyzed, but bias exists even when a variance-covariance or correlation matrix is used as input. Nevertheless, these imputed data sets can be useful for evaluating assumptions and for exploratory analyses that do not employ inferential statistics. Analysis of EM-imputed data sets also can provide insights when amounts of missing data are small if inferential statistics are interpreted with caution.

SPSS MVA performs EM to generate a data set with imputed values as well as variance-covariance and correlation matrices, and permits specification of some distributions other than normal. SPSS MVA also is extremely helpful for assessing patterns of missing data, providing *t* tests to predict missingness from other variables in the data set, and testing for MCAR, as seen in Table 4.1.

Structural equations modeling (SEM) programs (cf. AMOS, EQS, and LISREL in Chapter 14) typically have their own built-in imputation procedures which are based on EM. The programs do not produce data sets with imputed values but utilize appropriate standard errors in their analyses.

SAS, NORM, and SOLAS MDA can be used to create an EM-imputed data set by running the MI procedure with m (number of imputations) = 1, but analysis of the imputed data set is subject to the same cautions as noted for SPSS MVA. The EM variance-covariance matrix produced by NORM builds in appropriate standard errors, so that analyses based on those matrices are unbiased (Graham et al., 2003). Little and Rubin (1987) discuss EM and other methods in detail. EM through SPSS MVA is demonstrated in Section 10.7.1.1.

Multiple imputation also takes several steps to estimate missing data. First, logistic regression (Chapter 10) is used when cases with and without a missing value on a particular variable form the dichotomous DV. You determine which other variables are to be used as predictors in the logistic regression, which in turn provides an equation for estimating the missing values. Next, a random sample is taken (with replacement) from the cases with complete responses to identify the distribution of the variable with missing data.

Then several (m) random samples are taken (with replacement) from the distribution of the variable with missing data to provide estimates of that variable for each of m newly created (now complete) data sets. Rubin (1996) shows that five (or even three in some cases) such samples are adequate in many situations. You then perform your statistical analysis separately on the m new data sets and report the average parameter estimates (e.g., regression coefficients) from the multiple runs in your results.

Advantages of multiple imputation are that it can be applied to longitudinal data (e.g., for within-subjects IVs or time-series analysis) as well as data with single observations on each variable, and that it retains sampling variability (Statistical Solutions, Ltd., 1997). Another advantage is that it makes no assumptions about whether data are randomly missing. This is the method of choice for databases that are made available for analyses outside the agency that collected the data. That is, multiple data sets are generated, and other users may either make a choice of a single data set (with its inherent bias) or use the multiple data sets and report combined results. Reported results are the mean for each parameter estimate over the analyses of multiple data sets as well as the total variance estimate, which includes variance within imputations and between imputations—a measure of the true uncertainty in the data set caused by missing data (A. McDonnell, personal communication, August 24, 1999).

SOLAS MDA performs multiple imputation directly and provides a ROLLUP editor that combines the results from the newly created complete data sets (Statistical Solutions, Ltd., 1997). The editor shows the mean for each parameter and its total variance estimate, as well as within- and between-imputations variance estimates. The SOLAS MDA manual demonstrates multiple imputations with longitudinal data. Rubin (1996) provides further details about the procedure. With SPSS MVA, you apply your method m times via the EM procedure, using a random-number seed that changes for each new data set. Then you do your own averaging to establish the final parameter estimate.

NORM is a freely distributed program for multiple imputation available on the Internet (Schafer, 1999). The program currently is limited to normally distributed predictors and encompasses an EM procedure to estimate parameters, provide start values for the data augmentation step (multiple imputation), and help determine the proper number of imputations. A summary of the results of the multiple data sets produced by data augmentation is available as are the multiple data sets.

Newer versions of SAS use a three-step process to deal with multiple imputation of missing data. First, PROC MI provides an analysis of missing data patterns much like that of SPSS MVA (Table 4.1) but without the MCAR diagnosis or t tests to predict missingness from other variables. At the same time, a data set is generated with m subsets (default m = 5) with different imputations of missing data in each subset. A column indicating the imputation number (e.g., 1 to 5) is added to the data set. Next, the desired analysis module is run (e.g., REG, GLM, or MIXED) with syntax that requests a separate analysis for each imputation number and with some (but not all) results added to a data file. Finally, PROC MIANALYZE is run on the data file of results which combines the m sets of results into a single summary report.[3] Graham and colleagues (2003) report that m typically ranges from 5 to 20. Rubin (1996) suggests that often 3 to 5 imputations are adequate, as long as the amount of missing data is fairly small. He also asserts that any $m > 1$ is better than just one imputed data set. SAS MI and MIANALYZE are demonstrated in Section 5.7.4, where guidelines are given for choice of m.

SPSS MVA may be used to create multiple data sets with different imputations by running EM imputation m times with different random number seeds. However, you are then on your own for combining the results of the multiple analyses, and there is no provision for developing appropriate standard errors. Rubin (1987) discusses multiple imputation at length.

Other methods, such as *hot decking,* are available but they require specialized software and have few advantages in most situations over other imputation methods offered by SAS, SOLAS, and NORM.

4.1.3.3 *Using a Missing Data Correlation Matrix*

Another option with randomly missing data involves analysis of a missing data correlation matrix. In this option, all available pairs of values are used to calculate each of the correlations in **R**. A variable with 10 missing values has all its correlations with other variables based on 10 fewer pairs of numbers. If some of the other variables also have missing values, but in different cases, the number of complete pairs of variables is further reduced. Thus, each correlation in **R** can be based on a different number and a different subset of cases, depending on the pattern of missing values. Because the standard error of the sampling distribution for r is based on N, some correlations are less stable than others in the same correlation matrix.

But that is not the only problem. In a correlation matrix based on complete data, the sizes of some correlations place constraints on the sizes of others. In particular,

$$r_{13}r_{23} - \sqrt{(1 - r_{13}^2)(1 - r_{23}^2)} \le r_{12} \le r_{13}r_{23} + \sqrt{(1 - r_{13}^2)(1 - r_{23}^2)} \tag{4.2}$$

The correlation between variables 1 and 2, r_{12}, cannot be smaller than the value on the left or larger than the value on the right in a three-variable correlation matrix. If $r_{13} = .60$ and $r_{23} = .40$, then r_{12} cannot be smaller than $-.49$ or larger than .97. If, however, r_{12}, r_{23}, and r_{13} are all based on different subsets of cases because of missing data, the value for r_{12} can go out of range.

Most multivariate statistics involve calculation of eigenvalues and eigenvectors from a correlation matrix (see Appendix A). With loosened constraints on size of correlations in a missing data correlation matrix, eigenvalues sometimes become negative. Because eigenvalues represent variance, negative eigenvalues represent something akin to negative variance. Moreover, because the total variance that is partitioned in the analysis is a constant (usually equal to the number of variables), positive eigenvalues are inflated by the size of negative eigenvalues, resulting in inflation of variance. The statistics derived under these conditions can be quite distorted.

However, with a large sample and only a few missing values, eigenvalues are often all positive even if some correlations are based on slightly different pairs of cases. Under these conditions, a

---

[3]Output is especially sparse for procedures other than SAS REG.

missing data correlation matrix provides a reasonable multivariate solution and has the advantage of using all available data. Use of this option for the missing data problem should not be rejected out of hand but should be used cautiously with a wary eye to negative eigenvalues.

A missing value correlation matrix is prepared through the PAIRWISE deletion option in some of the SPSS programs. It is the default option for SAS CORR. If this is not an option of the program you want to run, then generate a missing data correlation matrix through another program for input to the one you are using.

4.1.3.4 Treating Missing Data as Data

It is possible that the fact that a value is missing is itself a very good predictor of the variable of interest in your research. If a dummy variable is created when cases with complete data are assigned 0 and cases with missing data 1, the liability of missing data could become an asset. The mean is inserted for missing values so that all cases are analyzed, and the dummy variable is used as simply another variable in analysis, as discussed by Cohen, Cohen, West, and Aiken (2003, pp. 431–451).

4.1.3.5 Repeating Analyses with and without Missing Data

If you use some method of estimating missing values or a missing data correlation matrix, consider repeating your analyses using only complete cases. This is particularly important if the data set is small, the proportion of missing values high, or data are missing in a nonrandom pattern. If the results are similar, you can have confidence in them. If they are different, however, you need to investigate the reasons for the difference, and either evaluate which result more nearly approximates "reality" or report both sets of results.

4.1.3.6 Choosing among Methods for Dealing with Missing Data

The first step in dealing with missing data is to observe their pattern to try to determine whether data are randomly missing. Deletion of cases is a reasonable choice if the pattern appears random and if only a very few cases have missing data, and those cases are missing data on different variables. However, if there is evidence of nonrandomness in the pattern of missing data, methods that preserve all cases for further analysis are preferred.

Deletion of a variable with a lot of missing data is also acceptable as long as that variable is not critical to the analysis. Or, if the variable is important, use a dummy variable that codes the fact that the scores are missing coupled with mean substitution to preserve the variable and make it possible to analyze all cases and variables.

It is best to avoid mean substitution unless the proportion of missing values is *very* small and there are no other options available to you. Using prior knowledge requires a great deal of confidence on the part of the researcher about the research area and expected results. Regression methods may be implemented (with some difficulty) without specialized software but are less desirable than EM methods.

EM methods sometimes offer the simplest and most reasonable approach to imputation of missing data, as long as your preliminary analysis provides evidence that scores are missing randomly (MCAR or MAR). Use of an EM covariance matrix, if the technique permits it as input, provides a less biased analysis a data set with imputed values. However, unless the EM program provides appropriate standard errors (as per the SEM programs of Chapter 14 or NORM), the strategy should be limited to data sets in which there is not a great deal of missing data, and inferential results (e.g., p values) are interpreted with caution. EM is especially appropriate for techniques that do not rely on inferential statistics, such as exploratory factor analysis (Chapter 13). Better yet is to incorporate EM methods into multiple imputation.

Multiple imputation is currently considered the most respectable method of dealing with missing data. It has the advantage of not requiring MCAR (and perhaps not even MAR) and can be used for any form of GLM analysis, such as regression, ANOVA, and logistic regression. The problem is that it is more difficult to implement and does not provide the full richness of output that is typical with other methods.

Using a missing data correlation matrix is tempting if your software offers it as an option for your analysis because it requires no extra steps. It makes most sense to use when missing data are scattered over variables, and there are no variables with a lot of missing values. The vagaries of missing data correlation matrices should be minimized as long as the data set is large and missing values are few.

Repeating analyses with and without missing data is highly recommended whenever any imputation method or a missing data correlation matrix is used and the proportion of missing values is high—especially if the data set is small.

4.1.4 Outliers

An outlier is a case with such an extreme value on one variable (a univariate outlier) or such a strange combination of scores on two or more variables (multivariate outlier) that it distorts statistics. Consider, for example, the bivariate scatterplot of Figure 4.1 in which several regression lines, all with slightly different slopes, provide a good fit to the data points inside the swarm. But when the data point labeled A in the upper right-hand portion of the scatterplot is also considered, the regression coefficient that is computed is the one from among the several good alternatives that provides the best fit to the extreme case. The case is an outlier because it has much more impact on the value of the regression coefficient than any of those inside the swarm.

Outliers are found in both univariate and multivariate situations, among both dichotomous and continuous variables, among both IVs and DVs, and in both data and results of analyses. They lead to

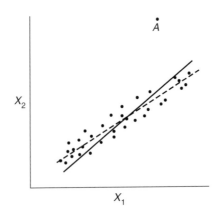

**FIGURE 4.1   Bivariate scatterplot for
showing impact of an outlier.**

both Type I and Type II errors, frequently with no clue as to which effect they have in a particular analysis. And they lead to results that do not generalize except to another sample with the same kind of outlier.

There are four reasons for the presence of an outlier. First is incorrect data entry. Cases that are extreme should be checked carefully to see that data are correctly entered. Second is failure to specify missing-value codes in computer syntax so that missing-value indicators are read as real data. Third is that the outlier is not a member of the population from which you intended to sample. If the case should not have been sampled, it is deleted once it is detected. Fourth is that the case is from the intended population but the distribution for the variable in the population has more extreme values than a normal distribution. In this event, the researcher retains the case but considers changing the value on the variable(s) so that the case no longer has as much impact. Although errors in data entry and missing values specification are easily found and remedied, deciding between alternatives three and four, between deletion and retention with alteration, is difficult.

4.1.4.1 *Detecting Univariate and Multivariate Outliers*

Univariate outliers are cases with an extreme value on one variable; multivariate outliers are cases with an unusual combination of scores on two or more variables. For example, a 15-year-old is perfectly within bounds regarding age, and someone who earns $45,000 a year is in bounds regarding income, but a 15-year-old who earns $45,000 a year is very unusual and is likely to be a multivariate outlier. Multivariate outliers can occur when several different populations are mixed in the same sample or when some important variables are omitted that, if included, would attach the outlier to the rest of the cases.

Univariate outliers are easier to spot. Among dichotomous variables, the cases on the "wrong" side of a very uneven split are likely univariate outliers. Rummel (1970) suggests deleting dichotomous variables with 90–10 splits between categories, or more, both because the correlation coefficients between these variables and others are truncated and because the scores for the cases in the small category are more influential than those in the category with numerous cases. Dichotomous variables with extreme splits are easily found in the programs for frequency distributions (SPSS FREQUENCIES, or SAS UNIVARIATE or Interactive Data Analysis) used during routine preliminary data screening.

Among continuous variables, the procedure for searching for outliers depends on whether data are grouped. If you are going to perform one of the analyses with ungrouped data (regression, canonical correlation, factor analysis, structural equation modeling, or some forms of time-series analysis), univariate and multivariate outliers are sought among all cases at once, as illustrated in Sections 4.2.1.1 (univariate) and 4.2.1.4 (multivariate). If you are going to perform one of the analyses with grouped data (ANCOVA, MANOVA or MANCOVA, profile analysis, discriminant analysis, logistic regression, survival analysis, or multilevel modeling) outliers are sought separately within each group, as illustrated in Sections 4.2.2.1 and 4.2.2.3.

Among continuous variables, univariate outliers are cases with very large standardized scores, z scores, on one or more variables, that are disconnected from the other z scores. Cases with standardized scores in excess of 3.29 ($p < .001$, two-tailed test) are potential outliers. However, the extremeness of a standardized score depends on the size of the sample; with a very large N, a few standardized scores in excess of 3.29 are expected. Z scores are available through SPSS EXPLORE or DESCRIPTIVES (where z scores are saved in the data file), and SAS STANDARD (with MEAN = 0 and STD = 1). Or you can hand-calculate z scores from any output that provides means, standard deviations, and maximum and minimum scores.

As an alternative or in addition to inspection of z scores, there are graphical methods for finding univariate outliers. Helpful plots are histograms, box plots, normal probability plots, or detrended normal probability plots. Histograms of variables are readily understood and available and may reveal one or more univariate outliers. There is usually a pileup of cases near the mean with cases trailing away in either direction. An outlier is a case (or a very few cases) that seems to be unattached to the rest of the distribution. Histograms for continuous variables are produced by SPSS FREQUENCIES (plus SORT and SPLIT for grouped data), and SAS UNIVARIATE or CHART (with BY for grouped data).

Box plots are simpler and literally box in observations that are around the median; cases that fall far away from the box are extreme. Normal probability plots and detrended normal probability plots are very useful for assessing normality of distributions of variables and are discussed in that context in Section 4.1.5.1. However, univariate outliers are visible in these plots as points that lie a considerable distance from others.

Once potential univariate outliers are located, the researcher decides whether transformations are acceptable. Transformations (Section 4.1.6) are undertaken both to improve the normality of distributions (Section 4.1.5.1) and to pull univariate outliers closer to the center of a distribution, thereby reducing their impact. Transformations, if acceptable, are undertaken prior to the search for multivariate outliers because the statistics used to reveal them (Mahalanobis distance and its variants) are also sensitive to failures of normality.

Mahalanobis distance is the distance of a case from the centroid of the remaining cases where the centroid is the point created at the intersection of the means of all the variables. In most data sets, the cases form a swarm around the centroid in multivariate space. Each case is represented in the swarm by a single point at its own peculiar combination of scores on all of the variables, just as each case is represented by a point at its own X, Y combination in a bivariate scatterplot. A case that is a multivariate outlier, however, lies outside the swarm, some distance from the other cases. Mahalanobis distance is one measure of that multivariate distance and it can be evaluated for each case using the χ^2 distribution.

Mahalanobis distance is tempered by the patterns of variances and covariances among the variables. It gives lower weight to variables with large variances and to groups of highly correlated variables. Under some conditions, Mahalanobis distance can either "mask" a real outlier (produce a false negative) or "swamp" a normal case (produce a false positive). Thus, it is not a perfectly reliable indicator of multivariate outliers and should be used with caution.

Mahalanobis distances are requested and interpreted in Sections 4.2.1.4 and 4.2.2.3, and numerous other places throughout the book. A very conservative probability estimate for a case being an outlier, say, $p < .001$ for the χ^2 value, is appropriate with Mahalanobis distance.

Other statistical measures used to identify multivariate outliers are leverage, discrepancy, and influence. Although developed in the context of multiple regression (Chapter 5), the three measures are now available for some of the other analyses. Leverage is related to Mahalanobis distance (or variations of it in the "hat" matrix) and is variously called HATDIAG, RHAT, or h_{ii}. Although leverage is related to Mahalanobis distance, it is measured on a different scale so that significance tests based on a χ^2 distribution do not apply.[4] Equation 4.3 shows the relationship between leverage—h_{ii}—and Mahalanobis distance.

$$\text{Mahalanobis distance} = (N - 1)(h_{ii} - 1/N) \tag{4.3}$$

[4]Lunneborg (1994) suggests that outliers be defined as cases with $h_{ii} \geq 2(k/N)$.

Or, as is sometimes more useful,

$$h_{ii} = \frac{\text{Mahalanobis distance}}{N-1} + \frac{1}{N}$$

The latter form is handy if you want to find a critical value for leverage at $\alpha = .001$ by translating the critical χ^2 value for Mahalanobis distance.

Cases with high leverage are far from the others, but they can be far out on basically the same line as the other cases, or far away and off the line. *Discrepancy* measures the extent to which a case is in line with the others. Figure 4.2(a) shows a case with high leverage and low discrepancy; Figure 4.2(b) shows a case with high leverage and high discrepancy. In Figure 4.2(c) is a case with low leverage and high discrepancy. In all of these figures, the outlier appears disconnected from the remaining scores.

Influence is a product of leverage and discrepancy (Fox, 1991). It assesses change in regression coefficients when a case is deleted; cases with influence scores larger than 1.00 are suspected of being outliers. Measures of influence are variations of Cook's distance and are identified in output as Cook's distance, modified Cook's distance, DFFITS, and DBETAS. For the interested reader, Fox (1991, pp. 29–30) describes these terms in more detail.

Leverage and/or Mahalanobis distance values are available as statistical methods of outlier detection in both statistical packages. However, recent research (e.g., Egan and Morgan, 1998; Hadi and Simonoff, 1993; Rousseeuw and van Zomeren, 1990) indicates that these methods are not perfectly reliable. Unfortunately, alternative methods are computationally challenging and not readily available in statistical packages. Therefore, multivariate outliers are currently most easily detected through Mahalanobis distance, or one of its cousins, but cautiously.

Statistics assessing the distance for each case, in turn, from all other cases, are available through SPSS REGRESSION by evoking Mahalanobis, Cook's, or Leverage values through the Save command in the Regression menu; these values are saved as separate columns in the data file and examined using standard descriptive procedures. To use the regression program just to find outliers, however, you must specify some variable (such as the case number) as DV, to find outliers

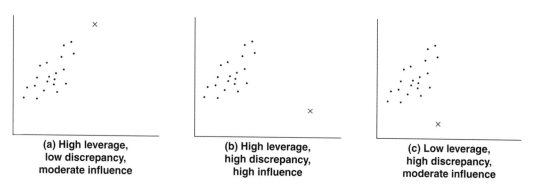

(a) High leverage,
low discrepancy,
moderate influence

(b) High leverage,
high discrepancy,
high influence

(c) Low leverage,
high discrepancy,
moderate influence

FIGURE 4.2   The relationships among leverage, discrepancy, and influence.

among the set of variables of interest, considered IVs. Alternatively, the 10 cases with largest Mahalanobis distance are printed out by SPSS REGRESSION using the **RESIDUALS** subcommand, as demonstrated in Section 4.2.1.4.

SAS regression programs provide a leverage, h_{ii}, value for each case that converts easily to Mahalanobis distance (Equation 4.3). These values are also saved to the data file and examined using standard statistical and graphical techniques.

When multivariate outliers are sought in grouped data, they are sought within each group separately. SPSS and SAS REGRESSION require separate runs for each group, each with its own error term. Programs in other packages, such as SYSTAT DISCRIM and BMDP7M, provide Mahalanobis distance for each case using a within-groups error term, so that outliers identified through those programs may be different from those identified by SPSS and SAS REGRESSION.

SPSS DISCRIMINANT provides outliers in the solution. These are not particularly helpful for screening (you would not want to delete cases just because the solution doesn't fit them very well), but are useful to evaluate generalizability of the results.

Frequently, some multivariate outliers hide behind other multivariate outliers—outliers are known to mask other outliers (Rousseeuw and van Zomeren, 1990). When the first few cases identified as outliers are deleted, the data set becomes more consistent and then other cases become extreme. Robust approaches to this problem have been proposed (e.g., Egan and Morgan, 1998; Hadi and Simonoff, 1993; Rousseeuw and van Zomeren, 1990), but these are not yet implemented in popular software packages. These methods can be approximated by screening for multivariate outliers several times, each time dealing with cases identified as outliers on the last run, until finally no new outliers are identified. But if the process of identifying ever more outliers seems to stretch into infinity, do a trial run with and without outliers to see if ones identified later are truly influencing results. If not, do not delete the later-identified outliers.

4.1.4.2 Describing Outliers

Once multivariate outliers are identified, you need to discover why the cases are extreme. (You already know why univariate outliers are extreme.) It is important to identify the variables on which the cases are deviant for three reasons. First, this procedure helps you decide whether the case is properly part of your sample. Second, if you are going to modify scores instead of delete cases, you have to know which scores to modify. Third, it provides an indication of the kinds of cases to which your results do not generalize.

If there are only a few multivariate outliers, it is reasonable to examine them individually. If there are several, you can examine them as a group to see if there are any variables that separate the group of outliers from the rest of the cases.

Whether you are trying to describe one or a group of outliers, the trick is to create a dummy grouping variable where the outlier(s) has one value and the rest of the cases another value. The dummy variable is then used as the grouping DV in discriminant analysis (Chapter 9) or logistic regression (Chapter 10), or as the DV in regression (Chapter 5). The goal is to identify the variables that distinguish outliers from the other cases. Variables on which the outlier(s) differs from the rest of the cases enter the equation; the remaining variables do not. Once those variables are identified, means on those variables for outlying and nonoutlying cases are found through any of the routine descriptive programs. Description of outliers is illustrated in Sections 4.2.1.4 and 4.2.2.3.

4.1.4.3   *Reducing the Influence of Outliers*

Once univariate outliers have been identified, there are several strategies for reducing their impact. But before you use one of them, check the data for the case to make sure that they are accurately entered into the data file. If the data are accurate, consider the possibility that one variable is responsible for most of the outliers. If so, elimination of the variable would reduce the number of outliers. If the variable is highly correlated with others or is not critical to the analysis, deletion of it is a good alternative.

If neither of these simple alternatives is reasonable, you must decide whether the cases that are outliers are properly part of the population from which you intended to sample. Cases with extreme scores, which are, nonetheless, apparently connected to the rest of the cases, are more likely to be a legitimate part of the sample. If the cases are not part of the population, they are deleted with no loss of generalizability of results to your intended population.

If you decide that the outliers are sampled from your target population, they remain in the analysis, but steps are taken to reduce their impact—variables are transformed or scores changed.

A first option for reducing impact of univariate outliers is variable transformation, undertaken to change the shape of the distribution to more nearly normal. In this case, outliers are considered part of a nonnormal distribution with tails that are too heavy so that too many cases fall at extreme values of the distribution. Cases that were outliers in the untransformed distribution are still on the tails of the transformed distribution, but their impact is reduced. Transformation of variables has other salutary effects, as described in Section 4.1.6.

A second option for univariate outliers is to change the score(s) on the variable(s) for the outlying case(s) so that they are deviant, but not as deviant as they were. For instance, assign the outlying case(s) a raw score on the offending variable that is one unit larger (or smaller) than the next most extreme score in the distribution. Because measurement of variables is sometimes rather arbitrary anyway, this is often an attractive alternative to reduce the impact of a univariate outlier.

Transformation or score alteration may not work for a truly multivariate outlier because the problem is with the combination of scores on two or more variables, not with the score on any one variable. The case is discrepant from the rest in its combinations of scores. Although the number of possible multivariate outliers is often substantially reduced after transformation or alteration of scores on variables, there are sometimes a few cases that are still far away from the others. These cases are usually deleted. If they are allowed to remain, it is with the knowledge that they may distort the results in almost any direction. Any transformations, changes of scores, and deletions are reported in the Results section together with the rationale.

4.1.4.4   *Outliers in a Solution*

Some cases may not fit well within a solution; the scores predicted for those cases by the selected model are very different from the actual scores for the cases. Such cases are identified *after* an analysis is completed, not as part of the screening process. To identify and eliminate or change scores for such cases before conducting the major analysis make the analysis look better than it should. Therefore, conducting the major analysis and then "retrofitting" is a procedure best limited to exploratory analysis. Chapters that describe techniques for ungrouped data deal with outliers in the solution when discussing the limitations of the technique.

4.1.5   Normality, Linearity, and Homoscedasticity

Underlying some multivariate procedures and most statistical tests of their outcomes is the assumption of multivariate normality. Multivariate normality is the assumption that each variable and all linear combinations of the variables are normally distributed. When the assumption is met, the residuals[5] of analysis are also normally distributed and independent. The assumption of multivariate normality is not readily tested because it is impractical to test an infinite number of linear combinations of variables for normality. Those tests that are available are overly sensitive.

The assumption of multivariate normality is made as part of derivation of many significance tests. Although it is tempting to conclude that most inferential statistics are robust[6] to violations of the assumption, that conclusion may not be warranted.[7] Bradley (1982) reports that statistical inference becomes less and less robust as distributions depart from normality, rapidly so under many conditions. And even when the statistics are used purely descriptively, normality, linearity, and homoscedasticity of variables enhance the analysis. The safest strategy, then, is to use transformations of variables to improve their normality unless there is some compelling reason not to.

The assumption of multivariate normality applies differently to different multivariate statistics. For analyses when subjects are not grouped, the assumption applies to the distributions of the variables themselves or to the residuals of the analyses; for analyses when subjects are grouped, the assumption applies to the sampling distributions[8] of means of variables.

If there is multivariate normality in ungrouped data, each variable is itself normally distributed and the relationships between pairs of variables, if present, are linear and *homoscedastic* (i.e., the variance of one variable is the same at all values of the other variable). The assumption of multivariate normality can be partially checked by examining the normality, linearity, and homoscedasticity of individual variables or through examination of residuals in analyses involving prediction.[9] The assumption is certainly violated, at least to some extent, if the individual variables (or the residuals) are not normally distributed or do not have pairwise linearity and homoscedasticity.

For grouped data, it is the sampling distributions of the means of variables that are to be normally distributed. The Central Limit Theorem reassures us that, with sufficiently large sample sizes, sampling distributions of means are normally distributed regardless of the distributions of variables. For example, if there are at least 20 degrees of freedom for error in a univariate ANOVA, the F test is said to be robust to violations of normality of variables (provided that there are no outliers).

---

[5]Residuals are leftovers. They are the segments of scores not accounted for by the multivariate analysis. They are also called "errors" between predicted and obtained scores where the analysis provides the predicted scores. Note that the practice of using a dummy DV such as case number to investigate multivariate outliers will *not* produce meaningful residuals plots.

[6]Robust means that the researcher is led to correctly reject the null hypothesis at a given alpha level the right number of times even if the distributions do not meet the assumptions of analysis. Often, Monte Carlo procedures are used where a distribution with some known properties is put into a computer, sampled from repeatedly, and repeatedly analyzed; the researcher studies the rates of retention and rejection of the null hypothesis against the known properties of the distribution in the computer.

[7]The univariate F test of mean differences, for example, is frequently said to be robust to violation of assumptions of normality and homogeneity of variance with large and equal samples, but Bradley (1984) questions this generalization.

[8]A sampling distribution is a distribution of statistics (not of raw scores) computed from random samples of a given size taken repeatedly from a population. For example, in univariate ANOVA, hypotheses are tested with respect to the sampling distribution of means (Chapter 3).

[9]Analysis of residuals to screen for normality, linearity, and homoscedasticity in multiple regression is discussed in Section 5.3.2.4.

These issues are discussed again in the third sections of Chapters 5 through 16 and 18 (online) as they apply directly to one or another of the multivariate procedures. For nonparametric procedures such as multiway frequency analysis (Chapter 16) and logistic regression (Chapter 10), there are no distributional assumptions. Instead, distributions of scores typically are hypothesized and observed distributions are tested against hypothesized distributions.

4.1.5.1 *Normality*

Screening continuous variables for normality is an important early step in almost every multivariate analysis, particularly when inference is a goal. Although normality of the variables is not always required for analysis, the solution is usually quite a bit better if the variables are all normally distributed. The solution is degraded, if the variables are not normally distributed, and particularly if they are nonnormal in very different ways (e.g., some positively and some negatively skewed).

Normality of variables is assessed by either statistical or graphical methods. Two components of normality are skewness and kurtosis. Skewness has to do with the symmetry of the distribution; a skewed variable is a variable whose mean is not in the center of the distribution. Kurtosis has to do with the peakedness of a distribution; a distribution is either too peaked (with short, thick tails) or too flat (with long, thin tails).[10] Figure 4.3 shows a normal distribution, distributions with skewness, and distributions with nonnormal kurtosis. A variable can have significant skewness, kurtosis, or both.

When a distribution is normal, the values of skewness and kurtosis are zero. If there is positive skewness, there is a pileup of cases to the left and the right tail is too long; with negative skewness, there is a pileup of cases to the right and the left tail is too long. Kurtosis values above zero indicate a distribution that is too peaked with short, thick tails, and kurtosis values below zero indicate a distribution that is too flat (also with too many cases in the tails).[11] Nonnormal kurtosis produces an underestimate of the variance of a variable.

There are significance tests for both skewness and kurtosis that test the obtained value against null hypotheses of zero. For instance, the standard error for skewness is approximately

$$s_s = \sqrt{\frac{6}{N}} \tag{4.4}$$

where N is the number of cases. The obtained skewness value is then compared with zero using the z distribution, where

$$z = \frac{S - 0}{s_s} \tag{4.5}$$

and S is the value reported for skewness. The standard error for kurtosis is approximately

$$s_k = \sqrt{\frac{24}{N}} \tag{4.6}$$

---

[10]If you decide that outliers are sampled from the intended population but that there are too many cases in the tails, you are saying that the distribution from which the outliers are sampled has kurtosis that departs from normal.

[11]The equation for kurtosis gives a value of 3 when the distribution is normal, but all of the statistical packages subtract 3 before printing kurtosis so that the expected value is zero.

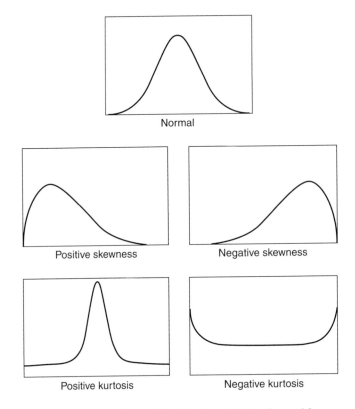

**FIGURE 4.3   Normal distribution, distributions with
skewness, and distributions with kurtoses.**

and the obtained kurtosis value is compared with zero using the z distribution, where

$$z = \frac{K - 0}{s_k} \tag{4.7}$$

and K is the value reported for kurtosis.

Conventional but conservative (.01 or .001) alpha levels are used to evaluate the significance of skewness and kurtosis with small to moderate samples, but if the sample is large, it is a good idea to look at the shape of the distribution instead of using formal inference tests. Because the standard errors for both skewness and kurtosis decrease with larger N, the null hypothesis is likely to be rejected with large samples when there are only minor deviations from normality.

In a large sample, a variable with statistically significant skewness often does not deviate enough from normality to make a substantive difference in the analysis. In other words, with large samples, the significance level of skewness is not as important as its actual size (worse the farther from zero) and the visual appearance of the distribution. In a large sample, the impact of departure from zero kurtosis also diminishes. For example, underestimates of variance associated with positive kurtosis (distributions with short, thick tails) disappear with samples of 100 or more cases; with negative kurtosis, underestimation of variance disappears with samples of 200 or more (Waternaux, 1976).

Values for skewness and kurtosis are available in several programs. SPSS FREQUENCIES, for instance, prints as options skewness, kurtosis, and their standard errors, and, in addition, superimposes a normal distribution over a frequency histogram for a variable if HISTOGRAM = NORMAL is specified. DESCRIPTIVES and EXPLORE also print skewness and kurtosis statistics. A histogram or stem-and-leaf plot is also available in SAS UNIVARIATE.[12]

Frequency histograms are an important graphical device for assessing normality, especially with the normal distribution as an overlay, but even more helpful than frequency histograms are expected normal probability plots and detrended expected normal probability plots. In these plots, the scores are ranked and sorted; then an expected normal value is computed and compared with the actual normal value for each case. The expected normal value is the z score that a case with that rank holds in a normal distribution; the normal value is the z score it has in the actual distribution. If the actual distribution is normal, then the points for the cases fall along the diagonal running from lower left to upper right, with some minor deviations due to random processes. Deviations from normality shift the points away from the diagonal.

Consider the expected normal probability plots for ATTDRUG and TIMEDRS through SPSS PPLOT in Figure 4.4. Syntax indicates the VARIABLES we are interested in are attdrug and timedrs. The remaining syntax is produced by default by the SPSS Windows menu system. As reported in Section 4.2.1.1, ATTDRUG is reasonably normally distributed (kurtosis = −0.447, skewness = −0.123) and TIMEDRS is too peaked and positively skewed (kurtosis = 13.101, skewness = 3.248, both significantly different from 0). The cases for ATTDRUG line up along the diagonal, whereas those for TIMEDRS do not. At low values of TIMEDRS, there are too many cases above the diagonal, and at high values, there are too many cases below the diagonal, reflecting the patterns of skewness and kurtosis.

Detrended normal probability plots for TIMEDRS and ATTDRUG are also in Figure 4.4. These plots are similar to expected normal probability plots except that deviations from the diagonal are plotted instead of values along the diagonal. In other words, the linear trend from lower left to upper right is removed. If the distribution of a variable is normal, as is ATTDRUG, the cases distribute themselves evenly above and below the horizontal line that intersects the Y axis at 0.0, the line of zero deviation from expected normal values. The skewness and kurtosis of TIMEDRS are again apparent from the cluster of points above the line at low values of TIMEDRS and below the line at high values of TIMEDRS. Normal probability plots for variables are also available in SAS UNIVARIATE and SPSS MANOVA. Many of these programs also produce detrended normal plots.

If you are going to perform an analysis with ungrouped data, an alternative to screening variables prior to analysis is conducting the analysis and then screening the residuals (the differences between the predicted and obtained DV values). If normality is present, the residuals are normally and independently distributed. That is, the differences between predicted and obtained scores—the errors—are symmetrically distributed around a mean value of zero and there are no contingencies among the errors. In multiple regression, residuals are also screened for normality through the expected normal probability plot and the detrended normal probability plot.[13] SPSS REGRESSION

---

[12]In structural equation modeling (Chapter 14), skewness and kurtosis for each variable are available in EQS and Mardia's Coefficient (the multivariate kurtosis measure) is available in EQS, PRELIS, and CALIS. In addition, PRELIS can be used to deal with nonnormality through alternative correlation coefficients, such as polyserial or polychoric (cf. Section 14.5.6).

[13]For grouped data, residuals have the same shape as within-group distributions because the predicted value is the mean, and subtracting a constant does not change the shape of the distribution. Many of the programs for grouped data plot the within-group distribution as an option, as discussed in the next few chapters when relevant.

```
PPLOT
 /VARIABLES = attdrug timedrs
 /NOLOG
 /NOSTANDARDIZE
 /TYPE = P – P
 /FRACTION = BLOM
 /TIES = MEAN
 /DIST = NORMAL.
```

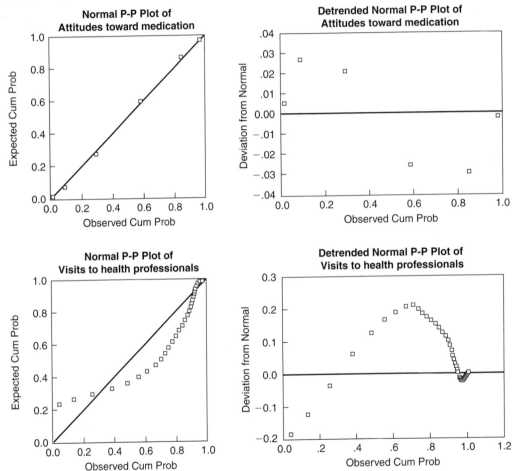

FIGURE 4.4 **Expected normal probability plot and detrended normal probability plot for ATTDRUG and TIMEDRS. SPSS PPLOT syntax and output.**

provides this diagnostic technique (and others, as discussed in Chapter 5). If the residuals are normally distributed, the expected normal probability plot and the detrended normal probability plot look just the same as they do if a variable is normally distributed. In regression, if the residuals plot looks normal, there is no reason to screen the individual variables for normality.

Although residuals will reveal departures from normality, the analyst has to resist temptation to look at the rest of the output to avoid "tinkering" with variables and cases to produce an anticipated result. Because screening the variables should lead to the same conclusions as screening residuals, it may be more objective to make one's decisions about transformations, deletion of outliers, and the like, on the basis of screening runs alone rather than screening through the outcome of analysis.[14]

With ungrouped data, if nonnormality is found, transformation of variables is considered. Common transformations are described in Section 4.1.6. Unless there are compelling reasons not to transform, it is probably better to do so. However, realize that even if each of the variables is normally distributed, or transformed to normal, there is no guarantee that all linear combinations of the variables are normally distributed. That is, if variables are each univariate normal, they do not necessarily have a multivariate normal distribution. However, it is more likely that the assumption of multivariate normality is met if all the variables are normally distributed.

4.1.5.2   *Linearity*

The assumption of linearity is that there is a straight-line relationship between two variables (where one or both of the variables can be combinations of several variables). Linearity is important in a practical sense because Pearson's *r* only captures the linear relationships among variables; if there are substantial nonlinear relationships among variables, they are ignored.

Nonlinearity is diagnosed either from residuals plots in analyses involving a predicted variable or from bivariate scatterplots between pairs of variables. In plots where standardized residuals are plotted against predicted values, nonlinearity is indicated when most of the residuals are above the zero line on the plot at some predicted values and below the zero line at other predicted values (see Chapter 5).

Linearity between two variables is assessed roughly by inspection of bivariate scatterplots. If both variables are normally distributed and linearly related, the scatterplot is oval-shaped. If one of the variables is nonnormal, then the scatterplot between this variable and the other is not oval. Examination of bivariate scatterplots is demonstrated in Section 4.2.1.2, along with transformation of a variable to enhance linearity.

However, sometimes the relationship between variables is simply not linear. Consider, for instance, number of symptoms and dosage of drug, as shown in Figure 4.5(a). It seems likely that there are lots of symptoms when the dosage is low, only a few symptoms when the dosage is moderate, and lots of symptoms again when the dosage is high. Number of symptoms and drug dosage are curvilinearly related. One alternative in this case is to use the square of number of symptoms to represent the curvilinear relationship instead of number of symptoms in the analysis. Another alternative is to recode dosage into two dummy variables (high vs. low on one dummy variable and a combination of high and low vs. medium on another dummy variable) and then use the dummy variables in place of dosage in analysis.[15] The dichotomous dummy variables can only have a linear relationship with other variables, if, indeed, there is any relationship at all after recoding.

Often, two variables have a mix of linear and curvilinear relationships, as shown in Figure 4.5(b). One variable generally gets smaller (or larger) as the other gets larger (or smaller) but there is

---

[14]We realize that others (e.g., Berry, 1993; Fox, 1991) have very different views about the wisdom of screening from residuals.

[15]A nonlinear analytic strategy is most appropriate here, such as nonlinear regression through SAS NLIN, but such strategies are beyond the scope of this book.

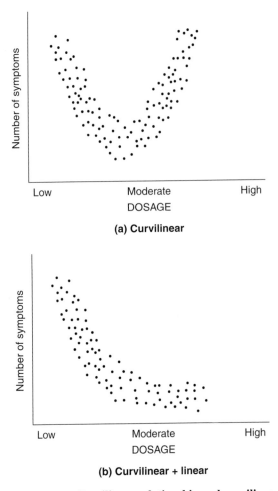

**FIGURE 4.5   Curvilinear relationship and curvilinear
plus linear relationship.**

also a curve to the relationship. For instance, symptoms might drop off with increasing dosage, but
only to a point; increasing dosage beyond the point does not result in further reduction or increase in
symptoms. In this case, the linear component may be strong enough that not much is lost by ignoring the curvilinear component unless it has important theoretical implications.

Assessing linearity through bivariate scatterplots is reminiscent of reading tea leaves, especially with small samples. And there are many cups of tea if there are several variables and all possible pairs are examined, especially when subjects are grouped and the analysis is done separately
within each group. If there are only a few variables, screening all possible pairs is not burdensome;
if there are numerous variables, you may want to use statistics on skewness to screen only pairs that
are likely to depart from linearity. Think, also, about pairs of variables that might have true nonlinearity and examine them through bivariate scatterplots. Bivariate scatterplots are produced by SPSS
GRAPH, and SAS PLOT, among other programs.

4.1.5.3   *Homoscedasticity, Homogeneity of Variance, and Homogeneity of Variance-Covariance Matrices*

For ungrouped data, the assumption of homoscedasticity is that the variability in scores for one continuous variable is roughly the same at all values of another continuous variable. For grouped data, this is the same as the assumption of homogeneity of variance when one of the variables is discrete (the grouping variable), the other is continuous (the DV); the variability in the DV is expected to be about the same at all levels of the grouping variable.

Homoscedasticity is related to the assumption of normality because when the assumption of multivariate normality is met, the relationships between variables are homoscedastic. The bivariate scatterplots between two variables are of roughly the same width all over with some bulging toward the middle. Homoscedasticity for a bivariate plot is illustrated in Figure 4.6(a).

Heteroscedasticity, the failure of homoscedasticity, is caused either by nonnormality of one of the variables or by the fact that one variable is related to some transformation of the other. Consider, for example, the relationship between age (X_1) and income (X_2), as depicted in Figure 4.6(b). People start out making about the same salaries, but with increasing age, people spread farther apart on income. The relationship is perfectly lawful, but it is not homoscedastic. In this example, income is likely to be positively skewed and transformation of income is likely to improve the homoscedasticity of its relationship with age.

Another source of heteroscedasticity is a greater error of measurement at some levels of an IV. For example, people in the age range 25 to 45 might be more concerned about their weight than people who are younger or older. Older and younger people would, as a result, give less reliable estimates of their weight, increasing the variance of weight scores at those ages.

It should be noted that heteroscedasticity is not fatal to an analysis of ungrouped data. The linear relationship between variables is captured by the analysis, but there is even more predictability if the heteroscedasticity is accounted for. If it is not, the analysis is weakened, but not invalidated.

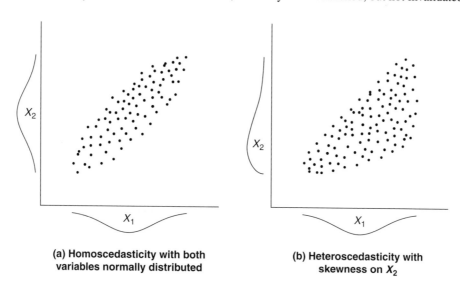

(a) Homoscedasticity with both
variables normally distributed

(b) Heteroscedasticity with
skewness on X_2

**FIGURE 4.6   Bivariate scatterplots under conditions of homoscedasticity
and heteroscedasticity.**

When data are grouped, homoscedasticity is known as homogeneity of variance. A great deal of research has assessed the robustness (or lack thereof) of ANOVA and ANOVA-like analyses to violation of homogeneity of variance. Recent guidelines have become more stringent than earlier, more cavalier ones. There are formal tests of homogeneity of variance but most too strict because they also assess normality. (An exception is Levene's test of homogeneity of variance, which is not typically sensitive to departures from normality.) Instead, once outliers are eliminated, homogeneity of variance is assessed with F_{max} in conjunction with sample-size ratios.

F_{max} is the ratio of the largest cell variance to the smallest. If sample sizes are relatively equal (within a ratio of 4 to 1 or less for largest to smallest cell size), an F_{max} as great as 10 is acceptable. As the cell size discrepancy increases (say, goes to 9 to 1 instead of 4 to 1), an F_{max} as small as 3 is associated with inflated Type I error if the larger variance is associated with the smaller cell size (Milligan, Wong, and Thompson, 1987).

Violations of homogeneity usually can be corrected by transformation of the DV scores. Interpretation, however, is then limited to the transformed scores. Another option is to use untransformed variables with a more stringent α level (for nominal $\alpha = .05$, use .025 with moderate violation and .01 with severe violation).

The multivariate analog of homogeneity of variance is homogeneity of variance-covariance matrices. As for univariate homogeneity of variance, inflated Type I error rate occurs when the greatest dispersion is associated with the smallest sample size. The formal test used by SPSS, Box's M, is too strict with the large sample sizes usually necessary for multivariate applications of ANOVA. Section 9.7.1.5 demonstrates an assessment of homogeneity of variance-covariance matrices through SAS DISCRIM using Bartlett's test. SAS DISCRIM permits a stringent α level for determining heterogeneity, and bases the discriminant analysis on separate variance-covariance matrices when the assumption of homogeneity is violated.

4.1.6   Common Data Transformations

Although data transformations are recommended as a remedy for outliers and for failures of normality, linearity, and homoscedasticity, they are not universally recommended. The reason is that an analysis is interpreted from the variables that are in it, and transformed variables are sometimes harder to interpret. For instance, although IQ scores are widely understood and meaningfully interpreted, the logarithm of IQ scores may be harder to explain.

Whether transformation increases difficulty of interpretation often depends on the scale in which the variable is measured. If the scale is meaningful or widely used, transformation often hinders interpretation, but if the scale is somewhat arbitrary anyway (as is often the case), transformation does not notably increase the difficulty of interpretation.

With ungrouped data, it is probably best to transform variables to normality unless interpretation is not feasible with the transformed scores. With grouped data, the assumption of normality is evaluated with respect to the sampling distribution of means (not the distribution of scores) and the Central Limit Theorem predicts normality with decently sized samples. However, transformations may improve the analysis and may have the further advantage of reducing the impact of outliers. Our recommendation, then, is to consider transformation of variables in all situations unless there is some reason not to.

If you decide to transform, it is important to check that the variable is normally or nearnormally distributed after transformation. Often you need to try first one transformation and then another until you find the transformation that produces the skewness and kurtosis values nearest zero, the prettiest picture, and/or the fewest outliers.

With almost every data set in which we have used transformations, the results of analysis have been substantially improved. This is particularly true when some variables are skewed and others are not, or variables are skewed very differently prior to transformation. However, if all the variables are skewed to about the same moderate extent, improvements of analysis with transformation are often marginal.

With grouped data, the test of mean differences after transformation is a test of differences between medians in the original data. After a distribution is normalized by transformation, the mean is equal to the median. The transformation affects the mean but not the median because the median depends only on rank order of cases. Therefore, conclusions about means of transformed distributions apply to medians of untransformed distributions. Transformation is undertaken because the distribution is skewed and the mean is not a good indicator of the central tendency of the scores in the distribution. For skewed distributions, the median is often a more appropriate measure of central tendency than the mean, anyway, so interpretation of differences in medians is appropriate.

Variables differ in the extent to which they diverge from normal. Figure 4.7 presents several distributions together with the transformations that are likely to render them normal. If the distribution differs moderately from normal, a square root transformation is tried first. If the distribution

TRANSFORMATION

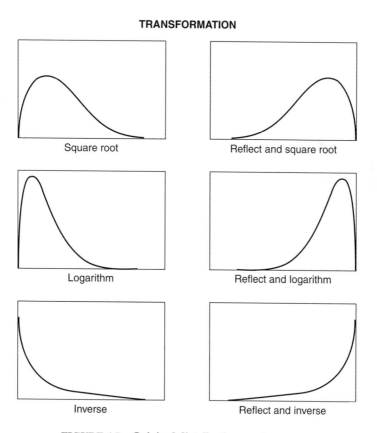

FIGURE 4.7   Original distributions and common
transformations to produce normality.

differs substantially, a log transformation is tried. If the distribution differs severely, the inverse is tried. According to Bradley (1982), the inverse is the best of several alternatives for J-shaped distributions, but even it may not render the distribution normal. Finally, if the departure from normality is severe and no transformation seems to help, you may want to try dichotomizing the variable.

The direction of the deviation is also considered. When distributions have positive skewness, as discussed earlier, the long tail is to the right. When they have negative skewness, the long tail is to the left. If there is negative skewness, the best strategy is to *reflect* the variable and then apply the appropriate transformation for positive skewness.[16] To reflect a variable, find the largest score in the distribution and add one to it to form a constant that is larger than any score in the distribution. Then create a new variable by subtracting each score from the constant. In this way, a variable with negative skewness is converted to one with positive skewness prior to transformation. When you interpret a reflected variable, be sure to reverse the direction of the interpretation as well (or consider rereflecting it after transformation).

Remember to check your transformations after applying them. If a variable is only moderately positively skewed, for instance, a square root transformation may make the variable moderately negatively skewed, and there is no advantage to transformation. Often you have to try several transformations before you find the most helpful one.

Syntax for transforming variables in SPSS and SAS is given in Table 4.3.[17] Notice that a constant is also added if the distribution contains a value less than one. A constant (to bring the smallest value to at least one) is added to each score to avoid taking the log, square root, or inverse of zero.

Different software packages handle missing data differently in various transformations. Be sure to check the manual to ensure that the program is treating missing data the way you want it to in the transformation.

It should be clearly understood that this section merely scratches the surface of the topic of transformations, about which a great deal more is known. The interested reader is referred to Box and Cox (1964) or Mosteller and Tukey (1977) for a more flexible and challenging approach to the problem of transformation.

4.1.7 Multicollinearity and Singularity

Multicollinearity and singularity are problems with a correlation matrix that occur when variables are too highly correlated. With multicollinearity, the variables are very highly correlated (say, .90 and above); with singularity, the variables are redundant; one of the variables is a combination of two or more of the other variables.

For example, scores on the Wechsler Adult Intelligence Scale (the WAIS) and scores on the Stanford-Binet Intelligence Scale are likely to be *multicollinear* because they are two similar measures of the same thing. But the total WAIS IQ score is *singular* with its subscales because the total score is found by combining subscale scores. When variables are multicollinear or singular, they

---

[16]Remember, however, that the interpretation of a reflected variable is just the opposite of what it was; if big numbers meant good things prior to reflecting the variable, big numbers mean bad things afterwards.

[17]Logarithmic (LO) and power (PO) transformations are also available in PRELIS for variables used in structural equation modeling (Chapter 14). A γ (GA) value is specified for power transformations; for example, $\gamma = 1/2$ provides a square root transform (PO GA = .5).

TABLE 4.3   Syntax for Common Data Transformations

	SPSS COMPUTE	SASᵃ DATA Procedure
Moderate positive skewness	NEWX=SQRT(X)	NEWX=SQRT(X)
Substantial positive skewness	NEWX=LG10(X)	NEWX=LOG10(X)
With zero	NEWX=LG10(X + C)	NEWX=LOG10(X + C)
Severe positive skewness	NEWX=1/X	NEWX=1/X
L-shaped With zero	NEWX=1/(X + C)	NEWX=1/(X+C)
Moderate negative skewness	NEWX=SQRT(K − X)	NEWX=SQRT(K−X)
Substantial negative skewness	NEWX=LG10(K − X)	NEWX=LOG10(K−X)
Severe negative skewness J-shaped	NEWX=1/(K − X)	NEWX=1/(K−X)

C = a constant added to each score so that the smallest score is 1.

K = a constant from which each score is subtracted so that the smallest score is 1; usually equal to the largest score + 1.

ᵃAlso may be done through SAS Interactive Data Analysis.

contain redundant information and they are not all needed in the same analysis. In other words, there are fewer variables than it appears and the correlation matrix is not of full rank because there are not really as many variables as columns.

Either bivariate or multivariate correlations can create multicollinearity or singularity. If a bivariate correlation is too high, it shows up in a correlation matrix as a correlation above .90, and, after deletion of one of the two redundant variables, the problem is solved. If it is a multivariate correlation that is too high, diagnosis is slightly more difficult because multivariate statistics are needed to find the offending variable. For example, although the WAIS IQ is a combination of its subscales, the bivariate correlations between total IQ and each of the subscale scores are not all that high. You would not know there was singularity by examination of the correlation matrix.

Multicollinearity and singularity cause both logical and statistical problems. The logical problem is that unless you are doing analysis of structure (factor analysis, principal components analysis, and structural-equation modeling), it is not a good idea to include redundant variables in the same analysis. They are not needed and, because they inflate the size of error terms, they actually weaken an analysis. Unless you are doing analysis of structure or are dealing with repeated measures of the same variable (as in various forms of ANOVA including profile analysis), think carefully before including two variables with a bivariate correlation of, say, .70 or more in the same analysis. You might omit one of the variables or you might create a composite score from the redundant variables.

The statistical problems created by singularity and multicollinearity occur at much higher correlations (.90 and higher). The problem is that singularity prohibits, and multicollinearity renders unstable, matrix inversion. Matrix inversion is the logical equivalent of division; calculations requiring division (and there are many of them—see the fourth sections of Chapters 5 through 16 and 18) cannot be performed on singular matrices because they produce determinants equal to zero that cannot be used as divisors (see Appendix A). Multicollinearity often occurs when you form cross-products or powers of variables and include them in the analysis along with the original variables, unless steps are taken to reduce the multicollinearity (Section 5.6.6).

With multicollinearity, the determinant is not exactly zero, but it is zero to several decimal places. Division by a near-zero determinant produces very large and unstable numbers in the inverted matrix. The sizes of numbers in the inverted matrix fluctuate wildly with only minor changes (say, in the second or third decimal place) in the sizes of the correlations in **R.** The portions of the multivariate solution that flow from an inverted matrix that is unstable are also unstable. In regression, for instance, error terms get so large that none of the coefficients is significant (Berry, 1993). For example, when r is .9, the precision of estimation of regression coefficients is halved (Fox, 1991).

Most programs protect against multicollinearity and singularity by computing SMCs for the variables. SMC is the squared multiple correlation of a variable where it serves as DV with the rest as IVs in multiple correlation (see Chapter 5). If the SMC is high, the variable is highly related to the others in the set and you have multicollinearity. If the SMC is 1, the variable is perfectly related to others in the set and you have singularity. Many programs convert the SMC values for each variable to tolerance $(1 - \text{SMC})$ and deal with tolerance instead of SMC.

Screening for singularity often takes the form of running your main analysis to see if the computer balks. Singularity aborts most runs except those for principal components analysis (see Chapter 13), where matrix inversion is not required. If the run aborts, you need to identify and delete the offending variable. A first step is to think about the variables. Did you create any of them from others of them; for instance, did you create one of them by adding together two others? If so, deletion of one removes singularity.

Screening for multicollinearity that causes statistical instability is also routine with most programs because they have tolerance criteria for inclusion of variables. If the tolerance $(1 - \text{SMC})$ is too low, the variable does not enter the analysis. Default tolerance levels range between .01 and .0001, so SMCs are .99 to .9999 before variables are excluded. You may wish to take control of this process, however, by adjusting the tolerance level (an option with many programs) or deciding yourself which variable(s) to delete instead of letting the program make the decision on purely statistical grounds. For this you need SMCs for each variable. Note that SMCs are *not* evaluated separately for each group if you are analyzing grouped data.

SMCs are available through factor analysis and regression programs in all packages. PRELIS provides SMCs for structural equation modeling. SAS and SPSS have incorporated *collinearity diagnostics* proposed by Belsely, Kuh, and Welsch (1980) in which a conditioning index is produced, as well as variance proportions associated with each variable, after standardization, for each root (see Chapters 12 and 13 and Appendix A for a discussion of roots and dimensions). Two or more variables with large variance proportions on the same dimension are those with problems.

Condition index is a measure of tightness or dependency of one variable on the others. The condition index is monotonic with SMC, but not linear with it. A high condition index is associated with variance inflation in the standard error of the parameter estimate for a variable. When its standard error becomes very large, the parameter estimate is highly uncertain. Each root (dimension) accounts for some proportion of the variance of each parameter estimated. A collinearity problem occurs when a root with a high condition index contributes strongly (has a high variance proportion) to the variance of two or more variables. Criteria for multicollinearity suggested by Belsely et al. (1980) are a conditioning index greater than 30 for a given dimension coupled with variance proportions greater than .50 for at least two different variables. Collinearity diagnostics are demonstrated in Section 4.2.1.6.

There are several options for dealing with collinearity if it is detected. First, if the only goal of analysis is prediction, you can ignore it. A second option is to delete the variable with the highest variance proportion. A third option is to sum or average the collinear variables. A fourth option is to compute principal components and use the components as the predictors instead of the original variables (see Chapter 13). A final alternative is to center one or more of the variables, as discussed in Chapters 5 and 15, if multicollinearity is caused by forming interactions or powers of continuous variables.

4.1.8 A Checklist and Some Practical Recommendations

Table 4.4 is a checklist for screening data. It is important to consider all the issues prior to the fundamental analysis lest you be tempted to make some of your decisions on the basis of how they influence the analysis. If you choose to screen through residuals, you cannot avoid doing an analysis at the same time; however, in these cases, you concentrate on the residuals and not on the other features of the analysis while making your screening decisions.

TABLE 4.4   Checklist for Screening Data

1. Inspect univariate descriptive statistics for accuracy of input
   a. Out-of-range values
   b. Plausible means and standard deviations
   c. Univariate outliers
2. Evaluate amount and distribution of missing data; deal with problem
3. Check pairwise plots for nonlinearity and heteroscedasticity
4. Identify and deal with nonnormal variables and univariate outliers
   a. Check skewness and kurtosis, probability plots
   b. Transform variables (if desirable)
   c. Check results of transformation
5. Identify and deal with multivariate outliers
   a. Variables causing multivariate outliers
   b. Description of multivariate outliers
6. Evaluate variables for multicollinearity and singularity

The order in which screening takes place is important because the decisions that you make at one step influence the outcomes of later steps. In a situation where you have both nonnormal variables and potential univariate outliers, a fundamental decision is whether you would prefer to transform variables, delete cases, or change scores on cases. If you transform variables first, you are likely to find fewer outliers. If you delete or modify the outliers first, you are likely to find fewer variables with nonnormality.

Of the two choices, transformation of variables is usually preferable. It typically reduces the number of outliers. It is likely to produce normality, linearity, and homoscedasticity among the variables. It increases the likelihood of multivariate normality to bring the data into conformity with one of the fundamental assumptions of most inferential tests. And on a very practical level, it usually enhances the analysis even if inference is not a goal. On the other hand, transformation may threaten interpretation, in which case all the statistical niceties are of little avail.

Or, if the impact of outliers is reduced first, you are less likely to find variables that are skewed because significant skewness is sometimes caused by extreme cases on the tails of the distributions. If you have cases that are univariate outliers because they are not part of the population from which you intended to sample, by all means delete them before checking distributions.

Last, as will become obvious in the next two sections, although the issues are different, the runs on which they are screened are not necessarily different. That is, the same run often provides you with information regarding two or more issues.

4.2   Complete Examples of Data Screening

Evaluation of assumptions is somewhat different for ungrouped and grouped data. That is, if you are going to perform multiple regression, canonical correlation, factor analysis, or structural equation modeling on ungrouped data, there is one approach to screening. If you are going to perform univariate or multivariate analysis of variance (including profile analysis), discriminant analysis, or multilevel modeling on grouped data, there is another approach to screening.[18]

Therefore, two complete examples are presented that use the same set of variables taken from the research described in Appendix B: number of visits to health professionals (TIMEDRS), attitudes toward drug use (ATTDRUG), attitudes toward housework (ATTHOUSE), INCOME, marital status (MSTATUS), and RACE. The grouping variable used in the analysis of grouped data is current employment status (EMPLMNT).[19] Data are in files labeled SCREEN.*.

Where possible in these examples, and for illustrative purposes, screening for ungrouped data is performed using SPSS, and screening for grouped data is performed using SAS programs.

4.2.1   Screening Ungrouped Data

A flow diagram for screening ungrouped data appears as Figure 4.8. The direction of flow assumes that data transformation is undertaken, as necessary. If transformation is not acceptable, then other procedures for handling outliers are used.

---

[18]If you are using multiway frequency analysis or logistic regression, there are far fewer assumptions than with these other analyses.

[19]This is a motley collection of variables chosen primarily for their statistical properties.

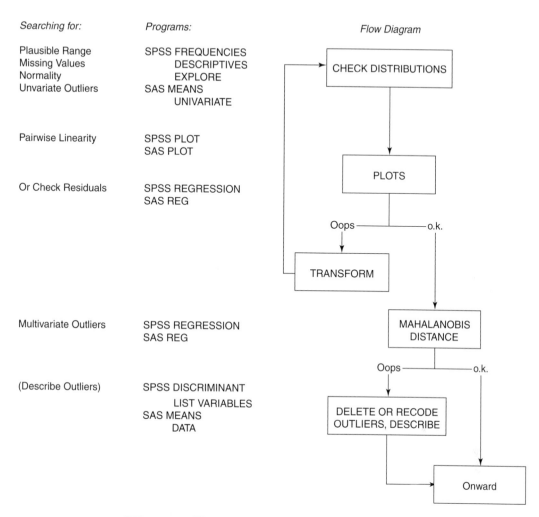

FIGURE 4.8   Flow diagram for screening ungrouped data.

4.2.1.1   Accuracy of Input, Missing Data, Distributions, and Univariate Outliers

A check on accuracy of data entry, missing data, skewness, and kurtosis for the data set is done through SPSS FREQUENCIES, as shown in Table 4.5.

The minimum and maximum values, means, and standard deviations of each of the variables are inspected for plausibility. For instance, the **Minimum** number of visits to health professionals (TIMEDRS) is 0 and the **Maximum** is 81, higher than expected but found to be accurate on checking the data sheets.[20] The **Mean** for the variable is 7.901, higher than the national average but not extremely so, and the standard deviation (**Std. Deviation**) is 10.948. These values are all reasonable, as are the values on the other variables. For instance, the ATTDRUG variable is constructed with a range of 5 to 10, so it is reassuring to find these values as **Minimum** and **Maximum**.

---

[20]The woman with this number of visits was terminally ill when she was interviewed.

TABLE 4.5 Syntax and SPSS FREQUENCIES Output Showing Descriptive Statistics and Histograms for Ungrouped Data

```
FREQUENCIES
  VARIABLES=timedrs attdrug atthouse income mstatus race
  /FORMAT=NOTABLE
  /STATISTICS=STDDEV VARIANCE MINIMUM MAXIMUM MEAN SKEWNESS SESKEW KURTOSIS
  SEKURT
  /HISTOGRAM NORMAL
  /ORDER= ANALYSIS
```

Statistics

		Visits to health professionals	Attitudes toward medication	Attitudes toward housework	Income	Whether currently married	Race
N	Valid	465	465	464	439	465	465
	Missing	0	0	1	26	0	0
Mean		7.90	7.69	23.54	4.21	1.78	1.09
Std. Deviation		10.948	1.156	4.484	2.419	.416	.284
Variance		119.870	1.337	20.102	5.851	.173	.081
Skewness		3.248	-.123	-.457	.582	-1.346	2.914
Std. Error of Skewness		.113	.113	.113	.117	.113	.113
Kurtosis		13.101	-.447	1.556	-.359	-.190	6.521
Std. Error of Kurtosis		.226	.226	.226	.233	.226	.226
Minimum		0	5	2	1	1	1
Maximum		81	10	35	10	2	2

TABLE 4.5   Continued

Histogram

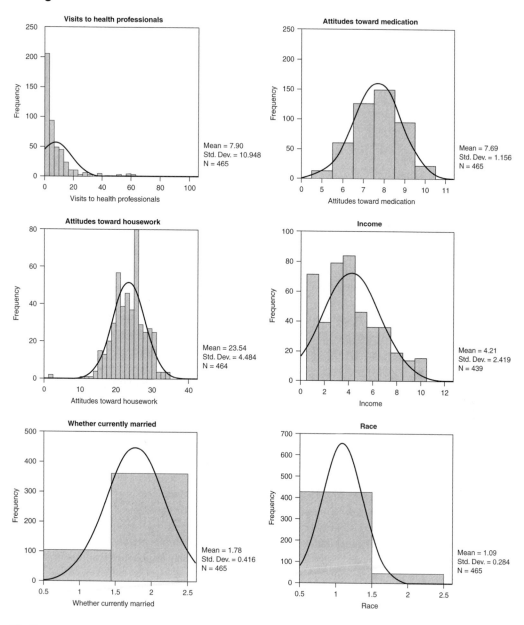

TIMEDRS shows no Missing cases but has strong positive Skewness (3.248). The significance of Skewness is evaluated by dividing it by Std. Error of Skewness, as in Equation 4.5,

$$z = \frac{3.248}{.113} = 28.74$$

to reveal a clear departure from symmetry. The distribution also has significant Kurtosis as evaluated by Equation 4.7,

$$z = \frac{13.101}{.226} = 57.97$$

The departures from normality are also obvious from inspection of the difference between frequencies expected under the normal distribution (the superimposed curve) and obtained frequencies. Because this variable is a candidate for transformation, evaluation of univariate outliers is deferred.

ATTDRUG, on the other hand, is well behaved. There are no Missing cases, and Skewness and Kurtosis are well within expected values. ATTHOUSE has a single missing value but is otherwise well distributed except for the two extremely low scores. The score of 2 is 4.8 standard deviations below the mean of ATTHOUSE (well beyond the $p = .001$ criterion of 3.29, two-tailed) and is disconnected from the other cases. It is not clear whether these are recording errors or if these two women actually enjoy housework that much. In any event, the decision is made to delete from further analysis the data from the two women with extremely favorable attitudes toward housework.

Information about these deletions is included in the report of results. The single missing value is replaced with the mean. (Section 10.7.1.1 illustrates a more sophisticated way of dealing with missing data in SPSS when the amount missing is greater than 5%.)

On INCOME, however, there are 26 cases with Missing values—more than 5% of the sample. If INCOME is not critical to the hypotheses, we delete it in subsequent analyses. If INCOME is important to the hypotheses, we could replace the missing values.

The two remaining variables are dichotomous and not evenly split. MSTATUS has a 362 to 103 split, roughly a 3.5 to 1 ratio, that is not particularly disturbing. But RACE, with a split greater than 10 to 1 is marginal. For this analysis, we choose to retain the variable, realizing that its association with other variables is deflated because of the uneven split.

Table 4.6 shows the distribution of ATTHOUSE with elimination of the univariate outliers. The mean for ATTHOUSE changes to 23.634, the value used to replace the missing ATTHOUSE score in subsequent analyses. The case with a missing value on ATTHOUSE becomes complete and available for use in all computations. The COMPUTE instructions filter out cases with values equal to or less than 2 on ATTHOUSE (univariate outliers) and the RECODE instruction sets the missing value to 23.63.

At this point, we have investigated the accuracy of data entry and the distributions of all variables, determined the number of missing values, found the mean for replacement of missing data, and found two univariate outliers that, when deleted, result in $N = 463$.

4.2.1.2 Linearity and Homoscedasticity

Because of nonnormality on at least one variable, SPSS GRAPH is run to check the bivariate plots for departures from linearity and homoscedasticity, as reproduced in Figure 4.9. The variables

TABLE 4.6   Syntax and SPSS FREQUENCIES Output Showing Descriptive Statistics and Histograms for ATTHOUSE with Univariate Outliers Deleted

```
USE ALL.
COMPUTE filter_$=(atthouse > 2).
VARIABLE LABEL filter_$ 'atthouse > 2 (FILTER)'.
VALUE LABELS filter_$ 0 'Not Selected' 1 'Selected'.
FORMAT filter_$ (f1.0).
FILTER BY filter_$.
EXECUTE.
RECODE
  atthouse (SYSMIS=23.63).
EXECUTE.
FREQUENCIES
  VARIABLES=atthouse /FORMAT=NOTABLE
  /STATISTICS=STDDEV VARIANCE MINIMUM MAXIMUM MEAN SKEWNESS SESKEW KURTOSIS
  SEKURT
  /HISTOGRAM NORMAL
  /ORDER= ANALYSIS.
```

Frequencies

Statistics

Attitudes toward housework

N	Valid	463
	Missing	0
Mean		23.63
Std. Deviation		4.258
Variance		18.128
Skewness		−.038
Std. Error of Skewness		.113
Kurtosis		−.254
Std. Error of Kurtosis		.226
Minimum		11
Maximum		35

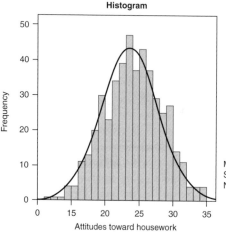

Histogram

Mean = 23.63
Std. Dev. = 4.258
N = 463

picked as *worst case* are those with the most discrepant distributions: TIMEDRS, which has the greatest departure from normality, and ATTDRUG, which is nicely distributed. (The SELECT IF instruction eliminates the univariate outliers on ATTHOUSE.)

In Figure 4.9, ATTDRUG is along the *Y* axis; turn the page so that the *Y* axis becomes the *X* axis and you can see the symmetry of the ATTDRUG distribution. TIMEDRS is along the *X* axis. The asymmetry of the distribution is apparent from the pileup of scores at low values of the variable. The overall shape of the scatterplot is not oval; the variables are not linearly related. Heteroscedasticity is evident in the greater variability in ATTDRUG scores for low than high values of TIMEDRS.

USEALL.
COMPUTE filter_$=(atthouse ~= 2).
VARIABLE LABEL filter_$ 'atthouse ~= 2 (FILTER)'
VALUE LABELS filter_$ 0 'Not Selected' 1 'Selected'.
FORMAT filter_$ (f1.0).
FILTER BY filter_$.
EXECUTE.
GRAPH
  /SCATTERPLOT(BIVAR)=timedrs WITH attdrug
  /MISSING=LISTWISE.

Graph

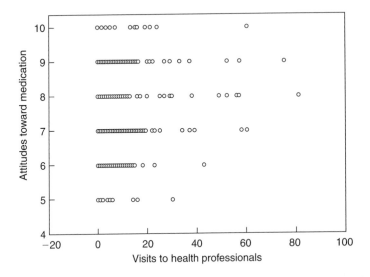

FIGURE 4.9   **Assessment of linearity through bivariate scatterplots, as produced by
SPSS GRAPH. This indicates ATTDRUG is normal; TIMEDRS is nonnormal.**

4.2.1.3   *Transformation*

Variables are transformed prior to searching for multivariate outliers. A logarithmic transformation is applied to TIMEDRS to overcome strong skewness. Because the smallest value on the variable is zero, one is added to each score as the transformation is performed, as indicated in the COMPUTE statement. Table 4.7 shows the distribution of TIMEDRS as transformed to LTIMEDRS.

Skewness is reduced from 3.248 to 0.221 and Kurtosis is reduced from 13.101 to -0.183 by the transformation. The frequency plot is not exactly pleasing (the frequencies are still too high for small scores), but the statistical evaluation of the distribution is much improved.

Figure 4.10 is a bivariate scatterplot between ATTDRUG and LTIMEDRS. Although still not perfect, the overall shape of the scatterplot is more nearly oval. The nonlinearity associated with nonnormality of one of the variables is "fixed" by transformation of the variable.

TABLE 4.7   Syntax and SPSS FREQUENCIES Output Showing Descriptive Statistics and Histograms for TIMEDRS after Logarithmic Transform

```
COMPUTE ltimedrs = lg10(timedrs+1).
EXECUTE.
FREQUENCIES
  VARIABLES=ltimedrs /FORMAT=NOTABLE
  /STATISTICS=STDDEV VARIANCE MINIMUM MAXIMUM MEAN SKEWNESS SESKEW KURTOSIS
  SEKURT
  /HISTOGRAM NORMAL
  /ORDER = ANALYSIS.
```

Frequencies

Statistics

ltimedrs

N	Valid	463
	Missing	0
Mean		.7424
Std. Deviation		.41579
Variance		.173
Skewness		.221
Std. Error of Skewness		.113
Kurtosis		−.183
Std. Error of Kurtosis		.226
Minimum		.00
Maximum		1.91

Histogram

Mean = 0.7424
Std. Dev. = 0.41579
N = 463

4.2.1.4   *Detecting Multivariate Outliers*

The 463 cases, with transformation applied to LTIMEDRS, are screened for multivariate outliers through SPSS REGRESSION (Table 4.8) using the RESIDUALS=OUTLIERS(MAHAL) syntax added to menu choices. Case labels (SUBNO) are used as the dummy DV, convenient because multivariate outliers among IVs are unaffected by the DV.[21] The remaining VARIABLES are considered independent variables.

The criterion for multivariate outliers is Mahalanobis distance at $p < .001$. Mahalanobis distance is evaluated as χ^2 with degrees of freedom equal to the number of variables, in this case five: LTIMEDRS, ATTDRUG, ATTHOUSE, MSTATUS, and RACE. Any case with a Mahal. Distance in Table 4.8 greater than $\chi^2(5) = 20.515$ (cf. Appendix C, Table C.4), then, is a multivariate outlier. As shown in Table 4.8, cases 117 and 193 are outliers among these variables in this data set.

---

[21]For a multiple-regression analysis, the actual DV would be used here rather than SUBNO as a dummy DV.

GRAPH
 /SCATTERPLOT(BIVAR) = ltimedrs WITH attdrug
 /MISSING=LISTWISE.

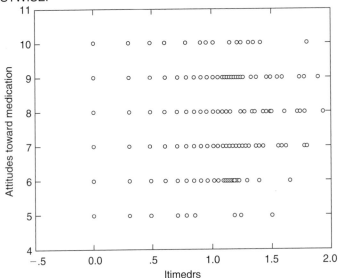

FIGURE 4.10 Assessment of linearity after log transformation of TIMEDRS, as produced by SPSS GRAPH.

There are 461 cases remaining if the two multivariate outliers are deleted. Little is lost by deleting the additional two outliers from the sample although transformation is an alternative because they are not particularly disconnected from the remaining cases. It is also necessary to determine why the two cases are multivariate outliers, to know how their deletion limits generalizability, and to include that information in the Results section.

4.2.1.5 Variables Causing Cases to Be Outliers

SPSS REGRESSION is used to identify the combination of variables on which case 117 (subject number 137 as found in the data editor) and case 193 (subject number 262) deviate from the remaining 462

TABLE 4.8 Syntax and Selected SPSS REGRESSION Output for Multivariate Outliers and Multicollinearity

REGRESSION
 /MISSING LISTWISE
 /STATISTICS COEFF OUTS R ANOVA COLLIN TOL
 /CRITERIA=PIN(.05) POUT(.10)
 /NOORIGIN
 /DEPENDENT subno
 /METHOD=ENTER attdrug atthouse mstatus race ltimedrs
 /RESIDUALS=OUTLIERS(MAHAL).

TABLE 4.8   Continued

Regression

Collinearity Diagnostics[a]

Model	Dimension	Eigenvalue	Condition Index	Variance Proportions					
				(Constant)	Attitudes toward medication	Attitudes toward housework	Whether currently married	Race	ltimedrs
1	1	5.656	1.000	.00	.00	.00	.00	.00	.01
	2	.210	5.193	.00	.00	.00	.01	.02	.92
	3	.060	9.688	.00	.00	.01	.29	.66	.01
	4	.043	11.508	.00	.03	.29	.46	.16	.06
	5	.025	15.113	.00	.53	.41	.06	.04	.00
	6	.007	28.872	.99	.43	.29	.18	.12	.00

[a]Dependent Variable: Subject number

Outlier Statistics[a]

		Case Number	Statistic
Mahal. Distance	1	117	21.837
	2	193	20.650
	3	435	19.968
	4	99	18.499
	5	335	18.469
	6	292	17.518
	7	58	17.373
	8	71	17.172
	9	102	16.942
	10	196	16.723

[a]Dependent Variable: Subject number

TABLE 4.9   SPSS REGRESSION Syntax and Partial Output Showing Variables Causing the 117th Case to Be an Outlier

```
COMPUTE dummy = 0.
EXECUTE.
IF (subno=137) dummy = 1.
EXECUTE.
REGRESSION
 /MISSING LISTWISE
 /STATISTICS COEFF OUTS
 /CRITERIA=PIN(.05) POUT(.10)
 /NOORIGIN
 /DEPENDENT dummy
 /METHOD=STEPWISE attdrug atthouse emplmnt mstatus race ltimedrs.
```

Coefficients[a]

Model		Unstandardized Coefficients		Standardized Coefficients	t	Sig.
		B	Std. Error	Beta		
1	(Constant)	−.024	.008		−2.881	.004
	race	.024	.008	.149	3.241	.001
2	(Constant)	.009	.016		.577	.564
	race	.025	.007	.151	3.304	.001
	Attitudes toward medication	−.004	.002	−.111	−2.419	.016
3	(Constant)	.003	.016		.169	.866
	race	.026	.007	.159	3.481	.001
	Attitudes toward medication	−.005	.002	−.123	−2.681	.008
	ltimedrs	.012	.005	.109	2.360	.019

[a]Dependent Variable: DUMMY

cases. Each outlying case is evaluated in a separate SPSS REGRESSION run after a dummy variable is created to separate the outlying case from the remaining cases. In Table 4.9, the dummy variable for subject 137 is created in the COMPUTE instruction with dummy = 0 and if (subno=137) dummy = 1. With the dummy variable as the DV and the remaining variables as IVs, you can find the variables that distinguish the outlier from the other cases.

For the 117th case (subno=137), RACE, ATTDRUG, and LTIMEDRS show up as significant predictors of the case (Table 4.10).

Variables separating case 193 (subno=262) from the other cases are RACE and LTIMEDRS. The final step in evaluating outlying cases is to determine how their scores on the variables that cause them to be outliers differ from the scores of the remaining sample. The SPSS LIST and DESCRIP-TIVES procedures are used, as shown in Table 4.11. The LIST procedure is run for each outlying

TABLE 4.10   SPSS REGRESSION Syntax and Partial Output Showing Variables Causing the 193rd Case to Be an Outlier

```
IF (subno=137) dummy = 0.
EXECUTE.
IF (subno=262) dummy = 1.
EXECUTE.
REGRESSION
 /MISSING LISTWISE
 /STATISTICS COEFF OUTS
 /CRITERIA=PIN(.05) POUT(.10)
 /NOORIGIN
 /DEPENDENT dummy
 /METHOD=STEPWISE attdrug atthouse emplmnt mstatus race ltimedrs.
```

Coefficients[a]

Model		Unstandardized Coefficients		Standardized Coefficients	t	Sig.
		B	Std. Error	Beta		
1	(Constant)	−.024	.008		−2.881	.004
	race	.024	.008	.149	3.241	.001
2	(Constant)	−.036	.009		−3.787	.000
	race	.026	.007	.158	3.436	.001
	ltimedrs	.014	.005	.121	2.634	.009

[a]Dependent Variable: DUMMY

TABLE 4.11   Syntax and SPSS Output Showing Variable Scores for Multivariate Outliers and Descriptive Statistics for All Cases

```
LIST VARIABLES=subno attdrug atthouse mstatus race ltimedrs
 /CASES FROM 117 TO 117.
LIST VARIABLES=subno attdrug atthouse mstatus race ltimedrs
 /CASES FROM 193 TO 193.
DESCRIPTIVES attdrug atthouse mstatus race ltimedrs.
```

List

```
subno    attdrug    atthouse    mstatus    race    ltimedrs
 137          5         24           2        2        1.49
Number of cases read: 117 Number of cases listed: 1
```

List

```
subno    attdrug    atthouse    mstatus    race    ltimedrs
 262          9         31           2        2        1.72
Number of cases read: 193 Number of cases listed: 1
```

(continued)

TABLE 4.11 Continued

Descriptives

Descriptive Statistics

	N	Minimum	Maximum	Mean	Std. Deviation
Attitudes toward medication	463	5	10	7.68	1.158
Attitudes toward housework	463	11	35	23.63	4.258
Whether currently married	463	1	2	1.78	.413
race	463	1	2	1.09	.284
ltimedrs	463	.00	1.91	.7424	.41579
Valid N (listwise)	463				

case to show its values on all the variables of interest. Then DESCRIPTIVES is used to show the average values for the remaining sample against which the outlying cases are compared.[22]

The 117th case is nonwhite on RACE, has very unfavorable attitudes regarding use of drugs (the lowest possible score on ATTDRUG), and has a high score on LTIMEDRS. The 193rd case is also nonwhite on RACE and has a very high score on LTIMEDRS. There is some question, then, about the generalizability of subsequent findings to nonwhite women who make numerous visits to physicians, especially in combination with unfavorable attitude toward use of drugs.

4.2.1.6 Multicollinearity

Evaluation of multicollinearity is produced in SPSS through the STATISTICS COLLIN instruction. As seen by the Collinearity Diagnostics output of Table 4.8, no multicollinearity is evident. Although the last root has a Condition Index that approaches 30, no dimension (row) has more than one Variance Proportion greater than .50.

Screening information as it might be described in a Results section of a journal article appears next.

Results

Prior to analysis, number of visits to health professionals, attitude toward drug use, attitude toward housework, income, marital status, and race were examined through various SPSS programs for accuracy of data entry, missing values, and fit between their

---

[22]These values are equal to those shown in the earlier FREQUENCIES runs but for deletion of the two univariate outliers.

distributions and the assumptions of multivariate analysis. The
single missing value on attitude toward housework was replaced by
the mean for all cases, while income, with missing values on more
than 5% of the cases, was deleted. The poor split on race (424 to
41) truncates its correlations with other variables, but it was
retained for analysis. To improve pairwise linearity and to reduce
the extreme skewness and kurtosis, visits to health professionals
was logarithmically transformed.

Two cases with extremely low z scores on attitude toward
housework were found to be univariate outliers; two other cases
were identified through Mahalanobis distance as multivariate outliers with $p < .001$.[23] All four outliers were deleted, leaving 461
cases for analysis.

4.2.2   Screening Grouped Data

For this example, the cases are divided into two groups according to the EMPLMNT (employment
status) variable; there are 246 cases who have paid work (EMPLMNT = 1) and 219 cases who are
housewives (EMPLMNT = 0). For illustrative purposes, variable transformation is considered inappropriate for this example, to be undertaken only if proved necessary. A flow diagram for screening
grouped data appears in Figure 4.11.

4.2.2.1   Accuracy of Input, Missing Data, Distributions, Homogeneity of Variance, and Univariate Outliers

SAS MEANS and Interactive Data Analysis provide descriptive statistics and histograms, respectively, for each group separately, as shown in Table 4.12. Menu choices are shown for SAS Interactive
Data Analysis because no SAS log file is provided. Note that data must be sorted by EMPLMNT
group before analyzing separately by groups. As with ungrouped data, accuracy of input is judged by
plausible Means and Std Devs and reasonable Maximum and Minimum values. The distributions are
judged by their overall shapes within each group. TIMEDRS is just as badly skewed when grouped as
when ungrouped, but this is of less concern when dealing with sampling distributions based on over

---

[23]Case 117 was nonwhite with very unfavorable attitudes regarding use of drugs but numerous visits to physicians. Case 193
was also nonwhite with numerous visits to physicians. Results of analysis may not generalize to nonwhite women with
numerous visits to physicians, particularly if they have very unfavorable attitudes toward use of drugs.

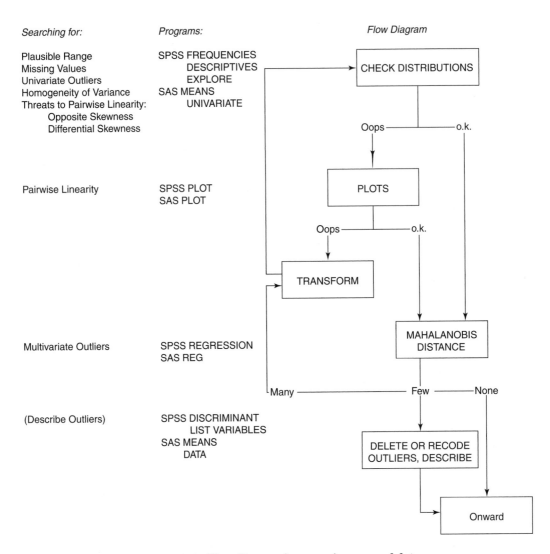

Searching for:

Plausible Range
Missing Values
Univariate Outliers
Homogeneity of Variance
Threats to Pairwise Linearity:
    Opposite Skewness
    Differential Skewness

Pairwise Linearity

Multivariate Outliers

(Describe Outliers)

Programs:

SPSS FREQUENCIES
    DESCRIPTIVES
    EXPLORE
SAS MEANS
    UNIVARIATE

SPSS PLOT
SAS PLOT

SPSS REGRESSION
SAS REG

SPSS DISCRIMINANT
    LIST VARIABLES
SAS MEANS
    DATA

Flow Diagram

CHECK DISTRIBUTIONS

Oops ——————— o.k.

PLOTS

Oops ——————— o.k.

TRANSFORM

MAHALANOBIS DISTANCE

Many ——————— Few ———— None

DELETE OR RECODE OUTLIERS, DESCRIBE

Onward

FIGURE 4.11   Flow diagram for screening grouped data.

200 cases unless the skewness causes nonlinearity among variables or there are outliers. ATTDRUG remains well distributed within each group. As shown in Table 4.12, the ATTHOUSE variable is nicely distributed as well, but the two cases in the employment status = 0 (paid workers) group with very low scores are outliers. With scores of 2, each case is 4.48 standard deviations below the mean for her group—beyond the $\alpha = .001$ criterion of 3.29 for a two-tailed test. Because there are more cases in the group of paid workers, it is decided to delete these two women with extremely favorable attitudes toward housework from further analysis and to report the deletion in the Results section. There is also a score missing within the group of women with paid work. ATTDRUG and most of the other variables have 246 cases in this group, but ATTHOUSE has only 245 cases. Because the case with the missing value is from the larger group, it is decided to delete the case from subsequent analyses.

TABLE 4.12 Syntax and Selected SAS MEANS and SAS Interactive Data Analysis Output Showing (a) Descriptive Statistics and (b) Histograms for Grouped Data

```
proc sort data = SASUSER.SCREEN;
  by EMPLMNT;
run;

proc means data=SASUSER.SCREEN      vardef=DF
  N NMISS MIN MAX MEAN VAR STD SKEWNESS KURTOSIS;
   var TIMDRS ATTDRUG ATTHOUSE INCOME MSTATUS RACE ;
by EMPLMNT;
run;
```

-----------------------------------Current employment status=0-----------------------------------

The MEANS Procedure

Variable	Label	N	N Miss	Minimum	Maximum	Mean
TIMEDRS	Visits to health professionals	246	0	0	81.0000000	7.2926829
ATTDRUG	Attitudes toward use of medication	246	0	5.0000000	10.0000000	7.5934959
ATTHOUSE	Attitudes toward housework	245	1	2.0000000	34.0000000	23.6408163
INCOME	Income code	235	11	1.0000000	10.0000000	4.2382979
MSTATUS	Current marital status	246	0	1.0000000	2.0000000	1.6869919
RACE	Ethnic affiliation	246	0	1.0000000	2.0000000	1.1097561

Variable	Variance	Std Dev	Skewness	Kurtosis
TIMEDRS	122.4527626	11.0658376	3.8716749	18.0765985
ATTDRUG	1.2381616	1.1127271	-0.1479593	-0.4724261
ATTHOUSE	23.3376715	4.8309079	-0.6828286	2.1614074
INCOME	5.9515185	2.4395734	0.5733054	-0.4287488
MSTATUS	0.2159117	0.4646630	-0.8114465	-1.3526182
RACE	0.0981085	0.3132228	2.5122223	4.3465331

-----------------------------------Current employment status=1-----------------------------------

Variable	Label	N	N Miss	Minimum	Maximum	Mean
TIMEDRS	Visits to health professionals	219	0	0	60.0000000	8.5844749
ATTDRUG	Attitudes toward use of medication	219	0	5.0000000	10.0000000	7.7899543
ATTHOUSE	Attitudes toward housework	219	0	11.0000000	35.0000000	23.4292237
INCOME	Income code	204	15	1.0000000	10.0000000	4.1764706
MSTATUS	Current marital status	219	0	1.0000000	2.0000000	1.8812785
RACE	Ethnic affiliation	219	0	1.0000000	2.0000000	1.0639269

(continued)

TABLE 4.12 Continued

Variable	Label	Variance	Std Dev	Skewness	Kurtosis
TIMEDRS	Visits to health professionals	116.6292991	10.7995046	2.5624008	7.8645362
ATTDRUG	Attitudes toward use of medication	1.4327427	1.1969723	-0.1380901	-0.4417282
ATTHOUSE	Attitudes toward housework	16.5488668	4.0680298	-0.0591932	0.0119403
INCOME	Income code	5.7618082	2.4003767	0.5948250	-0.2543926
MSTATUS	Current marital status	0.1051066	0.3242015	-2.3737868	3.6682817
RACE	Ethnic affiliation	0.0601148	0.2451832	3.5899053	10.9876845

(b) 1. Open SAS Interactive Data Analysis with appropriate data set (here Sasuser.Screen).
2. Choose Analyze and then Histogram/Bar Chart(Y).
3. Select Y variables: TIMEDRS ATTDRUG ATTHOUSE INCOME MSTATUS RACE.
4. In Group box, select EMPLMNT.
5. In Output dialog box, select Both, Vertical Axis at Left, and Horizontal Axis at Bottom.

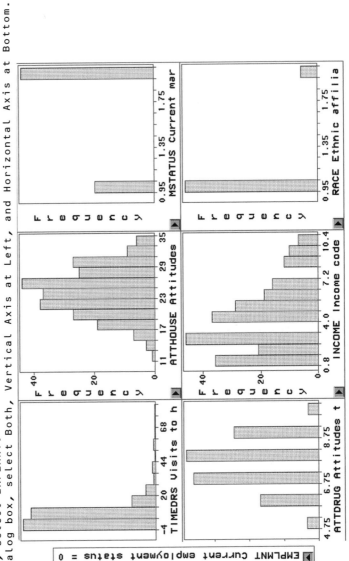

ED NOTE: REARRANGED TABLE AND FIGURE SLIGHTLY TO FIT IN SPACE ALLOTTED.

TABLE 4.12   Continued

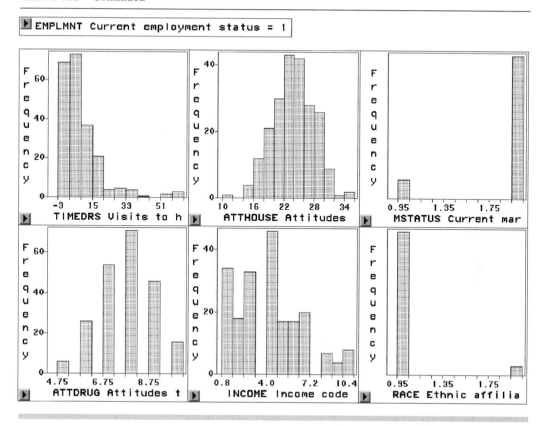

On INCOME, however, it is the smaller group, housewives, with the greater number of missing values; within that group almost 7% of the cases do not have INCOME scores. INCOME, then, is a good candidate for variable deletion, although other remedies are available should deletion seriously interfere with hypothesis testing.

The splits in the two dichotomous variables, MSTATUS and RACE, are about the same for grouped as for ungrouped data. The splits for both MSTATUS (for the housewife group) and for RACE (both groups) are disturbing, but we choose to retain them here.

For the remaining analyses, INCOME is deleted as a variable, and the case with the missing value as well as the two univariate outliers on ATTHOUSE are deleted, leaving a sample size of 462: 243 cases in the paid work group and 219 cases in the housewife group.

Because cell sample sizes are not very discrepant, variance ratios as great as 10 can be tolerated. All F_{max} are well below this criterion. As an example, for the two groups for ATTDRUG, $F_{max} = 1.197/1.113 = 1.16$. Thus, there is no concern about violation of homogeneity of variance nor of homogeneity of variance-covariance matrices.

```
data SASUSER.SCREENX;
  set SASUSER.SCREEN;
  if ATTHOUSE < 3 then delete;
run;
```

1. Open SAS Interactive Data Analysis with appropriate data set (here SASUSER.SCREENX).
2. Choose Analyze and the Scatter Plot (Y X).
3. Select Y variable: ATTDRUG.
4. Select X variable: TIMEDRS.
5. In Group box, select EMPLMNT.
6. In Output dialog box, select Names, Y Axis Vertical, Vertical Axis at Left, and Horizontal Axis at Bottom.

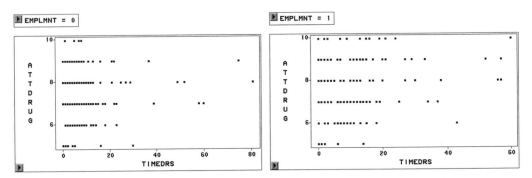

FIGURE 4.12   Syntax for creating reduced data set through SAS DATA and SAS Interactive Data Analysis setup and output showing within-group scatterplot of ATTDRUG vs. TIMEDRS.

4.2.2.2 Linearity

Because of the poor distribution on TIMEDRS, a check of scatterplots is warranted to see if TIME-DRS has a linear relationship with other variables. There is no need to check for linearity with MSTATUS and RACE because variables with two levels have only linear relationships with other variables. Of the two remaining variables, ATTHOUSE and ATTDRUG, the distribution of ATTDRUG differs most from that of TIMEDRS after univariate outliers are deleted.

Appropriately checked first, then, are within-group scatterplots of ATTDRUG versus TIME-DRS. Figure 4.12 shows syntax for creating the data set with extreme cases and missing data on ATTHOUSE deleted (SASUSER.SCREENX) and the setup and output for creating the scatterplots.

In the within-group scatterplots of Figure 4.12, there is ample evidence of skewness in the bunching up of scores at low values of TIMEDRS, but no suggestion of nonlinearity for these variables in the group of paid workers or housewives. Because the plots are acceptable, there is no evidence that the extreme skewness of TIMEDRS produces a harmful departure from linearity. Nor is there any reason to expect nonlinearity with the symmetrically distributed ATTHOUSE.

TABLE 4.13   Syntax for SAS REG and Selected Portion of Output File for Identification of Multivariate Outliers

```
proc reg data=SASUSER.SCREENX;
  by EMPLMNT
  model EMPLMNT = TIMEDRS ATTDRUG ATTHOUSE MSTATUS RACE;
  output out=SASUSER.SCRN_LEV H=H;
run;
```

SASUSER.SCRN_LEV

462 | 9

	SUBNO	TIMEDRS	ATTDRUG	ATTHOUSE	INCOME	EMPLMNT	MSTATUS	RACE	H
1	2	3	7	20	6	0	2	1	0.0106
2	3	0	8	23	3	0	2	1	0.0089
3	6	3	8	25	4	0	2	1	0.0080
4	10	4	8	30	8	0	1	1	0.0231
5	16	2	8	19	3	0	2	2	0.0440
6	22	2	6	21	7	0	1	1	0.0256
7	24	5	10	25	9	0	2	1	0.0269
8	25	3	6	19	4	0	2	1	0.0188
9	26	4	5	31	5	0	2	1	0.0418
10	27	2	8	25	2	0	2	1	0.0084
11	29	13	9	26	2	0	2	1	0.0150
12	30	7	9	33	1	0	2	1	0.0302
13	33	2	6	27	3	0	1	1	0.0273
14	34	5	9	30	1	0	2	1	0.0212
15	35	4	7	31	4	0	2	1	0.0197
16	36	6	6	25	5	0	2	1	0.0149
17	37	2	9	27	5	0	2	1	0.0165
18	40	7	7	32	3	0	2	1	0.0222
19	46	1	8	28	6	0	2	2	0.0478
20	47	3	8	16	1	0	2	1	0.0207
21	48	60	7	24	1	0	2	1	0.1036
22	49	5	8	19	1	0	2	1	0.0124
23	50	3	9	23	6	0	2	1	0.0144

4.2.2.3   Multivariate Outliers

Multivariate outliers within the groups are sought using SAS REG with EMPLMNT as the DV. Because outliers are sought only among IVs, the choice of the DV is simply one of convenience. Mahalanobis distance is unavailable directly but may be calculated from leverage values, which are added to the data file. Table 4.13 shows SAS REG syntax and a selected portion of the output data file (SASUSER.SCRN_LEV) that provides leverage (H) for each case from the centroid of each group. Missing data and univariate outliers have already been omitted (see syntax for Figure 4.12).

Critical values of Mahalanobis distance with 5 variables at $\alpha = .001$ is 20.515 (Appendix C, Table C.4). Equation 4.3 is used to convert this to a critical leverage value for each group.

For housewives:

$$h_{ii} = \frac{20.515}{219 - 1} + \frac{1}{219} = 0.103$$

TABLE 4.14 SAS DATA Syntax for Transformation of TIMEDRS;
Syntax for SAS REG and Selected Portion of Output File

```
data SASUSER.SCREENT;
  set SASUSER.SCREENX;
  LTIMEDRS = log10(LTIMEDRS + 1) ;
run;
proc reg data=SASUSER.SCREENT;
  by EMPLMNT;
  model EMPLMNT = LTIMEDRS ATTDRUG ATTHOUSE MSTATUS RACE;
  output out=sasuser.Scrnlev2 H=H;
run;
```

		Int	Int	Int	Int	Int	Int	Int	Int	Int	Int
	10	SUBNO	TIMEDRS	ATTDRUG	ATTHOUSE	INCOME	EMPLMNT	MSTATUS	RACE	LTIMEDR	H
462											
1		2	3	7	20	6	0	2	1	0.6021	0.0105
2		3	0	8	23	3	0	2	1	0.0000	0.0213
3		6	3	8	25	4	0	2	1	0.6021	0.0078
4		10	4	8	30	8	0	1	1	0.6990	0.0219
5		16	2	8	19	3	0	2	2	0.4771	0.0446
6		22	2	6	21	7	0	1	1	0.4771	0.0261
7		24	5	10	25	9	0	2	1	0.7782	0.0266
8		25	3	6	19	4	0	2	1	0.6021	0.0188
9		26	4	5	31	5	0	2	1	0.6990	0.0412
10		27	2	8	25	2	0	2	1	0.4771	0.0091
11		29	13	9	26	2	0	2	1	1.1461	0.0184
12		30	7	9	33	1	0	2	1	0.9031	0.0300
13		33	2	6	27	3	0	1	1	0.4771	0.0277
14		34	5	9	30	1	0	2	1	0.7782	0.0205
15		35	4	7	31	4	0	2	1	0.6990	0.0189
16		36	6	6	25	5	0	2	1	0.8451	0.0153
17		37	2	9	27	5	0	2	1	0.4771	0.0173
18		40	7	7	32	3	0	2	1	0.9031	0.0223
19		46	1	8	28	6	0	2	2	0.3010	0.0514
20		47	3	8	16	1	0	2	1	0.6021	0.0207
21		48	60	7	24	1	0	2	1	1.7853	0.0383
22		49	5	8	19	1	0	2	1	0.7782	0.0127
23		50	3	9	23	6	0	2	1	0.6021	0.0142

and for the paid work group:

$$h_{ii} = \frac{20.515}{243 - 1} + \frac{1}{243} = 0.089$$

The data set is shown for the first cases for the paid workers (group = 0). The last column, labeled H, shows leverage values. The 21st case, SUBNO #48 is a multivariate outlier among paid workers.

Altogether, 15 cases (about 3%) are identified as multivariate outliers: 3 paid workers and 10 housewives. Although this is not an exceptionally large number of cases to delete, it is worth investigating alternative strategies for dealing with outliers. The univariate summary statistics for TIMEDRS in Table 4.12 show a Maximum score of 81, converting to a standard score of $z = (81 - 7.293)/11.066 = 6.66$ among those with PAIDWORK and $z = 4.76$ among HOUSEWIFEs; the poorly distributed variable produces univariate outliers in both groups. The skewed histograms of Table 4.13 suggest a logarithmic transformation of TIMEDRS.

Table 4.14 shows output from a second run of SAS REG identical to the run in Table 4.13 except that TIMEDRS is replaced by LTIMEDR, its logarithmic transform (transformation syntax is

TABLE 4.15   SAS DATA Syntax for Forming the Dummy DV and Limiting Data to Housewives; Syntax and Partial SAS REG Output Showing Variables Causing SUBNO #262 to Be an Outlier among Housewives

```
data SASUSER.SCREEND;
  set SASUSER.SCREENT;
    DUMMY = 0;
  if SUBNO=262 then DUMMY=1;
  if EMPLMNT=0 then delete;
  run;

proc reg DATA=SASUSER.SCREEND;
model DUMMY = LTIMEDR ATTDRUG ATTHOUSE MSTATUS RACE/
  selection=FORWARD SLENTRY=0.05;;
run;
```

```
                     The REG Procedure
                      Model: MODEL1
                 Dependent Variable: DUMMY

                  Forward Selection: Step 2

             Parameter    Standard
Variable     Estimate      Error     Type II SS    F Value    Pr > F

Intercept    -0.09729     0.02167      0.08382       20.16     <.0001
LTIMEDR       0.02690     0.00995      0.03041        7.31     0.0074
RACE          0.07638     0.01791      0.07565       18.19     <.0001

           Bounds on condition number: 1.0105, 4.0422
----------------------------------------------------------------
No other variable met the 0.0500 significance level for entry into the model.
```

shown). The 21st case no longer is an outlier. With the transformed variable, the entire data set contains only five multivariate outliers, all of them in the housewife group. One of these, SUBNO #262 was also identified as a multivariate outlier in the ungrouped data set.[24]

4.2.2.4   Variables Causing Cases to Be Outliers

Identification of the variables causing outliers to be extreme proceeds in the same manner as for ungrouped data except that the values for the case are compared with the means for the group the case comes from. For subject number 262, a paid worker, the regression run is limited to paid workers, as shown in Table 4.15. First, Table 4.15 shows the SAS DATA syntax to create a dummy variable in which SUBNO #262 forms one code of the dummy variable and the remaining housewives form the other code. The dummy variable then serves as the DV in the SAS REG run. The table

[24]Note that these are different multivariate outliers than found by software used in earlier editions of this book.

TABLE 4.16   Syntax and SAS MEANS Output Showing Descriptive Statistics for Housewife Group

```
data SASUSER.SCREEND;
   set SASUSER.SCREENT;
   if EMPLMNT=0 then delete;
   run;

proc means data=SASUSER.SCREEND
   N   MEAN   STD;
   var LTIMEDR ATTDRUG MSTATUS RACE;
   by EMPLMNT;
run;
```
--------------------------Current employment status=1---------------------

The MEANS Procedure

Variable	Label	N	Mean	Std Dev
LTIMEDR		219	0.7657821	0.4414145
ATTDRUG	Attitudes toward use of medication	219	7.7899543	1.1969723
MSTATUS	Current marital status	219	1.8812785	0.3242015
RACE	Ethnic affiliation	219	1.0639269	0.2451832

shows that the same variables cause this woman to be an outlier from her group as from the entire sample: She differs on the combination of RACE and LTIMEDRS.

Regression runs for the other four cases are not shown, however the remaining four cases all differed on RACE. In addition, one case differed on ATTDRUG and another on MSTATUS.

As with ungrouped data, identification of variables on which cases are outliers is followed by an analysis of the scores on the variables for those cases. First, Table 4.16 shows the means on the three variables involved in outlying cases, separately by employment group. The data set is consulted for these values for the five outliers.

The data set shows that scores for subject number 262 are 2 for RACE and 1.763 for LTIME-DRS. For subject number 45, they are 2 for RACE and 1 (single) for MSTATUS. For subject number 119, they are 2 for RACE and 10 for ATTDRUG. For subject numbers 103 and 582, they are 2 for RACE. Thus all of the outliers are non-Caucasian housewives. A separate run of descriptive statistics for housewives (not shown) reveals that only 14 of the housewives in the sample of 219 are non-Caucasian. One of them also has an unusually large number of visits to health professionals, one is unmarried, and one has exceptionally unfavorable attitudes regarding use of drugs.

4.2.2.5 *Multicollinearity*

The collinearity diagnostics of a SAS REG run are used to assess multicollinearity for the two groups, combined (Table 4.17) after deleting the five multivariate outliers.

Screening information as it might be described in a Results section of a journal article appears next.

TABLE 4.17   Sytnax and Selected Multicollinearity Output from SAS REG.

```
data SASUSER.SCREENT;
   set SASUSER.SCREENT;
   if subno=45 or subno=103 or subno=19 or subno=262 or
      subno=582 then delete;
run;
proc reg data=SASUSER.SCREENF;
   model SUBNO= ATTDRUG ATTHOUSE MSTATUS RACE LTIMEDR/ COLLIN;
run;
```

Collinearity Diagnostics

Number	Eigenvalue	Condition Index
1	5.66323	1.00000
2	0.20682	5.23282
3	0.05618	10.03997
4	0.04246	11.54953
5	0.02463	15.16319
6	0.00668	29.12199

-----------------------Proportion of Variation-----------------------

Number	Intercept	ATTDRUG	ATTHOUSE	MSTATUS	RACE	LTIMEDR
1	0.00026447	0.00066681	0.00092323	0.00149	0.00172	0.00583
2	0.00090229	0.00163	0.00169	0.01208	0.01714	0.91984
3	0.00033382	0.00097697	0.00308	0.34932	0.61742	0.00424
4	0.00374	0.03562	0.30196	0.39771	0.19527	0.06458
5	0.00393	0.54441	0.40672	0.05798	0.04279	0.00416
6	0.99082	0.41670	0.28563	0.18142	0.12567	0.00136

Results

Prior to analysis, number of visits to health professionals, attitude toward drug use, attitude toward housework, income, marital status, and race were examined through various SAS programs for accuracy of data entry, missing values, and fit between their distributions and the assumptions of multivariate analysis. The variables were examined separately for the 246 employed women and the 219 housewives.

A case with a single missing value on attitude toward housework was deleted from the group of employed women, leaving 245 cases in that group. Income, with missing values on more than 5% of the cases, was deleted. Pairwise linearity was checked using within-group scatterplots and found to be satisfactory.

Two cases in the employed group were univariate outliers because of their extremely low z scores on attitude toward housework; these cases were deleted. By using Mahalanobis distance with $p < .001$, derived from leverage scores, 15 cases (about 3%) were identified as multivariate outliers in their own groups. Because several of these cases had extreme z scores on visits to health professionals and because that variable was severely skewed, a logarithmic transformation was applied. With the transformed variable in the variable set, only five cases were identified as multivariate outliers, all from the employed group.[25] With all seven outliers and the case with missing values deleted, 243 cases remained in the employed group and 214 in the group of housewives.

---

[25] All the outliers were non-Caucasian housewives. Thus, 36% (5/14) of the non-Caucasian housewives were outliers. One of them also had an unusually large number of visits to health professionals, one was unmarried, and one had exceptionally unfavorable attitudes regarding use of drugs. Thus, results may not generalize to non-Caucasian housewives, particularly those who are unmarried, make frequent visits to physicians, and have very unfavorable attitudes toward use of drugs.

CHAPTER 5

Multiple Regression

5.1 General Purpose and Description

Regression analyses are a set of statistical techniques that allow one to assess the relationship between one DV and several IVs. For example, is reading ability in primary grades (the DV) related to several IVs such as perceptual development, motor development, and age? The terms *regression* and *correlation* are used more or less interchangeably to label these procedures although the term regression is often used when the intent of the analysis is prediction, and the term correlation is used when the intent is simply to assess the relationship between the DV and the IVs.

Multiple regression is a popular technique in many disciplines. For example, Stefl-Mabry (2003) used standard multiple regression to study satisfaction derived from various sources of information (word-of-mouth, expert oral advice, internet, print news, nonfiction books, and radio/television news). Forty vignettes were developed with high and low levels of information and high and low consistency; individual regression analyses were performed for each of the 90 professional participants and then their standardized regression coefficients were averaged. The normative participant was most satisfied by expert oral advice, with nonfiction books and word-of-mouth next in order to produce satisfaction. Participants were consistent in the satisfaction they derived from various sources in various vignettes.

Baldry (2003) used sequential/hierarchical regression to study whether exposure to parental interpersonal violence contributed to bullying behavior above and beyond prediction afforded by demographic variables and parental child abuse. In the first step, gender (being a boy) and age (being older) were significantly related to bullying; in the second step, child abuse by the father (but not the mother) added to prediction of bullying behavior. In the final step, mother's violence against father significantly contributed to prediction, although father's violence against mother did not.[1] The full model was significant, but accounted for only 14% of the variance in bullying behaviors.

Regression techniques can be applied to a data set in which the IVs are correlated with one another and with the DV to varying degrees. One can, for instance, assess the relationship between a set of IVs such as education, income, and socioeconomic status with a DV such as occupational prestige. Because regression techniques can be used when the IVs are correlated, they are helpful both in experimental research when, for instance, correlation among IVs is created by unequal numbers of cases in cells, and in observational or survey research when nature has "manipulated" correlated variables. The flexibility of regression techniques is, then, especially useful to the researcher who is

---

[1]Without a report of the full correlations, it is difficult to know which variables in the same step of the analysis might have "knocked each other out."

interested in real-world or very complicated problems that cannot be meaningfully reduced to orthogonal designs in a laboratory setting.

Multiple regression is an extension of bivariate regression (see Section 3.5.2) in which several IVs instead of just one are combined to predict a value on a DV for each subject. The result of regression is a generalization of Equation 3.30 that represents the best prediction of a DV from several continuous (or dichotomous) IVs. The regression equation takes the following form:

$$Y' = A + B_1X_1 + B_2X_2 + \cdots + B_kX_k$$

where Y' is the predicted value on the DV, A is the Y intercept (the value of Y when all the X values are zero), the Xs represent the various IVs (of which there are k), and the Bs are the coefficients assigned to each of the IVs during regression. Although the same intercept and coefficients are used to predict the values on the DV for all cases in the sample, a different Y' value is predicted for each subject as a result of inserting the subject's own X values into the equation.

The goal of regression is to arrive at the set of B values, called *regression coefficients,* for the IVs that bring the Y values predicted from the equation as close as possible to the Y values obtained by measurement. The regression coefficients that are computed accomplish two intuitively appealing and highly desirable goals: they minimize (the sum of the squared) deviations between predicted and obtained Y values and they optimize the correlation between the predicted and obtained Y values for the data set. In fact, one of the important statistics derived from a regression analysis is the multiple-correlation coefficient, the Pearson product–moment correlation coefficient between the obtained and predicted Y values: $R = r_{yy'}$ (see Section 5.4.1).

Regression techniques consist of standard multiple regression, sequential (hierarchical) regression, and statistical (stepwise) regression. Differences between these techniques involve the way variables enter the equation: what happens to variance shared by variables and who determines the order in which variables enter the equation?

5.2   Kinds of Research Questions

The primary goal of regression analysis is usually to investigate the relationship between a DV and several IVs. As a preliminary step, one determines how strong the relationship is between the DV and IVs; then, with some ambiguity, one assesses the importance of each of the IVs to the relationship.

A more complicated goal might be to investigate the relationship between a DV and some IVs with the effect of other IVs statistically eliminated. Researchers often use regression to perform what is essentially a covariates analysis in which they ask if some critical variable (or variables) adds anything to a prediction equation for a DV after other IVs—the covariates—have already entered the equation. For example, does gender add to prediction of mathematical performance after statistical adjustment for extent and difficulty of mathematical training?

Another strategy is to compare the ability of several competing sets of IVs to predict a DV. Is use of Valium better predicted by a set of health variables or by a set of attitudinal variables?

All too often, regression is used to find the best prediction equation for some phenomenon regardless of the meaning of the variables in the equation, a goal met by statistical (stepwise) regression. In the several varieties of statistical regression, statistical criteria alone, computed from a single sample, determine which IVs enter the equation and the order in which they enter.

Regression analyses can be used with either continuous or dichotomous IVs. A variable that is initially discrete can be used if it is first converted into a set of dichotomous variables (numbering one fewer than the number of discrete categories) by dummy variable coding with 1s and 0s. For example, consider an initially discrete variable assessing religious affiliation in which 1 stands for Protestant, 2 for Catholic, 3 for Jewish, and 4 for none or other. The variable may be converted into a set of three new variables (Protestant = 1 vs. non-Protestant = 0, Catholic = 1 vs. non-Catholic = 0, Jewish = 1 vs. non-Jewish = 0), one variable for each degree of freedom. When the new variables are entered into regression as a group (as recommended by Fox, 1991), the variance due to the original discrete IV is analyzed, and, in addition, one can examine effects of the newly created dichotomous components. Dummy variable coding is covered in glorious detail in Cohen et al. (2003, Section 8.2).

ANOVA (Chapter 3) is a special case of regression in which main effects and interactions are a series of dichotomous IVs. The dichotomies are created by dummy-variable coding for the purpose of performing a statistical analysis. ANOVA problems can be handled through multiple regression, but multiple-regression problems often cannot readily be converted into ANOVA because of correlations among IVs and the presence of continuous IVs. If analyzed through ANOVA, continuous IVs have to be rendered discrete (e.g., high, medium, and low), a process that often results in loss of information and unequal cell sizes. In regression, the full range of continuous IVs is maintained. Simple ANOVA through regression is covered briefly in Section 5.6.5 and in detail for a variety of ANOVA models in Tabachnick and Fidell (2007).

As a statistical tool, regression is very helpful in answering a number of practical questions, as discussed in Sections 5.2.1 through 5.2.8.

5.2.1   Degree of Relationship

How good is the regression equation? Does the regression equation really provide better-than-chance prediction? Is the multiple correlation really any different from zero when allowances for naturally occurring fluctuations in such correlations are made? For example, can one reliably predict reading ability given knowledge of perceptual development, motor development, and age? The statistical procedures described in Section 5.6.2.1 allow you to determine if your multiple correlation is reliably different from zero.

5.2.2   Importance of IVs

If the multiple correlation is different from zero, you may want to ask which IVs are important in the equation and which IVs are not. For example, is knowledge of motor development helpful in predicting reading ability or can we do just as well with knowledge of only age and perceptual development? The methods described in Section 5.6.1 help you to evaluate the relative importance of various IVs to a regression solution.

5.2.3   Adding IVs

Suppose that you have just computed a regression equation and you want to know whether you can improve your prediction of the DV by adding one or more IVs to the equation. For example, is prediction of a child's reading ability enhanced by adding a variable reflecting parental interest in reading to the three IVs already included in the equation? A test for improvement of the multiple correlation after addition of one new variable is given in Section 5.6.1.2, and for improvement after addition of several new variables in Section 5.6.2.3.

5.2.4   Changing IVs

Although the regression equation is a linear equation (i.e., it does not contain squared values, cubed values, cross-products of variables, and the like), the researcher may include nonlinear relationships in the analysis by redefining IVs. Curvilinear relationships, for example, can be made available for analysis by squaring or raising original IVs to a higher power. Interaction can be made available for analysis by creating a new IV that is a cross-product of two or more original IVs and including it with the originals in the analysis. It is recommended that IVs be centered (replacing original scores with deviations from their mean) when including interactions or powers of IVs (cf. Section 5.6.6).

For an example of a curvilinear relationship, suppose a child's reading ability increases with increasing parental interest up to a point, and then levels off. Greater parental interest does not result in greater reading ability. If the square of parental interest is added as an IV, better prediction of a child's reading ability could be achieved.

Inspection of a scatterplot between predicted and obtained Y values (known as residuals analysis—see Section 5.3.2.4) may reveal that the relationship between the DV and the IVs also has more complicated components such as curvilinearity and interaction. To improve prediction or because of theoretical considerations, one may want to include some of these more complicated IVs. Procedures for using regression for nonlinear curve fitting are discussed in Cohen et al. (2003). There is danger, however, in too liberal use of powers or cross-products of IVs; the sample data may be overfit to the extent that results no longer generalize to a population.

5.2.5   Contingencies among IVs

You may be interested in the way that one IV behaves in the context of one, or a set, of other IVs. Sequential regression can be used to adjust statistically for the effects of some IVs while examining the relationship between an especially interesting IV and the DV. For example, after adjustment for differences in perceptual development and age, does motor development predict reading ability? This procedure is described in Section 5.5.2.

5.2.6   Comparing Sets of IVs

Is prediction of a DV from one set of IVs better than prediction from another set of IVs? For example, is prediction of reading ability based on perceptual and motor development and age as good as prediction from family income and parental educational attainments? Section 5.6.2.5 demonstrates a method for comparing the solutions given by two sets of predictors.

5.2.7   Predicting DV Scores for Members of a New Sample

One of the more important applications of regression involves predicting scores on a DV for subjects for whom only data on IVs are available. This application is fairly frequent in personnel selection for employment, selection of students for graduate training, and the like. Over a fairly long period, a researcher collects data on a DV, say, success in graduate school, and on several IVs, say, undergraduate GPA, GRE verbal scores, and GRE math scores. Regression analysis is performed and the regression equation obtained. If the IVs are strongly related to the DV, then, for a new sample of applicants to graduate school, regression coefficients are applied to IV scores to predict success in graduate school ahead of time. Admission to graduate school may, in fact, be based on prediction of success through regression.

The generalizability of a regression solution to a new sample is checked within a single large sample by a procedure called cross-validation. A regression equation is developed from a portion of a sample and then applied to the other portion of the sample. If the solution generalizes, the regression equation predicts DV scores better than chance for the new cases, as well. Section 5.5.3 demonstrates cross-validation with statistical regression.

5.2.8 Parameter Estimates

Parameter estimates in multiple regression are the unstandardized regression coefficients (B weights). A B weight for a particular IV represents the change in the DV associated with a one-unit change in that IV, all other IVs held constant. Suppose, for example, we want to predict graduate record exam (GRE) scores from grade point averages (GPA) and our analysis produces the following equation:

$$(GRE)' = 200 + 100(GPA)$$

$B = 100$ tells us that for each one-unit increase in GPA (e.g., from a GPA of 2.0 to one of 3.0), we expect a 100-point increase in GRE scores. Sometimes this is usefully expressed in terms of percentage of gain in the DV. For example, assuming the mean GRE is 500, an increase of one grade point represents a 20% $(100/500)$ average increase in GRE.

Accuracy of parameter estimates depends on agreement with the assumptions of multiple-regression analysis (cf. Section 5.3.2.4), including the assumption that IVs are measured without error. Therefore, interpretation has to be tempered by knowledge of the reliability of the IVs. You need to be cautious when interpreting regression coefficients with transformed variables, because the coefficients and interpretations of them apply only to the variable after transformation.

5.3 Limitations to Regression Analyses

Attention to issues surrounding assumptions of regression analysis has become a growth industry, partly because of the relative simplicity of regression compared to the multivariate techniques and partly because of the extensive use of multiple regression in all facets of science and commerce. A glance at the myriad of diagnostic tests available in regression programs confirms this view. However, it should also be noted that many of the popular diagnostic tests are concerned with poor fit of regression models to some cases—outliers in the solution—rather than tests conducted as part of screening.

This discussion barely skims the surface of the goodies available for screening your data and assessing the fit of cases to your solution, but should adequately cover most gross violations of assumptions. Berry (1993) and Fox (1991) offer some other interesting insights into regression assumptions and diagnostics.

5.3.1 Theoretical Issues

Regression analyses reveal relationships among variables but do not imply that the relationships are causal. Demonstration of causality is a logical and experimental, rather than statistical, problem. An apparently strong relationship between variables could stem from many sources, including the influence of other, currently unmeasured variables. One can make an airtight case for causal relationship among variables only by showing that manipulation of some of them is followed inexorably by change in others when all other variables are controlled.

Another problem for theory rather than statistics is inclusion of variables. Which DV should be used, and how is it to be measured? Which IVs should be examined, and how are they to be measured? If one already has some IVs in an equation, which IVs should be added to the equation for the most improvement in prediction? The answers to these questions can be provided by theory, astute observation, good hunches, or sometimes by careful examination of the distribution of residuals.

There are, however, some general considerations for choosing IVs. Regression will be best when each IV is strongly correlated with the DV but uncorrelated with other IVs. A general goal of regression, then, is to identify the fewest IVs necessary to predict a DV where each IV predicts a substantial and independent segment of the variability in the DV.

There are other considerations to selection of variables. If the goal of research is manipulation of some DV (say, body weight), it is strategic to include as IVs variables that can be manipulated (e.g., caloric intake and physical activity) as well as those that cannot (e.g., genetic predisposition). Or, if one is interested in predicting a variable such as annoyance caused by noise for a neighborhood, it is strategic to include cheaply obtained sets of IVs (e.g., neighborhood characteristics published by the Census Bureau) rather than expensively obtained ones (e.g., attitudes from in-depth interviews) if both sets of variables predict equally well.

It should be clearly understood that a regression solution is extremely sensitive to the combination of variables that is included in it. Whether an IV appears particularly important in a solution depends on the other IVs in the set. If the IV of interest is the only one that assesses some important facet of the DV, the IV will appear important; if the IV of interest is only one of several that assess the same important facet of the DV, it usually will appear less important. An optimal set of IVs is the smallest reliable, uncorrelated set that "covers the waterfront" with respect to the DV.[2]

Regression analysis assumes that IVs are measured without error, a clear impossibility in most social and behavioral science research. The best we can do is choose the most reliable IVs possible. More subtly, it is assumed that important unmeasured IVs, which contribute to error, are not correlated with any of the measured IVs. If, as Berry (1993) points out, an unmeasured IV is correlated with a measured IV, then the components of error are correlated with the measured IV, a violation of the assumption of independence of errors. Worse, the relationship between an unmeasured IV and a measured IV can change the estimates of the regression coefficients; if the relationship is positive, the coefficient for the measured IV is overestimated; if negative, underestimated. If the regression equation is to accurately reflect the contribution of each IV to prediction of the DV, then, all of the relevant IVs have to be included.

Analysis of residuals provides information important to both theoretical and practical issues in multiple regression analysis. Judicious inspection of residuals can help identify variables that are degrading rather than enhancing prediction. Plots of residuals identify failure to comply with distri-

---

[2]Knecht (personal communication, 2003) points to a potential problem of overfitting if the first few "good" IVs are selected from a very large pool of potential IVs; the regression can be too good due to chance factors alone.

butional assumptions. And residuals help identify cases that are outliers in the regression solution—cases poorly fit by the model. Procedures for examining residuals for normality, homoscedasticity, and the like are the same as those for examining any other variable, as discussed in Chapter 4 and in Section 5.3.2.4.

5.3.2 Practical Issues

In addition to theoretical considerations, use of multiple regression requires that several practical matters be attended to, as described in Sections 5.3.2.1 through 5.3.2.4.

5.3.2.1 *Ratio of Cases to IVs*

The cases-to-IVs ratio has to be substantial or the solution will be perfect—and meaningless. With more IVs than cases, one can find a regression solution that completely predicts the DV for each case, but only as an artifact of the cases-to-IV ratio.

Required sample size depends on a number of issues, including the desired power, alpha level, number of predictors, and expected effect sizes. Green (1991) provides a thorough discussion of these issues and some procedures to help decide how many cases are necessary. Some simple rules of thumb are $N \geq 50 + 8m$ (where m is the number of IVs) for testing the multiple correlation and $N \geq 104 + m$ for testing individual predictors. These rules of thumb assume a medium-size relationship between the IVs and the DV, $\alpha = .05$ and $\beta = .20$. For example, if you plan six predictors, you need $50 + (8)(6) = 98$ cases to test regression and $104 + 6 = 110$ cases for testing individual predictors. If you are interested in both the overall correlation and the individual IVs, *calculate N both ways and choose the larger number of cases.* Alternatively, you can consult one of the software programs that are available for estimating power in multiple regression, such as SAS POWER, SPSS Sample Power, or PASS (NCSS, 2002) or those available on the Internet (entering "statistical power" in your search engine or browser should produce a wealth of helpful programs, many of them free).

A higher cases-to-IV ratio is needed when the DV is skewed, a small effect size is anticipated, or substantial measurement error is expected from less reliable variables. That is, if the DV is not normally distributed and transformations are not undertaken, more cases are required. The size of anticipated effect is also relevant because more cases are needed to demonstrate a small effect than a large one. The following, more complex, rule of thumb that takes into account effect size is based on Green (1991): $N \geq (8/f^2) + (m - 1)$, where $f^2 = .02, .15$, and $.35$ for small, medium, and large effects, respectively, $f^2 = pr^2/(1 - pr^2)$, where pr^2 is the expected squared partial correlation for the IV with the smallest expected effect of interest. Finally, if variables are less reliable, measurement error is larger and more cases are needed.

It is also possible to have too many cases. As the number of cases becomes quite large, almost any multiple correlation will depart significantly from zero, even one that predicts negligible variance in the DV. For both statistical and practical reasons, then, one wants to measure the smallest number of cases that has a decent chance of revealing a relationship of a specified size.

If statistical (stepwise) regression is to be used, even more cases are needed. A cases-to-IV ratio of 40 to 1 is reasonable because statistical regression can produce a solution that does not generalize beyond the sample unless the sample is large. An even larger sample is needed in statistical regression if cross-validation (deriving the solution with some of the cases and testing it on the others) is used to test the generalizability of the solution. Cross-validation is illustrated in Section 5.5.3.

If you cannot measure as many cases as you would like, there are some strategies that may help. You can delete some IVs or create one (or more than one) IV that is a composite of several others. The new, composite IV is used in the analysis in place of the original IVs.

Be sure to verify that the analysis included as many cases as you think it should have. By default, regression programs delete cases for which there are missing values on any of the variables that can result in substantial loss of cases. Consult Chapter 4 if you have missing values and wish to estimate them rather than delete the cases.

5.3.2.2   *Absence of Outliers among the IVs and on the DV*

Extreme cases have too much impact on the regression solution and also affect the precision of estimation of the regression weights (Fox, 1991). With high leverage and low discrepancy (Figure 4.2), the standard errors of the regression coefficients are too small; with low leverage and high discrepancy, the standard errors of the regression coefficients are too large. Neither situation generalizes well to population values. Therefore, outliers should be deleted, rescored, or the variable transformed. See Chapter 4 for a summary of general procedures for detecting and dealing with univariate and multivariate outliers using both statistical tests and graphical methods, including evaluation of disconnectedness.

In regression, cases are evaluated for univariate extremeness with respect to the DV and each IV. Univariate outliers show up in initial screening runs (e.g., with SPSS FREQUENCIES or SAS Interactive Data Analysis) as cases far from the mean and unconnected with other cases on either plots or z-scores. Multivariate outliers among the IVs are sought using either statistical methods such as Mahalanobis distance (through SPSS REGRESSION or SAS GLM as described in Chapter 4) or by using graphical methods.

Screening for outliers can be performed either prior to a regression run (as recommended in Chapter 4) or through a residuals analysis after an initial regression run. The problems with an initial regression run are, first, the temptation to make screening decisions based on desired outcome, and, second, the overfitting that may occur if outliers in the solution are deleted along with outliers among the variables. It seems safer to deal with outliers among the variables in initial screening runs, and then determine the fit of the solution to the cases.

Regression programs offer more specialized tests for identifying outliers than most programs for the other techniques. SPSS REGRESSION provides Mahalanobis distance for multivariate outliers; SAS GLM provides leverage.

5.3.2.3   *Absence of Multicollinearity and Singularity*

Calculation of regression coefficients requires inversion of the matrix of correlations among the IVs (Equation 5.6), an inversion that is impossible if IVs are singular and unstable if they are multicollinear, as discussed in Chapter 4. Either of these problems can occur either because the IVs themselves are highly correlated, or because you have included interactions among IVs or powers of IVs in your analysis. In the latter case, the problem can be minimized by centering the variables, as discussed in Section 5.6.6.

Singularity and multicollinearity can be identified in screening runs through perfect or very high squared multiple correlations (SMC) among IVs (where each IV in turn serves as DV and the others are IVs), or very low tolerances (1 − SMC), or through multicollinearity diagnostics, as illustrated in Chapter 4.[3]

---

[3]For an extended discussion of the complicated relationship between outliers and collinearity, see Fox (1991).

In regression, multicollinearity is also signaled by very large (relative to the scale of the variables) standard errors for regression coefficients. Berry (1993) reports that when r is 0.9, the standard errors of the regression coefficients are doubled; when multicollinearity is present, none of the regression coefficients may be significant because of the large size of standard errors. Even tolerances as high as .5 or .6 may pose difficulties in testing and interpreting regression coefficients.

Most multiple-regression programs have default values for tolerance $(1 - SMC)$ that protect the user against inclusion of multicollinear IVs. If the default values for the programs are in place, IVs that are very highly correlated with IVs already in the equation are not entered. This makes sense both statistically and logically because the IVs threaten the analysis due to inflation of regression coefficients and because they are not needed due to their correlations with other IVs.

If variables are to be deleted, however, you probably want to make your own choice about which IV to delete on logical rather than statistical grounds by considering issues such as the reliability of the variables or the cost of measuring the variables. You may want to delete the least reliable variable, for instance, rather than the variable identified by the program with very low tolerance. With a less reliable IV deleted from the set of IVs, the tolerance for the IV in question may be sufficient for entry.

If multicollinearity is detected but you want to maintain your set of IVs anyway, ridge regression might be considered. Ridge regression is a controversial procedure that attempts to stabilize estimates of regression coefficients by inflating the variance that is analyzed. For a more thorough description of ridge regression, see Dillon and Goldstein (1984, Chapter 7). Although originally greeted with enthusiasm (cf. Price, 1977), serious questions about the procedure have been raised by Rozeboom (1979), Fox (1991), and others. If, after consulting this literature, you still want to employ ridge regression, it is available through SAS REG and as a macro in SPSS.

5.3.2.4   *Normality, Linearity, Homoscedasticity of Residuals*

Routine preanalysis screening procedures of Chapter 4 may be used to assess normality, linearity, and homoscedasticity. Regression programs, however, also offer an assessment of the three assumptions simultaneously through analysis of residuals produced by the programs.

Examination of residuals scatterplots provides a test of assumptions of normality, linearity, and homoscedasticity between predicted DV scores and errors of prediction. Assumptions of analysis are that the residuals (differences between obtained and predicted DV scores) are normally distributed about the predicted DV scores, that residuals have a straight-line relationship with predicted DV scores, and that the variance of the residuals about predicted DV scores is the same for all predicted scores.[4] When these assumptions are met, the residuals appear as in Figure 5.1(a).

Residuals scatterplots may be examined in lieu of or after initial screening runs. If residuals scatterplots are examined in lieu of initial screening, and the assumptions of analysis are deemed met, further screening of variables and cases is unnecessary. That is, if the residuals show normality, linearity, and homoscedasticity; if no outliers are evident; if the number of cases is sufficient; and if there is no evidence of multicollinearity or singularity, then regression requires only one run. (Parenthetically, we might note that we have never, in many years of multivariate analyses with many data sets, found this to be the case.) If, on the other hand, the residuals scatterplot from an initial run looks awful, further screening via the procedures in Chapter 4 is warranted.

---

[4]Note that there are no distributional assumptions about the IVs, other than their relationship with the DV. However, a prediction equation often is enhanced if IVs are normally distributed, primarily because linearity between the IV and DV is enhanced.

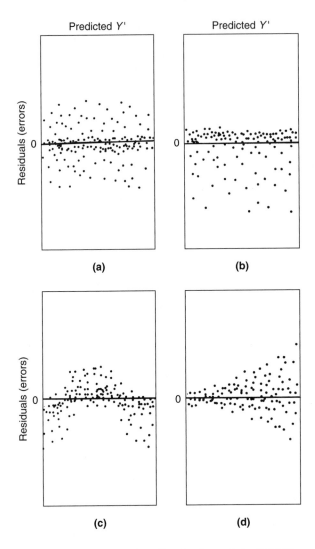

**FIGURE 5.1   Plots of predicted values of the DV (Y')
against residuals showing (a) assumptions met, (b) failure
of normality, (c) nonlinearity, and (d) heteroscedasticity.**

Residuals scatterplots are provided by all the statistical programs discussed in this chapter. All provide a scatterplot in which one axis is predicted scores and the other axis is errors of prediction. Which axis is which, however, and whether the predicted scores and residuals are standardized differ from program to program.

SPSS and SAS provide the plots directly in their regression programs. In SPSS, both predicted scores and errors of prediction are standardized; in SAS, they are not. In any event, it is the overall shape of the scatterplot that is of interest. If all assumptions are met, the residuals will be nearly rectangularly distributed with a concentration of scores along the center. As mentioned earlier, Figure 5.1(a) illustrates a distribution in which all assumptions are met.

The assumption of normality is that errors of prediction are normally distributed around each and every predicted DV score. The residuals scatterplot should reveal a pileup of residuals in the center of the plot at each value of predicted score and a normal distribution of residuals trailing off symmetrically from the center. Figure 5.1(b) illustrates a failure of normality, with a skewed distribution of residuals.

Linearity of relationship between predicted DV scores and errors of prediction is also assumed. If nonlinearity is present, the overall shape of the scatterplot is curved instead of rectangular, as shown in Figure 5.1(c). In this illustration, errors of prediction are generally in a negative direction for low and high predicted scores and in a positive direction for medium predicted scores. Typically, nonlinearity of residuals can be made linear by transforming IVs (or the DV) so that there is a linear relationship between each IV and the DV. If, however, there is a genuine curvilinear relationship between an IV and the DV, it may be necessary to include the square of the IV in the set of IVs.

Failure of linearity of residuals in regression does not invalidate an analysis so much as weaken it. A curvilinear relationship between the DV and an IV is a perfectly good relationship that is not completely captured by a linear correlation coefficient. The power of the analysis is reduced to the extent that the analysis cannot map the full extent of the relationships among the IVs and the DV.

The assumption of homoscedasticity is the assumption that the standard deviations of errors of prediction are approximately equal for all predicted DV scores. Heteroscedasticity also does not invalidate the analysis so much as weaken it. Homoscedasticity means that the band enclosing the residuals is approximately equal in width at all values of the predicted DV. Typical heteroscedasticity is a case in which the band becomes wider at larger predicted values, as illustrated in Figure 5.1(d). In this illustration, the errors of prediction increase as the size of the prediction increases. Serious heteroscedasticity occurs when the spread in standard deviations of residuals around predicted values is three times higher for the widest spread as for the most narrow spread (Fox, 1991). Heteroscedasticity may occur when some of the variables are skewed and others are not. Transformation of the variables may reduce or eliminate heteroscedasticity.

Heteroscedasticity can also result from interaction of an IV with another variable that is not part of the regression equation. For example, it may be that increasing variability in income with age is associated with education; for those with higher education, there is greater growth in income with age. Including education as well as age as predictors of income will strengthen the model as well as eliminate heteroscedasticity.

Another remedy is to use weighted (generalized) least squares regression, available as an option in all major regression programs. In this procedure, you weight the regression by the variance of the variable that produces the heteroscedasticity. For example, if you know that variance in the DV (e.g., income) increases with increasing values of an IV (e.g., age), you weight the regression by age. This latter remedy is less appealing than inclusion of the "interacting" variable (education), but may be more practical if you cannot identify or measure the interacting variable, or if the heteroscedasticity is a result of measurement error.

Under special and somewhat rare conditions, significance tests are available for linearity and homoscedasticity. Fox (1991, pp. 64–66) summarizes some of these significance tests for failure of linearity and heteroscedasticity, useful when some of the IVs are discrete with only a few categories.

5.3.2.5   *Independence of Errors*

Another assumption of regression, testable through residuals analysis, is that errors of prediction are independent of one another. In some instances, this assumption is violated as a function of something associated with the order of cases. Often "something" is time or distance. For example, time produces nonindependence of errors when subjects who are interviewed early in a survey exhibit more variability of response because of interviewer inexperience with a questionnaire. Distance produces nonindependence of errors when subjects who are farther away from a toxic source exhibit more variable reactions. Nonindependence of errors is, then, either a nuisance factor to be eliminated or of considerable research interest.

Nonindependence of errors associated with order of cases is assessed by entering cases in order and requesting a plot of residuals against sequence of cases. The associated Durbin-Watson statistic is a measure of autocorrelation of errors over the sequence of cases, and, if significant, indicates nonindependence of errors. Positive autocorrelation makes estimates of error variance too small, and results in inflation of the Type I error rate. Negative autocorrelation makes the estimates too large, and results in loss of power. Details on the use of this statistic and a test for its significance are given by Wesolowsky (1976). If nonindependence is found, consult Dillon and Goldstein (1984) for the options available to you.

5.3.2.6   *Absence of Outliers in the Solution*

Some cases may be poorly fit by the regression equation. These cases lower the multiple correlation. Examination of these cases is informative because they are the kinds of cases that are not well predicted by your solution.

Cases with large residuals are outliers in the solution. Residuals are available in raw or standardized form—with or without the outlying case deleted. A graphical method based on residuals uses leverage on the X axis and residuals on the Y axis. As in routine residuals plots, outlying cases in the solution fall outside the swarm of points produced by the remainder of the cases.

Examine the residuals plot. If outlying cases are evident, identify them through the list of standardized residuals for individual cases. The statistical criterion for identifying an outlier in the solution depends on the sample size; the larger the sample, the more likely that one or more residuals are present. When $N < 1000$, a criterion of $p = .001$ is appropriate; this p is associated with a standardized residuals in excess of about ± 3.3.

5.4   Fundamental Equations
for Multiple Regression

A data set appropriate for multiple regression consists of a sample of research units (e.g., graduate students) for whom scores are available on a number of IVs and on one DV. An unrealistically small sample of hypothetical data with three IVs and one DV is illustrated in Table 5.1.

Table 5.1 contains scores for six students on three IVs: a measure of professional motivation (MOTIV), a composite rating of qualifications for admissions to graduate training (QUAL), and a composite rating of performance in graduate courses (GRADE). The DV is a rating of performance on graduate comprehensive exams (COMPR). We ask how well we can predict COMPR from scores on MOTIV, QUAL, and GRADE.

TABLE 5.1   Small Sample of Hypothetical Data for Illustration of Multiple Correlation

Case No.	IVs MOTIV(X_1)	QUAL (X_2)	GRADE (X_3)	DV COMPR (Y)
1	14	19	19	18
2	11	11	8	9
3	8	10	14	8
4	13	5	10	8
5	10	9	8	5
6	10	7	9	12
Mean	11.00	10.17	11.33	10.00
Standard deviation	2.191	4.834	4.367	4.517

A sample of six cases is highly inadequate, of course, but the sample is sufficient to illustrate the calculation of multiple correlation and to demonstrate some analyses by canned computer programs. The reader is encouraged to work problems involving these data by hand as well as by available computer programs. Syntax and selected output for this example through SPSS REGRESSION and SAS REG appear in Section 5.4.3. A variety of ways are available to develop the "basic" equation for multiple correlation.

5.4.1   General Linear Equations

One way of developing multiple correlation is to obtain the prediction equation for Y' in order to compare the predicted value of the DV with obtained Y.

$$Y' = A + B_1X_1 + B_2X_2 + \cdots + B_kX_k \tag{5.1}$$

where Y' is the predicted value of Y, A is the value of Y' when all Xs are zero, B_1 to B_k represent regression coefficients, and X_1 to X_k represent the IVs.

The best-fitting regression coefficients produce a prediction equation for which squared differences between Y and Y' are at a minimum. Because squared errors of prediction—$(Y - Y')^2$—are minimized, this solution is called a least-squares solution.

In the sample problem, $k = 3$. That is, there are three IVs available to predict the DV, COMPR.

$$(\text{COMPR})' = A + B_M (\text{MOTIV}) + B_Q (\text{QUAL}) + B_G (\text{GRADE})$$

To predict a student's COMPR score, the available IV scores (MOTIV, QUAL, and GRADE) are multiplied by their respective regression coefficients. The coefficient-by-score products are summed and added to the intercept, or baseline, value (A).

Differences among the observed values of the DV (Y), the mean of Y ($\overline{Y}$), and the predicted values of Y (Y') are summed and squared, yielding estimates of variation attributable to different sources. Total sum of squares for Y is partitioned into a sum of squares due to regression and a sum of squares left over or residual.

$$SS_Y = SS_{reg} + SS_{res} \qquad (5.2)$$

Total sum of squares of Y:

$$SS_Y = \sum (Y - \overline{Y})^2$$

is, as usual, the sum of squared differences between each individual's observed Y score and the mean of Y over all N cases. The sum of squares for regression

$$SS_{reg} = \sum (Y' - \overline{Y})^2$$

is the portion of the variation in Y that is explained by use of the IVs as predictors. That is, it is the sum of squared differences between predicted Y' and the mean of Y because the mean of Y is the best prediction for the value of Y in the absence of any useful IVs. Sum-of-squares residual

$$SS_{res} = \sum (Y - Y')^2$$

is the sum of squared differences between observed Y and the predicted scores, Y', and represents errors in prediction.

The squared multiple correlation is

$$R^2 = \frac{SS_{reg}}{SS_Y} \qquad (5.3)$$

The squared multiple correlation, R^2, is the proportion of sum of squares for regression in the total sum of squares for Y.

The squared multiple correlation is, then, the proportion of variation in the DV that is predictable from the best linear combination of the IVs. The multiple correlation itself is the correlation between the obtained and predicted Y values; that is, $R = r_{yy'}$.

Total sum of squares (SS_Y) is calculated directly from the observed values of the DV. For example, in the sample problem, where the mean on the comprehensive examination is 10,

$$SS_C = (18 - 10)^2 + (9 - 10)^2 + (8 - 10)^2 + (8 - 10)^2 + (5 - 10)^2 + (12 - 10)^2 = 102$$

To find the remaining sources of variation, it is necessary to solve the prediction equation (Equation 5.1) for Y', which means finding the best-fitting A and B_i. The most direct method of deriving the equation involves thinking of multiple correlation in terms of individual correlations.

5.4.2   Matrix Equations

Another way of looking at R^2 is in terms of the correlations between each of the IVs and the DV. The squared multiple correlation is the sum across all IVs of the product of the correlation between the DV and IV and the (standardized) regression coefficient for the IV.

$$R^2 = \sum_{i=1}^{k} r_{yi} \beta_i \tag{5.4}$$

where each r_{yi} = correlation between the DV and the ith IV

β_i = standardized regression coefficient, or beta weight[5]

The standardized regression coefficient is the regression coefficient that would be applied to the standardized X_i value—the z-score of the X_i value—to predict standardized Y'.

Because r_{yi} are calculated directly from the data, computation of R^2 involves finding the standardized regression coefficients (β_i) for the k IVs. Derivation of the k equations in k unknowns is beyond the scope of this book. However, solution of these equations is easily illustrated using matrix algebra. For those who are not familiar with matrix algebra, the rudiments of it are available in Appendix A. Sections A.5 (matrix multiplication) and A.6 (matrix inversion or division) are the only portions of matrix algebra necessary to follow the next few steps. We encourage you to follow along using specialized matrix programs such as SAS IML, MATLAB, SYSTAT MATRIX, or SPSS MATRIX, or standard spreadsheet programs such as Quattro Pro or Excel.

In matrix form:

$$R^2 = \mathbf{R}_{yi} \mathbf{B}_i \tag{5.5}$$

where $\mathbf{R}_{yi}$ = row matrix of correlations between the DV and the k IVs

$\mathbf{B}_i$ = column matrix of standardized regression coefficients for the same k IVs

The standardized regression coefficients can be found by inverting the matrix of correlations among IVs and multiplying that inverse by the matrix of correlations between the DV and the IVs.

$$\mathbf{B}_i = \mathbf{R}_{ii}^{-1} \mathbf{R}_{iy} \tag{5.6}$$

where $\mathbf{B}_i$ = column matrix of standardized regression coefficients

$\mathbf{R}_{ii}^{-1}$ = inverse of the matrix of correlations among the IVs

$\mathbf{R}_{iy}$ = column matrix of correlations between the DV and the IVs

---

[5] β is used to indicate a sample standardized regression coefficient, rather than a population estimate of the unstandardized coefficient consistent with usage in software packages.

TABLE 5.2   Correlations among IVs and the DV for Sample Data in Table 5.1

		R_{ii}		R_{iy}
	MOTIV	*QUAL*	*GRADE*	*COMPR*
MOTIV	1.00000	.39658	.37631	.58613
QUAL	.39658	1.00000	.78329	.73284
GRADE	.37631	.78329	1.00000	.75043
R_{yi} COMPR	.58613	.73284	.75043	1.00000

TABLE 5.3   Inverse of Matrix of Intercorrelations among IVs for Sample Data in Table 5.1

	MOTIVE	**QUAL**	**GRADE**
MOTIVE	1.20255	−0.31684	−0.20435
QUAL	−0.31684	2.67113	−1.97305
GRADE	−0.20435	−1.97305	2.62238

Because multiplication by an inverse is the same as division, the column matrix of correlations between the IVs and the DV is divided by the correlation matrix of IVs.

These equations,[6] then, are used to calculate R^2 for the sample COMPR data from Table 5.1. All the required correlations are in Table 5.2.

Procedures for inverting a matrix are amply demonstrated elsewhere (e.g., Cooley & Lohnes, 1971; Harris, 2001) and are typically available in computer installations and spreadsheet programs. Because the procedure is extremely tedious by hand, and becomes increasingly so as the matrix becomes larger, the inverted matrix for the sample data is presented without calculation in Table 5.3.

From Equation 5.6, the $\mathbf{B}_i$ matrix is found by postmultiplying the $\mathbf{R}_{ii}^{-1}$ matrix by the $\mathbf{R}_{iy}$ matrix.

$$\mathbf{B}_i = \begin{bmatrix} 1.20255 & -0.31684 & -0.20435 \\ -0.31684 & 2.67113 & -1.97305 \\ -0.20435 & -1.97305 & 2.62238 \end{bmatrix} \begin{bmatrix} .58613 \\ .73284 \\ .75043 \end{bmatrix} = \begin{bmatrix} 0.31931 \\ 0.29117 \\ 0.40221 \end{bmatrix}$$

so that $\beta_M = 0.319$, $\beta_Q = 0.291$, and $\beta_G = 0.402$, Then, from Equation 5.5, we obtain

[6]Similar equations can be solved in terms of Σ (variance-covariance) matrices or $\mathbf{S}$ (sum-of-squares and cross-product) matrices as well as correlation matrices. If Σ or $\mathbf{S}$ matrices are used, the regression coefficients are unstandardized coefficients, as in Equation 5.1.

$$R^2 = [.58613 \quad .73284 \quad .75043] \begin{bmatrix} 0.31931 \\ 0.29117 \\ 0.40221 \end{bmatrix} = .70237$$

In this example, 70% of the variance in graduate comprehensive exam scores is predictable from knowledge of motivation, admission qualifications, and graduate course performance.

Once the standardized regression coefficients are available, they are used to write the equation for the predicted values of COMPR (Y'). If z-scores are used throughout, the beta weights (β_i) are used to set up the prediction equation. The equation is similar to Equation 5.1 except that there is no A (intercept) and both the IVs and predicted DV are in standardized form.

If, instead, the equation is needed in raw score form, the coefficients must first be transformed to unstandardized B_i coefficients.

$$B_i = \beta_i \left(\frac{S_Y}{S_i} \right) \tag{5.7}$$

Unstandardized coefficients (B_i) are found by multiplying standardized coefficients (beta weights—β_i) by the ratio of standard deviations of the DV and IV, where S_i is the standard deviation of the ith IV and S_Y is the standard deviation of the DV, and

$$A = \overline{Y} - \sum_{i=1}^{k} (B_i \overline{X}_i) \tag{5.8}$$

The intercept is the mean of the DV less the sum of the means of the IVs multiplied by their respective unstandardized coefficients.

For the sample problem of Table 5.1:

$$B_M = 0.319 \left(\frac{4.517}{2.191} \right) = 0.658$$

$$B_Q = 0.291 \left(\frac{4.517}{4.834} \right) = 0.272$$

$$B_G = 0.402 \left(\frac{4.517}{4.366} \right) = 0.416$$

$$A = 10 - [(0.658)(11.00) + (0.272)(10.17) + (0.416)(11.33)] = -4.72$$

The prediction equation for raw COMPR scores, once scores on MOTIV, QUAL, and GRADE are known, is

$$(COMPR)' = -4.72 + 0.658(MOTIV) + 0.272(QUAL) + 0.416(GRADE)$$

If a graduate student has ratings of 12, 14, and 15, respectively, on MOTIV, QUAL, and GRADE, the predicted rating on COMPR is

$$(COMPR)' = -4.72 + 0.658(12) + 0.272(14) + 0.416(15) = 13.22$$

The prediction equation also shows that for every one-unit change in GRADE, there is a change of about 0.4 point on COMPR if the values on the other IVs are held constant.

5.4.3   Computer Analyses of Small-Sample Example

Tables 5.4 and 5.5 demonstrate syntax and selected output for computer analyses of the data in Table 5.1, using default values. Table 5.4 illustrates SPSS REGRESSION. Table 5.5 shows a run through SAS REG.

TABLE 5.4   Syntax and Selected SPSS REGRESSION Output for Standard Multiple Regression on Sample Data in Table 5.1

```
REGRESSION
   /MISSING LISTWISE
   /STATISTICS COEFF OUTS R ANOVA
   /CRITERIA=PIN(.05) POUT(.10)
   /NOORIGIN
   /DEPENDENT compr
   /METHOD=ENTER motiv qual grade
```

Regression

Model Summary

Model	R	R Square	Adjusted R Square	Std. Error of the Estimate
1	.838[a]	.702	.256	3.8961

[a]Predictors: (Constant), GRADE, MOTIV, QUAL

ANOVA[b]

Model		Sum of Squares	df	Mean Square	F	Sig.
1	Regression	71.640	3	23.880	1.573	.411[a]
	Residual	30.360	2	15.180		
	Total	102.000	5			

[a]Predictors: (Constant), GRADE, MOTIV, QUAL

[b]Dependent Variable: COMPR

TABLE 5.4   Continued

Coefficients[a]

Model		Unstandardized Coefficients		Standardized Coefficients	t	Sig.
		B	Std. Error	Beta		
1	(Constant)	−4.722	9.066		−.521	.654
	MOTIV	.658	.872	.319	.755	.529
	QUAL	.272	.589	.291	.462	.690
	GRADE	.416	.646	.402	.644	.586

[a]Dependent Variable: COMPR

TABLE 5.5   Syntax and SAS REG Output for Standard Multiple Regression on Sample Data of Table 5.1

```
proc reg data=SASUSER.SS_REG;
    model COMPR= SUBJNO MOTIV QUAL GRADE;
run;
```

                    The REG Procedure
                      Model: MODEL1
                  Dependent Variable: COMPR

    Number of Observations Read   7
    Number of Observations Used   6
    Number of Observations with Missing Values   1

                    Analysis of Variance

                           Sum of   Mean
    Source   DF   Squares   Square   F Value   Pr > F

    Model   3   71.64007   23.88002   1.57   0.4114
    Error   2   30.35993   25.17997
    Corrected Total 5   102.00000

            Root MSE   3.89615   R-Square   0.7024
            Dependent Mean   10.00000   Adj R-Sq   0.2559
            Coeff Var   38.96148

                    Parameter Estimates

                     Parameter   Standard
    Variable   DF   Estimate   Error   t Value   Pr > |t|

    Intercept   1   -4.72180   9.06565   -0.52   0.6544
    MOTIV   1   0.65827   0.87213   0.75   0.5292
    QUAL   1   0.27205   0.58911   0.46   0.6896
    GRADE   1   0.41603   0.64619   0.64   0.5857

In SPSS REGRESSION, the DV is specified as compr. METHOD=ENTER, followed by the list of IVs, is the instruction that specifies standard multiple regression.

In standard multiple regression, results are given for only one step in the Model Summary table. The table includes R, R^2, adjusted R^2 (see Section 5.6.3) and Std. Error of the Estimate, the standard error of the predicted score, Y'. Then the ANOVA table shows details of the F test of the hypothesis that multiple regression is zero (see Section 5.6.2.1). The following are the regression coefficients and their significance tests, including B weights, the standard error of B (Std. Error), β weights (Beta), t tests for the coefficients (see Section 5.6.2.2), and their significance levels (Sig.). The term (Constant) refers to the intercept (A).

In using SAS REG for standard multiple regression, the variables for the regression equation are specified in the MODEL statement, with the DV on the left side of the equation and the IVs on the right.

In the ANOVA table, the sum of squares for regression is called Model and residual is called Error. Total SS and df are in the row labeled C Total. Below the ANOVA is the standard error of the estimate, shown as the square root of the error term, MS_{res} (Root MSE). Also printed are the mean of the DV (Dep Mean), R^2, adjusted R^2, and the coefficient of variation (Coeff Var)—defined here as 100 times the standard error of the estimate divided by the mean of the DV. The section labeled Parameter Estimates includes the usual B coefficients in the Parameter Estimate column, their standard errors, t tests for those coefficients and significance levels (Pr > |t|). Standardized regression coefficients are not printed unless requested.

Additional features of these programs are discussed in Section 5.8.

5.5   Major Types of Multiple Regression

There are three major analytic strategies in multiple regression: standard multiple regression, sequential (hierarchical) regression, and statistical (stepwise) regression. Differences among the strategies involve what happens to overlapping variability due to correlated IVs and who determines the order of entry of IVs into the equation.

Consider the Venn diagram in Figure 5.2(a) in which there are three IVs and one DV. IV_1 and IV_2 both correlate substantially with the DV and with each other. IV_3 correlates to a lesser extent with the DV and to a negligible extent with IV_2. R^2 for this situation is the area $a + b + c + d + e$. Area a comes unequivocally from IV_1; area c unequivocally from IV_2; area e from IV_3. However, there is ambiguity regarding areas b and d. Both areas could be predicted from either of two IVs; area b from either IV_1 or IV_2, area d from either IV_2 or IV_3. To which IV should the contested area be assigned? The interpretation of analysis can be drastically affected by choice of strategy because the apparent importance of the various IVs to the solution changes.

5.5.1   Standard Multiple Regression

The standard model is the one used in the solution for the small-sample graduate student data in Table 5.1. In the standard, or simultaneous, model, all IVs enter into the regression equation at once; each one is assessed as if it had entered the regression after all other IVs had entered. Each IV is evaluated in terms of what it adds to prediction of the DV that is different from the predictability afforded by all the other IVs.

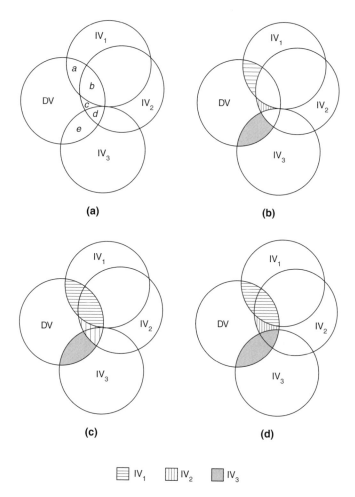

IV_1 IV_2 IV_3

**FIGURE 5.2   Venn diagrams illustrating (a) overlapping
variance sections; and allocation of overlapping variance in
(b) standard multiple regression, (c) sequential regression,
and (d) statistical (stepwise) regression.**

Consider Figure 5.2(b). The darkened areas of the figure indicate the variability accorded each of the IVs when the procedure of Section 5.6.1.1 is used. IV_1 "gets credit" for area a, IV_2 for area c, and IV_3 for area e. That is, each IV is assigned only the area of its unique contribution. The overlapping areas, b and d, contribute to R^2, but are not assigned to any of the individual IVs.

In standard multiple regression, it is possible for a variable like IV_2 to appear unimportant in the solution when it actually is highly correlated with the DV. If the area of that correlation is whittled away by other IVs, the unique contribution of the IV is often very small despite a substantial correlation with the DV. For this reason, both the full correlation and the unique contribution of the IV need to be considered in interpretation.

Standard multiple regression is handled in the SPSS package by the REGRESSION program, as are all other types of multiple regression. A selected part of output is given in Table 5.4 for the sample problem of Table 5.1. (Full interpretation of program output is available in substantive examples presented later in this chapter.) SAS REG is used for standard analyses, as illustrated in Table 5.6.

5.5.2 Sequential Multiple Regression

In sequential regression (sometimes called hierarchical regression), IVs enter the equation in an order specified by the researcher. Each IV (or set of IVs) is assessed in terms of what it adds to the equation at its own point of entry. Consider the example in Figure 5.2(c). Assume that the researcher assigns IV_1 first entry, IV_2 second entry, and IV_3 third entry. In assessing importance of variables by the procedure of Section 5.6.1.2, IV_1 "gets credit" for areas a and b, IV_2 for areas c and d, and IV_3 for area e. Each IV is assigned the variability, unique and overlapping, left to it at its own point of entry. Notice that the apparent importance of IV_2 would increase dramatically if it were assigned first entry and, therefore, "got credit" for b, c, and d.

The researcher normally assigns order of entry of variables according to logical or theoretical considerations. For example, IVs that are presumed (or manipulated) to be causally prior are given higher priority of entry. For instance, height might be considered prior to amount of training in assessing success as a basketball player and accorded a higher priority of entry. Variables with greater theoretical importance could also be given early entry.

Or the opposite tack could be taken. The research could enter manipulated or other variables of major importance on later steps, with "nuisance" variables given higher priority for entry. The lesser, or nuisance, set is entered first; then the major set is evaluated for what it adds to the prediction over and above the lesser set. For example, we might want to see how well we can predict reading speed (the DV) from intensity and length of a speed-reading course (the major IVs) while holding constant initial differences in reading speed (the nuisance IV). This is the basic analysis of covariance problem in regression format.

IVs can be entered one at a time or in blocks. The analysis proceeds in steps, with information about variables both in and out of the equation given in computer output at each step. Finally, after all variables are entered, summary statistics are provided along with the information available at the last step.

In the SPSS package, sequential regression is performed by the REGRESSION program. SAS REG also has interactive modes in which individual IVs can be entered sequentially.

Syntax and selected output are shown for the sample problem of Table 5.1 using SPSS REGRESSION. In Table 5.6, with higher priority given to admission qualifications and course performance, and lower priority given to motivation, the sequence is indicated by having two ENTER instructions, one for each step of the model. CHANGE is added to STATISTICS to provide a table that shows the gain in prediction afforded by motiv, once qual and grade are in the model. SPSS REGRESSION sequential analysis is interpreted in detail in Section 5.7.3.

5.5.3 Statistical (Stepwise) Regression

Statistical regression (sometimes generically called stepwise regression) is a controversial procedure, in which order of entry of variables is based solely on statistical criteria. The meaning or

TABLE 5.6   Syntax and Selected SPSS REGRESSION Output for Sequential Multiple Regression on Sample Data in Table 5.1

```
REGRESSION
  /MISSING LISTWISE
  /STATISTICS COEFF OUTS R ANOVA CHANGE
  /CRITERIA=PIN(.05) POUT(.10)
  /NOORIGIN
  /DEPENDENT compr
  /METHOD=ENTER qual grade /METHOD=ENTER motiv.
```

Regression

Model Summary

Model	R	R Square	Adjusted R Square	Std. Error of the Estimate	R Square Change	F Change	df1	df2	Sig. F Change
					Change Statistics				
1	.786[a]	.618	.363	3.6059	.618	2.422	2	3	.236
2	.838[b]	.702	.256	3.8961	.085	.570	1	2	.529

[a]Predictors: (Constant), GRADE, QUAL
[b]Predictors: (Constant), GRADE, QUAL, MOTIV

ANOVA[c]

Model		Sum of Squares	df	Mean Square	F	Sig.
1	Regression	62.992	2	31.496	2.422	.236[a]
	Residual	39.008	3	13.003		
	Total	102.000	5			
2	Regression	71.640	3	23.880	1.573	.411[b]
	Residual	30.360	2	15.180		
	Total	102.000	5			

[a]Predictors: (Constant), GRADE, QUAL
[b]Predictors: (Constant), GRADE, QUAL, MOTIV
[c]Dependent Variable: COMPR

Coefficients[a]

Model		Unstandardized Coefficients B	Unstandardized Coefficients Std. Error	Standardized Coefficients Beta	t	Sig.
1	(Constant)	1.084	4.441		.244	.823
	QUAL	.351	.537	.375	.653	.560
	GRADE	.472	.594	.456	.795	.485
2	(Constant)	-4.722	9.066		-.521	.654
	QUAL	.272	.589	.291	.462	.690
	GRADE	.416	.646	.402	.644	.586
	MOTIV	.658	.872	.319	.755	.529

[a]Dependent Variable: COMPR

139

interpretation of the variables is not relevant. Decisions about which variables are included and which omitted from the equation are based solely on statistics computed from the particular sample drawn; minor differences in these statistics can have a profound effect on the apparent importance of an IV.

Consider the example in Figure 5.2(d). IV_1 and IV_2 both correlate substantially with the DV; IV_3 correlates less strongly. The choice between IV_1 and IV_2 for first entry is based on which of the two IVs has the higher full correlation with the DV, even if the higher correlation shows up in the second or third decimal place. Let's say that IV_1 has the higher correlation with the DV and enters first. It "gets credit" for areas a and b. At the second step, IV_2 and IV_3 are compared, where IV_2 has areas c and d available to add to prediction, and IV_3 has areas e and d. At this point, IV_3 contributes more strongly to R^2 and enters the equation. IV_2 is now assessed for whether or not its remaining area, c, contributes significantly to R^2. If it does, IV_2 enters the equation; otherwise, it does not despite the fact that it is almost as highly correlated with the DV as the variable that entered first. For this reason, interpretation of a statistical regression equation is hazardous unless the researcher takes special care to remember the message of the initial DV–IV correlations.

There are actually three versions of statistical regression: forward selection, backward deletion, and stepwise regression. In forward selection, the equation starts out empty and IVs are added one at a time provided they meet the statistical criteria for entry. Once in the equation, an IV stays in. Bendel and Afifi (1977) recommend a liberal criterion for entry of predictors in forward regression. Important variables are less likely to be excluded from the model with a more liberal probability level for entry of .15 to .20 rather than .05. In backward deletion, the equation starts out with all IVs entered and they are deleted one at a time if they do not contribute significantly to regression. Stepwise regression is a compromise between the two other procedures in which the equation starts out empty and IVs are added one at a time if they meet statistical criteria, but they may also be deleted at any step where they no longer contribute significantly to regression.

Statistical regression is typically used to develop a subset of IVs that is useful in predicting the DV, and to eliminate those IVs that do not provide additional prediction to the IVs already in the equation. For this reason, statistical regression may have some utility if the only aim of the researcher is a prediction equation.

Even so, the sample from which the equation is drawn should be large and representative because statistical regression tends to capitalize on chance and overfit data. It capitalizes on chance because decisions about which variables to include are dependent on potentially minor differences in statistics computed from a single sample, where some variability in the statistics from sample to sample is expected. It overfits data because the equation derived from a single sample is too close to the sample and may not generalize well to the population.

An additional problem is that statistical analysis may not lead to the optimum solution in terms of R^2. Several IVs considered together may increase R^2, whereas any one of them considered alone does not. In simple statistical regression, none of the IVs enters. By specifying that IVs enter in blocks, one can set up combinations of sequential and statistical regression. A block of high-priority IVs is set up to compete among themselves stepwise for order of entry; then a second block of IVs compete among themselves for order of entry. The regression is sequential over blocks, but statistical within blocks.

Cross-validation with a second sample is highly recommended for statistical regression and is accomplished through several steps. First, the data set is divided into two random samples; a recommended split is 80% for the statistical regression analysis and the remaining 20% as the cross-validation sample (SYSTAT Software Inc., 2004, p. II-16). After the statistical regression on the larger subsample is run, predicted scores are created for the smaller cross-validation sample using the regression coefficients produced by the analysis. Finally, predicted scores and actual scores are correlated to find R^2 for the smaller sample. A large discrepancy between R^2 for the smaller and larger samples indicates overfitting and lack of generalizability of the results of the analysis.

Tables 5.7 and 5.8 show cross-validation with statistical regression through SAS. There is forward selection on an 80% sample of a hypothetical data set with the same variables as in Section 5.4, but with 100 cases, followed by cross-validation using the remaining 20% of the cases. Syntax in Table 5.7 first creates the two random samples, with the 80% sample coded 1 and the 20% sample coded 0. Then the two samples are formed, SAMP80 and SAMP20. Then the 80% sample (SAMP80) is selected for the statistical regression run.

The output shows that only MOTIV and GRADE enter the equation. QUAL does not reliably add to prediction of COMPR over and above that produced by the other two IVs when order of entry is chosen by a statistical criterion for this sample of 77 cases. (Note that the 80% random sample actually produced 77 rather than 80 cases.)

Syntax in Table 5.8 shows the creation of predicted scores for the 20% cross-validation sample, followed by a request for a correlation between predicted (PREDCOMP) and actual (COMPR) scores for that sample. The prediction equation is taken from the last section of Table 5.7. The first two lines turn off the selection of the 80% sample and turn on the selected of the cross-validation sample.

The correlation between predicted and actual scores is squared ($R^2 = .90417^2 = .81752$) to compare it with $R^2 = .726$ for the larger sample. In this case, the cross-validation sample is better predicted by the regression equation than the sample that generated the equation. This is an unusual result, but one that would make a researcher breathe a sigh of relief after using statistical regression.

SPSS REGRESSION provides statistical regression in a manner similar to that of sequential regression. However, STEPWISE is chosen as the METHOD, rather than ENTER.

Another option to avoid overfitting is bootstrapping, available in SPSS using a macro available when the pack is installed: oms_bootstrapping.sps. Instructions are available within the macro. Bootstrapping is a process by which statistics (e.g., regression weights) are generated over a very large number of replications, with samples drawn with replacement from a data set. For example, there might be 1,000 bootstrap samples of 6 cases drawn from the small-sample data set of 6 cases. Each case may be drawn more than once, or not at all, because of replacement. In a given sample, for instance, case 1 might be drawn twice, case 2 drawn twice, case 3 drawn once, case 4 drawn once, and cases 5 and 6 not drawn. Descriptive statistics and histograms are then viewed for the requested statistics. For example, a 1,000-replication bootstrap of the small-sample data set yielded an average intercept of -4.74 and average B weights of 0.68, -0.12, and 1.07. Values for QUAL and GRADE are very different from the values of Section 5.4 for such a small sample.

At the very least, separate analyses of two halves of an available sample should be conducted to avoid overfitting, with conclusions limited to results that are consistent for both analyses.

On the other hand, statistical regression is a handy (and acceptable) procedure for determining which variables are associated with the difference between an outlier and remaining cases, as shown in Sections 4.2.1 and 4.2.2. Here, there is no intent to generalize to any population—use of statistical regression is just to describe some characteristics of the sample.

TABLE 5.7 **Forward Statistical Regression on an 80% Subsample. Syntax and Selected SAS REG Output**

```
data SASUSER.REGRESSX;
        set SASUSER.CROSSVAL;
            samp = 0;
            if uniform(13068) < .80 then samp = 1;
run;
data SASUSER.SAMP80;
            set SASUSER.REGRESSX;
            where samp = 1;
data SASUSER.SAMP20;
            set SASUSER.REGRESSX;
            where samp = 0;
run;
proc reg data=SASUSER.SAMP80;
        model COMPR= MOTIV QUAL GRADE/ selection= FORWARD;
run;
```

                        The REG Procedure
                          Model: MODEL1
                    Dependent Variable: COMPR

            Number of Observations Read   77
            Number of Observations Used   77

                    Forward Selection: Step 1

    Variable GRADE Entered: R-Square = 0.5828 and C(p) = 52.7179

                        Analysis of Variance

                                Sum of   Mean
        Source   DF   Squares   Square   F Value   Pr > F

        Model   1   797.72241   797.72241   104.77   <.0001
        Error   75   571.04564   7.61394
        Corrected Total 76   1368.76805

                    Parameter   Standard
        Variable   Estimate   Error   Type II SS   F Value   Pr > F

        Intercept   1.19240   0.95137   11.96056   1.57   0.2140
        GRADE   0.79919   0.07808   797.72241   104.77   <.0001

                    Bounds on condition number: 1, 1

                    Forward Selection: Step 2

    Variable MOTIV Entered: R-Square = 0.7568 and C(p) = 2.2839

TABLE 5.7   Continued

Analysis of Variance

Source	DF	Sum of Squares	Mean Square	F Value	Pr > F
Model	2	1035.89216	517.94608	115.14	<.0001
Error	74	332.87589	4.49832		
Corrected Total	76	1368.76805			

Variable	Parameter Estimate	Standard Error	Type II SS	F Value	Pr > F
Intercept	-5.86448	1.21462	104.86431	23.31	<.0001
MOTIV	0.78067	0.10729	238.16975	52.95	<.0001
GRADE	0.65712	0.06311	487.68516	108.41	<.0001

5.5.4   Choosing among Regression Strategies

To simply assess relationships among variables and answer the basic question of multiple correlation, the method of choice is standard multiple regression. However, standard multiple regression is atheoretical—a shotgun approach. Reasons for using sequential regression are theoretical or for testing explicit hypotheses.

TABLE 5.8   Correlation between Predicted and Actual Scores on Comprehension. Syntax and Selected SAS CORR Output

```
data SASUSER.PRED20;
   set SASUSER.SAMP20
  PREDCOMP = -5.86448 + 0.78067*MOTIV + 0.65712*GRADE
run;
proc corr data=SASUSER.PRED20 PEARSON;
     var COMPR   PREDCOMP;
run;
```

The CORR Procedure
Pearson Correlation Coefficients, N = 23
Prob > |r| under H0: Rho=0

	COMPR	PREDCOMP
COMPR	1.00000	0.90417 <.0001
PREDCOMP	0.90417 <.0001	1.00000

Sequential regression allows the researcher to control the advancement of the regression process. Importance of IVs in the prediction equation is determined by the researcher according to logic or theory. Explicit hypotheses are tested about proportion of variance attributable to some IVs after variance due to IVs already in the equation is accounted for.

Although there are similarities in programs used and output produced for sequential and statistical regression, there are fundamental differences in the way that IVs enter the prediction equation and in the interpretations that can be made from the results. In sequential regression, the researcher controls entry of variables, whereas in statistical regression, statistics computed from sample data control order of entry. Statistical regression is, therefore, a model-building rather than model-testing procedure. As an exploratory technique, it may be useful for such purposes as eliminating variables that are clearly superfluous in order to tighten up future research. However, clearly superfluous IVs will show up in any of the procedures. Also, results of statistical regression can be very misleading unless based on samples that are large and highly representative of the population of interest. When multicollinearity or singularity is present, statistical regression may be helpful in identifying multicollinear variables, as indicated in Chapter 4.

For the example of Section 5.4, in which performance on graduate comprehensive exam (COMPR) is predicted from professional motivation (MOTIV), qualifications for graduate training (QUAL), and performance in graduate courses (GRADE), the differences among regression strategies might be phrased as follows. If standard multiple regression is used, two fundamental questions are asked: (1) What is the size of the overall relationship between COMPR and the set of IVs: MOTIV, QUAL, and GRADE? (2) How much of the relationship is contributed uniquely by each IV? If sequential regression is used, with QUAL and GRADE entered before MOTIV, the question is: Does MOTIV significantly add to prediction of COMPR after differences among students in QUAL and GRADE have been statistically eliminated? If statistical regression is used, one asks: What is the best linear combination of IVs to predict the DV in this sample?

5.6   Some Important Issues

5.6.1   Importance of IVs

If the IVs are uncorrelated with each other, assessment of the contribution of each of them to multiple regression is straightforward. IVs with bigger correlations or higher standardized regression coefficients are more important to the solution than those with lower (absolute) values. (Because unstandardized regression coefficients are in a metric that depends on the metric of the original variables, their sizes are harder to interpret. A large regression coefficient for an IV with a low correlation with the DV can also be misleading because the IV predicts the DV well only after another IV suppresses irrelevant variance, as shown in Section 5.6.4.)

If the IVs are correlated with each other, assessment of the importance of each of them to regression is more ambiguous. The correlation between an IV and the DV reflects variance shared with the DV, but some of that variance may be predictable from other IVs.

To get the most straightforward answer regarding the importance of an IV to regression, one needs to consider the type of regression it is, and both the full and unique relationship between the IV and the DV. This section reviews several of the issues to be considered when assessing the importance of an IV to standard multiple, sequential, or statistical regression. In all cases, one needs to compare the total relationship of the IV with the DV, the unique relationship of the IV with the DV,

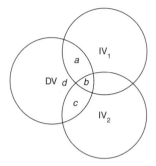

	Standard Multiple	**Sequential**
r_i^2	IV_1 $(a+b)/(a+b+c+d)$	$(a+b)/(a+b+c+d)$
	IV_2 $(c+b)/(a+b+c+d)$	$(c+b)/(a+b+c+d)$
sr_i^2	IV_1 $a/(a+b+c+d)$	$(a+b)/(a+b+c+d)$
	IV_2 $c/(a+b+c+d)$	$c/(a+b+c+d)$
pr_i^2	IV_1 $a/(a+d)$	$(a+b)/(a+b+d)$
	IV_2 $c/(c+d)$	$c/(c+d)$

FIGURE 5.3 Areas representing squared correlation, squared semipartial correlation, and squared partial correlation in standard multiple and sequential regression (where IV_1 is given priority over IV_2).

and the correlations of the IVs with each other in order to get a complete picture of the function of an IV in regression. The total relationship of the IV with the DV (correlation) and the correlations of the IVs with each other are given in the correlation matrix. The unique contribution of an IV to predicting a DV is generally assessed by either partial or semipartial correlation.

For standard multiple and sequential regression, the relationships between correlation, partial correlation, and semipartial correlation are given in Figure 5.3 for a simple case of one DV and two IVs. In the figure, squared correlation, partial correlation, and semipartial correlation coefficients are defined as areas created by overlapping circles. Area $a + b + c + d$ is the total area of the DV and reduces to a value of 1 in many equations. Area b is the segment of the variability of the DV that can be explained by either IV_1 or IV_2 and is the segment that creates the ambiguity. Notice that it is the denominators that change between squared semipartial and partial correlation.

In a partial correlation, the contribution of the other IVs is taken out of both the IV and the DV. In a semipartial correlation, the contribution of other IVs is taken out of only the IV. Thus, squared semipartial correlation expresses the unique contribution of the IV to the total variance of the DV. Squared semipartial correlation (sr_i^2) is the more useful measure of importance of an IV.[7] The interpretation of sr_i^2 differs, however, depending on the type of multiple regression employed.

---

[7]Procedures for obtaining partial correlations are available in earlier editions of this book but are omitted in this edition.

5.6.1.1 Standard Multiple Regression

In standard multiple regression, sr_i^2 for an IV is the amount by which R^2 is reduced if that IV is deleted from the regression equation. That is, sr_i^2 represents the unique contribution of the IV to R^2 in that set of IVs.

When the IVs are correlated, squared semipartial correlations do not necessarily sum to multiple R^2. The sum of sr_i^2 is usually smaller than R^2 (although under some rather extreme circumstances, the sum can be larger than R^2). When the sum is smaller, the difference between R^2 and the sum of sr_i^2 for all IVs represents shared variance, variance that is contributed to R^2 by two or more IVs. It is rather common to find substantial R^2, with sr_i^2 for all IVs quite small.

Table 5.9 summarizes procedures for finding sr_i^2 (and pr_i^2) for both standard multiple and sequential regression through SPSS and SAS.

SPSS and SAS provide versions of sr_i^2 as part of their output. If you use SPSS REGRESSION, sr_i is optionally available as **Part Correlations** by requesting **STATISTICS = ZPP** (part and partial correlations on the Statistics menu). SAS REG provides squared semipartial correlations when **SCORR2** is requested (from the parameter estimates menu: Print Type II squared semipartial correlations).

In all standard multiple-regression programs, F_i (or T_i) is the significance test for sr_i^2, pr_i^2, B_i,[5] and β_i, as discussed in Section 5.6.2.

5.6.1.2 Sequential or Statistical Regression

In these two forms of regression, sr_i^2 is interpreted as the amount of variance added to R^2 by each IV at the point that it enters the equation. The research question is, How much does this IV add to multiple R^2 after IVs with higher priority have contributed their share to prediction of the DV? Thus, the apparent importance of an IV is very likely to depend on its point of entry into the equation. In sequential and statistical regression, the sr_i^2 do, indeed, sum to R^2; consult Figure 5.2 if you want to review this point.

As reviewed in Table 5.9, SPSS REGRESSION provides squared semipartial correlations as part of output for sequential and statistical regression when **CHANGE** statistics are requested.[8] For SPSS, sr_i^2 is **R Square Change** for each IV in the **Model Summary** table (see Table 5.6).

SAS REG (for sequential regression) provides R^2 for each step. You can calculate sr_i^2 by subtraction between subsequent steps.

5.6.2 Statistical Inference

This section covers significance tests for multiple regression and for regression coefficients for individual IVs. A test, F_{inc}, is also described for evaluating the statistical significance of adding two or more IVs to a prediction equation in sequential or statistical analysis. Calculations of confidence limits for unstandardized regression coefficients and procedures for comparing the predictive capacity of two different sets of IVs conclude the section.

When the researcher is using statistical regression as an exploratory tool, inferential procedures of any kind may be inappropriate. Inferential procedures require that the researcher have a

---

[8]If you request one of the statistical regression options in PROC REG, squared semipartial correlations will be printed out in a summary table as Partial R**2.

TABLE 5.9   Procedures for Finding sr_i^2 and pr_i^2 through SPSS and SAS for Standard Multiple and Sequential Regression

	sr_i^2	pr_i^2
Standard Multiple Regression		
SPSS REGRESSION	STATISTICS ZPP Part	STATISTICS ZPP Partial
SAS REG	SCORR2	PCORR2
Sequential Regression		
SPSS REGRESSION	R Square Change in Model Summary table	Not available
SAS REG	SCORR1	PCORR1

hypothesis to test. When statistical regression is used to snoop data, there may be no hypothesis, even though the statistics themselves are available.

5.6.2.1   *Test for Multiple* R

The overall inferential test in multiple regression is whether the sample of scores is drawn from a population in which multiple R is zero. This is equivalent to the null hypothesis that all correlations between DV and IVs and all regression coefficients are zero. With large N, the test of this hypothesis becomes trivial because it is almost certain to be rejected.

For standard multiple and sequential regression, the test of this hypothesis is presented in all computer outputs as analysis of variance. For sequential regression (and for standard multiple regression performed through stepwise programs), the analysis of variance table at the last step gives the relevant information. The F ratio for mean square regression over mean square residual tests the significance of multiple R. Mean square regression is the sum of squares for regression in Equation 5.2 divided by k degrees of freedom; mean square residual is the sum of squares for residual in the same equation divided by $(N - k - 1)$ degrees of freedom.

If you insist on inference in statistical regression, adjustments are necessary because all potential IVs do not enter the equation and the test for R^2 is not distributed as F. Therefore, the analysis of variance table at the last step (or for the "best" equation) is misleading; the reported F is biased so that the F ratio actually reflects a Type I error rate in excess of α.

Wilkinson and Dallal (1981) have developed tables for critical R^2 when forward selection procedures are used for statistical addition of variables and the selection stops when the F-to-enter for the next variable falls below some preset value. Appendix C contains Table C.5 that shows how large multiple R^2 must be to be statistically significant at .05 or .01 levels, given N, k, and F, where N is sample size, k is the number of potential IVs, and F is the minimum F-to-enter that is specified. F-to-enter values that can be chosen are 2, 3, or 4.

For example, for a statistical regression in which there are 100 subjects, 20 potential IVs, and an F-to-enter value of 3 chosen for the solution, a multiple R^2 of approximately .19 is required to be considered significantly different from zero at $\alpha = .05$ (and approximately .26 at $\alpha = .01$). Wilkinson and Dallal report that linear interpolation on N and k works well. However, they caution against extensive extrapolation for values of N and k beyond those given in the table. In sequential regression, this table is also used to find critical R^2 values if a post hoc decision is made to terminate regression as soon as R^2 reaches statistical significance, if the appropriate F-to-enter is substituted for R^2 probability value as the stopping rule.

Wilkinson and Dallal recommend forward selection procedures in favor of other selection methods (e.g., stepwise selection). They argue that, in practice, results using different procedures are not likely to be substantially different. Further, forward selection is computationally simple and efficient, and allows straightforward specification of stopping rules. If you are able to specify in advance the number of variables you wish to select, an alternative set of tables is provided by Wilkinson (1979) to evaluate significance of multiple R^2 with forward selection.

After looking at the data, you may wish to test the significance of some subsets of IVs in predicting the DV where a subset may even consist of a single IV. If several post hoc tests like this are desired, Type I errors become increasingly likely. Larzelere and Mulaik (1977) recommend the following conservative F test to keep Type I error rate below alpha for all combinations of IVs:

$$F = \frac{R_s^2/k}{(1 - R_s^2)/(N - k - 1)} \tag{5.9}$$

where R_s^2 is the squared multiple (or bivariate) correlation to be tested for significance, and k is the total number of IVs. Obtained F is compared with tabled F, with k and $(N - k - 1)$ degrees of freedom (Table C.3). That is, the critical value of F for each subset is the same as the critical value for the overall multiple R.

In the sample problem of Section 5.4, the bivariate correlation between MOTIV and COMPR (from Table 5.2) is tested post hoc as follows:

$$F = \frac{.58613^2/3}{(1 - .58613^2)/2} = 0.349, \quad \text{with df} = 3, 2$$

which is obviously not significant. (Note that results can be nonsensical with very small samples.)

5.6.2.2   Test of Regression Components

In standard multiple regression, for each IV, the same significance test evaluates B_i, β_i, pr_i, and sr_i. The test is straightforward, and results are given in computer output. In SPSS, t_i tests the unique contribution of each IV and appears in the output section labeled Coefficients (see Table 5.4). Degrees of freedom are 1 and df_{res}, which appears in the accompanying ANOVA table. In SAS REG, T_i or F_i values are given for each IV, tested with df_{res} from the analysis of variance table (see Tables 5.5 and 5.6).

Recall the limitations of these significance tests. The significance tests are sensitive only to the unique variance an IV adds to R^2. A very important IV that shares variance with another IV in the analysis may be nonsignificant although the two IVs in combination are responsible in large part for the size of R^2. An IV that is highly correlated with the DV but has a nonsignificant regression coefficient may have suffered just such a fate. For this reason, it is important to report and interpret

r_{iy} in addition to F_i for each IV, as shown later in Table 5.13, summarizing the results of a complete example.

For statistical and sequential regression, assessment of contribution of variables is more complex, and appropriate significance tests may not appear in the computer output. First, there is inherent ambiguity in the testing of each variable. In statistical and sequential regression, tests of sr_i^2 are not the same as tests of the regression coefficients (B_i and β_i). Regression coefficients are independent of order of entry of the IVs, whereas sr_i^2 depend directly on order of entry. Because sr_i^2 reflects "importance" as typically of interest in sequential or statistical regression, tests based on sr_i^2 are discussed here.[9]

SPSS and SAS provide significance tests for sr_i^2 in summary tables. For SPSS, the test is F Change—F ratio for change in R^2—that is accompanied by a significance value, Sig F Change. If you use SAS REG, you need to calculate F for sr_i^2 as found by subtraction (cf. Section 5.6.1.2) using the following equation:

$$F_i = \frac{sr_i^2}{(1 - R^2)/\text{df}_{res}} \tag{5.10}$$

The F_i for each IV is based on sr_i^2 (the squared semipartial correlation), multiple R^2 at the final step, and residual degrees of freedom from the analysis of variance table for the final step.

Note that these are incremental F ratios, F_{inc}, because they test the incremental change in R^2 as variables in each step are added to prediction.

5.6.2.3   Test of Added Subset of IVs

For sequential and statistical regression, one can test whether a block of two or more variables significantly increases R^2 above the R^2 predicted by a set of variables already in the equation.

$$F_{inc} = \frac{(R_{wi}^2 - R_{wo}^2)/m}{(1 - R^2)/\text{df}_{res}} \tag{5.11}$$

where F_{inc} is the incremental F ratio; R_{wi}^2 is the multiple R^2 achieved with the new block of IVs in the equation; R_{wo}^2 is the multiple R^2 without the new block of IVs in the equation; m is the number of IVs in the new block; and $\text{df}_{res} = (N - k - 1)$ is residual degrees of freedom in the final analysis of variance table.

Both R_{wi}^2 and R_{wo}^2 are found in the summary table of any program that produces one. The null hypothesis of no increase in R^2 is tested as F with m and df_{res} degrees of freedom. If the null hypothesis is rejected, the new block of IVs does significantly increase the explained variance.

---

[9]For combined standard-sequential regression, it might be desirable to use the "standard" method for all IVs simply to maintain consistency. If so, be sure to report that the F test is for regression coefficients.

Although this is a poor example because only one variable is in the new block, we can use the sequential example in Table 5.6 to test whether MOTIV adds significantly to the variance contributed by the first two variables to enter the equation, QUAL and GRADE.

$$F_{inc} = \frac{(.70235 - .61757)/1}{(1 - .70235)/2} = 0.570 \qquad \text{with df} = 1, 2$$

Because only one variable was entered, this test is the same as **F Change** for Model 2 in Table 5.6 in the **Model Summary** output. Thus, F_{inc} can be used when there is only one variable in the block, but the information is already provided in the output. Indeed, as noted previously, any test of a step in a sequential model is a form of F_{inc}.

5.6.2.4   *Confidence Limits around* **B** *and Multiple* **R**^{2}

To estimate population values, confidence limits for unstandardized regression coefficients (B_i) are calculated. Standard errors of unstandardized regression coefficients, unstandardized regression coefficients, and the critical two-tailed value of t for the desired confidence level (based on $N - 2$ degrees of freedom, where N is the sample size) are used in Equation 5.12.

$$CL_{B_i} = B_i \pm SE_{B_i}(t_{\alpha/2}) \tag{5.12}$$

The $1 - \alpha$ confidence limits for the unstandardized regression coefficient for the ith IV (CL_{B_i}) are the regression coefficient (B_i) plus or minus the standard error of the regression coefficient (SE_{B_i}) times the critical value of t, with ($N - 2$) degrees of freedom at the desired level of α.

If 95% confidence limits are requested, they are given in SPSS REGRESSION output in the segment of output titled **Coefficients**. With other output or when 99% confidence limits are desired, Equation 5.12 is used. Unstandardized regression coefficients and the standard errors of those coefficients appear in the sections labeled Coefficients or Parameter Estimates.

For the example in Table 5.4, the 95% confidence limits for GRADE, with df = 4, are

$$CL_{BG} = 0.416 \pm 0.646(2.78) = 0.416 \pm 1.796 = -1.380 \leftrightarrow 2.212$$

If the confidence interval contains zero, one retains the null hypothesis that the population regression coefficient is zero.

Confidence limits around R^2 also are calculated to estimate population values. Steiger and Fouladi (1992) have provided a computer program R2 (included with data sets for this book) to find them. The **Confidence Interval** is chosen as the **Option**, and **Maximize Accuracy** is chosen as the **Algorithm**. Using the R^2 value of .702, Figure 5.4a shows setup values of 6 observations, 4 variables (including 3 IVs plus the DV) and probability value of .95. As seen in Figure 5.4b, the R2 program provides 95% confidence limits from .00 to .89. Again, inclusion of zero suggests no statistically significant effect.

(a)
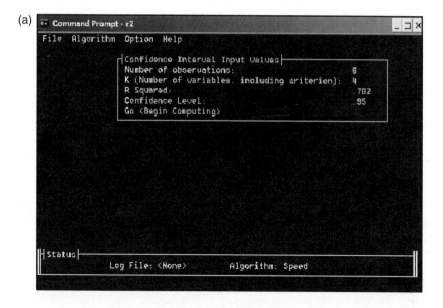

(b)
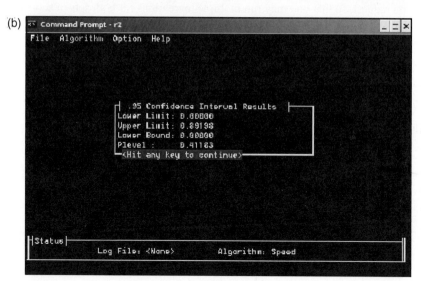

**FIGURE 5.4   Confidence limits around R^2 using Steiger and Fouladi's
(1992) software: (a) setup and (b) results.**

Software also is available in SAS and SPSS to find confidence limits around R^2 using values
of the F ratio and degrees of freedom as input. These programs by Smithson (2003), also included
with the data sets for this book, are demonstrated in Sections 6.6.2, 7.6.2, and elsewhere.

5.6.2.5 *Comparing Two Sets of Predictors*

It is sometimes of interest to know whether one set of IVs predicts a DV better than another set of IVs. For example, can ratings of current belly dancing ability be better predicted by personality tests or by past dance and musical training?

The procedure for finding out is fairly convoluted, but if you have a large sample and you are willing to develop a data file with a pair of predicted scores for each subject in your sample, a test for the significance of the difference between two "correlated correlations" (both correlations are based on the same sample and share a variable) is available (Steiger, 1980). (If sample size is small, non-independence among predicted scores for cases can result in serious violation of the assumptions underlying the test.)

As suggested in Section 5.4.1, a multiple correlation can be thought of as a simple correlation between obtained DVs and predicted DVs; that is, $R = r_{yy'}$. If there are two sets of predictors, $Y_{a'}$ and $Y_{b'}$ (where, for example, $Y_{a'}$ is prediction from personality scores and $Y_{b'}$ is prediction from past training), a comparison of their relative effectiveness in predicting Y is made by testing for the significance of the difference between $r_{yy'_a}$ and $r_{yy'_b}$. For simplicity, let's call these r_{ya} and r_{yb}.

To test the difference, we need to know the correlation between the predicted scores from set A (personality) and those from set B (training), that is, $r_{ya'yb'}$ or, simplified, r_{ab}. This is where file manipulation procedures or hand entering become necessary. (SPSS REGRESSION saves predicted scores to the data file on request, so that you may run multiple regressions for both sets of scores as IVs and save them to the same data file.)

The z test for the difference between r_{ya} and r_{yb} is

$$\overline{Z}^* = (z_{ya} - z_{yb})\sqrt{\frac{N-3}{2 - 2s_{ya,yb}}} \tag{5.13}$$

where N is, as usual, the sample size,

$$z_{ya} = (1/2)\ln\left(\frac{1 + r_{ya}}{1 - r_{ya}}\right) \quad \text{and} \quad z_{yb} = (1/2)\ln\left(\frac{1 + r_{yb}}{1 - r_{yb}}\right)$$

and

$$s_{ya,yb} = \frac{[(r_{ab})(1 - 2\bar{r}^2)] - [(1/2)(\bar{r}^2)(1 - 2\bar{r}^2 - r_{ab}^2)]}{(1 - \bar{r}^2)^2}$$

where $\bar{r} = (1/2)(r_{ya} + r_{yb})$.

So, for the example, if the correlation between currently measured ability and ability as predicted from personality scores is .40 ($R_a = r_{ya} = .40$), the correlation between currently measured ability and ability as predicted from past training is .50 ($R_b = r_{yb} = .50$), and the correlation between ability as predicted from personality and ability as predicted from training is .10 ($r_{ab} = .10$), and $N = 103$,

$$\bar{r} = (1/2)(.40 + .50) = .45$$

$$\overline{s_{ya,yb}} = \frac{[(.10)(1 - 2(.45)^2)] - [(1/2)(.45^2)(1 - 2(.45^2) - .10^2)]}{(1 - .45^2)^2} = .0004226$$

$$z_{ya} = (1/2)\ln\left(\frac{1 + .40}{1 - .40}\right) = .42365$$

$$z_{yb} = (1/2)\ln\left(\frac{1 + .50}{1 - .50}\right) = .54931$$

and, finally,

$$\overline{Z^*} = (.42365 - .54931)\sqrt{\frac{103 - 3}{2 - .000845}} = -0.88874$$

Because $\overline{Z^*}$ is within the critical values of ±1.96 for a two-tailed test, there is no statistically significant difference between multiple R when predicting Y from Y'_a or Y'_b. That is, there is no statistically significant difference in predicting current belly dancing ability from past training versus personality tests.

Steiger (1980) and Steiger and Browne (1984) present additional significance tests for situations where both the DV and the IVs are different, but from the same sample, and for comparing the difference between any two correlations within a correlation matrix.

5.6.3 Adjustment of R^2

Just as simple r_{xy} from a sample is expected to fluctuate around the value of the correlation in the population, sample R is expected to fluctuate around the population value. But multiple R never takes on a negative value, so all chance fluctuations are in the positive direction and add to the magnitude of R. As in any sampling distribution, the magnitude of chance fluctuations is larger with smaller sample sizes. Therefore, R tends to be overestimated, and the smaller the sample the greater the overestimation. For this reason, in estimating the population value of R, adjustment is made for expected inflation in sample R.

All the programs discussed in Section 5.8 routinely provide adjusted R^2. Wherry (1931) provides a simple equation for this adjustment, which is called $\tilde{R}^2$:

$$\tilde{R}^2 = 1 - (1 - R^2)\left(\frac{N - 1}{N - k - 1}\right) \tag{5.14}$$

where N = sample size
$\quad\quad k$ = number of IVs
$\quad\quad R^2$ = squared multiple correlation

For the small sample problem,

$$\tilde{R}^2 = 1 - (1 - .70235)(5/2) = .25588$$

as printed out for SPSS, Table 5.4.

For statistical regression, Cohen et al. (2003) recommend k based on the number of IVs considered for inclusion, rather than on the number of IVs selected by the program. They also suggest the convention of reporting $\tilde{R}^2 = 0$ when the value spuriously becomes negative.

When the number of subjects is 60 or fewer and there are numerous IVs (say, more than 20), Equation 5.15 may provide inadequate adjustment for R^2. The adjusted value may be off by as much as .10 (Cattin, 1980). In these situations of small N and numerous IVs,

$$R_s^2 = \frac{(N - k - 3)\tilde{R}^4 + \tilde{R}^2}{(N - 2k - 2)\tilde{R}^2 + k} \tag{5.15}$$

Equation 5.16 (Browne, 1975) provides further adjustment:

$$\tilde{R}^4 = (\tilde{R}^2)^2 - \frac{2k(1 - \tilde{R}^2)^2}{(N - 1)(N - k + 1)} \tag{5.16}$$

The adjusted R^2 for small samples is a function of the number of cases, N, the number of IVs, k, and the $\tilde{R}^2$ value as found from Equation 5.15.

When N is less than 50, Cattin (1980) provides an equation that produces even less bias but requires far more computation.

5.6.4   Suppressor Variables

Sometimes you may find an IV that is useful in predicting the DV and in increasing the multiple R^2 by virtue of its correlations with other IVs. This IV is called a suppressor variable because it *suppresses* variance that is irrelevant to prediction of the DV. Thus, a suppressor variable is defined not by its own regression weight, but by its enhancement of the effects of other variables in the set of IVs. It is a suppressor only for those variables whose regression weights are increased (Conger, 1974). In a full discussion of suppressor variables, Cohen et al. (2003) describe and provide examples of several varieties of suppression.

For instance, one might administer as IVs two paper-and-pencil tests, a test of ability to list dance patterns and a test of test-taking ability. By itself, the first test poorly predicts the DV (say, belly dancing ability) and the second test does not predict the DV at all. However, in the context of the test of test taking, the relationship between ability to list dance patterns and belly dancing ability improves. The second IV serves as a suppressor variable because by removing variance due to ability in taking tests, prediction of the DV by the first IV is enhanced.

The foregoing is an example of classical suppression (called traditional by Conger, 1974). Another type is cooperative or reciprocal suppression, in which IVs correlate positively with the DV and correlate negatively with each other (or vice versa). Both IVs end up with higher correlations with the DV after each IV is adjusted for the other. For example, the ability to list dance patterns and prior musical training might be negatively correlated, although both predict belly dancing ability somewhat. In the context of both of the predictors, belly dancing ability is predicted more fully than expected on the basis of adding the separate predictive ability of the two IVs.

A third type of suppression occurs when the sign of a regression weight of an IV is the opposite of what would be expected on the basis of its correlation with the DV. This is negative or net suppression. Prediction still is enhanced because the magnitude of the effect of the IV is greater (although the sign is opposite) in the presence of the suppressor. Suppose that belly dance ability is positively predicted by both knowledge of dance steps and previous dance training, and that the IVs are positively correlated. The regression weight of previous dance training might turn out to be negative, but stronger than would be expected on the basis of its bivariate correlation with belly dance ability. Thus, knowledge of dance steps is a negative suppressor for previous dance training.

In output, the presence of a suppressor variable is identified by the pattern of regression coefficients and correlations of each IV with the DV. Compare the simple correlation between each IV and the DV in the correlation matrix with the standardized regression coefficient (beta weight) for the IV. If the beta weight is significantly different from zero, either one of the following two conditions signals the presence of a suppressor variable: (1) the absolute value of the simple correlation between IV and DV is substantially smaller than the beta weight for the IV, or (2) the simple correlation and beta weight have opposite signs. There is as yet no statistical test available to assess how different a regression weight and a simple correlation need to be to identify suppression (Smith, Ager, and Williams, 1992).

It is often difficult to identify which variable is doing the suppression if there are more than two or three IVs. If you know that a suppressor variable is present, you need to search for it among the regression coefficients and correlations of the IVs. The suppressor is among the ones that are congruent, where the correlation with the DV and the regression coefficients are consistent in size and direction. One strategy is to systematically leave each congruent IV out of the equation and examine the changes in regression coefficients for the IV(s) with inconsistent regression coefficients and correlations in the original equation.

If a suppressor variable is identified, it is properly interpreted as a variable that enhances the importance of other IVs by virtue of suppression of irrelevant variance in them. If the suppressor is not identified, Tzelgov and Henik (1991) suggest an approach in which the focus is on suppression situations rather than on specific suppressor variables.

5.6.5   Regression Approach to ANOVA

Analysis of variance is a part of the general linear model, as is regression. Indeed, ANOVA can be viewed as a form of multiple regression in which the IVs are levels of discrete variables rather than the more usual continuous variables of regression. This approach is briefly reviewed here. Interested readers are referred to Tabachnick and Fidell (2007) for a fuller description and demonstrations of the regression approach to a variety of ANOVA models.

Chapter 2 reviews the traditional approach to calculation of sums of squares for analysis of variance. The regression approach to calculating the same sums of squares involves creating a variable for each df of the IVs that separates the levels of IVs. For example, a one-way between-subjects ANOVA with two levels requires only one variable, X, to separate its levels. If contrast coding is used, cases in a_1 are coded as 1 and cases in a_2 as -1, as shown in Table 5.10. (There are several other forms of coding, but for many applications, contrast coding works best.)

TABLE 5.10 Contrast Coding for a One-Way Between-Subjects ANOVA, Where A Has Two Levels

Level of A	Case	X	Y (DV Scores)
	s_1	1	
a_1	s_2	1	
	s_3	1	
	s_4	-1	
a_2	s_5	-1	
	s_6	-1	

When a column for DV scores, Y, is added, the bivariate correlation between X and Y, the intercept, and the regression coefficient, B, can be calculated as per Equations 3.31 and 3.32. Thus, the data set fits within the model of bivariate regression (Equation 3.30).

To convert this to ANOVA, total sum of squares is calculated for Y using Equation 3.5, where $\overline{Y}$ is the grand mean of the DV column (equivalent to SS_{total} of Equation 3.9). Sum of squares for regression (effect of A) and residual (error) are also calculated and they correspond to the SS_{bg} and SS_{wg} of Equation 3.10, respectively.[10] Degrees of freedom for A is the number of X columns needed to code the levels of A (1 in this example). Degrees of freedom total is the number of cases minus 1 ($6 - 1 = 5$ here). And degrees of freedom for error is the difference between the other two df ($5 - 1 = 4$ here). Mean squares and the F ratio are found in the usual manner.

Additional levels of A make the coding more interesting because more X columns are needed to separate the levels of A and the problem moves from bivariate to multiple regression. For example, with three levels of A and contrast coding, you need two X columns, as shown in Table 5.11.

The X_1 column codes the difference between a_1 and a_2, with all cases in a_3 assigned a code of zero. The X_2 column combines a_1 and a_2 (by giving all cases in both levels the same code of 1) against a_3 (with a code of -2).

In addition to a vector for Y, then, you have two X vectors, and a data set appropriate for multiple regression, Equation 5.1. Total sum of squares is calculated in the usual way, ignoring group membership. Separate sums of squares are calculated for X_1 and X_2 and are added together to form SS_A (as long as orthogonal coding is used—Section 3.2.6.2—and there are equal sample sizes in each cell). Sum of squares for error is found by subtraction, and mean squares and the F ratio for the omnibus effect of A are found in the usual manner.

The regression approach is especially handy when orthogonal comparisons are desired because those are the codes in the X columns. The difference between this procedure and traditional approach to specific comparisons is that the weighting coefficients are applied to each case rather than to group means. The sums of squares for the X_1 and X_2 columns can be evaluated separately to test the comparisons represented in those columns.

---

[10]Details for calculating ANOVA sums of squares using the regression approach are provided by Tabachnick and Fidell (2007).

TABLE 5.11   Contrast Coding for a One-Way Between-Subjects ANOVA, Where A Has Three Levels

Level of A	Case	X_1	X_2	Y (DV Scores)
a_1	s_1	1	1	
	s_2	1	1	
	s_3	1	1	
a_2	s_4	-1	1	
	s_5	-1	1	
	s_6	-1	1	
a_3	s_7	0	-2	
	s_8	0	-2	
	s_9	0	-2	

As the ANOVA problem grows, the number of columns needed to code for various IVs and their interactions[11] also grows, as does the complexity of the multiple regression, but the general principles remain the same. In several situations, the regression approach to ANOVA is easier to understand, if not compute, than the traditional approach. This fascinating topic is explored in great (excruciating?) detail in Tabachnick and Fidell (2007).

5.6.6   Centering When Interactions and Powers of IVs Are Included

Interactions between discrete IVs are common and are discussed in any standard ANOVA text. Interactions between continuous IVs are less common, but are of interest if we want to test whether the regression coefficient or importance of one IV (X_1) varies over the range of another IV (X_2). If so, X_2 is said to moderate the relationship between X_1 and the DV. For example, is the importance of education in predicting occupational prestige the same over the range of income? If there is interaction, the regression coefficient (B) for education in the regression equation depends on income. A different regression coefficient for education is needed for different incomes. Think about what the interaction would look like if income and education were discrete variables. If income has three levels (low, middle, and high) and so does education (high school grad, college grad, postgrad degree), you could plot a separate line for education at each level of income (or vice versa) and each line would have a different slope. The same plot could be generated for continuous variables, except that distinct values for income must be used.

When you want to include interactions of IVs or power of IVs in the prediction equation, they can cause problems of multicollinearity unless they have been centered: converted to deviation scores so that each variable has a mean of zero (Aiken and West, 1991). Note that centering does not

[11]Problems in interpreting parameter estimates may occur if interactions are listed in the regression equation before main effects.

require that scores be standardized, because it is not necessary to divide the score's deviation from its mean by its standard deviation. Centering an IV does not affect its simple correlation with other variables, but it does affect regression coefficients for interactions or powers of IVs included in the regression equation. (There is no advantage to centering the DV.)

Recall from Chapter 4 that computational problems arise when IVs are highly correlated. If the IVs with interactions are not centered, their product (as well as higher-order polynomial terms such as X_1^2) is highly correlated with the component IVs. That is, X_1X_2 is highly correlated with both X_1 and with X_2; X_1^2 is also highly correlated with X_1. Note that the problem with multicollinearity in this case is strictly statistical; the logical problems sometimes associated multicollinearity among supposedly different predictors are not at issue. In the case of interaction between IVs or powers of an IV, multicollinearity is caused by the measurement scales of the component IVs and can be ameliorated by centering them.

Analyses with centered variables lead to the same unstandardized regression coefficients for simple terms in the equation (e.g., B_1 for X_1 and B_2 for X_2) as when uncentered. The significance test for the interaction also is the same, although the unstandardized regression coefficient is not (e.g., B_3 for X_1X_2). However, the standardized regression coefficients (β) are different for all effects. If a standardized solution is desired, the strategy suggested by Friedrich (1982) is to convert all scores to z-scores, including the DV, and apply the usual solution. The computer output column showing "unstandardized" regression coefficients (Parameter Estimate, Coefficient, B) actually shows standardized regressions coefficients, β. Ignore any output referring to standardized regression coefficients. However, the intercept for the standardized solution for centered data is not necessarily zero, as it always is for noncentered data.

When interaction terms are statistically significant, plots are useful for interpretation. Plots are generated by solving the regression equation at chosen levels of X_2, typically high, medium, and low levels. In the absence of theoretical reasons for choice of levels, Cohen et al. (2003) suggest levels corresponding to the mean of X_2, one standard deviation above and one standard deviation below the mean as the medium, high, and low levels, respectively. Then, for each slope, you substitute the chosen value of X_2 in the rearranged regression equation:

$$Y' = (A + B_2X_2) + (B_1 + B_3X_2)X_1 \qquad (5.17)$$

where B_3 is the regression coefficient for the interaction.

Suppose, for example, that $A = 2$, $B_1 = 3$, $B_2 = 3.5$, $B_3 = 4$, and the X_2 value at one standard deviation below the mean is -2.5. The regression line for the DV at the low value of X_2 is

$$Y' = [2 + (3.5)(-2.5)] + [3 + (4)(-2.5)]X_1$$
$$= -6.75 - 7.00X_1$$

If X_2 at one standard deviation above the mean is 2.5 and the regression line for the DV at the high value of X_2 is

$$Y' = [2 + (3.5)(2.5)] + [3 + (4)(2.5)]X_1$$
$$= 10.00 + 13.00X_1$$

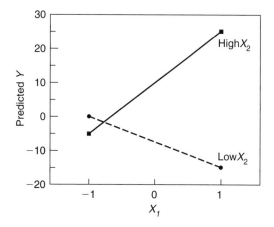

FIGURE 5.5   Interaction between two continuous IVs: X_1 and X_2.

Figure 5.5 is the resulting plot. Each regression equation is solved for two values of X_1 within a reasonable range of values and the resulting DV values plotted. For this example, the very simplest are chosen: $X_1 = -1$ and $X_1 = 1$. For the low value of X_2 (-2.5) then, $Y' = 0.25$ when $X_1 = -1$ and $Y' = -13.75$ when $X_1 = 1$. For the high value of X_2 (2.5), $Y' = -2.25$ when $X_1 = -1$ and $Y' = 23.15$ when $X_1 = 1$.

The finding of a significant interaction in multiple regression often is followed up by a simple effects analysis, just as it is in ANOVA (cf. Chapter 8). In multiple regression, this means that the relationship between Y and X_1 is tested separately at the chosen levels of X_2. Aiken and West (1991) call this a simple slope analysis and provide techniques for testing the significance of each of the slopes, both as planned and post hoc comparisons.

Aiken and West (1991) provide a wealth of information about interactions among continuous variables, including higher-order interactions, relationships taken to higher-order powers, and dealing with interactions between discrete and continuous variables. Their book is highly recommended if you plan to do serious work with interactions in multiple regression. Alternatively, Holmbeck (1997) suggests use of structural equation modeling (the topic of Chapter 14) when interactions between continuous variables (and presumably their powers) are included and sample size is large.

5.6.7   Mediation in Causal Sequence

If you have a hypothetical causal sequence of three (or more) variables, the middle variable is considered a mediator (indirect effect) that represents at least part of the chain of events leading to changes in the DV. For example, there is a relationship between gender and number of visits to health care professionals, but what mechanism underlies the relationship? You might propose that the linkage is through some aspect of personality. That is, you might hypothesize that gender "causes" some differences in personality which, in turn, "cause" women to make more visits to health care professionals. Gender, personality, and visits are in causal sequence, with gender the IV, personality the mediator, and visits the DV.

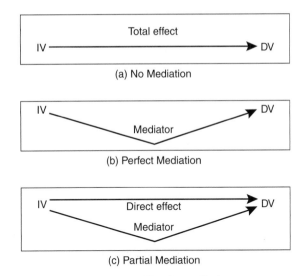

FIGURE 5.6   Simple mediation.

As seen in Figure 5.6, the relationship between the IV and the DV is called the *total effect*. The *direct effect* is the relationship between the IV and the DV after "controlling for" the mediator. According to Baron and Kinney (1986) a variable is confirmed as a mediator if 1) there is a significant relationship between the IV and the DV, 2) there is a significant relationship between the IV and the mediator, 3) the mediator still predicts the DV after controlling for the IV, and 4) the relationship between the IV and the DV is reduced when the mediator is in the equation. If the relationship between the IV and the DV goes to zero when the mediator is in the equation, mediation is said to be perfect (or full, or complete, Figure 5.5(b)); if the relationship is diminished, but not to zero, mediation is said to be partial (Figure 5.5(c)).

In the example, personality is a mediator if there is a relationship between gender and visits, there is a relationship between gender and personality, personality predicts visits even after controlling for gender, and the relationship between gender and visits is smaller when personality is in the equation. If the relationship between gender and visits is plausibly zero when personality is in the equation, the mediation is perfect. If the relationship is smaller, but not zero, mediation is partial. In this example, you might expect the relationship to be reduced, but not to zero, due to childbearing.

Note that the three variables (the IV, the mediator, and the DV) are hypothesized to occur in a causal sequence. In the example, gender is presumed to "cause" personality which, in turn "causes" visits to health care professionals. There are other types of relationships between three variables (e.g., interaction, Section 5.6.6) that do not involve a sequence of causal relationships. Note also that this discussion is of simple mediation with three variables. As seen in Chapter 14 (Structural Equation Modeling), there are many other forms of mediation. For example, there may be more than one mediator in a sequence, or mediators may be operating in parallel instead of in sequence. Further, in SEM, mediators may be directly measured or latent.

Sobel (1982), among others, presents a method for testing the significance of a simple mediator by testing the difference between the total effect and the direct effect. In the example, the mediating effects of personality are tested as the difference between the relationship of gender and visits with and without consideration of personality. If the relationship between gender and visits is not reduced by adding personality to the equation, personality is not a mediator of the relationship. The Sobel method requires just one significance test for mediation rather than several as proposed by Baron and Kinney and is thus less susceptible to familywise alpha errors. Preacher and Hayes (2004) provide both SPSS and SAS macros for following the Baron and Kinney procedure and formally testing mediation as recommended by Sobel. They also discuss the assumption of the formal test (normality of sampling distribution) and bootstrapping methods for circumventing it. Tests of indirect effects in SEM are demonstrated in Section 14.6.2.

5.7   Complete Examples of Regression Analysis

To illustrate applications of regression analysis, variables are chosen from among those measured in the research described in Appendix B, Section B.1. Two analyses are reported here, both with number of visits to health professionals (TIMEDRS) as the DV and both using the SPSS REGRESSION program. Files are REGRESS.*.

The first example is a standard multiple regression between the DV and number of physical health symptoms (PHYHEAL), number of mental health symptoms (MENHEAL), and stress from acute life changes (STRESS). From this analysis, one can assess the degree of relationship between the DV and IVs, the proportion of variance in the DV predicted by regression, and the relative importance of the various IVs to the solution.

The second example demonstrates sequential regression with the same DV and IVs. The first step of the analysis is entry of PHYHEAL to determine how much variance in number of visits to health professionals can be accounted for by differences in physical health. The second step is entry of STRESS to determine if there is a significant increase in R^2 when differences in stress are added to the equation. The final step is entry of MENHEAL to determine if differences in mental health are related to number of visits to health professionals after differences in physical health and stress are statistically accounted for.

5.7.1   Evaluation of Assumptions

Because both analyses use the same variables, this screening is appropriate for both.

5.7.1.1   *Ratio of Cases to IVs*

With 465 respondents and 3 IVs, the number of cases is well above the minimum requirement of 107 (104 + 3) for testing individual predictors in standard multiple regression. There are no missing data.

5.7.1.2   *Normality, Linearity, Homoscedasticity, and Independence of Residuals*

We choose for didactic purposes to conduct preliminary screening through residuals. The initial run through SPSS REGRESSION uses untransformed variables in a standard multiple regression to produce the scatterplot of residuals against predicted DV scores that appears in Figure 5.7.

Notice the execrable overall shape of the scatterplot that indicates violation of many of the assumptions of regression. Comparison of Figure 5.7 with Figure 5.1(a) (in Section 5.3.2.4) suggests further analysis of the distributions of the variables. (It was noticed in passing, although we tried not to look, that R^2 for this analysis was significant, but only .22.)

```
REGRESSION
  /MISSING LISTWISE
  /STATISTICS COEFF OUTS R ANOVA
  /CRITERIA=PIN(.05) POUT(.10)
  /NOORIGIN
  /DEPENDENT timedrs
  METHOD=ENTER phyheal menheal stress
  /SCATTERPLOT=(*ZRESID ,*ZPRED).
```

Scatterplot

Dependent Variable: Visits to health professionals

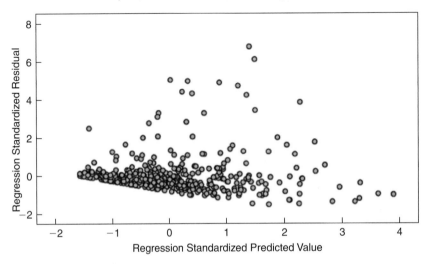

FIGURE 5.7   SPSS REGRESSION syntax and residuals scatterplot for original variables.

SPSS EXPLORE is used to examine the distributions of the variables, as shown in Table 5.12. All the variables have significant positive skewness (see Chapter 4), which explains, at least in part, the problems in the residuals scatterplot. Logarithmic and square root transformations are applied as appropriate, and the transformed distributions checked once again for skewness. Thus, TIMEDRS and PHYHEAL (with logarithmic transformations) become LTIMEDRS and LPHYHEAL, whereas

TABLE 5.12   Syntax and Output for Examining Distributions of Variables through SPSS EXPLORE

```
EXAMINE
  VARIABLES=timedrs phyheal menheal stress
  /PLOT BOXPLOT HISTOGRAM NPPLOT
  /COMPARE GROUP
  /STATISTICS DESCRIPTIVES EXTREME
  /CINTERVAL 95
  /MISSING LISTWISE
  /NOTOTAL.
```

TABLE 5.12   Continued

Descriptives

			Statistic	Std. Error
Visits to health professionals	Mean		7.90	.508
	95% Confidence Interval for Mean	Lower Bound	6.90	
		Upper Bound	8.90	
	5% Trimmed Mean		6.20	
	Median		4.00	
	Variance		119.870	
	Std. Deviation		10.948	
	Minimum		0	
	Maximum		81	
	Range		81	
	Interquartile Range		8	
	Skewness		3.248	.113
	Kurtosis		13.101	.226
Physical health symptoms	Mean		4.97	.111
	95% Confidence Interval for Mean	Lower Bound	4.75	
		Upper Bound	5.19	
	5% Trimmed Mean		4.79	
	Median		5.00	
	Variance		5.704	
	Std. Deviation		2.388	
	Minimum		2	
	Maximum		15	
	Range		13	
	Interquartile Range		3	
	Skewness		1.031	.113
	Kurtosis		1.124	.226
Mental health symptoms	Mean		6.12	.194
	95% Confidence Interval for Mean	Lower Bound	5.74	
		Upper Bound	6.50	
	5% Trimmed Mean		5.93	
	Median		6.00	
	Variance		17.586	
	Std. Deviation		4.194	
	Minimum		0	
	Maximum		18	
	Range		18	
	Interquartile Range		6	
	Skewness		.602	.113
	Kurtosis		−.292	.226
Stressful life events	Mean		204.22	6.297
	95% Confidence Interval for Mean	Lower Bound	191.84	
		Upper Bound	216.59	
	5% Trimmed Mean		195.60	
	Median		178.00	
	Variance		18439.662	
	Std. Deviation		135.793	
	Minimum		0	
	Maximum		920	
	Range		920	
	Interquartile Range		180	
	Skewness		1.043	.113
	Kurtosis		1.801	.226

(continued)

TABLE 5.12   Continued

Extreme Values

			Case Number	Value
Visits to health professionals	Highest	1	405	81
		2	290	75
		3	40	60
		4	168	60
		5	249	58
	Lowest	1	437	0
		2	435	0
		3	428	0
		4	385	0
		5	376	0[a]
Physical health symptoms	Highest	1	277	15
		2	373	14
		3	381	13
		4	391	13
		5	64	12[b]
	Lowest	1	454	2
		2	449	2
		3	440	2
		4	419	2
		5	418	2[c]
Mental health symptoms	Highest	1	52	18
		2	103	18
		3	113	18
		4	144	17
		5	198	17
	Lowest	1	462	0
		2	454	0
		3	352	0
		4	344	0
		5	340	0[a]
Stressful life events	Highest	1	403	920
		2	405	731
		3	444	643
		4	195	597
		5	304	594
	Lowest	1	446	0
		2	401	0
		3	387	0
		4	339	0
		5	328	0[a]

[a]Only a partial list of cases with the value 0 are shown in the table of lower extremes.

[b]Only a partial list of cases with the value 12 are shown in the table of upper extremes.

[c]Only a partial list of cases with the value 2 are shown in the table of lower extremes.

TABLE 5.12 Continued

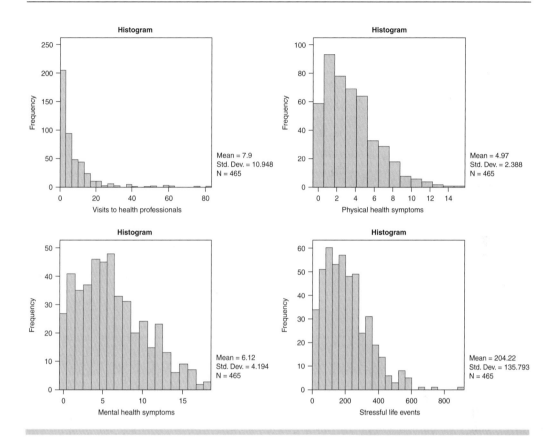

STRESS (with a square root transformation) becomes SSTRESS.[12] In the case of MENHEAL, application of the milder square root transformation makes the variable significantly negatively skewed, so no transformation is undertaken.

Table 5.13 shows output from FREQUENCIES for one of the transformed variables, LTIMEDRS, the worst prior to transformation. Transformations similarly reduce skewness in the other two transformed variables.

The residuals scatterplot from SPSS REGRESSION following regression with the transformed variables appears as Figure 5.8. Notice that, although the scatterplot is still not perfectly rectangular, its shape is considerably improved over that in Figure 5.7.

5.7.1.3 Outliers

Univariate outliers in the DV and in the IVs are sought using output from Table 5.12. The highest values in the histograms appear disconnected from the next highest scores for TIMEDRS and STRESS;

---

[12]Note the DV (TIMEDRS) is transformed to meet the assumptions of multiple regression. Transformation of the IVs is undertaken to enhance prediction.

TABLE 5.13   Syntax and Output for Examining Distribution of Transformed Variable through SPSS FREQUENCIES

```
FREQUENCIES
   VARIABLES=ltimedrs/
   /STATISTICS=STDDEV VARIANCE MINIMUM MAXIMUM MEAN SKEWNESS SESKEW
   KURTOSIS
   SEKURT
   /HISTOGRAM NORMAL
   /ORDER = ANALYSIS.
```

Statistics

ltimedrs

N	Valid	465
	Missing	0
Mean		.7413
Median		.6990
Std. Deviation		.41525
Skewness		.228
Std. Error of Skewness		.113
Kurtosis		−.177
Std. Error of Kurtosis		.226
Minimum		.00
Maximum		1.91

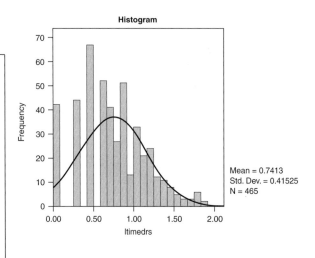

Histogram

Mean = 0.7413
Std. Dev. = 0.41525
N = 465

from the table of **Extreme Values,** the two highest values for TIMEDRS (81 and 75) have z-scores of 6.68 and 6.13, respectively. The three highest values for STRESS (920, 731, and 643) have z-scores of 5.27, 3.88, and 3.23, respectively. The highest value for PHYHEAL (15) has a z-score of 4.20, but it does not appear disconnected from the rest of the distribution. Because there is no reason not to, the decision is to transform TIMEDRS and STRESS because of the presence of outliers and PHYHEAL because of failure of normality.

Once variables are transformed, the highest scores for LTIMEDRS (see Table 5.13), LPHY-HEAL, and SSTRESS (not shown) no longer appear disconnected from the rest of their distributions; the z-scores associated with the highest scores are now 2.81, 2.58, and 3.41 respectively. In a sample of this size, these values seem reasonable.

Multivariate outliers are sought using the transformed IVs as part of an SPSS REGRESSION run in which the Mahalanobis distance of each case to the centroid of all cases is computed. The ten cases with the largest distance are printed (see Table 5.14). Mahalanobis distance is distributed as a chi-square (χ^2) variable, with degrees of freedom equal to the number of IVs. To determine which cases are multivariate outliers, one looks up critical χ^2 at the desired alpha level (Table C.4). In this case, critical χ^2 at $\alpha = .001$ for 3 df is 16.266. Any case with a value larger than 16.266 in the **Sta-tistic** column of the **Outlier Statistics** table is a multivariate outlier among the IVs. None of the

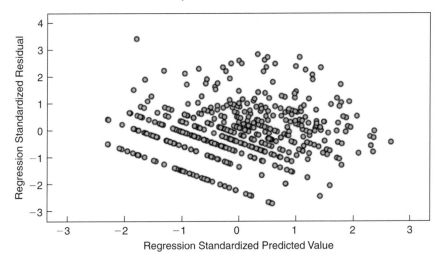

FIGURE 5.8 **Residuals scatterplot following regression with transformed variables. Output from SPSS REGRESSION. See Table 5.15 for syntax.**

cases has a value in excess of 16.266. (If outliers are found, the procedures detailed in Chapter 4 are followed to reduce their influence.)

Note that Figure 5.8 shows no outliers in the solution; none of the standardized residuals exceeds 3.29.

5.7.1.4 *Multicollinearity and Singularity*

None of the tolerances $(1 - SMC)$ listed in Table 5.14 approaches zero. Collinearity diagnostics indicate no cause for concern using the criteria of Section 4.1.7. The REGRESSION run of Table 5.15 additionally resolves doubts about possible multicollinearity and singularity among the transformed IVs. All variables enter the equation without violating the default value for tolerance (cf. Chapter 4). Further, the highest correlation among the IVs, between MENHEAL and LPHYHEAL, is .511. (If multicollinearity is indicated, redundant IVs are dealt with as discussed in Chapter 4.)

5.7.2 Standard Multiple Regression

SPSS REGRESSION is used to compute a standard multiple regression between LTIMEDRS (the transformed DV), and MENHEAL, LPHYHEAL, and SSTRESS (the IVs), as shown in Table 5.15.

Included in the REGRESSION output are descriptive statistics, including a correlation table, the values of R, R^2, and adjusted R^2, and a summary of the analysis of variance for regression. The significance level for R is found in the **ANOVA** table with $F(3, 461) = 92.90, p < .001$. In the table labeled **Coefficients** are printed unstandardized and standardized regression coefficients with their

TABLE 5.14   Syntax and Output from SPSS REGRESSION Showing Multivariate Outliers and Collinearity Diagnostics

```
REGRESSION
   /MISSING LISTWISE
   /STATISTICS COEFF OUTS R ANOVA COLLIN TOL
   /CRITERIA=PIN(.05) POUT(.10)
   /NOORIGIN
   /DEPENDENT ltimedrs
   /METHOD=ENTER lphyheal menheal sstress
   /SCATTERPLOT=(*ZRESID,*ZPRED)
   /RESIDUALS = outliers(mahal).
```

Collinearity Diagnostics[a]

Model	Dimension	Eigenvalue	Condition Index	Variance Proportions			
				(Constant)	lphyheal	Mental health symptoms	sstress
1	1	3.685	1.000	.00	.00	.01	.01
	2	.201	4.286	.06	.01	.80	.03
	3	.076	6.971	.07	.25	.00	.85
	4	.039	9.747	.87	.74	.19	.11

[a]Dependent Variable: LTIMEDRS

Outlier Statistics[a]

		Case Number	Statistic
Mahal. Distance	1	403	14.135
	2	125	11.649
	3	198	10.569
	4	52	10.548
	5	446	10.225
	6	159	9.351
	7	33	8.628
	8	280	8.587
	9	405	8.431
	10	113	8.353

[a]Dependent Variable: LTIMEDRS

significance levels, 95% confidence intervals, and three correlations: Zero-order (matching the IV–DV values of the correlation table), semipartial (Part), and partial.

The significance levels for the regression coefficients are assessed through t statistics, which are evaluated against 461 df, or through confidence intervals. Only two of the IVs, SSTRESS and LPHYHEAL, contribute significantly to regression with ts of 4.67 and 11.928, respectively. The significance of SSTRESS and LPHYHEAL is confirmed by their 95% confidence intervals that do not include zero as a possible value.

TABLE 5.15   Standard Multiple Regression Analysis of LTIMEDRS (the SPSS REGRESSION DV) with MENHEAL, SSTRESS, and LPHYHEAL (the IVs). SPSS REGRESSION Syntax and Selected Output

```
REGRESSION
  /DESCRIPTIVES MEAN STDDEV CORR SIG N
  /MISSING LISTWISE
  /STATISTICS COEFF OUTS CI R ANOVA ZPP
  /CRITERIA=PIN(.05) POUT(.10)
  /NOORIGIN
  /DEPENDENT ltimedrs
  /METHOD=ENTER lphyheal menheal sstress.
```

Regression

Descriptive Statistics

	Mean	Std. Deviation	N
ltimedrs	.7413	.41525	465
lphyheal	.6484	.20620	465
Mental health symptoms	6.12	4.194	465
sstress	13.3995	4.97217	465

Correlations

		ltimedrs	lphyheal	Mental health symptoms	sstress
Pearson Correlation	ltimedrs	1.000	.586	.355	.359
	lphyheal	.586	1.000	.511	.317
	Mental health symptoms	.355	.511	1.000	.383
	sstress	.359	.317	.383	1.000
Sig. (1-tailed)	ltimedrs	.	.000	.000	.000
	lphyheal	.000	.	.000	.000
	Mental health symptoms	.000	.000	.	.000
	sstress	.000	.000	.000	.
N	ltimedrs	465	465	465	465
	lphyheal	465	465	465	465
	Mental health symptoms	465	465	465	465
	sstress	465	465	465	465

Model Summary

Model	R	R Square	Adjusted R Square	Std. Error of the Estimate
1	.614[a]	.377	.373	.3289

[a]Predictors: (Constant), sstress, lphyheal, Mental health symptoms

(continued)

TABLE 5.15   Continued

ANOVA[b]

Model		Sum of Squares	df	Mean Square	F	Sig.
1	Regression	30.146	3	10.049	92.901	.000[a]
	Residual	49.864	461	.108		
	Total	80.010	464			

[a]Predictors: (Constant), sstress, lphyheal, Mental health symptoms
[b]Dependent Variable: ltimedrs

Coefficients[a]

Model		Unstandardized Coefficients		Standardized Coefficients	t	Sig.	95% Confidence Interval for B		Correlations		
		B	Std. Error	Beta			Lower Bound	Upper Bound	Zero-order	Partial	Part
1	(Constant)	-.155	.058		-2.661	.008	-.270	-.041			
	lphyheal	1.040	.087	.516	11.928	.000	.869	1.211	.586	.486	.439
	Mental health symptoms	.002	.004	.019	.428	.669	-.007	.011	.355	.020	.016
	sstress	.016	.003	.188	4.671	.000	.009	.022	.359	.213	.172

[a]Dependent Variable: ltimedrs

Semipartial correlations are labeled Part in the Coefficients section. These values, when squared, indicate the amount by which R^2 would be reduced if an IV were omitted from the equation. The sum for the two significant IVs ($.172^2 + .439^2 = .222$) is the amount of R^2 attributable to unique sources. The difference between R^2 and unique variance ($.377 - .222 = .155$) represents variance that SSTRESS, LPHYHEAL, and MENHEAL jointly contribute to R^2.

Information from this analysis is summarized in Table 5.16 in a form that might be appropriate for publication in a professional journal. Confidence limits around R^2 are found using Steiger and Fouladi's (1992) software, as per Section 5.6.2.4.

It is noted from the correlation matrix in Table 5.15 that MENHEAL correlates with LTIMEDRS $r = .355$) but does not contribute significantly to regression. If Equation 5.10 is used post hoc to evaluate the significance of the correlation coefficient,

$$ F = \frac{(.355)^2/3}{1 - (.355)^2/(465 - 3 - 1)} = 22.16 $$

the correlation between MENHEAL and LTIMEDRS differs reliably from zero; $F(3, 461) = 22.16$, $p < .01$.

Thus, although the bivariate correlation between MENHEAL and LTIMEDRS is reliably different from zero, the relationship seems to be mediated by, or redundant to, the relationship between LTIMEDRS and other IVs in the set. Had the researcher measured only MENHEAL and LTIMEDRS, however, the significant correlation might have led to stronger conclusions than are warranted about the relationship between mental health and number of visits to health professionals.

Table 5.17 contains a checklist of analyses performed with standard multiple regression. An example of a Results section in journal format appears next.

TABLE 5.16   Standard Multiple Regression of Health and Stress Variables on Number of Visits to Health Professionals

Variables	Visits to Dr. (log) (DV)	Physical Health (log)	Stress (Sq. root)	Mental health	B	β	sr^2 (unique)
Physical health (log)	.59				1.040**	0.52	.19
Stress (sq. root)	.36	.32	.38		0.016**	0.19	.03
Mental health	.36	.51			0.002	0.02	
				Intercept $= -0.155$			
Means	0.74	0.65	13.40	6.12			
Standard deviations	0.42	0.21	4.97	4.19		$R^2 = .38$[a]	
						Adjusted $R^2 = .37$	
						$R = .61$**	

**p < .01

[a]Unique variability $= .22$; shared variability $= .16$, 95% confidence limits from .30 to .44.

TABLE 5.17   Checklist for Standard Multiple Regression

1. Issues
   a. Ratio of cases to IVs and missing data
   b. Normality, linearity, and homoscedasticity of residuals
   c. Outliers
   d. Multicollinearity and singularity
   e. Outliers in the solution
2. Major analyses
   a. Multiple R^2 and its confidence limits, F ratio
   b. Adjusted multiple R^2, overall proportion of variance accounted for
   c. Significance of regression coefficients
   d. Squared semipartial correlations
3. Additional analyses
   a. Post hoc significance of correlations
   b. Unstandardized (B) weights, confidence limits
   c. Standardized (β) weights
   d. Unique versus shared variability
   e. Suppressor variables
   f. Prediction equation

Results

A standard multiple regression was performed between number of visits to health professionals as the dependent variable and physical health, mental health, and stress as independent variables. Analysis was performed using SPSS REGRESSION and SPSS EXPLORE for evaluation of assumptions.

Results of evaluation of assumptions led to transformation of the variables to reduce skewness, reduce the number of outliers, and improve the normality, linearity, and homoscedasticity of residuals. A square root transformation was used on the measure of stress. Logarithmic transformations were used on number of visits to health professionals and on physical health. One IV, mental health, was positively skewed without transformation and negatively skewed with it; it was not transformed. With the use of a $p < .001$ criterion for Mahalanobis distance no outliers among the cases were found. No cases had missing data and no suppressor variables were found, $N = 465$.

Table 5.16 displays the correlations between the variables, the unstandardized regression coefficients (B) and intercept, the standardized regression coefficients (β), the semipartial correlations (sr_i^2), R^2, and adjusted R^2. R for regression was significantly different from zero, $F(3, 461) = 92.90$, $p < .001$, with R^2 at .38 and 95% confidence limits from .30 to .44. The adjusted R^2 value of .37 indicates that more than a third of the variability in visits to health professionals is predicted by number of physical health symptoms, stress, and mental health symptoms. For the two regression coefficients that differed significantly from zero, 95% confidence limits were calculated. The confidence limits for (square root of) stress were 0.0091 to 0.0223, and those for (log of) physical health were 0.8686 to 1.2113.

The three IVs in combination contributed another .15 in shared variability. Altogether, 38% (37% adjusted) of the variability in visits to health professionals was predicted by knowing scores on these three IVs. The size and direction of the relationships suggest that more visits to health professionals are made among women with a large number of physical health symptoms and higher stress. Between those two, however, number of physical health symptoms is much more important, as indicated by the squared semipartial correlations.

Although the bivariate correlation between (log of) visits to health professionals and mental health was statistically different from zero using a post hoc correction, $r = .36$, $F(3, 461) = 22.16$, $p < .01$, mental health did not contribute significantly to regression. Apparently, the relationship between the number of visits to health professionals and mental health is mediated by the relationships between physical health, stress, and visits to health professionals.

5.7.3   Sequential Regression

The second example involves the same three IVs entered one at a time in an order determined by the researcher. LPHYHEAL is the first IV to enter, followed by SSTRESS and then MENHEAL. The main research question is whether information regarding differences in mental health can be used to predict visits to health professionals after differences in physical health and in acute stress are statistically eliminated. In other words, do people go to health professionals for more numerous mental health symptoms if they have physical health and stress similar to other people?

Table 5.18 shows syntax and selected portions of the output for sequential analysis using the SPSS REGRESSION program. Notice that a complete regression solution is provided at the end of

TABLE 5.18   Syntax and Selected Output for SPSS Sequential Regression

```
REGRESSION
  /MISSING LISTWISE
  /STATISTICS COEFF OUTS CI R ANOVA CHANGE
  /CRITERIA=PIN(.05) POUT(.10)
  /NOORIGIN
  /DEPENDENT ltimedrs
  /METHOD=ENTER lphyheal /METHOD=ENTER sstress /METHOD=ENTER menheal.
```

Regression

Variables Entered/Removed[b]

Model	Variables Entered	Variables Removed	Method
1	lphyheal[a]	.	Enter
2	sstress[a]	.	Enter
3	Mental health symptoms[a]	.	Enter

[a]All requested variables entered.
[b]Dependent Variable: ltimedrs

Model Summary

Model	R	R Square	Adjusted R Square	Std. Error of the Estimate	Change Statistics R Square Change	F Change	df1	df2	Sig. F Change
1	.586[a]	.343	.342	.3369	.343	241.826	1	463	.000
2	.614[b]	.377	.374	.3286	.033	24.772	1	462	.000
3	.614[c]	.377	.373	.3289	.000	.183	1	461	.669

[a]Predictors: (Constant), lphyheal
[b]Predictors: (Constant), lphyheal, sstress
[c]Predictors: (Constant), lphyheal, sstress, Mental health symptoms

TABLE 5.18 Continued

ANOVA[d]

Model		Sum of Squares	df	Mean Square	F	Sig.
1	Regression	27.452	1	27.452	241.826	.000[a]
	Residual	52.559	463	.114		
	Total	80.010	464			
2	Regression	30.126	2	15.063	139.507	.000[b]
	Residual	49.884	462	.108		
	Total	80.010	464			
3	Regression	30.146	3	10.049	92.901	.000[c]
	Residual	49.864	461	.108		
	Total	80.010	464			

[a]Predictors: (Constant), lphyheal
[b]Predictors: (Constant), lphyheal, sstress
[c]Predictors: (Constant), lphyheal, sstress, mental health symptoms
[d]Dependent Variable: ltimedrs

Coefficients[a]

Model		Unstandardized Coefficients		Standardized Coefficients	t	Sig.	95% Confidence Interval for B	
		B	Std. Error	Beta			Lower Bound	Upper Bound
1	(Constant)	-2.4	.052		-.456	.648	-.125	.078
	lphyheal	1.180	.076	.586	15.551	.000	1.031	1.329
2	(Contstant)	-.160	.057		-2.785	.006	-.272	-.047
	lphyheal	1.057	.078	.525	13.546	.000	.903	1.210
	sstress	1.61	.003	.193	4.977	.000	.010	.022
3	(Constant)	-.155	.058		-2.661	.008	-.270	-.041
	lphyheal	1.040	.087	.516	11.928	.000	.869	1.211
	sstress	1.57	.003	.188	4.671	.000	.009	.022
	Mental health symptoms	1.88	.004	.019	.428	.669	-.007	.011

[a]Dependent Variable: ltimedrs

(continued)

175

TABLE 5.18   Continued

Excluded Variables[c]

Model		Beta In	t	Sig.	Partial Correlation	Collinearity Statistics Tolerance
1	sstress	.193[a]	4.977	.000	.226	.900
	Mental health symptoms	.075[a]	1.721	.086	.080	.739
2	Mental health symptoms	.019[b]	.428	.669	.020	.684

[a]Predictors in the Model: (Constant), lphyheal
[b]Predictors in the Model: (Constant), lphyheal, sstress
[c]Dependent Variable: ltimedrs

each step. The significance of the bivariate relationship between LTIMEDRS and LPHYHEAL is assessed at the end of step 1, $F(1, 463) = 241.83, p < .001$. The bivariate correlation is .59, accounting for 34% of the variance. After step 2, with both LPHYHEAL and SSTRESS in the equation, $F(2, 462) = 139.51$, $p < .01$, $R = .61$, and $R^2 = .38$. With the addition of MENHEAL, $F(3, 461) = 92.90, R = .61$, and $R^2 = .38$. Increments in R^2 at each step are read directly from the R Square Change column of the Model Summary table. Thus, $sr^2_{LPHYHEAL} = .34, sr^2_{SSTRESS} = .03$, and $sr^2_{MENHEAL} = .00$.

By using procedures of Section 5.6.2.3, significance of the addition of SSTRESS to the equation is indicated in the output at the second step (Model 2, where SSTRESS entered) as F for SSTRESS in the segment labeled Change Statistics. Because the F value of 24.772 exceeds critical F with 1 and 461 df (df_{res} at the end of analysis), SSTRESS is making a significant contribution to the equation at this step.

Similarly, significance of the addition of MENHEAL to the equation is indicated for Model 3, where the F for MENHEAL is .183. Because this F value does not exceed critical F with 1 and 461 df, MENHEAL is not significantly improving R^2 at its point of entry.

The significance levels of the squared semipartial correlations are also available in the Model Summary table as F Change, with probability value Sig F Change for evaluating the significance of the added IV.

Thus, there is no significant increase in prediction of LTIMEDRS by addition of MENHEAL to the equation if differences in LPHYHEAL and SSTRESS are already accounted for. Apparently, the answer is "no" to the question: Do people with numerous mental health symptoms go to health professionals more often if they have physical health and stress similar to others? A summary of information from this output appears in Table 5.19.

Table 5.20 is a checklist of items to consider in sequential regression. An example of a Results section in journal format appears next.

TABLE 5.19   Sequential Regression of Health and Stress Variables on Number of Visits to Health Professionals

Variables	Visits to Dr. (log) DV	Physical Health (log)	Stress (sq. root)	Mental Health	B	SE B	β	sr^2 (incremental)
Physical health (log)	.59				1.04	1.04**	0.52	.34**
Stress (sq. root)	.36	.32			0.02	0.016**	0.19	.03**
Mental health	.36	.51	.38		0.00	0.002	0.02	.00
Intercept					−0.16	−0.155		
Means	0.74	0.65	13.40	6.12				
Standard deviations	0.42	0.21	4.97	4.19			$R^2 = .38$[a]	
							Adjusted $R^2 = .37$	
							$R = .61$**	

[a]95% confidence limits from .30 to .44.

*p < .05.

**p < .01.

TABLE 5.20   Checklist for Sequential Regression Analysis

1. Issues
   a. Ratio of cases to IVs and missing data
   b. Normality, linearity, and homoscedasticity of residuals
   c. Outliers
   d. Multicollinearity and singularity
   e. Outliers in the solution
2. Major analyses
   a. Multiple R^2, and its confidence limits, F ratio
   b. Adjusted R^2, proportion of variance accounted for
   c. Squared semipartial correlations
   d. Significance of regression coefficients
   e. Incremental F
3. Additional analyses
   a. Unstandardized (B) weights, confidence limits
   b. Standardized ($β$) weights
   c. Prediction equation from stepwise analysis
   d. Post hoc significance of correlations
   e. Suppressor variables
   f. Cross-validation (stepwise)

Results

Sequential regression was employed to determine if addition of information regarding stress and then mental health symptoms improved prediction of visits to health professionals beyond that afforded by differences in physical health. Analysis was performed using SPSS REGRESSION and SPSS EXPLORE for evaluation of assumptions.

These results led to transformation of the variables to reduce skewness, reduce the number of outliers, and improve the normality, linearity, and homoscedasticity of residuals. A square root transformation was used on the measure of stress. Logarithmic transformations were used on the number of visits to health professionals and physical health. One IV, mental health, was positively skewed without transformation and negatively skewed with it; it was not transformed. With the use of a $p < .001$ criterion for Mahalanobis distance, no outliers among the cases were identified. No cases had missing data and no suppressor variables were found, $N = 465$.

Table 5.19 displays the correlations between the variables, the unstandardized regression coefficients (B) and intercept, the standardized regression coefficients (β), the semipartial correlations (sr_i^2), and R, R^2, and adjusted R^2 after entry of all three IVs. R was significantly different from zero at the end of each step. After step 3, with all IVs in the equation, $R^2 = .38$ with 95% confidence limits from .30 to .44, $F(3, 461) = 92.90$, $p < .01$. The adjusted R^2 value of .37 indicates that more than a third of the variability in visits to health professionals is predicted by number of physical health symptoms and stress.

After step 1, with log of physical health in the equation, $R^2 = .34$, F_{inc} (1, 461) = 241.83, $p < .001$. After step 2, with square root of stress added to prediction of (log of) visits to health professionals by (log of) physical health, $R^2 = .38$, F_{inc} (1, 461) = 24.77, $p < .01$. Addition of square root of stress to the equation with physical health results in a significant increment in R^2. After step 3, with mental health added to prediction of visits by (log of) physical health and square root of stress, $R^2 = .38$ (adjusted $R^2 = .37$), F_{inc} (1, 461) = 0.18. Addition of mental health to the equation did not reliably improve R^2. This pattern of results suggests that over a third of the variability in number of visits to health professionals is predicted by number of physical health symptoms. Level of stress contributes modestly to that prediction; number of mental health symptoms adds no further prediction.

5.7.4   Example of Standard Multiple Regression with Missing Values Multiply Imputed

A number of cases were randomly deleted from the SASUSER.REGRESS file after transformations to create a new file, SASUSER.REGRESSMI, with a fair amount of missing data. Table 5.21 shows the first 15 cases of the data set. There are no missing data on the DV, LTIMEDRS.

Multiple imputation through SAS is a three-step procedure.

1. Run PROC MI to create the multiply-imputation data set, with *m* imputations (subsets) in which missing values are imputed from a distribution of missing values.
2. Run the analysis for each of the imputations (e.g., PROC REG) on the file with *m* subsets and save parameter estimates (regression coefficients) in a second file.
3. Run PROC MIANALYZE to combine the results of the *m* analyses into a single set of parameter estimates.

Table 5.22 shows SAS MI syntax and selected output to create 5 imputations (the default *m*) and save the resulting data set in SASUSER.ANCOUTMI. The DV is not included in the var list; including LTIMEDRS could artificially inflate prediction by letting the DV influence the imputed values. Instead, only cases with nonmissing data on the DV are included in the data set.

The output first shows the missing data patterns, of which there are seven. The first, and most common pattern, is complete data with 284 cases (61.08% of the 465 cases). The second most common pattern is one in which MENHEAL is missing, with 69 cases (14.84%), and so on. The table

TABLE 5.21   Partial View of SASUSER.REGRESSMI with Missing Data

SUBJNO	MENHEAL	LTIMEDRS	LPHYHEAL	SSTRESS
1	.	0.3010	0.6990	.
2	.	0.6021	0.6021	20.3715
3	4	0.0000	.	9.5917
4	2	1.1461	0.3010	15.5242
5	6	1.2041	0.4771	9.2736
6	5	0.6021	0.6990	15.7162
7	6	0.4771	0.6990	.
8	5	0.0000	.	3.4641
9	4	0.9031	0.6990	16.4012
10	9	0.6990	0.4771	19.7737
11	.	1.2041	0.7782	15.3948
12	.	0.0000	0.4771	3.6056
13	10	0.4771	0.4771	9.1652
14	9	1.1461	.	12.0000
15	2	0.4771	0.4771	11.6190

also shows means on each of the unmissing variables for each pattern. For example, the mean for MENHEAL when both SSTRESS and LPHYHEAL are missing is 10, as opposed to the mean for MENHEAL when data are complete (6.05).

The table labeled **EM (Posterior Mode) Estimates** shows the means and covariances (cf. Section 1.6.3) for the first step of multiple imputation, formation of the EM covariance matrix (Section 4.1.3.2). The next table, **Multiple Imputation Variance Information**, shows the partition of total variance for the variables with missing data into variance between imputations and variance within imputations. That is, how much do the predictors individually vary, on average, within each imputed data subset and how much do their means vary among the five imputed data subsets. The **Relative Increase in Variance** is a measure of increased uncertainty due to missing data. Notice that the variable with the most missing data, MENHEAL, has the greatest relative increase in variance. Relative efficiency is related to power; greater relative efficiency is associated with a smaller standard error for testing a parameter estimate. Again, the fewer the missing data, the greater the relative efficiency. Choice of m also affects relative efficiency.[13] Finally, the **Multiple Imputation Parameter Estimates** are the average, minimum, and maximum values for means of the variables with missing data averaged over the 5 imputed data sets. The **t for H0:** is uninteresting, testing whether the mean differs from zero. Thus, the mean for MENHEAL varies from 5.95 to 6.16 over the five data subsets, with an average of 6.05 with 95% confidence limits from 5.63 to 6.48.

[13]Relative efficiency of a parameter estimate depends on m and the amount of missing data. Rubin (1987) provides the following equation for relative efficiency:

$$\%Efficiency = \left(1 + \frac{\gamma}{m}\right)^{-1}$$

where γ is the rate of missing data. For example, with $m = 5$ and 10% of data missing, relative efficiency is 98%.

TABLE 5.22 Syntax and Selected SAS MI Output to Create Multiply-Imputed Data Set

```
proc mi data=SASUSER.REGRESSMI seed=45792
  out=SASUSER.REGOUTMI; var LPHYHEAL MENHEAL SSTRESS;
run:
```

Missing Data Patterns

						--------Group Means--------		
Group	LPHYHEAL	MENHEAL	SSTRESS	Freq	Percent	LPHYHEAL	MENHEAL	SSTRESS
1	X	X	X	284	61.08	0.647020	6.049296	13.454842
2	X	X	.	26	5.59	0.630585	4.884615	.
3	X	.	X	69	14.84	0.646678	.	12.948310
4	X	.	.	12	2.58	0.774916	.	.
5	.	X	X	56	12.04	.	6.375000	13.820692
6	.	X	.	4	0.86	.	10.000000	.
7	.	.	X	14	3.01	.	.	14.053803

EM (Posterior Mode) Estimates

TYPE	_NAME_	LPHYHEAL	MENHEAL	SSTRESS
MEAN		0.652349	6.079942	13.451916
COV	LPHYHEAL	0.041785	0.441157	0.310229
COV	MENHEAL	0.441157	17.314557	7.770443
COV	SSTRESS	0.310229	7.770443	24.269465

Multiple Imputation Variance Information

----------------Variance----------------

Variable	Between	Within	Total	DF
LPHYHEAL	0.000006673	0.000092988	0.000101	254.93
MENHEAL	0.007131	0.037496	0.046054	88.567
SSTRESS	0.002474	0.052313	0.055281	332.44

Multiple Imputation Variance Information

Variable	Relative Increase in Variance	Fraction Missing Information	Relative Efficiency
LPHYHEAL	0.086119	0.082171	0.983832
MENHEAL	0.228232	0.199523	0.961627
SSTRESS	0.056744	0.055058	0.989108

Multiple Imputation Parameter Estimates

Variable	Mean	Std Error	95% Confidence	Limits	DF
LPHYHEAL	0.651101	0.010050	0.63131	0.67089	254.93
MENHEAL	6.054922	0.214601	5.62849	6.48136	88.567
SSTRESS	13.423400	0.235120	12.96089	13.88591	332.44

Multiple Imputation Parameter Estimates

Variable	Minimum	Maximum	Mu0	t for H0: Mean=Mu0	Pr > |t|
LPHYHEAL	0.646559	0.652602	0	64.79	<.0001
MENHEAL	5.953237	6.160381	0	28.21	<.0001
SSTRESS	13.380532	13.502912	0	57.09	<.0001

Table 5.23 shows a portion of the imputed data set, with the imputation variable (1 to 5) as well as missing values filled in. The latter portion of imputation 1 is shown, along with the first 15 cases of imputation 2.

The next step is to run SAS REG on the 5 imputations. This is done by including the by _Imputation_ instruction in the syntax, as seen in Table 5.24. The results of the analyses in terms of parameter estimates (regression coefficients, B) and variance-covariance matrices are sent to the output file: REGOUT.

All of the imputations show similar results, with F values ranging from 62.55 to 72.39 (all $p < .0001$), but these are considerably smaller than the $F = 92.90$ of the full-data standard multiple regression analysis of Section 5.7.2. Adjusted R^2 ranges from .28 to .32, as compared with .37 with the full data reported in Table 5.16.

TABLE 5.23   Partial View of SASUSER.REGOUTMI with Imputation Variable and Missing Data Imputed

SASUSER.REGOUTMI						
6	Int	Int	Int	Int	Int	Int
2325	_Imputation_	SUBJNO	MENHEAL	LTIMEDRS	LPHYHEAL	SSTRESS
459	1	717	11.1393	1.2788	0.7782	12.8841
460	1	724	9.0000	1.1761	0.7709	18.1934
461	1	754	6.0000	0.6021	0.4771	8.5440
462	1	755	5.3851	0.6990	0.6021	8.1240
463	1	756	6.0000	1.2041	0.9542	11.9164
464	1	757	4.0000	0.6990	0.7782	9.3274
465	1	758	2.0000	0.6021	0.6990	12.2066
466	2	1	7.7236	0.3010	0.6990	3.6336
467	2	2	9.4462	0.6021	0.6021	20.3715
468	2	3	4.0000	0.0000	0.6812	9.5917
469	2	4	2.0000	1.1461	0.3010	15.5242
470	2	5	6.0000	1.2041	0.4771	9.2736
471	2	6	5.0000	0.6021	0.6990	15.7162
472	2	7	6.0000	0.4771	0.6990	9.1690
473	2	8	5.0000	0.0000	0.7080	3.4641
474	2	9	4.0000	0.9031	0.6990	16.4012
475	2	10	9.0000	0.6990	0.4771	19.7737
476	2	11	3.6350	1.2041	0.7782	15.3948
477	2	12	5.3116	0.0000	0.4771	3.6056
478	2	13	10.0000	0.4771	0.4771	9.1652
479	2	14	9.0000	1.1461	0.7556	12.0000
480	2	15	2.0000	0.4771	0.4771	11.6190
481	2	16	4.0000	0.4771	0.4771	17.0587
482	2	21	-0.2698	0.3010	0.4771	12.1751
483	2	22	3.8717	0.4771	0.8451	15.2643
484	2	23	6.0000	0.7782	0.6021	12.1244
485	2	24	9.4105	0.7782	1.1047	17.5499
486	2	25	12.0000	0.6021	0.6021	11.0454
487	2	26	3.0000	0.6990	0.3010	17.5214
488	2	27	4.0000	0.4771	0.4771	15.7480
489	2	28	8.9444	0.0000	0.6990	11.0454
490	2	29	8.2556	1.1461	0.8451	19.5959

TABLE 5.24 SAS REG Syntax and Selected Output for Multiple Regression on All Five Imputations

```
proc reg data=SASUSER.REGOUTMI outest=REGOUT covout;
    model LTIMEDRS = LPHYHEAL MENHEAL SSTRESS;
    by_Imputation_;
run;
```

------------------------Imputation Number=1------------------------

The REG Procedure
Model: MODEL 1
Dependent Variable: LTIMEDRS LTIMEDRS

| Number of Observations Read | 465 |
| Number of Observations Used | 465 |

Analysis of Variance

Source	DF	Sum of Squares	Mean Square	F Value	Pr > F
Model	3	25.62241	8.54080	72.39	<.0001
Error	461	54.38776	0.11798		
Corrected Total	464	80.01017			

Root MSE	0.34348	R-Square	0.3203	
Dependent Mean	0.74129	Adj R-Sq	0.3158	
Coeff Var	46.33561			

Parameter Estimates

Variable	Label	DF	Parameter Estimate	Standard Error	t Value	Pr > |t|
Intercept	Intercept	1	−0.10954	0.06239	−1.76	0.0798
LPHYHEAL	LPHYHEAL	1	0.91117	0.09324	9.77	<.0001
MENHEAL	MENHEAL	1	0.00275	0.00467	0.59	0.5566
SSTRESS	SSTRESS	1	0.01800	0.00361	4.98	<.0001

------------------------Imputation Number=2------------------------

The REG Procedure
Model: MODEL1
Dependent Variable: LTIMEDRS LTIMEDRS

| Number of Observations Read | 465 |
| Number of Observations Used | 465 |

Analysis of Variance

Source	DF	Sum of Squares	Mean Square	F Value	Pr > F
Model	3	23.14719	7.71573	62.55	<.0001
Error	461	56.86297	0.12335		
Corrected Total	464	80.01017			

(continued)

183

TABLE 5.24 Continued

```
               Root MSE            0.35121    R-Square    0.2893
               Dependent Mean      0.74129    Adj R-Sq    0.2847
               Coeff Var          47.37826
```

Parameter Estimates

Variable	Label	DF	Parameter Estimate	Standard Error	t Value	Pr > |t|
Intercept	Intercept	1	−0.03392	0.06067	−0.56	0.5764
LPHYHEAL	LPHYHEAL	1	0.79920	0.09270	8.62	<.0001
MENHEAL	MENHEAL	1	0.00259	0.00481	0.54	0.5902
SSTRESS	SSTRESS	1	0.01770	0.00358	4.94	<.0001

```
------------------------Imputation Number=3-------------------------
```

The REG Procedure
Model: MODEL 1
Dependent Variable: LTIMEDRS LTIMEDRS

```
        Number of Observations Read              465
        Number of Observations Used              465
```

Analysis of Variance

Source	DF	Sum of Squares	Mean Square	F Value	Pr > F
Model	3	24.22616	8.07539	66.74	<.0001
Error	461	55.78401	0.12101		
Corrected Total	464	80.01017			

```
               Root MSE            0.34786    R-Square    0.3028
               Dependent Mean      0.74129    Adj R-Sq    0.2983
               Coeff Var          46.92661
```

Parameter Estimates

Variable	Label	DF	Parameter Estimate	Standard Error	t Value	Pr > |t|
Intercept	Intercept	1	−0.09996	0.06381	−1.57	0.1179
LPHYHEAL	LPHYHEAL	1	0.86373	0.09238	9.35	<.0001
MENHEAL	MENHEAL	1	0.00255	0.00464	0.55	0.5826
SSTRESS	SSTRESS	1	0.01954	0.00354	5.51	<.0001

```
------------------------Imputation Number=4-------------------------
```

The REG Procedure
Model: MODEL 1
Dependent Variable: LTIMEDRS LTIMEDRS

```
        Number of Observations Read              465
        Number of Observations Used              465
```

TABLE 5.24   Continued

```
                         Analysis of Variance

Source            DF    Sum of Squares    Mean Square    F Value    Pr > F
Model              3          24.27240        8.09080      66.92    <.0001
Error            461          55.73777        0.12091
Corrected Total  464          80.01017

              Root MSE              0.34772    R-Square     0.3034
              Dependent Mean        0.74129    Adj R-Sq     0.2988
              Coeff Var            46.90716

                         Parameter Estimates

                              Parameter    Standard
Variable    Label       DF     Estimate       Error    t Value    Pr > |t|
Intercept   Intercept    1     -0.07371     0.06260      -1.18      0.2396
LPHYHEAL    LPHYHEAL     1      0.87685     0.09194       9.54      <.0001
MENHEAL     MENHEAL      1      0.00138     0.00475       0.29      0.7725
SSTRESS     SSTRESS      1      0.01790     0.00362       4.95      <.0001

------------------------Imputation Number=5-------------------------

                       The REG Procedure
                        Model: MODEL1
               Dependent Variable: LTIMEDRS LTIMEDRS

          Number of Observations Read:          465
          Number of Observations Used:          465

                       Analysis of Variance

Source            DF    Sum of Squares    Mean Square    F Value    Pr > F
Model              3          25.12355        8.37452      70.34    <.0001
Error            461          54.88662        0.11906
Corrected Total  464          80.01017

              Root MSE              0.34505    R-Square     0.3140
              Dependent Mean        0.74129    Adj R-Sq     0.3095
              Coeff Var            46.54763

                       Parameter Estimates

Variable    Label       DF    Parameter     Standard    t Value    Pr > |t|
Intercept   Intercept    1    Estimate         Error      -1.42      0.1569
LPHYHEAL    LPHYHEAL     1     -0.08577       0.06050       9.72      <.0001
MENHEAL     MENHEAL      1      0.88661       0.09123      -0.18      0.8605
SSTRESS     SSTRESS      1     -0.00083306    0.00474       5.37      <.0001
                                0.01879       0.00350
```

Table 5.25 shows syntax and output for combining the results of the five imputations through SAS MIANALYZE. Data from the five imputations are used. Variables are the ones for which Parameter Estimates were sent to the REGOUT data set.

TABLE 5.25   SAS MIANALYZE Syntax and Output for Combining Parameter Estimates from the Results of Five Multiple Regression Analyses

```
proc mianalyze data=REGOUT:
     var Intercept LPHYHEAL MENHEAL SSTRESS
run;
```

The MIANALYZE Procedure

Model Information

Data Set	WORK.REGOUT
Number of Imputations	5

Multiple Imputation Variance Information

	--------------	Variance	--------------	
Parameter	Between	Within	Total	DF
Intercept	0.000866	0.003845	0.004884	88.288
LPHYHEAL	0.001760	0.008520	0.010631	101.37
MENHEAL	0.000002286	0.000022309	0.000025052	333.65
SSTRESS	0.000000589	0.000012756	0.000013462	1453.4

Multiple Imputation Variance Information

Parameter	Relative Increase in Variance	Fraction Missing Information	Relative Efficiency
Intercept	0.270411	0.230099	0.956005
LPHYHEAL	0.247882	0.213998	0.958957
MENHEAL	0.122956	0.114783	0.977559
SSTRESS	0.055366	0.053762	0.989362

Multiple Imputation Parameter Estimates

Parameter	Estimate	Std Error	95% Confidence	Limits	DF
Intercept	−0.080582	0.069888	−0.21946	0.058300	88.288
LPHYHEAL	0.867511	0.103108	0.66298	1.072042	101.37
MENHEAL	0.001687	0.005005	−0.00816	0.011533	333.65
SSTRESS	0.018384	0.003669	0.01119	0.025581	1453.4

Multiple Imputation Parameter Estimates

Parameter	Minimum	Maximum	Theta0	t for H0: Parameter=Theta0	Pr > |t|
Intercept	−0.109545	−0.033922	0	−1.15	0.2520
LPHYHEAL	0.799195	0.911169	0	8.41	<.0001
MENHEAL	−0.000833	0.002750	0	0.34	0.7363
SSTRESS	0.017697	0.019539	0	5.01	<.0001

Parameter estimates differ from those of the standard multiple regression of Section 5.7.2 but conclusions do not. Again, only LPHYHEAL and SSTRESS contribute to prediction of LTIMEDRS, MENHEAL does not. Note that DF are based on the fraction of missing data and *m*: the more data missing, the smaller the DF. If a DF value is close to 1, you need greater *m* because estimates are unstable.

MIANALYZE does not provide an ANOVA table to assess the results of overall prediction or any form of multiple R^2. Instead, ranges of results are reported from the PROC REG runs of Table 5.24. The outcome of this sample multiple imputation analysis is that some power is lost relative to the complete data of SASUSER.REGRESS.

As another comparison, Table 5.26 shows the results of a standard multiple regression using only the 284 complete cases of SASUSER.REGRESSMI (partially shown in Table 5.22). SAS REG uses listwise deletion as default, so that cases with missing data on any of the variables are deleted before analysis.

**TABLE 5.26   SAS REG Syntax and Output for Standard Multiple
Regression with Listwise Deletion of Missing Values**

```
proc reg data=SASUSER.REGRESSMI
   model LTIMEDRS = LPHYHEAL MENHEAL SSTRESS;
run;
```

The REG Procedure
Model: MODEL 1
Dependent Variable: LTIMEDRS LTIMEDRS

Number of Observations Read	465
Number of Observations Used	284
Number of Observations with Missing Values	181

Analysis of Variance

Source	DF	Sum of Squares	Mean Square	F Value	Pr > F
Model	3	18.43408	6.14469	57.36	<.0001
Error	280	29.99510	0.10713		
Corrected Total	283	48.42918			

Root MSE	0.32730	R-Square	0.3806
Dependent Mean	0.78354	Adj R-Sq	0.3740
Coeff Var	41.77218		

Parameter Estimates

Variable	Label	DF	Parameter Estimate	Standard Error	t Value	Pr > |t|
Intercept	Intercept	1	-0.09014	0.07529	-1.20	0.2322
LPHYHEAL	LPHYHEAL	1	1.05972	0.11127	9.52	<.0001
MENHEAL	MENHEAL	1	0.00398	0.00555	0.72	0.4741
SSTRESS	SSTRESS	1	0.01219	0.00435	2.80	0.0055

Additional power in terms of statistical significance is lost by deleting all cases with any missing data. Overall F has been reduced to 57.36 (although p remains $< .0001$) and there is a larger p value for the test of SSTRESS. However, adjusted R^2 has not been reduced by this strategy, unlike that of multiple imputation. Also, in this case, parameter estimates appear to more closely resemble those of the analysis of the full data set.

5.8   Comparison of Programs

The popularity of multiple regression is reflected in the abundance of applicable programs. SPSS, SYSTAT, and SAS each have a single, highly flexible program for the various types of multiple regression. The packages also have programs for the more exotic forms of multiple regression, such as nonlinear regression, probit regression, logistic regression (cf. Chapter 10), and the like.

Direct comparisons of programs for standard regression are summarized in Table 5.27. Additional features for statistical and sequential regression are summarized in Table 5.28. Features include those that are available only through syntax. Some of these features are elaborated in Sections 5.7.1 through 5.7.4.

5.8.1   SPSS Package

The distinctive feature of SPSS REGRESSION, summarized in Tables 5.27 and 5.28, is flexibility. SPSS REGRESSION offers four options for treatment of missing data (described in the on-disk help system). Data can be input raw or as correlation or covariance matrices. Data can be limited to a subset of cases, with residuals statistics and plots reported separately for selected and unselected cases.

A special option is available so that correlation matrices are printed only when one or more of the correlations cannot be calculated. Also convenient is the optional printing of semipartial correlations for standard multiple regression and 95% confidence intervals for regression coefficients.

The statistical procedure offers forward, backward, and stepwise selection of variables, with several user-modifiable statistical criteria for variable selection.

A series of **METHOD=ENTER** subcommands are used for sequential regression. Each **ENTER** subcommand is evaluated in turn; the IV or IVs listed after each **ENTER** subcommand are evaluated in that order. Within a single subcommand, SPSS enters the IVs in order of decreasing tolerance. If there is more than one IV in the subcommand, they are treated as a block in the Model Summary table where changes in the equation are evaluated.

Extensive analysis of residuals is available. For example, a table of predicted scores and residuals can be requested and accompanied by a plot of standardized residuals against standardized predicted values of the DV (z-scores of the Y' values). Plots of standardized residuals against sequenced cases are also available. For a sequenced file, one can request a Durbin-Watson statistic, which is used for a test of autocorrelation between adjacent cases. In addition, you can request Mahalanobis distance for cases as a convenient way of evaluating outliers. This is the only program within the SPSS packages that offers Mahalanobis distance. A case labeling variable may be specified, so that subject numbers for outliers can be easily identified. Partial residual plots (partialing out all but one of the IVs) are available in SPSS REGRESSION.

The flexibility of input for the SPSS REGRESSION program does not carry over into output. The only difference between standard and statistical or sequential regression is in the printing of each

TABLE 5.27   Comparison of Programs for Standard Multiple Regression

Feature	SPSS REGRESSION	SAS REG	SYSTAT REGRESS
Input			
Correlation matrix input	Yes	Yes	Yes
Covariance matrix input	Yes	Yes	Yes
SSCP matrix input	No	Yes	Yes
Missing data options	Yes	No	No
Regression through the origin	ORIGIN	NOINT	Yes[c]
Tolerance option	TOLERANCE	SINGULAR	Tolerance
Post hoc hypotheses[a]	TEST	Yes	Yes
Optional error terms	No	Yes	No
Collinearity diagnostics	COLLIN	COLLIN	PRINT=MEDIUM
Select subset of cases	Yes	WEIGHT	WEIGHT
Weighted least squares	REGWGT	Yes	WEIGHT
Multivariate multiple regression	No	MTEST	No
Setwise regression	No	Yes	No
Ridge regression	No	RIDGE	Yes
Identify case labeling variable	RESIDUALS ID	No	No
Bayesian regression	No	No	Yes
Resampling	No	No	Yes
Regression output			
Analysis of variance for regression	ANOVA	Analysis of Variance	Analysis of Variance
Multiple R	R	No	Multiple R
R^2	R Square	R-square	Squared multiple R
Adjusted R^2	Adjusted R Square	Adj R-sq	Adjusted Squared Multiple R
Standard error of Y'	Std. Error of the Estimate	Root MSE	Std. error of estimate
Coefficient of variation	No	Coeff Var	No
Correlation matrix	Yes	CORR	No
Significance levels of correlation matrix	Yes[a]	No	No
Sum-of-squares and cross-products (SSCP) matrix	Yes	USSCP	No
Covariance matrix	Yes	No	No
Means and standard deviations	Yes	Yes	No

(continued)

TABLE 5.27   Continued

Feature	SPSS REGRESSION	SAS REG	SYSTAT REGRESS
Regression output *(continued)*			
Matrix of correlation coefficients if some not computed	Yes	N.A.	N.A.
N for each correlation coefficient	Yes	N.A.	N.A.
Sum of squares for each variable	No	Uncorrected SS	No
Unstandardized regression coefficients	B	Parameter Estimate	Coefficient
Standard error of regression coefficient	Std. Error	Standard Error	Std. Error
F or t test of regression coefficient	t (F optional)	t Value	t
Significance for regression coefficient	Sig.	$Pr > |t|$	P
Intercept (constant)	(Constant)	Intercept	CONSTANT
Standardized regression coefficient	Beta	Standardized Estimate	Std. Coef
Approx. standard error of β	Std. Error	No	No
Partial correlation	Partial	Squared Partial Corr Type II	No
Semipartial correlation or sr_i^2	Part	Squared Semi-partial Corr Type II	No
Bivariate correlation with DV	No	Corr	Zero Order
Tolerance	Yes	Yes	Yes
Variance-covariance matrix for unstandardized B coefficients	Yes	Yes	No
Correlation matrix of B coefficients	No	No	Yes
Correlation matrix for unstandardized B coefficients	Yes	Yes	Yes (PRINT=LONG)
95% confidence interval for B	Yes	Yes	Yes (PRINT=MEDIUM)
Specify alternative variation for CI for B	No	Yes	No
Hypothesis matrices	No	Yes	No
Collinearity diagnostics	Yes	Yes	Yes
Residuals			
Predicted scores, residuals and standardized residuals	Yes	Yes	Data file
Partial residuals	No	No	Data file

TABLE 5.27   Continued

Feature	SPSS REGRESSION	SAS REG	SYSTAT REGRESS
Residuals *(continued)*			
95% confidence interval for predicted value	No	95% CL Predict	No
Plot of standardized residuals against predicted scores	Yes	Yes	Yes
Normal plot of residuals	Yes	No	No
Durbin-Watson statistic	Yes	Yes	Yes
Leverage diagnostics (e.g., Mahalanobis distance)	Mahal. Distance	Hat Diag H	Data file (Leverage)
Influence diagnostics (e.g., Cook's distance)	Cook's Distance	Cook's D	Data file
Histograms	Yes	No	No
Casewise plots	Yes	No	No
Partial plots	Yes	Yes	No
Other plots available	Yes	Yes	No
Summary statistics for residuals	Yes	Yes	No
Save predicted values/residuals	Yes	Yes	Yes

Note: SAS and SYSTAT GLM can also be used for standard multiple regression.

[a]Does *not* use Larzelere and Mulaik (1977) correction.

[b]Use Type II for standard MR.

[c]Omit CONSTANT from MODEL.

step (Model) in each table, and in an extension of the model summary table available through the CHANGE instruction. Otherwise, the statistics and parameter estimates are identical. These values, however, have different meanings, depending on the type of analysis. For example, you can request semipartial correlations (called Part) through the ZPP statistics. But these apply only to standard multiple regression. As pointed out in Section 5.6.1, semipartial correlations for statistical or sequential analysis appear in the Model Summary table (obtained through the CHANGE statistic) as R Square Change.

5.8.2   SAS System

Currently, SAS REG is the all-purpose regression program in the SAS system. In addition, GLM can be used for regression analysis; it is more flexible and powerful than SAS but also more difficult to use.

SAS REG handles correlation, covariance, or SSCP matrix input but has no options for dealing with missing data. A case is deleted if it contains any missing values.

TABLE 5.28 Comparison of Additional Features for Stepwise and/or Sequential Regression

Feature	SPSS REGRESSION	SAS REG	SYSTAT REGRESS
Input			
Specify stepping algorithm	Yes	Yes	Yes
Specify F to enter and/or remove	FIN/FOUT	No	FEnter/FRemove
Specify probability of F to enter and/or remove	PIN/POUT	SLE/SLS	Enter/Remove
Specify maximum number of steps	MAXSTEPS	MAXSTEP	Max step
Specify maximum number of variables	No	STOP	No
Request selection statistics (e.g., AIC, Mallow's C_p)	SELECTION	Yes	No
Force variables into equation	ENTER	INCLUDE	FORCE
Specify order of entry (hierarchy)	ENTER	No	No
IV sets for entry in single step	ENTER	GROUPNAMES	No
Interactive processing	No	Yes	Yes
Regression Output			
Analysis of variance for regression, each step	ANOVA	Analysis of Variance	No[a]
Multiple R, each step	R	No	R
R^2, each step	R Square	R-Square	R-Square
Mallow's C_p, each step	Mallow's Prediction Criterion	C(p)	No
Standard error of Y', each step	Std. Error of the Estimate	No	No[a]
Adjusted R^2, each step	Adjusted R Square	No	No[a]
Variables in equation/Coefficients (each step)			
Unstandardized regression coefficients	B	Parameter Estimate	Coefficient
Standard error of regression coefficient	Std. Error	Std. Error	Std. Error
95% confidence interval for B	Yes	No	Yes
Standard regression coefficient	Beta	Standardized Estimate	Std. Coef
F (or T) to remove	t (F optional)	F Value	F
p to remove	Sig.	Pr > F	'P'
Intercept	(Constant)	Intercept	Constant
Tolerance	No	No	Tol.

TABLE 5.28 Continued

Feature	SPSS REGRESSION	SAS REG	SYSTAT REGRESS
Variables not in equation /Excluded Variables (each step)			
Standardized regression coefficient for entering	Beta In	No	No
Partial correlation coefficient for entering	Partial Correlation	No	Part. corr.
Tolerance	Yes	No	Yes
F (or T) to enter	t (F optional)	No	F
p to enter	Sig.	Yes	'P'
Summary table/Change Statistics			
Multiple R	R	No	No
R^2	R Square	Model R-Square	No
Adjusted R^2	Adjusted R Square	No	No
Change in R^2 (squared semipartial correlation)	R Square Change	Partial R-Square	No
F_{inc}	F Change	F Value	No
Degrees of freedom for F_{inc}	df1, df2		No
p for F_{inc}	Sig. F Change	Pr > .F	No
Standard regression coefficient	Beta	No	No
Mallow's C_p	Mallow's Prediction Criterion	C(p)	No
Number of variables in the equation	No	Number Vars In	No

[a]Available by running separate standard multiple regressions for each step.

In SAS REG, two types of semipartial correlations are available. The sr_i^2 appropriate for standard multiple regression is the one that uses TYPE II (partial) sums of squares (that can also be printed). SAS REG also does multivariate multiple regression as a form of canonical correlation analysis where specific hypotheses can be tested.

For statistical regression using SAS REG, the usual forward, backward, and stepwise criteria for selection are available, in addition to five others. The statistical criteria available are the probability of F to enter and F to remove an IV from the equation as well as maximum number of variables. Interactive processing can be used to build sequential models.

Several criteria are available for setwise regression. You make the choice of the best subset by comparing values on your chosen criterion or on R^2, which is printed for all criteria. No information is given about individual IVs in the various subsets.

Sequential regression is handled interactively, in which an initial model statement is followed by instructions to add one or more variables at each subsequent step. Full output is available at the end of each step, but there is no summary table.

Tables of residuals and other diagnostics, including Cook's distance and hat diagonal (a measure of leverage, cf. Equation 4.3), are extensive, but are saved to file rather than printed. However, a plotting facility within SAS REG allows extensive plotting of residuals. SAS REG also provides partial residual plots where all IVs except one are partialed out. Recent enhancements to SAS REG greatly increase plotting capabilities.

5.8.3 SYSTAT System

Multiple regression in SYSTAT Version 11 is most easily done through REGRESS, although GLM also may be used. Statistical regression options include forward and backward stepping. Options are also available to modify α level to enter and remove, change tolerance, and force the first k variables into the equation. By choosing the interactive mode for stepwise regression, you can specify individual variables to enter the equation at each step, allowing a simple form of sequential regression. This is the only program reviewed that permits standard multiple regression with resampling and Bayesian multiple regression.

Matrix input is accepted in SYSTAT REGRESS, but processing of output of matrices or descriptive statistics requires the use of other programs in the SYSTAT package. Residuals and other diagnostics are handled by saving values to a file, which can then be printed out or plotted through SYSTAT PLOT (a plot of unstandardized residuals against predicted scores is shown in the output). In this way, you can take a look at Cook's value or leverage, from which Mahalanobis distance can be computed (Equation 4.3), for each case. Or, you can find summary values for residuals and diagnostics through SYSTAT STATS, the program for descriptive statistics.

6

Analysis of Covariance

6.1 General Purpose and Description

Analysis of covariance is an extension of analysis of variance in which main effects and interactions of IVs are assessed after DV scores are adjusted for differences associated with one or more covariates (CVs), variables that are measured before the DV and are correlated with it.[1] The major question for ANCOVA (analysis of covariance) is essentially the same as for ANOVA: Are mean differences among groups on the adjusted DV likely to have occurred by chance? For example, is there a mean difference between a treated group and a control group on a posttest (the DV) after posttest scores are adjusted for differences in pretest scores (the CV)?

Analysis of covariance is used for three major purposes. The first purpose is to increase the sensitivity of the test of main effects and interactions by reducing the error term; the error term is adjusted for, and hopefully reduced by, the relationship between the DV and the CV(s). The second purpose is to adjust the means on the DV themselves to what they would be if all subjects scored equally on the CV(s). The third use of ANCOVA occurs in MANOVA (Chapter 7) where the researcher assesses one DV after adjustment for other DVs that are treated as CVs.

The first use of ANCOVA is the most common. In an experimental setting, ANCOVA increases the power of an F test for a main effect or interaction by removing predictable variance associated with CV(s) from the error term. That is, CVs are used to assess the "noise" where "noise" is undesirable variance in the DV (e.g., individual differences) that is estimated by scores on CVs (e.g., pretests). Use of ANCOVA in experiments is discussed by Tabachnick and Fidell (2007).

An experimental ANCOVA strategy was taken by Copeland, Blow, and Barry (2003), who used repeated-measures ANCOVA to investigate the effect of a brief intervention program to reduce at-risk drinking on health-care utilization among veterans. Covariates were age, race/ethnicity, living alone and educational, as well as pre-intervention utilization. Effects of intervention were examined 9 and 18 months after intervention. Veterans were randomly assigned to be presented with either a General Health Advice booklet or a Brief Alcohol Intervention booklet. Thus, this was a 2×2 between-within-subjects ANCOVA with two levels of the between-subjects IV, intervention, and two

---

[1]Strictly speaking, ANCOVA, like multiple regression, is not a multivariate technique because it involves a single DV. For the purposes of this book, however, it is convenient to consider it along with multivariate analyses.

levels of the within-subjects IV, time of assessment. The intervention booklet was found to increase utilization of outpatient care services in the short term, but there was no noticeable effect on long term utilization of inpatient/outpatient services.

The second use of ANCOVA commonly occurs in nonexperimental situations when subjects cannot be randomly assigned to treatments. ANCOVA is used as a statistical matching procedure, although interpretation is fraught with difficulty, as discussed in Section 6.3.1. ANCOVA is used primarily to adjust group means to what they would be if all subjects scored identically on the CV(s). Differences between subjects on CVs are removed so that, presumably, the only differences that remain are related to the effects of the grouping IV(s). (Differences could also, of course, be due to attributes that have not been used as CVs.)

This second application of ANCOVA is primarily for descriptive model building: the CV enhances prediction of the DV, but there is no implication of causality. If the research question to be answered involves causality, ANCOVA is no substitute for running an experiment.

As an example, suppose we are looking at regional differences in political attitudes where the DV is some measure of liberalism-conservatism. Regions of the United States form the IV, say, Northeast, South, Midwest, and West. Two variables that are expected to vary with political attitude and with geographical region are socioeconomic status and age. These two variables serve as CVs. The statistical analysis tests the null hypothesis that political attitudes do not differ with geographical region after adjusting for socioeconomic status and age. However, if age and socioeconomic differences are inextricably tied to geography, adjustment for them is not realistic. And, of course, there is no implication that political attitudes are caused in any way by geographic region. Further, unreliability in measurement of the CV and the DV–CV relationship may lead to over- or underadjustment of scores and means and, therefore, to misleading results. These issues are discussed in greater detail throughout the chapter.

This nonexperimental ANCOVA strategy was taken by Brambilla et al. (2003), who looked at differences in amygdala volumes between 24 adults diagnosed with bipolar disorder and 36 healthy controls.[2] Volumetric measures of amygdala and other structures were performed blindly, with semi-automated software. Covariates were age, gender and intracranial volume. The researchers found significantly larger left amygdala volumes for bipolar patients compared with controls, but no significant differences between the two groups in the other temporal lobe structures.

Serum cholesterol concentrations were examined in violent and nonviolent female suicide attempters by Vevera, Žukor, Morcinek, and Papežeová (2003). Their sample consisted of three groups of women admitted to a psychiatric department in a retrospective case-control design: 19 women with a history of violent suicide attempts, 51 with a history of nonviolent attempts, and 70 nonsuicidal "controls" matched by psychiatric diagnosis and actual age. Scheffé-adjusted post-hoc tests found significantly lower cholesterol levels in violent-suicide attempters than nonviolent attempters or nonsuicidal women. No significant difference was found between the latter two groups.

In the third major application of ANCOVA, discussed more fully in Chapter 7, ANCOVA is used to interpret IV differences when several DVs are used in MANOVA. After a multivariate analysis of variance, it is frequently desirable to assess the contribution of the various DVs to significant differences among IVs. One way to do this is to test DVs, in turn, with the effects of other DVs

---

[2]We cringe at this common but misleading use of "control". We prefer using "comparison" group here and reserving the term control for those randomly assigned to a control group.

removed. Removal of the effects of other DVs is accomplished by treating them as CVs in a procedure called a *stepdown analysis.*

The statistical operations are identical in all three major applications of ANCOVA. As in ANOVA, variance in scores is partitioned into variance due to differences between groups and variance due to differences within groups. Squared differences between scores and various means are summed (see Chapter 3) and these sums of squares, when divided by appropriate degrees of freedom, provide estimates of variance attributable to different sources (main effects of IVs, interactions between IVs, and error). Ratios of variances then provide tests of hypotheses about the effects of IVs on the DV.

However, in ANCOVA, the regression of one or more CVs on the DV is estimated first. Then DV scores and means are adjusted to remove the linear effects of the CV(s) before analysis of variance is performed on these adjusted values.

Lee (1975) presents an intuitively appealing illustration of the manner in which ANCOVA reduces error variance in a one-way between-subjects design with three levels of the IV (Figure 6.1). Note that the vertical axis on the right-hand side of the figure illustrates scores and group means in ANOVA. The error term is computed from the sum of squared deviations of DV scores around their associated group means. In this case, the error term is substantial because there is considerable spread in scores within each group.

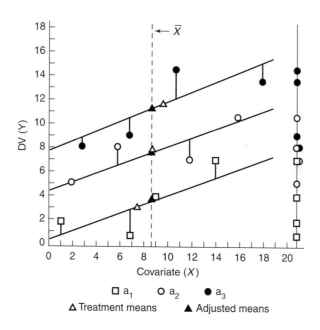

FIGURE 6.1 Plot of hypothetical data. The straight lines with common slope are those that best fit the data for the three treatments. The data points are also plotted along the single vertical line on the right as they would be analyzed in ANOVA.

Source: From *Experimental Design and Analysis* by Wayne Lee. Copyright © 1975 by W.H. Freeman and Company. Used with permission.

When the same scores are analyzed in ANCOVA, a regression line is found first that relates the DV to the CV. The error term is based on the (sum of squared) deviations of the DV scores from the regression line running through each group mean instead of from the means themselves. Consider the score in the lower left-hand corner of Figure 6.1. The score is near the regression line (a small deviation for error in ANCOVA) but far from the mean for its own group (a large deviation for error in ANOVA). As long as the slope of the regression lines is not zero, ANCOVA produces a smaller sum of squares for error than ANOVA. If the slope is zero, error sum of squares is the same as in ANOVA but error mean square is larger because CVs use up degrees of freedom.

CVs can be used in all ANOVA designs—factorial between-subjects, within-subjects, mixed within-between, nonorthogonal, and so on. Analyses of these more complex designs are readily available in only a few programs, however. Similarly, specific comparisons and trend analysis of adjusted means are possible in ANCOVA but not always readily available through the programs.

6.2   Kinds of Research Questions

As with ANOVA, the question in ANCOVA is whether mean differences in the DV between groups are larger than expected by chance. In ANCOVA, however, one gets a more precise look at the IV–DV relationship after removal of the effect of CV(s).

6.2.1   Main Effects of IVs

Holding all else constant, are changes in behavior associated with different levels of an IV larger than expected through random fluctuations occurring by chance? For example, is test anxiety affected by treatment, after holding constant prior individual differences in test anxiety? Does political attitude vary with geographical region, after holding constant differences in socioeconomic status and age? The procedures described in Section 6.4 answer this question by testing the null hypothesis that the IV has no systematic effect on the DV.

With more than one IV, separate statistical tests are available for each one. Suppose there is a second IV in the political attitude example, for example, religious affiliation, with four groups: Protestant, Catholic, Jewish, and none-or-other. In addition to the test of geographic region, there is also a test of differences in attitudes associated with religious affiliation after adjustment for differences in socioeconomic status and age.

6.2.2   Interactions among IVs

Holding all else constant, does change in behavior over levels of one IV depend on levels of another IV? That is, do IVs interact in their effect on behavior? (See Chapter 3 for a discussion of interaction.) For the political attitude example where religious affiliation is added as a second IV, are differences in attitudes over geographic region the same for all religions, after adjusting for socioeconomic status and age?

Tests of interactions, while interpreted differently from main effects, are statistically similar, as demonstrated in Section 6.6. With more than two IVs, numerous interactions are generated.

Except for common error terms, each interaction is tested separately from other interactions and from main effects. All tests are independent when sample sizes in all groups are equal and the design is balanced.

6.2.3   Specific Comparisons and Trend Analysis

When statistically significant effects are found in a design with more than two levels of a single IV, it is often desirable to pinpoint the nature of the differences. Which groups differ significantly from each other? Or, is there a simple trend over sequential levels of an IV? For the test anxiety example, we ask whether (1) the two treatment groups are more effective in reducing test anxiety than the waiting-list control, after adjusting for individual differences in test anxiety; and if (2) among the two treatment groups, desensitization is more effective than relaxation training in reducing test anxiety, again after adjusting for preexisting differences in test anxiety?

These two questions could be asked in planned comparisons instead of asking, through routine ANCOVA, the omnibus question of whether means are the same for all three levels of the IV. Or, with some loss in sensitivity, these two questions could be asked post hoc after finding a main effect of the IV in ANCOVA. Planned and post hoc comparisons are discussed in Section 6.5.4.3.

6.2.4   Effects of Covariates

Analysis of covariance is based on a linear regression (Chapter 5) between CV(s) and the DV, but there is no guarantee that the regression is statistically significant. The regression can be evaluated statistically by testing the CV(s) as a source of variance in DV scores, as discussed in Section 6.5.2. For instance, consider the test anxiety example where the CV is a pretest and the DV a posttest. To what extent is it possible to predict posttest anxiety from pretest anxiety, ignoring effects of differential treatment?

6.2.5   Effect Size

If a main effect or interaction of IVs is reliably associated with changes in the DV, the next logical question is: How much? How much of the variance in the adjusted DV scores—adjusted for the CV(s)—is associated with the IV(s)? In the test anxiety example, if a main effect is found between the means for desensitization, relaxation training, and control group, one then asks: What proportion of variance in the adjusted test anxiety scores is attributed to the IV? Effect sizes and their confidence intervals are demonstrated in Sections 6.4.2, 6.5.4.4, and 6.6.2.1.

6.2.6   Parameter Estimates

If any main effects or interactions are statistically significant, what are the estimated population parameters (adjusted mean and standard deviation or confidence interval) for each level of the IV or combination of levels of the IVs? How do group scores differ, on the average, on the DV, after adjustment for CVs? For the test anxiety example, if there is a main effect of treatment, what is the average adjusted posttest anxiety score in each of the three groups? The reporting of parameter estimates is demonstrated in Section 6.6.

6.3 Limitations to Analysis of Covariance

6.3.1 Theoretical Issues

As with ANOVA, the statistical test in no way assures that changes in the DV were caused by the IV. The inference of causality is a logical rather than a statistical problem that depends on the manner in which subjects are assigned to levels of the IV(s), manipulation of levels of the IV(s) by the researcher, and the controls used in the research. The statistical test is available to test hypotheses from both nonexperimental and experimental research, but only in the latter case is attribution of causality justified.

Choice of CVs is a logical exercise as well. As a general rule, one wants a very small number of CVs, all correlated with the DV and none correlated with each other. The goal is to obtain maximum adjustment of the DV with minimum loss of degrees of freedom for error. Calculation of the regression of the DV on the CV(s) results in the loss of one degree of freedom for error for each CV. Thus *the gain in power from decreased sum of squares for error may be offset by the loss in degrees of freedom.* When there is a substantial correlation between the DV and a CV, increased sensitivity due to reduced error variance offsets the loss of a degree of freedom for error. With multiple CVs, however, a point of diminishing returns is quickly reached, especially if the CVs correlate with one another (see Section 6.5.1).

In experimental work, a frequent caution is that the CVs must be independent of treatment. It is suggested that data on CVs be gathered before treatment is administered. Violation of this precept results in removal of some portion of the effect of the IV on the DV—that portion of the effect that is associated with the CV. In this situation, adjusted group means may be closer together than unadjusted means. Further, the adjusted means may be difficult to interpret.

In nonexperimental work, adjustment for prior differences in means associated with CVs is appropriate. If the adjustment reduces mean differences on the DV, so be it—unadjusted differences reflect unwanted influences (other than the IV) on the DV. In other words, mean differences on a CV associated with an IV are quite legitimately corrected for as long as the CV differences are not caused by the IV (Overall & Woodward, 1977).

When ANCOVA is used to evaluate a series of DVs after MANOVA, independence of the "CVs" and the IV is not required. Because CVs are actually DVs, it is expected that they be dependent on the IV.

In all uses of ANCOVA, however, adjusted means must be interpreted with great caution because the adjusted mean DV score may not correspond to any situation in the real world. Adjusted means are the means that would have occurred if all subjects had the same scores on the CVs. Especially in non-experimental work, such a situation may be so unrealistic as to make the adjusted values meaningless.

Sources of bias in ANCOVA are many and subtle and can produce either under- or overadjustment of the DV. At best, the nonexperimental use of ANCOVA allows you to look at IV–DV relationships (noncausal) adjusted for the effects of CVs, as measured. If causal inference regarding effects is desired, there is no substitute for random assignment of subjects. Don't expect ANCOVA to permit causal inference of treatment effects with nonrandomly assigned groups. If random assignment is absolutely impossible, or if it breaks down because of nonrandom loss of subjects, be sure to thoroughly ground yourself in the literature regarding use of ANCOVA in such cases, starting with Cook and Campbell (1979).

Limitations to generalizability apply to ANCOVA as they do to ANOVA or any other statistical test. One can generalize only to those populations from which a random sample is taken.

ANCOVA may, in some limited sense, sometimes adjust for a failure to randomly assign the sample to groups, but it does not affect the relationship between the sample and the population to which one can generalize.

6.3.2   Practical Issues

The ANCOVA model assumes reliability of CVs, linearity between pairs of CVs and between CVs and the DV, and homogeneity of regression, in addition to the usual ANOVA assumptions of normality and homogeneity of variance.

6.3.2.1   Unequal Sample Sizes, Missing Data, and Ratio of Cases to IVs

If scores on the DV are missing in a between-subjects ANCOVA, this is reflected as the problem of unequal n because all IV levels or combinations of IV levels do not contain equal numbers of cases. Consult Section 6.5.4.2 for strategies for dealing with unequal sample sizes. If some subjects are missing scores on CV(s), or if, in within-subjects ANCOVA, some DV scores are missing for some subjects, this is more clearly a missing-data problem. Consult Chapter 4 for methods of dealing with missing data.

Sample sizes in each cell must be sufficient to ensure adequate power. Indeed, the point of including covariates in an analysis often is to increase power. There are many software programs available to calculate required sample sizes depending on desired power and anticipated means and standard deviations in an ANOVA. Try a "statistical power" search on the Web to find some of them. These are easily applied to ANCOVA by substituting anticipated adjusted means or expected differences between adjusted means.

6.3.2.2   Absence of Outliers

Within each group, univariate outliers can occur in the DV or any one of the CVs. Multivariate outliers can occur in the space of the DV and CV(s). Multivariate outliers among DV and CV(s) can produce heterogeneity of regression (Section 6.3.2.7), leading to rejection of ANCOVA or at least unreasonable adjustment of the DV. If the CVs are serving as a convenience in most analyses, rejection of ANCOVA because there are multivariate outliers is hardly convenient.

Consult Chapter 4 for methods dealing with univariate outliers in the DV or CV(s) and multivariate outliers among the DV and CV(s). Tests for univariate and multivariate outliers within each group are demonstrated in Section 6.6.1.

6.3.2.3   Absence of Multicollinearity and Singularity

If there are multiple CVs, they should not be highly correlated with each other. *Highly correlated CVs should be eliminated*, both because they add no adjustment to the DV over that of other CVs and because of potential computational difficulties if they are singular or multicollinear. Most programs for ANCOVA automatically guard against statistical multicollinearity or singularity of CVs, however logical problems with redundancy among CVs occur far short of that criterion. For purposes of ANCOVA, *any CV with a squared multiple correlation (SMC) in excess of .50 may be considered redundant and deleted from further analysis*. Calculation of SMCs among CVs is demonstrated in Section 6.6.1.5.

6.3.2.4   *Normality of Sampling Distributions*

As in all ANOVA, it is assumed that the sampling distributions of means, as described in Chapter 3, are normal within each group. Note that it is the sampling distributions of means and not the raw scores within each cell that need to be normally distributed. Without knowledge of population values, or production of actual sampling distributions of means, there is no way to test this assumption. However, the central limit theorem suggests that, with large samples, sampling distributions are normal even if raw scores are not. *With relatively equal sample sizes in groups, no outliers, and two-tailed tests, robustness is expected with 20 degrees of freedom for error.* (See Chapter 3 for calculation of error degrees of freedom.)

Larger samples are necessary for one-tailed tests. With small, unequal samples or with outliers present, it may be necessary to consider data transformation (cf. Chapter 4).

6.3.2.5   *Homogeneity of Variance*

It is assumed in ANCOVA that the variance of DV scores within each cell of the design is a separate estimate of the same population variance. In ANCOVA the covariances are also evaluated for homogeneity of variance. If a CV fails the test, either a more stringent test of main effects and interactions is required (e.g., $\alpha = .025$ instead of .05) or the CV is dropped from the analysis. Section 4.1.5.3 provides guidelines and formal tests for evaluating homogeneity of variance, and remedies for violation of the assumption.

6.3.2.6   *Linearity*

The ANCOVA model is based on the assumption that the relationship between each CV and the DV and the relationships among pairs of CVs are linear. As with multiple regression (Chapter 5), violation of this assumption reduces the power of the statistical test; errors in statistical decision making are in a conservative direction. Error terms are not reduced as fully as they might be, optimum matching of groups is not achieved, and group means are incompletely adjusted. Section 4.5.1.2 discusses methods for assessing linearity.

Where curvilinearity is indicated, it may be corrected by transforming some of the CVs. Or, because of the difficulties in interpreting transformed variables, you may consider eliminating a CV that produces nonlinearity. Or a higher-order power of the CV can be used to produce an alternative CV that incorporates nonlinear influences.

6.3.2.7   *Homogeneity of Regression*

Adjustment of scores in ANCOVA is made on the basis of an average within-cell regression coefficient. The assumption is that the slope of the regression between the DV and the CV(s) within each cell is an estimate of the same population regression coefficient, that is, that the slopes are equal for all cells.

Heterogeneity of regression implies that there is a different DV–CV(s) slope in some cells of the design, or, that there is an interaction between IV(s) and CV(s). If IV(s) and CV(s) interact, the relationship between the CV(s) and the DV is different at different levels of the IV(s), and the CV adjustment that is needed for various cells is different. Figure 6.2 illustrates, for three groups, perfect homogeneity of regression (equality of slopes) and extreme heterogeneity of regression (inequality of slopes).

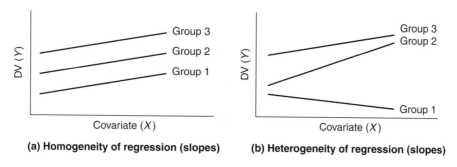

FIGURE 6.2   DV–CV regression lines for three groups plotted on the same coordinates for conditions of (a) homogeneity and (b) heterogeneity of regression.

If a between-subjects design is used, test the assumption of homogeneity of regression according to procedures described in Section 6.5.3. If any other design is used, and interaction between IVs and CVs is suspected, ANCOVA is inappropriate. If there is no reason to suspect an IV–CV interaction with complex designs, it is probably safe to proceed with ANCOVA on the basis of the robustness of the model. Alternatives to ANCOVA are discussed in Section 6.5.5.

6.3.2.8   Reliability of Covariates

It is assumed in ANCOVA that CVs are measured without error; that they are perfectly reliable. In the case of such variables as sex and age, the assumption can usually be justified. With self-report of demographic variables, and with variables measured psychometrically, such assumptions are not so easily made. And variables such as attitude may be reliable at the point of measurement, but fluctuate over short periods.

In experimental research, unreliable CVs lead to loss of power and a conservative statistical test through underadjustment of the error term. In nonexperimental applications, however, unreliable CVs can lead to either under- or overadjustment of the means. Group means may be either spread too far apart (Type I error) or compressed too closely together (Type II error). The degree of error depends on how unreliable the CVs are. *In nonexperimental research, limit CVs to those that can be measured reliably ($r_{xx} > .8$).*

If fallible CVs are absolutely unavoidable, they can be adjusted for unreliability. However, there is no one procedure that produces appropriate adjustment under all conditions nor is there even agreement about which procedure is most appropriate for which application. Because of this disagreement, and because procedures for correction require use of sophisticated programs, they are not covered in this book. The interested reader is referred to Cohen et al. (2003) or to procedures discussed by St. Pierre (1978).

6.4   Fundamental Equations for Analysis of Covariance

In the simplest application of analysis of covariance there is a DV score, a grouping variable (IV), and a CV score for each subject. An example of such a small hypothetical data set is in Table 6.1. The

TABLE 6.1   Small-Sample Data for Illustration of Analysis of Covariance

	Groups					
	Treatment 1		Treatment 2		Control	
	Pre	Post	Pre	Post	Pre	Post
	85	100	86	92	90	95
	80	98	82	99	87	80
	92	105	95	108	78	82
Sums	257	303	263	299	255	257

IV is type of treatment given to a sample of nine learning-disabled children. Three children are assigned to one of two treatment groups or to a control group, so that sample size of each group is three. For each of the nine children, two scores are measured, CV and DV. The CV is a pretest score on the reading subtest of the Wide Range Achievement Test (WRAT-A), measured before the study begins. The DV is a posttest score on the same test measured at the end of the study. The pretest score on the WRAT-A is the CV because the goal of ANCOVA here is to adjust reading achievement scores after treatment (the DV) for individual differences in achievement before the scores can be affected by the treatment. This adjustment is expected to reduce variability in posttest scores within each of the treatment groups, resulting in a more powerful test of treatment differences.

The research question is: Does differential treatment of learning-disabled children affect reading scores, after adjusting for differences in the children's prior reading ability? The sample size is, of course, inadequate for a realistic test of the research question but is convenient for illustration of the techniques in ANCOVA. The reader is encouraged to follow this example with hand calculations. Computer analyses using two popular programs follow this section.

6.4.1   Sums of Squares and Cross Products

Equations for ANCOVA are an extension of those for ANOVA, as discussed in Chapter 3. Averaged squared deviations from means—variances—are partitioned into variance associated with different levels of the IV (between-groups variance) and variance associated with differences in scores within groups (unaccounted for or error variance). Variance is partitioned by summing and squaring differences between scores and various means.

$$\sum_i \sum_j (Y_{ij} - GM)^2 = n \sum_j (\overline{Y}_j - GM)^2 + \sum_i \sum_j (Y_{ij} - \overline{Y}_j)^2 \tag{6.1}$$

or

$$SS_{total} = SS_{bg} + SS_{wg}$$

The total sum of squared differences between scores on Y (the DV) and the grand mean (GM) is partitioned into two components: sum of squared differences between group means ($\overline{Y}_j$) and the grand mean (i.e., systematic or between-groups variability); and sum of squared differences between individual scores (Y_{ij}) and their respective group means (i.e., error).

In ANCOVA, there are two additional partitions. First, the differences between CV scores and their GM are partitioned into between- and within-groups sums of squares:

$$SS_{total(x)} = SS_{bg(x)} + SS_{wg(x)}$$ (6.2)

The total sum of squared differences on the CV (X) is partitioned into differences between groups and differences within groups.

Similarly, the covariance (the linear relationship between the DV and the CV) is partitioned into sums of products associated with covariance between groups and sums of products associated with covariance within groups.

$$SP_{total} = SP_{bg} + SP_{wg}$$ (6.3)

A *sum of squares* involves taking deviations of scores from means (e.g., $X_{ij} - \overline{X}_j$ or $Y_{ij} - \overline{Y}_j$), squaring them, and then summing the squares over all subjects; a *sum of products* involves taking two deviations from the same subject (e.g., both $X_{ij} - X_j$ and $Y_{ij} - Y_j$), multiplying them together (instead of squaring), and then summing the products over all subjects (Section 1.6.4). As discussed in Chapter 3, the means that are used to produce the deviations are different for the different sources of variance in the research design.

The partitions for the CV (Equation 6.2) and the partitions for the association between the CV and DV (Equation 6.3) are used to adjust the sums of squares for the DV according to the following equations:

$$SS'_{bg} = SS_{bg} - \left[\frac{(SP_{bg} + SP_{wg})^2}{SS_{bg(x)} + SS_{wg(x)}} - \frac{(SP_{wg})^2}{SS_{wg(x)}} \right]$$ (6.4)

The adjusted between-groups sum of squares (SS'_{bg}) is found by subtracting from the unadjusted between-groups sum of squares a term based on sums of squares associated with the CV, X, and sums of products for the linear relationship between the DV and the CV.

$$SS'_{wg} = SS_{wg} - \frac{(SP_{wg})^2}{SS_{wg(x)}}$$ (6.5)

The adjusted within-groups sum of squares (SS'_{wg}) is found by subtracting from the unadjusted within-groups sum of squares a term based on within-groups sums of squares and products associated with the CV and with the linear relationship between the DV and the CV.

This can be expressed in an alternate form. The adjustment for each score consists of subtracting from the deviation of that score from the grand mean a value that is based on the deviation of the corresponding CV from the grand mean on the CV, weighted by the regression coefficient for predicting the DV from the CV. Symbolically, for an individual score:

$$(Y - Y') = (Y - GM_y) - B_{y \cdot x}(X - GM_x) \tag{6.6}$$

The adjustment for any subject's score $(Y - Y')$ is obtained by subtracting from the unadjusted deviation score $(Y - GM_y)$ the individual's deviation on the CV $(X - GM_x)$ weighted by the regression coefficient, $B_{y \cdot x}$.

Once adjusted sums of squares are found, mean squares are found as usual by dividing by appropriate degrees of freedom. The only difference in degrees of freedom between ANOVA and ANCOVA is that in ANCOVA the error degrees of freedom are reduced by one for each CV because a degree of freedom is used up in estimating each regression coefficient.

For computational convenience, raw score equations rather than deviation equations are provided in Table 6.2 for Equations 6.4 and 6.5. Note that these equations apply only to equal-n designs.

When applied to the data in Table 6.1, the six sums of squares and products are as follows:

$$SS_{bg} = \frac{(303)^2 + (299)^2 + (257)^2}{3} - \frac{(859)^2}{(3)(3)} = 432.889$$

$$SS_{wg} = (100)^2 + (98)^2 + (105)^2 + (92)^2 + (99)^2 + (108)^2 + (95)^2 + (80)^2 + (82)^2$$
$$- \frac{(303)^2 + (299)^2 + (257)^2}{3} = 287.333$$

$$SS_{bg(x)} = \frac{(257)^2 + (263)^2 + (255)^2}{3} - \frac{(775)^2}{(3)(3)} = 11.556$$

$$SS_{wg(x)} = (85)^2 + (80)^2 + (92)^2 + (86)^2 + (82)^2 + (95)^2$$
$$+ (90)^2 + (87)^2 + (78)^2 - \frac{(257)^2 + (263)^2 + (255)^2}{3} = 239.333$$

$$SP_{bg} = \frac{(257)(303) + (263)(299) + (255)(257)}{3} - \frac{(775)(859)}{(3)(3)} = 44.889$$

$$SP_{wg} = (85)(100) + (80)(98) + (92)(105) + (86)(92) + (82)(99)$$
$$+ (95)(108) + (90)(105) + (87)(80) + (78)(82)$$
$$- \frac{(257)(303) + (263)(299) + (255)(257)}{3} = 181.667$$

TABLE 6.2 Computation Equations for Sums of Squares and Cross-Products in One-Way Between-Subjects Analysis of Covariance

Source	Sum of Squares for Y (DV)	Sum of Squares for X (covariate)	Sum of Products
Between groups	$SS_{bg} = \dfrac{\sum\limits^{k}\left(\sum\limits^{n} Y\right)^2}{n} - \dfrac{\left(\sum\limits^{k}\sum\limits^{n} Y\right)^2}{kn}$	$SS_{bg(x)} = \dfrac{\sum\limits^{k}\left(\sum\limits^{n} X\right)^2}{n} - \dfrac{\left(\sum\limits^{k}\sum\limits^{n} X\right)^2}{kn}$	$SP_{bg} = \dfrac{\sum\limits^{k}\left(\sum\limits^{n} Y\right)\left(\sum\limits^{n} X\right)}{n} - \dfrac{\left(\sum\limits^{k}\sum\limits^{n} Y\right)\left(\sum\limits^{k}\sum\limits^{n} X\right)}{kn}$
Within groups	$SS_{wg} = \sum\limits^{k}\sum\limits^{n} Y^2 - \dfrac{\sum\limits^{k}\left(\sum\limits^{n} Y\right)^2}{n}$	$SS_{wg(x)} = \sum\limits^{k}\sum\limits^{n} X^2 - \dfrac{\sum\limits^{k}\left(\sum\limits^{n} X\right)^2}{n}$	$SP_{wg} = \sum\limits^{k}\sum\limits^{n} (XY) - \dfrac{\sum\limits^{k}\left(\sum\limits^{n} Y\right)\left(\sum\limits^{n} X\right)}{n}$

Note: k = number of groups; n = number of subjects per group.

These values are conveniently summarized in a sum-of-squares and cross-products matrix (cf. Chapter 1). For the between-groups sums of squares and cross-products,

$$\mathbf{S}_{bg} = \begin{bmatrix} 11.556 & 44.889 \\ 44.889 & 432.889 \end{bmatrix}$$

The first entry (first row and first column) is the sum of squares for the CV and the last (second row, second column) is the sum of squares for the DV; the sum of products is shown in the off-diagonal portion of the matrix. For the within-groups sums of squares and cross-products, arranged similarly,

$$\mathbf{S}_{wg} = \begin{bmatrix} 239.333 & 181.667 \\ 181.667 & 287.333 \end{bmatrix}$$

From these values, the adjusted sums of squares are found as per Equations 6.4 and 6.5.

$$SS'_{bg} = 432.889 - \left[\frac{(44.889 + 181.667)^2}{11.556 + 239.333} - \frac{(181.667)^2}{239.333} \right] = 366.202$$

$$SS'_{wg} = 287.333 - \frac{(181.667)^2}{239.333} = 149.438$$

6.4.2 Significance Test and Effect Size

These values are entered into a source table such as Table 6.3. Degrees of freedom for between-groups variance are $k - 1$, and for within-groups variance $N - k - c$. (N = total sample size, k = number of levels of the IV, and c = number of CVs.)

As usual, mean squares are found by dividing sums of squares by appropriate degrees of freedom. The hypothesis that there are no differences among groups is tested by the F ratio formed by dividing the adjusted mean square between groups by the adjusted mean square within groups.

$$F = \frac{183.101}{29.888} = 6.13$$

From a standard F table, we find that the obtained F of 6.13 exceeds the critical F of 5.79 at $\alpha = .05$ with 2 and 5 df. We therefore reject the null hypothesis of no change in WRAT reading scores associated with the three treatment levels, after adjustment for pretest reading scores.

TABLE 6.3 Analysis of Covariance for Data of Table 6.1

Source of Variance	Adjusted SS	df	MS	F
Between groups	366.202	2	183.101	6.13*
Within groups	149.439	5	29.888	

*$p < .05$.

TABLE 6.4   Analysis of Variance for Data of Table 6.1

Source of Variance	SS	df	MS	F
Between groups	432.889	2	216.444	4.52
Within groups	287.333	6	47.889	

The effect size then is assessed using η^2.

$$\eta^2 = \frac{SS'_{bg}}{SS'_{bg} + SS'_{wg}} \tag{6.7}$$

For the sample data,

$$\eta^2 = \frac{366.202}{366.202 + 149.438} = .71$$

We conclude that 71% of the variance in the adjusted DV scores (WRAT-R) is associated with treatment. Confidence intervals around η^2, a form of R^2, are available through the Smithson (2003) SPSS or SAS files. Section 6.6.2 demonstrates use of the Smithson's SAS program for finding effect sizes and their confidence limits. Section 7.6.2 demonstrates use of Smithson's SPSS program for finding effect sizes and their confidence limits. Using NoncF3.sps, the 95% confidence interval for partial η^2 ranges from 0 to .83.

ANOVA for the same data appears in Table 6.4. Compare the results with those of ANCOVA in Table 6.3. ANOVA produces larger sums of squares, especially for the error term. There is also one more degree of freedom for error because there is no CV. However, in ANOVA the null hypothesis is retained while in ANCOVA it is rejected. Thus, use of the CV has reduced the "noise" in the error term for this example.

ANCOVA extends to factorial and within-subjects designs (Section 6.5.4.1), unequal n (Section 6.5.4.2), and multiple CVs (Section 6.5.2). In all cases, the analysis is done on adjusted, rather than raw, DV scores.

6.4.3   Computer Analyses of Small-Sample Example

Tables 6.5 and 6.6 demonstrate analyses of covariance of this small data set using SPSS GLM and SAS GLM. Minimal output is requested for each of the programs, although much more is available upon request.

In SPSS GLM UNIANOVA (General Factorial in the menu), the DV (POST) is specified in the ANOVA paragraph followed BY the IV (TREATMNT) WITH the CV (PRE) as seen in Table 6.5. Some default syntax generated by the menu system (but not necessary for analysis) is not shown here.

The output is an ANOVA source table with some extraneous sources of variation. The only sources of interest are PRE, TREATMNT, and Error (S/A). Sums of squares for PRE and

TABLE 6.5   Syntax and Selected SPSS GLM Output for Analysis of Covariance on Sample Data in Table 6.1

```
UNIANOVA
 POST BY TREATMNT WITH PRE
 /METHOD = SSTYPE(3)
 /INTERCEPT = INCLUDE
 /CRITERIA = ALPHA(.05)
 /DESIGN = PRE TREATMNT
```

Univariate Analysis of Variance

Tests of Between-Subjects Effects

Dependent Variable: POST

Source	Type III Sum of Squares	df	Mean Square	F	Sig.
Corrected Model	570.784[a]	3	190.261	6.366	.037
Intercept	29.103	1	29.103	.974	.369
PRE	137.895	1	137.895	4.614	.084
TREATMNT	366.201	2	183.101	6.126	.045
Error	149.439	5	29.888		
Total	82707.000	9			
Corrected Total	720.222	8			

[a]R Squared = .793 (Adjusted R Squared = .668)

TREATMNT are adjusted for each other when Type III Sums of Squares (the default in GLM) are used. The Error sum of squares also is adjusted (compare with Table 6.3). The Corrected Total is inappropriate for calculating η^2; instead form an adjusted total by summing the adjusted SS for TREATMNT and Error (Eq. 6.7). The R Squared and Adjusted R Squared shown here are *not* measures of effect size for treatment.

SAS GLM requires the IV to be specified in the class instruction. Then a model instruction is set up, with the DV on the left side of the equation, and the IV and CV on the right side as seen in Table 6.6. Two source tables are provided, one for the problem as a regression and the other for the problem as a more standard ANOVA, both with the same Error term.

The first test in the regression table asks if there is significant prediction of the DV by the combination of the IV and the CV. The output resembles that of standard multiple regression (Chapter 5) and includes R-Square (also inappropriate), the unadjusted Mean on POST (the DV), the Root MSE (square root of the error mean square), and Coeff Var the coefficient of variation (100 times the Root MSE divided by the mean of the DV).

In the ANOVA-like table, two forms of tests are given by default, labeled Type I SS and Type III SS. The sums of squares for TREATMNT are the same in both forms because both adjust treatment for the CV. In Type III SS the sum of squares for the CV (PRE) is adjusted

**TABLE 6.6   Syntax and SAS GLM Output for Analysis of Covariance
on Sample Data in Table 6.1**

```
proc glm data=SASUSER.SS_ANCOV;
    class TREATMNT;
    model POST = PRE TREATMNT
run;
```

General Linear Models Procedure
Class Level Information

Class	Levels	Values
TREATMNT	3	1 2 3

Number of Observations Read   9
Number of Observations Used   9

Dependent Variable: POST

Source	DF	Sum of Squares	Mean Square	F Value	Pr > F
Model	3	570.7835036	190.2611679	6.37	0.0369
Error	5	149.4387187	29.8877437		
Corrected Total	8	720.2222222			

R-Square	Coeff Var	Root MSE	POST Mean
0.792510	5.727906	5.466968	95.444444

Source	DF	Type I SS	Mean Square	F Value	Pr > F
PRE	1	204.5822754	204.5822754	6.85	0.0473
TREATMNT	2	366.2012282	183.1006141	6.13	0.0452

Source	DF	Type III SS	Mean Square	F Value	Pr > F
PRE	1	137.8946147	137.8946147	4.61	0.0845
TREATMNT	2	366.2012282	183.1006141	6.13	0.0452

for treatment, but the sum of squares for the CV is not adjusted for the effect of treatment in `Type I SS`.

6.5   Some Important Issues

6.5.1   Choosing Covariates

One wants to use an optimal set of CVs if several are available. When too many CVs are used and they are correlated with each other, a point of diminishing returns in adjustment of the DV is quickly reached. Power is reduced because numerous correlated CVs subtract degrees of freedom from the error term while not removing commensurate sums of squares for error. Preliminary analysis of the CVs improves chances of picking a good set.

Statistically, the goal is to identify a small set of CVs that are uncorrelated with each other but correlated with the DV. Conceptually, one wants to select CVs that adjust the DV for predictable but unwanted sources of variability. It may be possible to pick the CVs on theoretical grounds or on the basis of knowledge of the literature regarding important sources of variability that should be controlled.

If theory is unavailable or the literature is insufficiently developed to provide a guide to important sources of variability in the DV, statistical considerations assist the selection of CVs. One strategy is to look at correlations among CVs and select one from among each of those groups of potential CVs that are substantially correlated with each other, perhaps by choosing the one with the highest correlation with the DV. Alternatively, stepwise regression may be used to pick an optimal set.

If N is large and power is not a problem, it may still be worthwhile to find a small set of CVs for the sake of parsimony. Useless CVs are identified in the first ANCOVA run. Then further runs are made, each time eliminating CVs, until a small set of useful CVs is found. The analysis with the smallest set of CVs is reported, but mention is made in the Results section of the discarded CV(s) and the fact that the pattern of results did not change when they were eliminated.

6.5.2   Evaluation of Covariates

CVs in ANCOVA can themselves be interpreted as predictors of the DV. From a sequential regression perspective (Chapter 5), each CV is a high-priority, continuous IV with remaining IVs (main effects and interactions) evaluated after the relationship between the CV and the DV is removed.

Significance tests for CVs assess their utility in adjusting the DV. If a CV is significant, it provides adjustment of the DV scores. For the example in Table 6.5, the CV, PRE, does not provide significant adjustment to the DV, POST, with $F(1, 5) = 4.61$, $p > .05$. PRE is interpreted in the same way as any IV in multiple regression (Chapter 5).

With multiple CVs, all CVs enter the multiple regression equation at once and, as a set, are treated as a standard multiple regression (Section 5.5.1). Within the set of CVs, the significance of each CV is assessed as if it entered the equation last; only the unique relationship between the CV and the DV is tested for significance after overlapping variability with other CVs, in their relationship with the DV, is removed. Therefore, although a CV may be significantly correlated with the DV when considered individually, it may add no significant adjustment to the DV when considered last. When interpreting the utility of a CV, it is necessary to consider correlations among CVs, correlations between each CV and the DV, and significance levels for each CV as reported in ANCOVA source tables. Evaluation of CVs is demonstrated in Section 6.6.

Unstandardized regression coefficients, provided by most canned computer programs on request, have the same meaning as regression coefficients described in Chapter 5. However, with unequal n, interpretation of the coefficients depends on the method used for adjustment. When Method 1 (standard multiple regression—see Table 6.10 and Section 6.5.4.2) is used, the significance of the regression coefficients for CVs is assessed as if the CV entered the regression equation after all main effects and interactions. With other methods, however, CVs enter the equation first, or after main effects but before interactions. The coefficients are evaluated at whatever point the CVs enter the equation.

6.5.3   Test for Homogeneity of Regression

The assumption of homogeneity of regression is that the slopes of the regression of the DV on the CV(s) (the regression coefficients or *B* weights as described in Chapter 5) are the same for all cells of a design. Both homogeneity and heterogeneity of regression are illustrated in Figure 6.2. Because the average of the slopes for all cells is used to adjust the DV, it is assumed that the slopes do not differ significantly either from one another or from a single estimate of the population value. If the null hypothesis of equality among slopes is rejected, the analysis of covariance is inappropriate and an alternative strategy as described in Sections 6.3.2.7 and 6.5.5 should be used.

Hand calculation of the test of homogeneity of regression (see, for instance, Keppel & Wickens, 2004, or Tabachnick and Fidell, 2007) is extremely tedious. The most straightforward program for testing homogeneity of regression is SPSS MANOVA (available only in syntax mode.)

Special language is provided for the test in SPSS MANOVA, as shown in Table 6.7. The inclusion of the IV by CV interaction (**PRE BY TREATMNT**) as the last effect in the **DESIGN** instruction, after the CV and the IV, provides the test for homogeneity of regression. Placing this effect last and requesting **METHOD=SEQUENTIAL** ensures that the test for the interaction is adjusted for the CV and the IV. The **ANALYSIS** instruction identifies **POST** as the DV. The test for **PRE BY TREATMNT** shows that there is no violation of homogeneity of regression: $p = .967$.

Programs based on the general linear model (SPSS and SAS GLM) test for homogeneity of regression by evaluating the CV(s) by IV(s) interaction (e.g., PRE by TREATMNT) as the last effect entering a model (Chapter 5).

6.5.4   Design Complexity

Extension of ANCOVA to factorial between-subjects designs is straightforward as long as sample sizes within cells are equal. Partitioning of sources of variance follows ANOVA (cf. Chapter 3) with

TABLE 6.7   SPSS MANOVA Syntax and Selected Output for Homogeneity of Regression

```
MANOVA
 POST BY TREATMNT(1, 3) WITH PRE
 /PRINT=SIGNIF(BRIEF)
 /ANALYSIS = POST
 /METHOD=SEQUENTIAL
 /DESIGN PRE TREATMNT PRE BY TREATMNT.
```

Manova

```
* * * * * * A n a l y s i s   o f   V a r i a n c e -- design 1 * * * * * *
```

Tests of Significance for POST using SEQUENTIAL Sums of Squares

Source of Variation	SS	DF	MS	F	Sig of F
WITHIN+RESIDUAL	146.15	3	48.72		
PRE	204.58	1	204.58	4.20	.133
TREATMNT	366.20	2	183.10	3.76	.152
PRE BY TREATMNT	3.29	2	1.64	.03	.967

"subjects nested within cells" as the simple error term. Sums of squares are adjusted for the average association between the CV and the DV in each cell, just as they are for the one-way design demonstrated in Section 6.4.

There are, however, two major design complexities that arise: within-subjects IVs, and unequal sample sizes in the cells of a factorial design. And, just as in ANOVA, contrasts are appropriate for a significant IV with more than two levels and assessment of effect size is appropriate for all effects.

6.5.4.1 Within-Subjects and Mixed Within-Between Designs

Just as a DV can be measured once or repeatedly, so also a CV can be measured once or repeatedly. In fact, the same design may contain one or more CVs measured once and other CVs measured repeatedly.

A CV that is measured only once does not provide adjustment to a within-subjects effect because it provides the same adjustment (equivalent to no adjustment) for each level of the effect. The CV does, however, adjust any between-subjects effects in the same design. Thus ANCOVA with one or more CVs measured once is useful in a design with both between- and within-subjects effects for increasing the power of the test of between-subjects IVs. Both between- and within-subjects effects are adjusted for CVs measured repeatedly.

ANOVA and ANCOVA with repeated measures (within-subjects effects) are more complicated than designs with only between-subjects effects. One complication is that some programs cannot handle repeatedly measured CVs. A second complication is the assumption of sphericity, as discussed in Chapter 8 (which also discusses alternatives in the event of violation of the assumption). A third complication (more computational than conceptual) is development of separate error terms for various segments of within-subjects effects.

6.5.4.1.1 Same Covariate(s) for all Cells

There are two approaches to analysis of designs with at least one within-subjects IV, the traditional approach and the general linear model approach. In the traditional approach, the within-subjects effects in the designs are not adjusted for CVs. In the GLM approach, the repeated measures are adjusted by the interaction of the CV(s) with the within-subjects effects. SPSS MANOVA takes the traditional approach to ANCOVA.[3] Programs labeled GLM in SAS and SPSS use the general linear model approach. All programs provide adjusted marginal and cell means. Tabachnick and Fidell (2007) show examples for both these approaches. It is not clear that the GLM strategy generally makes much sense or enhances power in ANCOVA. There is usually no *a priori* reason to expect a different relationship between the DV and CV for different levels of the within-subjects IV(s). If no such relationship is present, the loss of degrees of freedom for estimating this effect could more than offset the reduction in sum of squares for error and result in a less powerful test of the within-subjects effects. When using a GLM program, you could rerun the analysis without the CV to obtain the traditional within-subjects portion of the source table.

6.5.4.1.2 Varying Covariate(s) over Cells

There are two common designs where covariates differ for cells: matched randomized blocks designs where cases in the cells of a within-subjects IV actually are different cases (cf. Section

---

[3]Covariates that do not vary over trials in within-subjects designs are listed in parentheses after WITH (see Table 6.7).

3.2.3), and designs where the covariate is reassessed prior to administration of each level of the within-subjects IV(s). In addition, a stepdown analysis uses higher priority DVs as covariates, which change for each level of the within-subjects IV. When the covariates differ for levels of the within-subjects IV(s), they are potentially useful in enhancing power for all effects.

In SPSS MANOVA, covariates that vary over within-subjects levels are listed without parentheses after the WITH instruction (see Table 6.7). However, GLM programs and SPSS MIXED MODELS have no special syntax for specifying a covariate that changes with each level of the within-subjects IV. Instead, the problem is set up as a randomized-groups design with one IV representing cases where measurements for each trial are on a separate line. For this example, each case has two lines, one for each trial. Trial number, covariate score, and DV score are on each line. The within-subjects design is simulated by considering both cases and trials randomized-groups IVs. Table 6.8 shows a hypothetical data set arranged for this analysis.

Table 6.9 shows SAS GLM setup and output for this analysis. Because no interactions are requested, the CASE by T interaction is the error term. Type III sums of squares (s s3) are requested to limit output.

Used this way, the SAS GLM source table provides a test of CASE as an IV, usually not available in a within-subjects design. However, there is no test for sphericity using this setup, relevant if there are more than two levels of the within-subjects IV. Therefore, some alternative to a univariate within-subjects analysis, such as trend analysis, is warranted if there is reason to believe that sphericity might be violated. This issue is discussed in detail in Section 8.5.1.

TABLE 6.8   Hypothetical Data for Within-Subjects ANCOVA with Varying Covariate, One-Line-Per-Trial Setup

CASE	T	X	Y
1	1	4	9
1	2	3	15
2	1	8	10
2	2	6	16
3	1	13	14
3	2	10	20
4	1	1	6
4	2	3	9
5	1	8	11
5	2	9	15
6	1	10	10
6	2	9	9
7	1	5	7
7	2	8	12
8	1	9	12
8	2	9	20
9	1	11	14
9	2	10	20

TABLE 6.9 Within-Subjects ANCOVA: Varying Covariate through SAS GLM

```
proc glm data=SASUSER.SPLTPLOT;
  class T CASE;
  model Y = T CASE X / ss3;
  lsmeans T/;
  means T;
run;
```

Class Level Information

Class	Levels	Values
CASE	9	1 2 3 4 5 6 7 8 9
T	2	1 2

Number of Observations Read   18
Number of Observations Used   18

The GLM Procedure

Dependent Variable: Y

Source	DF	Sum of Squares	Mean Square	F Value	Pr > F
Model	10	295.5359231	29.5535923	7.93	0.0059
Error	7	26.07518880	3.7250269		
Corrected Total	17	321.6111111			

R-Square	Coeff Var	Root MSE	Y Mean
0.918923	15.17056	1.930033	12.72222

Source	DF	Type III SS	Mean Square	F Value	Pr > F
T	1	99.1581454	99.1581454	26.62	0.0013
CASE	8	105.1901182	13.1487648	3.53	0.0569
X	1	0.7025898	0.7025898	0.19	0.6771

TABLE 6.9 Continued

```
               The GLM Procedure
             Least Squares Means

            T       Y LSMEAN
            1      10.3575606
            2      15.0868839
             The GLM Procedure
```

Level of		------------Y------------		------------X------------	
T	N	Mean	Std Dev	Mean	Std Dev
1	9	10.3333333	2.78388218	7.66666667	3.74165739
2	9	15.1111111	4.42844342	7.44444444	2.78886676

6.5.4.2 Unequal Sample Sizes

Two problems arise in a factorial design if cells have unequal numbers of scores. First, there is ambiguity regarding a marginal mean from cells with unequal n. Is the marginal mean the mean of the means, or is the marginal mean the mean of the scores? Second, the total sums of squares for all effects is greater than SS_{total} and there is ambiguity regarding assignment of overlapping sums of squares to sources. The design has become nonorthogonal and tests for main effects and interactions are no longer independent (cf. Section 3.2.5.3). The problem generalizes directly to ANCOVA.

If equalizing cell sizes by random deletion of cases is undesirable, there are a number of strategies for dealing with unequal n. The choice among strategies depends on the type of research. Of the three major methods described by Overall and Spiegel (1969), Method 1 is usually appropriate for experimental research, Method 2 for survey or nonexperimental research, and Method 3 for research in which the researcher has clear priorities for effects.

Table 6.10 summarizes research situations calling for different methods and notes some of the jargon used by various sources. As Table 6.10 reveals, there are a number of ways of viewing these methods; the terminology associated with these viewpoints by different authors is quite different and, sometimes, seemingly contradictory. Choice of method affects adjusted (estimated) means as well as significance tests of effects.

Differences in these methods are easiest to understand from the perspective of multiple regression (Chapter 5). Method 1 is like standard multiple regression with each main effect and interaction assessed after adjustment is made for all other main effects and interactions, as well as for CVs. The same hypotheses are tested as in the unweighted-means approach where each cell mean is given equal weight regardless of its sample size. Interactions are listed after their constituent main effects even in Method 1, because order of listing may affect parameter estimates. This is the recommended approach for experimental research unless there are reasons for doing otherwise.

Reasons include a desire to give heavier weighting to some effects than others because of importance, or because unequal population sizes have resulted from treatments that occur naturally. (If a design intended to be equal-n ends up grossly unequal, the problem is not type of adjustment but, more seriously, differential dropout.)

Method 2 imposes a hierarchy of testing effects where main effects are adjusted for each other and for CVs, while interactions are adjusted for main effects, for CVs, and for same- and lower-level

TABLE 6.10   Terminology for Strategies for Adjustment for Unequal Cell Size

Research Type	Overall and Spiegel (1969)	SPSS GLM	SPSS MANOVA	SAS
1. Experiments designed to be equal-*n*, with random dropout. All cells equally important.	Method 1	METHOD= SSTYPE(3) —(default) METHOD= SSTYPE(4)[a]	METHOD= UNIQUE— (default)	TYPE III and TYPE IV[a]
2. Nonexperimental research in which sample sizes reflect importance of cells. Main effects have equal priority.[b]	Method 2	METHOD= SSTYPE(2)	METHOD= EXPERIMENTAL	TYPE II
3. Like number 2 above, except all effects have unequal priority.	Method 3	METHOD= SSTYPE(1)	METHOD= SEQUENTIAL	TYPE I

[a]Types III and IV differ only if there are missing cells.

[b]The programs take different approaches to adjustment for interaction effects.

interactions. (The SAS implementation also makes adjustments for some higher order effects.) The order of priority for adjustment emphasizes main effects over interactions and lower-order interactions over higher-order interactions. This is labeled METHOD=EXPERIMENTAL in SPSS MANOVA although it is normally used in nonexperimental work when there is a desire to weight marginal means by the sizes of samples in cells from which they are computed. The adjustment assigns heavier weighting to cells with larger sample sizes when computing marginal means and lower-order interactions. Method 3 allows the researcher to set up the sequence of adjustment of CVs, main effects, and interactions.

All programs in the reviewed packages perform ANCOVA with unequal sample sizes. SAS GLM and SPSS MANOVA and GLM provide for design complexity and flexibility in adjustment for unequal *n*.

Some researchers advocate use of Method 1 always. Because Method 1 is the most conservative, you are unlikely to draw criticism by using it. On the other hand, you risk loss of power with a nonexperimental design and perhaps interpretability and generalizability by treating all cells as if they had equal sample sizes.

6.5.4.3   Specific Comparisons and Trend Analysis

If there are more than two levels of an IV, the finding of a significant main effect or interaction in ANOVA or ANCOVA is often insufficient for full interpretation of the effects of IV(s) on the DV. The omnibus *F* test of a main effect or interaction gives no information as to which means are significantly different from which other means. With a qualitative IV (whose levels differ in kind) the researcher generates contrast coefficients to compare some adjusted mean(s) against other adjusted means. With a quantitative IV (whose levels differ in amount rather than in kind), trend coefficients are used to see if adjusted means of the DV follow a linear or quadratic pattern, say, over increasing levels of the IV.

As with ANOVA (Chapter 3), specific comparisons or trends can be either planned as part of the research design, or tested post hoc as part of a data-snooping procedure after omnibus analyses are completed. For planned comparisons, protection against inflated Type I error is achieved by running a small number of comparisons instead of omnibus F (where the number does not exceed the available degrees of freedom) and by working with an orthogonal set of coefficients. For post hoc comparisons, the probability of Type I error increases with the number of possible comparisons, so adjustment is made for inflated α error.

Comparisons are achieved by specifying coefficients and running analyses based on these coefficients. The comparisons can be simple (between two marginal or cell means with the other means left out) or complicated (where means for some cells or margins are pooled and contrasted with means for other cells or margins, or where coefficients for trend—linear, quadratic, etc.—are used). The difficulty of conducting comparisons depends on the complexity of the design and the effect to be analyzed. Comparisons are more difficult if the design has within-subjects IVs, either alone or in combination with between-subjects IVs, where problems arise from the need to develop a separate error term for each comparison. Comparisons are more difficult for interactions than for main effects because there are several approaches to comparisons for interactions. Some of these issues are reviewed in Section 8.5.2.

Pairwise tests of adjusted means are also available through options in SPSS MANOVA and SAS GLM. In addition, SAS GLM provides several tests that incorporate post hoc adjustments, such as Bonferroni.

Table 6.11 shows syntax and location of output for orthogonal contrasts and pairwise comparisons with a request for Tukey adjustment though SPSS MANOVA and GLM, and SAS GLM.

The orthogonal comparisons are based on coefficients for testing the linear and quadratic trends of TREATMNT, respectively.[4] (Because the IV in this sample is not quantitative, trend analysis is inappropriate; trend coefficients are used here for illustration only.) Note that the first (linear) comparison also is a "pairwise" comparison; the first (treatment 1) group is compared with the third (control) group.

The printed output of all the programs assumes that orthogonal comparisons are planned. Otherwise, some adjustment needs to be made by hand for inflation of Type I error rate when post hoc tests are done. Equation 3.24 shows the Scheffé adjustment for a single IV. Obtained F is compared with an adjusted critical F produced by multiplying the tabled F value (in this case 5.79 for 2 and 5 df, $\alpha = .05$) by the degrees of freedom associated with the number of cells, or $k - 1$. For this example, the adjusted critical F value is $2(5.79) = 11.58$, and the difference between the first treatment group and the control group ($F = 11.00$) fails to reach statistical significance, although it did so as a planned comparison.

In designs with more than one IV, the size of the Scheffé adjustment to critical F for post hoc comparisons depends on the degrees of freedom for the effect being analyzed. For a two-way design, for example, with IVs A and B, adjusted critical F for A is tabled critical F for A multiplied by $(a - 1)$ (where a is the number of levels of A); adjusted critical F for B is tabled critical F for B multiplied

---

[4]Coefficients for orthogonal polynomials are available in most standard ANOVA texts such as Tabachnick and Fidell (2007), Keppel & Wickens (2004), or Brown et al. (1991).

TABLE 6.11 Syntax for Orthogonal Comparisons and Pairwise Comparisons with Tukey Adjustment

Type of Comparison	Program	Syntax	Section of Output	Name of Effect
Orthogonal	SPSS GLM	UNIANOVA POST BY TREATMNT WITH PRE /METHOD = SSTYPE(3) /INTERCEPT = INCLUDE /CRITERIA = ALPHA(.05) /LMATRIX "LINEAR" TREATMNT 1 0 -1 /LMATRIX "QUADRATIC" TREATMNT 1 -2 1 /DESIGN = PRE TREATMNT.	**Custom Hypothesis Tests:** **Test Results**	Contrast

	SPSS MANOVA	MANOVA
		POST BY TREATMNT(1,3) WITH PRE
		/METHOD = UNIQUE
		/PARTITION (TREATMNT)
		/CONTRAST(TREATMNT)=SPECIAL(1 1 1,
		                                         1 0 -1,
		                                         1 -2 1)
		
		/ANALYSIS POST
		/DESIGN = PRE TREATMNT(1) TREATMNT(2).
			Tests of significance for POST using UNIQUE sums of squares	TREATMNT(1), TREATMNT(2)

	SAS GLM	PROC GLM DATA=SASUSER.SS_ANCOV ;
		    CLASS TREATMNT;
		    MODEL POST = TREATMNT PRE;
		    CONTRAST 'LINEAR' TREATMNT 1 0 -1;
		    CONTRAST 'QUADRATIC' TREATMNT 1 -2 1;
		run;
			Contrast	linear, quadratic

Pairwise with Tukey	PSS GLM	Tukey adjustment not available with covariates[a]		
	SPSS MANOVA	Pairwise comparisons with adjustments not available.		

	SAS GLM	PROC GLM DATA=SASUSER.SS_ANCOV ;
		    CLASS TREATMNT;
		    MODEL POST = TREATMNT PRE ;
		    LSMEANS TREATMNT / ADJUST=TUKEY P ;
		RUN;
			Least Squares Means for effect TREATMNT	i/j

[a]LSD, Sidak, and Bonferroni are available as adjustments for post hoc pairwise comparisons in the EMMEANS instruction.

by $(b - 1)$; adjusted critical F for the AB interaction is tabled critical F for the interaction multiplied by $(a - 1)(b - 1)$.[5]

If appropriate programs are unavailable, hand calculations for specific comparisons (including pairwise comparisons) are not particularly difficult, as long as sample sizes are equal for each cell. Equation 3.23 is for hand calculation of comparisons; to apply it, obtain the adjusted cell or marginal means and the error mean square from an omnibus ANCOVA program.

6.5.4.4 Effect Size

Once an effect is found to be statistically significant, the next logical question is: How important is the effect? It is becoming common now to report effect sizes and their confidence limits even when effects are *not* statistically significant. Importance is usually assessed as the percentage of variance in the DV that is associated with the IV. For one-way designs, the strength of association between an effect and DV (i.e., effect size) for adjusted sums of squares is found using Equation 6.7. For factorial designs, one uses an extension of Equation 6.7.

The numerator for η^2 is the adjusted sum of squares for the main effect or interaction being evaluated; the denominator is the total adjusted sum of squares. The total adjusted sum of squares includes adjusted sums of squares for all main effects, interactions, and error terms but does not include components for CVs or the mean, which are typically printed out by computer programs. To find the strength of association between an effect and the adjusted DV scores, then,

$$\eta^2 = \frac{SS'_{effect}}{SS'_{total}} \tag{6.8}$$

In multifactorial designs, the size of η^2 for a particular effect is, in part, dependent on the strength of other effects in the design. In a design where there are several main effects and interactions, η^2 for a particular effect is diminished because other effects increase the size of the denominator. An alternative method of computing η^2 (partial η^2) uses in the denominator only the adjusted sum of squares for the effect being tested and the adjusted sum of squares for the appropriate error term for that effect (see Chapter 3 for appropriate error terms).

$$\text{partial } \eta^2 = \frac{SS'_{effect}}{SS'_{effect} + SS'_{error}} \tag{6.9}$$

Confidence intervals around partial η^2 (equivalent to η^2 in a one-way design) are demonstrated in Section 6.6.2.1.

6.5.5 Alternatives to ANCOVA

Because of the stringent limitations to ANCOVA and potential ambiguity in interpreting results of ANCOVA, alternative analytical strategies are often sought. The availability of alternatives depends

---

[5]The adjusted critical F for an interaction is insufficient if a great many post hoc comparisons are undertaken; it should suffice, however, if a moderate number are performed. If a great many are undertaken, multiply by $ab - 1$ instead of $(a - 1)(b - 1)$.

on such issues as the scale of measurement of the CV(s) and the DV, the time that elapses between measurement of the CV and assignment to treatment, and the difficulty of interpreting results.

When the CV(s) and the DV are measured on the same scale, two alternatives are available: use of difference (change) scores and conversion of the pretest and posttest scores into a within-subjects IV. In the first alternative, the difference between a pretest score (the previous CV) and a posttest score (the previous DV) is computed for each subject and used as the DV in ANOVA. If the research question is phrased in terms of "change," then difference scores provide the answer. For example, suppose self-esteem is measured both before and after a year of either belly dance or aerobic dance classes. If difference scores are used, the research question is: Does one year of belly dance training change self-esteem scores more than participation in aerobic dance classes? If ANCOVA is used, the research question is: Do belly dance classes produce greater self-esteem than aerobic dance classes after adjustment for pretreatment differences in self-esteem?

A problem with change scores and mixed ANOVA is ceiling and floor effects (or, more generally, skewness). A change score (or interaction) may be small because the pretest score is very near the end of the scale and no treatment effect can change it very much, or it may be small because the effect of treatment is small—the result is the same in either case and the researcher is hard pressed to decide between them. Further, when the DV is skewed, a change in a mean also produces a change in the shape of the distribution and a potentially misleading significance test (Jamieson and Howk, 1992). Another problem with difference scores is their potential unreliability. They are less reliable than either the pre- or posttest scores, so they are not to be recommended unless used with a highly reliable test (Harlow, 2002). If either ANCOVA or ANOVA with change scores is possible, then, ANCOVA is usually the better approach when the data are skewed and transformations are not undertaken (Jamieson, 1999).

When CVs are measured on any continuous scale, other alternatives are available: randomized blocks and blocking. In the randomized-block design, subjects are matched into blocks—equated—on the basis of scores on what would have been the CV(s). Each block has as many subjects as the number of levels of the IV in a one-way design or number of cells in a larger design (cf. Section 3.2.3). Subjects within each block are randomly assigned to levels or cells of the IV(s) for treatment. In the analytic phase, subjects in the same block are treated as if they were the same person, in a within-subjects analysis.

Disadvantages to this approach are the strong assumption of sphericity of a within-subjects analysis and the loss of degrees of freedom for error without commensurate loss of sums of squares for error if the variables used to block are not highly related to the DV. In addition, implementation of the randomized-block design requires the added step of equating subjects before randomly assigning them to treatment, a step that may be inconvenient, if not impossible, in some applications.

Another alternative is use of blocking. Subjects are measured on potential CV(s) and then grouped according to their scores (e.g., into groups of high, medium, and low self-esteem on the basis of pretest scores). The groups of subjects become the levels of another full-scale IV that are crossed with the levels of the IV(s) of interest in factorial design. Interpretation of the main effect of the IV of interest is straightforward and variation due to the potential CV(s) is removed from the estimate of error variance and assessed as a separate main effect. Furthermore, if the assumption of homogeneity of regression would have been violated in ANCOVA, it shows up as an interaction between the blocking IV and the IV of interest.

Blocking has several advantages over ANCOVA and the other alternatives listed here. First, it has none of the assumptions of ANCOVA or within-subjects ANOVA. Second, the relationship between the potential CV(s) and the DV need not be linear (blocking is less powerful when the CV–DV relationship *is* linear); curvilinear relationships can be captured in ANOVA when three (or more) levels of an IV are analyzed. Blocking, then, is preferable to ANCOVA in many situations, and particularly for experimental, rather than correlational, research.

Blocking can also be expanded to multiple CVs. That is, several new IVs, one per CV, can be developed through blocking and, with some difficulty, crossed in factorial design. However, as the number of IVs increases, the design rapidly becomes very large and cumbersome to implement.

For some applications, however, ANCOVA is preferable to blocking. When the relationship between the DV and the CV is linear, ANCOVA is more powerful than blocking. And, if the assumptions of ANCOVA are met, conversion of a continuous CV to a discrete IV can result in loss of information. Finally, practical limitations may prevent measurement of potential CV(s) sufficiently in advance of treatment to accomplish random assignment of equal numbers of subjects to the cells of the design. When blocking is attempted after treatment, sample sizes within cells are likely to be highly discrepant, leading to the problems of unequal *n*.

In some applications, a combination of blocking and ANCOVA may turn out to be best. Some potential CVs are used to create new IVs, while others are analyzed as CVs.

A final alternative is multilevel modeling (MLM) (Chapter 15). MLM has no assumption of homogeneity of regression; heterogeneity is dealt with by creating a second level of analysis consisting of groups and specifying that groups may have different slopes (relationships between the DV and DVs) as well as different intercepts (means on the DV).

6.6 Complete Example of Analysis of Covariance

The research described in Appendix B, Section B.1, provides the data for this illustration of ANCOVA. The research question is whether attitudes toward drugs are associated with current employment status and/or religious affiliation. Files are ANCOVA.*.

Attitude toward drugs (ATTDRUG) serves as the DV, with increasingly high scores reflecting more favorable attitudes. The two IVs, factorially combined, are: current employment status (EMPLMNT) with two levels, (1) employed and (2) unemployed; and religious affiliation (RELIGION) with four levels, (1) none-or-other, (2) Catholic, (3) Protestant, and (4) Jewish.

In examining other data for this sample of women, three variables stand out that could be expected to relate to attitudes toward drugs and might obscure effects of employment status and religion. These variables are general state of physical health, mental health, and the use of psychotropic drugs. In order to control for the effects of these three variables on attitudes toward drugs, they are treated as CVs. CVs, then, are physical health (PHYHEAL), mental health (MENHEAL), and sum of all psychotropic drug uses, prescription and over-the-counter (PSYDRUG). For all three CVs, larger scores reflect increasingly poor health or more use of drugs.

The 2 × 4 analysis of covariance, then, provides a test of the effects of employment status, religion, and their interaction on attitudes toward drugs after adjustment for differences in physical health, mental health, and use of psychotropic drugs. Note that this is a form of ANCOVA in which no causal inference can be made.

6.6.1 Evaluation of Assumptions

These variables are examined with respect to practical limitations of ANCOVA as described in Section 6.3.2.

6.6.1.1   Unequal n and Missing Data

SAS MEANS provides an initial screening run to look at descriptive statistics for DV and CVs for the eight groups. Three women out of 465 failed to provide information on religious affiliation (not shown). Because RELIGION is one of the IVs for which cell sizes are unequal in any event, the three cases are dropped from analysis. Output for the DV and the three CVs is shown in Table 6.12 for the first two groups: RELIGION = 1 EMPLMNT = 1(employed women who report other or no religious affiliation) and RELIGION = 3 EMPLMNT = 2 (unemployed Protestant women). Sample sizes for the eight groups are in Table 6.13.

The cell-size approach (Method 3 of Section 6.5.4.2) for dealing with unequal *n* is chosen for this study. This method weights cells by their sample sizes, which, in this study, are meaningful because they represent population sizes for the groups. Religion is given higher priority to reflect its temporally prior status to employment.

6.6.1.2   Normality

Table 6.12 shows positive skewness for some variables. Because the assumption of normality applies to the sampling distribution of means, and not to the raw scores, skewness by itself poses no problem. With the large sample size and use of two-tailed tests, normality of sampling distributions of means is anticipated.

6.6.1.3   Linearity

There is no reason to expect curvilinearity considering the variables used and the fact that the variables, when skewed, are all skewed in the same direction. Had there been reason to suspect curvilinearity, within-group scattergrams would have been examined through a SAS PLOT run.

6.6.1.4   Outliers

The maximum values in the SAS MEANS run of Table 6.13 show that, although no outliers are evident for the DV, several cases are univariate outliers for two of the CVs, PHYHEAL and PSYDRUG. Note, for example, that $z = (43 - 5.096)/8.641 = 4.39$ for the largest PSYDRUG score among unemployed protestant women. Positive skewness is also visible for these variables.

To facilitate the decision between transformation of variables and deletion of outliers, separate regression analyses are run on the eight groups through SAS REG, with a request for the h (leverage) statistic. Critical χ^2 at $\alpha = .001$ with 3 covariates is 16.266. This is translated into critical values for leverage for each of the eight groups based on their sample sizes (Table 6.13), using Equation 4.3, as seen in Table 6.14. For example,

$$h_{ii} = \frac{\text{Mahalanobis distance}}{N-1} + \frac{1}{N} = \frac{16.266}{45} + \frac{1}{46} = 0.3832$$

for the first group, employed women with none-or-other affiliation.

Table 6.15 shows syntax and a portion of the output data set with leverage values as produced by SAS REG. The DATA step insures that the cases with missing data on the IVs are not included in the analysis. Only the three CVs are included in the calculation of leverage values; use of ATTDRUG as a DV has no effect on the calculations.

TABLE 6.12 Syntax and Partial Output of Screening Run for Distributions and Univariate Outliers Using SAS MEANS

```
proc sort data = SASUSER.ANCOVA;
    by RELIGION EMPLMNT;
run;

proc means data = SASUSER.ANCOVA    vardef=DF
    N NMISS MIN MAX MEAN VAR STD SKEWNESS KURTOSIS;
    var ATTDRUG PHYHEAL MENHEAL PSYDRUG;
    by RELIGION EMPLMNT;
run;
```

-------- Religious affiliation=1 Current employment status=1 --------

			N			
Variable	Label	N	Miss	Minimum	Maximum
ATTDRUG	Attitudes toward use of medication	46	0	5.0000000	10.0000000
PHYHEAL	Physical health symptoms	46	0	2.0000000	9.0000000
MENHEAL	Mental health symptoms	46	0	0	17.0000000
PSYDRUG	Use of psychotropic drugs	46	0	0	32.0000000

Variable	Label	Mean	Variance	Std Dev
ATTDRUG	Attitudes toward use of medication	7.6739130	1.8246377	1.3507915
PHYHEAL	Physical health symptoms	5.0652174	3.5734300	1.8903518
MENHEAL	Mental health symptoms	6.5434783	16.4314010	4.0535665
PSYDRUG	Use of psychotropic drugs	5.3478261	58.6318841	7.6571459

Variable	Label	Skewness	Kurtosis
ATTDRUG	Attitudes toward use of medication	-0.2741204	-0.6773951
PHYHEAL	Physical health symptoms	0.3560510	-0.9003343
MENHEAL	Mental health symptoms	0.6048430	0.1683358
PSYDRUG	Use of psychotropic drugs	1.6761585	2.6270392

(continued)

225

TABLE 6.12 Continued

------------------- Religious affiliation=3 Current employment status=2 -------------------

Variable	Label	N	N Miss	Minimum	Maximum
ATTDRUG	Attitudes toward use of medication	83	0	5.0000000	10.0000000
PHYHEAL	Physical health symptoms	83	0	2.0000000	13.0000000
MENHEAL	Mental health symptoms	83	0	0	16.0000000
PSYDRUG	Use of psychotropic drugs	83	0	0	43.0000000

Variable	Label	Mean	Variance	Std Dev
ATTDRUG	Attitudes toward use of medication	7.8433735	1.4263885	1.1943151
PHYHEAL	Physical health symptoms	5.3734940	7.7246547	2.7793263
MENHEAL	Mental health symptoms	6.3132530	20.2177490	4.4964151
PSYDRUG	Use of psychotropic drugs	5.0963855	74.6735234	8.6413843

Variable	Label	Skewness	Kurtosis
ATTDRUG	Attitudes toward use of medication	-0.1299988	-0.3616168
PHYHEAL	Physical health symptoms	0.8526678	0.2040353
MENHEAL	Mental health symptoms	0.4068114	-0.8266558
PSYDRUG	Use of psychotropic drugs	2.3466630	5.8717433

TABLE 6.13   Sample Sizes for Eight Groups

Employment Status	Religious Affiliation			
	None-or-Other	*Catholic*	*Protestant*	*Jewish*
Employed	46	63	92	44
Unemployed	30	56	83	48

TABLE 6.14   Critical Leverage Values for Each Group

Employment Status	Religious Affiliation			
	None-or-Other	*Catholic*	*Protestant*	*Jewish*
Employed	0.3832	0.2782	0.1896	0.4010
Unemployed	0.5942	0.3136	0.2104	0.3669

Table 6.15 shows that subject number 213 is a multivariate outlier with a leverage value of 0.3264 in the unemployed Catholic group, which has a critical value of 0.3136. (Note that case numbers are for the sorted data file with 3 cases deleted.) Altogether, four cases are multivariate outliers in four different groups.

This is a borderline case in terms of whether to transform variables or delete outliers—somewhere between "few" and "many." A log transform of the two skewed variables is undertaken to see if outliers remain after transformation. LPSYDRUG is created as the logarithm of PSYDRUG (incremented by 1 since many of the values are at zero) and LPHYHEAL as the logarithm of PHYHEAL. See Table 4.3 for SAS DATA syntax for transforming variables. Transformed as well as original variables are saved into a file labeled ANC_LEVT.

A second run of SAS REG (not shown) with the three CVs (two of them transformed) revealed no outliers at $\alpha = .001$. All four former outliers are within acceptable distance from their groups once the two CVs are transformed. The decision is to proceed with the analysis using the two transformed CVs rather than to delete cases, although the alternative decision is also acceptable in this situation.

6.6.1.5   Multicollinearity and Singularity

SAS FACTOR provides squared multiple correlations for each variable as a DV with the remaining variables acting as IVs. This is helpful for detecting the presence of multicollinearity and singularity among the CVs, as seen in Table 6.16 for the transformed variables. There is no danger of multicollinearity or singularity because the largest **SMC** (R^2) = .30. In any event, SAS GLM guards against statistical problems due to multicollinearity.

**TABLE 6.15   Test for Multivariate Outliers. SAS REG Syntax
and Selected Portion of Output Data Set**

```
data SASUSER.ANC_LEV;
  set SASUSER.ANCOVA;
  if RELIGION =. or EMPLMNT =. then delete;
run;
proc reg;
     by RELIGION EMPLMNT;
   model ATTDRUG= PHYHEAL MENHEAL PSYDRUG;
   output out=SASUSER.ANC_LEV h=LEVERAGE;
run;
```

SASUSER.ANC_LEV1								
► 8	Int	Int	Int	Int	Int	Int	Int	Int
462	SUBNO	ATTDRUG	PHYHEAL	MENHEAL	PSYDRUG	EMPLMNT	RELIGION	LEVERAGE
148	97	8	3	6	0	2	2	0.0367
149	119	10	4	0	0	2	2	0.0559
150	126	8	5	10	0	2	2	0.0483
151	130	8	8	15	13	2	2	0.1013
152	131	8	5	12	0	2	2	0.0742
153	138	8	6	11	10	2	2	0.0510
154	148	10	8	7	3	2	2	0.0701
155	183	10	9	12	10	2	2	0.0847
156	205	9	5	3	0	2	2	0.0358
157	213	10	7	10	25	2	2	0.3264
158	228	8	4	16	6	2	2	0.1487
159	235	10	6	11	9	2	2	0.0448
160	238	9	6	8	8	2	2	0.0314
161	250	6	3	4	4	2	2	0.0420
162	251	8	3	4	0	2	2	0.0331
163	252	9	4	1	7	2	2	0.0799

TABLE 6.16   Check for Multicollinearity Through SAS FACTOR. Syntax and Selected Output

```
proc factor data=SASUSER.ANC_LEVT priors = smc;
   var LPHYHEAL MENHEAL LPSYDRUG;
run;
```

                         The FACTOR Procedure
              Initial Factor Method: Principal Factors
                  Prior Communality Estimates: SMC

            LPHYHEAL   MENHEAL   LPSYDRUG
          0.30276222   0.28483712   0.16531267

6.6.1.6 Homogeneity of Variance

Sample variances are available from a SAS MEANS run with the transformed variables, requesting only sample sizes, means, and variances, partially shown in Table 6.17. For the DV, find the largest and smallest variances over the groups. For example, the variance for ATTDRUG in the employed

TABLE 6.17 Sample Sizes, Means, and Variances for Transformed Variables Through SAS MEANS. Syntax and Selected Output

```
proc means data=SASUSER.ANC_LEVT vardef=DF
  N MEAN VAR;
    var ATTDRUG LPHYHEAL MENHEAL LPSYDRUG;
  by RELIGION EMPLMNT;
run;
```

------- Religious affiliation=1 Current employment status=1 -------

The MEANS Procedure

Variable	Label	N	Mean	Variance
ATTDRUG	Attitudes toward use of medication	46	7.6739130	1.8246377
LPHYHEAL		46	0.6733215	0.0289652
MENHEAL	Mental health symptoms	46	6.5434783	16.4314010
LPSYDRUG		46	0.4881946	0.2867148

------- Religious affiliation=2 Current employment status=1 -------

Variable	Label	N	Mean	Variance
ATTDRUG	Attitudes toward use of medication	63	7.6666667	0.9677419
LPHYHEAL		63	0.6095463	0.0477688
MENHEAL	Mental health symptoms	63	5.8412698	22.5873016
LPSYDRUG		63	0.3275272	0.1631570

Catholic group = .968 (the smallest variance). The largest variance is for the first group (employed with no or other religious affiliation), with $s^2 = 1.825$. The variance ratio $(F_{max}) = 1.89$, well below the criterion of 10:1. There is no need for a formal test of homogeneity of variance with this variance ratio since the ratio of sample sizes is less than 4:1 ($92/30 = 3.07$), and there are no outliers. Tests for homogeneity of variance among CVs likewise show no concern for heterogeneity of variance.

6.6.1.7 *Homogeneity of Regression*

Homogeneity of regression is not tested automatically in any analysis of variance programs in SAS. However, it can be evaluated by forming interactions between effects (main effects and interactions) and covariates through the ANCOVA procedure demonstrated in the next section. Although covariates cannot be pooled, each covariate can be evaluated separately. Thus, the tests for homogeneity of regression will be demonstrated in Section 6.6.2 after the main analysis of covariance.

6.6.1.8 *Reliability of Covariates*

The three CVs, MENHEAL, PHYHEAL, and PSYDRUG, were measured as counts of symptoms or drug use—"have you ever . . . ?" It is assumed that people are reasonably consistent in reporting the presence or absence of symptoms and that high reliability is likely. Therefore no adjustment in ANCOVA is made for unreliability of CVs.

6.6.2 Analysis of Covariance

6.6.2.1 *Main Analysis*

The program chosen for the major two-way analysis of covariance is SAS GLM. The cell size weights (Table 6.10, number 3, Method 3, SSTYPE I) approach to adjustment of unequal *n* is chosen for this set of survey data. Ease of use, then, makes SAS GLM a convenient program for this unequal-*n* data set.

Syntax and selected output from application of SAS GLM to these data appear in Table 6.18. The **model** instruction shows the DV on the left of the equation and the CVs and IVs on the right. Order of entry for Type I sums of squares follows the order on the right side of the equation.

Source tables for both Type I and Type III sums of squares are shown in Table 6.18. Type III sums of squares, which are adjusted for all other effects, are used to evaluate covariates. Type I sums of squares, which are adjusted for all previous but not following effects, are used to evaluate main effects of religious affiliation and employment status and their interaction.

In this example, the main effect of religion reaches statistical significance, $F(3, 451) = 2.86$, $p = .0366$. The only CV that reaches statistical reliability is LPSYDRUG, $F(1, 451) = 39.09$, $p < .0001$. The source table is summarized in Table 6.19.

TABLE 6.18 Syntax and Selected Output from SAS GLM Analysis of Covariance Run

```
proc glm data=SASUSER.ANC_LEVT;
  class RELIGION EMPLMNT;
  model ATTDRUG = LPHYHEAL MENHEAL LPSYDRUG RELIGION EMPLMNT RELIGION*EMPLMNT;
run;
```

The GLM Procedure

Dependent Variable: ATTDRUG   Attitudes toward use of medication

Source	DF	Sum of Squares	Mean Square	F Value	Pr > F
Model	10	78.6827786	7.8682779	6.58	<.0001
Error	451	539.1786932	1.1955182		
Corrected Total	461	617.8614719			

R-Square	Coeff Var	Root MSE	ATTDRUG Mean
0.127347	14.22957	1.093398	7.683983

Source	DF	Type I SS	Mean Square	F Value	Pr > F
LPHYHEAL	1	9.90840138	9.90840138	8.29	0.0042
MENHEAL	1	0.13429906	0.13429906	0.11	0.7377
LPSYDRUG	1	45.72530836	45.72530836	38.25	<.0001
RELIGION	3	10.25921450	3.41973817	2.86	0.0366
EMPLMNT	1	3.78463591	3.78463591	3.17	0.0759
RELIGION*EMPLMNT	3	8.87091941	2.95697314	2.47	0.0611

Source	DF	Type III SS	Mean Square	F Value	Pr > F
LPHYHEAL	1	0.62999012	0.62999012	0.53	0.4683
MENHEAL	1	1.42886911	1.42886911	1.20	0.2749
LPSYDRUG	1	46.73742309	46.73742309	39.09	<.0001
RELIGION	3	12.19374680	4.06458227	3.40	0.0178
EMPLMNT	1	1.07961032	1.07961032	0.90	0.3425
RELIGION*EMPLMNT	3	8.87091941	2.95697314	2.47	0.0611

TABLE 6.19   Analysis of Covariance of Attitude Toward Drugs

Source of Variance	Adjusted SS	df	MS	F
Religious affiliation	10.26	3	3.42	2.86*
Employment status (adjusted for religious affiliation)	3.78	1	3.78	3.17
Interaction	8.87	3	2.95	2.47
Covariates (adjusted for all effects)				
Physical health (log)	0.63	1	0.63	0.42
Mental health	1.43	1	1.43	1.20
Drug uses (log)	46.74	1	46.74	39.09
Error	539.18	451	1.20	

Entries in the sum of squares column of the source table are used to calculate partial η^2 as a measure of effect size for main effects and the interaction (Sections 6.4 and 6.5.4.4). For RELIGION:

$$\text{partial } \eta^2 = \frac{10.259}{10.259 + 539.179} = .02$$

Table 6.20 shows the use of the Smithson (2003) procedure to find partial η^2 and its confidence limits for all the effects, whether statistically significant or not. Confidence limits for effect sizes as well as partial η^2 are found by adding values of the syntax file: NoncF2.sas. Table 6.20 shows a portion of the syntax file with added values shaded. Filled in values (from Table 6.19) are, respectively, *F,* numerator df, denominator df, and the proportion for the desired confidence interval, here .95, respectively, These replace the default values filled into NoncF2.sas. Effects appear in Table 6.20 in the following order: RELIGION, EMPLMNT, and the interaction. The output column labeled rsq is partial η^2; lower and upper confidence limits are labeled rsqlow and rsqupp, respectively.

Thus, the 95% confidence interval for the size of the RELIGION effect using the Smithson (2003) SAS procedure ranges from .00 to .04. Effect size for employment status is .01 with a 95% confidence interval ranging from .00 to .03. For the interaction, partial $\eta^2 = .02$ also, with a 95% confidence interval ranging from .00 to .04.

Unadjusted and adjusted (**Least Squares**) marginal means and confidence intervals for RELIGION are shown in Table 6.21, provided by SAS GLM. These could be requested in the main ANCOVA run.

Note that unadjusted means are provided for CVs as well as the DV. Because no a priori hypotheses about differences among religious groups were generated, planned comparisons are not appropriate. A glance at the four adjusted means in Table 6.21 suggests a straightforward interpretation; the none-or-other group has the least favorable attitude toward use of psychotropic drugs, the Catholic group the most favorable attitude, and the Protestant and Jewish groups an intermediate attitude. In the absence of specific questions about differences between means, there is no compelling reason to evaluate post hoc the significance of these differences, although they certainly provide a rich source of speculation for future research. Relevant means are summarized in Table 6.22. Means with 95% confidence limits are in Figure 6.3.

TABLE 6.20 Data Set Output from NoncF2.sas for Effect Size (rsq) with 95% Confidence Limits (rsqlow and rsqupp)

```
.
.
.
rsqlow = ncplow / (ncplow + df1 + df2 + 1);
rsqupp = ncpupp / (ncpupp + df1 + df2 + 1);
cards;
2.860   3   451   0.95   0.95
3.170   1   451   0.95   0.95
2.470   3   451   0.95   0.95
;
proc print;
run;
```

Obs	F	df1	df2	conf	prlow	prupp	ncplow	ncpupp	rsq	rsqlow	rsqupp
1	2.860	3	451	0.95	0.975	0.025	0.000	21.385	0.01867	0.00000	0.04489
2	3.170	1	451	0.95	0.975	0.025	0.000	14.009	0.00698	0.00000	0.03000
3	2.470	3	451	0.95	0.975	0.025	0.000	19.347	0.01616	0.00000	0.04079

233

TABLE 6.21   Adjusted and Unadjusted Mean Attitude toward Drugs for Four Categories of Religion. SAS GLM Syntax and Selected Output

```
proc glm data=SASUSER.ANC_LEVT;
   class RELIGION EMPLMNT;
   model ATTDRUG = LPHYHEAL MENHEAL LPSYDRUG RELIGION|EMPLMNT;
   means RELIGION;
   lsmeans RELIGION;
run;
```

Level of RELIGION	N	----------ATTDRUG---------- Mean	Std Dev	---------LPHYHEAL--------- Mean	Std Dev
1	76	7.44736842	1.31041710	0.65639689	0.16746166
2	119	7.84033613	1.04948085	0.63034302	0.20673876
3	175	7.66857143	1.14665216	0.66175666	0.22091000
4	92	7.70652174	1.16296366	0.64575134	0.20602015

Level of RELIGION	N	---------MENHEAL---------- Mean	Std Dev	---------LPSYDRUG--------- Mean	Std Dev
1	76	5.93421053	3.99444790	0.40110624	0.49754148
2	119	6.13445378	4.65764247	0.32712918	0.43952791
3	175	6.08571429	4.10054264	0.43477410	0.51643966
4	92	6.40217391	3.97259785	0.50695061	0.53592293

Least Squares Means

RELIGION	ATTDRUG LSMEAN
1	7.40744399
2	7.91776374
3	7.66041635
4	7.64441304

RELIGION	ATTDRUG LSMEAN	95% Confidence Limits	
1	7.407444	7.154984	7.659903
2	7.917764	7.719421	8.116106
3	7.660416	7.497559	7.823274
4	7.644413	7.419188	7.869639

TABLE 6.22   Adjusted and Unadjusted Mean Attitude Toward Drugs for Four Categories of Religious Affiliation

Religion	Adjusted Mean	Unadjusted Mean
None-or-other	7.41	7.45
Catholic	7.92	7.84
Protestant	7.66	7.67
Jewish	7.64	7.71

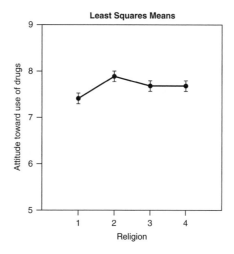

FIGURE 6.3   Adjusted means for attitude toward drug use with error bars representing the 95% confidence interval for the mean. Graph produced through SYSTAT 11.

6.6.2.2   *Evaluation of Covariates*

Information about utility of covariates is provided in Table 6.18 where only LPSYDRUG was seen to adjust the DV, ATTDRUG, after adjustment for all other covariates and effects is taken into account. Table 6.23 shows the pooled within-cell correlations among the DV and CVs as produced by SAS GLM. These correlations are adjusted for differences among cells—the bivariate correlations are found within each cell and then averaged (pooled). However, the correlations are not adjusted for each other. The run is done as a MANOVA, with the DV and CVs all treated as multiple DVs. This way, relationships among all four variables are shown. The `printe` instruction requests the pooled within-cell correlation table. The `nouni` instruction limits the output.

TABLE 6.23   Pooled Within-Cell Correlations among the DV and CVs. SAS GLM Syntax and Selected Output

```
proc glm data=SASUSER.ANC_LEVT;
   class RELIGION EMPLMNT;
   model ATTDRUG LPHYHEAL MENHEAL LPSYDRUG =
         RELIGION|EMPLMNT /nouni;
   manova h=_all_ / printe;
run;
```

Partial Correlation Coefficients from the Error SSCP Matrix / Prob > |r|

DF = 454	ATTDRUG	LPHYHEAL	MENHEAL	LPSYDRUG
ATTDRUG	1.000000	0.121087	0.064048	0.301193
		0.0097	0.1726	<.0001
LPHYHEAL	0.121087	1.000000	0.509539	0.364506
	0.0097		<.0001	<.0001
MENHEAL	0.064048	0.509539	1.000000	0.333499
	0.1726	<.0001		<.0001
LPSYDRUG	0.301193	0.364506	0.333499	1.000000
	<.0001	<.0001	<.0001	

TABLE 6.24   Pooled Within-Cell Correlations among Three Covariates and the Dependent Variable, Attitude Toward Drugs

	Physical Health (LOG)	Mental Health	Drug Uses (LOG)
Attitude toward drugs	.121*	.064	.301*
Physical health (LOG)		.510*	.365*
Mental health			.333*

*$p < .01$

    Table 6.23 shows that both LPHYHEAL and LPSYDRUG are related to the DV, ATTDRUG. However, only LPSYDRUG is effective as a covariate once adjustment is made for the other CVs and effects, as seen in Table 6.18. Table 6.23 shows the reason why; LPHYHEAL and LPSYDRUG are themselves related. According to the criteria of Section 6.5.1, then, use of MENHEAL as a covariate in future research is not warranted (it has, in fact, lowered the power of this analysis) and use of LPHYHEAL is questionable. Table 6.24 summarizes the pooled within-cell correlations.

TABLE 6.25   Analysis of Covariance for Evaluating Homogeneity of Regression. SAS GLM Syntax and Selected Output

```
proc glm data=SASUSER.ANC_LEVT;
   class RELIGION EMPLMNT;
   model ATTDRUG = RELIGION|EMPLMNT LPHYHEAL|RELIGION|EMPLMNT
                   LPSYDRUG|RELIGION|EMPLMNT MENHEAL|RELIGION|EMPLMNT;
run;
```

Source	DF	Type III SS	Mean Square	F Value	Pr > F
RELIGION	3	1.87454235	0.62484745	0.52	0.6663
EMPLMNT	1	0.80314068	0.80314068	0.67	0.4125
RELIGION*EMPLMNT	3	1.83822082	0.61274027	0.51	0.6732
LPHYHEAL	1	1.18143631	1.18143631	0.99	0.3203
LPHYHEAL*RELIGION	3	0.49791878	0.16597293	0.14	0.9366
LPHYHEAL*EMPLMNT	1	0.62917073	0.62917073	0.53	0.4682
LPHYHE*RELIGI*EMPLMN	3	2.96516036	0.98838679	0.83	0.4789
LPSYDRUG	1	37.24237057	37.24237057	31.20	<.0001
LPSYDRUG*RELIGION	3	5.14751739	1.71583913	1.44	0.2312
LPSYDRUG*EMPLMNT	1	2.09112695	2.09112695	1.75	0.1863
LPSYDR*RELIGI*EMPLMN	3	9.69187356	3.23062452	2.71	0.0449
MENHEAL	1	0.44710058	0.44710058	0.37	0.5408
MENHEAL*RELIGION	3	3.92657123	1.30885708	1.10	0.3503
MENHEAL*EMPLMNT	1	0.05659585	0.05659585	0.05	0.8277
MENHEA*RELIGI*EMPLMN	3	4.80558448	1.60186149	1.34	0.2601

6.6.2.3   *Homogeneity of Regression Run*

The SAS GLM run to test for homogeneity of regression (Table 6.25) adds all interactions between effects and CVs to the analysis of Table 6.18. A hierarchical notation is used in which effects separated by a "|" include interactions and all lower order effects. For example MENHEAL|RELIGION|EMPLMNT includes MENHEAL, RELIGION, EMPLMNT, MENHEAL*RELIGION, MENHEAL*EMPLMNT, RELIGION*EMPLMNT, and MENHEAL*RELIGION*EMPLMNT.

Effects of interest here are those which interact with CVs. Only one of them, LPSYDRUG*RELIGION*EMPLOYMENT is statistically significant at $\alpha = .05$, suggesting different relationships between ATTDRUG and LPSYDRUG among the eight groups of women. However, a more stringent alpha criterion than .05 is advisable for the multitude of tests produced by this method of evaluating homogeneity of regression. (Indeed, an SPSS MANOVA run, which pools the covariates into a single test, showed no violation of the assumption.)

A checklist for analysis of covariance appears as Table 6.26. An example of a Results section, in journal format, follows for the analysis described above.

TABLE 6.26   Checklist for Analysis of Covariance

1. Issues
   a. Unequal sample size and missing data
   b. Within-cell outliers
   c. Normality
   d. Homogeneity of variance
   e. Within-cell linearity
   f. Homogeneity of regression
   g. Reliability of CVs
2. Major analyses
   a. Main effect(s) or planned comparison. If significant: Adjusted marginal means and standard deviations or standard errors or confidence intervals
   b. Interactions or planned comparisons. If significant: Adjusted cell means and standard deviations or standard errors or confidence intervals (in table or interaction graph)
   c. Effect sizes with confidence intervals for all effects
3. Additional analyses
   a. Evaluation of CV effects
   b. Evaluation of intercorrelations
   c. Post hoc comparisons (if appropriate)
   d. Unadjusted marginal and/or cell means (if significant main effect and/or interaction) if nonexperimental application

---

Results

   A 2 × 4 between-subjects analysis of covariance was performed on attitude toward drugs. Independent variables consisted of current employment status (employed and unemployed) and religious identification (None-or-other, Catholic, Protestant, and Jewish), factorially combined. Covariates were physical health, mental health, and the sum of psychotropic drug uses. Analyses were performed by SAS GLM, weighting cells by their sample sizes to adjust for unequal *n*.

   Results of evaluation of the assumptions of normality of sampling distributions, linearity, homogeneity of variance, homogeneity of regression, and reliability of covariates were satisfactory. Presence of outliers led to transformation of two of the covariates. Logarithmic transforms were made of physical health and the

sum of psychotropic drug uses. No outliers remained after transformation. The original sample of 465 was reduced to 462 by three
women who did not provide information as to religious affiliation.

After adjustment by covariates, attitude toward drugs varied
significantly with religious affiliation, as summarized in .19, with
$F(3, 451) = 2.86$, $p < .05$. The strength of the relationship between
adjusted attitudes toward drugs and religion was weak, however, with
partial $\eta^2 = .02$, 95% confidence limits from .00 to .04. The adjusted
marginal means, as displayed in Table 6.22 and, with 95% confidence
interval, in Figure 6.2, show that the most favorable attitudes
toward drugs were held by Catholic women, and least favorable attitudes by women who either were unaffiliated with a religion or identified with some religion other than the major three. Attitudes
among Protestant and Jewish women were almost identical, on average,
for this sample, and fell between those of the two other groups.

No statistically significant main effect of current employment
status was found. Nor was there a significant interaction between
employment status and religion after adjustment for covariates. For
employment, partial $\eta^2 = .01$ with 95% confidence limits from .00 to
.03. For the interaction, partial $\eta^2 = .02$ with 95% confidence limits
from .00 to .04.

Pooled within-cell correlations among covariates and attitude
toward drugs are shown in Table 6.24. Two of the covariates, logarithm of physical health and logarithm of drug use, were significantly associated with the dependent variable. However, only
logarithm of drug use uniquely adjusted the attitude scores,
$F(1, 451) = 39.09$, $p < .01$, after covariates were adjusted for
other covariates, main effects, and interaction. The remaining
two covariates, mental health and logarithm of physical health,
provided no statistically significant unique adjustment.

6.7 Comparison of Programs

For the novice, there is a bewildering array of canned computer programs in SPSS (REGRESSION, GLM, and MANOVA), and SYSTAT (ANOVA, GLM, and REGRESS) packages for ANCOVA. For our purposes, the programs based on regression (SYSTAT REGRESS and SPSS REGRESSION) are not discussed because they offer little advantage over the other, more easily used programs.

SAS has a single general linear model program designed for use with both discrete and continuous variables. This program deals well with ANCOVA. Features of eight programs are described in Table 6.27.

6.7.1 SPSS Package

Two SPSS programs perform ANCOVA: GLM and MANOVA. Both programs are rich and highly flexible. Both provide a great deal of information about adjusted and unadjusted statistics, and have alternatives for dealing with unequal n. They are the only programs that offer power analyses and effect sizes in the form of partial η^2. Also available in GLM are plots of means. MANOVA shines in its ability to test assumptions such as homogeneity of regression (see Section 6.5.3). It provides adjusted cell and marginal means; specific comparisons and trend analysis are readily available. However, SPSS does not facilitate the search for multivariate outliers among the DV and CV(s) in each group. Only a measure of influence, Cook's distance, is available in GLM (leverage values produced do not differ within cells).

6.7.2 SAS System

SAS GLM is a program for univariate and multivariate analysis of variance and covariance. SAS GLM offers analysis of complex designs, several adjustments for unequal n, a test for sphericity for within-subjects IVs, a full array of descriptive statistics (upon request), and a wide variety of post hoc tests in addition to user-specified comparisons and trend analysis. Although there is no example of a test for homogeneity of regression in the SAS manual, the procedures described in Section 6.6.2.3 can be followed.

6.7.3 SYSTAT System

SYSTAT ANOVA Version 11 is an easily used program which handles all types of ANOVA and ANCOVA. In addition, SYSTAT GLM is a multivariate general linear program that does almost everything that the ANOVA program does, and more, but is not always quite as easy to set up. MANOVA is a recent addition to the menu, but brings forth the same dialog box as GLM. All three programs handle repeated measures and post hoc comparisons. For some unknown reason, sphericity tests are not available for repeated measures in "long" output. The programs provide a great deal of control over error terms and comparisons. Cell means adjusted for CVs are produced and plotted. The program also provides leverage values which may be converted to Mahalanobis distance, as per Equation 4.3. The program provides adjustment for violation of sphericity in within-subject designs.

The major advantage in using SYSTAT GLM over ANOVA is the greater flexibility with unequal-n designs. Only method 1 is available in the ANOVA program. GLM allows specification of a

TABLE 6.27   Comparison of Selected Programs for Analysis of Covariance

Feature	SPSS GLM	SPSS MANOVA[a]	SAS GLM[a]	SYSTAT ANOVA,[a] GLM,[a] and MANOVA[a]
Input				
Maximum number of IVs	No limit	10	No limit	No limit
Choice of unequal-n adjustment	Yes	Yes	Yes	No[d]
Within-subjects IVs	Yes	Yes	Yes	Yes
Specify tolerance	EPS	No	SINGULAR	Yes
Specify separate variance error term for contrasts	No	No	Yes	Yes
Resampling	No	No	No	Yes
Output				
Source table	Yes	No	Yes	Yes
Unadjusted cell means	Yes	Yes	Yes	No
Confidence interval for unadjusted cell means	No	Yes	No	No
Unadjusted marginal means	Yes	Yes	Yes	No
Cell standard deviations	Yes	Yes	Yes	No
Adjusted cell means	EMMEANS	PMEANS	LSMEAN	PRINT MEDIUM
Standard errors or SDs for adjusted cell means	Yes	No	STDERR	PRINT MEDIUM
Adjusted marginal means	Yes	Yes[b]	LSMEAN	No
Standard error for adjusted marginal means	Yes	Yes	STDERR	No
Power analysis	OPOWER	POWER	No[e]	No
Effect sizes	ETASQ	POWER	No	No
Test for equality of slope (homogeneity of regression) for multiple covariates	No	Yes	No	Yes
Post hoc tests with adjustment	Yes	Yes[c]	Yes	Yes
User-specified contrasts	Yes	Yes	Yes	Yes
Hypothesis SSCP matrices	No	Yes	Yes	No
Pooled within-cell error SSCP matrices	No	Yes	Yes	No
Hypothesis covariance matrices	No	Yes	No	No
Pooled within-cell covariance matrix	No	Yes	No	No
Group covariance matrices	No	Yes	No	No

(continued)

TABLE 6.27   Continued

Feature	SPSS GLM	SPSS MANOVA[a]	SAS GLM[a]	SYSTAT ANOVA[a] GLM,[a] and MANOVA[a]
Output *(continued)*				
Pooled within-cell correlation matrix	No	Yes	Yes	No
Group correlation matrices	No	Yes	No	No
Covariance matrix for adjusted group means	No	No	Data file	No
Regression coefficient for each CV	Yes	Yes	Yes	No
Regression coefficient for each cell	No	No	Yes	No
Multiple R and/or R^2	Yes	Yes	Yes	Yes
Test for homogeneity of variance	Yes	Yes	No	No
Test for sphericity	Yes	Yes	Yes	No
Adjustment for heterogeneity of covariance	Yes	Yes	Yes	Yes[f]
Predicted values and residuals	Yes	Yes	Yes	Data file
Plots of means	Yes	No	No	Yes
Multivariate influence and/or leverage statistics by cell	No	No	No	Data file

[a]Additional features described in Chapter 7 (MANOVA).

[b]Available through the CONSPLUS procedure.

[c]Bonferroni and Scheffé confidence intervals.

[d]Some flexibility is possible in GLM.

[e]Power analysis for ANCOVA is in a separate program: GLMPOWER.

[f]Not available in "long" output.

MEANS model, which, when WEIGHTS are applied to cell means, provides a weighted means analysis. The manual also describes the multiple models that can be estimated to provide sums of squares that correspond to the four SS types of adjustments for unequal n. GLM is also the only SYSTAT program set up for simple effects and such designs as Latin square, nesting, and incomplete blocks.

7

Multivariate Analysis of Variance and Covariance

7.1 General Purpose and Description

Multivariate analysis of variance (MANOVA) is a generalization of ANOVA to a situation in which there are several DVs. For example, suppose a researcher is interested in the effect of different types of treatments on several types of anxieties: test anxiety, anxiety in reaction to minor life stresses, and so-called free-floating anxiety. The IV is different treatment with three levels (desensitization, relaxation training, and a waiting-list control). After random assignment of subjects to treatments and a subsequent period of treatment, subjects are measured for test anxiety, stress anxiety, and free-floating anxiety. Scores on all three measures for each subject serve as DVs. MANOVA is used to ask whether a combination of the three anxiety measures varies as a function of treatment. MANOVA is statistically identical to discriminant analysis, the subject of Chapter 9. The difference between the techniques is one of emphasis only. MANOVA emphasizes the mean differences and statistical significance of differences among groups. Discriminant analysis emphasizes prediction of group membership and the dimensions on which groups differ.

ANOVA tests whether mean differences among groups on a single DV are likely to have occurred by chance. MANOVA tests whether mean differences among groups on a combination of DVs are likely to have occurred by chance. In MANOVA, a new DV that maximizes group differences is created from the set of DVs. The new DV is a linear combination of measured DVs, combined so as to separate the groups as much as possible. ANOVA is then performed on the newly created DV. As in ANOVA, hypotheses about means in MANOVA are tested by comparing variances—hence multivariate analysis of variance.

In factorial or more complicated MANOVA, a different linear combination of DVs is formed for each main effect and interaction. If gender of subject is added to the example as a second IV, one combination of the three DVs maximizes the separation of the three treatment groups, a second combination maximizes separation of women and men, and a third combination maximizes separation of the cells of the interaction. Further, if the treatment IV has more than two levels, the DVs can be recombined in yet other ways to maximize the separation of groups formed by comparisons.[1]

MANOVA has a number of advantages over ANOVA. First, by measuring several DVs instead of only one, the researcher improves the chance of discovering what it is that changes as a result of different treatments and their interactions. For instance, desensitization may have an advantage over

[1]The linear combinations themselves are of interest in discriminant analysis (Chapter 9).

relaxation training or waiting-list control, but only on test anxiety; the effect is missing if test anxiety isn't one of your DVs. A second advantage of MANOVA over a series of ANOVAs when there are several DVs is protection against inflated Type I error due to multiple tests of (likely) correlated DVs.

Another advantage of MANOVA is that, under certain, probably rare conditions, it may reveal differences not shown in separate ANOVAs. Such a situation is shown in Figure 7.1 for a one-way design with two levels. In this figure, the axes represent frequency distributions for each of two DVs, Y_1 and Y_2. Notice that from the point of view of either axis, the distributions are sufficiently overlapping that a mean difference might not be found in ANOVA. The ellipses in the quadrant, however, represent the distributions of Y_1 and Y_2 for each group separately. When responses to two DVs are considered in combination, group differences become apparent. Thus, MANOVA, which considers DVs in combination, may occasionally be more powerful than separate ANOVAs.

But there are no free lunches in statistics, either. MANOVA is a substantially more complicated analysis than ANOVA. There are several important assumptions to consider, and there is often some ambiguity in interpretation of the effects of IVs on any single DV. Further, the situations in which MANOVA is more powerful than ANOVA are quite limited; often MANOVA is considerably less powerful than ANOVA, particularly in finding significant group differences for a particular DV. Thus, our recommendation is to think very carefully about the need for more than one DV in light of the added complexity and ambiguity of analysis and the likelihood that multiple DVs may be redundant (see also Section 7.5.3). Even moderately correlated DVs diminish the power of MANOVA. Figure 7.2 shows a set of hypothetical relationships between a single IV and four DVs. DV1 is highly related to the IV and shares some variance with DV2 and DV3. DV2 is related to both DV1 and DV3 and shares very little unique variance with the IV, although by itself in a univariate ANOVA might be related to the IV. DV3 is somewhat related to the IV, but also to all of the other DVs. DV4 is highly

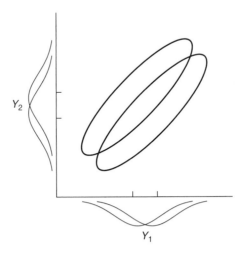

FIGURE 7.1 Advantage of MANOVA, which combines DVs, over ANOVA. Each axis represents a DV; frequency distributions projected to axes show considerable overlap, while ellipses, showing DVs in combination, do not.

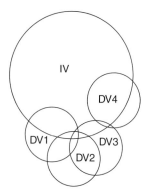

**FIGURE 7.2   Hypothetical relationships
among a single IV and four DVs.**

related to the IV and shares only a little bit of variance with DV3. Thus, DV2 is completely redundant with the other DVs, and DV3 adds only a bit of unique variance to the set. However, DV2 would be useful as a CV if that use made sense conceptually. DV2 reduces the total variance in DV1 and DV2, and most of the variance reduced is *not* related to the IV. Therefore, DV2 reduces the error variance in DV1 and DV3 (the variance that is not overlapping with the IV).

Multivariate analysis of covariance (MANCOVA) is the multivariate extension of ANCOVA (Chapter 6). MANCOVA asks if there are statistically significant mean differences among groups after adjusting the newly created DV for differences on one or more covariates. For the example, suppose that before treatment subjects are pretested on test anxiety, minor stress anxiety, and free-floating anxiety. When pretest scores are used as covariates, MANCOVA asks if mean anxiety on the composite score differs in the three treatment groups, after adjusting for preexisting differences in the three types of anxieties.

MANCOVA is useful in the same ways as ANCOVA. First, in experimental work, it serves as a noise-reducing device where variance associated with the covariate(s) is removed from error variance; smaller error variance provides a more powerful test of mean differences among groups. Second, in nonexperimental work, MANCOVA provides statistical matching of groups when random assignment to groups is not possible. Prior differences among groups are accounted for by adjusting DVs as if all subjects scored the same on the covariate(s). (But review Chapter 6 for a discussion of the logical difficulties of using covariates this way.)

ANCOVA is used after MANOVA (or MANCOVA) in Roy-Bargmann stepdown analysis where the goal is to assess the contributions of the various DVs to a significant effect. One asks whether, after adjusting for differences on higher-priority DVs serving as covariates, there is any significant mean difference among groups on a lower-priority DV. That is, does a lower-priority DV provide additional separation of groups beyond that of the DVs already used? In this sense, ANCOVA is used as a tool in interpreting MANOVA results.

Although computing procedures and programs for MANOVA and MANCOVA are not as well developed as for ANOVA and ANCOVA, there is in theory no limit to the generalization of the model, despite complications that arise. There is no reason why all types of designs—one-way, factorial, repeated measures, nonorthogonal, and so on—cannot be extended to research with several DVs. Questions of effect size, specific comparisons, and trend analysis are equally interesting with

MANOVA. In addition, there is the question of importance of DVs—that is, which DVs are affected by the IVs and which are not.

MANOVA developed in the tradition of ANOVA. Traditionally, MANOVA was applied to experimental situations where all, or at least some, IVs are manipulated and subjects are randomly assigned to groups, usually with equal cell sizes. Discriminant analysis (Chapter 9) developed in the context of nonexperimental research where groups are formed naturally and are not usually the same size. MANOVA asks if mean differences among groups on the combined DV are larger than expected by chance; discriminant analysis asks if there is some combination of variables that reliably separates groups. But there is no mathematical distinction between MANOVA and discriminant analysis. At a practical level, computer programs for discriminant analysis are more informative but are also, for the most part, limited to one-way designs. Therefore, analysis of one-way MANOVA is deferred to Chapter 9 and the present chapter covers factorial MANOVA and MANCOVA.

Mason (2003) used a 2 × 5 between-subjects MANOVA to investigate male and female high school students' beliefs about math. The six scales serving as DVs were in agreement with items concerning ability to solve difficult math problems, need for complex procedures for word problems, importance of understanding concepts, importance of word problems, effect of effort, and usefulness of math in everyday life. Multivariate tests of both main effects were statistically significant, but the interaction was not. Post hoc Tukey HSD tests were used to investigate the individual DVs. Belief in usefulness of math and need for complex procedures increased over the grades; belief in ability to solve difficult problems increased from the first to second year and then decreased. Girls were found to be more likely to believe in the importance of understanding concepts than boys.

A more complex MANOVA design was employed by Pisula (2003) who studied responses to novelty in high- and low-avoidance rats. IVs were sex of rat, subline (high vs. low avoidance), and 8 time intervals. Thus, this was a 2 × 2 × 8 mixed between-between-within MANOVA. DVs were four durations spent inside various zones, duration of object contact, duration of floor sniffing, and number of walking onsets. Multivariate results were not reported, but the table of (presumably) univariate F tests suggests significant results for all effects except the sex by subline interaction. All DVs showed significant differences over trials. All DVs associated with time spent inside various zones also showed significant differences between high- and low-avoidance sublines, as did number of walking onsets. Duration of object contact and number of walking onsets showed sex differences. All DVs except walking onsets also showed significant two-way interactions, and duration of object contact showed a significant three-way interaction.

A MANCOVA approach was taken by Hay (2003) to investigate quality of life variables in bulimic eating disorders. Two types of disorders were identified: regular binge eating and extreme weight control. These each were compared with a non–eating-disordered group in separate MANCOVAs. It is not clear why (or if) these were not combined into a single three-group one-way MANCOVA with planned comparisons between each eating-disorder group and the comparison group. Covariates were age, gender, income level, and BMI (body mass index). Three sets of DVs (physical and mental health components of SF-36 scores, eight SF-36 subscale scores, and six utility AqoL scores) were entered into three separate MANCOVAs for each of the comparisons, resulting in a total of 6 MANCOVAs. The emphasis in interpretation was on variance explained (η^2) for each analysis. For example, 23% of the variance in mental and physical scores was associated with regular binge eating after adjusting for CVs, but only 5% of the variance was associated with extreme weight control behaviors. Similarly, binge eating was associated with greater variance in SF-36 subscale scores and in AqoL scores than were extreme weight control behavior.

7.2   Kinds of Research Questions

The goal of research using MANOVA is to discover whether behavior, as reflected by the DVs, is changed by manipulation (or other action) of the IVs. Statistical techniques are currently available for answering the types of questions posed in Sections 7.2.1 through 7.2.8.

7.2.1   Main Effects of IVs

Holding all else constant, are mean differences in the composite DV among groups at different levels of an IV larger than expected by chance? The statistical procedures described in Sections 7.4.1 and 7.4.3 are designed to answer this question, by testing the null hypothesis that the IV has no systematic effect on the optimal linear combination of DVs.

As in ANOVA, "holding all else constant" refers to a variety of procedures: (1) controlling the effects of other IVs by "crossing over" them in a factorial arrangement, (2) controlling extraneous variables by holding them constant (e.g., running only women as subjects), counterbalancing their effects, or randomizing their effects, or (3) using covariates to produce an "as if constant" state by statistically adjusting for differences on covariates.

In the anxiety-reduction example, the test of main effect asks: Are there mean differences in anxiety—measured by test anxiety, stress anxiety, and free-floating anxiety—associated with differences in treatment? With addition of covariates, the question is: Are there differences in anxiety associated with treatment, after adjustment for individual differences in anxiety prior to treatment?

When there are two or more IVs, separate tests are made for each IV. Further, when sample sizes are equal in all cells, the separate tests are independent of one another (except for use of a common error term) so that the test of one IV in no way predicts the outcome of the test of another IV. If the example is extended to include gender of subject as an IV, and if there are equal numbers of subjects in all cells, the design produces tests of the main effect of treatment and of gender of subject, the two tests independent of each other.

7.2.2   Interactions among IVs

Holding all else constant, does change in the DV over levels of one IV depend on the level of another IV? The test of interaction is similar to the test of main effect, but interpreted differently, as discussed more fully in Chapter 3 and in Sections 7.4.1 and 7.4.3. In the example, the test of interaction asks: Is the pattern of response to the three types of treatments the same for men as it is for women? If the interaction is significant, it indicates that one type of treatment "works better" for women while another type "works better" for men.

With more than two IVs, there are multiple interactions. Each interaction is tested separately from tests of other main effects and interactions, and these tests (but for a common error term) are independent when sample sizes in all cells are equal.

7.2.3   Importance of DVs

If there are significant differences for one or more of the main effects or interactions, the researcher usually asks which of the DVs are changed and which are unaffected by the IVs. If the main effect of treatment is significant, it may be that only test anxiety is changed while stress anxiety and free-floating anxiety do not differ with treatment. As mentioned in Section 7.1, Roy-Bargmann stepdown analysis is often used where each DV is assessed in ANCOVA with higher-priority DVs serving as covariates. Stepdown analysis and other procedures for assessing importance of DVs appear in Section 7.5.3.

7.2.4   Parameter Estimates

Ordinarily, marginal means are the best estimates of population parameters for main effects and cell means are the best estimates of population parameters for interactions. But when Roy-Bargmann stepdown analysis is used to test the importance of the DVs, the means that are tested are adjusted means rather than sample means. In the example, suppose free-floating anxiety is given first, stress anxiety second, and test anxiety third priority. Now suppose that a stepdown analysis shows that only test anxiety is affected by differential treatment. The means that are tested for test anxiety are not sample means, but sample means adjusted for stress anxiety and free-floating anxiety. In MANCOVA, additional adjustment is made for covariates. Interpretation and reporting of results are based on both adjusted and sample means, as illustrated in Section 7.6. In any event, means are accompanied by some measure of variability: standard deviations, standard errors, and/or confidence intervals.

7.2.5   Specific Comparisons and Trend Analysis

If an interaction or a main effect for an IV with more than two levels is significant, you probably want to ask which levels of main effect or cells of interaction are different from which others. If, in the example, treatment with three levels is significant, the researcher would be likely to want to ask if the pooled average for the two treated groups is different from the average for the waiting-list control, and if the average for relaxation training is different from the average for desensitization. Indeed, the researcher may have planned to ask these questions instead of the omnibus F questions about treatment. Similarly, if the interaction of gender of subject and treatment is significant, you may want to ask if there is a significant difference in the average response of women and men to, for instance, desensitization.

Specific comparisons and trend analysis are discussed more fully in Sections 7.5.4, 3.2.6, 6.5.4.3, and 8.5.2.

7.2.6   Effect Size

If a main effect or interaction reliably affects behavior, the next logical question is: How much? What proportion of variance of the linear combination of DV scores is attributable to the effect? You can determine, for instance, the proportion of the variance in the linear combination of anxiety scores that is associated with differences in treatment. These procedures are described in Section 7.4.1. Procedures are also available for finding the effect sizes for individually significant DVs as demonstrated in Section 7.6, along with confidence intervals for effect sizes.

7.2.7   Effects of Covariates

When covariates are used, the researcher normally wants to assess their utility. Do the covariates provide statistically significant adjustment and what is the nature of the DV-covariate relationship? For example, when pretests of test, stress, and free-floating anxiety are used as covariates, to what degree does each covariate adjust the composite DV? Assessment of covariates is demonstrated in Section 7.6.3.1.

7.2.8   Repeated-Measures Analysis of Variance

MANOVA is an alternative to repeated-measures ANOVA in which responses to the levels of the within-subjects IV are simply viewed as separate DVs. Suppose, in the example, that measures of test anxiety are taken three times (instead of measuring three different kinds of anxiety once), before, immediately after, and 6 months after treatment. Results could be analyzed as a two-way ANOVA, with treatment as a between-subjects IV and tests as a within-subject IV, or as a one-way MANOVA, with treatment as a between-subjects IV and the three testing occasions as three DVs.

As discussed in Sections 3.2.3 and 8.5.1, repeated measures ANOVA has the often-violated assumption of sphericity. When the assumption is violated, significance tests are too liberal and some alternative to ANOVA is necessary. Other alternatives are adjusted tests of the significance of the within-subjects IV (e.g., Huynh-Feldt), decomposition of the repeated-measures IV into an orthogonal series of single degree of freedom tests (e.g., trend analysis), and profile analysis of repeated measures (Chapter 8).

7.3   Limitations to Multivariate Analysis of Variance and Covariance

7.3.1   Theoretical Issues

As with all other procedures, attribution of causality to IVs is in no way assured by the statistical test. This caution is especially relevant because MANOVA, as an extension of ANOVA, stems from experimental research where IVs are typically manipulated by the experimenter and desire for causal inference provides the reason behind elaborate controls. But the statistical test is available whether or not IVs are manipulated, subjects randomly assigned, and controls implemented. Therefore, the inference that significant changes in the DVs are caused by concomitant changes in the IVs is a logical exercise, not a statistical one.

Choice of variables is also a question of logic and research design rather than of statistics. Skill is required in choosing IVs and levels of IVs, as well as DVs that have some chance of showing effects of the IVs. A further consideration in choice of DVs is the extent of likely correlation among them. The best choice is a set of DVs that are uncorrelated with each other because they each measure a separate aspect of the influence of the IVs. When DVs are correlated, they measure the same or similar facets of behavior in slightly different ways. What is gained by inclusion of several measures of the same thing? Might there be some way of combining DVs or deleting some of them so that the analysis is simpler?

In addition to choice of number and type of DVs is choice of the order in which DVs enter a stepdown analysis if Roy-Bargmann stepdown F is the method chosen to assess the importance of DVs (see Section 7.5.3.2). Priority is usually given to more important DVs or to DVs that are considered causally prior to others in theory. The choice is not trivial because the significance of a DV may well depend on how high a priority it is given, just as in sequential multiple regression the significance of an IV is likely to depend on its position in the sequence.

When MANCOVA is used, the same limitations apply as in ANCOVA. Consult Sections 6.3.1 and 6.5 for a review of some of the hazards associated with interpretation of designs that include covariates.

Finally, the usual limits to generalizability apply. The results of MANOVA and MANCOVA generalize only to those populations from which the researcher has randomly sampled. And although MANCOVA may, in some very limited situations, adjust for failure to randomly assign subjects to groups, MANCOVA does not adjust for failure to sample from segments of the population to which one wishes to generalize.

7.3.2 Practical Issues

In addition to the theoretical and logical issues discussed above, the statistical procedure demands consideration of some practical matters.

7.3.2.1 *Unequal Sample Sizes, Missing Data, and Power*

Problems associated with unequal cell sizes are discussed in Section 6.5.4.2. Problems caused by incomplete data (and solutions to them) are discussed in Chapters 4 and 6 (particularly Section 6.3.2.1). The discussion applies to MANOVA and, in fact, may be even more relevant because, as experiments are complicated by numerous DVs and, perhaps, covariates, the probability of missing data increases.

In addition, when using MANOVA, it is necessary to have more cases than DVs in every cell. With numerous DVs this requirement can become burdensome, especially when the design is complicated and there are numerous cells. There are two reasons for the requirement. The first is associated with the assumption of homogeneity of variance-covariance matrices (see Section 7.3.2.4). If a cell has more DVs than cases, the cell becomes singular and the assumption is untestable. If the cell has only one or two more cases than DVs, the assumption is likely to be rejected. Thus MANOVA as an analytic strategy may be discarded because of a failed assumption when the assumption failed because the cases-to-DVs ratio is too low.

Second, the power of the analysis is lowered unless there are more cases than DVs in every cell because of reduced degrees of freedom for error. One likely outcome of reduced power is a nonsignificant multivariate F, but one or more significant univariate Fs (and a very unhappy researcher). Sample sizes in each cell must be sufficient in any event to ensure adequate power. There are many software programs available to calculate required sample sizes depending on desired power and anticipated means and standard deviations in an ANOVA. An Internet search for "statistical power" reveals a number of them, some of which are free. One quick-and-dirty way to apply these is to pick the DV with the smallest expected difference that you want to show statistical significance—your minimum significant DV. One program specifically designed to assess power in MANOVA is GANOVA (Woodward, Bonett, & Brecht, 1990). Another is NCSS PASS (2002), which now includes power analysis for between-subjects MANOVA. Required sample size also may be estimated through SPSS MANOVA by a process of successive approximation. For post hoc estimates of power at a given sample size, you compute a constant weighting variable, weight cases by that variable, and rerun the analysis until desired power is achieved (David P. Nichols, SPSS, personal communication, April 19, 2005). Matrix input is useful for a priori estimates of sample size using SPSS MANOVA (D'Amico, Neilands, & Zambarano, 2001).

Power in MANOVA also depends on the relationships among DVs. Power for the multivariate test is highest when the pooled within-cell correlation among two DVs is high and negative. The multivariate test has much less power when the correlation is positive, zero, or moderately negative. An interesting thing happens, however, when one of two DVs is affected by the treatment and the other is not. The higher the absolute value of the correlation between the two DVs, the greater the power of the multivariate test (Woodward et al., 1990).

7.3.2.2   *Multivariate Normality*

Significance tests for MANOVA, MANCOVA, and other multivariate techniques are based on the multivariate normal distribution. Multivariate normality implies that the sampling distributions of means of the various DVs in each cell and all linear combinations of them are normally distributed. With univariate F and large samples, the central limit theorem suggests that the sampling distribution of means approaches normality even when raw scores do not. Univariate F is robust to modest violations of normality as long as there are at least 20 degrees of freedom for error in a univariate ANOVA and the violations are not due to outliers (Section 4.1.5). Even with unequal n and only a few DVs, a sample size of about 20 in the smallest cell should ensure robustness (Mardia, 1971). In Monte Carlo studies, Seo, Kanda, and Fujikoshi (1995) have shown robustness to nonnormality in MANOVA with overall $N = 40$ ($n = 10$ per group).

With small, unequal samples, normality of DVs is assessed by reliance on judgment. Are the individual DVs expected to be fairly normally distributed in the population? If not, is some transformation likely to produce normality? With a nonnormally distributed covariate consider transformation or deletion. Covariates are often included as a convenience in reducing error, but it is hardly a convenience if it reduces power.

7.3.2.3   *Absence of Outliers*

One of the more serious limitations of MANOVA (and ANOVA) is its sensitivity to outliers. Especially worrisome is that an outlier can produce either a Type I or a Type II error, with no clue in the analysis as to which is occurring. Therefore, it is highly recommended that a test for outliers accompany any use of MANOVA.

Several programs are available for screening for univariate and multivariate outliers (cf. Chapter 4). *Run tests for univariate and multivariate outliers for each cell of the design separately and change, transform, or eliminate them.* Report the change, transformation, or deletion of outlying cases. Screening runs for within-cell univariate and multivariate outliers are shown in Sections 6.6.1.4 and 7.6.1.4.

7.3.2.4   *Homogeneity of Variance-Covariance Matrices*

The multivariate generalization of homogeneity of variance for individual DVs is homogeneity of variance-covariance matrices as discussed in Section 4.1.5.3.[2] The assumption is that variance-covariance matrices within each cell of the design are sampled from the same population variance-

---

[2]In MANOVA, homogeneity of variance for each of the DVs is also assumed. See Section 8.3.2.4 for discussion and recommendations.

covariance matrix and can reasonably be pooled to create a single estimate of error.[3] If the within-cell error matrices are heterogeneous, the pooled matrix is misleading as an estimate of error variance.

The following guidelines for testing this assumption in MANOVA are based on a generalization of a Monte Carlo test of robustness for T^2 (Hakstian, Roed, & Lind, 1979). If sample sizes are equal, robustness of significance tests is expected; disregard the outcome of Box's M test, a notoriously sensitive test of homogeneity of variance-covariance matrices available through SPSS MANOVA.

However, if sample sizes are unequal and Box's M test is significant at $p < .001$, then robustness is not guaranteed. The more numerous the DVs and the greater the discrepancy in cell sample sizes, the greater the potential distortion of alpha levels. Look at both sample sizes and the sizes of the variances and covariances for the cells. If cells with larger samples produce larger variances and covariances, the alpha level is conservative so that null hypotheses can be rejected with confidence. If, however, cells with smaller samples produce larger variances and covariances, the significance test is too liberal. Null hypotheses are retained with confidence but indications of mean differences are suspect. Use Pillai's criterion instead of Wilks' lambda (see Section 7.5.2) to evaluate multivariate significance (Olson, 1979); or equalize sample sizes by random deletion of cases, if power can be maintained at reasonable levels.

7.3.2.5 Linearity

MANOVA and MANCOVA assume linear relationships among all pairs of DVs, all pairs of covariates, and all DV–covariate pairs in each cell. Deviations from linearity reduce the power of the statistical tests because (1) the linear combinations of DVs do not maximize the separation of groups for the IVs, and (2) covariates do not maximize adjustment for error. Section 4.1.5.2 provides guidelines for checking for and dealing with nonlinearity. If serious curvilinearity is found with a covariate, consider deletion; if curvilinearity is found with a DV, consider transformation—provided, of course, that increased difficulty in interpretation of a transformed DV is worth the increase in power.

7.3.2.6 Homogeneity of Regression

In Roy-Bargmann stepdown analysis (Section 7.5.3.2) and in MANCOVA (Section 7.4.3) it is assumed that the regression between covariates and DVs in one group is the same as the regression in other groups so that using the average regression to adjust for covariates in all groups is reasonable.

In both MANOVA and MANCOVA, if Roy-Bargmann stepdown analysis is used, the importance of a DV in a hierarchy of DVs is assessed in ANCOVA with higher-priority DVs serving as covariates. Homogeneity of regression is required for each step of the analysis, as each DV, in turn, joins the list of covariates. If heterogeneity of regression is found at a step, the rest of the stepdown analysis is uninterpretable. Once violation occurs, the IV-"covariate" interaction is itself interpreted and the DV causing violation is eliminated from further steps.

In MANCOVA (like ANCOVA) heterogeneity of regression implies that there is interaction between the IV(s) and the covariates and that a different adjustment of DVs for covariates is needed in different groups. If interaction between IVs and covariates is suspected, MANCOVA is an inappropriate analytic strategy, both statistically and logically. Consult Sections 6.3.2.7 and 6.5.5 for alternatives to MANCOVA where heterogeneity of regression is found.

---

[3]Don't confuse this assumption with the assumption of sphericity that is relevant to repeated-measures ANOVA or MANOVA, as discussed in Section 6.5.4.1 and 8.5.1.

For MANOVA, test for stepdown homogeneity of regression, and for MANCOVA, test for over-all and stepdown homogeneity of regression. These procedures are demonstrated in Section 7.6.1.6.

7.3.2.7   Reliability of Covariates

In MANCOVA as in ANCOVA, the *F* test for mean differences is more powerful if covariates are reliable. If covariates are not reliable, either increased Type I or Type II errors can occur. Reliability of covariates is discussed more fully in Section 6.3.2.8.

In Roy-Bargmann stepdown analysis where all but the lowest-priority DV act as covariates in assessing other DVs, unreliability of any of the DVs (say, $r_{yy} < .8$) raises questions about stepdown analysis as well as about the rest of the research effort. When DVs are unreliable, use another method for assessing the importance of DVs (Section 7.5.3) and report known or suspected unreliability of covariates and high-priority DVs in your Results section.

7.3.2.8   Absence of Multicollinearity and Singularity

When correlations among DVs are high, one DV is a near-linear combination of other DVs; the DV provides information that is redundant to the information available in one or more of the other DVs. It is both statistically and logically suspect to include all the DVs in analysis and *the usual solution is deletion of the redundant DV.* However, if there is some compelling theoretical reason to retain all DVs, a principal components analysis (cf. Chapter 13) is done on the pooled within-cell correlation matrix, and component scores are entered as an alternative set of DVs.

SAS and SPSS GLM protect against multicollinearity and singularity through computation of pooled within-cell tolerance $(1 - SMC)$ for each DV; DVs with insufficient tolerance are deleted from analysis. In SPSS MANOVA, singularity or multicollinearity may be present when the determinant of the within-cell correlation matrix is near zero (say, less than .0001). Section 4.1.7 discusses multi-collinearity and singularity and has suggestions for identifying the redundant variable(s).

7.4   Fundamental Equations for Multivariate Analysis of Variance and Covariance

7.4.1   Multivariate Analysis of Variance

A minimum data set for MANOVA has one or more IVs, each with two or more levels, and two or more DVs for each subject within each combination of IVs. A fictitious small sample with two DVs and two IVs is illustrated in Table 7.1. The first IV is degree of disability with three levels—mild, moderate, and severe—and the second is treatment with two levels—treatment and no treatment. These two IVs in factorial arrangement produce six cells; three children are assigned to each cell so there are 3×6 or 18 children in the study. Each child produces two DVs: score on the reading subtest of the Wide Range Achievement Test (WRAT-R) and score on the arithmetic subtest (WRAT-A). In addition an IQ score is given in parentheses for each child to be used as a covariate in Section 7.4.3.

The test of the main effect of treatment asks: Disregarding degree of disability, does treatment affect the composite score created from the two subtests of the WRAT? The test of interaction asks: Does the effect of treatment on a difference composite score from the two subtests differ as a function of degree of disability?

TABLE 7.1 Small-Sample Data for Illustration of Multivariate Analysis of Variance

	Mild			Moderate			Severe		
	WRAT-R	*WRAT-A*	*(IQ)*	*WRAT-R*	*WRAT-A*	*(IQ)*	*WRAT-R*	*WRAT-A*	*(IQ)*
Treatment	115	108	(110)	100	105	(115)	89	78	(99)
	98	105	(102)	105	95	(98)	100	85	(102)
	107	98	(100)	95	98	(100)	90	95	(100)
Control	90	92	(108)	70	80	(100)	65	62	(101)
	85	95	(115)	85	68	(99)	80	70	(95)
	80	81	(95)	78	82	(105)	72	73	(102)

The test of the main effect of disability is automatically provided in the analysis but is trivial in this example. The question is: Are scores on the WRAT affected by degree of disability? Because degree of disability is at least partially defined by difficulty in reading and/or arithmetic, a significant effect provides no useful information. On the other hand, the absence of this effect would lead us to question the adequacy of classification.

The sample size of three children per cell is highly inadequate for a realistic test but serves to illustrate the techniques of MANOVA. Additionally, if causal inference is intended, the researcher should randomly assign children to the levels of treatment. The reader is encouraged to analyze these data by hand and by computer. Syntax and selected output for this example appear in Section 7.4.2 for several appropriate programs.

MANOVA follows the model of ANOVA where variance in scores is partitioned into variance attributable to difference among scores within groups and to differences among groups. Squared differences between scores and various means are summed (see Chapter 3); these sums of squares, when divided by appropriate degrees of freedom, provide estimates of variance attributable to different sources (main effects of IVs, interactions among IVs, and error). Ratios of variances provide tests of hypotheses about the effects of IVs on the DV.

In MANOVA, however, each subject has a score on each of several DVs. When several DVs for each subject are measured, there is a matrix of scores (subjects by DVs) rather than a simple set of DVs within each group. Matrices of difference scores are formed by subtracting from each score an appropriate mean; then the matrix of differences is squared. When the squared differences are summed, a sum-of-squares-and-cross-products matrix, an **S** matrix, is formed, analogous to a sum of squares in ANOVA (Section 16.4). Determinants[4] of the various **S** matrices are found, and ratios between them provide tests of hypotheses about the effects of the IVs on the linear combination of DVs. In MANCOVA, the sums of squares and cross products in the **S** matrix are adjusted for covariates, just as sums of squares are adjusted in ANCOVA (Chapter 6).

The MANOVA equation for equal *n* is developed below through extension of ANOVA. The simplest partition apportions variance to systematic sources (variance attributable to differences

---

[4]A determinant, as described in Appendix A, can be viewed as a measure of generalized variance for a matrix.

between groups) and to unknown sources of error (variance attributable to differences in scores within groups). To do this, differences between scores and various means are squared and summed.

$$\sum_i \sum_j (Y_{ij} - GM)^2 = n \sum_j (\bar{Y}_j - GM)^2 + \sum_i \sum_j (Y_{ij} - \bar{Y}_j)^2 \tag{7.1}$$

The total sum of squared differences between scores on Y (the DV) and the grand mean (GM) is partitioned into sum of squared differences between group means $(\bar{Y}_j)$ and the grand mean (i.e., systematic or between-groups variability), and sum of squared difference between individual scores (Y_{ij}) and their respective group means.

or

$$SS_{total} = SS_{bg} - SS_{wg}$$

For designs with more than one IV, SS_{bg} is further partitioned into variance associated with the first IV (e.g., degree of disability, abbreviated D), variance associated with the second IV (treatment, or T), and variance associated with the interaction between degree of disability and treatment (or DT).

$$n_{km} \sum_k \sum_m (DT_{km} - GM_{km})^2 = n_k \sum_k (D_k - GM)^2 + n_m \sum_m (T_m - GM)^2$$

$$+ \left[n_{km} \sum_k \sum_m (DT_{km} - GM)^2 - n_k \sum_k (D_k - GM)^2 - n_m \sum_m (T_m - GM)^2 \right] \tag{7.2}$$

The sum of squared differences between cell (DT_{km}) means and the grand mean is partitioned into (1) sum of squared differences between means associated with different levels of disability (D_k) and the grand mean; (2) sum of squared differences between means associated with different levels of treatment (T_m) and the grand mean; and (3) sum of squared differences associated with combinations of treatment and disability (DT_{km}) and the grand mean, from which differences associated with D_k and T_m are subtracted. Each n is the number of scores composing the relevant marginal or cell mean.

or

$$SS_{bg} = SS_D + SS_T + SS_{DT}$$

The full partition for this factorial between-subjects design is

$$\sum_i \sum_k \sum_m (Y_{ikm} - GM)^2 = n_k \sum_k (D_k - GM)^2 + n_m \sum_m (T_m - GM)^2$$

$$+ \left[n_{km} \sum_k \sum_m (DT_{km} - GM)^2 - n_k \sum_k (D_k - GM)^2 - n_m \sum_m (T_m - GM)^2 \right] \tag{7.3}$$

$$+ \sum_i \sum_k \sum_m (Y_{ikm} - DT_{km})^2$$

For MANOVA, there is no single DV but rather a column matrix (or vector) of Y_{ikm} values of scores on each DV. For the example in Table 7.1, column matrices of Y scores for the three children in the first cell of the design (mild disability with treatment) are

$$Y_{i11} = \begin{bmatrix} 115 \\ 108 \end{bmatrix} \begin{bmatrix} 98 \\ 105 \end{bmatrix} \begin{bmatrix} 107 \\ 98 \end{bmatrix}$$

Similarly, there is a column matrix of disability—D_k—means for mild, moderate, and severe levels of D, with one mean in each matrix for each DV.

$$D_1 = \begin{bmatrix} 95.83 \\ 96.50 \end{bmatrix} \quad D_2 = \begin{bmatrix} 88.83 \\ 88.00 \end{bmatrix} \quad D_3 = \begin{bmatrix} 82.67 \\ 77.17 \end{bmatrix}$$

where 95.83 is the mean on WRAT-R and 96.50 is the mean on WRAT-A for children with mild disability, averaged over treatment and control groups.

Matrices for treatment—T_m—means, averaged over children with all levels of disability are

$$T_1 = \begin{bmatrix} 99.89 \\ 96.33 \end{bmatrix} \quad T_2 = \begin{bmatrix} 78.33 \\ 78.11 \end{bmatrix}$$

Similarly, there are six matrices of cell means (DT_{km}) averaged over the three children in each group.

Finally, there is a single matrix of grand means (**GM**), one for each DV, averaged over all children in the experiment.

$$GM = \begin{bmatrix} 89.11 \\ 87.22 \end{bmatrix}$$

As illustrated in Appendix A, differences are found by simply subtracting one matrix from another, to produce difference matrices. The matrix counterpart of a difference score, then, is a difference matrix. To produce the error term for this example, the matrix of grand means (**GM**) is subtracted from each of the matrixes of individual scores (Y_{ikm}). Thus for the first child in the example:

$$(Y_{111} - GM) = \begin{bmatrix} 115 \\ 108 \end{bmatrix} - \begin{bmatrix} 89.11 \\ 87.22 \end{bmatrix} = \begin{bmatrix} 25.89 \\ 20.75 \end{bmatrix}$$

In ANOVA, difference scores are squared. The matrix counterpart of squaring is multiplication by a transpose. That is, each column matrix is multiplied by its corresponding row matrix (see Appendix A for matrix transposition and multiplication) to produce a sum-of-squares and cross-products matrix. For example, for the first child in the first group of the design:

$$(Y_{111} - GM)(Y_{111} - GM)' = \begin{bmatrix} 25.89 \\ 20.78 \end{bmatrix} [25.89 \quad 20.78] = \begin{bmatrix} 670.29 & 537.99 \\ 537.99 & 431.81 \end{bmatrix}$$

These matrices are then summed over subjects and over groups, just as squared differences are summed in univariate ANOVA.[5] The order of summing and squaring is the same in MANOVA as in ANOVA for a comparable design. The resulting matrix (**S**) is called by various names: sum-of-squares and cross-products, cross-products, or sum-of-products. The MANOVA partition of sums-of-squares and cross-products for our factorial example is represented below in a matrix form of Equation 7.3:

$$\sum_i \sum_k \sum_m (\mathbf{Y}_{ikm} - \mathbf{GM})(\mathbf{Y}_{ikm} - \mathbf{GM})'$$

$$= n_k \sum_k (\mathbf{D}_k - \mathbf{GM})(\mathbf{D}_k - \mathbf{GM})' + n_m \sum_m (\mathbf{T}_m - \mathbf{GM})(\mathbf{T}_m - \mathbf{GM})'$$

$$+ \left[n_{km} \sum_k \sum_m (\mathbf{DT}_{km} - \mathbf{GM})(\mathbf{DT}_{km} - \mathbf{GM})' - n_k \sum_k (\mathbf{D}_k - \mathbf{GM})(\mathbf{D}_k - \mathbf{GM})' \right.$$

$$\left. - n_m \sum_m (\mathbf{T}_m - \mathbf{GM})(\mathbf{T}_m - \mathbf{GM})' \right] + \sum_i \sum_k \sum_m (\mathbf{Y}_{ikm} - \mathbf{DT}_{km})(\mathbf{Y}_{ikm} - \mathbf{DT}_{km})'$$

or

$$\mathbf{S}_{\text{total}} = \mathbf{S}_D + \mathbf{S}_T + \mathbf{S}_{DT} + \mathbf{S}_{S(DT)}$$

The total cross-products matrix ($\mathbf{S}_{\text{total}}$) is partitioned into cross-products matrices for differences associated with degree of disability, with treatment, with the interaction between disability and treatment, and for error–subjects within groups ($\mathbf{S}_{S(DT)}$).

For the example in Table 7.1, the four resulting cross-products matrices[6] are

$$\mathbf{S}_D = \begin{bmatrix} 570.29 & 761.72 \\ 761.72 & 1126.78 \end{bmatrix} \qquad \mathbf{S}_T = \begin{bmatrix} 2090.89 & 1767.56 \\ 1767.56 & 1494.22 \end{bmatrix}$$

$$\mathbf{S}_{DT} = \begin{bmatrix} 2.11 & 5.28 \\ 5.28 & 52.78 \end{bmatrix} \qquad \mathbf{S}_{S(DT)} = \begin{bmatrix} 544.00 & 31.00 \\ 31.00 & 539.33 \end{bmatrix}$$

Notice that all these matrices are symmetrical, with the elements top left to bottom right diagonal representing sums of squares (that, when divided by degrees of freedom, produce variances), and with the off-diagonal elements representing sums of cross products (that, when divided by degrees of freedom, produce covariances). In this example, the first element in the major diagonal (top left to bottom right) is the sum of squares for the first DV, WRAT-R, and the second element is the sum of

---

[5]We highly recommend using a matrix algebra program, such as a spreadsheet or SPSS MATRIX, MATLAB, or SAS IML, to follow the more complex matrix equations to come.

[6]Numbers producing these matrices were carried to 8 digits before rounding.

squares for the second DV, WRAT-A. The off-diagonal elements are the sums of cross-products between WRAT-R and WRAT-A.

In ANOVA, sums of squares are divided by degrees of freedom to produce variances, or mean squares. In MANOVA, the matrix analog of variance is a determinant (see Appendix A); the determinant is found for each cross-products matrix. In ANOVA, ratios of variances are formed to test main effects and interactions. In MANOVA, ratios of determinants are formed to test main effects and interactions when using Wilks' lambda (see Section 7.5.2 for additional criteria). These ratios follow the general form

$$\Lambda = \frac{|\mathbf{S}_{error}|}{|\mathbf{S}_{effect} + \mathbf{S}_{error}|} \tag{7.4}$$

Wilks' lambda (Λ) is the ratio of the determinant of the error cross-products matrix to the determinant of the sum of the error and effect cross-products matrices.

To find Wilks' lambda, the within-groups matrix is added to matrices corresponding to main effects and interactions before determinants are found. For the example, the matrix produced by adding the $\mathbf{S}_{DT}$ matrix for interaction to the $\mathbf{S}_{S(DT)}$ matrix for subjects within groups (error) is

$$\mathbf{S}_{DT} + \mathbf{S}_{S(DT)} = \begin{bmatrix} 2.11 & 5.28 \\ 5.28 & 52.78 \end{bmatrix} + \begin{bmatrix} 544.00 & 31.00 \\ 31.00 & 539.33 \end{bmatrix}$$

$$= \begin{bmatrix} 546.11 & 36.28 \\ 36.28 & 592.11 \end{bmatrix}$$

For the four matrices needed to test main effect of disability, main effect of treatment, and the treatment-disability interaction, the determinants are

$$|\mathbf{S}_{S(DT)}| = 292436.52$$

$$|\mathbf{S}_D + \mathbf{S}_{S(DT)}| = 1228124.71$$

$$|\mathbf{S}_T + \mathbf{S}_{S(DT)}| = 2123362.49$$

$$|\mathbf{S}_{DT} + \mathbf{S}_{S(DT)}| = 322040.95$$

At this point a source table, similar to the source table for ANOVA, is useful, as presented in Table 7.2. The first column lists sources of variance; in this case the two main effects and the interaction. The error term does not appear. The second column contains the value of Wilks' lambda.

Wilks' lambda is a ratio of determinants, as described in Equation 7.4. For example, for the interaction between disability and treatment, Wilks' lambda is

$$\Lambda = \frac{|\mathbf{S}_{S(DT)}|}{|\mathbf{S}_{DT} + \mathbf{S}_{S(DT)}|} = \frac{292436.52}{322040.95} = .908068$$

Tables for evaluating Wilks' lambda directly are rare, however, an approximation to F has been derived that closely fits Λ. The last three columns of Table 7.2, then, represent the approximate F values and their associated degrees of freedom.

TABLE 7.2 Multivariate Analysis of Variance of WRAT-R and WRAT-A Scores

Source of Variance	Wilks' Lambda	df_1	df_2	Multivariate F
Treatment	.13772	2.00	11.00	34.43570**
Disability	.25526	4.00	22.00	5.38602*
Treatment by disability	.90807	4.00	22.00	0.27170

*$p < .01$.

**$p < .001$.

The following procedure for calculating approximate F (Rao, 1952) is based on Wilks' lambda and the various degrees of freedom associated with it.

$$\text{Approximate } F(df_1, df_2) = \left(\frac{1-y}{y}\right)\left(\frac{df_2}{df_1}\right) \tag{7.5}$$

where df_1 and df_2 are defined below as the degrees of freedom for testing the F ratio, and y is

$$y = \Lambda^{1/s} \tag{7.6}$$

Λ is defined in Equation 7.4, and s is[7]

$$s = \min(p, df_{effect}) \tag{7.7}$$

where p is the number of DVs, and df_{effect} is the degrees of freedom for the effect being tested. And

$$df_1 = p(df_{effect})$$

and

$$df_2 = s\left[(df_{error}) - \frac{p - df_{effect} + 1}{2}\right] - \left[\frac{p(df_{effect}) - 2}{2}\right]$$

where df_{error} is the degrees of freedom associated with the error term.

For the test of interaction in the sample problem, we have

$p = 2$ the number of DVs

$df_{effect} = 2$ the number of treatment levels minus 1 times the number of disability levels minus 1 or $(t-1)(d-1)$

$df_{error} = 12$ the number of treatment levels times the number of disability levels times the quantity $n-1$ (where n is the number of scores per cell for each DV)—that is, $df_{error} = dt(n-1)$

---

[7]When $p = 1$, we have univariate ANOVA.

Thus

$$s = \min(p, \text{df}_{\text{effect}}) = 2$$

$$y = .908068^{1/2} = .952926$$

$$\text{df}_1 = 2(2) = 4$$

$$\text{df}_2 = 2\left[12 - \frac{2 - 2 + 1}{2}\right] - \left[\frac{2(2) - 2}{2}\right] = 22$$

$$\text{Approximate } F(4, 22) = \left(\frac{.047074}{.952926}\right)\left(\frac{22}{4}\right) = 0.2717$$

This approximate F value is tested for significance by using the usual tables of F at selected α. In this example, the interaction between disability and treatment is not statistically significant with 4 and 22 df, because the observed value of 0.2717 does not exceed the critical value of 2.82 at $\alpha = .05$.

Following the same procedures, the effect of treatment is statistically significant, with the observed value of 34.44 exceeding the critical value of 3.98 with 2 and 11 df, $\alpha = .05$. The effect of degree of disability is also statistically significant, with the observed value of 5.39 exceeding the critical value of 2.82 with 4 and 22 df, $\alpha = .05$. (As noted previously, this main effect is not of research interest, but does serve to validate the classification procedure.) In Table 7.2, significance is indicated at the highest level of α reached, following standard practice.

A measure of effect size is readily available from Wilks' lambda.[8] For MANOVA:

$$\eta^2 = 1 - \Lambda \tag{7.8}$$

This equation represents the variance accounted for by the best linear combination of DVs as explained below.

In a one-way analysis, according to Equation 7.4, Wilks' lambda is the ratio of (the determinant of) the error matrix and (the determinant of) the total sum-of-squares and cross-products matrix. The determinant of the error matrix—Λ—is the variance not accounted for by the combined DVs so $1 - \Lambda$ is the variance that is accounted for.

Thus, for each statistically significant effect, the proportion of variance accounted for is easily calculated using Equation 7.8. For example, the main effect of treatment:

$$\eta_T^2 = 1 - \Lambda_T = 1 - .137721 = .862279$$

In the example, 86% of the variance in the best linear combination of WRAT-R and WRAT-A scores is accounted for by assignment to levels of treatment. The square root of η^2 ($\eta = .93$) is a form of correlation between WRAT scores and assignment to treatment.

However, unlike η^2 in the analogous ANOVA design, the sum of η^2 for all effects in MANOVA may be greater than 1.0 because DVs are recombined for each effect. This lessens the appeal of an interpretation in terms of proportion of variance accounted for, although the size of η^2 is still a measure of the relative importance of an effect.

---

[8]An alternative measure of effect size is canonical correlation, printed out by some computer programs. Canonical correlation is the correlation between the optimal linear combination of IV levels and the optimal linear combination of DVs where optimal is chosen to maximize the correlation between combined IVs and DVs. Canonical correlation as a general procedure is discussed in Chapter 12, and the relation between canonical correlation and MANOVA is discussed briefly in Chapter 17.

Another difficulty in using this form of η^2 is that effects tend to be much larger in the multivariate than in the univariate case. Therefore, a recommended alternative, when $s > 1$ is

$$\text{partial } \eta^2 = 1 - \Lambda^{1/s} \tag{7.9}$$

Estimated effect size is reduced to .63 with the use of partial η^2 for the current data, a more reasonable assessment. Confidence limits around effect sizes are in Section 7.6.

7.4.2 Computer Analyses of Small-Sample Example

Tables 7.3 through 7.5 show syntax and selected minimal output for SPSS MANOVA, SPSS GLM, and SAS GLM, respectively.

In SPSS MANOVA (Table 7.3) simple MANOVA source tables, resembling those of ANOVA, are printed out when PRINT=SIGNIF(BRIEF) is requested. After interpretive material is printed (not shown), the source table is shown, labeled Tests using UNIQUE sums of squares and WITHIN+RESIDUAL. WITHIN+RESIDUAL refers to the pooled within-cell error SSCP matrix (Section 7.4.1) plus any effects not tested, the error term chosen by default for MANOVA.

For the example, the two-way MANOVA source table consists of the two main effects and the interaction. For each source, you are given Wilks' lambda, Approximate (multivariate) F with numerator and denominator degrees of freedom (Hyp. DF and Error DF, respectively), and the probability level achieved for the significance test.

Syntax for SPSS GLM is similar to that of MANOVA, except that levels of IVs are not shown in parentheses. METHOD, INTERCEPT, and CRITERIA instructions are produced by the menu system by default.

Output consists of a source table that includes four tests of the multivariate effects, Pillai's, Wilks', Hotelling's, and Roy's (see Section 7.5.2 for a discussion of these tests). All are identical when there are only two levels of a between-subjects IV. The results of Wilks' Lambda test match those of SPSS MANOVA in Table 7.3. This is followed by univariate tests on each of the DVs, in the

TABLE 7.3   MANOVA on Small-Sample Example through SPSS MANOVA (Syntax and Output)

```
MANOVA
 WRATR WRATA BY TREATMNT(1,2) DISABLTY(1,3)
 /PRINT=SIGNIF(BRIEF)
 /DESIGN = TREATMNT DISABLTY TREATMNT*DISABLTY.

* * * * * * A n a l y s i s   o f   V a r i a n c e—design 1 * * * * * *

Multivariate Tests of Significance
Tests using UNIQUE sums of squares and WITHIN+RESIDUAL error term
Source of Variation     Wilks    Approx F   Hyp. DF    Error DF    Sig of F

TREATMNT                .138     34.436      2.00      11.000       .000
DISABLTY                .255      5.386      4.00      22.000       .004
TREATMNT * DISABLTY     .908       .272      4.00      22.000       .893
```

TABLE 7.4 MANOVA on Small-Sample Example through SPSS GLM (Syntax and Selected Output)

```
GLM
 wratr wrata BY treatmnt disablty
 /METHOD = SSTYPE(3)
 /INTERCEPT = INCLUDE
 /CRITERIA = ALPHA(.05)
 /DESIGN = treatmnt disablty treatmnt*disablty.
```

General Linear Model

Between-Subjects Factors

		Value Label	N
Treatment type	1.00	Treatment	9
	2.00	Control	9
Degree of disability	1.00	Mild	6
	2.00	Moderate	6
	3.00	Severe	6

Multivariate Tests[c]

Effect		Value	F	Hypothesis df	Error df	Sig.
Intercept	Pillai's Trace	.998	2687.779[a]	2.000	11.000	.000
	Wilks' Lambda	.002	2687.779[a]	2.000	11.000	.000
	Hotelling's Trace	488.687	2687.779[a]	2.000	11.000	.000
	Roy's Largest Root	488.687	2687.779[a]	2.000	11.000	.000
Treatmnt	Pillai's Trace	.862	34.436[a]	2.000	11.000	.000
	Wilks' Lambda	.138	34.436[a]	2.000	11.000	.000
	Hotelling's Trace	6.261	34.436[a]	2.000	11.000	.000
	Roy's Largest Root	6.261	34.436[a]	2.000	11.000	.000
Disablty	Pillai's Trace	.750	3.604	4.000	24.000	.019
	Wilks' Lambda	.255	5.386[a]	4.000	22.000	.004
	Hotelling's Trace	2.895	7.238	4.000	20.000	.001
	Roy's Largest Root	2.887	17.323[b]	2.000	12.000	.000
Treatmnt * disablty	Pillai's Trace	.092	.290	4.000	24.000	.882
	Wilks' Lambda	.908	.272[a]	4.000	22.000	.893
	Hotelling's Trace	.101	.252	4.000	20.000	.905
	Roy's Largest Root	.098	.588[b]	2.000	12.000	.571

[a]Exact statistic
[b]The statistic is an upper bound on F that yields a lower bound on the significance level.
[c]Design: Intercept+Treatmnt+Disablty+Treatmnt * Disablty

TABLE 7.4   Continued

Tests of Between-Subjects Effects

Source	Dependent Variable	Type III Sum of Squares	df	Mean Square	F	Sig.
Corrected Model	WRAT - Reading	2613.778[a]	5	522.756	11.531	.000
	WRAT - Arithmetic	2673.778[b]	5	534.756	11.898	.000
Intercept	WRAT - Reading	142934.222	1	142934.222	3152.961	.000
	WRAT - Arithmetic	136938.889	1	136938.999	3046.848	.000
Treatmnt	WRAT - Reading	2090.889	1	2090.889	46.123	.000
	WRAT - Arithmetic	1494.222	1	1494.222	33.246	.000
Disablty	WRAT - Reading	520.778	2	260.389	5.744	.018
	WRAT - Arithmetic	1126.778	2	563.389	12.535	.001
Treatmnt * Disablty	WRAT - Reading	2.111	2	1.056	.023	.977
	WRAT - Arithmetic	52.778	2	26.389	.587	.571
Error	WRAT - Reading	544.000	12	45.333		
	WRAT - Arithmetic	539.333	12	44.944		
Total	WRAT - Reading	146092.000	18			
	WRAT - Arithmetic	140152.000	18			
Corrected Total	WRAT - Reading	3157.778	17			
	WRAT - Arithmetic	3213.111	17			

[a]R Squared = .828 (Adjusted R Squared = .756)
[b]R Squared = .832 (Adjusted R Squared = .762)

table labeled Tests of Between-Subjects Effects. The format of the table follows that of univariate ANOVA (see Table 6.5). Note that interpretation of MANOVA through univariate ANOVAs is *not* recommended (cf. Section 7.5.3.1).

In SAS GLM (Table 7.5) IVs are defined in a class instruction and the model instruction defines the DVs and the effects to be considered. The nouni instruction suppresses printing of descriptive statistics and univariate *F* tests. The manova h = _all_ instruction requests tests of all main effects and interactions listed in the model instruction, and short condenses the printout.

The output begins with some interpretative information (not shown), followed by separate sections for TREATMNT, DISABLTY, and TREATMNT*DISABLTY. Each source table is preceded by information about characteristic roots and vectors of the error SSCP matrix (not shown—these are discussed in Chapters 9, 12, and 13), and the three df parameters (Section 7.4.1). Each source table shows results of four multivariate tests, fully labeled (cf. Section 7.5.2).

7.4.3   Multivariate Analysis of Covariance

In MANCOVA, the linear combination of DVs is adjusted for differences in the covariates. The adjusted linear combination of DVs is the combination that would be obtained if all participants had the same scores on the covariates. For this example, pre-experimental IQ scores (listed in parentheses in Table 7.1) are used as covariates.

In MANCOVA the basic partition of variance is the same as in MANOVA. However, all the matrices—$\mathbf{Y}_{ikm}$, $\mathbf{D}_k$, $\mathbf{T}_m$, $\mathbf{DT}_{km}$, and **GM**—have three entries in our example; the first entry is the covariate (IQ score) and the second two entries are the two DV scores (WRAT-R and WRAT-A). For example, for the first child with mild disability and treatment, the column matrix of covariate and DV scores is

$$\mathbf{Y}_{111} = \begin{bmatrix} 110 \\ 115 \\ 108 \end{bmatrix} \quad \begin{matrix} \text{(IQ)} \\ \text{(WRAT-R)} \\ \text{(WRAT-A)} \end{matrix}$$

As in MANOVA, difference matrices are found by subtraction, and then the squares and cross-products matrices are found by multiplying each difference matrix by its transpose to form the **S** matrices.

At this point another departure from MANOVA occurs. The **S** matrices are partitioned into sections corresponding to the covariates, the DVs, and the cross-products of covariates and DVs. For the example, the cross-products matrix for the main effect of treatment is

$$\mathbf{S}_T = \begin{bmatrix} [2.00] & [64.67 \quad 54.67] \\ \begin{bmatrix} 64.67 \\ 54.67 \end{bmatrix} & \begin{bmatrix} 2090.89 & 1767.56 \\ 1767.56 & 1494.22 \end{bmatrix} \end{bmatrix}$$

The lower right-hand partition is the $\mathbf{S}_T$ matrix for the DVs (or $\mathbf{S}_T^{(Y)}$) and is the same as the $\mathbf{S}_T$ matrix developed in Section 7.4.1. The upper left matrix is the sum of squares for the covariate (or $\mathbf{S}_T^{(X)}$). (With additional covariates, this segment becomes a full sum-of-squares and cross-products matrix.) Finally, the two off-diagonal segments contain cross-products of covariates and DVs (or $\mathbf{S}_T^{(XY)}$).

TABLE 7.5   MANOVA on Small-Sample Example through SAS GLM (Syntax and Selected Output)

```
proc glm data=SASUSER.SS_MANOV;
   class TREATMNT DISABLTY;
   model WRATR WRATA=TREATMNT DISABLTY TREATMNT*DISABLTY / nouni;
   manova h=_all_ / short;
run;
```

```
             MANOVA Test Criteria and Exact F Statistics for
               the Hypothesis of NO Overall TREATMNT Effect
                  H = Type III SSCP Matrix for TREATMNT
                       E = Error SSCP Matrix

                       S=1    M=0    N=4.5

Statistic                      Value    F Value   Num DF   Den DF   Pr > F

Wilks' lambda               0.13772139    34.44       2       11    <.0001
Pillai's Trace              0.86227861    34.44       2       11    <.0001
Hotelling-Lawley Trace      6.26103637    34.44       2       11    <.0001
Roy's Greatest Root         6.26103637    34.44       2       11    <.0001


          Characteristic Roots and Vectors of: E Inverse * H, where
                   H = Type III SSCP Matrix for DISABLTY
                        E = Error SSCP Matrix


Characteristic                        Characteristic Vector V'EV=1
      Root       Percent                    WRATR            WRATA

  2.88724085      99.73                  0.02260839       0.03531017
  0.00779322       0.27                 -0.03651215       0.02476743


             MANOVA Test Criteria and F Approximations for
               the Hypothesis of NO Overall DISABLTY Effect
                  H = Type III SSCP Matrix for DISABLTY
                       E = Error SSCP Matrix

                      S=2    M=-0.5    N=4.5

Statistic                      Value    F Value   Num DF   Den DF   Pr > F

Wilks' lambda               0.25526256     5.39       4       22    0.0035
Pillai's Trace              0.75048108     3.60       4       24    0.0195
Hotelling-Lawley Trace      2.89503407     7.79       4   12.235    0.0023
Roy's Greatest Root         2.88724085    17.32       2       12    0.0003

      NOTE: F Statistic for Roy's Greatest Root is an upper bound.
         NOTE: F Statistic for Wilks' lambda is exact.
```

(continued)

TABLE 7.5 **Continued**

```
     Characteristic Roots and Vectors of: E Inverse * H, where
           H = Type III SSCP Matrix for TREATMNT*DISABLTY
                    E = Error SSCP Matrix

Characteristic                        Characteristic Vector V'EV=1
         Root     Percent                    WRATR              WRATA

    0.09803883     97.11                   0.00187535        0.04291087
    0.00291470      2.89                   0.04290407       -0.00434641

            MANOVA Test Criteria and F Approximations for
         the Hypothesis of NO Overall TREATMNT*DISABLTY Effect
            H = Type III SSCP Matrix for TREATMNT*DISABLTY
                    E = Error SSCP Matrix

                S=2    M=-0.5    N=4.5

Statistic                    Value        F   Num DF   Den DF  Pr > F

Wilks' lambda             0.90806786    0.27      4       22   0.8930
Pillai's Trace            0.09219163    0.29      4       24   0.8816
Hotelling-Lawley Trace    0.10095353    0.27      4   12.235   0.8908
Roy's Greatest Root       0.09803883    0.59      2       12   0.5706

   NOTE: F Statistic for Roy's Greatest Root is an upper bound.
         NOTE: F Statistic for Wilks' lambda is exact.
```

Adjusted or S^* matrices are formed from these segments. The S^* matrix is the sums-of-squares and the cross-products of DVs adjusted for effects of covariates. Each sum of squares and each cross-product is adjusted by a value that reflects variance due to differences in the covariate.

In matrix terms, the adjustment is

$$S^* = S^{(Y)} - S^{(YX)}(S^{(X)})^{-1}S^{(XY)} \tag{7.10}$$

The adjusted cross-products matrix S^* is found by subtracting from the unadjusted cross-products matrix of DVs ($S^{(Y)}$) a product based on the cross-products matrix for covariate(s) ($S^{(X)}$) and cross-products matrices for the relation between the covariates and the DVs ($S^{(YX)}$ and $S^{(XY)}$).

The adjustment is made for the regression of the DVs (Y) on the covariates (X). Because $S^{(XY)}$ is the transpose of $S^{(YX)}$, their multiplication is analogous to a squaring operation. Multiplying by the inverse of $S^{(X)}$ is analogous to division. As shown in Chapter 3 for simple scalar numbers, the regression coefficient is the sum of cross-products between X and Y, divided by the sum of squares for X.

An adjustment is made to each $\mathbf{S}$ matrix to produce $\mathbf{S}^*$ matrices. The $\mathbf{S}^*$ matrices are 2×2 matrices, but their entries are usually smaller than those in the original MANOVA $\mathbf{S}$ matrices. For the example, the reduced $\mathbf{S}^*$ matrices are

$$\mathbf{S}_D^* = \begin{bmatrix} 388.18 & 500.49 \\ 500.49 & 654.57 \end{bmatrix} \qquad \mathbf{S}_T^* = \begin{bmatrix} 2059.50 & 1708.24 \\ 1708.24 & 1416.88 \end{bmatrix}$$

$$\mathbf{S}_{DT}^* = \begin{bmatrix} 2.06 & 0.87 \\ 0.87 & 19.61 \end{bmatrix} \qquad \mathbf{S}_{S(DT)}^* = \begin{bmatrix} 528.41 & -26.62 \\ -26.62 & 324.95 \end{bmatrix}$$

Note that, as in the lower right-hand partition, cross-products matrices may have negative values for entries other than the major diagonal which contains sums of squares.

Tests appropriate for MANOVA are applied to the adjusted $\mathbf{S}^*$ matrices. Ratios of determinants are formed to test hypotheses about main effects and interactions by using Wilks' lambda criterion (Equation 7.4). For the example, the determinants of the four matrices needed to test the three hypotheses (two main effects and the interaction) are

$$|\mathbf{S}_{S(DT)}^*| = 171032.69$$

$$|\mathbf{S}_D^* + \mathbf{S}_{S(DT)}^*| = 673383.31$$

$$|\mathbf{S}_T^* + \mathbf{S}_{S(DT)}^*| = 1680076.69$$

$$|\mathbf{S}_{DT}^* + \mathbf{S}_{S(DT)}^*| = 182152.59$$

The source table for MANCOVA, analogous to that produced for MANOVA, for the sample data is in Table 7.6.

One new item in this source table that is not in the MANOVA table of Section 7.4.1 is the variance in the DVs due to the covariate. (With more than one covariate, there is a line for combined covariates and a line for each of the individual covariates.) As in ANCOVA, one degree of freedom

TABLE 7.6 Multivariate Analysis of Covariance of WRAT-R and WRAT-A Scores

Source of Variance	Wilks' Lambda	df_1	df_2	Multivariate F
Covariate	.58485	2.00	10.00	3.54913
Treatment	.13772	2.00	10.00	44.11554**
Disability	.25526	4.00	22.00	4.92112*
Treatment by Disability	.90807	4.00	22.00	0.15997

$*p < .01.$

$**p < .001.$

for error is used for each covariate so that df_2 and s of Equation 7.5 are modified. For MANCOVA, then,

$$s = \min(p + q, df_{effect}) \tag{7.11}$$

where q is the number of covariates and all other terms are defined as in Equation 7.7.

$$df_2 = s\left[(df_{error}) - \frac{(p + q) - df_{effect} + 1}{2}\right] - \left[\frac{(p + q)(df_{effect}) - 2}{2}\right]$$

Approximate F is used to test the significance of the covariate–DV relationship as well as main effects and interactions. If a significant relationship is found, Wilks' lambda is used to find the effect size as shown in Equations 7.6 or 7.9.

7.5   Some Important Issues

7.5.1   MANOVA vs. ANOVAs

MANOVA works best with highly negatively correlated DVs and acceptably well with moderately correlated DVs in either direction (about $|.6|$). For example, two DVs, such as time to complete a task and number of errors, might be expected to have a moderate negative correlation and are best analyzed through MANOVA. MANOVA is less attractive if correlations among DVs are very highly positive or near zero (Woodward et al., 1990).

Using very highly positively correlated DVs in MANOVA is wasteful. For example, the effects of the Head Start program might be tested in a MANOVA with the WISC and Stanford-Binet as DVs. The overall multivariate test works acceptably well, but after the highest priority DV is entered in stepdown analysis, tests of remaining DVs are ambiguous. Once that DV becomes a covariate, there is no variance remaining in the lower priority DVs to be related to IV main effects or interactions. Univariate tests also are highly misleading, because they suggest effects on different behaviors when actually there is one behavior being measured repeatedly. Better strategies are to pick a single DV (preferably the most reliable) or to create a composite score (an average if the DVs are commensurate or a principal component score if they are not) for use in ANOVA.

MANOVA also is wasteful if DVs are uncorrelated—naturally, or if they are factor or component scores. The multivariate test has lower power than the univariate and there is little difference between univariate and stepdown results. The only advantage to MANOVA over separate ANOVAs on each DV is control of familywise Type I error. However, this error rate can be controlled by applying a Bonferroni correction (cf. Equation 7.12) to each test in a set of separate ANOVAs on each DV, although that could potentially result in a more conservative analysis than MANOVA.

Sometimes there is a mix of correlated and uncorrelated DVs. For example, there may be a set of moderately correlated DVs related to performance on a task and another set of moderately correlated DVs related to attitudes. Separate MANOVAs on each of the two sets of moderately correlated DVs are likely to produce the most interesting interpretations as long as appropriate adjustments are made for familywise error rate for the multiple MANOVAs. Or one set might serve as covariates in a single MANCOVA.

7.5.2 Criteria for Statistical Inference

Several multivariate statistics are available in MANOVA programs to test significance of main effects and interactions: Wilks' lambda, Hotelling's trace criterion, Pillai's criterion, as well as Roy's gcr criterion. When an effect has only two levels ($s = 1$, 1 df in the univariate sense), the F tests for Wilks' lambda, Hotelling's trace, and Pillai's criterion are identical. And usually when an effect has more than two levels ($s > 1$ and df > 1 in the univariate sense), the F values are slightly different, but either all three statistics are significant or all are nonsignificant. Occasionally, however, some of the statistics are significant while others are not, and the researcher is left wondering which result to believe.

When there is only one degree of freedom for effect, there is only one way to combine the DVs to separate the two groups from each other. However, when there is more than one degree of freedom for effect, there is more than one way to combine DVs to separate groups. For example, with three groups, one way of combining DVs may separate the first group from the other two while the second way of combining DVs separates the second group from the third. Each way of combining DVs is a dimension along which groups differ (as described in gory detail in Chapter 9) and each generates a statistic.

When there is more than one degree of freedom for effect, Wilks' lambda, Hotelling's trace criterion, and Pillai's criterion pool the statistics from each dimension to test the effect; Roy's gcr criterion uses only the first dimension (in our example, the way of combining DVs that separates the first group from the other two) and is the preferred test statistic for a few researchers (Harris, 2001). Most researchers, however, use one of the pooled statistics to test the effect (Olson, 1976).

Wilks' lambda, defined in Equation 7.4 and Section 7.4.1, is a likelihood ratio statistic that tests the likelihood of the data under the assumption of equal population mean vectors for all groups against the likelihood under the assumption that population mean vectors are identical to those of the sample mean vectors for the different groups. Wilks' lambda is the pooled ratio of error variance to effect variance plus error variance. Hotelling's trace is the pooled ratio of effect variance to error variance. Pillai's criterion is simply the pooled effect variances.

Wilks' lambda, Hotelling's trace, and Roy's gcr criterion are often more powerful than Pillai's criterion when there is more than one dimension but the first dimension provides most of the separation of groups; they are less powerful when separation of groups is distributed over dimensions. But Pillai's criterion is said to be more robust than the other three (Olson, 1979). As sample size decreases, unequal n's appear, and the assumption of homogeneity of variance-covariance matrices is violated (Section 7.3.2.2), the advantage of Pillai's criterion in terms of robustness is more important. When the research design is less than ideal, then Pillai's criterion is the criterion of choice.

In terms of availability, all the MANOVA programs reviewed here provide Wilks' lambda, as do most research reports, so that Wilks' lambda is the criterion of choice unless there is reason to use Pillai's criterion. Programs differ in the other statistics provided (see Section 7.7).

In addition to potentially conflicting significance tests for multivariate F is the irritation of a nonsignificant multivariate F but a significant univariate F for one of the DVs. If the researcher measures only one DV—the right one—the effect is significant, but because more DVs are measured, it is not. Why doesn't MANOVA combine DVs with a weight of 1 for the significant DV and a weight of zero for the rest? In fact, MANOVA comes close to doing just that, but multivariate F is often not as powerful as univariate or stepdown F and significance can be lost. If this happens, about the best one can do is report the nonsignificant multivariate F and offer the univariate and/or stepdown result as a guide to future research.

7.5.3 Assessing DVs

When a main effect or interaction is significant in MANOVA, the researcher has usually planned to pursue the finding to discover which DVs are affected. But the problems of assessing DVs in significant multivariate effects are similar to the problems of assigning importance to IVs in multiple regression (Chapter 5). First, there are multiple significance tests so some adjustment is necessary for inflated Type I error. Second, if DVs are uncorrelated, there is no ambiguity in assignment of variance to them, but if DVs are correlated, assignment of overlapping variance to DVs is problematical.

7.5.3.1 Univariate F

If pooled within-group correlations among DVs are zero (and they never are unless they are formed by principal components analysis), univariate ANOVAs, one per DV, give the relevant information about their importance. Using ANOVA for uncorrelated DVs is analogous to assessing importance of IVs in multiple regression by the magnitude of their individual correlations with the DV. The DVs that have significant univariate Fs are the important ones, and they can be ranked in importance by effect size. However, because of inflated Type I error rate due to multiple testing, more stringent alpha levels are required.

Because there are multiple ANOVAs, a Bonferroni type adjustment is made for inflated Type I error. The researcher assigns alpha for each DV so that alpha for the set of DVs does not exceed some critical value.

$$\alpha = 1 - (1 - \alpha_1)(1 - \alpha_2)\ldots(1 - \alpha_p) \tag{7.12}$$

The Type I error rate (α) is based on the error rate for testing the first DV (α_1), the second DV (α_2), and all other DVs to the pth, or last, DV (α_p).

All the alphas can be set at the same level, or more important DVs can be given more liberal alphas. For example, if there are four DVs and α for each DV is set at .01, the overall alpha level according to Equation 7.12 is .039, acceptably below .05 overall. Or if α is set at .02 for 2 DVs, and at .001 for the other 2 DVs, overall α is .042, also below .05. A close approximation if all α_i are to be the same is:

$$\alpha_i = \alpha_{fw}/p$$

where α_{fw} is the family-wise error rate (e.g., .05) and p is the number of tests

Correlated DVs pose two problems with univariate Fs. First, correlated DVs measure overlapping aspects of the same behavior. To say that two of them are both "significant" mistakenly suggests that the IV affects two different behaviors. For example, if the two DVs are Stanford-Binet IQ and WISC IQ, they are so highly correlated that an IV that affects one surely affects the other. The second problem with reporting univariate Fs for correlated DVs is inflation of Type I error rate; with correlated DVs, the univariate Fs are not independent and no straightforward adjustment of the error rate is possible. In this situation, reporting univariate ANOVAs violates the spirit of MANOVA. However, this is still the most common method of interpreting the results of a MANOVA.

Although reporting univariate F for each DV is a simple tactic, the report should also contain the pooled within-group correlations among DVs so the reader can make necessary interpretive adjustments. The pooled within-group correlation matrix is provided by SPSS MANOVA and SAS GLM.

In the example of Table 7.2, there is a significant multivariate effect of treatment (and of disability, although, as previously noted, it is not interesting in this example). It is appropriate to ask which of the two DVs is affected by treatment. Univariate ANOVAs for WRAT-R and WRAT-A are in Tables 7.7 and 7.8, respectively. The pooled within-group correlation between WRAT-R and WRAT-A is .057 with 12 df. Because the DVs are relatively uncorrelated, univariate F with adjustment of α for multiple tests might be considered appropriate (but note the stepdown results in the following section). There are two DVs, so each is set at alpha .025.[9] With 2 and 12 df, critical F is 5.10; with 1 and 12 df, critical F is 6.55. There is a main effect of treatment (and disability) for both WRAT-R and WRAT-A.

7.5.3.2 Roy-Bargmann Stepdown Analysis[10]

The problem of correlated univariate F tests with correlated DVs is resolved by stepdown analysis (Bock, 1966; Bock & Haggard, 1968). Stepdown analysis of DVs is analogous to testing the importance of IVs in multiple regression by sequential analysis. Priorities are assigned to DVs according to theoretical or practical considerations.[11] The highest-priority DV is tested in univariate ANOVA,

TABLE 7.7   Univariate Analysis of Variance of WRAT-R Scores

Source	SS	df	MS	F
D	520.7778	2	260.3889	5.7439
T	2090.8889	1	2090.8889	46.1225
DT	2.1111	2	1.0556	0.0233
S(DT)	544.0000	12	45.3333	

TABLE 7.8   Univariate Analysis of Variance of WRAT-A Scores

Source	SS	df	MS	F
D	1126.7778	2	563.3889	12.5352
T	1494.2222	1	1494.2222	33.2460
DT	52.7778	2	26.3889	0.5871
S(DT)	539.5668	12	44.9444	

[9]When the design is very complicated and generates many main effects and interactions, further adjustment of α is necessary in order to keep overall α under .15 or so, across the ANOVAs for the DVs.

[10]Stepdown analysis can be run in lieu of MANOVA where a significant stepdown F is interpreted as a significant multivariate effect for the main effect or interaction.

[11]It is also possible to assign priority on the basis of statistical criteria such as univariate F, but the analysis suffers all the problems inherent in stepwise regression, discussed in Chapter 5.

with appropriate adjustment of alpha. The rest of the DVs are tested in a series of ANCOVAs; each successive DV is tested with higher-priority DVs as covariates to see what, if anything, it adds to the combination of DVs already tested. Because successive ANCOVAs are independent, adjustment for inflated Type I error due to multiple testing is the same as in Section 7.5.3.1.

For the example, we assign WRAT-R scores higher priority since reading problems represent the most common presenting symptoms for learning disabled children. To keep overall alpha below .05, individual alpha levels are set at .025 for each of the two DVs. WRAT-R scores are analyzed through univariate ANOVA, as displayed in Table 7.7. Because the main effect of disability is not interesting and the interaction is not statistically significant in MANOVA (Table 7.2), the only effect of interest is treatment. The critical value for testing the treatment effect (6.55 with 1 and 12 df at $\alpha = .025$) is clearly exceeded by the obtained F of 46.1225.

WRAT-A scores are analyzed in ANCOVA with WRAT-R scores as covariate. The results of this analysis appear in Table 7.9.[12] For the treatment effect, critical F with 1 and 11 df at $\alpha = .025$ is 6.72. This exceeds the obtained F of 5.49. Thus, according to stepdown analysis, the significant effect of treatment is represented in WRAT-R scores, with nothing added by WRAT-A scores.

Note that WRAT-A scores show significant univariate but not stepdown F. Because WRAT-A scores are not significant in stepdown analysis does not mean they are unaffected by treatment but rather that no unique variability is shared with treatment after adjustment for differences in WRAT-R. This result occurs despite the relatively low correlation between the DVs.

This procedure can be extended to sets of DVs through MANCOVA. If the DVs fall into categories, such as scholastic variables and attitudinal variables, one can ask whether there is any change in attitudinal variables as a result of an IV, after adjustment for differences in scholastic variables. The attitudinal variables serve as DVs in MANCOVA while the scholastic variables serve as covariates.

7.5.3.3   Using Discriminant Analysis

Discriminant analysis, as discussed more fully in Chapter 9, provides information useful in assessing DVs (DVs are predictors in the context of discriminant analysis). A structure (loading) matrix is produced which contains correlations between the linear combination of DVs that maximizes treatment differences and the DVs themselves. DVs that correlate highly with the combination are more important to discrimination among groups.

TABLE 7.9   Analysis of Covariance of WRAT-A Scores, with WRAT-R Scores as the Covariate

Source	SS	df	MS	F
Covariate	1.7665	1	1.7665	0.0361
D	538.3662	2	269.1831	5.5082
T	268.3081	1	268.3081	5.4903
DT	52.1344	2	26.0672	0.5334
S(DT)	537.5668	11	48.8679	

[12]A full stepdown analysis is produced as an option through SPSS MANOVA. For illustration, however, it is helpful to show how the analysis develops.

Discriminant analysis also can be used to test each of the DVs in the standard multiple regression sense; the effect on each DV is assessed after adjustment for all other DVs. That is, each DV is assessed as if it were the last one to enter an equation. This is demonstrated in Section 9.6.4.

7.5.3.4   *Choosing among Strategies for Assessing DVs*

You may find the procedures of Sections 9.6.3 and 9.6.4 more useful than univariate or stepdown F for assessing DVs when you have a significant multivariate main effect with more than two levels. Similarly, you may find the procedures described in Section 8.5.2 helpful for assessment of DVs if you have a significant multivariate interaction.

The choice between univariate and stepdown F is not always easy, and often you want to use both. When there is no correlation among the DVs, univariate F with adjustment for Type I error is acceptable. When DVs are correlated, as they almost always are, stepdown F is preferable on grounds of statistical purity, but you have to prioritize the DVs and the results can be difficult to interpret.

If DVs are correlated and there is some compelling priority ordering of them, stepdown analysis is clearly called for, with univariate Fs and pooled within-cell correlations reported simply as supplemental information. For significant lower-priority DVs, marginal and/or cell means adjusted for higher-priority DVs are reported and interpreted.

If the DVs are correlated but the ordering is somewhat arbitrary, an initial decision in favor of stepdown analysis is made. If the pattern of results from stepdown analysis makes sense in the light of the pattern of univariate results, interpretation takes both patterns into account with emphasis on DVs that are significant in stepdown analysis. If, for example, a DV has a significant univariate F but a nonsignificant stepdown F, interpretation is straightforward: The variance the DV shares with the IV is already accounted for through overlapping variance with one or more higher-priority DVs. This is the interpretation of WRAT-A in the preceding section and the strategy followed in Section 7.6.

But if a DV has a nonsignificant univariate F and a significant stepdown F, interpretation is much more difficult. In the presence of higher-order DVs as covariates, the DV suddenly takes on "importance." In this case, interpretation is tied to the context in which the DVs entered the stepdown analysis. It may be worthwhile at this point, especially if there is only a weak basis for ordering DVs, to forgo evaluation of statistical significance of DVs and resort to simple description. After finding a significant multivariate effect, unadjusted marginal and/or cell means are reported for DVs with high univariate Fs but significance levels are not given.

An alternative to attempting interpretation of either univariate or stepdown F is interpretation of loading matrices in discriminant analysis, as discussed in Section 9.6.3.2. This process is facilitated when SPSS MANOVA or SAS GLM is used because information about the discriminant functions is provided as a routine part of the output. Alternatively, a discriminant analysis may be run on the data.

Another perspective is whether DVs differ significantly in the effects of IVs on them. For example: Does treatment affect reading significantly more than it affects arithmetic? Tests for contrasts among DVs have been developed in the context of meta-analysis with its emphasis on comparing effect sizes. Rosenthal (2001) demonstrates these techniques.

7.5.4   Specific Comparisons and Trend Analysis

When there are more than two levels in a significant multivariate main effect and when a DV is important to the main effect, the researcher often wants to perform specific comparisons or trend analysis of the DV to pinpoint the source of the significant difference. Similarly, when there is a significant multivariate interaction and a DV is important to the interaction, the researcher follows up the finding with comparisons on the DV. Specific comparisons may also be done on multivariate effects. These are often less interpretable than comparisons on individual DVs, unless DVs are all scaled in the same direction, or are based on factor or principal component scores. Review Sections 3.2.6, 6.5.4.3, and 8.5.2 for examples and discussions of comparisons. The issues and procedures are the same for individual DVs in MANOVA as in ANOVA.

Comparisons are either planned (performed in lieu of omnibus F) or post hoc (performed after omnibus F to snoop the data). When comparisons are post hoc, an extension of the Scheffé procedure is used to protect against inflated Type I error due to multiple tests. The procedure is very conservative but allows for an unlimited number of comparisons. Following Scheffé for ANOVA (see Section 3.2.6), the tabled critical value of F is multiplied by the degrees of freedom for the effect being tested to produce an adjusted, and much more stringent, F. If marginal means for a main effect are being contrasted, the degrees of freedom are those associated with the main effect. If cell means are being contrasted, our recommendation is to use the degrees of freedom associated with the interaction.

Various types of contrasts on individual DVs are demonstrated in Sections 8.5.2.1 and 8.5.2.3. The difference between setting up contrasts on individual DVs and setting up contrasts on the combination is that all DVs are included in the syntax. Table 7.10 shows syntax for trend analysis and user-specified orthogonal contrasts on the main effect of DISABLTY for the small-sample example. The coefficients illustrated for the orthogonal contrasts actually are the trend coefficients. Note that SPSS GLM requires fractions in part of the **LMATRIX** command to produce the right answers.

Use of this syntax also provides univariate tests of contrasts for each DV. None of these contrasts are adjusted for post hoc analysis. The usual corrections are to be applied to minimize inflated Type I error rate unless comparisons are planned (cf. Sections 3.2.6.5, 6.5.4.3, and 8.5.2).

7.5.5   Design Complexity

When between-subjects designs have more than two IVs, extension of MANOVA is straightforward as long as sample sizes are equal within each cell of the design. The partition of variance continues to follow ANOVA, with a variance component computed for each main effect and interaction. The pooled variance-covariance matrix due to differences among subjects within cells serves as the single error term. Assessment of DVs and comparisons proceed as described in Sections 7.5.3 and 7.5.4.

Two major design complexities that arise, however, are inclusion of within-subjects IVs and unequal sample sizes in cells.

7.5.5.1   *Within-Subjects and Between-Within Designs*

The simplest design with repeated measures is a one-way within-subjects design where the same subjects are measured on a single DV on several different occasions. The design can be complicated by addition of between-subjects IVs or more within-subjects IVs. Consult Chapters 3 and 6 for discussion of some of the problems that arise in ANOVA with repeated measures.

Repeated measures is extended to MANOVA when the researcher measures several DVs on several different occasions. The occasions can be viewed in two ways. In the traditional sense, occasions is a within-subjects IV with as many levels as occasions (Chapter 3). Alternatively, each occasion can

TABLE 7.10 Syntax for Orthogonal Comparisons and Trend Analysis

Type of Comparison	Program	Syntax	Section of Output	Name of Effect

Orthogonal	SPSS GLM	GLM
		WRATR WRATA BY TREATMNT DISABLTY
		/METHOD = SSTYPE(3)
		/INTERCEPT = INCLUDE
		/CRITERIA = ALPHA(.05)
		/LMATRIX "LINEAR" DISABLTY 1 0 -1
		             TREATMNT*DISABLTY 1/2 0 -1/2 1/2 0 -1/2
		/LMATRIX "QUADRATIC" DISABLTY 1 -2 1
		             TREATMNT*DISABLTY 1/2 -2/2 1/2 1/2 -2/2 1/2
		/DESIGN = TREATMNT DISABLTY TREATMNT*DISABLTY.
			Custom Hypothesis Tests:
			
			Multivariate Test Results
				Wilks' Lambda

	SPSS MANOVA	MANOVA
		WRATR WRATA BY TREATMNT (1, 2) DISABLTY (1, 3)
		/METHOD = UNIQUE
		/PARTITION (DISABLTY)
		/CONTRAST(DISABLTY)=SPECIAL(1 1 1,
		                        1 0 -1,
		                        1 -2 1)
		
		/DESIGN = TREATMNT DISABLTY(1)
		DISABLTY(2) TREATMNT BY DISABLTY.
			EFFECT...
			DISABLTY (2)
			
			EFFECT...
			DISABLTY (1)
				Wilks' Lambda

	SAS GLM	PROC GLM DATA=SASUSER.SS_MANOV; CLASS TREATMNT DISABLTY; MODEL WRATR WRATA = TREATMNT DISABLTY TREATMNT*DISABLTY; CONTRAST 'LINEAR' DISABLTY 1 0 -1; CONTRAST 'QUADRATIC' DISABLTY 1 -2 1; manova h=_all_/short; run;	MANOVA Test Criteria... No Overall linear (quadratic) Effect	Wilks' Lambda

(continued)

275

TABLE 7.10   Continued

Type of Comparison	Program	Syntax	Section of Output	Name of Effect
Trend Analysis	SPSS GLM	No special syntax; done as any other user-specified contrasts.	EFFECT... DISABLTY(2)	Wilks' Lambda

	SPSS MANOVA	MANOVA
		WRATR WRATA BY TREATMNT(1,2)
		    DISABLTY(1,3)
		/METHOD = UNIQUE
		/PARTITION (DISABLTY)
		/CONTRAST(DISABLTY)=
		    POLYNOMIAL (1,2,3)
		/DESIGN = TREATMNT DISABLTY(1)
		    DISABLTY(2)
		    TREATMNT BY DISABLTY.
			`EFFECT...`
			`DISABLTY(1)`	

	SAS GLM	No special syntax for between-subjects IVs.		

be treated as a separate DV—one DV per occasion (Section 7.2.8). In this latter view, if there is more than one DV measured on each occasion, the design is said to be doubly multivariate—multiple DVs are measured on multiple occasions. (There is no distinction between the two views when there are only two levels of the within-subjects IV.)

Section 8.5.3 discusses a doubly-multivariate analysis of a small data set with a between-subjects IV (PROGRAM), a within-subjects IV (MONTH), and two DVs (WTLOSS and ESTEEM), both measured three times. A complete example of a doubly-multivariate design is in Section 8.6.2.

It also is possible to have multiple DVs, but treat the within-subjects IV univariately. This is useful when (1) there are only two levels of the within-subjects IV, (2) there is no concern with violation of sphericity (Sections 3.2.3 and 8.5.1), or (3) a trend analysis is planned to replace the omnibus tests of the within-subjects IV and any interactions with the within-subjects IV. All programs that do doubly-multivariate analysis also show univariate results, therefore the syntax is the same as that used in Section 8.5.3.

7.5.5.2   Unequal Sample Sizes

When cells in a factorial ANOVA have an unequal number of scores, the sum of squares for effect plus error no longer equals the total sum of squares, and tests of main effects and interactions are correlated. There are a number of ways to adjust for overlap in sums of squares (cf. Woodward & Overall, 1975), as discussed in some detail in Section 6.5.4.2, particularly Table 6.10. Both the problem and the solutions generalize to MANOVA.

All the MANOVA programs described in Section 7.7 adjust for unequal *n*. SPSS MANOVA offers both Method 1 adjustment (METHOD = UNIQUE), which is default, and Method 3 adjustment (METHOD = SEQUENTIAL). Method 3 adjustment with survey data through SPSS MANOVA is shown in Section 7.6.2. Method 1—called SSTYPE(3)—is the default among four options in SPSS GLM. In SAS GLM, Method 1 (called `TYPE III` or `TYPE IV`) also is the default among four options available.

7.6 Complete Examples of Multivariate Analysis of Variance and Covariance

In the research described in Appendix B, Section B.1, there is interest in whether the means of several of the variables differ as a function of sex role identification. Are there differences in self-esteem, introversion-extraversion, neuroticism, and so on associated with a woman's masculinity and femininity? Files are MANOVA.*.

Sex role identification is defined by the masculinity and femininity scales of the Bem Sex Role Inventory (Bem, 1974). Each scale is divided at its median to produce two levels of masculinity (high and low), two levels of femininity (high and low), and four groups: Undifferentiated (low femininity, low masculinity), Feminine (high femininity, low masculinity), Masculine (low femininity, high masculinity), and Androgynous (high femininity, high masculinity). The design produces a main effect of masculinity, a main effect of femininity, and a masculinity–femininity interaction.[13]

DVs for this analysis are self-esteem (ESTEEM), internal versus external locus of control (CONTROL), attitudes toward women's role (ATTROLE), socioeconomic level (SEL2), introversion-extraversion (INTEXT), and neuroticism (NEUROTIC). Scales are coded so that higher scores generally represent the more "negative" trait: low self-esteem, greater neuroticism, etc.

Omnibus MANOVA (Section 7.6.2) asks whether these DVs are associated with the two IVs (femininity and masculinity) or their interaction. The Roy-Bargmann stepdown analysis, in conjunction with the univariate F values, allows us to examine the pattern of relationships between DVs and each IV.

In a second example (Section 7.6.3), MANCOVA is performed with SEL2, CONTROL, and ATTROLE used as covariates and ESTEEM, INTEXT, and NEUROTIC used as DVs. The research question is whether the three personality DVs vary as a function of sex role identification (the two IVs and their interaction) after adjusting for differences in socioeconomic status, attitudes toward women's role, and beliefs regarding locus of control of reinforcements.

7.6.1 Evaluation of Assumptions

Before proceeding with MANOVA and MANCOVA, we must assess the variables with respect to practical limitations of the techniques.

7.6.1.1 Unequal Sample Sizes and Missing Data

SPSS FREQUENCIES is run with SORT and SPLIT FILE to divide cases into the four groups. Data and distributions for each DV within each group are inspected for missing values, shape, and variance (see Table 7.11 for output on the CONTROL variable for the Feminine group). The run reveals the presence of a case for which the CONTROL score is missing. No datum is missing on any of the

---

[13]Some would argue with the wisdom of considering masculinity and femininity separate IVs, and of performing a median split on them to create groups. This example is used for didactic purposes.

TABLE 7.11 Syntax and Selected SPSS FREQUENCIES Output for MANOVA Variables Split by Group

```
MISSING VALUES CONTROL (0)
SORT CASES BY ANDRM.
SPLIT FILE
    SEPARATE BY ANDRM.
FREQUENCIES
    VARIABLES=ESTEEM CONTROL ATTROLE SEL2 INTEXT NEUROTIC /FORMAT NOTABLE
    /STATISTICS=STDDEV VARIANCE MINIMUM MAXIMUM MEAN SKEWNESS SESKEW KURTOSIS
    SEKURT
    HISTOGRAM NORMAL
    /ORDER=ANALYSIS.
```

Frequencies   Groups-4 = Feminine

Statistics[a]

		Self esteem	Locus of control	Attitude toward role of women	Socio-economic level	Introversion-extroversion	Neuroticism
N	Valid	173	172	173	173	173	173
	Missing	0	1	0	0	0	0
Mean		16.4913	6.7733	37.0520	40.402643	11.3266	8.9653
Std. Deviation		3.48688	1.26620	6.28145	24.659579	3.66219	5.10688
Variance		12.158	1.603	39.457	608.095	13.412	26.080
Skewness		.471	.541	.076	-.235	-.327	.238
Std. Error of Skewness		.185	.185	.185	.185	.185	.185
Kurtosis		.651	-.381	-.204	-1.284	-.335	-.689
Std. Error of Kurtosis		.367	.368	.367	.367	.367	.367
Minimum		9.00	5.00	22.00	.00000	2.00	.00
Maximum		28.00	10.00	55.00	81.00000	20.00	23.00

[a]Groups-4 = Feminine

TABLE 7.11 Continued

Histogram

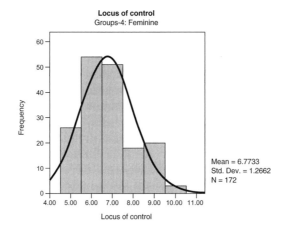

Locus of control
Groups-4: Feminine

Mean = 6.7733
Std. Dev. = 1.2662
N = 172

other DVs for the 369 women who were administered the Bem Sex Role Inventory. Deletion of the case with the missing value, then, reduces the available sample size to 368.

Sample sizes are quite different in the four groups: There are 71 Undifferentiated, 172 Feminine, 36 Masculine, and 89 Androgynous women in the sample. Because it is assumed that these differences in sample size reflect real processes in the population, the sequential approach to adjustment for unequal n is used with FEM (femininity) given priority over MASC (masculinity), and FEM by MASC (interaction between femininity and masculinity).

7.6.1.2 Multivariate Normality

The sample size of 368 includes over 35 cases for each cell of the 2×2 between-subjects design, more than the 20 df for error suggested to assure multivariate normality of the sampling distribution of means, even with unequal sample sizes; there are far more cases than DVs in the smallest cell. Further, the distributions for the full run (of which CONTROL in Table 7.11 is a part) produce no cause for alarm. Skewness is not extreme and, when present, is roughly the same for the DVs.

Two-tailed tests are automatically performed by the computer programs used. That is, the F test looks for differences between means in either direction.

7.6.1.3 Linearity

The full output for the run of Table 7.11 reveals no cause for worry about linearity. All DVs in each group have reasonably balanced distributions so there is no need to examine scatterplots for each pair of DVs within each group. Had scatterplots been necessary, SPSS PLOT would have been used with the SORT and SPLIT FILE syntax in Table 7.11.

7.6.1.4 Outliers

No univariate outliers were found using a criterion $z = |3.3|$ ($\alpha = .001$) with the minimum and maximum values in the full output of Table 7.11. SPSS REGRESSION is used with the split file in place to check for multivariate outliers within each of the four groups (Table 7.12). The RESIDUALS= OUTLIERS(MAHAL) instruction produces the 10 most outlying cases for each of the groups. With six variables and a criterion $\alpha = .001$, critical $\chi^2 = 22.458$; no multivariate outliers are found.

7.6.1.5   *Homogeneity of Variance-Covariance Matrices*

As a preliminary check for robustness, sample variances (in the full run of Table 7.11) for each DV are compared across the four groups. For no DV does the ratio of largest to smallest variance approach 10:1. As a matter of fact, the largest ratio is about 1.5:1 for the Undifferentiated versus Androgynous groups on CONTROL.

Sample sizes are widely discrepant, with a ratio of almost 5:1 for the Feminine to Masculine groups. However, with very small differences in variance and two-tailed tests, the discrepancy in sam-

---

TABLE 7.12   Mahalanobis Distance Values for Assessing Multivariate Outliers (Syntax and Selected Portion of Output from SPSS REGRESSION)

```
REGRESSION
 /MISSING LISTWISE
 /STATISTICS COEFF OUTS R ANOVA
 /CRITERIA=PIN(.05) POUT(.10)
 /NOORIGIN
 /DEPENDENT CASENO
 /METHOD=ENTER ESTEEM CONTROL ATTROLE SEL2 INTEXT NEUROTIC
 /RESIDUALS=OUTLIERS(MAHAL).
```

Regression

Groups-4 = Undifferentiated

Outlier Statistics[a,b]

		Case Number	Statistic
Mahal. Distance	1	32	14.975
	2	71	14.229
	3	64	11.777
	4	5	11.577
	5	41	11.371
	6	37	10.042
	7	55	9.378
	8	3	9.352
	9	1	9.318
	10	25	8.704

[a]Dependent Variable: CASENO
[b]Groups-4 = Undifferentiated

Groups-4 = Masculine

Outlier Statistics[a,b]

		Case Number	Statistic
Mahal. Distance	1	277	14.294
	2	276	11.773
	3	249	11.609
	4	267	10.993
	5	251	9.175
	6	253	8.276
	7	246	7.984
	8	271	7.917
	9	278	7.406
	10	273	7.101

[a]Dependent Variable: CASENO
[b]Groups-4 = Masculine

TABLE 7.12 Continued

Groups-4 = Feminine

Outlier Statistics[a,b]

		Case Number	Statistic
Mahal. Distance	1	209	19.348
	2	208	15.888
	3	116	14.813
	4	92	14.633
	5	167	14.314
	6	233	13.464
	7	150	13.165
	8	79	12.794
	9	138	12.181
	10	179	11.916

[a]Dependent Variable: CASENO
[b]Groups-4 = Feminine

Groups-4 = Androgynous

Outlier Statistics[a,b]

		Case Number	Statistic
Mahal. Distance	1	326	19.622
	2	301	15.804
	3	315	15.025
	4	312	10.498
	5	288	10.381
	6	347	10.101
	7	285	9.792
	8	338	9.659
	9	318	9.203
	10	302	8.912

[a]Dependent Variable: CASENO
[b]Groups-4 = Androgynous

ple sizes does not invalidate use of MANOVA. The very sensitive Box's M test for homogeneity of dispersion matrices (performed through SPSS MANOVA as part of the major analysis in Table 7.15) produces $F(63, 63020) = 1.07$, $p > .05$, supporting the conclusion of homogeneity of variance-covariance matrices.

7.6.1.6 Homogeneity of Regression

Because Roy-Bargmann stepdown analysis is planned to assess the importance of DVs after MANOVA, a test of homogeneity of regression is necessary for each step of the stepdown analysis. Table 7.13 shows the SPSS MANOVA syntax for tests of homogeneity of regression where each DV, in turn, serves as DV on one step and then becomes a covariate on the next and all remaining steps (the split file instruction first is turned off).

Table 7.13 also contains output for the last two steps where CONTROL serves as DV with ESTEEM, ATTROLE, NEUROTIC, and INTEXT as covariates, and then SEL2 is the DV with ESTEEM, ATTROLE, NEUROTIC, INTEXT, and CONTROL as covariates. At each step, the relevant effect is the one appearing last in the column labeled `Source of Variation`, so that for SEL2 the F value for homogeneity of regression is $F(15, 344) = 1.46$, $p > .01$. (The more stringent cutoff is used here because robustness is expected.) Homogeneity of regression is established for all steps.

For MANCOVA, an overall test of homogeneity of regression is required, in addition to stepdown tests. Syntax for all tests is shown in Table 7.14. The ANALYSIS sentence with three DVs specifies the overall test, while the ANALYSIS sentences with one DV each are for stepdown analysis. Output for the overall test and the last stepdown test is also shown in Table 7.14. Multivariate output is printed for the overall test because there are three DVs; univariate results are given for the stepdown tests. All runs show sufficient homogeneity of regression for this analysis.

TABLE 7.13   Test for Homogeneity of Regression for MANOVA Stepdown Analysis (Syntax and Selected Output for Last Two Tests from SPSS MANOVA)

```
SPLIT FILE
 OFF.
MANOVA ESTEEM,ATTROLE,NEUROTIC,INTEXT,CONTROL,SEL2 BY FEM, MASC(1,2)
    /PRINT=SIGNIF(BRIEF)
    /ANALYSIS=ATTROLE
    /DESIGN=ESTEEM,FEM,MASC,FEM BY MASC, ESTEEM BY FEM BY MASC
    /ANALYSIS=NEUROTIC
    /DESIGN=ATTROLE,ESTEEM,FEM,MASC,FEM BY MASC, POOL(ATTROLE,ESTEEM)
        BY FEM + POOL(ATTROLE,ESTEEM) BY MASC + POOL(ATTROLE,
        ESTEEM)BY FEM BY MASC/
    /ANALYSIS=INTEXT
    /DESIGN=NEUROTIC,ATTROLE,ESTEEM,FEM,MASC,FEM BY MASC, POOL(NEUROTIC,
        ATTROLE,ESTEEM) BY FEM + POOL(NEUROTIC,ATTROLE,ESTEEM)
        BY MASC + POOL(NEUROTIC,ATTROLE,ESTEEM) BY FEM BY MASC
    /ANALYSIS=CONTROL
    /DESIGN=INTEXT,NEUROTIC,ATTROLE,ESTEEM FEM,MASC FEM BY MASC,
        POOL(INTEXT,NEUROTIC,ATTROLE,ESTEEM) BY FEM +
        POOL(INTEXT,NEUROTIC,ATTROLE,ESTEEM) BY MASC +
        POOL(INTEXT,NEUROTIC,ATTROLE,ESTEEM) BY FEM BY MASC
    /ANALYSIS=SEL2
    /DESIGN=CONTROL,INTEXT,NEUROTIC,ATTROLE,ESTEEM,FEM,MASC,FEM BY MASC,
        POOL(CONTROL,INTEXT,NEUROTIC,ATTROLE,ESTEEM) BY FEM +
        POOL(CONTROL,INTEXT,NEUROTIC,ATTROLE,ESTEEM) BY MASC +
        POOL(CONTROL,INTEXT,NEUROTIC,ATTROLE,ESTEEM) BY FEM BY MASC.
```

```
Tests of Significance for CONTROL using UNIQUE sums of squares
```

Source of Variation	SS	DF	MS	F	Sig of F
WITHIN+RESIDUAL	442.61	348	1.27		
INTEXT	2.19	1	2.19	1.72	.190
NEUROTIC	42.16	1	42.16	33.15	.000
ATTROLE	.67	1	.67	.52	.470
ESTEEM	14.52	1	14.52	11.42	.001
FEM	2.80	1	2.80	2.20	.139
MASC	3.02	1	3.02	2.38	.124
FEM BY MASC	.00	1	.00	.00	.995
POOL(INTEXT NEUROTIC ATTROLE ESTEEM) BY FEM + POOL(INTEXT NEUROTIC ATTROLE ESTEEM) BY MASC + POOL(INTEXT NEUROTIC ATTROLE ESTEEM) BY FEM BY MASC	19.78	12	1.65	1.30	.219

TABLE 7.13　Continued

```
Tests of Significance of SEL2 using UNIQUE sums of squares
```

Source of Variation	SS	DF	MS	F	Sig of F
WITHIN+RESIDUAL	220340.10	344	640.52		
CONTROL	1525.23	1	1525.23	2.38	.124
INTEXT	.99	1	.99	.00	.969
NEUR	262.94	1	262.94	.41	.522
ATT	182.98	1	182.98	.29	.593
EST	157.77	1	157.77	.25	.620
FEM	1069.23	1	1069.23	1.67	.197
MASC	37.34	1	37.34	.06	.809
FEM BY MASC	1530.73	1	1530.73	2.39	.123
POOL (CONTROL INTEXT NEUROTIC ATTROLE EST EEM) BY FEM + POOL(C ONTROL INTEXT NEUROT IC ATTROLE ESTEEM) B Y MASC + POOL(CONTRO L INTEXT NEUROTIC AT TROLE ESTEEM) BY FEM　BY MASC)	14017.22	15	934.48	1.46	.118

TABLE 7.14　Tests of Homogeneity of Regression for MANCOVA and Stepdown Analysis (Syntax and Partial Output for Overall Tests and Last Stepdown Test from SPSS MANOVA)

```
MANOVA   ESTEEM,ATTROLE,NEUROTIC,INTEXT,CONTROL,SEL2 BY FEM MASC(1,2)
    /PRINT=SIGNIF(BRIEF)
    /ANALYSIS=ESTEEM,INTEXT,NEUROTIC
    /DESIGN=CONTROL,ATTROLE,SEL2,FEM,MASC,FEM BY MASC,
       POOL(CONTROL,ATTROLE,SEL2) BY FEM +
       POOL(CONTROL,ATTROLE,SEL2) BY MASC +
       POOL(CONTROL,ATTROLE,SEL2) BY FEM BY MASC
    /ANALYSIS=ESTEEM
    /DESIGN=CONTROL,ATTROLE,SEL2,FEM,MASC,FEM BY MASC,
       POOL(CONTROL,ATTROLE,SEL2) BY FEM +
       POOL(CONTROL,ATTROLE,SEL2) BY MASC +
       POOL(CONTROL,ATTROLE,SEL2) BY FEM BY MASC
    /ANALYSIS=INTEXT
    /DESIGN=ESTEEM,CONTROL,ATTROLE,SEL2,FEM,MASC,FEM BY MASC,
       POOL(ESTEEM,CONTROL,ATTROLE,SEL2) BY FEM +
       POOL(ESTEEM,CONTROL,ATTROLE,SEL2) BY MASC +
       POOL(ESTEEM,CONTROL,ATTROLE,SEL2) BY FEM BY MASC
    /ANALYSIS=NEUROTIC
    /DESIGN=INTEXT,ESTEEM,CONTROL,ATTROLE,SEL2,FEM,MASC,FEM BY MASC,
       POOL(INTEXT,ESTEEM,CONTROL,ATTROLE,SEL2) BY FEM+
       POOL(INTEXT,ESTEEM,CONTROL,ATTROLE,SEL2) BY MASC +
       POOL(INTEXT,ESTEEM,CONTROL,ATTROLE,SEL2) BY FEM BY MASC.
```

(continued)

TABLE 7.14   Continued

```
* * * * * * A n a l y s i s   o f   V a r i a n c e — design 1 * * * * * *
```

Multivariate Tests of Significance
Tests using UNIQUE sums of squares and WITHIN+RESIDUAL error term

Source of Variation	Wilks	Approx F	Hyp. DF	Error DF	Sig of F
CONTROL	.814	26.656	3.00	350.000	.000
ATTROLE	.973	3.221	3.00	350.000	.023
SEL2	.999	.105	3.00	350.000	.957
FEM	.992	.949	3.00	350.000	.417
MASC	.993	.824	3.00	350.000	.481
FEM BY MASC	.988	1.414	3.00	350.000	.238
POOL(CONTROL ATTROL E SEL2) BY FEM + POO L(CONTROL ATTROLE SE L2) BY MASC + POOL(C ONTROL ATTROLE SEL2) BY FEM BY MASC	.933	.911	27.00	1022.823	.596

```
* * * * * * A n a l y s i s   o f   V a r i a n c e — design 1 * * * * * *
```

Tests of Significance for NEUROTIC using UNIQUE sums of squares

Source of Variation	SS	DF	MS	F	Sig of F
WITHIN+RESIDUAL	6662.67	344	19.37		
INTEXT	82.19	1	82.19	4.24	.040
ESTEEM	308.32	1	308.32	15.92	.000
CONTROL	699.04	1	699.04	36.09	.000
ATTROLE	1.18	1	1.18	.06	.805
SEL2	2.24	1	2.24	.12	.734
FEM	.07	1	.07	.00	.952
MASC	74.66	1	74.66	3.85	.050
FEM BY MASC	1.65	1	1.65	.09	.770
POOL(INTEXT ESTEEM C ONTROL ATTROLE SEL2) BY FEM + POOL(INTEX T ESTEEM CONTROL ATT ROLE SEL2) BY MASC + POOL(INTEXT ESTEEM CONTROL ATTROLE SEL2) BY FEM BY MASC	420.19	15	28.01	1.45	.124

7.6.1.7   Reliability of Covariates

For the stepdown analysis in MANOVA, all DVs except ESTEEM must be reliable because all act as covariates. Based on the nature of scale development and data collection procedures, there is no reason to expect unreliability of a magnitude harmful to covariance analysis for ATTROLE, NEU-ROTIC, INTEXT, CONTROL, and SEL2. These same variables act as true or stepdown covariates in the MANCOVA analysis.

7.6.1.8 Multicollinearity and Singularity

The log-determinant of the pooled within-cells correlation matrix is found (through SPSS MANOVA syntax in Table 7.15) to be $-.4336$, yielding a determinant of 2.71. This is sufficiently different from zero that multicollinearity is not judged to be a problem.

7.6.2 Multivariate Analysis of Variance

Syntax and partial output of omnibus MANOVA produced by SPSS MANOVA appear in Table 7.15. The order of IVs listed in the MANOVA statement together with METHOD=SEQUENTIAL sets up

TABLE 7.15 Multivariate Analysis of Variance of Composite of DVs (ESTEEM, CONTROL, ATTROLE, SEL2, INTEXT, and NEUROTIC), as a Function of (Top to Bottom) FEMININITY by MASCULINITY Interaction, MASCULINITY, and FEMININITY (Syntax and Selected Output from SPSS MANOVA)

```
MANOVA      ESTEEM,ATTROLE,NEUROTIC,INTEXT,CONTROL,SEL2 BY FEM,MASC(1,2)
            /PRINT=SIGNIF(STEPDOWN), ERROR(COR),
              HOMOGENEITY(BARTLETT,COCHRAN,BOXM)
            /METHOD=SEQUENTIAL
            /DESIGN FEM MASC FEM BY MASC.
```

```
EFFECT.. FEM BY MASC
Multivariate Tests of Significance (S = 1, M = 2 , N = 178 1/2)
```

Test Name	Value	Exact F	Hypoth. DF	Error DF	Sig. of F
Pillais	.00816	.49230	6.00	359.00	.814
Hotellings	.00823	.49230	6.00	359.00	.814
Wilks	.99184	.49230	6.00	359.00	.814
Roys	.00816				

```
Note.. F statistics are exact.
EFFECT.. MASC
Multivariate Tests of Significance (S = 1, M = 2 , N = 178 1/2)
```

Test Name	Value	Exact F	Hypoth. DF	Error DF	Sig. of F
Pillais	.24363	19.27301	6.00	359.00	.000
Hotellings	.32211	19.27301	6.00	359.00	.000
Wilks	.75637	19.27301	6.00	359.00	.000
Roys	.24363				

```
Note.. F statistics are exact.

EFFECT.. FEM
Multivariate Tests of Significance (S = 1, M = 2 , N = 178 1/2)
```

Test Name	Value	Exact F	Hypoth. DF	Error DF	Sig. of F
Pillais	.08101	5.27423	6.00	359.00	.000
Hotellings	.08815	5.27423	6.00	359.00	.000
Wilks	.91899	5.27423	6.00	359.00	.000
Roys	.08101				

```
Note.. F statistics are exact.
```

the priority for testing FEM before MASC in this unequal-n design. Results are reported for FEM by MASC, MASC, and FEM, in turn. Tests are reported out in order of adjustment where FEM by MASC is adjusted for both MASC and FEM, and MASC is adjusted for FEM.

Four multivariate statistics are reported for each effect. Because there is only one degree of freedom for each effect, three of the tests—Pillai's, Hotelling's, and Wilks'—produce the same F.[14] Both main effects are highly significant, but there is no statistically significant interaction. If desired, effect size for the composite DV for each main effect is found using Equation 7.8 (shown in SPSS MANOVA as Pillai's value) or 7.9. In this case, full and partial η^2 are the same for each of the three effects because $s = 1$ for all of them. Confidence limits for effect sizes are found by entering values from Table 7.15 (**Exact F, Hypoth. DF, Error DF**, and the percentage for the desired confidence interval) into Smithson's (2003) NoncF.sav and running it through NoncF3.sps. Results are added to NoncF.sav, as seen in Table 7.16. (Note that partial η^2 also is reported as **r2**.) Thus, for the main effect of FEM, partial $\eta^2 = .08$ with 95% confidence limits from .02 to .13. For the main effect of MASC, partial $\eta^2 = .24$ with 95% confidence limits from .16 to .30. For the interaction, partial $\eta^2 = .01$ with 95% confidence limits from .00 to .02.

Because omnibus MANOVA shows significant main effects, it is appropriate to investigate further the nature of the relationships among the IVs and DVs. Correlations, univariate Fs, and stepdown Fs help clarify the relationships.

The degree to which DVs are correlated provides information as to the independence of behaviors. Pooled within-cell correlations, adjusted for IVs, as produced by SPSS MANOVA through PRINT = ERROR(COR), appear in Table 7.17. (Diagonal elements are pooled standard deviations.) Correlations among ESTEEM, NEUROTIC, and CONTROL are in excess of .30 so stepdown analysis is appropriate.

Even if stepdown analysis is the primary procedure, knowledge of univariate Fs is required to correctly interpret the pattern of stepdown Fs. And, although the statistical significance of these F values is misleading, investigators frequently are interested in the ANOVA that would have been produced if each DV had been investigated in isolation. These univariate analyses are produced automatically by SPSS MANOVA and shown in Table 7.18 for the three effects in turn: FEM by MASC, MASC, and FEM. F values are substantial for all DVs except SEL2 for MASC and ESTEEM, ATT-ROLE, and INTEXT for FEM.

Finally, Roy-Bargmann stepdown analysis, produced by PRINT=SIGNIF(STEPDOWN), allows a statistically pure look at the significance of DVs, in context, with Type I error rate controlled.

TABLE 7.16   Data Set Output from NoncF3.sps for Effect Size (r2) with 95% Confidence Limits (lr2 and ur2) for Interaction, MASC, and FEM, Respectively

	fval	df1	df2	conf	lc2	ucdf	uc2	lcdf	power	r2	lr2	ur2
1	.4923	6	359	.950	.000	.186	5.875	.02	.1306	.01	.00	.02
2	19.2730	6	359	.950	70.0	.975	160.0	.02	1.000	.24	.16	.30
3	5.2742	6	359	.950	9.08	.975	52.34	.02	.9910	.08	.02	.13

[14]For more complex designs, a single source table containing all effects can be obtained through PRINT=SIGNIF(BRIEF) but the table displays only Wilks' lambda.

TABLE 7.17   Pooled Within-Cell Correlations among Six DVs (Selected Output from SPSS MANOVA—See Table 7.15 for Syntax)

```
WITHIN+RESIDUAL Correlations with Std. Devs. on Diagonal

             ESTEEM    ATTROLE    NEUROTIC    INTEXT    CONTROL    SEL2

ESTEEM        3.533
ATTROLE        .145     6.227
NEUROTIC       .358      .051       4.965
INTEXT        -.164      .011       -.009      3.587
CONTROL        .348     -.031        .387      -.083     1.267
SEL2          -.035      .016       -.015       .055     -.084     25.501
```

TABLE 7.18   Univariate Analyses of Variance of Six DVs for Effects of (Top to Bottom) FEM by MASC Interaction, Masculinity, and Femininity (Selected Output from SPSS MANOVA—See Table 7.15 for Syntax)

```
EFFECT.. FEM BY MASC (Cont.)
Univariate F-tests with (1,364) D. F.
```

Variable	Hypoth. SS	Error SS	Hypoth. MS	Error MS	F	Sig. of F
ESTEEM	17.48685	4544.44694	17.48685	12.48474	1.40066	.237
ATTROLE	36.79594	14115.1212	36.79594	38.77781	.94889	.331
NEUROTIC	.20239	8973.67662	.20239	24.65296	.00821	.928
INTEXT	.02264	4684.17900	.02264	12.86862	.00176	.967
CONTROL	.89539	584.14258	.89539	1.60479	.55795	.456
SEL2	353.58143	236708.966	353.58143	650.29936	.54372	.461

```
EFFECT.. MASC (Cont.)
Univariate F-tests with (1,364) D. F.
```

Variable	Hypoth. SS	Error SS	Hypoth. MS	Error MS	F	Sig. of F
ESTEEM	979.60086	4544.44694	979.60086	12.48474	78.46383	.000
ATTROLE	1426.75675	14115.1212	1426.75675	38.77781	36.79313	.000
NEUROTIC	179.53396	8973.67662	179.53396	24.65296	7.28245	.007
INTEXT	327.40797	4684.17900	327.40797	12.86862	25.44235	.000
CONTROL	11.85923	584.14258	11.85923	1.60479	7.38991	.007
SEL2	1105.38196	236708.966	1105.38196	650.29936	1.69980	.193

```
EFFECT.. FEM (Cont.)
Univariate F-tests with (1,364) D. F.
```

Variable	Hypoth. SS	Error SS	Hypoth. MS	Error MS	F	Sig. of F
ESTEEM	101.46536	4544.44694	101.46536	12.48474	8.12715	.005
ATTROLE	610.88860	14115.1212	610.88860	38.77781	15.75356	.000
NEUROTIC	44.05442	8973.67662	44.05442	24.65296	1.78698	.182
INTEXT	87.75996	4684.17900	87.75996	12.86862	6.81968	.009
CONTROL	2.83106	584.14258	2.83106	1.60479	1.76414	.185
SEL2	9.00691	236708.966	9.00691	650.29936	.01385	.906

For this study, the following priority order of DVs is developed, from most to least important: ESTEEM, ATTROLE, NEUROTIC, INTEXT, CONTROL, SEL2. Following the procedures for stepdown analysis (Section 7.5.3.2), the highest-priority DV, ESTEEM, is tested in univariate ANOVA. The second-priority DV, ATTROLE, is assessed in ANCOVA with ESTEEM as the covariate. The third-priority DV, NEUROTIC, is tested with ESTEEM and ATTROLE as covariates, and so on, until all DVs are analyzed. Stepdown analyses for the interaction and both main effects are in Table 7.19.

For purposes of journal reporting, critical information from Tables 7.18 and 7.19 is consolidated into a single table with both univariate and stepdown analyses, as shown in Table 7.20. The alpha level established for each DV is reported along with the significance levels for stepdown F. The final three columns show partial η^2 with 95% confidence limits for all stepdown effects, described later.

For the main effect of FEM, ESTEEM and ATTROLE are significant. (INTEXT would be significant in ANOVA but its variance is already accounted for through overlap with ESTEEM, as noted

TABLE 7.19   Stepdown Analyses of Six Ordered DVs for (Top to Bottom) FEM by MASC Interaction, Masculinity, and Femininity (Selected Output from SPSS MANOVA— see Table 7.16 for Syntax)

Roy-Bargman Stepdown F-tests

Variable	Hypoth. MS	Error MS	Stepdown F	Hypoth. DF	Error DF	Sig. of F
ESTEEM	17.48685	12.48474	1.40066	1	364	.237
ATTROLE	24.85653	38.06383	.65302	1	363	.420
NEUROTIC	2.69735	21.61699	.12478	1	362	.724
INTEXT	.26110	12.57182	.02077	1	361	.885
CONTROL	.41040	1.28441	.31952	1	360	.572
SEL2	297.09000	652.80588	.45510	1	359	.500

Roy-Bargman Stepdown F-tests

Variable	Hypoth. MS	Error MS	Stepdown F	Hypoth. DF	Error DF	Sig. of F
ESTEEM	979.60086	12.48474	78.46383	1	364	.000
ATTROLE	728.51682	38.06383	19.13935	1	363	.000
NEUROTIC	4.14529	21.61699	.19176	1	362	.662
INTEXT	139.98354	12.57182	11.13471	1	361	.001
CONTROL	.00082	1.28441	.00064	1	360	.980
SEL2	406.59619	652.80588	.62284	1	359	.431

Roy-Bargman Stepdown F-tests

Variable	Hypoth. MS	Error MS	Stepdown F	Hypoth. DF	Error DF	Sig. of F
ESTEEM	101.46536	12.48474	8.12715	1	364	.005
ATTROLE	728.76735	38.06383	19.14593	1	363	.000
NEUROTIC	2.21946	21.61699	.10267	1	362	.749
INTEXT	47.98941	12.57182	3.81722	1	361	.052
CONTROL	.05836	1.28441	.04543	1	360	.831
SEL2	15.94930	652.80588	.02443	1	359	.876

TABLE 7.20   Tests of Femininity, Masculinity, and Their Interaction

IV	DV	Univariate F	df	Stepdown F	df	α	Partial η^2	CL around Partial η^2 per α Lower	Upper
Femininity	ESTEEM	8.13[a]	1/364	8.13**	1/364	.01	.02	.00	.07
	ATTROLE	15.75[a]	1/364	19.15**	1/363	.01	.05	.01	.12
	NEUROTIC	1.79	1/364	0.10	1/362	.01	.00	.00	.01
	INTEXT	6.82[a]	1/364	3.82	1/361	.01	.01	.00	.05
	CONTROL	1.76	1/364	0.05	1/360	.01	.00	.00	.01
	SEL2	0.01	1/364	0.02	1/359	.001	.00	.00	.00
Masculinity	ESTEEM	78.46[a]	1/364	78.46**	1/364	.01	.18	.09	.27
	ATTROLE	36.79[a]	1/364	19.14**	1/363	.01	.05	.01	.12
	NEUROTIC	7.28[a]	1/364	0.19	1/362	.01	.00	.00	.02
	INTEXT	25.44[a]	1/364	11.13**	1/361	.01	.03	.00	.09
	CONTROL	7.39[a]	1/364	0.00	1/360	.01	.00	.00	.00
	SEL2	1.70	1/364	0.62	1/359	.001	.00	.00	.04
Femininity by masculinity interaction	ESTEEM	1.40	1/364	1.40	1/364	.01	.00	.00	.04
	ATTROLE	0.95	1/364	0.65	1/363	.01	.00	.00	.03
	NEUROTIC	0.01	1/364	0.12	1/362	.01	.00	.00	.01
	INTEXT	0.00	1/364	0.02	1/361	.01	.00	.00	.00
	CONTROL	0.56	1/364	0.32	1/360	.01	.00	.00	.03
	SEL2	0.54	1/364	0.46	1/359	.001	.00	.00	.04

[a]Significance level cannot be evaluated but would reach $p < .01$ in univariate context.

**$p < .01$.

in the pooled within-cell correlation matrix.) For the main effect of MASC, ESTEEM, ATTROLE, and INTEXT are significant. (NEUROTIC and CONTROL would be significant in ANOVA, but their variance is also already accounted for through overlap with ESTEEM, ATTROLE, and, in the case of CONTROL, NEUROTIC and INTEXT.)

For the DVs significant in stepdown analysis, the relevant adjusted marginal means are needed for interpretation. Marginal means are needed for ESTEEM for FEM and for MASC adjusted for FEM. Also needed are marginal means for ATTROLE with ESTEEM as a covariate for both FEM, and MASC adjusted for FEM; lastly, marginal means are needed for INTEXT with ESTEEM, ATTROLE, and NEUROTIC as covariates for MASC adjusted for FEM. Table 7.21 contains syntax and selected output for these marginal means as produced through SPSS MANOVA. In the table, level of effect is identified under PARAMETER and mean is under Coeff. Thus, the mean for ESTEEM at level 1 of FEM is 16.57. Marginal means for effects with univariate, but not stepdown, differences are shown in Table 7.22 where means for NEUROTIC and CONTROL are found for the main effect of MASC adjusted for FEM.

Effect size for each DV is evaluated as partial η^2 (Equation 3.25, 3.26, 6.7, 6.8, or 6.9). The information you need for calculation of η^2 is available in SPSS MANOVA stepdown tables (see Table 7.19) but not in a convenient form; mean squares are given in the tables but you need sums of squares for calculation of η^2. Smithson's (2003) program (NoncF3.sps) calculates confidence limits for effect sizes

TABLE 7.21   Adjusted Marginal Means for ESTEEM; ATTROLE with ESTEEM as a Covariate; and INTEXT with ESTEEM, ATTROLE, and NEUROTIC as Covariates (Syntax and Selected Output from SPSS MANOVA)

```
MANOVA      ESTEEM,ATTROLE,NEUROTIC,INTEXT,CONTROL,SEL2 BY FEM,MASC(1,2)
      /PRINT=PARAMETERS(ESTIM)
    /ANALYSIS=ESTEEM /DESIGN=CONSPLUS FEM
            /DESIGN=FEM,CONSPLUS MASC
    /ANALYSIS=ATTROLE WITH ESTEEM /DESIGN=CONSPLUS FEM
            /DESIGN=FEM, CONSPLUS MASC
    /ANALYSIS=INTEXT WITH ESTEEM,ATTROLE,NEUROTIC
            /DESIGN=FEM, CONSPLUS MASC.
```

```
Estimates for ESTEEM
--- Individual univariate .9500 confidence intervals

CONSPLUS FEM

Parameter      Coeff.    Std. Err.     t-Value   Sig. t Lower -95%  CL- Upper

      1    16.5700935     .37617     44.04964    .00000    15.83037   17.30982
      2    15.4137931     .24085     63.99636    .00000    14.94016   15.88742


Estimates for ESTEEM
--- Individual univariate .9500 confidence intervals

CONSPLUS MASC

Parameter      Coeff.    Std. Err.     t-Value   Sig. t Lower -95%  CL- Upper

      2    17.1588560     .24196     70.91464    .00000    16.68304   17.63468
      3    13.7138144     .32770     41.84820    .00000    13.06939   14.35824


Estimates for ATTROLE adjusted for 1 covariate
--- Individual univariate .9500 confidence intervals

CONSPLUS FEM

Parameter      Coeff.    Std. Err.     t-Value   Sig. t Lower -95%  CL- Upper

      1    32.5743167     .61462     52.99941    .00000    31.36568   33.78295
      2    35.9063146     .39204     91.58908    .00000    35.13538   36.67725


Estimates for ATTROLE adjusted for 1 covariate
--- Individual univariate .9500 confidence intervals

CONSPLUS MASC

Parameter      Coeff.    Std. Err.     t-Value   Sig. t Lower -95%  CL- Upper

      2    35.3849271     .44123     80.19697    .00000    34.51726   36.25260
      3    32.1251995     .60108     53.44537    .00000    30.94316   33.30723
```

TABLE 7.21 Continued

```
Estimates for INTEXT adjusted for 3 covariates
— Individual univariate .9500 confidence intervals

CONSPLUS MASC
```

Parameter	Coeff.	Std. Err.	t-Value	Sig. t	Lower -95%	CL- Upper
2	11.0013930	.25372	43.36122	.00000	10.50245	11.50033
3	12.4772029	.35546	35.10172	.00000	11.77818	13.17623

Note: Coeff. = adjusted marginal mean; first parameter = low, second parameter = high.

TABLE 7.22 Unadjusted Marginal Means for Neurotic and Control (Syntax and Selected Output from SPSS MANOVA)

```
MANOVA     ESTEEM,ATTROLE,NEUROTIC,INTEXT,CONTROL,SEL2 BY FEM,MASC(1,2)
       /PRINT=PARAMETERS(ESTIM)
       /ANALYSIS=NEUROTIC /DESIGN=FEM, CONSPLUS MASC
       /ANALYSIS=CONTROL /DESIGN=FEM, CONSPLUS MASC.

Estimates for NEUROTIC
--- Individual univariate .9500 confidence intervals

CONSPLUS MASC
```

Parameter	Coeff.	Std. Err.	t-Value	Sig. t	Lower -95%	CL- Upper
2	9.37093830	.33937	27.61309	.00000	8.70358	10.03830
3	7.89610411	.45962	17.17971	.00000	6.99227	8.79994

```
Estimates for CONTROL
--- Individual univariate .9500 confidence intervals

CONSPLUS MASC
```

Parameter	Coeff.	Std. Err.	t-Value	Sig. t	Lower -95%	CL- Upper
2	6.89163310	.08665	79.53388	.00000	6.72124	7.06203
3	6.51258160	.11735	55.49504	.00000	6.28181	6.74336

Note: Coeff. = unadjusted marginal mean; first parameter = low, second parameter = high.

and also calculates the effect size itself from F (stepdown or otherwise), df for effect (df1) and error (df2), and the percentage associated with the desired confidence limits. These four values are entered into the data sheet (NoncF.sav). The remaining columns of NoncF.sav are filled in when NoncF3.sps is run. The relevant output columns are r2 (equivalent to partial η^2 of Equation 6.9), lr2 and ur2, the lower and upper confidence limits, respectively, for the effect size. Table 7.23 shows the input/output data set for all of the stepdown effects following the order in Table 7.20, e.g., 1 = ESTEEM for FEM, 2 = ATTROLE for FEM, 3 = NEUROTIC for FEM and so on. Values filled into the first three columns are from Table 7.20. The value of .99 or .999 filled in for the confidence limits reflects the chosen α level for each effect.

A checklist for MANOVA appears in Table 7.24. An example of a Results section, in journal format, follows for the study just described.

TABLE 7.23   Data Set Output for Stepdown Effects from NoncF3.sps for Effect Size (r2) with 95% Confidence Limits (lr2 and ur2)

	fval	df1	df2	conf	lc2	ucdf	uc2	lcdf	power	r2	lr2	ur2
1	8.1300	1	364	.990	.0208	.9950	29.5824	.0050	.5146	.02	.00	.07
2	19.1500	1	363	.990	3.1051	.9950	48.7540	.0050	.9403	.05	.01	.12
3	.0100	1	362	.990	.0000	.0796	.4300	.0642	.0052	.00	.00	.00
4	3.8200	1	361	.990	.0000	.9486	20.5773	.0050	.1955	.01	.00	.05
5	.0500	1	360	.990	.0000	.1768	2.1500	.0614	.0061	.00	.00	.01
6	.0200	1	359	.999	.0000	.1124	.8600	.0733	.0006	.00	.00	.00
7	78.4600	1	364	.990	37.6976	.9950	133.7230	.0050	1.0000	.18	.09	.27
8	19.1400	1	363	.990	3.1034	.9950	48.7472	.0050	.9402	.05	.01	.12
9	.1900	1	362	.990	.0000	.3368	8.1700	.0072	.0094	.00	.00	.02
10	11.1300	1	361	.990	.5444	.9950	35.1617	.0050	.6985	.03	.00	.09
11	.0000	1	360	.990	.0000	.1587	.0000	.1587	.0015	.00	.00	.00
12	.6200	1	359	.999	.0000	.5684	16.6625	.0005	.0035	.00	.00	.04
13	1.4000	1	364	.990	.0000	.7625	14.1422	.0050	.0519	.00	.00	.04
14	.6500	1	363	.990	.0000	.5794	11.4359	.0050	.0227	.00	.00	.03
15	.1200	1	362	.990	.0000	.2708	5.1600	.0227	.0077	.00	.00	.01
16	.0200	1	361	.990	.0000	.1124	.8600	.0733	.0054	.00	.00	.00
17	.3200	1	360	.990	.0000	.4280	9.8200	.0050	.0128	.00	.00	.03
18	.4600	1	359	.999	.0000	.5019	15.7550	.0005	.0025	.00	.00	.04

TABLE 7.24   Checklist for Multivariate Analysis of Variance

1. Issues
   a. Unequal sample sizes and missing data
   b. Normality of sampling distributions
   c. Outliers
   d. Homogeneity of variance-covariance matrices
   e. Linearity
   f. In stepdown, when DVs act as covariates
      (1) Homogeneity of regression
      (2) Reliability of DVs
   g. Multicollinearity and singularity
2. Major analyses: Planned comparisons or omnibus F, when significant. Importance of DVs
   a. Within-cell correlations, stepdown F, univariate F
   b. Effect sizes with confidence interval for significant stepdown F
   c. Means or adjusted marginal and/or cell means for significant F, with standard deviations, standard errors, or confidence intervals
3. Multivariate effect size(s) with confidence interval(s) for planned comparisons or omnibus F
4. Additional analyses
   a. Post hoc comparisons
   b. Interpretation of IV–covariates interaction (if homogeneity of regression violated)

Results

A 2 × 2 between-subjects multivariate analysis of variance was performed on six dependent variables: Self-esteem, attitude toward the role of women, neuroticism, introversion-extraversion, locus of control, and socioeconomic level. Independent variables were masculinity (low and high) and femininity (low and high).

SPSS MANOVA was used for the analyses with the sequential adjustment for nonorthogonality. Order of entry of IVs was femininity, then masculinity. Total N of 369 was reduced to 368 with the deletion of a case missing a score on locus of control. There were no univariate or multivariate within-cell outliers at $p < .001$. Results of evaluation of assumptions of normality, homogeneity of variance-covariance matrices, linearity, and multicollinearity were satisfactory.

With the use of Wilks' criterion, the combined DVs were significantly affected by both masculinity, $F(6, 359) = 19.27$, $p < .001$, and femininity, $F(6, 359) = 5.27$, $p < .001$, but not by their interaction, $F(6, 359) = 0.49$, $p > .05$. The results reflected a modest association between masculinity scores (low vs. high) and the combined DVs, partial $\eta^2 = .24$ with 95% confidence limits from .16 to .30. The association was even less substantial between femininity and the DVs, partial $\eta^2 = .08$ with 95% confidence limits from .02 to .13. For the nonsignificant interaction, partial $\eta^2 = .01$ with 95% confidence limits from .00 to .02. [*F and Pillai's value (partial η^2) are from Table 7.15; confidence limits for partial η^2 are found through NoncF3.sps.*]

To investigate the impact of each main effect on the individual DVs, a Roy-Bargmann stepdown analysis was performed on the prioritized DVs. All DVs were judged to be sufficiently reliable to warrant stepdown analysis. In stepdown analysis each DV was analyzed, in turn, with higher-priority DVs treated as covariates and with the highest-priority DV tested in a univariate ANOVA. Homogeneity of regression was achieved for all components of the stepdown analysis.

Results of this analysis are summarized in Table 7.20. An experimentwise error rate of 5% was achieved by the apportionment of alpha as shown in the last column of Table 7.20 for each of the DVs.

A unique contribution to predicting differences between those low and high on femininity was made by self-esteem, stepdown $F(1, 364)$ = 8.13, $p < .01$, partial η^2 = .02 with 99% confidence limits from .00 to .07. Self-esteem was scored inversely, so women with higher femininity scores showed greater self-esteem (mean self-esteem = 15.41, SE = 0.24) than those with lower femininity (mean self-esteem = 16.57, SE = 0.38). After the pattern of differences measured by self-esteem was entered, a difference was also found on attitude toward the role of women, stepdown $F(1, 363)$ = 19.15, $p < .01$, partial η^2 = .05 with confidence limits from .01 to .12. Women with higher femininity scores had more conservative attitudes toward women's role (adjusted mean attitude = 35.90, SE = 0.35) than those lower in femininity (adjusted mean attitude = 32.57, SE = 0.61). Although a univariate comparison revealed that those higher in femininity also were more extroverted, univariate $F(1, 364)$ = 6.82, this difference was already represented in the stepdown analysis by higher-priority DVs.

Three DVs—self-esteem, attitude toward role of women, and introvert-extrovert—made unique contributions to the composite DV that best distinguished between those high and low in masculinity. The greatest contribution was made by self-esteem, the highest-priority DV, stepdown $F(1, 364)$ = 78.46, $p < .01$, partial η^2 = .18 with confidence limits from .09 to .27. Women scoring high in masculinity had higher self-esteem (mean self-esteem = 13.71, SE = 0.33) than those scoring low (mean self-esteem = 17.16, SE = 0.24). With differences due to self-esteem already entered, attitudes toward the role of women made a unique contribution, stepdown $F(1, 363)$ = 19.14, $p < .01$, partial η^2 = .05 with confidence limits from .01 to .12. Women scoring lower in masculinity had more conservative attitudes toward the proper role of women (adjusted mean attitude = 35.39, SE = 0.44) than those scoring higher (adjusted mean attitude = 32.13, SE = 0.60). Introversion-extraversion, adjusted by self-esteem, attitudes toward women's role, and neuroticism also made a unique contribution to the composite DV, stepdown $F(1, 361) = 11.13$, $p < .01$, partial $\eta^2 = .03$ with confidence limits from .00 to .09. Women with higher masculinity were more extroverted (mean adjusted introversion-extraversion score = 12.48) than lower masculinity women (mean adjusted introversion-extraversion score = 11.00). Univariate analyses revealed that women with higher masculinity scores were also less neurotic, univariate $F(1, 364) = 7.28$, and had a more internal locus of control, univariate $F(1, 364)$ 7.39, differences that were already accounted for in the composite DV by higher-priority DVs. [*Means adjusted for main effects and for other DVs for stepdown interpretation are from Table 7.21, partial η^2 values and confidence limits are from Table 7.23. Means adjusted for main effects but not other DVs for univariate interpretation are in Table 7.22.*]

High-masculinity women, then, have greater self-esteem, less conservative attitudes toward the role of women, and more extraversion than women scoring low on masculinity. High femininity is associated with greater self-esteem and more conservative attitudes toward women's role than low femininity. Of the five effects, however, only the association between masculinity and self-esteem shows even a moderate proportion of shared variance.

Pooled within-cell correlations among DVs are shown in Table 7.17. The only relationships accounting for more than 10% of variance are between self-esteem and neuroticism ($r = .36$), locus of control and self-esteem ($r = .35$), and between neuroticism and locus of control ($r = .39$). Women who are high in neuroticism tend to have lower self-esteem and more external locus of control.

7.6.3 Multivariate Analysis of Covariance

For MANCOVA the same six variables are used as for MANOVA but ESTEEM, INTEXT, and NEUROTIC are used as DVs and CONTROL, ATTROLE, and SEL2 are used as covariates. The research question is whether there are personality differences associated with femininity, masculinity, and their interaction after adjustment for differences in attitudes and socioeconomic status.

Syntax and partial output of omnibus MANCOVA as produced by SPSS MANOVA appear in Table 7.25. As in MANOVA, Method 3 adjustment for unequal *n* is used with MASC adjusted for FEM and the interaction is adjusted for FEM and MASC. And, as in MANOVA, both main effects are highly significant but there is no interaction. Effect sizes for the three effects are Pillai's values. Entering `Approx. F` and appropriate df and percentage values into the NoncF.sav program and running NoncF3.sps, 95% confidence limits for these effect sizes are .00 to 08 for FEM, .08 to .21 for MASC, and .00 to .01 for the interaction.

7.6.3.1 Assessing Covariates

Under `EFFECT..WITHIN+RESIDUAL Regression` is the multivariate significance test for the relationship between the set of DVs (ESTEEM, INTEXT, and NEUROTIC) and the set of covariates (CONTROL, ATTROLE, and SEL2) after adjustment for IVs. Partial η^2 is calculated through the NoncF3.sps algorithm (Pillai's criterion is inappropriate unless $s = 1$) using `Approx. F` and appropriate df and is found to be .10 with 95% confidence limits from .06 to .13.

Because there is multivariate significance, it is useful to look at the three multiple regression analyses of each DV in turn, with covariates acting as IVs (see Chapter 5). The syntax of Table 7.25 automatically produces these regressions. They are done on the pooled within-cell correlation matrix, so that effects of the IVs are eliminated.

The results of the DV-covariate multiple regressions are shown in Table 7.26. At the top of Table 7.26 are the results of the univariate and stepdown analysis, summarizing the results of multiple regressions for the three DVs independently and then in priority order (see Section 7.6.3.2). At the bottom of Table 7.26 under `Regression analysis for WITHIN+RESIDUAL error term` are the separate regressions for each DV with covariates as IVs. For ESTEEM, two covariates, CONTROL and ATTROLE, are significantly related but SEL2 is not. None of the three covariates is related to INTEXT. Finally, for NEUROTIC, only CONTROL is significantly related. Because SEL2 provides no adjustment to any of the DVs, it could be omitted from future analyses.

7.6.3.2 Assessing DVs

Procedures for evaluating DVs, now adjusted for covariates, follow those specified in Section 7.6.2 for MANOVA. Correlations among all DVs, among covariates, and between DVs and covariates are informative so all the correlations in Table 7.17 are still relevant.[15]

Univariate *F*s are now adjusted for covariates. The univariate ANCOVAs produced by the SPSS MANOVA run specified in Table 7.25 are shown in Table 7.27. Although significance levels are misleading, there are substantial *F* values for ESTEEM and INTEXT for MASC (adjusted for FEM) and for FEM.

For interpretation of effects of IVs on DVs adjusted for covariates, comparison of stepdown *F*s with univariate *F*s again provides the best information. The priority order of DVs for this analysis is

---

[15]For MANCOVA, SPSS MANOVA prints pooled within-cell correlations among DVs (called criteria) adjusted for covariates. To get a pooled within-cell correlation matrix for covariates as well as DVs, you need a run in which covariates are included in the set of DVs.

TABLE 7.25   Multivariate Analysis of Covariance of Composite of DVs (ESTEEM, INTEXT, and NEUROTIC) as a Function of (Top to Bottom) FEM by MASC Interaction, Masculinity, and Femininity; Covariates are ATTROLE, CONTROL, and SEL2 (Syntax and Selected Output from SPSS MANOVA)

```
MANOVA      ESTEEM,ATTROLE,NEUROTIC,INTEXT,CONTROL,SEL2 BY FEM,MASC(1,2)
      /ANALYSIS=ESTEEM,INTEXT,NEUROTIC WITH CONTROL,ATTROLE,SEL2
      /PRINT=SIGNIF(STEPDOWN), ERROR(COR),
          HOMOGENEITY(BARTLETT,COCHRAN,BOXM)
      /METHOD=SEQUENTIAL
      /DESIGN FEM MASC FEM BY MASC.
```

```
EFFECT .. WITHIN+RESIDUAL Regression
Multivariate Tests of Significance (S = 3, M = -1/2, N = 178 1/2)
```

Test Name	Value	Approx. F	Hypoth. DF	Error DF	Sig. of F
Pillais	.23026	10.00372	9.00	1083.00	.000
Hotellings	.29094	11.56236	9.00	1073.00	.000
Wilks	.77250	10.86414	9.00	873.86	.000
Roys	.21770				

```
EFFECT.. FEM BY MASC
Multivariate Tests of Significance (S = 1, M = 1/2, N = 178 1/2)
```

Test Name	Value	Approx. F	Hypoth. DF	Error DF	Sig. of F
Pillais	.00263	.31551	3.00	359.00	.814
Hotellings	.00264	.31551	3.00	359.00	.814
Wilks	.99737	.31551	3.00	359.00	.814
Roys	.00263				

Note.. F statistics are exact.

```
EFFECT.. MASC
Multivariate Tests of Significance (S = 1, M = 1/2, N = 178 1/2)
```

Test Name	Value	Approx. F	Hypoth. DF	Error DF	Sig. of F
Pillais	.14683	20.59478	3.00	359.00	.000
Hotellings	.17210	20.59478	3.00	359.00	.000
Wilks	.85317	20.59478	3.00	359.00	.000
Roys	.14683				

Note.. F statistics are exact.

```
EFFECT.. FEM
Multivariate Tests of Significance (S = 1, M = 1/2, N = 178 1/2)
```

Test Name	Value	Approx. F	Hypoth. DF	Error DF	Sig. of F
Pillais	.03755	4.66837	3.00	359.00	.003
Hotellings	.03901	4.66837	3.00	359.00	.003
Wilks	.96245	4.66837	3.00	359.00	.003
Roys	.03755				

Note.. F statistics are exact.

TABLE 7.26 Assessment of Covariates: Univariate, Stepdown, and Multiple Regression Analyses for Three DVs with Three Covariates (Selected Output from SPSS MANOVA—see Table 7.25 for Syntax)

```
EFFECT.. WITHIN+RESIDUAL Regression (Cont.)
Univariate F-tests with (3,361) D. F.
```

Variable	Hypoth. SS	Error SS	Hypoth. MS	Error MS	F
ESTEEM	660.84204	3883.60490	220.28068	10.75791	20.47616
INTEXT	43.66605	4640.51295	14.55535	12.85461	1.13231
NEUROTIC	1384.16059	7589.51604	461.38686	21.02359	21.94615

Variable	Sig. of F
ESTEEM	.000
INTEXT	.336
NEUROTIC	.000

```
------------------------------------------------------------------
Roy-Bargman Stepdown F-tests
```

Variable	Hypoth. MS	Error MS	Stepdown F	Hypoth. DF	Error DF	Sig. of F
ESTEEM	220.28068	10.75791	20.47616	3	361	.000
INTEXT	6.35936	12.60679	.50444	3	360	.679
NEUROTIC	239.94209	19.72942	12.16164	3	359	.000

```
------------------------------------------------------------------
Regression analysis for WITHIN+RESIDUAL error term
--- Individual Univariate .9500 confidence intervals
Dependent variable .. ESTEEM          Self-esteem
```

COVARIATE	B	Beta	Std. Err.	t-Value	Sig. of t
CONTROL	.98173	.32005	.136	7.205	.000
ATTROLE	.08861	.15008	.028	3.208	.001
SEL2	-.00111	-.00723	.007	-.164	.869

```
Regression analysis for WITHIN+RESIDUAL error term
Dependent variable .. ESTEEM          Self-esteem
```

COVARIATE	Lower -95%	CL- Upper
CONTROL	.714	1.250
ATTROLE	.034	.143
SEL2	-.014	.012

```
Dependent variable .. INTEXT          Introversion-extroversion
```

COVARIATE	B	Beta	Std. Err.	t-Value	Sig. of t
CONTROL	-.22322	-.07655	.149	-1.499	.135
ATTROLE	.00456	.00812	.030	.151	.880
SEL2	.00682	.04662	.007	.922	.357

TABLE 7.26   Continued

COVARIATE	Lower -95%	CL- Upper
CONTROL	-.516	.070
ATTROLE	-.055	.064
SEL2	-.008	.021

Dependent variable .. NEUROTIC   Neuroticism

COVARIATE	B	Beta	Std. Err.	t-Value	Sig. of t
CONTROL	1.53128	.39102	.190	8.040	.000
ATTROLE	.04971	.06595	.039	1.287	.199
SEL2	.00328	.01670	.009	.347	.729

COVARIATE	Lower -95%	CL- Upper
CONTROL	1.157	1.906
ATTROLE	-.026	.126
SEL2	-.015	.022

TABLE 7.27   Univariate Analyses of Covariance of Three DVs Adjusted for Three Covariates for (Top to Bottom) FEM by MASC Interaction, Masculinity, and Femininity (Selected Output from SPSS MANOVA—see Table 7.25 for Syntax)

EFFECT.. FEM BY MASC (Cont.)
 Univariate F-tests with (1,361) D. F.

Variable	Hypoth. SS	Error SS	Hypoth. MS	Error MS	F	Sig. of F
ESTEEM	7.21931	3883.60490	7.21931	10.75791	.67107	.413
INTEXT	2.59032	464051.295	2.59032	1285.46065	.00202	.964
NEUROTIC	1.52636	7589.51604	1.52636	21.02359	.07260	.788

EFFECT.. MASC (Cont.)
Univariate F-tests with (1,361) D. F.

Variable	Hypoth. SS	Error SS	Hypoth. MS	Error MS	F	Sig. of F
ESTEEM	533.61774	3883.60490	533.61774	10.75791	49.60237	.000
INTEXT	26444.5451	464051.295	26444.5451	1285.46065	20.57204	.000
NEUROTIC	35.82929	7589.51604	35.82929	21.02359	1.70424	.193

EFFECT.. FEM (Cont.)
Univariate F-tests with (1,361) D. F.

Variable	Hypoth. SS	Error SS	Hypoth. MS	Error MS	F	Sig. of F
ESTEEM	107.44454	3883.60490	107.44454	10.75791	9.98749	.002
INTEXT	7494.31182	464051.295	7494.31182	1285.46065	5.83006	.016
NEUROTIC	26.81431	7589.51604	26.81431	21.02359	1.27544	.259

ESTEEM, INTEXT, and NEUROTIC. ESTEEM is evaluated after adjustment only for the three covariates. INTEXT is adjusted for effects of ESTEEM and the three covariates; NEUROTIC is adjusted for ESTEEM and INTEXT and the three covariates. In effect, then, INTEXT is adjusted for four covariates and NEUROTIC is adjusted for five.

Stepdown analysis for the interaction and two main effects is in Table 7.28. The results are the same as those in MANOVA except that there is no longer a main effect of FEM on INTEXT after adjustment for four covariates. The relationship between FEM and INTEXT is already represented by the relationship between FEM and ESTEEM. Consolidation of information from Tables 7.27 and 7.28, as well as some information from Table 7.26, appears in Table 7.29, along with apportionment of the .05 alpha error to the various tests and effect sizes with their confidence limits based on the α error chosen.

For the DVs associated with significant main effects, interpretation requires associated marginal means. Table 7.30 contains syntax and adjusted marginal means for ESTEEM and for INTEXT (which is adjusted for ESTEEM as well as covariates) for FEM and for MASC adjusted for FEM. Syntax and marginal means for the main effect of FEM on INTEXT (univariate but not stepdown effect) appear in Table 7.31.

Effect sizes and their confidence limits for stepdown effects are found through Smithson's (2003) program as for MANOVA. Table 7.32 shows the input/output for that analysis using values from Table 7.28. Values chosen for confidence limits reflect apportionment of α. A checklist for MANCOVA appears in Table 7.33. An example of a Results section, as might be appropriate for journal presentation, follows.

TABLE 7.28   Stepdown Analyses of Three Ordered DVs Adjusted for Three Covariates for (Top to Bottom) FEM by MASC Interaction, Masculinity, and Femininity (Selected Output from SPSS MANOVA—see Table 7.25 for Syntax)

Roy-Bargman Stepdown F-tests

Variable	Hypoth. MS	Error MS	StepDown F	Hypoth. DF	Error DF	Sig. of F
ESTEEM	7.21931	10.75791	.67107	1	361	.413
INTEXT	.35520	12.60679	.02817	1	360	.867
NEUROTIC	4.94321	19.72942	.25055	1	359	.617

Roy-Bargman Stepdown F-tests

Variable	Hypoth. MS	Error MS	StepDown F	Hypoth. DF	Error DF	Sig. of F
ESTEEM	533.61774	10.75791	49.60237	1	361	.000
INTEXT	137.74436	12.60679	10.92621	1	360	.001
NEUROTIC	1.07421	19.72942	.05445	1	359	.816

Roy-Bargman Stepdown F-tests

Variable	Hypoth. MS	Error MS	StepDown F	Hypoth. DF	Error DF	Sig. of F
ESTEEM	107.44454	10.75791	9.98749	1	361	.002
INTEXT	47.36159	12.60679	3.75683	1	360	.053
NEUROTIC	4.23502	19.72942	.21466	1	359	.643

TABLE 7.29   Tests of Covariates, Femininity, Masculinity (Adjusted for Femininity), and Interaction

IV	DV	Univariate F	df	Stepdown F	df	α	Partial η^2	CL around Partial η^2 per α Lower	Upper
Covariates	ESTEEM	20.48[a]	3/361	20.48**	3/361	.02	.15	.07	.22
	INTEXT	1.13	3/361	0.50	3/360	.02	.00	.00	.02
	NEUROTIC	21.95[a]	3/361	12.16**	3/359	.01	.09	.03	.17
Femininity	ESTEEM	9.99[a]	1/361	9.99**	1/361	.02	.03	.00	.08
	INTEXT	5.83[a]	1/361	3.76	1/360	.02	.01	.00	.05
	NEUROTIC	1.28	1/361	0.21	1/359	.01	.00	.00	.02
Masculinity	ESTEEM	49.60[a]	1/361	49.60**	1/361	.02	.12	.06	.20
	INTEXT	20.57[a]	1/361	10.93**	1/360	.02	.03	.00	.08
	NEUROTIC	1.70	1/361	0.05	1/359	.01	.00	.00	.01
Femininity by	ESTEEM	0.67	1/361	0.67	1/361	.02	.00	.00	.03
masculinity	INTEXT	0.00	1/361	0.03	1/360	.01	.00	.00	.00
interaction	NEUROTIC	0.07	1/361	0.25	1/359	.01	.00	.00	.03

[a]Significance level cannot be evaluated but would reach $p < .02$ in univariate context.

**$p < .01$.

TABLE 7.30   Adjusted Marginal Means for Esteem Adjusted for Three Covariates and INTEXT Adjusted for ESTEEM Plus Three Covariates (Syntax and Selected Output from SPSS MANOVA)

```
MANOVA      ESTEEM,ATTROLE,NEUROTIC,INTEXT,CONTROL,SEL2 BY FEM,MASC(1,2)
    /PRINT=PARAMETERS(ESTIM)
    /ANALYSIS=ESTEEM WITH CONTROL,ATTROLE,SEL2
    /DESIGN=CONSPLUS FEM /DESIGN=FEM,CONSPLUS MASC
    /ANALYSIS=INTEXT WITH CONTROL,ATTROLE,SEL2,ESTEEM
    /DESIGN=FEM, CONSPLUS,MASC.
```

```
Estimates for ESTEEM adjusted for 3 covariates
--- Individual univariate .9500 confidence intervals

CONSPLUS FEM

 Parameter     Coeff.   Std. Err.     t-Value    Sig. t Lower -95%   CL- Upper

        1   16.7151721     .34268    48.77769    .00000   16.04128    17.38906
        2   15.3543164     .21743    70.61815    .00000   14.92674    15.78189

Estimates for ESTEEM adjusted for 3 covariates
--- Individual univariate .9500 confidence intervals

CONSPLUS MASC

 Parameter     Coeff.   Std. Err.     t-Value    Sig. t Lower -95%   CL- Upper

        2   16.9175545     .22684    74.57875    .00000   16.47146    17.36365
        3   14.2243700     .31940    44.53415    .00000   13.59625    14.85249
```

(continued)

TABLE 7.30 Continued

```
Estimates for INTEXT adjusted for 4 covariates
--- Individual univariate .9500 confidence intervals

CONSPLUS MASC
```

Parameter	Coeff.	Std. Err.	t-Value	Sig. t	Lower -95% CL-	Upper
2	11.0058841	.25416	43.30276	.00000	10.50606	11.50571
3	12.4718467	.35617	35.01620	.00000	11.77141	13.17228

Note: Coeff. = adjusted marginal mean; first parameter = low, second parameter = high.

TABLE 7.31 Marginal Means for INTEXT Adjusted for Three Covariates Only (Syntax and Selected Output from SPSS MANOVA)

```
MANOVA ESTEEM,ATTROLE,NEUROTIC,INTEXT,CONTROL,SEL2 BY FEM,MASC(1,2)
        /PRINT=PARAMETERS(ESTIM)
        /ANALYSIS=INTEXT WITH CONTROL,ATTROLE,SEL2, ESTEEM
        /DESIGN=CONSPLUS FEM.
```

```
Estimates for INTEXT adjusted for 4 covariates
--- Individual univariate .9500 confidence intervals

CONSPLUS FEM
```

Parameter	Coeff.	Std. Err.	t-Value	Sig. t	Lower -95% CL-	Upper
1	11.1014711	.35676	31.11753	.00000	10.39989	11.80305
2	11.9085923	.22495	52.93984	.00000	11.46623	12.35096

Note: Coeff. = adjusted marginal mean; first parameter = low, second parameter = high.

TABLE 7.32 Data Set Output from NoncF3.sps for Effect Size (r2) with 95% Confidence Limits.

	fval	df1	df2	conf	lc2	ucdf	uc2	lcdf	power	r2	lr2	ur2
1	20.4800	3	361	.980	27.6000	.9900	102.9400	.0100	1.0000	.15	.07	.22
2	.5000	3	360	.980	.0000	.3175	9.1211	.0100	.0485	.00	.00	.02
3	12.1600	3	359	.990	10.0225	.9950	72.9125	.0050	.9960	.09	.03	.17
4	9.9900	1	361	.980	.6643	.9900	30.2627	.0100	.7187	.03	.00	.08
5	3.7600	1	360	.980	.0000	.9467	18.2345	.0100	.2610	.01	.00	.05
6	.2100	1	359	.990	.0000	.3530	9.0300	.0052	.0099	.00	.00	.02
7	49.6000	1	361	.980	21.4579	.9900	89.1734	.0100	1.0000	.12	.06	.20
8	10.9300	1	360	.980	.9188	.9900	31.8934	.0100	.7655	.03	.00	.08
9	.0500	1	359	.999	.0000	.1768	2.1500	.0614	.0007	.00	.00	.01
10	.6700	1	361	.980	.0000	.5864	9.8825	.0100	.0397	.00	.00	.03
11	.0300	1	360	.980	.0000	.1374	1.2900	.0726	.0111	.00	.00	.00
12	.2500	1	359	.999	.0000	.3826	10.7500	.0027	.0015	.00	.00	.03

TABLE 7.33   Checklist for Multivariate Analysis of Covariance

1. Issues
   a. Unequal sample sizes and missing data
   b. Normality of sampling distributions
   c. Outliers
   d. Homogeneity of variance-covariance matrices
   e. Linearity
   f. Homogeneity of regression
      (1) Covariates
      (2) DVs for stepdown analysis
   g. Reliability of covariates (and DVs for stepdown)
   h. Multicollinearity and singularity
2. Major analyses: Planned comparisons or omnibus F; when significant: Importance of DVs
   a. Within-cell correlations, stepdown F, univariate F
   b. Effect size with its confidence interval for significant stepdown F
   c. Adjusted marginal and/or cell means for significant F, and standard deviations or standard errors or confidence intervals
3. Multivariate effect size(s) with confidence interval(s) for planned comparisons or omnibus F.
4. Additional analyses
   a. Assessment of covariates
   b. Interpretation of IV-covariates interaction (if homogeneity of regression violated for stepdown analysis)
   c. Post hoc comparisons

Results

   A 2 × 2 between-subjects multivariate analysis of covariance was performed on three dependent variables associated with personality of respondents: self-esteem, introversion-extraversion, and neuroticism. Adjustment was made for three covariates: attitude toward role of women, locus of control, and socioeconomic status. Independent variables were masculinity (high and low) and femininity (high and low).

   SPSS MANOVA was used for the analyses with the sequential adjustment for nonorthogonality. Order of entry of IVs was femininity, then masculinity. Total N = 369 was reduced to 368 with the deletion of a case missing a score on locus of control. There were no univariate or multivariate within-cell outliers at α = .001. Results of evaluation of assumptions of normality, homogeneity of variance-covariance matrices, linearity, and multicollinearity were satisfactory. Covariates were judged to be adequately reliable for covariance analysis.

With the use of Wilks' criterion, the combined DVs were significantly related to the combined covariates, approximate $F(9, 873)$ = 10.86, p < .01, to femininity, $F(3, 359)$ = 4.67, p < .01, and to masculinity, $F(3, 359)$ = 20.59, p < .001 but not to the interaction, $F(3, 359)$ = 0.31, p > .05. There was a modest association between DVs and covariates, partial η^2 = .10 with confidence limits from .06 to .29. A somewhat larger association was found between combined DVs and the main effect of masculinity, partial η^2 = .15 with confidence limits from .08 to .21, but the association between the main effect of femininity and the combined DVs was smaller, partial η^2 = .04 with confidence limits from .00 to .08. Effect size for the nonsignificant interaction was .00 with confidence limits from .00 to .01. [*F is from Table 7.25; partial* h^2 *and their confidence limits are found through Smithson's NoncF3.sps for main effects, interaction, and covariates.*]

To investigate more specifically the power of the covariates to adjust dependent variables, multiple regressions were run for each DV in turn, with covariates acting as multiple predictors. Two of the three covariates, locus of control and attitudes toward women's role, provided significant adjustment to self-esteem. The B value of .98 (confidence interval from .71 to 1.25) for locus of control was significantly different from zero, $t(361)$ = 7.21, p < .001, as was the B value of .09 (confidence interval from .03 to .14) for attitudes toward women's role, $t(361) = 3.21$, $p < .01$. None of the covariates provided adjustment to the introversion-extraversion scale. For neuroticism, only locus of control reached statistical significance, with $B = 1.53$ (confidence interval from 1.16 to 1.91), $t(361) = 8.04$, $p < .001$. For none of the DVs did socioeconomic status provide significant adjustment. [*Information about relationships for individual DVs and CVs is from Table 7.26.*]

Effects of masculinity and femininity on the DVs after adjustment for covariates were investigated in univariate and Roy-Bargmann step-down analysis, in which self-esteem was given the highest priority, introversion-extraversion second priority (so that adjustment was made for self-esteem as well as for the three covariates), and neuroticism third priority (so that adjustment was made for self-esteem and introversion-extraversion as well as for the three covariates). Homogeneity of regression was satisfactory for this analysis, and DVs were judged to be sufficiently reliable to act as covariates. Results of this analysis are summarized in Table 7.29. An experimentwise error rate of 5% for each effect was achieved by apportioning alpha according to the values shown in the last column of the table.

After adjusting for differences on the covariates, self-esteem made a significant contribution to the composite of the DVs that best distinguishes between women who were high or low in femininity, step-down $F(1, 361) = 9.99$, $p < .01$, partial $\eta^2 = .03$ with confidence limits from .00 to .08. With self-esteem scored inversely, women with higher femininity scores showed greater self-esteem after adjustment for covariates (adjusted mean self-esteem = 15.35, SE = 0.22) than those scoring lower on femininity (adjusted mean self-esteem = 16.72, SE = 0.34). Univariate analysis revealed that a statistically significant difference was also present on the introversion-extraversion measure, with higher-femininity women more extraverted, univariate $F(1, 361) =$ 5.83, a difference already accounted for by covariates and the higher-priority DV. [*Adjusted means are from Tables 7.30 and 7.31; partial η^2 and confidence limits are from Table 7.32.*]

Lower- versus higher-masculinity women differed in self-esteem, the highest-priority DV, after adjustment for covariates, stepdown $F(1, 361) = 49.60$, $p < .01$, partial $\eta^2 = .12$ with confidence limits from .06 to .20. Greater self-esteem was found among higher-masculinity women (adjusted mean = 14.22, SE = 0.32) than among lower-masculinity women (adjusted mean = 16.92, SE = 0.23). The measure of introversion and extraversion, adjusted for covariates and self-esteem, was also related to differences in masculinity, stepdown $F(1, 360) = 10.93$, $p < .01$, partial $\eta^2 = .03$ with confidence limits from .00 to .08. Women scoring higher on the masculinity scale were more extraverted (adjusted mean extraversion 12.47, SE = 0.36) than those showing lower masculinity (adjusted mean extraversion = 11.01, SE = 0.25).

High-masculinity women, then, are characterized by greater self-esteem and extraversion than low-masculinity women when adjustments are made for differences in socioeconomic status, attitudes toward women's role, and locus of control. High-femininity women show greater self-esteem than low-femininity women with adjustment for those covariates.

Pooled within-cell correlations among dependent variables and covariates are shown in Table 7.17. The only relationships accounting for more than 10% of variance are between self-esteem and neuroticism ($r = .36$), locus of control and self-esteem ($r = .35$), and between neuroticism and locus of control ($r = .39$). Women who are high in neuroticism tend to have lower self-esteem and are more likely to attribute reinforcements to external sources.

7.7   Comparison of Programs

SPSS, SAS, and SYSTAT all have highly flexible and full-featured MANOVA programs, as seen in Table 7.34. One-way between-subjects MANOVA is also available through discriminant function programs, as discussed in Chapter 9.

7.7.1   SPSS Package

SPSS has two programs, MANOVA (available only through syntax) and GLM. Features of the two programs are quite different, so that you may want to use both programs for an analysis.

Both programs offer several methods of adjustment for unequal n and several statistical criteria for multivariate effects. In repeated-measures designs, the sphericity test offered by both programs evaluates the sphericity assumption; if the assumption is rejected (that is, if the test is significant), one of the alternatives to repeated-measures ANOVA—MANOVA, for instance—is appropriate. There are also the Greenhouse-Geisser, Huynh-Feldt and lower-bound epsilons for adjustment of df for sphericity. SPSS MANOVA and GLM do the adjustment and provide significance levels for the effects with adjusted df.

SPSS MANOVA has several features that make it superior to any of the other programs reviewed here. It is the only program that performs Roy-Bargmann stepdown analysis as an option (Section 7.5.3.2). Use of other programs requires a separate ANCOVA run for each DV after the one of highest priority. SPSS MANOVA also is the only program that has special syntax for pooling covariates to test homogeneity of regression for MANCOVA and stepdown analysis (Section 7.6.1.6). If the assumption is violated, the manuals describe procedures for ANCOVA with separate regression estimates, if that is your choice. Full simple effects analyses are easily specified using the MWITHIN instruction (Section 8.5.2). SPSS MANOVA also is easier to use for user-specified comparisons. Bivariate collinearity and homogeneity of variance-covariance matrices are readily tested in SPSS MANOVA through within-cell correlations and homogeneity of dispersion matrices, respectively. Multicollinearity is assessed through the determinant of the within-cells correlation matrix (cf. Section 4.1.7).

Both programs provide complete descriptive statistics for unadjusted means and standard deviations, however, adjusted means for marginal and cell effects are more easily specified in SPSS GLM through the EMMEANS instruction. SPSS MANOVA provides adjusted cell means easily, but marginal means require rather convoluted CONSPLUS instructions, as seen in Section 7.6. SPSS GLM provides leverage values (that are easily converted to Mahalanobis distance) to assess multivariate outliers.

For between-subjects designs, both programs offer Bartlett's test of sphericity, which tests the null hypothesis that correlations among DVs are zero; if they are, univariate F (with Bonferroni adjustment) is used instead of stepdown F to test the importance of DVs (Section 7.5.3.1).

A principal components analysis can be performed on the DVs through SPSS MANOVA, as described in the manuals. In the case of multicollinearity or singularity among DVs (see Chapter 4), principal components analysis can be used to produce composite variables that are orthogonal to one another. However, the program still performs MANOVA on the raw DV scores, not the component scores. If MANOVA for component scores is desired, use the results of PCA and the COMPUTE facility to generate component scores for use as DVs.

TABLE 7.34   Comparison of Programs for Multivariate Analysis of Variance and Covariance[a]

Feature	SPSS GLM	SPSS MANOVA	SAS GLM	SYSTAT ANOVA, GLM, and MANOVA[g]
Input				
Variety of strategies for unequal n	Yes	Yes	Yes	Yes
Specify tolerance	EPS	No	SINGULAR	Yes
Specify exact tests for multivariate effects	No	No	MSTAT= EXACT	No
Output				
Standard source table for Wilks' lambda	No	PRINT = SIGNIF (BRIEF)	No	No
Cell covariance matrices	No	Yes	No	No
Cell covariance matrix determinants	No	Yes	No	No
Cell correlation matrices	No	Yes	No	No
Cell SSCP matrices	No	Yes	No	No
Cell SSCP determinants	No	Yes	No	No
Unadjusted marginal means for factorial design	Yes	Yes	Yes	No
Unadjusted cell means	Yes	Yes	Yes	No
Unadjusted cell standard deviations	Yes	Yes	Yes	No
Confidence interval around unadjusted cell means	No	Yes	No	No
Adjusted cell means	EMMEANS	PMEANS	LSMEANS	PRINT MEDIUM
Standard errors for adjusted cell means	EMMEANS	No	LSMEANS	PRINT MEDIUM
Adjusted marginal means	EMMEANS	Yes[b]	LSMEANS	No
Standard errors for adjusted marginal means	EMMEANS	No	LSMEANS	No
Wilks' lambda with approximate F statistic	Yes	Yes	Yes	Yes
Criteria other than Wilks'	Yes	Yes	Yes	Yes
Multivariate influence/leverage statistics by cell	No	No	No	Data file
Canonical (discriminant function) statistics[c]	No	Yes	Yes	Yes
Univariate F tests	Yes	Yes	Yes	Yes
Averaged univariate F tests	No	Yes	No	No
Stepdown F tests (DVs)	No	Yes	No	No
Sphericity test	Yes	Yes	No	Yes[h]
Adjustment for failure of sphericity	Yes	Yes	Yes	No
Tests for univariate homogeneity of variance	Yes	Yes	Yes[c]	No
Test for homogeneity of covariance matrices	Box's M	Box's M	No	No

TABLE 7.34   Continued

Feature	SPSS GLM	SPSS MANOVA	SAS GLM	SYSTAT ANOVA, GLM, and MANOVA[g]
Output *(continued)*				
Principal component analysis of residuals	No	Yes	No	Yes
Hypothesis SSCP matrices	Yes	No	Yes	Yes
Hypothesis covariance matrices	No	Yes	No	No
Inverse of hypothesis SSCP matrices	No	No	Yes	Yes
Pooled within-cell error SSCP matrix	Yes	Yes	Yes	Yes
Pooled within-cell covariance matrix	No	Yes	No	Yes
Pooled within-cell correlation matrix	No	Yes	Yes	Yes
Total SSCP matrix	No	No	No	Yes
Determinants of pooled within-cell correlation matrix	No	Yes	No	No
Covariance matrix for adjusted cell means	No	No	Data file	No
Effect size for univariate tests	ETASQ	POWER	No	No
Power analysis	OPOWER	POWER	No[f]	No
SMCs with effects for each DV	No	No	No	Yes
Confidence intervals for multivariate tests	No	Yes	No	No
Post hoc tests with adjustment	POSTHOC	Yes[d]	Yes	Yes
Specific comparisons	Yes	Yes	Yes	Yes
Tests of simple effects (complete)	No	Yes	No	No
Homogeneity of regression	No	Yes	No	Yes
ANCOVA with separate regression estimates	No	Yes	No	No
Regression coefficient for each covariate	PARAMETER	Yes	No	PRINT LONG
Regression coefficient for each cell	Yes	No	Yes	No
R^2 for model	Yes	No	Yes	No
Coefficient of variation	No	No	Yes	No
Normalized plots for each DV and covariate	No	Yes	No	No
Predicted values and residuals for each case	Data file	Yes	Yes	Data file
Confidence limits for predicted values	No	No	Yes	No
Residuals plots	No	Yes	No	No

[a]Additional features are discussed in Chapter 6 (ANCOVA).

[b]Available through CONSPLUS procedure, see Section 7.6.

[c]Discussed more fully in Chapter 9.

[d]Bonferroni and Scheffé confidence intervals.

[e]One-way design only.

[f]Available in a separate program: GLMPOWER.

[g]MANOVA, added to SYSTAT in Version 11, differs from GLM only in its menu access.

[h]Not available in "long" output.

7.7.2   SAS System

MANOVA in SAS is done through the PROC GLM. This general linear model program, like SPSS GLM, has great flexibility in testing models and specific comparisons. Four types of adjustments for unequal-n are available, called **TYPE I** through **TYPE IV** estimable functions (cf. 6.5.4.2); this program is considered by some to have provided the archetypes of the choices available for unequal-n adjustment. Adjusted cell and marginal means are printed out with the **LSMEANS** instruction. SAS tests multivariate outliers by adding leverage values (which may be converted to Mahalanobis distance) to the data set (cf. Section 6.6.1.4). Exact tests of multivariate effects may be requested in place of the usual F approximation.

SAS GLM provides Greenhouse-Geisser and Huynh-Feldt adjustments to degrees of freedom and significance tests for effects using adjusted df. There is no explicit test for homogeneity of regression, but because this program can be used for any form of multiple regression, the assumption can be tested as a regression problem where the interaction between the covariate(s) and IV(s) is an explicit term in the regression equation (Section 6.5.3).

There is abundant information about residuals, as expected from a program that can be used for multiple regression. Should you want to plot residuals, however, a run through the PLOT procedure is required. As with most SAS programs, the output requires a fair amount of effort to decode until you become accustomed to the style.

7.7.3   SYSTAT System

In SYSTAT, the GLM, ANOVA, and MANOVA programs may be used for simple, fully factorial MANOVA, however GLM and MANOVA are recommended for more complex designs for their numerous features and flexibility, and because they are not much more difficult to set up.

Model 1 adjustment for unequal n is provided by default, along with a strong argument as to its benefits. Other options are available, however, by specification of error terms or a series of sequential regression analyses. Several criteria are provided for tests of multivariate hypotheses, along with a great deal of flexibility in specifying these hypotheses. Leverage values are saved in a data set by request, and may be converted to Mahalanobis distance as per Equation 4.3 to assess multivariate outliers.

The program provides cell least squares means and their standard errors, adjusted for covariates, if any. Other univariate statistics are not provided in the program, but they can be obtained through the STATS module.

Like SPSS MANOVA, principal components analysis can be done on the pooled within-cell correlation matrix. But also like the SPSS program, the MANOVA is performed on the original scores.

8
Profile Analysis: The Multivariate Approach to Repeated Measures

8.1 General Purpose and Description

Profile analysis is a special application of multivariate analysis of variance (MANOVA) to a situation where there are several DVs, all measured on the same scale.[1] The set of DVs can either come from one DV measured several different times, or several different DVs all measured at one time. There is also a popular extension of the analysis where several different DVs are measured at several different times, called the *doubly-multivariate design.*

The more common application is in research where subjects are measured repeatedly on the same DV. For example, math achievement tests are given at various points during a semester to test the effects of alternative educational programs such as traditional classroom vs. computer-assisted instruction. Used this way, profile analysis offers a multivariate alternative to the univariate *F* test for the within-subjects effect and its interactions (see Chapter 3).[2] The choice between profile analysis and univariate repeated-measures ANOVA depends on sample size, power, and whether statistical assumptions of repeated-measured ANOVA are met. These issues are discussed fully in Section 8.5.1.

Less commonly, profile analysis is used to compare profiles of two or more groups measured on several different scales, all at one time. For example, psychoanalysts and behavior therapists are both given, say, the Profile of Mood States (POMS). The DVs are the various scales of the POMS, tension-anxiety, vigor, anger-hostility, and so on, all measured on the same scale. The analysis asks if the two groups have the same pattern of means on the subscales.

Rapidly growing in popularity is use of repeated-measures MANOVA for doubly-multivariate designs where several DVs, not all measured on the same scale, are measured repeatedly. For example, math competence is measured several times during a semester, each time by both a grade on a math test (one DV) and a scale of math anxiety (a second DV). A discussion of doubly-multivariate analysis appears in Section 8.5.3. A complete example appears in Section 8.6.2.

Current computer programs allow the application of profile analysis to complex designs where, for instance, groups are classified along two or more dimensions to create two or more

---

[1]The term *profile analysis* is also applied to techniques for measuring resemblance among profile patterns through cluster analysis rather than the MANOVA strategy described in this chapter.

[2]The term profile analysis is used for convenience here as synonymous with "taking the multivariate approach to repeated measures."

between-subjects IVs, as in ANOVA. For example, POMS profiles are examined for both male and female psychoanalysts and behavior therapists; or changes in math competence during a semester as a result of either traditional education or computer-assisted instruction are evaluated for both elementary and junior high school students.

Rangaswamy and colleagues (2002) used profile analysis to examine beta power in the EEGs of alcoholics. Separate repeated-measures MANOVAs were run on men and women and three beta bands. The within-subjects IV consisted of the 22 pairs of electrodes in various frontal, central, and parietal locations. The between-subjects IV was alcohol-dependent or not, with 307 participants in each group. Post hoc analyses examined the three sets of locations. Beta power was found to be elevated in alcoholics in all three bands, more so for men than women, and with greater increase in the central region. Additional univariate follow-up analyses were conducted.

A repeated-measures MANCOVA was used by Martis et al. (2003) to examine changes in neurocognitive functioning with repetitive transcranial magnetic stimulation (rTMS) in 15 severely depressed patients. A $2 \times 4 \times 2$ within-within-between design was employed with two time periods (baseline and post-treatment), four neuropsychological domains, and two groups (rTMS responder or nonresponder in terms of depression reduction). Domains were created by combining a series of tests into a single z-score for each domain. Covariates presumably were HDRS (a depression scale) scores and length of treatment (it is not clear why baseline scores were not also used as covariates). No significant main effects were observed, however a couple of significant interactions were found. Interactions were examined as simple effects at each domain. The interaction between time and neuropsychological domain indicated improvement in scores across 3 of the 4 domains between baseline and post-treatment times: working memory-executive function, objective memory, and fine motor speed. The interaction between response and neuropsycholgical domain showed significant differences between responders and nonresponders for one of the domains: attention and mental speed.

8.2   Kinds of Research Questions

The major question for profile analysis is whether groups have different profiles on a set of measures. To apply profile analysis, all measures must have the same range of possible scores, with the same score value having the same meaning on all the measures. There is restriction on the scaling of the measures because in two of the major tests of profile analysis (parallelism and flatness) the numbers that are actually tested are difference[3] scores between DVs measured on adjacent occasions or some other transformation of the set of DVs. Difference scores are called segments in profile analysis.

8.2.1   Parallelism of Profiles

Do different groups have parallel profiles? This is commonly known as the test of parallelism and is the primary question addressed by profile analysis. When using the profile approach to univariate

---

[3]Although difference scores are notoriously unreliable, their use here is just as a statistical convenience (as any other transformation would be). The difference scores are not interpreted per se.

repeated-measures ANOVA, the parallelism test is the test of interaction. For example, do traditional and computer assisted instruction lead to the same pattern of gains in achievement over the course of testing? Or do changes in achievement depend on which method of instruction is used? Using the therapist example, do psychoanalysts and behavior therapists have the same pattern of highs and lows on the various mood states measured by the POMS?

8.2.2   Overall Difference among Groups

Whether or not groups produce parallel profiles, does one group, on average, score higher on the collected set of measures than another? For example, does one method of instruction lead to greater overall math achievement than the other method? Or does one type of therapist have reliably higher scores on the set of states measured by the POMS than the other?

In garden-variety univariate ANOVA, this question is answered by test of the "groups" hypothesis; in profile analysis jargon, this is the "levels" hypothesis. It addresses the same question as the between-subjects main effect in repeated-measures ANOVA.

8.2.3   Flatness of Profiles

The third question addressed by profile analysis concerns the similarity of response to all DVs, independent of groups. Do all the DVs elicit the same average response? In profile jargon, this tests the "flatness" hypothesis. This question is typically relevant only if the profiles are parallel. If the profiles are not parallel, then at least one of them is necessarily not flat. Although it is conceivable that non-flat profiles from two or more groups could cancel each other out to produce, on average, a flat profile, this result is often not of research interest.

In the instructional example, the flatness test evaluates whether achievement changes over the period of testing. In this context the flatness test evaluates the same hypothesis as the within-subjects main effect in repeated-measures ANOVA. Using the therapist example, if psychoanalysts and behavior therapists have the same pattern of mood states on the POMS (that is, they have parallel profiles), one might ask whether therapists, as a whole, are notably high or low on any of the states.

8.2.4   Contrasts Following Profile Analysis

With more than two groups or more than two measures, differences in parallelism, flatness, and/or level can result from a variety of sources. For example, if a group of client-centered therapists is added to the therapist study, and the parallelism or levels hypothesis is rejected, it is not obvious whether it is the behavior therapists who differ from the other two groups, the psychoanalysts who differ from the client-centered therapists, or exactly which group differs from which other group or groups. Contrasts following profile analysis are discussed in Section 8.5.2.

8.2.5   Parameter Estimates

Parameters are estimated whenever statistically significant differences are found between groups or measures. For profile analysis, the major description of results is typically a plot of profiles in which the means for each of the DVs are plotted for each of the groups. In addition, if the null hypothesis regarding levels is rejected, assessment of group means and group standard deviations, standard errors, or confidence intervals is helpful. And if the null hypothesis regarding flatness is rejected and the finding is of interest, a plot of scores combined over groups is instructive, along with standard deviations or standard errors. Profile plots are demonstrated in Sections 8.4 and 8.6.

8.2.6   Effect Size

As with all statistical techniques, estimates of effect size are appropriate for all effects of interest. Such effect size measures and confidence intervals around them are demonstrated in Sections 8.4 and 8.6.

8.3   Limitations to Profile Analysis

8.3.1   Theoretical Issues

Choice of DVs is more limited in profile analysis than in usual applications of multivariate statistics because DVs must be commensurate except in the doubly-multivariate application. That is, they must all have been subjected to the same scaling techniques.

In applications where profile analysis is used as an alternative to univariate repeated-measures ANOVA, this requirement is met because all DVs are literally the same measure. In other applications of profile analysis, however, careful consideration of the measurement of the DVs is required to assure that the units of measurement are the same. One way to produce commensurability is to use standardized scores, such as z-scores, instead of raw scores for the DVs. In this case, each DV is standardized using the pooled within-groups standard deviation (the square root of the error mean square for the DV) provided by univariate one-way between-subjects ANOVA for the DV. There is some danger in generalizing results with this approach, however, because sample standard deviations are used to form z-scores. Similar problems arise when factor or component scores are used as measures, based on factor or component analysis of sample data. More commonly, commensurate DVs are subscales of standardized tests, in which subtests are scaled in the same manner.

Differences among profiles are causally attributed to differences in treatments among groups if, and only if, groups are formed by random assignment, levels of IVs manipulated, and proper experimental controls maintained. Causality, as usual, is not addressed by the statistical test. Generalizability, as well, is influenced by sampling strategy, not by choice of statistical tests. That is, the results of profile analysis generalize only to the populations from which cases are randomly sampled.

As described in Section 8.2 and derived in Section 8.4, the DVs in profile analysis are differences (segments) between the scores for adjacent levels of the within-subjects IV, or some other transformation of them. Creating difference scores is one of the ways to equate the number of DVs and the degrees of freedom for the within-subjects IV. Although different programs use different transformations, the resulting omnibus analysis is insensitive to them. Technically, then, limitations should be assessed with regard to segments or otherwise transformed DVs; however, it is reasonable (and a lot simpler) to assess the DVs in their original form. That is, for purposes of assessing limitations, scores for the levels of the within-subjects IV are treated as the set of DVs.

8.3.2   Practical Issues

8.3.2.1   *Sample Size, Missing Data, and Power*

The sample size in each group is an important issue in profile analysis, as in MANOVA, because *there should be more research units in the smallest group than there are DVs.* This is recommended both because of considerations of power and for evaluation of the assumption of homogeneity of variance-covariance matrices (cf. Sections 7.3.2.1 and 7.3.2.4). In the choice between univariate repeated-measures ANOVA and profile analysis, sample size per group is often the deciding factor.

Unequal sample sizes typically provide no special difficulty in profile analysis because each hypothesis is tested as if in a one-way design and, as discussed in Section 6.5.4.2, unequal *n* creates difficulties in interpretation only in designs with more than one between-subjects IV. However, unequal *n* sometimes has implications for evaluating homogeneity of variance-covariance matrices, as discussed in Section 8.3.2.4. If some measures are missing from some cases, the usual problems and solutions discussed in Chapter 4 are modified for profile analysis, because all measures are commensurate and indeed may be the same measure. Imputation of missing values is discussed in Section 8.5.5.

As always, larger sample sizes produce greater power, all else being equal. There also are power implications in the choice between the univariate and multivariate approaches to repeated measures. Generally, there is greater power in the multivariate approach, given the adjustment for violation of sphericity that is often required for the univariate approach, but surprises do occur. Section 7.3.2.1 discusses issues of power and software to estimate sample sizes.

8.3.2.2   *Multivariate Normality*

Profile analysis is as robust to violation of normality as other forms of MANOVA (cf. Section 7.3.2.2). So, *unless there are fewer cases than DVs in the smallest group and highly unequal* n, *deviation from normality of sampling distributions is not expected.* In the unhappy event of small, unequal samples, however, a look at the distributions of DVs for each group is in order. If distributions of DVs show marked, highly-significant skewness, some normalizing transformations might be investigated (cf. Chapter 4).

8.3.2.3   *Absence of Outliers*

As in all MANOVA, profile analysis is extremely sensitive to outliers. *Tests for univariate and multivariate outliers*, detailed in Chapter 4, *are applied to DVs.* These tests are demonstrated in Section 8.6.1.

8.3.2.4   *Homogeneity of Variance-Covariance Matrices*

If sample sizes are equal, evaluation of homogeneity of variance-covariance matrices is not necessary. However, if sample sizes are notably discrepant, Box's *M* test is available through SPSS MANOVA as a preliminary test of the homogeneity of the variance-covariance matrices. Box's *M* is too sensitive for use at routine α levels, but if the test of homogeneity is rejected at highly significant levels, the guidelines in Section 7.3.2.4 are appropriate.

Univariate homogeneity of variance is also assumed, but the robustness of ANOVA generalizes to profile analysis. Unless sample sizes are highly divergent or there is evidence of strong heterogeneity (variance ratio of 10:1 or larger) of the DVs (cf. Section 6.3.2.5), this assumption is probably safely ignored.

8.3.2.5 Linearity

For the parallelism and flatness tests, linearity of the relationships among DVs is assumed. This assumption is evaluated by examining scatterplots between all pairs of DVs through SPSS PLOT, or SAS CORR or PLOT. Because the major consequence of failure of linearity is loss of power in the parallelism test, violation is somewhat mitigated by large sample sizes. Therefore, *with many symmetrically distributed DVs and large sample sizes, the issue may be ignored.* On the other hand, if distributions are notably skewed in different directions, check a few scatterplots for the variables with the most discrepant distributions to assure that the assumption is not too badly violated.

8.3.2.6 Absence of Multicollinearity and Singularity

Section 7.3.2.8 discusses multicollinearity and singularity for MANOVA. Criteria for logical multicollinearity are quite different, however, for the multivariate approach to repeated measures. Correlations among DVs are expected to be quite high when they are the same measure taken from the same cases over time. Therefore, only statistical multicollinearity poses difficulties, and even then only if tolerance (1 − SMC) is less than .001 for the measures combined over groups.

8.4 Fundamental Equations for Profile Analysis

Table 8.1 is a hypothetical data set appropriate for using profile analysis as an alternative to repeated-measures ANOVA. The three groups (the IV) whose profiles are compared are belly dancers, politicians, and administrators (or substitute your favorite scapegoat). The five respondents in each of these occupational groups participate in four leisure activities (the DVs) and, during each, are asked to rate their satisfaction on a 10-point scale. The leisure activities are reading, dancing, watching TV, and skiing. The profiles are illustrated in Figure 8.1, where mean ratings of each group for each activity are plotted.

Profile analysis tests of parallelism and flatness are multivariate and involve sum-of-squares and cross-products matrices. But the levels test is a univariate test, equivalent to the between-subjects main effect in repeated-measures ANOVA.

8.4.1 Differences in Levels

For the example, the levels test examines differences between the means of the three occupational groups combined over the four activities. Are the group means of 7.30, 5.00, and 3.15 significantly different from each other?

The relevant equation for partitioning variance is adapted from Equation 3.8 as follows:

$$\sum_i \sum_j (Y_{ij} - \text{GM})^2 = np \sum_j (\overline{Y}_j - \text{GM})^2 + p \sum_i \sum_j (Y_{ij} - \overline{Y}_j)^2 \tag{8.1}$$

where n is the number of subjects in each group and p is the number of measures, in this case the number of ratings made by each respondent.

TABLE 8.1 Small Sample of Hypothetical Data for Illustration of Profile Analysis

Group		Case No.	Read	Dance	TV	Ski	Combined Activities
				Activity			
Belly dancers		1	7	10	6	5	7.00
		2	8	9	5	7	7.25
		3	5	10	5	8	7.00
		4	6	10	6	8	7.50
		5	7	8	7	9	7.75
	Mean		6.60	9.40	5.80	7.40	7.30
Politicians		6	4	4	4	4	4.00
		7	6	4	5	3	4.50
		8	5	5	5	6	5.25
		9	6	6	6	7	6.25
		10	4	5	6	5	5.00
	Mean		5.00	4.80	5.20	5.00	5.00
Administrators		11	3	1	1	2	1.75
		12	5	3	1	5	3.50
		13	4	2	2	5	3.25
		14	7	1	2	4	3.50
		15	6	3	3	3	3.75
	Mean		5.00	2.00	1.80	3.80	3.15
Grand mean			5.53	5.40	4.27	5.40	5.15

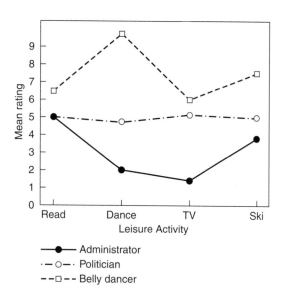

**FIGURE 8.1 Profiles of leisure-time ratings
for three occupations.**

The partition of variance in Equation 8.1 produces the total, within-groups, and between-groups sums of squares, respectively, as in Equation 3.9. Because the score for each subject in the levels test is the subject's average score over the four activities, degrees of freedom follow Equations 3.10 through 3.13, with N equal to the total number of subjects and k equal to the number of groups.

For the hypothetical data of Table 8.1:

$$SS_{bg} = (5)(4)[(7.30 - 5.15)^2 + (5.00 - 5.15)^2 + (3.15 - 5.15)^2]$$

and

$$SS_{wg} = (4)[(7.00 - 7.30)^2 + (7.25 - 7.30)^2 + \cdots + (3.75 - 3.15)^2]$$

$$df_{bg} = k - 1 = 2$$

$$df_{wg} = N - k = 12$$

The levels test for the example produces a standard ANOVA source table for a one-way univariate test, as summarized in Table 8.2. There is a statistically significant difference between occupational groups in average rating of satisfaction during the four leisure activities.

Standard univariate η^2 is used to evaluate the effect size for occupational groups:

$$\eta^2 = \frac{SS_{bg}}{SS_{bg} + SS_{wg}} = \frac{172.90}{172.90 + 23.50} = .88 \tag{8.2}$$

The confidence interval, found through Smithson's (2003) procedure described earlier, ranges from .61 to .92.

8.4.2 Parallelism

Tests of parallelism and flatness are conducted through hypotheses about adjacent segments of the profiles. The test of parallelism, for example, asks if the difference (segment) between reading and dancing is the same for belly dancers, politicians, and administrators. How about the difference between dancing and watching TV?

The most straightforward demonstration of the parallelism test begins by converting the data matrix into difference scores.[4] For the example, the four DVs are turned into three differences, as

TABLE 8.2   ANOVA Summary Table for Test of Levels Effect for Small-Sample Example of Table 8.1

Source of Variance	SS	df	MS	F
Between groups	172.90	2	86.45	44.14*
Within groups	23.50	12	1.96	

*$p < .001$.

[4]Other transformations are equally valid but generally more complex.

shown in Table 8.3. The difference scores are created from adjacent pairs of activities, but in this example, as in many uses of profile analysis, the order of the DVs is arbitrary. In profile analysis it is often true that segments are formed from arbitrarily transformed DVs and have no intrinsic meaning. This sometimes creates difficulty in interpreting statistical findings of computer software and may lead to the decision to apply some other transformation, such as polynomial (to produce a trend analysis).

In Table 8.3 the first entry for the first case is the difference between READ and DANCE scores, that is, $7 - 10 = -3$. Her second score is the difference in ratings between DANCE and TV: $10 - 6 = 4$, and so on.

A one-way MANOVA on the segments tests the parallelism hypothesis. Because each segment represents a slope between two original DVs, if there is a multivariate difference between groups then one or more slopes are different and the profiles are not parallel.

Using procedures developed in Chapter 7, the total sum-of-squares and cross-products matrix $(\mathbf{S}_{total})$ is partitioned into the between-groups matrix $(\mathbf{S}_{bg})$ and the within-groups or error matrix $(\mathbf{S}_{wg})$.[5] To produce the within-groups matrix, each person's score matrix, $\mathbf{Y}_{ikm}$, has subtracted from it

TABLE 8.3 Scores for Adjacent Segments for Small-Sample Hypothetical Data

Group		Case No.	Read vs. Dance	Dance vs. TV	TV vs. Ski
				Segment	
Belly dancers		1	−3	4	1
		2	−1	4	−2
		3	−5	5	−3
		4	−4	4	−2
		5	−1	1	−2
	Mean		−2.8	3.6	−1.6
Politicians		6	0	0	0
		7	2	−1	2
		8	0	0	−1
		9	0	0	−1
		10	−1	−1	1
	Mean		0.2	−0.4	0.2
Administrators		11	2	0	−1
		12	2	2	−4
		13	2	0	−3
		14	6	−1	−2
		15	3	0	0
	Mean		3.0	0.2	−2.0
Grand mean			0.13	1.13	−1.13

[5]Other methods of forming **S** matrices can be used to produce the same result.

the mean matrix for that group, $\mathbf{M}_m$. The resulting difference matrix is multiplied by its transpose to create the sum-of-squares and cross-products matrix. For the first belly dancer:

$$(\mathbf{Y}_{111} - \mathbf{M}_1) = \begin{bmatrix} -3 \\ 4 \\ 1 \end{bmatrix} - \begin{bmatrix} -2.8 \\ 3.6 \\ -1.6 \end{bmatrix} = \begin{bmatrix} -0.2 \\ 0.4 \\ 2.6 \end{bmatrix}$$

and

$$(\mathbf{Y}_{111} - \mathbf{M}_1)(\mathbf{Y}_{111} - \mathbf{M}_1)' = \begin{bmatrix} -0.2 \\ 0.4 \\ 2.6 \end{bmatrix} [-0.2 \quad 0.4 \quad 2.6]$$

$$= \begin{bmatrix} 0.04 & -0.08 & -0.52 \\ -0.08 & 0.16 & 1.04 \\ -0.52 & 1.04 & 6.76 \end{bmatrix}$$

This is the sum-of-squares and cross-products matrix for the first case. When these matrices are added over all cases and groups, the result is the error matrix, $\mathbf{S}_{wg}$:

$$\mathbf{S}_{wg} = \begin{bmatrix} 29.6 & -13.2 & 6.4 \\ -13.2 & 15.2 & -6.8 \\ 6.4 & -6.8 & 26.0 \end{bmatrix}$$

To produce the between-groups matrix, $\mathbf{S}_{bg}$, the grand matrix, **GM**, is subtracted from each mean matrix, $\mathbf{M}_k$, to form a difference matrix for each group. The mean matrix for each group in the example is

$$\mathbf{M}_1 = \begin{bmatrix} -2.8 \\ 3.6 \\ -1.6 \end{bmatrix} \quad \mathbf{M}_2 = \begin{bmatrix} 0.2 \\ -0.4 \\ 0.2 \end{bmatrix} \quad \mathbf{M}_3 = \begin{bmatrix} 3.0 \\ 0.2 \\ -2.0 \end{bmatrix}$$

and the grand mean matrix is

$$\mathbf{GM} = \begin{bmatrix} 0.13 \\ 1.13 \\ -1.13 \end{bmatrix}$$

The between-groups sum-of-squares and cross-products matrix $\mathbf{S}_{bg}$ is formed by multiplying each group difference matrix by its transpose, and then adding the three resulting matrices. After multiplying each entry by $n = 5$, to provide for summation over subjects,

$$\mathbf{S}_{bg} = \begin{bmatrix} 84.133 & -50.067 & -5.133 \\ -50.067 & 46.533 & 11.933 \\ -5.133 & -11.933 & 13.733 \end{bmatrix}$$

Wilks' Lambda (Λ) tests the hypothesis of parallelism by evaluating the ratio of the determinant of the within-groups cross-products matrix to the determinant of the matrix formed by the sum of the within- and between-groups cross-products matrices:

$$\Lambda = \frac{|\mathbf{S}_{wg}|}{|\mathbf{S}_{wg} + \mathbf{S}_{bg}|} \tag{8.3}$$

For the example, Wilks' Lambda for testing parallelism is

$$\Lambda = \frac{6325.2826}{6325.2826 + 76598.7334} = .076279$$

By applying the procedures of Section 7.4.1, one finds an approximate $F(6, 20) = 8.74$, $p < .001$, leading to rejection of the hypothesis of parallelism. That is, the three profiles of Figure 8.1 are not parallel. Effect size is measured as partial η^2:[6]

$$\text{partial } \eta^2 = 1 - \Lambda^{1/2} \tag{8.4}$$

For this example, then,

$$\text{partial } \eta^2 = 1 - .076279^{1/2} = .72$$

Seventy-two percent of the variance in the segments as combined for this test is accounted for by the difference in shape of the profiles for the three groups. Confidence limits (per Smithson, 2003) are .33 to .78. Recall from Chapter 7 that segments are combined here to maximize group differences for parallelism. A different combination of segments is used for the test of flatness.

8.4.3 Flatness

Because the hypothesis of parallelism is rejected for this example, the test of flatness is irrelevant; the question of flatness of combined profiles of Figure 8.1 makes no sense because at least one of them (and in this case probably two) is not flat. The flatness test is computed here to conclude the demonstration of this example.

---

[6]Partial η^2 is available through SPSS GLM and through Smithson's (2003) procedures discussed earlier.

Statistically, the test is whether, with groups combined, the three segments of Table 8.3 deviate from zero. That is, if segments are interpreted as slopes in Figure 8.1, are any of the slopes for the combined groups different from zero (nonhorizontal)? The test subtracts a set of hypothesized grand means, representing the null hypothesis, from the matrix of actual grand means:

$$(\mathbf{GM} - 0) = \begin{bmatrix} 0.13 \\ 1.13 \\ -1.13 \end{bmatrix} - \begin{bmatrix} 0 \\ 0 \\ 0 \end{bmatrix} = \begin{bmatrix} 0.13 \\ 1.13 \\ -1.13 \end{bmatrix}$$

The test of flatness is a multivariate generalization of the one-sample t test demonstrated in Chapter 3. Because it is a one-sample test, it is most conveniently evaluated through Hotelling's T^2, or trace:[7]

$$T^2 = N(\mathbf{GM} - 0)'\mathbf{S}_{wg}^{-1}(\mathbf{GM} - 0) \tag{8.5}$$

where N is the total number of cases and $\mathbf{S}_{wg}^{-1}$ is the inverse of the within-groups sum-of-squares and cross-products matrix developed in Section 8.4.2.

For the example:

$$T^2 = (15)[0.13 \quad 1.13 \quad -1.13] \begin{bmatrix} .05517 & .04738 & -.00119 \\ .04738 & .11520 & .01847 \\ -.00119 & .01847 & .04358 \end{bmatrix} \begin{bmatrix} 0.13 \\ 1.13 \\ -1.13 \end{bmatrix}$$

$$= 2.5825$$

From this is found F, with $p - 1$ and $N - k - p + 2$ degrees of freedom, where p is the number of original DVs (in this case 4), and k is the number of groups (3).

$$F = \frac{N - k - p + 2}{p - 1}(T^2) \tag{8.6}$$

so that

$$F = \frac{15 - 3 - 4 + 2}{4 - 1}(2.5825) = 8.608$$

with 3 and 10 degrees of freedom, $p < .01$ and the test shows significant deviation from flatness.

A measure of effect size is found through Hotelling's T^2 that bears a simple relationship to lambda.

$$\Lambda = \frac{1}{1 + T^2} = \frac{1}{1 + 2.5825} = .27913$$

[7]This is sometimes referred to as Hotelling's T.

Lambda, in turn, is used to find η^2 (note that there is no difference between η^2 and partial η^2 because $s = 1$):

$$\eta^2 = 1 - \Lambda = 1 - .27913 = .72$$

showing that 72% of the variance in this combination of segments is accounted for by nonflatness of the profile collapsed over groups. Smithson's (2003) procedure finds confidence limits from .15 to .81.

8.4.4 Computer Analyses of Small-Sample Example

Tables 8.4 through 8.6 show syntax and selected output for computer analyses of the data in Table 8.1. Table 8.4 illustrates SPSS MANOVA with brief output. SPSS GLM is illustrated in Table 8.5. Table 8.6 demonstrates profile analysis through SAS GLM, with short printout requested. All programs are set up as repeated-measures ANOVA, which automatically produces both univariate and multivariate results.

The three programs differ substantially in syntax and presentation of the three tests. To set up SPSS MANOVA for profile analysis, the DVs (levels of the within-subject effect) READ TO SKI are followed in the MANOVA statement by the keyword BY and the grouping variable with its

TABLE 8.4 Profile Analysis of Small-Sample Example through SPSS MANOVA (Syntax and Partial Output)

```
MANOVA   READ TO SKI BY OCCUP(1,3)
            /WSFACTOR=ACTIVITY(4)
            /WSDESIGN=ACTIVITY
            /PRINT=SIGNIF(BRIEF)
            /DESIGN.
```

Tests of Between-Subjects Effects.

Tests of Significance for T1 using UNIQUE sums of squares

Source of Variation	SS	DF	MS	F	Sig of F
WITHIN+RESIDUAL	23.50	12	1.96		
occup	172.90	2	86.45	44.14	.000

Multivariate Tests of Significance

Tests using UNIQUE sums of squares and WITHIN+RESIDUAL error term

Source of Variation	Wilks	Approx F	Hyp. DF	Error DF	Sig of F
ACTIVITY	.279	8.608	3.00	10.000	.004
occup BY ACTIVITY	.076	8.736	6.00	20.000	.000

levels—OCCUP(1,3). The DVs are combined for profile analysis in the WSFACTOR instruction and labeled ACTIVITY(4) to indicate four levels for the within-subjects factor.

In the SPSS MANOVA output, the levels test for differences among groups is the test of OCCUP in the section labeled Tests of Significance for T1.... This is followed by information about tests and adjustments for sphericity (not shown, cf. Section 8.5.1). The flatness and parallelism tests appear in the section labeled Tests using UNIQUE sums of squares and WITHIN+RESIDUAL for ACTIVITY and OCCUP by ACTIVITY, respectively. Output is limited in this example by the PRINT=SIGNIF(BRIEF) instruction. Without this statement, separate source tables are printed for flatness and parallelism, each containing several multivariate tests (demonstrated in Section 8.6.1). Univariate tests for repeated-measures factors (ACTIVITY and OCCUP by ACTIVITY) as well as tests for sphericity are also printed, but omitted here.

SPSS GLM has a similar setup for repeated-measures ANOVA. The EMMEANS instructions request tables of adjusted means as parameter estimates.

Multivariate tests for all four criteria of Section 7.5.2 are shown for ACTIVITY (flatness) and ACTIVITY by OCCUP (parallelism). The test for levels, OCCUP, is shown in the section labeled Tests of Between-Subjects Effects. Parameter estimate tables for the three effects include means, standard errors, and 95% confidence intervals. Information on univariate tests of OCCUP and OCCUP by ACTIVITY, including trend analysis and sphericity tests, has been omitted here.

In SAS GLM, the class instruction identifies OCCUP as the grouping variable. The model instruction shows the DVs on the left of an equation and the IV on the right. Profile analysis is distinguished from ordinary MANOVA by the instructions in the line beginning repeated, as seen in Table 8.6.

The results are presented in two multivariate tables and a univariate table. The first table, labeled ...Hypothesis of no ACTIVITY Effect, shows four fully-labeled multivariate tests of flatness. The second table shows the same four multivariate tests of parallelism, labeled ...Hypothesis of no ACTIVITY*OCCUP Effect. The test of levels is the test for OCCUP in the third table, labeled Test of Hypotheses for Between Subjects Effects. Univariate tests of ACTIVITY and OCCUP by ACTIVITY are omitted here. More extensive output is available if the SHORT instruction is omitted.

TABLE 8.5   Profile Analysis of Small-Sample Example through SPSS GLM (Syntax and Output)

```
GLM
  read dance tv ski BY occup
  /WSFACTOR = activity 4 Polynomial
  /MEASURE = rating
  /METHOD = SSTYPE(3)
  /EMMEANS = TABLES(occup)
  /EMMEANS = TABLES(activity)
  /EMMEANS = TABLES(occup*activity)
  /CRITERIA = ALPHA(.05)
  /WSDESIGN = activity
  /DESIGN = occup.
```

TABLE 8.5 Continued

General Linear Model

Multivariate Tests[c]

Effect		Value	F	Hypothesis df	Error df	Sig.
activity	Pillai's Trace	.721	8.608[a]	3.000	10.000	.004
	Wilks' Lambda	.279	8.608[a]	3.000	10.000	.004
	Hotelling's Trace	2.582	8.608[a]	3.000	10.000	.004
	Roy's Largest Root	2.582	8.608[a]	3.000	10.000	.004
activity * occup	Pillai's Trace	1.433	9.276	6.000	22.000	.000
	Wilks' Lambda	.076	8.736[a]	6.000	20.000	.000
	Hotelling's Trace	5.428	8.142	6.000	18.000	.000
	Roy's Largest Root	3.541	12.982[b]	3.000	11.000	.001

[a] Exact statistic

[b] The statistic is an upper bound on F that yields a lower bound on the significance level.

[c]
Design: Intercept+occup
Within Subjects Design: activity

Tests of Between-Subjects Effects

Measure: rating
Transformed Variable: Average

Source	Type III Sum of Squares	df	Mean Square	F	Sig.
Intercept	1591.350	1	1591.350	812.604	.000
occup	172.900	2	86.450	44.145	.000
Error	23.500	12	1.958		

(continued)

TABLE 8.5   Continued

Estimated Marginal Means

1. Occupation

Measure: rating

Occupation	Mean	Std. Error	95% Confidence Interval	
			Lower Bound	Upper Bound
Belly dancer	7.300	.313	6.618	7.982
Politician	5.000	.313	4.318	5.682
Administrator	3.150	.313	2.468	3.832

2. activity

Measure: rating

activity	Mean	Std. Error	95% Confidence Interval	
			Lower Bound	Upper Bound
1	5.533	.327	4.822	6.245
2	5.400	.236	4.886	5.914
3	4.267	.216	3.796	4.737
4	5.400	.380	4.572	6.228

3. Occupation * activity

Measure: rating

Occupation	activity	Mean	Std. Error	95% Confidence Interval	
				Lower Bound	Upper Bound
Belly dancer	1	6.600	.566	5.367	7.833
	2	9.400	.408	8.511	10.289
	3	5.800	.374	4.985	6.615
	4	7.400	.658	5.966	8.834
Politician	1	5.000	.566	3.767	6.233
	2	4.800	.408	3.911	5.689
	3	5.200	.374	4.385	6.015
	4	5.000	.658	3.566	6.434
Administrator	1	5.000	.566	3.767	6.233
	2	2.000	.408	1.111	2.889
	3	1.800	.374	.985	2.615
	4	3.800	.658	2.366	5.234

TABLE 8.6 Profile Analysis of Small-Sample Example through SAS GLM (Syntax and Selected Output)

```
proc glm data=SASUSER.SSPROFIL;
  class OCCUP;
  model READ DANCE TV SKI = OCCUP/NOUNI;
  repeated ACTIVITY 4 profile/short;
run;
```

Manova Test Criteria and Exact F Statistics for the Hypothesis of no ACTIVITY Effect
H = Type III SSCP Matrix for ACTIVITY
E = Error SSCP Matrix

S=1   M=0.5   N=4

Statistic	Value	F Value	Num DF	Den DF	Pr > F
Wilks' Lambda	0.27913735	8.61	3	10	0.0040
Pillai's Trace	0.72086265	8.61	3	10	0.0040
Hotelling-Lawley Trace	2.58246571	8.61	3	10	0.0040
Roy's Greatest Root	2.58246571	8.61	3	10	0.0040

(continued)

TABLE 8.6 Continued

Manova Test Criteria and F Approximations for the Hypothesis of no ACTIVITY*OCCUP Effect
H = Type III SSCP Matrix for ACTIVITY*OCCUP
E = Error SSCP Matrix

S=2   M=0   N=4

Statistic	Value	F Value	Num DF	Den DF	Pr > F
Wilks' Lambda	0.07627855	8.74	6	20	<.0001
Pillai's Trace	1.43341443	9.28	6	22	<.0001
Hotelling-Lawley Trace	5.42784967	8.73	6	11.714	0.0009
Roy's Greatest Root	3.54059987	12.98	3	11	0.0006

NOTE: F Statistic for Roy's Greatest Root is an upper bound.
NOTE: F Statistic for Wilks' Lambda is exact.

The GLM Procedure
Repeated Measures Analysis of Variance
Tests of Hypotheses for Between Subjects Effects

Source	DF	Type III SS	Mean Square	F Value	Pr > F
OCCUP	2	172.9000000	86.4500000	44.14	<.0001
Error	12	23.5000000	1.9583333		

8.5   Some Important Issues

Issues discussed here are unique to profile analysis, or at least they affect profile analysis differently from traditional MANOVA. Issues such as choice among statistical criteria, for instance, are identical whether the DVs are analyzed directly (as in MANOVA) or converted into segments or some other transformation (as in profile analysis). Therefore, the reader is referred to Section 7.5 for consideration of these matters.

8.5.1   Univariate vs. Multivariate Approach
to Repeated Measures

Research where the same cases are repeatedly measured with the same instrument is common in many sciences. Longitudinal or developmental studies, research that requires follow-up, studies where changes in time are of interest—all involve repeated measurement. Further, many studies of short-term phenomena have repeated measurement of the same subjects under several experimental conditions, resulting in an economical research design.

When there are repeated measures, a variety of analytical strategies are available, each with advantages and disadvantages. Choice among the strategies depends upon details of research design and conformity between the data and the assumptions of analysis.

Univariate repeated-measures ANOVA with more than 1 df for the repeated-measure IV requires sphericity. Although the test for homogeneity of covariance, a component of sphericity, is fairly complicated, the notion is conceptually simple. All pairs of levels of the within-subjects variable need to have equivalent correlations. For example, consider a longitudinal study in which children are measured yearly from ages 5 to 10. If there is homogeneity of covariance, the correlation between scores on the DV for ages 5 and 6 should be about the same as the correlation between scores between ages 5 and 7, or 5 and 8, or 6 and 10, etc. In applications like these, however, the assumption is almost surely violated. Things measured closer in time tend to be more highly correlated than things measured farther away in time; the correlation between scores measured at ages 5 and 6 is likely to be much higher than the correlation between scores measured at ages 5 and 10. Thus, whenever time is a within-subjects IV, the assumption of homogeneity of covariance is likely to be violated, leading to increased Type I error. Both packages routinely provide information about sphericity directly in their output: SPSS GLM and MANOVA each show a sphericity test for the significance of departure from the assumption. The issue is moot when there are only two levels of the within-subjects IV. In that case, sphericity is not an issue and univariate results match multivariate results.

If there is violation of sphericity, several alternatives are available, as also discussed in Section 6.5.4.1. The first is to use one of the significance tests that is adjusted for violation of the assumption, such as Greenhouse-Geisser or Huynh-Feldt.[8] In all applicable programs in the three packages both Greenhouse-Geisser (G-G) and Huynh-Feldt (H-F) values are provided, along with adjusted significance levels.

---

[8]See Keppel and Wickens (2004, pp. 378–379) for a discussion of the differences between the two types of adjustments (referring to the Huynh-Feldt procedure as the Box correction). Even greater insights are available through consultation with the original sources: Greenhouse and Geisser (1959) as well as Huynh and Feldt (1976).

A second strategy, available through all three programs, is a more stringent adjustment of the statistical criterion leading to a more honest Type I error rate, but lower power. This strategy has the advantage of simplicity of interpretation (because familiar main effects and interactions are evaluated) and simplicity of decision-making (you decide on one of the strategies before performing the analysis and then take your chances with respect to power).

For all of the multivariate programs, however, results of profile analysis are also printed out, and you have availed yourself of the third strategy, whether you meant to or not. Profile analysis, called the multivariate approach to repeated measures, is a statistically acceptable alternative to repeated-measures ANOVA. Other requirements such as homogeneity of variance-covariance matrices and absence of multicollinearity and singularity must be met, but they are less likely to be violated.

Profile analysis requires more cases than univariate repeated-measures ANOVA—more cases than DVs in the smallest group. If the sample is too small, the choice between multivariate and univariate approaches is automatically resolved in favor of the univariate approach, with adjustment for failure of sphericity, as necessary.

Sometimes, however, the choice is not so simple and you find yourself with two sets of results. If the conclusions from both sets of results are the same, it often is easier to report the univariate solution, while noting that the multivariate solution is similar. But if conclusions differ between the two sets of results, you have a dilemma. Choice between conflicting results requires attention to the details of the research design. Clean, counterbalanced experimental designs "fit" better within the univariate model, while nonexperimental or contaminated designs often require the multivariate model that is more forgiving statistically, but more ambiguous to interpret.

The best solution, the fourth alternative, often is to perform trend analysis (or some other set of single df contrasts) instead of either profile analysis or repeated-measures ANOVA if it makes conceptual sense within the context of the research design. Many longitudinal, follow-up, and other time-related studies lend themselves beautifully to interpretation in terms of trends. Because statistical tests of trends and other contrasts each use a single degree of freedom of the within-subjects IV, there is no possibility of violation of sphericity. Furthermore, none of the assumptions of the multivariate approach needs be met. SPSS GLM automatically prints out a full trend analysis for a repeated-measures analysis.

A fifth alternative is straightforward MANOVA where DVs are treated directly (cf. Chapter 7), without transformation. The design becomes a one-way between-subjects analysis of the grouping variable with the repeated measures used simply as multiple DVs. There are two problems with conversion of repeated measures to MANOVA. First, because the design is now one-way between-subjects, MANOVA does not produce the interaction (parallelism) test most often of interest in a repeated-measures design. Second, MANOVA allows a Roy-Bargmann stepdown analysis, but not a trend analysis of DVs after finding a multivariate effect.

A final alternative is to use multilevel modeling (Chapter 15) in which the repeated measures form the first level in a hierarchical analysis. This approach, although fraught with complexities of random effects and maximum likelihood analysis, is highly flexible in dealing with missing data and varying intervals between measurements. Sphericity is not an issue because each analysis deals with only one comparison, e.g., linear trend.

In summary then, if the levels of the IV differ along a single dimension such as time or dosage and trend analysis makes sense, use it. Or, if the design is a clean experiment where cases have been randomly assigned to treatment and there are expected to be no carryover effects, the univariate repeated-measures approach is probably justified. (But just to be on the safe side, use a program that tests and adjusts for violation of sphericity.) If, however, the levels of the IV do not vary along a single dimension but violation of sphericity is likely, and if there are lots more cases than DVs, it is probably a good idea to choose either profile analysis or MANOVA.

8.5.2   Contrasts in Profile Analysis

When there are more than two levels of a significant effect in profile analysis, it is often desirable to perform contrasts to pinpoint sources of variability. For instance, because there is an overall difference between administrators, belly dancers, and politicians in their ratings of satisfaction with leisure time activities (see Section 8.4), contrasts are needed to discover which groups differ from which other groups. Are belly dancers the same as administrators? Politicians? Neither?

It is probably easiest to think of contrasts following profile analysis as coming from a regular ANOVA design with (at least) one grouping variable and one repeated measure, even when the application of the technique is to multiple, commensurate DVs. That is, the most interpretable contrasts following profile analysis are likely to be the ones that would also be appropriate after a mixed within- and between-subjects ANOVA.

There are, of course, numerous contrast procedures and the choice among them depends on what makes most sense in a given research setting. With a single control group, Dunnett's procedure often makes most sense. Or if all pairwise comparisons are desired, the Tukey test is most appropriate. Or if there are numerous repeated measures and/or normative data are available, a confidence interval procedure, such as that used in Section 8.6.1, may make the most sense. With relatively few repeated measures, a Scheffé type procedure is probably the most general (if also the most conservative) and is the procedure illustrated in this section.

It is important to remember that the contrasts recommended here explore differences in original DV scores while the significance tests in profile analysis for parallelism and flatness typically evaluate segments. Although there is a logical problem with following up a significance test based on segments with a contrast based on the original scores, performing contrasts on segments or some other transformation of the variables seems even worse because of difficulty in interpreting the results.

Contrasts in repeated-measures ANOVA with both grouping variables and repeated measures is not the easiest of topics, as you probably recall. First, when parallelism (interaction) is significant, there is the choice between a simple-effects analysis and an interaction-contrasts analysis. Second, there is a need in some cases to develop separate error terms for some of the contrasts. Third is the need to apply an adjustment such as Scheffé to the F test to avoid too liberal a rejection of the null hypothesis. The researcher who is fascinated by these topics is referred to Tabachnick and Fidell (2007) for a detailed discussion of them. The present effort is to illustrate several possible approaches and to recommend guidelines for when each is likely to be appropriate.

The most appropriate contrast to perform depends on which effect or combination of effects—levels, flatness, or parallelism—is significant. If either levels or flatness is significant, but parallelism (interaction) is not, contrasts are performed on marginal means. If the test for levels is significant, contrasts are formed on marginal values for the grouping variable. If the test for flatness is significant, contrasts are formed on the repeated-measures marginal values. Because contrasts formed on marginal values "fall out" of computer runs for interaction contrasts, they are illustrated in Section 8.5.2.3.

Sections 8.5.2.1 and 8.5.2.2 describe simple-effects analyses, appropriate if parallelism is significant. In simple-effects analysis, one variable is held constant at some value while mean differences are examined on the levels of the other variable, as seen in Figure 8.2. For instance, the level

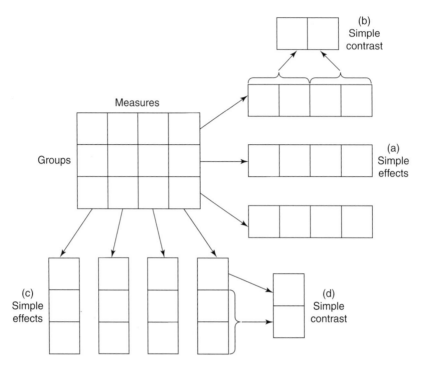

FIGURE 8.2 **Simple-effects analysis exploring: (a) differences among measures for each group, followed by (b) a simple contrast between measures for one group: and (c) differences among groups for each measure, followed by (d) a simple contrast between groups for one measure.**

of group is held constant at belly dancer while mean differences are examined among the leisure time activities [Figure 8.2(a)]. The researcher asks if belly dancers have mean differences in satisfaction with different leisure activities. Or leisure activity is held constant at dance while mean differences are explored between administrators, politicians, and belly dancers [Figure 8.2(c)]. The researcher asks whether the three groups have different mean satisfaction while dancing.

Section 8.5.2.1 illustrates a simple-effects analysis followed by simple contrasts [Figures 8.2(c) and (d)] for the case where parallelism and flatness effects are both significant, but the levels effect is not. Section 8.5.2.2 illustrates a simple-effects analysis followed by simple contrasts [Figures 8.2(a) and (b)] for the case where parallelism and levels are both significant, but the flatness effect is not. This particular pattern of simple-effects analysis is recommended because of the confounding inherent in analyzing simple effects.

The analysis is confounded because when the groups (levels) effect is held constant to analyze the repeated measure in a one-way within-subjects ANOVA, both the sum of squares for interaction and the sum of squares for the repeated measure are partitioned. When the repeated measure is held constant so the groups (levels) effect is analyzed in a one-way between-subjects ANOVA, both the sum of squares for interaction and the sum of squares for the group effect are partitioned. Because in simple-effects analyses the interaction sum of squares is confounded with one or the other of the main effects, it seems best to confound it with a nonsignificant main effect where possible. This recommendation is followed in Sections 8.5.2.1 and 8.5.2.2.

Measures

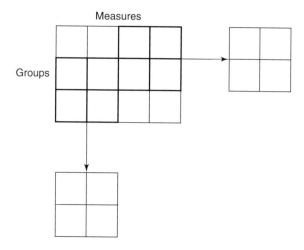

Groups

**FIGURE 8.3   Interaction-contrasts analyses exploring
small (2 × 2) interactions formed by partitioning a large
(3 × 4) interaction.**

Section 8.5.2.3 describes an interaction-contrasts analysis. In such an analysis, an interaction between two IVs is examined through one or more smaller interactions (Figure 8.3). For instance, the significant interaction between the three groups on four leisure activities in the example might be reduced to examination of the difference between two groups on only two of the activities. One could, for instance, ask if there is a significant interaction in satisfaction between belly dancers and administrators while watching TV vs. dancing. Or one could pool the results for administrator and politician and contrast them with belly dancer for one side of the interaction, while pooling the results for the two sedentary activities (watching TV and reading) against the results for the two active activities (dancing and skiing) as the other side of the interaction. The researcher asks whether there is an interaction between dancers and the other professionals in their satisfaction while engaged in sedentary vs. active leisure time activities.

An interaction-contrasts analysis is not a confounded analysis; only the sum of squares for interaction is partitioned. Thus, it is appropriate whenever the interaction is significant and regardless of the significance of the other two effects. However, because simple effects are generally easier to understand and explain, it seems better to perform them when possible. For this reason, we recommend an interaction-contrasts analysis to explore the parallelism effect only when both the levels and flatness effects are also significant.

8.5.2.1   *Parallelism and Flatness Significant, Levels Not Significant (Simple-effects Analysis)*

When parallelism and flatness are both significant, a simple-effects analysis is recommended where differences among means for groups are examined separately at each level of the repeated measure [Figure 8.2(c)]. For the example, differences in means among politicians, administrators, and belly dancers are sought first in the reading variable, then in dance, then in TV, and finally in skiing. (Not all these effects need to be examined, of course, if they are not of interest.)

Table 8.7 shows syntax and location of output for SPSS ONEWAY, GLM, and MANOVA, and SAS GLM for performing simple-effects analysis on groups with repeated measures held constant.

TABLE 8.7 Syntax for Simple-Effects Analysis of Occupation, Holding Activity Constant

Program	Syntax	Section of Output	Name of Effect
SPSS Compare Means	ONEWAY dance BY occup /MISSING ANALYSIS.	**ANOVA**	Between Groups

SPSS MANOVA	MANOVA read TO ski BY occup(1,3)
	    /WSFACTOR=activity(4)
	    /WSDESIGN=MWITHIN activity(1)
	    MWITHIN activity(2)
	    MWITHIN activity(3)
	    MWITHIN activity(4)
	    /RENAME=READ, DANCE, TV, SKI
	    /DESIGN.
		Tests involving `'MWITHIN ACTIVITY(2)..` etc.	`occup BY MWITHIN ACTIVITY(2),` etc.

SAS GLM	`proc glm data=SASUSER.SSPROFIL;`
	    `class OCCUP;`
	    `model DANCE = OCCUP;`
	`run;`
		`Dependent Variable: DANCE Source Type III SS`	`OCCUP`

The syntax of Table 8.7 shows simple effects of occupation only for one DV, DANCE (except for SPSS MANOVA, which is set up to do all simple effects at once). Parallel syntax produces simple effects for the other DVs. To evaluate the significance of the simple effects, a Scheffé adjustment (see Section 3.2.6) is applied to unadjusted critical F under the assumptions that these tests are performed post hoc and that the researcher wants to control for familywise Type I error. For these contrasts the Scheffé adjustment is

$$F_s = (k - 1)F_{(k-1),k(n-1)} \tag{8.7}$$

where k is the number of groups and n is the number of subjects in each group. For the example, using $\alpha = 0.05$

$$F_s = (3 - 1)F_{(2,12)} = 7.76$$

By this criterion, there are not statistically significant mean differences between the groups when the DV is READ, but there are statistically significant differences when the DV is DANCE, TV, or SKI.

Because there are three groups, these findings are still ambiguous. Which group or groups are different from which other group or groups? To pursue the analysis, simple contrasts are performed [Figure 8.2(d)]. Contrast coefficients are applied to the levels of the grouping variable to determine the source of the difference. For the example, contrast coefficients compare the mean for belly dancers with the mean for the other two groups combined for DANCE. Syntax and location of output from the three programs are shown in Table 8.8.

TABLE 8.8 Syntax for Simple Comparisons on Occupations, Holding Activity Constant

Program	Syntax	Section of Output	Name of Effect
SPSS Compare Means	ONEWAY dance BY occup /CONTRAST= 2 –1 –1 /MISSING ANALYSIS.	**Contrast Tests**	Assume equal variances[a]

SPSS MANOVA	MANOVA dance BY occup(1,3)
	   /PARTITION(occup)
	   /CONTRAST(occup)=SPECIAL (1 1 1,
	                                2 –1 –1,
	                                0 1 –1)
	
	   /DESIGN=occup(1).
		Tests of Significance for dance using...	OCCUP(1)

SAS GLM	proc glm data=SASUSER.SSPROFIL;
	    class OCCUP;
	    model DANCE = OCCUP;
	    contrast 'BD VS. OTHERS' OCCUP 2 –1 –1;
	run;
		Dependent Variable: DANCE
		Contrast
			BD VS. OTHERS

[a]t is given rather than F; recall that $t^2 = F$.

The CONTRAST procedure is used for both programs. For this analysis, the sum of squares and mean square for the contrast is 120.000, error mean square is .83333, and F is 144.00 (with $t = 12.00$). This F exceeds the F_s adjusted critical value of 7.76; it is no surprise to find there is a statistically significant difference between belly dancers and others in their satisfaction while engaging in DANCE.

8.5.2.2   *Parallelism and Levels Significant, Flatness Not Significant (Simple-effects Analysis)*

This combination of findings occurs rarely because if parallelism and levels are significant, flatness is nonsignificant only if profiles for different groups are mirror images that cancel each other out.

The simple-effects analysis recommended here examines mean differences among the various DVs in series of one-way within-subjects ANOVAs with each group in turn held constant [Figure 8.2(a)]. For the example, mean differences between READ, DANCE, TV, and SKI are sought first for belly dancers, then for politicians, and then for administrators. The researcher inquires whether each group, in turn, is more satisfied during some activities than during others.

Table 8.9 shows the syntax and location of output for Belly Dancers (OCCUP = 1) for the three programs.

SPSS GLM requires separate runs for each occupation. SAS GLM and SPSS MANOVA permit analyses by OCCUP, so that results are printed out for simple effects for all levels of occupation at one time.

For these simple effects, the Scheffé adjustment to critical F is

$$F_s = (p - 1)F_{(p-1),k(p-1)(n-1)} \tag{8.8}$$

where p is the number of repeated measures, n is the number of subjects in each group, and k is the number of groups. For the example

$$F_s = (4 - 1)F_{(3, 36)} = 8.76$$

The F value for simple effects of activity for belly dancers (7.66) does not exceed adjusted critical F in the output of SPSS and SAS GLM, which use an error term based only on a single OCCUP (df = 12). However, SPSS MANOVA uses an error term based on all occupations (df = 36), producing $F = 10.71$, leading to ambiguous conclusions. In any event, the effect is significant as a planned comparison, or probably with a less stringent adjustment for family Type I error rate.

A statistically significant finding is also ambiguous in this case because there are more than two activities. Contrast coefficients are therefore applied to the levels of the repeated measure to examine the pattern of differences in greater detail [Figure 8.2(b)].

Table 8.10 shows syntax and location of output for a simple contrast for the three programs. The contrast that is illustrated compares the pooled mean for the two sedentary activities (READ and TV) against the pooled mean for the two active activities (DANCE and SKI) for (you guessed it) belly dancers.

The F value of 15.365 produced by SAS and SPSS GLM exceeds F_s of 8.76 as well as unadjusted critical F and indicates that belly dancers have statistically significant mean differences in their satisfaction during active vs. sedentary activities. The F value of 16.98 produced by SPSS MANOVA also exceeds critical F. The difference in F values again is produced by different error

TABLE 8.9 Syntax for Simple Effects of Activity, Holding Occupation Constant

Program	Syntax	Section of Output	Name of Effect

SPSS GLM	SELECT IF (occup = 1).
	GLM
	
	  read dance tv ski
	  /WSFACTOR = activity 4 Polynomial
	  /METHOD = SSTYPE(3)
	  /CRITERIA = ALPHA(.05)
	  /WSDESIGN = activity.
		Tests of Within-Subjects Effects	ACTIVITY

SPSS MANOVA	MANOVA read TO ski BY occup(1,3)
	  /WSFACTOR=activity(4)
	  /PRINT=SIGNIF(BRIEF)
	  /DESIGN=MWITHIN occup(1), MWITHIN occup(2),
	               MWITHIN occup(3).
		AVERAGED Tests
		of Signficance
		for MEAS.1
		using ....
			MWITHIN
			OCCUP(1),
			etc.

SAS GLM	proc glm data=SASUSER.SSPROFIL;
	  by OCCUP;
	  class OCCUP;
	  model READ DANCE TV SKI = /nouni;
	  repeated ACTIVITY 4 /short;
	run;
		OCCUP=1
		Source
			ACTIVITY

337

TABLE 8.10   Syntax for Simple Comparisons on Activity, Holding Occupation Constant

Program	Syntax	Section of Output	Name of Effect
SPSS GLM	SELECT IF (occup = 1). GLM read dance tv ski /WSFACTOR activity 4 SPECIAL(1 1 1 1 -1 1 -1 1 -1 0 1 0 0 -1 0 1) /METHOD = SSTYPE(3) /CRITERIA = ALPHA(.05) /WSDESIGN = activity.	**Tests of Within- Subjects Contrasts**	activity L1
SPSS MANOVA	MANOVA read TO ski BY occup(1,3) /WSFACTOR=activity(4) /PARTITION(activity) /CONTRAST(activity)=SPECIAL (1 1 1 1, -1 1 -1 1, -1 0 1 0, 0 -1 0 1) /WSDESIGN=activity(1) /RENAME=overall, sedvsact, readvstv, danvssk/ /PRINT=SIGNIF(BRIEF) /DESIGN=MWITHIN occup(1).	Tests of Significance for SEDVSACT using. . .	MWITHIN OCCUP (1) BY ACTIVITY (1)
SAS GLM	proc glm data=SASUSER.SSPROFIL; where OCCUP=1: model READ DANCE TV SKI =; manova m = –1*READ + 1*DANCE – 1*TV + 1*ski H=INTERCEPT; run;	MANOVA Test...no Overall Intercept Effect	Wilks' Lambda

terms. Only SPSS MANOVA uses an error term based on all three occupations, with $df_{error} = 12$ rather than 4.

8.5.2.3 *Parallelism, Levels, and Flatness Significant (Interaction Contrasts)*

When all three effects are significant, an interaction-contrasts analysis is often most appropriate. This analysis partitions the sum of squares for interaction into a series of smaller interactions (Figure 8.3). Smaller interactions are obtained by deleting or combining groups or measures with use of appropriate contrast coefficients.

For the example, illustrated in Table 8.11, the means for administrators and politicians are combined and compared with the mean of belly dancers, while the means for TV and READ are combined and compared with the combined mean of DANCE and SKI. The researcher asks whether belly dancers and others have the same pattern of satisfaction during sedentary vs. active leisure activities.

The F value for the contrast is 15.37 for SPSS and SAS GLM and 15.21 for SPSS MANOVA due to minor differences in algorithms.

Interaction contrasts also need Scheffé adjustment to critical F to hold down the rate of familywise error. For an interaction, the Scheffé adjustment is

$$F_s = (p - 1)(k - 1)F_{(p-1)(k-1),k(p-1)(n-1)} \tag{8.9}$$

where p is the number of repeated measures, k is the number of groups, and n is the number of subjects in each group. For the example

$$F_s = (4 - 1)(3 - 1)F_{(6, 36)} = 14.52$$

Because the F value for the interaction contrast exceeds F_s, there is an interaction between belly dancers vs. others in their satisfaction during sedentary vs. active leisure time activities. A look at the means in Table 8.1 reveals that belly dancers favor active leisure time activities to a greater extent than others.

8.5.2.4 *Only Parallelism Significant*

If the only significance is in the interaction of groups with repeated measures, any of the analyses in Section 8.5.2 is appropriate. The decision between simple-effects analysis and interaction contrasts is based on which is more informative and easier to explain. Writers and readers seem likely to have an easier time explaining results of procedures in Section 8.5.2.2.[9]

8.5.3 Doubly-Multivariate Designs

In a doubly-multivariate design, noncommensurate DVs are repeatedly measured. For example, children in classrooms with either traditional or computer assisted instruction are measured at several

---

[9]If you managed to read this far, go have a beer.

TABLE 8.11 **Syntax for Interaction Contrasts, Belly Dancers *vs.* Others and Active *vs.* Sedentary Activities**

Program	Syntax	Location of Output	Name of Effect

SPSS GLM	```GLM```
	 ``  read dance tv ski BY occup``
	 ``  /METHOD = SSTYPE(3)``
	 ``  /CRITERIA = ALPHA(.05)``
	 ``  /INTERCEPT = INCLUDE``
	 ``  /DESIGN = occup``
	 ``  /MMATRIX = "sed vs. act" read -1 dance 1 tv -1 ski 1``
	 ``  /LMATRIX = "bd vs. other" occup 2 -1 -1.``
		Test Results	Contrast

SPSS MANOVA	```MANOVA read TO ski BY occup(1,3)```
	 ``  /WSFACTOR=activity(4)``
	 ``  /WSDESIGN=activity``
	 ``  /CONTRAST(activity)=SPECIAL (1 1 1 1,``
	 ``                               -1 1-1 1,``
	 ``                               -1 0 1 0,``
	 ``                               0-1 0 1)/``
	
	 ``  /WSDESIGN=activity(1)/``
	 ``  /RENAME=OVERALL, SEDVSACT, READVSTV, DANVSSK/``
	 ``  /PARTITION(occup)/``
	 ``  /CONTRAST(occup)=SPECIAL (1 1 1,``
	 ``                              2-1-1,``
	 ``                              0 1-1)/``
	
	 ``  /PRINT=SIGNIF(BRIEF)/ERROR=WITHIN/``
	 ``  /DESIGN=occup(1) VS WITHIN.``
		Tests involving 'ACTIVITY(1)' Within-Subject Effect	OCCUP(1) BY ACTIVITY(1)

SAS GLM	```proc glm data=SASUSER.SSPROFIL;```
	 ``  class OCCUP;``
	 ``  model READ DANCE TV SKI = OCCUP;``
	 ``  contrast 'BD VS. OTHERS' OCCUP 2 -1 -1;``
	 ``  manova m = -1*READ + 1*DANCE -1*TV + 1*SKI;``
	 ``run;``
		Manova Test .... no Overall BD VS. OTHERS Effect	Wilks' Lambda[a]

[a] This is a multivariate test, but produces the same *F* ratio and df as other programs.

points over the semester on reading achievement, general information, and math achievement. There are two ways to conceptualize the analysis. If treated in a singly multivariate fashion, this is a between-within design (groups by time) with multiple DVs. The time effect, however, has the assumption of sphericity. To circumvent the assumption, the analysis becomes doubly multivariate where both the within-subjects part of the design and the multiple DVs are analyzed multivariately. The between-subject effect is singly multivariate; the within-subject effects and interactions are doubly multivariate.

The number of cases needed is determined by the between-subjects effect, using the same criteria as in MANOVA (Section 7.3.2.1). Because of the within-subjects IV, however, it is probably wise to have a few additional subjects in each group, especially if there is reason to suspect heterogeneity of the variance-covariance matrices.

For SPSS GLM and MANOVA, the procedure is not difficult. Syntax for the two programs is similar, but output looks quite different (a complete example of a doubly-multivariate analysis through SPSS MANOVA is in Section 8.6.2). The SAS manual has an example of a doubly-multivariate design in the GLM chapter.

Consider a study with repeatedly measured noncommensurate DVs. The between-subjects (levels) IV is three weight-loss programs (PROGRAM): a control group (CONTROL), a group that diets (DIET), and a group that both diets and exercises (DIET-EX). The major DV is weight loss (WTLOSS) and a secondary DV is self-esteem (ESTEEM). The DVs are measured at the end of the first, second, and third months of treatment. The within-subject IV (flatness) treated multivariately, then, is MONTH that the measures are taken. That is, the commensurate DVs are MONTH1, MONTH2, and MONTH3.

Table 8.12 shows syntax and location of output for both multivariate and univariate tests of effects in SPSS GLM and MANOVA, as well as SAS GLM. Syntax is fairly simple through the SPSS programs; SAS requires special syntax in the form of matrices or combinations of DVs for each of the three effects: parallelism, flatness, and levels. SPSS GLM sets up the repeated-measures effects as a univariate trend analysis for each DV by default. The syntax in Table 8.12 requests univariate (and for SPSS MANOVA stepdown) trend analysis for each DV for the remaining programs as well. Other options for coding univariate effects may be used if trend analysis is inappropriate for the within-subjects IV and interaction. Decomposing univariate effects into trend analysis or other specific comparisons avoids the need to assume sphericity.

All three programs provide identical multivariate tests of the three effects: doubly multivariate for parallelism and flatness and singly multivariate for levels. All programs also show a full trend analysis of the flatness (trend of marginal means of month) and parallelism (trend of month by program) effects for each DV using the syntax of Table 8.12. SPSS GLM shows cell and marginal means adjusted for unequal n (but not for stepdown analysis). SAS shows adjusted cell means, but marginal means must be found by averaging cell means.

SPSS MANOVA provides stepdown analysis, as well as univariate tests; for the levels effect (program) as well as the trend analysis for the flatness and parallelism effects. Means adjusted for unequal n and/or stepdown analysis require separate CONSPLUS runs, as per Section 7.6.

Separate runs are required for each DV, except the first, to create a stepdown analysis, if any program other than SPSS MANOVA is used. This is done by declaring the higher priority DV(s) to be covariates in a mixed within-between ANOVA. Because the covariate as well as the DV is measured at each time period, this is the case of a covariate that varies over levels of the within-subjects IV (see Section 6.5.4.1). However, as seen in Table 6.8, this requires a rearrangement of the data set so that there are as many lines per case as there are levels of the within-subjects IV.

TABLE 8.12 Syntax and Location of Output for Doubly-Multivariate ANOVA

a. MULTIVARIATE EFFECTS

Program	Syntax	Parallelism		Flatness		Levels	
		Section of Output	Name of Effect	Section of Output	Name of Effect	Section of Output	Name of Effect
SPSS GLM	GLM wtloss1 wtloss2 wtloss3 esteem1 esteem2 esteem3 BY program /WSFACTOR = month 3 Polynomial /MEASURE = wtloss esteem /METHOD = SSTYPE(3) /EMMEANS = TABLES(program) /EMMEANS = TABLES(month) /EMMEANS = TABLES(program*month) /CRITERIA = ALPHA(.05) /WSDESIGN = month /DESIGN = program.	**Multivariate Tests**	Within Subjects: month * program	**Multivariate Tests**	Within Subjects: month	**Multivariate Tests**	Between Subjects: program
SPSS MANOVA	MANOVA wtloss1 TO esteem3 BY program(1,3) /WSFACTOR=month(3) /MEASURES= wtloss, esteem /TRANSFORM(wtloss1 to esteem3) = polynomial /RENAME = WTLOSS WTLIN WTQUAD ESTEEM ESTLIN ESTQUAD /WSDESIGN=MONTH /PRINT=SIGNIF(UNIV, STEPDOWN) /DESIGN=PROGRAM.	EFFECT... PROGRAM BY MONTH	Multivariate Tests of Significance	EFFECT... MONTH	Multivariate Tests of Significance	EFFECT... PROGRAM	Multivariate Tests of Significance

TABLE 8.12 Continued

a. MULTIVARIATE EFFECTS (continued)

Program	Syntax	Parallelism		Flatness		Levels	
		Section of Output	Name of Effect	Section of Output	Name of Effect	Section of Output	Name of Effect

SAS GLM	`proc glm data=SASUSER.SSDOUBLE;`
	`class PROGRAM;`
	`model WTLOSS1 WTLOSS2 WTLOSS3`
	`      ESTEEM1 ESTEEM2 ESTEEM3=PROGRAM;`
	`/*Test for LEVELS effect */`
	`manova h=PROGRAM`
	`m=WTLOSS1+WTLOSS2+WTLOSS3,`
	`ESTEEM1+ESTEEM2+ESTEEM3/summary;`
	`/*Test for FLATNESS effect*/`
	`manova h=intercept`
	`   m = (-1 0 1 0 0 0,`
	`         1 -2 1 0 0 0,`
	`         0 0 0 -1 0 1,`
	`         0 0 0 1 -2 1)/summary;`
	`/*Test for PARALLELISM effect */`
	`manova h=PROGRAM`
	`   m =(-1 0 1 0 0 0,`
	`        1 -2 1 0 0 0,`
	`        0 0 0 -1 0 1,`
	`        0 0 0 1 -2 1)/summary;`
	`lsmeans PROGRAM;`
	`run;`
		MANOVA Test Criteria... No Overall PROGRAM Effect
		
		(Final portion of output)
			MANOVA Test Statistic	MANOVA Test Criteria... No Overall Intercept Effect	MANOVA Test Statistic	MANOVA Test Criteria... No Overall PROGRAM Effect
						
						(First portion of output)
							MANOVA Test Statistic

(continued)

343

TABLE 8.12 Continued

b. UNIVARIATE AND TREND EFFECTS AND MEANS

Program	Parallelism		Flatness		Levels		Means (Unadjusted for Stepdown Analysis)
	Section of Output	Name of Effect	Section of Output	Name of Effect	Section of Output	Name of Effect	
SPSS GLM	Univariate Tests	month * program	Univariate Tests	month	Tests of Between-Subjects Effects	program	Estimated Marginal Means
SPSS MANOVA	EFFECT... PROGRAM BY MONTH	Univariate F-tests... WTLIN WTQUAD ESTLIN ESTQUAD	EFFECT... MONTH	Univariate F-tests... WTLIN WTQUAD ESTLIN ESTQUAD	EFFECT... PROGRAM	Univariate F-tests... WTLOSS ESTEEM	See CONSPLUS procedure of Section 7.6, Table 7.21.
SAS GLM	MANOVA Test Criteria... No Overall PROGRAM Effect (*Final portion of output*)	Dependent Variable: MVAR1 (*Wtloss linear*) MVAR2 (*Wtloss quadratic*) MVAR3 (*Esteem linear*) MVAR4 (*Esteem quadratic*)	MANOVA Test Criteria... No Overall Intercept Effect	Dependent Variable: MVAR1 (*Wtloss linear*) MVAR2 (*Wtloss quadratic*) MVAR3 (*Esteem linear*) MVAR4 (*Esteem quadratic*)	MANOVA Test Criteria... No Overall PROGRAM Effect (*First portion of output*)	Dependent Variable: MVAR1 (*Wtloss*) MVAR2 (*Esteem*)	Least Squares Means

[a]Note: Italicized labels do not appear in output.

8.5.4 Classifying Profiles

A procedure typically available in programs designed for discriminant analysis is the classification of cases into groups on the basis of a best-fit statistical function. The principle of classification is often of interest in research where profile analysis is appropriate. If it is found that groups differ on their profiles, it could be useful to classify new cases into groups according to their profiles.

For example, given a profile of scores for different groups on a standardized test such as the Illinois Test of Psycholinguistic Abilities, one might use the profile of a new child to see if that child more closely resembles a group of children who have difficulty reading or a group who does not show such difficulty. If statistically significant profile differences were available before the age at which children are taught to read, classification according to profiles could provide a powerful diagnostic tool.

Note that this is no different from using classification procedures in discriminant analysis (Chapter 9). It is simply mentioned here because choice of profile analysis as the initial vehicle for testing group differences does not preclude use of classification. To use a discriminant program for classification, one simply defines the levels of the IV as "groups" and the DVs as "predictors."

8.5.5 Imputation of Missing Values

Issues of Section 4.1.3.2 apply to repeated-measures MANOVA. However, many of the procedures for imputing missing values described in that section do not take into account the commensurate nature of measures in profile analysis or, for that matter, any design with repeated measures. Multiple imputation through SOLAS MDA is applicable to longitudinal data (or any other repeated measures) but is difficult to implement. Or, if you happen to have BMDP5V (Dixon, 1992), the program imputes and prints out missing values for univariate repeated-measures analysis, which may then be added to the data set for multivariate analysis. None of the other procedures of Table 4.2 takes advantage of commensurate measurement.

A popular method (e.g., Myers & Well, 2002) is to replace the missing value with a value estimated from the mean for that level of the repeated factor and for that case. The following equation takes into account both the mean for the case and the mean for the level of A, the commensurate factor, as well as the mean for the group, B.

$$Y_{ij}^* = \frac{sS_i' + aA_j' - B'}{(a-1)(s-1)} \tag{8.10}$$

where Y_{ij}^* = predicted score to replace missing score,
s = the number of cases in the group,
S_i' = the sum of the known values for that case,
a = the number of levels of A, the within-subjects factor,
A_j' = the sum of the known values of A, and
B' = the sum of all known values for the group.

Say that the final score, s_{15} in a_4, is missing from Table 8.1. The remaining scores for that case (in a_1, a_2, and a_3) sum to $6 + 3 + 3 = 12$. The remaining scores for a_4 in b_3 (administrators) sum to

$2 + 5 + 5 + 4 = 16$. The remaining scores for the entire table sum to 60. Plugging these values into Equation 8.10,

$$Y^*_{15,4} = \frac{5(12) + 4(16) - 60}{(3)(4)} = 5.33$$

This procedure may produce an error term that is a bit too small because the estimated value is often too consistent with the other values. A more conservative α level is recommended for all tests if the proportion of missing values imputed is greater than 5%.

8.6   Complete Examples of Profile Analysis

Two complete examples of profile analysis are presented. The first is an analysis of subtests of the WISC (the commensurate measure) for three types of learning-disabled children. The second is a study of mental rotation of either a letter or a symbol over five sessions, using as DVs the slope and intercept of reaction time calculated over four angles of rotation.

8.6.1   Profile Analysis of Subscales of the WISC

Variables are chosen from among those in the learning disabilities data bank described in Appendix B, Section B.2 to illustrate the application of profile analysis. Three groups are formed on the basis of the preference of learning-disabled children for age of playmates (AGEMATE): children whose parents report that they have (1) preference for playmates younger than themselves, (2) preference for playmates older than themselves, and (3) preference for playmates the same age as themselves or no preference. Data are in PROFILE.*.

DVs are the 11 subtests of the Wechsler Intelligence Scale for Children given either in its original or revised (WISC-R) form, depending on the date of administration of the test. The subtests are information (INFO), comprehension (COMP), arithmetic (ARITH), similarities (SIMIL), vocabulary (VOCAB), digit span (DIGIT), picture completion (PICTCOMP), picture arrangement (PARANG), block design (BLOCK), object assembly (OBJECT), and CODING.

The primary question is whether profiles of learning-disabled children on the WISC subscales differ if the children are grouped on the basis of their choice of age of playmates (the parallelism test). Secondary questions are whether preference for age of playmates is associated with overall IQ (the levels test), and whether the subtest pattern of the combined group of learning-disabled children is flat (the flatness test), as it is for the population on which the WISC was standardized.

8.6.1.1   *Evaluation of Assumptions*

Assumptions and limitations of profile analysis are evaluated as described in Section 8.3.2.

8.6.1.1.1   *Unequal Sample Sizes and Missing Data*
From the sample of 177 learning-disabled children given the WISC or WISC-R, a preliminary run of SAS MEANS (Table 8.13) is used to reveal the extent and pattern of missing data. Missing data

TABLE 8.13   Identification of Missing Data (Syntax and Output from SAS MEANS)

```
proc sort data=SASUSER.PROFILE;
   by AGEMATE;
run;
proc means vardef=DF
   N NMISS;
   var INFO COMP ARITH SIMIL VOCAB DIGIT PICTCOMP PARANG BLOCK OBJECT CODING;
   by AGEMATE;
run;
```

---------------- Preferred age of playmates=. ----------------

The MEANS Procedure

Variable	Label	N	N Miss
INFO	Information	9	0
COMP	Comprehension	9	0
ARITH	Arithmetic	9	0
SIMIL	Similarities	9	0
VOCAB	Vocabulary	9	0
DIGIT	Digit Span	9	0
PICTCOMP	Picture Completion	9	0
PARANG	Picture Arrangement	9	0
BLOCK	Block Design	9	0
OBJECT	Object Assembly	9	0
CODING	Coding	9	0

---------------- Preferred age of playmates=1 ----------------

Variable	Label	N	N Miss
INFO	Information	46	0
COMP	Comprehension	46	0
ARITH	Arithmetic	46	0
SIMIL	Similarities	46	0
VOCAB	Vocabulary	46	0
DIGIT	Digit Span	45	1
PICTCOMP	Picture Completion	46	0
PARANG	Picture Arrangement	46	0
BLOCK	Block Design	46	0
OBJECT	Object Assembly	46	0
CODING	Coding	46	0

(continued)

TABLE 8.13 Continued

```
---------------- Preferred age of playmates=2 ----------------

                    The MEANS Procedure

                                               N
          Variable   Label                N   Miss
          --------------------------------------------
          INFO       Information          55    0
          COMP       Comprehension        55    0
          ARITH      Arithmetic           55    0
          SIMIL      Similarities         55    0
          VOCAB      Vocabulary           55    0
          DIGIT      Digit Span           55    0
          PICTCOMP   Picture Completion   55    0
          PARANG     Picture Arrangement  55    0
          BLOCK      Block Design         55    0
          OBJECT     Object Assembly      55    0
          CODING     Coding               54    1
          --------------------------------------------


---------------- Preferred age of playmates=3 ----------------

                                               N
          Variable   Label                N   Miss
          --------------------------------------------
          INFO       Information          67    0
          COMP       Comprehension        65    2
          ARITH      Arithmetic           67    0
          SIMIL      Similarities         67    0
          VOCAB      Vocabulary           67    0
          DIGIT      Digit Span           67    0
          PICTCOMP   Picture Completion   67    0
          PARANG     Picture Arrangement  67    0
          BLOCK      Block Design         67    0
          OBJECT     Object Assembly      67    0
          CODING     Coding               67    0
          --------------------------------------------
```

are sought among the DVs (subtests, levels of the within-subjects IV) for cases grouped by AGEMATE as indicated in the **by AGEMATE** instruction. Nine cases cannot be grouped according to preferred age of playmates, leaving 168 cases with group identification. Four children are identified as missing data through the SAS MEANS run. Because so few cases have missing data, and the missing variables are scattered over groups and DVs, it is decided to delete them from analysis, leaving $N = 164$. Other strategies for dealing with missing data are discussed in Chapter 4 and in Section 8.5.5.

Of the remaining 164 children, 45 are in the group preferring younger playmates, 54 older playmates, and 65 same age playmates or no preference. This leaves 4.5 times as many cases as DVs in the smallest group, posing no problems for multivariate analysis.

8.6.1.1.2 *Multivariate Normality*

Groups are large and not notably discrepant in size. Therefore, the central limit theorem should assure acceptably normal sampling distributions of means for use in profile analysis. The **data** step in Table 8.14 deletes cases with missing data and no group identification and provides a new, complete data file labeled PROFILEC to be used in all subsequent analyses. SAS MEANS output for those data shows all of the DVs to be well behaved; summary statistics for the first group, the one that prefers younger playmates, for example, are in Table 8.14. Skewness and kurtosis values are acceptable for all DVs in all groups.

The levels test is based on the average of the DVs. However, this should pose no problem; since the individual DVs are so well behaved, there is no reason to expect problems with the average of them. Had the DVs shown serious departures from normality, an "average" variable could have been created through a transformation and tested through the usual procedures of Section 4.2.2.1.

8.6.1.1.3 *Linearity*

Considering the well-behaved nature of these DVs and the known linear relationship among subtests of the WISC, no threats to linearity are anticipated.

8.6.1.1.4 *Outliers*

As seen in the univariate summary statistics of Table 8.14 for the first group, one standard score (ARITH) has $z = (19 - 9.22)/2.713 = 3.6$, suggesting a univariate outlier. No other standard scores are greater than $|3.3|$. A SAS REG run with leverage values saved is done as per Table 6.15 and reveals no multivariate outliers with a criterion of $p = .001$ (not shown). The decision is made to retain the univariate outlier since the subtest score of 19 is acceptable, and trial analyses with and without the outlier removed made no difference in the results (cf. Section 4.1.4.3).

TABLE 8.14 Univariate Summary Statistics Through SAS MEANS for Complete Data (Syntax and Selected Output)

```
data SASUSER.PROFILEC;
  set SASUSER.PROFILE;
  if AGEMATE=. or DIGIT=. or COMP=. or CODING=. then delete;
run;

proc means data=SASUSER.PROFILEC vardef=DF
    N MIN MAX MEAN VAR STD SKEWNESS KURTOSIS; ;
    var INFO COMP ARITH SIMIL VOCAB DIGIT PICTCOMP PARANG BLOCK OBJECT CODING;
    by AGEMATE;
run;
```

(continued)

TABLE 8.14 Continued

```
------------------- Preferred age of playmates=1 -------------------

                              The MEANS Procedure

Variable   Label                   N    Minimum         Maximum         Mean         Variance

INFO       Information            45    4.0000000      19.0000000     9.0666667     11.0636364
COMP       Comprehension          45    3.0000000      18.0000000     9.5111111      8.3919192
ARITH      Arithmetic             45    5.0000000      19.0000000     9.2222222      7.3585859
SIMIL      Similarities           45    5.0000000      19.0000000     9.8666667     10.6181818
VOCAB      Vocabulary             45    2.0000000      19.0000000    10.2888889     12.1191919
DIGIT      Digit Span             45    3.0000000      16.0000000     8.5333333      7.2090909
PICTCOMP   Picture Completion     45    5.0000000      17.0000000    11.2000000      7.1181818
PARANG     Picture Arrangement    45    5.0000000      15.0000000    10.0888889      5.6282828
BLOCK      Block Design           45    3.0000000      19.0000000    10.0444444      8.8616162
OBJECT     Object Assembly        45    3.0000000      14.0000000    10.4666667      6.8000000
CODING     Coding                 45    4.0000000      14.0000000     8.6444444      6.3707071


Variable   Label                  Std Dev        Skewness       Kurtosis

INFO       Information           3.3262045      0.5904910      0.4696812
COMP       Comprehension         2.8968809      0.8421902      1.6737862
ARITH      Arithmetic            2.7126714      1.0052275      2.6589394
SIMIL      Similarities          3.2585552      0.7760773      0.4300328
VOCAB      Vocabulary            3.4812630      0.5545930      0.8634583
DIGIT      Digit Span            2.6849750      0.5549781      0.5300003
PICTCOMP   Picture Completion    2.6679921     -0.2545848     -0.1464723
PARANG     Picture Arrangement   2.3724002      0.2314390     -0.1781970
BLOCK      Block Design          2.9768467      0.3011471      1.3983036
OBJECT     Object Assembly       2.6076810     -0.8214876      0.3214474
CODING     Coding                2.5240260      0.2440899     -0.5540206
```

8.6.1.1.5 Homogeneity of Variance-Covariance Matrices

Evidence for relatively equal variances is available from the full SAS MEANS run of Table 8.14, where variances are given for each variable within each group. All the variances are quite close in value across groups; for no variable is there a between-group ratio of largest to smallest variance approaching 10:1.

8.6.1.1.6 Multicollinearity and Singularity

Standardization of the WISC subtests indicates that although subtests are correlated, particularly within the two sets comprising verbal and performance IQ, there is no concern that SMCs would be so large as to create statistical multicollinearity or singularity. In any event, SAS GLM prevents such variables from entering an analysis.

8.6.1.2　Profile Analysis

Syntax and major output for profile analysis of the 11 WISC subtests for the three groups as produced by SAS GLM appear in Table 8.15. Significance tests are shown, in turn, for flatness (**SUB-TEST**), parallelism (**SUBTEST*AGEMATE**), and levels (**AGEMATE**).

The parallelism test, called the test of the **SUBTEST*AGEMATE** effect, shows significantly different profiles for the three AGEMATE groups. The various multivariate tests of parallelism produce slightly different probability levels for α, all less than 0.05. The test shows that there are statistically significant differences among the three AGEMATE groups in their profiles on the WISC. The profiles are illustrated in Figure 8.4. Mean values for the plots are found in the cell means portion of the output in Table 8.15, produced by the statement **means AGEMATE**. Effect sizes for all three tests—SUBTEST, SUBTEST*AGEMATE, and AGEMATE, respectively—are found through Smithson's (2003) NoncF2.sas procedures; partial syntax and results are in Table 8.16.

For interpretation of the nonparallel profiles, a contrast procedure is needed to determine which WISC subtests separate the three groups of children. Because there are so many subtests, however, the procedure of Section 8.5.2 is unwieldy. The decision is made to evaluate profiles in terms of subtests on which group averages fall outside the confidence interval of the pooled profile. Table 8.17 shows marginal means and standard deviations for each subtest to derive these confidence intervals.

In order to compensate for multiple testing, a wider confidence interval is developed for each test to reflect an experimentwise 95% confidence interval. Alpha rate is set at .0015 for each test to account for the 33 comparisons available—3 groups at each of 11 subtests—generating a 99.85% confidence interval. Because an N of 164 produces a t distribution similar to z, it is appropriate to base the confidence interval on $z = 3.19$.

For the first subtest, INFO,

$$P(\overline{Y} - zs_m < \mu < \overline{Y} + zs_m) = 99.85 \tag{8.11}$$

$$P(9.55488 - 3.19(3.03609)/\sqrt{164}) < \mu < 9.55488 + 3.19(3.03609)/\sqrt{164} = 99.85$$

$$P(8.79860 < \mu < 10.31116) = 99.85$$

Because none of the group means on INFO falls outside this interval for the INFO subtest, profiles are not differentiated on the basis of the information subtest of the WISC. It is not necessary to

TABLE 8.15 Syntax and Selected Output from SAS GLM Profile Analysis of 11 WISC Subtests

```
proc glm data=SASUSER.PROFILEC;
  class AGEMATE;
  model INFO COMP ARITH SIMIL VOCAB DIGIT
    PICTCOMP PARANG BLOCK OBJECT CODING = AGEMATE/nouni;
  repeated SUBTEST 11 / summary;
  means AGEMATE;
run;
```

Manova Test Criteria and Exact F Statistics for the Hypothesis of no SUBTEST Effect
H = Type III SSCP Matrix for SUBTEST
E = Error SSCP Matrix

S=1   M=4   N=75

Statistic	Value	F Value	Num DF	Den DF	Pr > F
Wilks' Lambda	0.53556008	13.18	10	152	<.0001
Pillai's Trace	0.46443992	13.18	10	152	<.0001
Hotelling-Lawley Trace	0.86720415	13.18	10	152	<.0001
Roy's Greatest Root	0.86720415	13.18	10	152	<.0001

Manova Test Criteria and F Approximations for the Hypothesis of no SUBTEST*AGEMATE Effect
H = Type III SSCP Matrix for SUBTEST*AGEMATE
E = Error SSCP Matrix

S=2   M=3.5   N=75

Statistic	Value	F Value	Num DF	Den DF	Pr > F
Wilks' Lambda	0.78398427	1.97	20	304	0.0087
Pillai's Trace	0.22243093	1.91	20	306	0.0113
Hotelling-Lawley Trace	0.26735297	2.02	20	253.32	0.0070
Roy's Greatest Root	0.23209691	3.55	10	153	0.0003

NOTE: F Statistic for Roy's Greatest Root is an upper bound.
NOTE: F Statistic for Wilks' Lambda is exact.

352

TABLE 8.15 Continued

The GLM Procedure
Repeated Measures Analysis of Variance
Tests of Hypotheses for Between Subjects Effects

Source	DF	Type III SS	Mean Square	F Value	Pr > F
AGEMATE	2	49.611331	24.805665	0.81	0.4456
Error	161	4916.226807	30.535570		

The GLM Procedure

Level of AGEMATE	N	INFO Mean	INFO Std Dev	COMP Mean	COMP Std Dev	ARITH Mean	ARITH Std Dev
1	45	9.0666667	3.32620450	9.5111111	2.89688094	9.22222222	2.71267135
2	54	10.1851852	3.27410696	10.4259259	2.87212926	8.79629630	2.25187243
3	65	9.3692308	2.54072597	10.1230769	2.88047138	9.13846154	2.49297089

Level of AGEMATE	N	SIMIL Mean	SIMIL Std Dev	VOCAB Mean	VOCAB Std Dev	DIGIT Mean	DIGIT Std Dev
1	45	9.8666667	3.25855517	10.2888889	3.48126298	8.53333333	2.68497503
2	54	11.2037037	2.98031042	11.4629630	2.80641424	9.01851852	2.53645675
3	65	10.7538462	3.44615170	10.3692308	2.75323237	8.73846154	2.62367183

Level of AGEMATE	N	PICTCOMP Mean	PICTCOMP Std Dev	PARANG Mean	PARANG Std Dev	BLOCK Mean	BLOCK Std Dev
1	45	11.2000000	2.66799209	10.0888889	2.37240023	10.0444444	2.97684668
2	54	9.7962963	3.33296644	10.7037037	2.95008557	10.2962963	2.89195129
3	65	11.1538462	2.77956139	10.3846154	2.59622507	10.6461538	2.45859951

Level of AGEMATE	N	OBJECT Mean	OBJECT Std Dev	CODING Mean	CODING Std Dev
1	45	10.4666667	2.60768096	8.64444444	2.52402596
2	54	10.9074074	2.87650547	8.81481481	2.97203363
3	65	11.0769231	2.92781751	8.20000000	2.80735641

TABLE 8.16 Effect Sizes (rsq) with Lower and Upper Confidence Limits (rsqlow and rsqupp) for Profile Analysis of WISC-R Subtests: Partial Syntax and Output

```
                    .
                    .
                    .
rsq = df1 * F / (df2 + df1 * F);
rsqlow = ncplow / (ncplow + df1 + df2 + 1);
rsqupp = ncpupp / (ncpupp + df1 + df2 + 1)
cards;
```

```
13.180    10    152    .95
1.970     20    304    .95
0.810      2    151    .95
```

```
;
proc print;
run;
```

The SAS System

Obs	F	df1	df2	conf	prlow	prupp	ncplow	ncpupp	rsq	rsqlow	rsqupp
1	13.180	10	152	0.95	0.975	0.025	74.900	181.205	0.46441	0.31484	0.52644
2	1.970	20	304	0.95	0.975	0.025	2.534	47.014	0.11474	0.00774	0.12638
3	0.810	2	151	0.95	0.975	0.025	0.000	8.836	0.01061	0.00000	0.05427

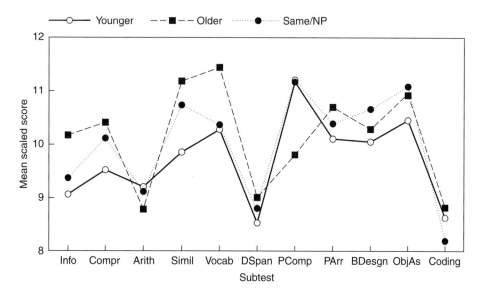

FIGURE 8.4 **Profiles of WISC scores for three AGEMATE groups.**

TABLE 8.17 **Syntax and Output for Marginal Means and Standard Deviations for Each Subtest: All Groups Combined**

```
proc means data=SASUSER.PROFILEC vardef=DF MEAN STD;
 var INFO COMP ARITH SIMIL VOCAB DIGIT PICTCOMP PARANG BLOCK
     OBJECT CODING ;
run;
```

The Means Procedure

Variable	Label	Mean	Std Dev
INFO	Information	9.5548780	3.0360877
COMP	Comprehension	10.0548780	2.8869349
ARITH	Arithmetic	9.0487805	2.4714440
SIMIL	Similarities	10.6585366	3.2699394
VOCAB	Vocabulary	10.7073171	3.0152488
DIGIT	Digit Span	8.7743902	2.6032689
PICTCOMP	Picture Completion	10.7195122	2.9980541
PARANG	Picture Arrangement	10.4085366	2.6557358
BLOCK	Block Design	10.3658537	2.7470551
OBJECT	Object Assembly	10.8536585	2.8202682
CODING	Coding	8.5243902	2.7857024

calculate intervals for any variable for which none of the groups deviates from the 95% confidence interval, because they cannot deviate from a wider interval. Therefore, intervals are calculated only for SIMIL, COMP, VOCAB, and PICTCOMP. Applying Equation 8.11 to these variables, significant profile deviation is found for vocabulary and picture completion. (The direction of differences is given in the Results section that follows.)

The first omnibus test produced by SAS GLM is the **SUBTEST** effect, for which the flatness hypothesis is rejected. All multivariate criteria show essentially the same result, but Hotelling's criterion, with approximate $F(10, 152) = 13.18, p < .001$, is most appropriately reported because it is a test of a single group (all groups combined).

Although not usually of interest when the hypothesis of parallelism is rejected, the flatness test, labeled **SUBTEST**, is interesting in this case because it reveals differences between learning-disabled children (our three groups combined) and the sample used for standardizing the WISC, for which the profile is necessarily flat. (The WISC was standardized so that all subtests produce the same mean value.) Any sample that differs from a flat profile, that is, has different mean values on various subtests, diverges from the standard profile of the WISC.

Appropriate contrasts for the flatness test in this example are simple one-sample z tests (cf. Section 3.1.1) against the standardized population values for each subtest with mean = 10.0 and standard deviation = 3.0. In this case we are less interested in how the subtests differ from one another than in how they differ from the normative population. (Had we been interested in differences among subtests for this sample, the contrasts procedures of Section 8.5.2 could have been applied.)

As a correction for post hoc inflation of experimentwise Type I error rate, individual alpha for each of the 11 z tests is set at .0045, meeting the requirements of

$$\alpha_{ew} = 1 - (1 - .0045)^{11} < .05$$

TABLE 8.18 Results of z Tests Comparing Each Subtest with WISC Population Mean (Alpha = .0045, Two-Tailed Test)

Subtest	Mean for Entire Sample	z for Comparison with Population Mean
Information	9.55488	−1.90
Similarities	10.65854	2.81
Arithmetic	9.04878	−4.06*
Comprehension	10.05488	0.23
Vocabulary	10.70732	3.02*
Digit span	8.77439	−5.23*
Picture completion	10.71951	3.07*
Picture arrangement	10.40854	1.74
Block design	10.36585	1.56
Object assembly	10.85366	3.64*
Coding	8.52439	−6.30*

*$p < .0045$.

as per Equation 7.12. Because most z tables (cf. Table C.1) are set up for testing one-sided hypotheses, critical α is divided in half to find critical z for rejecting the hypothesis of no difference between our sample and the population; the resulting criterion z is ± 2.845.

For the first subtest, INFO, the mean for our entire sample (from Table 8.17) is 9.55488. Application of the z test results in

$$z = \frac{\overline{Y} - \mu}{\sigma/\sqrt{N}} = \frac{9.55488 - 10}{3.0/\sqrt{164}} = -1.900$$

For INFO, then, there is no significant difference between the learning disabled group and the normative population. Results of these individual z tests appear in Table 8.18.

The final significance test is for **AGEMATE** (levels), in the section labeled **Tests of Hypotheses for Between Subjects Effects** and shows no statistically significant differences among groups on the average of the subtests, $F(2, 161) = 0.81, p = .4456$. This, also, typically is of no interest when parallelism is rejected.

A checklist for profile analysis appears in Table 8.19. Following is an example of a Results section in APA journal format.

TABLE 8.19   Checklist for Profile Analysis

1. Issues
   a. Unequal sample sizes and missing data
   b. Normality of sampling distributions
   c. Outliers
   d. Homogeneity of variance-covariance matrices
   e. Linearity
   f. Multicollinearity and singularity
2. Major analysis
   a. Tests for parallelism. If significant: Figure showing profile for deviation from parallelism
   b. Test for differences among levels, if appropriate. If significant: Marginal means for groups and standard deviations or standard errors or confidence intervals
   c. Test for deviation from flatness, if appropriate. If significant: Means for measures and standard deviations or standard errors or confidence intervals
   d. Effect sizes with confidence limits for all three tests
3. Additional analyses
   a. Planned comparisons
   b. Post hoc comparisons appropriate for significant effect(s)
      (1) Comparisons among groups
      (2) Comparisons among measures
      (3) Comparisons among measures within groups
   c. Power analysis for nonsignificant effects

Results

A profile analysis was performed on 11 subtests of the Wechsler Intelligence Scale for Children (WISC): information, similarities, arithmetic, comprehension, vocabulary, digit span, picture completion, picture arrangement, block design, object assembly, and coding. The grouping variable was preference for age of playmates, divided into children who (1) prefer younger playmates, (2) prefer older playmates, and (3) those who have no preference or prefer playmates the same age as themselves.

SAS MEANS and REG were used for data screening. Four children in the original sample, scattered through groups and DVs, had missing data on one or more subtest, reducing the sample size to 164. No univariate or multivariate outliers were detected among these children, with $p = .001$. After deletion of cases with missing data, assumptions regarding normality of sampling distributions, homogeneity of variance-covariance matrices, linearity, and multicollinearity were met.

SAS GLM was used for the major analysis. Using Wilks' criterion, the profiles, seen in Figure 8.4, deviated significantly from parallelism, $F(20, 304) = 1.97$, $p = .009$, partial $\eta^2 = .11$ with confidence limits from .01 to .13. For the levels test, no statistically significant differences were found among groups when scores were averaged over all subtests, $F(2, 161) = 0.81$, $p = .45$, partial $\eta^2 = .01$ with confidence limits from 0 to 0. When averaged over groups, however, subtests were found by Hotelling's criterion to deviate significantly from flatness, $F(10, 152) = 13.18$, $p < .001$, partial $\eta^2 = .46$, with confidence limits from .31 to .53.

To evaluate deviation from parallelism of the profiles, confidence limits were calculated around the mean of the profile for the three groups combined. Alpha error for each confidence interval was set at .0015 to achieve an experimentwise error rate of 5%. Therefore, 99.85% limits were evaluated for the pooled profile. For two of the subtests, one or more groups had means that fell outside these limits. Children who preferred older playmates had a significantly higher mean on the vocabulary subtest (mean = 11.46) than that of the pooled groups (where the 99.85% confidence limits were 9.956 to 11.458); children who preferred older playmates had significantly lower scores on the picture completion subtest (mean = 9.80) than that of the pooled groups (99.85% confidence limits were 9.973 to 11.466).

Deviation from flatness was evaluated by identifying which subtests differed from those of the standardization population of the WISC, with mean = 10 and standard deviation = 3 for each subtest. Experimentwise α = .05 was achieved by setting α for each test at .0045. As seen in Table 8.18, learning-disabled children had significantly lower scores than the WISC normative population in arithmetic, digit span, and coding. On the other hand, these children had significantly higher than normal performance on vocabulary, picture completion, and object assembly.

Thus, learning disabled children who prefer older playmates are characterized by having higher vocabulary and lower picture completion scores than the average of learning disabled children. As a group, the learning disabled children in this sample had lower scores on arithmetic, digit span, and coding than children-at-large, but higher scores on vocabulary, picture completion, and object assembly.

8.6.2   Doubly-Multivariate Analysis of Reaction Time

This analysis is on a data set from mental rotation experiments conducted by Damos (1989), described in greater detail in Appendix B. The between-subjects IV (levels) is whether the target object was the letter G or a symbol. The within-subjects IV treated multivariately consists of the first four testing sessions. The two noncommensurate DVs are the (1) slope and (2) intercept calculated from reaction times over four angles of rotation. Thus, intercept is the average reaction time and slope is the change in reaction time as a function of angle of rotation. Data files are DBLMULT.*. The major question is whether practice effects over the four sessions are different for the two target objects.

8.6.2.1   Evaluation of Assumptions

8.6.2.1.1   Unequal Sample Sizes, Missing Data, Multivariate Normality, and Linearity
Sample sizes in this data set are equal: 10 cases per group and there are no missing data among the eight DVs (2 measures over four occasions). Groups are small but equal in size and there are a few more cases than DVs in each group. Therefore, there is no concern about deviation from multivariate normality. Indeed, the SPSS DESCRIPTIVES output of Table 8.20 shows very small skewness and kurtosis values. These well-behaved variables also pose no threat to linearity.

8.6.2.1.2   Outliers
The **SAVE** instruction in the syntax of Table 8.20 adds standardized scores for each variable and each case to the SPSS data file. This provides a convenient way to look for univariate outliers, particularly when sample sizes are small. A criterion $\alpha = .01$ is used, so that any case with a z-score $> |2.58|$ is considered an outlier on that variable. Only one score approaches that criterion: case number 13 has a z-score of 2.58 on the slope measure for the second session.

Multivariate outliers are sought through SPSS REGRESSION, as per Table 7.12. Criterion χ^2 with 8 df at $\alpha = .01$ is 20.09. By this criterion, none of the cases is a multivariate outlier; the largest Mahalanobis distance in either group is 8.09. Therefore, all cases are retained for analysis.

8.6.2.1.3   Homogeneity of Variance-Covariance Matrices
Table 8.20 shows the ratio of variances for all eight variables to be well within acceptable limits, particularly for this equal-n data set. Indeed, all variance ratios are 2.5:1 or less.

8.6.2.1.4   Homogeneity of Regression
SPSS MANOVA is used to test homogeneity of regression for the stepdown analysis, in which the second DV, slope, is adjusted for the first, intercept. Table 8.21 shows the syntax and final portion of output for the test. The last source of variance is the one that tests homogeneity of regression. The assumption is supported with $p = .138$ because it is $> .05$.

8.6.2.1.5   Reliability of DVs
Intercept acts as a covariate for slope in the stepdown analysis. There is no reason to doubt the reliability of intercept as a measure, because it is a derived value based on a measure (response time) recorded electronically on equipment checked periodically.

8.6.2.1.6   Multicollinearity and Singularity
Correlations among DVs are expected to be high, particularly within slope and intercept sets, but not so high as to threaten statistical multicollinearity. Correlations between slopes and intercepts

TABLE 8.20 Descriptive Statistics for the Eight DVs (SPSS DESCRIPTIVES Syntax and Output)

```
SPLIT FILE
  SEPARATE BY group.
DESCRIPTIVES
  VARIABLES=slope1 intrcpt1 slope2 intrcpt2 slope3 intrcpt3 slope4 intrcpt4
  /SAVE
  /STATISTICS=MEAN STDDEV VARIANCE KURTOSIS SKEWNESS.
```

Descriptives

Group identification = Letter G

Descriptive Statistics[a]

	N	Mean	Std.	Variance	Skewness		Kurtosis	
	Statistic	Statistic	Statistic	Statistic	Statistic	Std. Error	Statistic	Std. Error
SLOPE1	10	642.62	129.889	16871.090	1.017	.687	.632	1.334
INTRCPT1	10	200.70	42.260	1785.917	.160	.687	-.077	1.334
SLOPE2	10	581.13	97.983	9600.605	.541	.687	-.034	1.334
INTRCPT2	10	133.75	46.921	2201.560	.644	.687	-.940	1.334
SLOPE3	10	516.87	64.834	4203.427	.159	.687	-.354	1.334
INTRCPT3	10	90.46	30.111	906.651	.421	.687	-1.114	1.334
SLOPE4	10	505.39	67.218	4518.283	-.129	.687	-.512	1.334
INTRCPT4	10	72.39	26.132	682.892	.517	.687	-.694	1.334
Valid N (listwise)	10							

aGroup identification = Letter G

(continued)

TABLE 8.20 Continued

Group identification = Symbol

Descriptive Statistics[a]

	N	Mean	Std.	Variance	Skewness		Kurtosis	
	Statistic	Statistic	Statistic	Statistic	Statistic	Std. Error	Statistic	Std. Error
SLOPE1	10	654.50	375.559	141044.9	−.245	.687	1.015	1.334
INTRCPT1	10	40.85	49.698	2469.870	.514	.687	−1.065	1.334
SLOPE2	10	647.53	153.991	23713.180	2.113	.687	5.706	1.334
INTRCPT2	10	24.30	29.707	882.521	1.305	.687	1.152	1.334
SLOPE3	10	568.22	59.952	3594.241	−.490	.687	.269	1.334
INTRCPT3	10	22.95	29.120	847.957	.960	.687	−.471	1.334
SLOPE4	10	535.82	50.101	2510.106	−.441	.687	−.781	1.334
INTRCPT4	10	22.46	26.397	696.804	1.036	.687	.003	1.334
Valid N (listwise)	10							

[a]Group identification = Symbol

TABLE 8.21 Syntax and Selected SPSS MANOVA Output for Test of Homogeneity of Regression

```
SPLIT FILE
  OFF.
MANOVA
  INTRCPT1 INTRCPT2 INTRCPT3 INTRCPT4 SLOPE1 SLOPE2 SLOPE3 SLOPE4
      BY GROUP(1, 2)
  /PRINT=SIGNIF(BRIEF)
  /ANALYSIS = SLOPE1 SLOPE2 SLOPE3 SLOPE4
  /DESIGN = POOL(INTRCPT1 INTRCPT2 INTRCPT3 INTRCPT4) GROUP
          POOL(INTRCPT1 INTRCPT2 INTRCPT3 INTRCPT4) BY GROUP.
```

```
Multivariate Tests of Significance
Tests using UNIQUE sums of squares and WITHIN+RESIDUAL error term
```

Source of Variation	Wilks	Approx F	Hyp. DF	Error DF	Sig of F
POOL(INTRCPT1 INTRCPT2 INTRCPT3 INTRCPT4)	.054	2.204	16.00	22.023	.043
GROUP	.570	1.320	4.00	7.000	.350
POOL(INTRCPT1 INTRCPT2 INTRCPT3 INTRCPT4) BY GROUP	.091	1.644	16.00	22.023	.138

are not expected to be substantial. The determinant of the variance-covariance matrix in the main analysis provides assurance that there is no statistical multicollinearity ($p > .00001$), as seen in the main analysis in Table 8.21.

8.6.2.2 Doubly-Multivariate Analysis of Slope and Intercept

SPSS MANOVA is chosen for the main analysis, which includes a full trend analysis on the repeated measures effects: flatness (the main effect of session) and parallelism (the session by group interaction). The program also provides a stepdown analysis without the need for reconfiguring the data set (cf. Section 6.5.4.1).

Table 8.22 shows syntax and output for the omnibus analysis and stepdown trend analyses. ERROR(COR) requests residual (pooled within-cell) correlation matrix; HOMOGENEITY (BOXM) provides the determinant of the pooled within-cell variance-covariance matrix. The RENAME instruction makes the output easier to read. INT_LIN is the linear trend of the group by session interaction for intercept, INT_QUAD is the quadratic trend of the interaction for intercept, SL_CUBIC is the cubic trend of the group by session interaction for slope, and so on. EFSIZE in the PRINT paragraph requests effect sizes along with univariate and stepdown results.

The three sections labeled Multivariate Tests of Significance (for GROUP, GROUP BY SESSION, and SESSION) show that all three effects are statistically significant, $p < .0005$. Because parallelism is rejected, with a strong SESSION BY GROUP interaction, multivariate $F(6, 13) = 9.92$, flatness and levels effects are not interpreted. Table 8.23 shows effect sizes

```
MANOVA
INTRCPT1 INTRCPT2 INTRCPT3 INTRCPT4 SLOPE1 SLOPE2 SLOPE3 SLOPE4
    BY GROUP(1, 2)
/WSFACTOR = SESSION(4)
/MEASURES = INTERCPT, SLOPE
/TRANSFORM(INTRCPT1 TO SLOPE4) = POLYNOMIAL
/RENAME=INTERCPT INT_LIN INT_QUAD INT_CUBIC
        SLOPE SL_LIN SL_QUAD SL_CUBIC
/WSDESIGN = SESSION
/PRINT=SIGNIF(UNIV, STEPDOWN, EFSIZE) ERROR(CORR) HOMOGENEITY(BOXM)
/DESIGN = GROUP.
```

```
Determinant of pooled Covariance matrix of dependent vars. = 7.06128136E+23
LOG(Determinant) =                                                   54.91408
```

WITHIN+RESIDUAL Correlations with Std. Devs. on Diagonal

	INTERCPT	SLOPE
INTERCPT	64.765	
SLOPE	.187	232.856

EFFECT.. GROUP
Multivariate Tests of Significance (S = 1, M = 0, N = 7 1/2)

Test Name	Value	Exact F	Hypoth. DF	Error DF	Sig. of F
Pillais	.73049	23.03842	2.00	17.00	.000
Hotellings	2.71040	23.03842	2.00	17.00	.000
Wilks	.26951	23.03842	2.00	17.00	.000
Roys	.73049				

Note.. F statistics are exact.

Multivariate Effect Size

TEST NAME	Effect Size
(All)	.730

Univariate F-tests with (1,18) D. F.

Variable	Hypoth. SS	Error SS	Hypoth. MS	Error MS	F	Sig. of F
INTERCPT	186963.468	75501.1333	186963.468	4194.50740	44.57340	.000
SLOPE	32024.7168	975990.893	32024.7168	54221.7163	.59063	.452

Variable	ETA Square
INTERCPT	.71234
SLOPE	.03177

Roy-Bargman Stepdown F-tests

Variable	Hypoth. MS	Error MS	Stepdown F	Hypoth. DF	Error DF	Sig. of F
INTERCPT	186963.468	4194.50740	44.57340	1	18	.000
SLOPE	63428.5669	55404.9667	1.14482	1	17	.300

TABLE 8.22   Continued

EFFECT.. GROUP (Cont.)

Tests involving 'SESSION' Within-Subject Effect.

EFFECT.. GROUP BY SESSION
Multivariate Tests of Significance (S = 1, M = 2 , N = 5 1/2)

Test Name	Value	Exact F	Hypoth. DF	Error DF	Sig. of F
Pillais	.82070	9.91711	6.00	13.00	.000
Hotellings	4.57713	9.91711	6.00	13.00	.000
Wilks	.17930	9.91711	6.00	13.00	.000
Roys	.82070				

Note.. F statistics are exact.

Multivariate Effect Size

TEST NAME	Effect Size
(All)	.821

Univariate F-tests with (1,18) D. F.

Variable	Hypoth. SS	Error SS	Hypoth. MS	Error MS	F	Sig. of F
INT_LIN	34539.9233	11822.7068	34539.9233	656.81704	52.58683	.000
INT_QUAD	1345.75055	4783.30620	1345.75055	265.73923	5.06418	.037
INT_CUBI	63.15616	2160.40148	63.15616	120.02230	.52620	.478
SL_LIN	412.00434	677335.981	412.00434	37629.7767	.01095	.918
SL_QUAD	7115.88944	165948.577	7115.88944	9219.36538	.77184	.391
SL_CUBIC	1013.73020	35227.3443	1013.73020	1957.07469	.51798	.481

Variable	ETA Square
INT_LIN	.74499
INT_QUAD	.21957
INT_CUBI	.02840
SL_LIN	.00061
SL_QUAD	.04112
SL_CUBIC	.02797

Roy-Bargman Stepdown F-tests

Variable	Hypoth. MS	Error MS	Stepdown F	Hypoth. DF	Error DF	Sig. of F
INT_LIN	34539.9233	656.81704	52.58683	1	18	.000
INT_QUAD	209.21632	195.44139	1.07048	1	17	.315
INT_CUBI	2.36753	45.37427	.05218	1	16	.822
SL_LIN	6198.06067	39897.8795	.15535	1	15	.699
SL_QUAD	710.44434	1472.26563	.48255	1	14	.499
SL_CUBIC	917.69207	255.81569	3.58732	1	13	.081

(continued)

TABLE 8.22 Continued

```
EFFECT.. SESSION
Multivariate Tests of Significance (S = 1, M = 2, N = 5 1/2)
```

Test Name	Value	Exact F	Hypoth. DF	Error DF	Sig. of F
Pillais	.88957	17.45295	6.00	13.00	.000
Hotellings	8.05521	17.45295	6.00	13.00	.000
Wilks	.11043	17.45295	6.00	13.00	.000
Roys	.88957				

```
Note.. F statistics are exact.
```

```
Multivariate Effect Size
```

TEST NAME	Effect Size
(All)	.890

```
Univariate F-tests with (1,18) D. F.
```

Variable	Hypoth. SS	Error SS	Hypoth. MS	Error MS	F	Sig. of F
INT_LIN	58748.6922	11822.7068	58748.6922	656.81704	89.44453	.000
INT_QUAD	5269.58732	4783.30620	5269.58732	265.73923	19.82992	.000
INT_CUBI	40.83996	2160.40148	40.83996	120.02230	.34027	.567
SL_LIN	207614.672	677335.981	207614.672	37629.7767	5.51730	.030
SL_QUAD	755.44788	165948.577	755.44788	9219.36538	.08194	.778
SL_CUBIC	7637.13270	35227.3443	7637.13270	1957.07469	3.90232	.064

Variable	ETA Square
INT_LIN	.83247
INT_QUAD	.52419
INT_CUBI	.01855
SL_LIN	.23461
SL_QUAD	.00453
SL_CUBIC	.17817

```
Roy-Bargman Stepdown F-tests
```

Variable	Hypoth. MS	Error MS	Stepdown F	Hypoth. DF	Error DF	Sig. of F
INT_LIN	58748.6922	656.81704	89.44453	1	18	.000
INT_QUAD	26.62826	195.44139	.13625	1	17	.717
INT_CUBI	1.74540	45.37427	.03847	1	16	.847
SL_LIN	70401.2193	39897.8795	1.76454	1	15	.204
SL_QUAD	6294.30308	1472.26563	4.27525	1	14	.058
SL_CUBIC	96.64789	255.81569	.37780	1	13	.549

TABLE 8.23   Effect Sizes (r2) with Lower and Upper Confidence Limits (lr2 and ur2) for Doubly-Multivariate Analysis of Reaction Time

	fval	df1	df2	conf	lc2	ucdf	uc2	lcdf	power	r2	lr2	ur2
1	9.9171	6	13	.950	11.6604	.9750	120.0513	.0250	.9999	.82	.37	.86
2	17.4530	6	13	.950	27.1680	.9750	206.4357	.0250	1.0000	.89	.58	.91
3	23.0384	2	17	.950	13.2066	.9750	93.7622	.0250	1.0000	.73	.40	.82

and their confidence limits through Smithson's (2003) procedure for all three effects: parallelism, flatness, and levels, respectively.

The trend analysis for the two DVs, is in the section labeled `Roy-Bargman Stepdown F-tests in the EFFECT ... GROUP BY SESSION` section of Table 8.22. With $\alpha = .0083$ to compensate for inflated Type I error rate with the six DVs, only the linear trend of the interaction for intercept is statistically reliable, $F(1, 18) = 52.59$. Effect size is calculated using Equation 3.26:

$$\text{partial } \eta^2 = \frac{34539.92}{34539.92 + 11822.71} = .74$$

Note that the univariate effect size can be used because this trend is in the first-priority DV. Table 8.24 shows a summary of the stepdown trend analysis of the group by session interaction, in a form suitable for publication.

Figure 8.5 plots the profiles over the four sessions for the two groups. Figure 8.5 shows that reaction time intercept is much longer for the symbol than for the letter G, but rather rapidly declines over the four sessions, while reaction time intercept for the letter G stays low and fairly stable over the four sessions. Thus, the linear trend of the interaction indicates that the linear trend is greater for the symbol than for the letter.

Table 8.25 summarizes cell means and standard deviations from Table 8.20.

TABLE 8.24   Stepdown Tests of the Trend Analysis of the Group by Session Interaction

| | | Univariate | | Stepdown | | Partial | 99.17% CL around Partial η^2 | |
IV	Trend	F	df	F	df	η^2	Lower	Upper
Intercept	Linear	52.59[a]	1/18	52.59**	1/18	.75	.32	.87
	Quadratic	5.06	1/18	1.07	1/17	.06	.00	.42
	Cubic	0.53	1/18	0.05	1/16	.00	.00	.11
Slope	Linear	0.01	1/18	0.16	1/15	.01	.00	.29
	Quadratic	0.77	1/18	0.48	1/14	.03	.00	.41
	Cubic	0.52	1/18	3.59	1/13	.22	.00	.59

[a]Significance level cannot be evaluated but would reach $p < .0083$ in univariate context.

**$p < .0083$.

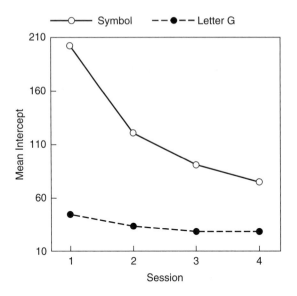

FIGURE 8.5 **Intercept of reaction time over
four angles of rotation as a function of
session and target object.**

TABLE 8.25 **Intercept and Slope over Four Angles of
Rotation of Reaction Time for Two Target Objects for
the First Four Sessions**

Target Object	Session			
	1	*2*	*3*	*4*
Intercept				
Letter G				
M	200.70	133.75	90.46	72.39
SD	42.46	46.92	30.11	26.13
Symbol				
M	40.85	24.30	22.95	22.46
SD	49.70	29.71	29.12	24.12
Slope				
Letter G				
M	642.62	581.13	516.87	505.39
SD	129.89	97.98	64.83	67.22
Symbol				
M	654.50	647.53	568.22	535.82
SD	375.56	153.99	59. 95	50.10

The pooled-within cell correlation matrix (Correlation in the section labeled `WITHIN+RESIDUAL Correlations with Std. Devs. On Diagonal`) shows that, indeed, the correlation between measures of intercept and slope is small.

Table 8.26 is a checklist for the doubly-multivariate analysis. It is followed by a Results section, in journal format, for the analysis just described.

TABLE 8.26   Checklist for Doubly-Multivariate Analysis of Variance

1. Issues
    a. Unequal sample sizes and missing data
    b. Normality of sampling distributions
    c. Outliers
    d. Homogeneity of variance-covariance matrices
    e. Linearity
    f. In stepdown analysis, when DVs act as covariates
        (1) Homogeneity of regression
        (2) Reliability of DVs
    g. Multicollinearity and singularity
2. Major analyses: Planned comparisons or omnibus F, when significant
    a. Parallelism. If significant: Importance of DVs
        (1) Within-cell correlations, stepdown F, univariate F
        (2) Effect size with confidence limits for significant stepdown F
        (3) Profile plot and table of cell means or adjusted cell means and standard deviations, standard errors, or confidence intervals.
    b. Test for differences among levels, if appropriate. If significant: Importance of DVs
        (1) Within-cell correlations, stepdown F, univariate F
        (2) Effect size with confidence limits for significant stepdown F
        (3) Marginal or adjusted marginal means and standard deviations, standard errors, or confidence intervals
    c. Test for deviation from flatness, if appropriate. If significant: Importance of DVs
        (1) Within-cell correlations, stepdown F, univariate F
        (2) Effect size with confidence limits for significant F
        (3) Marginal or adjusted marginal means and standard deviations, standard errors, or confidence intervals
    d. Effect sizes and confidence intervals for tests of parallelism, levels, and flatness
3. Additional analyses
    a. Post hoc comparisons appropriate for significant effect(s)
        (1) Comparisons among groups
        (2) Comparisons among measures
        (3) Comparisons among measures within groups
    b. Power analysis for nonsignificant effects

Results

A doubly-multivariate analysis of variance was performed on two measures of reaction time: intercept and slope of the regression line over four angles of rotation. Intercept represents overall reaction time; slope represents change in reaction time as a function of angle of rotation. Two target objects formed the between-subjects IV: the letter G and a symbol. The within-subjects IV treated multivariately was the first four sessions of the entire set of 20 sessions. Trend analysis was planned for the main effect of sessions as well as the group by session interaction. $N = 10$ for each of the groups.

No data were missing, nor were there univariate or multivariate outliers at $\alpha = .01$. Results of evaluation of assumptions of doubly-multivariate analysis of variance were satisfactory. Cell means and standard deviations for the two DVs over all combinations of group and session are in Table 8.25.

The group by session interaction (deviation from parallelism) was strong and statistically significant multivariate $F(6, 13) = 9.92$, $p < .005$, partial $\eta^2 = .82$ with confidence limits from .37 to .86. Reaction-time changes over the four sessions differed for the two target types. Although group and session main effects also were statistically significant they are not interpreted in the presence of the strong interaction. Partial η^2 for the group main effect was .73 with confidence limits from .40 to .82. For the session main effect, partial $\eta^2 = .89$ with confidence limits from .58 to .91.

A Roy-Bargmann stepdown analysis was performed on the trend analysis of the DVs, with the three trends of intercept as the first DV and the three trends of slope adjusted for intercept as the second. Intercept was judged sufficiently reliable as a covariate for slope to warrant stepdown analysis. Homogeneity of regression was achieved for the stepdown analysis. An experiment-wise error rate of 5% was achieved by setting α = .083 for each of the six components (three trends each of intercept and slope). Table 8.24 shows the results of the trend analysis.

The only significant stepdown effect for the group by session interaction was the linear trend of intercept, $F(1, 18)$ = 52.59, p < .005, partial η^2 = .74 with confidence limits from .32 to .87. Figure 8.5 plots the profiles for the two groups (letter G and symbol) over the four sessions. Mean reaction time (intercept) is much longer for the symbol than for the letter G, but rapidly declines over the four sessions. Reaction time for the letter G stays low and fairly stable over the four sessions. There is no evidence that change in reaction time as a function of angle of rotation (slope) is different for the two target objects.

Thus, reaction time to the letter G is quite fast, and does not change with practice. Reaction time is much slower for the symbol, but improves with practice. The change in reaction time associated with differing rotation of the target object does not depend on type of object or practice.

8.7   Comparison of Programs

Programs for MANOVA are covered in detail in Chapter 7. Therefore, this section is limited to those features of particular relevance to profile analysis. SPSS and SYSTAT each have two programs useful for profile analysis. SAS has an additional program that can be used for profile analysis, but it is limited to equal-n designs. The SAS manual shows by example how to set up doubly-multivariate designs. SYSTAT has an example available in online help files. SPSS GLM syntax for doubly-multivariate designs is shown in the manual, but no such help is available for SPSS MANOVA, unless you happen to have an old (1986) SPSS[x] manual handy. All programs provide output for

TABLE 8.27   Comparison of Programs for Profile Analysis[a]

	SPSS MANOVA	SPSS GLM	SAS GLM and ANOVA	SYSTAT GLM and ANOVA
Input				
Variety of strategies for unequal n	Yes	Yes	Yes[b]	No
Special specification for doubly-multivariate analysis	Yes	Yes	Yes	No
Special specification for simple effects	Yes	No	No	No
Output				
Single source table for Wilks' lambda	PRINT= SIGNIF (BRIEF)	Yes	No	No
Specific comparisons	Yes	Yes	Yes	Yes
Within-cells correlation matrix	Yes	Yes	Yes	Yes
Determinant of within-cells variance-covariance matrix	Yes	No	No	No
Cell means and standard deviations	PRINT= CELLINFO (MEANS)	EMMEANS	LSMEANS	PRINT MEDIUM
Marginal means	OMEANS	EMMEANS	LSMEANS	No
Marginal standard deviations or standard errors	No	Yes	STDERR	No
Confidence intervals around cell means	Yes	Yes	No	No
Wilks' lambda and F for parallelism	Yes	Yes	Yes	Yes
Pillai's criterion	Yes	Yes	Yes	Yes
Additional statistical criteria	Yes	Yes	Yes	Yes
Test for homogeneity of covariance/sphericity	Yes	No	No	No
Greenhouse-Geisser epsilon and adjusted p	Yes	Yes	Yes	Yes
Huynh-Feldt epsilon and adjusted p	Yes	Yes	Yes	Yes
Predicted values and residuals for each case	Yes	Yes	Yes	Data file
Residuals plot	Yes	No	No	No
Homogeneity of variance-covariance matrices	Box's M	Box's M	No	No
Test for multivariate outliers	No	No	Data file	Data file
Effect sizes (strength of association)	EFSIZE	ETASQ	No	No

[a]Additional features of these programs appear in Chapter 9 (MANOVA).

[b]SAS ANOVA requires equal n.

within-subjects effects in both univariate and multivariate form. All programs also provide information useful for deciding between multivariate and univariate approaches, with Greenhouse-Geisser and Huynh-Feldt adjustments for violation of sphericity in the univariate approach. Features of the programs appear in Table 8.27.

8.7.1 SPSS Package

Profile analysis in SPSS MANOVA or GLM is run like any other repeated-measures design. Output includes both univariate and multivariate tests for all within-subjects effects (flatness) and mixed interactions (parallelism).

The **MEASURES** command allows the multiple DVs to each be given a generic name in doubly-multivariate designs, making the output more readable. Also, several repeated-measures analyses can be specified within a single run.

The **MWITHIN** feature in SPSS MANOVA provides for testing simple effects when repeated measures are analyzed. The full output for SPSS MANOVA consists of three separate source tables, one each for parallelism, levels, and flatness. Specification of **PRINT= SIGNIF(BRIEF)** is used to simplify the multivariate output. The determinant of the within-cells correlation matrix is used as an aid to deciding whether or not further investigation of multicollinearity is needed. This is the only program that provides stepdown analysis for a doubly-multivariate design, along with straightforward syntax for testing homogeneity of regression. Adjusted means are not easily obtained, however, as demonstrated in Section 7.6.

SPSS GLM provides more easily interpreted output, with means adjusted for unequal-n (but not stepdown analysis in a doubly-multivariate design) readily available. Although the determinant of the variance-covariance matrix is not available, the program protects against statistical multicollinearity and singularity.

Both SPSS programs permit the residuals plots, which provide evidence of skewness and outliers. And Box's M test is available in both of them as an ultrasensitive test of homogeneity of variance-covariance matrices. But tests of linearity and homoscedasticity are not directly available within SPSS GLM or MANOVA and a test for multivariate outliers requires use of SPSS REGRESSION separately for each group.

8.7.2 SAS System

Profile analysis is available through the GLM (general linear model) procedure of SAS or through ANOVA if the groups have equal sample sizes. The two procedures use very similar syntax conventions and provide the same output. In GLM and ANOVA, profile analysis is treated as a special case of repeated-measures ANOVA. An explicit statement that generates the segments between adjacent levels of the within-subjects factor is available: `profile`.

Both univariate and multivariate results are provided by default, with the multivariate output providing the tests of parallelism and flatness. However, the output is not particularly easy to read. Each multivariate effect appears separately but instead of a single table for each effect, there are multiple sections filled with cryptic symbols. These symbols are defined above the output for that effect. Only the test of levels appears in a familiar source table. Specification of `short` within the `repeated` statement provides condensed output for the multivariate tests of flatness and parallelism, but the tests for each effect are still separated.

8.7.3   SYSTAT System

The GLM and MANOVA programs in SYSTAT handle profile analysis through the REPEAT format. (SYSTAT ANOVA also does profile analysis, but with few features and less flexibility.) There are three forms for printing output—long, medium, and short. The short form is the default option. The long form provides such extras as error correlation matrices and canonical analysis, as well as cell means and standard errors for the DVs. Additional statistics are available through the STATS program but the data set must be sorted by groups. SYSTAT GLM automatically prints out a full trend analysis on the within-subjects variable. There is no example in the manual to follow for a doubly-multivariate analysis.

Multivariate outliers are found by applying the discriminant procedure detailed in the SYSTAT manual. Leverage values for each group, which can be converted to Mahalanobis distances, are saved to a file. Additional assumptions are not directly tested through the GLM procedure although assumptions such as linearity and homogeneity of variance can be evaluated through STATS and GRAPH. No test for homogeneity of variance-covariance matrices is available.

9

Discriminant Analysis

9.1 General Purpose and Description

The goal of discriminant analysis is to predict group membership from a set of predictors. For example, can a differential diagnosis among a group of nondisabled children, a group of children with learning disability, and a group with emotional disorder be made reliably from a set of psychological test scores? The three groups are nondisabled children, children with learning disabilities, and children with emotional disorders. The predictors are a set of psychological test scores such as the Illinois Test of Psycholinguistic Ability, subtests of the Wide Range Achievement Test, Figure Drawing tests, and the Wechsler Intelligence Scale for Children.

Discriminant analysis (DISCRIM) is MANOVA turned around. In MANOVA, we ask whether group membership is associated with statistically significant mean differences on a combination of DVs. If the answer to that question is yes, then the combination of variables can be used to predict group membership—the DISCRIM perspective. In univariate terms, a significant difference among groups implies that, given a score, you can predict (imperfectly, no doubt) which group it comes from.

Semantically, however, confusion arises between MANOVA and DISCRIM because in MANOVA the IVs are the groups and the DVs predictors while in DISCRIM the IVs are the predictors and the DVs are the groups. We have tried to avoid confusion here by always referring to IVs as *predictors* and to DVs as *groups* or *grouping variables*.[1]

Mathematically, MANOVA and DISCRIM are the same, although the emphases often differ. The major question in MANOVA is whether group membership is associated with statistically significant mean differences in combined DV scores, analogous in DISCRIM to the question of whether predictors can be combined to predict group membership reliably. In many cases, DISCRIM is carried to the point of actually putting cases into groups in a process called classification.

Classification is a major extension of DISCRIM over MANOVA. Most computer programs for DISCRIM evaluate the adequacy of classification. How well does the classification procedure do? How many learning-disabled kids in the original sample, or a cross-validation sample, are classified

---

[1] Many texts also refer to IVs or predictors as discriminating variables and to DVs or groups as classification variables. However, there are also discriminant functions and classification functions to contend with, so the terminology becomes quite confusing. We have tried to simplify it by using only the terms predictors and groups.

correctly? When errors occur, what is their nature? Are learning-disabled kids more often confused with nondisabled kids or with kids suffering emotional disorders?

A second difference involves interpretation of differences among the predictors. In MANOVA, there is frequently an effort to decide which DVs are associated with group differences, but rarely an effort to interpret the pattern of differences among the DVs as a whole. In DISCRIM, there is often an effort to interpret the pattern of differences among the predictors as a whole in an attempt to understand the dimensions along which groups differ.

Complexity arises with this attempt, however, because with more than two groups there may be more than one way to combine the predictors to differentiate among groups. There may, in fact, be as many dimensions that discriminate among groups as there are degrees of freedom for the groups or the number of predictors (whichever is smaller). For example, if there are only two groups, there is only one linear combination of predictors that best separates them. Figure 9.1(a) illustrates the separation of two group *centroids* (multivariate version of means), $\overline{Y}_1$ and $\overline{Y}_2$, on a single axis, X, which represents the best linear combination of predictors that separate groups 1 and 2. A line parallel to the imaginary line that connects the two centroids represents the linear combination of Xs, or the first *discriminant function*[2] of X. Once a third group is added, however, it may not fall along that line. To maximally separate the three groups, it may be necessary to add a second linear combination of Xs, or second discriminant function. In the example in Figure 9.1(b), the first discriminant function separates the centroids of the first group from the centroids of the other two groups but does not distinguish the centroids for the second and third groups. The second discriminant function separates the centroid of the third group from the other two but does not distinguish between the first two groups.

On the other hand, the group means *might* fall along a single straight line even with three (or more) groups. If that is the case, only the first discriminant function is necessary to describe the dif-

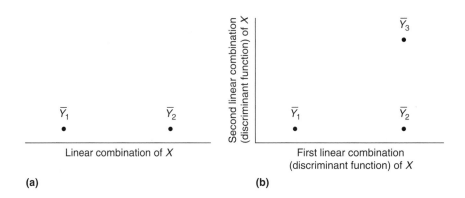

(a) Linear combination of X

(b) First linear combination (discriminant function) of X

FIGURE 9.1   **(a) Plot of two group centroids, $\overline{Y}_1$ and $\overline{Y}_2$, on a scale representing the linear combinations of X. (b) Plot of two linear combinations of X required to distinguish among three group centroids, $\overline{Y}_1$, $\overline{Y}_2$, and $\overline{Y}_3$.**

[2]Discriminant functions are also known as roots, canonical variates, principal components, dimensions, etc., depending on the statistical technique in which they are developed.

ferences among groups. The number of discriminant functions necessary to describe the group separation may be smaller than the maximum number available (which is the number of predictors or the number of groups minus 1, whichever is fewer). With more than two groups, then, discriminant analysis is a truly multivariate technique which is interpreted as such, with multiple *Y*s representing groups and multiple *X*s representing predictors.

In our example of three groups of children (nondisabled, learning-disabled, and emotionally disordered) given a variety of psychological measures, one way of combining the psychological test scores may tend to separate the nondisabled group from the two groups with disorders, while a second way of combining the test scores may tend to separate the group with learning disabilities from the group with emotional disorders. The researcher attempts to understand the "message" in the two ways of combining test scores to separate groups differently. What is the meaning of the combination of scores that separates nondisabled from disabled kids, and what is the meaning of the different combination of scores that separates kids with one kind of disorder from kids with another? This attempt is facilitated by the statistics available in many of the canned computer programs for DISCRIM that are not printed in some programs for MANOVA.

Thus there are two facets of DISCRIM, and one or both may be emphasized in any given research application. The researcher may simply be interested in a decision rule for classifying cases where the number of dimensions and their meaning is irrelevant. Or the emphasis may be on interpreting the results of DISCRIM in terms of the combinations of predictors—called discriminant functions—that separate various groups from each other.

A DISCRIM version of covariate analysis (MANCOVA) is available, because DISCRIM can be set up in a sequential manner. When sequential DISCRIM is used, the covariate is simply a predictor that is given top priority. For example, a researcher might consider the score on the Wechsler Intelligence Scale for Children a covariate and ask how well the Wide Range Achievement Test, the Illinois Test of Psycholinguistic Ability, and Figure Drawings differentiate between nondisabled, learning-disabled, and emotionally disordered children after differences in IQ are accounted for.

If groups are arranged in a factorial design, it is frequently best to rephrase research questions so that they are answered within the framework of MANOVA. (However, DISCRIM can in some circumstances be directly applied to factorial designs as discussed in Section 9.6.6.) Similarly, DISCRIM programs make no provision for within-subjects variables. If a within-subjects analysis is desired, the question is also rephrased in terms of MANOVA or profile analysis. For this reason the emphasis in this chapter is on one-way between-subjects DISCRIM.

Honigsfeld and Dunn (2003) used a two-group discriminant analysis (in addition to other analyses) to examine gender differences in learning styles internationally. The 1,637 7th to 13th grade participants were from five countries. A significant multivariate difference was found between boys and girls, indicating that the linear combination of learning-style elements discriminated between them, with an effect size (R^2) of about .08. Girls were found to score higher on a discriminant function comprised of responsibility, self-motivation, teacher motivation, persistence, learning in several ways, and parent motivation.

GAD (generalized anxiety disorder) in older adults was investigated by Diefenbach et al. (2003) in a three-group stepwise discriminant analysis (diagnosable GAD, subsyndromal minor GAD, and normal volunteers). The original set of predictors consisted of four self-report measures: the Penn State Worry Questionnaire (PSWQ), State-Trait Anxiety Inventory-Trait Scale (STAI-Trait), Beck Depression Inventory (BDI), and Quality of Life Inventory (QOLI). The final set of two predictors,

PSWQ and STAI-Trait, formed a single discriminant function that significantly discriminated among the three groups, accounting for 99% of the variance. That is, the three groups varied along a single dimension, with the GAD group centroid higher than the subsyndromal GAD centroid, which in turn was higher than the normal centroid. The solution classified 73% of the cases into their correct groups (replicated with cross-validation). However, there was relatively poor classification of the subsyndromal GAD group, with 38% of them classified into the GAD group. That is, subsyndromal GAD participants were more similar to GAD participants than normals on worry/anxiety symptoms.

9.2   Kinds of Research Questions

The primary goals of DISCRIM are to find the dimension or dimensions along which groups differ, and to find classification functions to predict group membership. The degree to which these goals are met depends, of course, on choice of predictors. Typically, the choice is made either on the basis of theory about which variables should provide information about group membership, or on the basis of pragmatic considerations such as expense, convenience, or unobtrusiveness.

It should be emphasized that the same data are profitably analyzed through either MANOVA or DISCRIM programs, and frequently through both, depending on the kinds of questions you want to ask. If group sizes are very unequal, and/or distributional assumptions are untenable, logistic regression also answers most of the same questions. In any event, statistical procedures are readily available within canned computer programs for answering the following types of questions generally associated with DISCRIM.

9.2.1   Significance of Prediction

Can group membership be predicted reliably from the set of predictors? For example, can we do better than chance in predicting whether children are learning-disabled, emotionally disordered, or nondisabled on the basis of the set of psychological test scores? This is the major question of DISCRIM that the statistical procedures described in Section 9.6.1 are designed to answer. The question is identical to the question about "main effects of IVs" for a one-way MANOVA.

9.2.2   Number of Significant Discriminant Functions

Along how many dimensions do groups differ reliably? For the three groups of children in our example, two discriminant functions are possible, and neither, one, or both may be statistically significant. For example, the first function may separate the nondisabled group from the other two while the second, which would separate the group with learning disability from the group with emotional disorders, is not statistically significant. This pattern of results indicates the predictors can differentiate nondisabled from disabled kids, but cannot separate learning-disabled kids from kids with emotional disorders.

In DISCRIM, the first discriminant function provides the best separation among groups. Then a second discriminant function, orthogonal to the first, is found that best separates groups on the basis of associations not used in the first discriminant function. This procedure of finding successive orthogonal discriminant functions continues until all possible dimensions are evaluated. The number of possible dimensions is either one fewer than the number of groups or equal to the number of predictor variables, whichever is smaller. Typically, only the first one or two discriminant functions reliably discriminate among groups; remaining functions provide no additional information about group membership and are better ignored. Tests of significance for discriminant functions are discussed in Section 9.6.2.

9.2.3 Dimensions of Discrimination

How can the dimensions along which groups are separated be interpreted? Where are groups located along the discriminant functions, and how do predictors correlate with the discriminant functions? In our example, if two significant discriminant functions are found, which predictors correlate highly with each function? What pattern of test scores discriminates between nondisabled children and the other two groups (first discriminant function)? And what pattern of scores discriminates between children with learning disabilities and children with emotional disorders (second discriminant function)? These questions are discussed in Section 9.6.3.

9.2.4 Classification Functions

What linear equation(s) can be used to classify new cases into groups? For example, suppose we have the battery of psychological test scores for a group of new, undiagnosed children. How can we combine (weight) their scores to achieve the most reliable diagnosis? Procedures for deriving and using classification functions are discussed in Sections 9.4.2 and 9.6.7.[3]

9.2.5 Adequacy of Classification

Given classification functions, what proportion of cases is correctly classified? When errors occur, how are cases misclassified? For instance, what proportion of learning-disabled children is correctly classified as learning-disabled, and, among those who are incorrectly classified, are they more often put into the group of nondisabled children or into the group of emotionally disordered children?

Classification functions are used to predict group membership for new cases and to check the adequacy of classification for cases in the same sample through cross-validation. If the researcher knows that some groups are more likely to occur, or if some kinds of misclassification are especially undesirable, the classification procedure can be modified. Procedures for deriving classification functions and modifying them are discussed in Section 9.4.2; procedures for testing them are discussed in Section 9.6.7.

9.2.6 Effect Size

What is the degree of relationship between group membership and the set of predictors? If the first discriminant function separates the nondisabled group from the other two groups, how much does the variance for groups overlap the variance in combined test scores? If the second discriminant

---

[3]Discriminant analysis provides classification of cases into groups where group membership is known, at least for the sample from whom the classification equations are derived. Cluster analysis is a similar procedure except that group membership is not known. Instead, the analysis develops groups on the basis of similarities among cases.

function separates learning-disabled from emotionally disordered children, how much does the variance for these groups overlap the combined test scores for this discriminant function? This is basically a question of percent of variance accounted for and, as seen in Section 9.4.1, is answered through canonical correlation (Chapter 12). A canonical correlation is a multiple-multiple correlation because there are multiple variables on both sides of the regression equation. A multiple correlation has multiple predictor variables (IVs) and a single criterion (DV). A canonical correlation has multiple criteria as well—the df for the groups provide the multiple criteria in discriminant analysis. A canonical correlation is found for each discriminant function that, when squared, indicates the proportion of variance shared between groups and predictors on that function. Confidence limits can be found for these effect size measures. An effect size and its confidence limits also are available for the overall discriminant analysis, identical to that available for omnibus MANOVA. Finally, an effect size and associated confidence interval may be found for group contrasts. Section 9.6.5 discusses all of these measures of effect size.

9.2.7   Importance of Predictor Variables

Which predictors are most important in predicting group membership? Which test scores are helpful for separating nondisabled children from children with disorders, and which are helpful for separating learning-disabled from emotionally disordered children?

Questions about importance of predictors are analogous to those of importance of DVs in MANOVA, to those of IVs in multiple regression, and to those of IVs and DVs in canonical correlation. One procedure in DISCRIM is to interpret the correlations between the predictors and the discriminant functions, as discussed in Section 9.6.3.2. A second procedure is to evaluate predictors by how well they separate each group from all the others, as discussed in Section 9.6.4. (Or importance can be evaluated as in MANOVA, Section 7.5.3.)

9.2.8   Significance of Prediction with Covariates

After statistically removing the effects of one or more covariates, can one reliably predict group membership from a set of predictors? In DISCRIM, as in MANOVA, the ability of some predictors to promote group separation can be assessed after adjustment for prior variables. If scores on the Wechsler Intelligence Scale for Children (WISC) are considered the covariate and given first entry in DISCRIM, do scores on the Illinois Test of Psycholinguistic Ability (ITPA), the Wide Range Achievement Test, and Figure Drawings contribute to prediction of group membership when they are added to the equation?

Rephrased in terms of sequential discriminant analysis, the question becomes, Do scores on the ITPA, the Wide Range Achievement Test, and Figure Drawings provide significantly better classification among the three groups than that afforded by scores on the WISC alone? Sequential DISCRIM is discussed in Section 9.5.2. Tests for contribution of added predictors are given in Section 9.6.7.3.

9.2.9   Estimation of Group Means

If predictors discriminate among groups, it is important to report just how the groups differ on those variables. The best estimate of central tendency in a population is the sample mean. If, for example, the ITPA discriminates between groups with learning disabilities and emotional disorders, it is worthwhile to compare and report the mean ITPA score for learning-disabled children and the mean ITPA score for emotionally disordered children.

9.3 Limitations to Discriminant Analysis

9.3.1 Theoretical Issues

Because DISCRIM is typically used to predict membership in naturally occurring groups rather than groups formed by random assignment, questions such as why we can reliably predict group membership, or what causes differential membership are often not asked. If, however, group membership has occurred by random assignment, inferences of causality are justifiable as long as proper experimental controls have been instituted. The DISCRIM question then becomes: Does treatment following random assignment to groups produce enough difference in the predictors that we can now reliably separate groups on the basis of those variables?

As implied, limitations to DISCRIM are the same as limitations to MANOVA. The usual difficulties of generalizability apply to DISCRIM. But the cross-validation procedure described in Section 9.6.7.1 gives some indication of the generalizability of a solution.

9.3.2 Practical Issues

Practical issues for DISCRIM are basically the same as for MANOVA. Therefore, they are discussed here only to the extent of identifying the similarities between MANOVA and DISCRIM and identifying the situations in which assumptions for MANOVA and DISCRIM differ.

Classification makes fewer statistical demands than does inference. If classification is the primary goal, then most of the following requirements (except for outliers and homogeneity of variance-covariance matrices) are relaxed. If, for example, you achieve 95% accuracy in classification, you hardly worry about the shape of distributions. Nevertheless, DISCRIM is optimal under the same conditions where MANOVA is optimal; and, if the classification rate is unsatisfactory, it may be because of violation of assumptions or limitations. And, of course, deviation from assumptions may distort tests of statistical significance just as in MANOVA.

9.3.2.1 Unequal Sample Sizes, Missing Data, and Power

As DISCRIM is typically a one-way analysis, *no special problems are posed by unequal sample sizes in groups.*[4] In classification, however, a decision is required as to whether you want the a priori probabilities of assignment to groups to be influenced by sample size. That is, do you want the probability with which a case is assigned to a group to reflect the fact that the group itself is more (or less) probable in the sample? Section 9.4.2 discusses this issue, and use of unequal a priori probabilities is demonstrated in Section 9.7. Regarding missing data (absence of scores on predictors for some cases), consult Section 6.3.2.1 and Chapter 4 for a review of problems and potential solutions.

As discussed in Section 7.3.2.1, *the sample size of the smallest group should exceed the number of predictor variables.* Although sequential and stepwise DISCRIM avoid the problems of

---

[4]Actually a problem does occur if rotation is desired because discriminant functions may be nonorthogonal with unequal n (cf. Chapter 13), but rotation of axes is uncommon in discriminant analysis.

multicollinearity and singularity by a tolerance test at each step, overfitting (producing results so close to the sample they don't generalize to other samples) occurs with all forms of DISCRIM if the number of cases does not notably exceed the number of predictors in the smallest group.[5]

Issues of power, also, are the same as for MANOVA if tests of statistical significance are to be applied. Section 7.3.2.1 discusses these issues and methods for determining sample size to obtain desired power.

9.3.2.2   Multivariate Normality

When using statistical inference in DISCRIM, the assumption of multivariate normality is that scores on predictors are independently and randomly sampled from a population, and that the sampling distribution of any linear combination of predictors is normally distributed. No tests are currently feasible for testing the normality of all linear combinations of sampling distributions of means of predictors.

However, DISCRIM, like MANOVA, is robust to failures of normality if violation is caused by skewness rather than outliers. Recall that *a sample size that would produce 20 df for error in the univariate ANOVA case should ensure robustness with respect to multivariate normality, as long as sample sizes are equal and two-tailed tests are used.* (Calculation of df for error in the univariate case is discussed in Section 3.2.1.)

Because tests for DISCRIM typically are two-tailed, this requirement poses no difficulty. Sample sizes, however, are often not equal for applications of DISCRIM because naturally occurring groups rarely occur or are sampled with equal numbers of cases in groups. As differences in sample size among groups increase, larger overall sample sizes are necessary to assure robustness. As a conservative recommendation, robustness is expected with 20 cases in the smallest group if there are only a few predictors (say, five or fewer).

If samples are both small and unequal in size, assessment of normality is a matter of judgment. Are predictors expected to have normal sampling distributions in the population being sampled? If not, the transformation of one or more predictors (cf. Chapter 4) may be worthwhile.

9.3.2.3   Absence of Outliers

DISCRIM, like MANOVA, is highly sensitive to inclusion of outliers. Therefore, *run a test for univariate and multivariate outliers for each group separately, and transform or eliminate significant outliers before DISCRIM* (see Chapter 4).

9.3.2.4   Homogeneity of Variance-Covariance Matrices

In inference, when sample sizes are equal or large, DISCRIM, like MANOVA (Section 7.3.2.4), is robust to violation of the assumption of equality of within-group variance-covariance (dispersion) matrices. However, when sample sizes are unequal and small, results of significance testing may be misleading if there is heterogeneity of the variance-covariance matrices.

Although inference is usually robust with respect to heterogeneity of variance-covariance matrices with decently sized samples, classification is not. Cases tend to be overclassified into

---

[5]Also, highly unequal sample sizes are better handled by logistic regression (Chapter 10) than by discriminant analysis.

groups with greater dispersion. If classification is an important goal of analysis, test for homogeneity of variance-covariance matrices.

Homogeneity of variance-covariance matrices is assessed through procedures of Section 7.3.2.4 or by *inspection of scatterplots of scores on the first two discriminant functions produced separately for each group.* These scatterplots are available through SPSS DISCRIMINANT. *Rough equality in overall size of the scatterplots is evidence of homogeneity of variance-covariance matrices.* Anderson's test, available in SAS DISCRIM (`pool=test`) assesses homogeneity of variance-covariance matrices, but is also sensitive to nonnormality. This test is demonstrated in Section 9.7.1.5. Another overly sensitive test, Box's *M*, is available in SPSS MANOVA and DISCRIMINANT.

If heterogeneity is found, one can transform predictors, use separate covariance matrices during classification, use quadratic discriminant analysis (shown in Section 9.7), or use nonparametric classification. Transformation of predictors follows procedures of Chapter 4. Classification on the basis of separate covariance matrices, the second remedy, is available through SPSS DISCRIMINANT and SAS DISCRIM. Because this procedure often leads to overfitting, it should be used only when the sample is large enough to permit cross-validation (Section 9.6.7.1). Quadratic discrimination analysis, the third remedy, is available in SAS DISCRIM (cf. Section 9.7). This procedure avoids overclassification into groups with greater dispersion, but performs poorly with small samples (Norušis, 1990).

SAS DISCRIM uses separate matrices and computes quadratic discriminant functions with the instruction `pool=no`. With the instruction `pool=test`, SAS DISCRIM uses the pooled variance-covariance matrix only if heterogeneity of variance-covariance matrices is not significant (Section 9.7.2). With small samples, nonnormal predictors, and heterogeneity of variance-covariance matrices, SAS DISCRIM offers a fourth remedy—nonparametric classification methods—which avoid overclassification into groups with greater dispersion and are robust to nonnormality.

Therefore, *transform variables if there is significant departure from homogeneity, samples are small and unequal, and inference is the major goal. If the emphasis is on classification and dispersions are unequal, use (1) separate covariance matrices and/or quadratic discriminant analysis if samples are large and variables are normal and (2) nonparametric classification methods if variables are nonnormal and/or samples are small.*

9.3.2.5 *Linearity*

The DISCRIM model assumes linear relationships among all pairs of predictors within each group. The assumption is less serious (from some points of view) than others, however, in that violation leads to reduced power rather than increased Type I error. The procedures in Section 6.3.2.6 may be applied to test for and improve linearity and to increase power.

9.3.2.6 *Absence of Multicollinearity and Singularity*

Multicollinearity or singularity may occur with highly redundant predictors, making matrix inversion unreliable. Fortunately, most computer programs for DISCRIM protect against this possibility by testing tolerance. Predictors with insufficient tolerance are excluded.

Guidelines for assessing multicollinearity and singularity for programs that do not include tolerance tests, and for dealing with multicollinearity or singularity when it occurs, are in Section 7.3.2.8. Note that analysis is done on predictors, not "DVs" in DISCRIM.

9.4   Fundamental Equations for Discriminant Analysis

Hypothetical scores on four predictors are given for three groups of learning-disabled children for demonstration of DISCRIM. Scores for three cases in each of the three groups are shown in Table 9.1.

The three groups are MEMORY (children whose major difficulty seems to be with tasks related to memory), PERCEPTION (children who show difficulty in visual perception), and COMMUNI-CATION (children with language difficulty). The four predictors are PERF (Performance Scale IQ of the WISC), INFO (Information subtest of the WISC), VERBEXP (Verbal Expression subtest of the ITPA), and AGE (chronological age in years). The grouping variable, then, is type of learning disability, and the predictors are selected scores from psychodiagnostic instruments and age.

Fundamental equations are presented for two major parts of DISCRIM: discriminant functions and classification equations. Syntax and selected output for this example appear in Section 9.4.3 for SPSS DISCRIMINANT and SAS DISCRIM.

9.4.1   Derivation and Test of Discriminant Functions

The fundamental equations for testing the significance of a set of discriminant functions are the same as for MANOVA, discussed in Chapter 7. Variance in the set of predictors is partitioned into two sources: variance attributable to differences between groups and variance attributable to differences within groups. Through procedures shown in Equations 7.1 to 7.3, cross-products matrices are formed.

$$\mathbf{S}_{\text{total}} = \mathbf{S}_{bg} + \mathbf{S}_{wg} \tag{9.1}$$

The total cross-products matrix ($\mathbf{S}_{\text{total}}$) is partitioned into a cross-products matrix associated with differences between groups ($\mathbf{S}_{bg}$) and a cross-products matrix of differences within groups ($\mathbf{S}_{wg}$).

TABLE 9.1   Hypothetical Small Data Set for Illustration of Discriminant Analysis

	Predictors			
Group	PERF	INFO	VERBEXP	AGE
	87	5	31	6.4
MEMORY	97	7	36	8.3
	112	9	42	7.2
	102	16	45	7.0
PERCEPTION	85	10	38	7.6
	76	9	32	6.2
	120	12	30	8.4
COMMUNICATION	85	8	28	6.3
	99	9	27	8.2

For the example in Table 9.1, the resulting cross-products matrices are

$$\mathbf{S}_{bg} = \begin{bmatrix} 314.89 & -71.56 & -180.00 & 14.49 \\ -71.56 & 32.89 & 8.00 & -2.22 \\ -180.00 & 8.00 & 168.00 & -10.40 \\ 14.49 & -2.22 & -10.40 & 0.74 \end{bmatrix}$$

$$\mathbf{S}_{wg} = \begin{bmatrix} 1186.00 & 220.00 & 348.33 & 50.00 \\ 220.00 & 45.33 & 73.67 & 6.37 \\ 348.33 & 73.67 & 150.00 & 9.73 \\ 50.00 & 6.37 & 9.73 & 5.49 \end{bmatrix}$$

Determinants[6] for these matrices are

$$|\mathbf{S}_{wg}| = 4.70034789 \times 10^{13}$$

$$|\mathbf{S}_{bg} + \mathbf{S}_{wg}| = 448.63489 \times 10^{13}$$

Following procedures in Equation 7.4, Wilks' lambda[7] for these matrices is

$$\Lambda = \frac{|\mathbf{S}_{wg}|}{|\mathbf{S}_{bg} + \mathbf{S}_{wg}|} = .010477$$

To find the approximate F ratio, as per Equation 7.5, the following values are used:

$p = 4$ the number of predictor variables

$df_{bg} = 2$ the number of groups minus one, or $k - 1$

$df_{wg} = 6$ the number of groups times the quantity $n - 1$, where n is the number of cases per group. Because n is often not equal for all groups in DISCRIM, an alternative equation for df_{wg} is $N - k$, where N is the total number of cases in all groups—9 in this case.

Thus we obtain

$$s = \min (p, k - 1) = \min (4, 2) = 2$$

$$y = (.010477)^{1/2} = .102357$$

$$df_2 = (2) \left[6 - \frac{4 - 2 + 1}{2} \right] - \left[\frac{4(2) - 2}{2} \right] = 6$$

$$df_1 = 4(2) = 8$$

$$\text{Approximate } F(8, 6) = \left(\frac{1 - .102357}{.102357} \right) \left(\frac{6}{8} \right) = 6.58$$

---

[6]A determinant, as described in Appendix A, can be viewed as a measure of generalized variance of a matrix.

[7]Alternative statistical criteria are discussed in Section 9.6.1.1. Note that bg and wg are used in place of effect and error, respectively, in these equations.

Critical F with 8 and 6 df at $\alpha = 0.05$ is 4.15. Because obtained F exceeds critical F, we conclude that the three groups of children can be distinguished on the basis of the combination of the four predictors.

This is a test of overall relationship between groups and predictors. It is the same as the overall test of a main effect in MANOVA. In MANOVA, this result is followed by an assessment of the importance of the various DVs to the main effect. In DISCRIM, however, when an overall relationship is found between groups and predictors, the next step is to examine the discriminant functions that compose the overall relationship.

The maximum number of discriminant functions is either (1) the number of predictors or (2) the degrees of freedom for groups, whichever is smaller. Because there are three groups (and four predictors) in this example, there are potentially two discriminant functions contributing to the overall relationship. And, because the overall relationship is statistically significant, at least the first discriminant function is very likely to be significant, and both may be significant.

Discriminant functions are like regression equations; a discriminant function score for a case is predicted from the sum of the series of predictors, each weighted by a coefficient. There is one set of discriminant function coefficients for the first discriminant function, a second set of coefficients for the second discriminant function, and so forth. Subjects get separate discriminant function scores for each discriminant function when their own scores on predictors are inserted into the equations.

To solve for the (standardized) discriminant function score for the ith function, Equation 9.2 is used.

$$D_i = d_{i1}z_1 + d_{i2}z_2 + \cdots + d_{ip}z_p \tag{9.2}$$

A child's standardized score on the ith discriminant function (D_i) is found by multiplying the standardized score on each predictor (z) by its standardized discriminant function coefficient (d_i) and then adding the products for all predictors.

Discriminant function coefficients are found in the same manner as are coefficients for canonical variates (to be described in Section 12.4.2). In fact, DISCRIM is basically a problem in canonical correlation with group membership on one side of the equation and predictors on the other, where successive canonical variates (here called discriminant functions) are computed. In DISCRIM, d_i are chosen to maximize differences between groups relative to differences within groups.

Just as in multiple regression, Equation 9.2 can be written either for raw scores or for standardized scores. A discriminant function score for a case, then, can also be produced by multiplying the raw score on each predictor by its associated unstandardized discriminant function coefficient, adding the products over all predictors, and adding a constant to adjust for the means. The score produced in this way is the same D_i as produced in Equation 9.2. The mean of each discriminant function over all cases is zero, because the mean of each predictor, when standardized, is zero. The standard deviation of each D_i is 1.

Just as D_i can be calculated for each case, a mean value of D_i can be calculated for each group. The members of each group considered together have a mean score on a discriminant function that is the distance of the group, in standard deviation units, from the zero mean of the discriminant function. Group means on D_i are typically called centroids in reduced space, the space having been reduced from that of the p predictors to a single dimension, or discriminant function.

A canonical correlation is found for each discriminant function. Canonical correlations are found by solving for the eigenvalues and eigenvectors of a correlation matrix, in a process described in Chapters 12 and 13. An eigenvalue is a form of a squared canonical correlation which, as is usual for squared correlation coefficients, represents overlapping variance among variables, in this case, between predictors and groups. Successive discriminant functions are evaluated for significance, as discussed in Section 9.6.2. Also discussed in subsequent sections are structure matrices of loadings and group centroids.

If there are only two groups, discriminant function scores can be used to classify cases into groups. A case is classified into one group if its D_i score is above zero, and into the other group if the D_i score is below zero. With numerous groups, classification is possible from the discriminant functions, but it is simpler to use the procedure in the following section.

9.4.2 Classification

To assign cases into groups, a classification equation is developed for each group. Three classification equations are developed for the example in Table 9.1, where there are three groups. Data for each case are inserted into each classification equation to develop a classification score for each group for the case. The case is assigned to the group for which it has the highest classification score.

In its simplest form, the basic classification equation for the jth group ($j = 1, 2, ..., k$) is

$$C_j = c_{j0} + c_{j1}X_1 + c_{j2}X_2 + \cdots + c_{jp}X_p \tag{9.3}$$

A score on the classification function for group j (C_j) is found by multiplying the raw score on each predictor (X) by its associated classification function coefficient (c_j), summing over all predictors, and adding a constant c_{j0}.

Classification coefficients, c_j, are found from the means of the p predictors and the pooled within-group variance-covariance matrix, $\mathbf{W}$. The within-group covariance matrix is produced by dividing each element in the cross-products matrix, $\mathbf{S}_{wg}$, by the within-group degrees of freedom, $N - k$. In matrix form,

$$\mathbf{C}_j = \mathbf{W}^{-1}\mathbf{M}_j \tag{9.4}$$

The column matrix of classification coefficients for group j ($\mathbf{C}_j = c_{j1}, c_{j2}, ..., c_{jp}$) is found by multiplying the inverse of the within-group variance-covariance matrix ($\mathbf{W}^{-1}$) by a column matrix of means for group j on the p variables ($\mathbf{M}_j = X_{j1}, X_{j2}, ..., X_{jp}$).

The constant for group j, c_{j0}, is found as follows:

$$c_{j0} = \left(-\frac{1}{2}\right)\mathbf{C}'_j\mathbf{M}_j \tag{9.5}$$

The constant for the classification function for group j (c_{j0}) is formed by multiplying $-1/2$ times the transpose of the column matrix of classification coefficients for group j ($\mathbf{C}'_j$) times the column matrix of means for group j ($\mathbf{M}_j$).

For the sample data, each element in the $\mathbf{S}_{wg}$ matrix from Section 9.4.1 is divided by $df_{wg} = df_{error} = 6$ to produce the within-group variance-covariance matrix:

$$\mathbf{W}_{bg} = \begin{bmatrix} 214.33 & 36.67 & 58.06 & 8.33 \\ 36.67 & 7.56 & 12.28 & 1.06 \\ 58.06 & 12.28 & 25.00 & 1.62 \\ 8.33 & 1.06 & 1.62 & 0.92 \end{bmatrix}$$

The inverse of the within-group variance-covariance matrix is

$$\mathbf{W}^{-1} = \begin{bmatrix} 0.04362 & -0.20195 & 0.00956 & -0.17990 \\ -0.21095 & 1.62970 & -0.37037 & 0.60623 \\ 0.00956 & -0.37037 & 0.20071 & -0.01299 \\ -0.17990 & 0.60623 & -0.01299 & 2.05006 \end{bmatrix}$$

Multiplying $\mathbf{W}^{-1}$ by the column matrix of means for the first group gives the matrix of classification coefficients for that group, as per Equation 9.4.

$$\mathbf{C}_1 = \mathbf{W}^{-1} \begin{bmatrix} 98.67 \\ 7.00 \\ 36.33 \\ 7.30 \end{bmatrix} = \begin{bmatrix} 1.92 \\ -17.56 \\ 5.55 \\ 0.99 \end{bmatrix}$$

The constant for group 1, then, according to Equation 9.5, is

$$c_{1,0} = (-1/2)[1.92, \quad -17.56, \quad 5.55, \quad 0.99] \begin{bmatrix} 98.67 \\ 7.00 \\ 36.33 \\ 7.30 \end{bmatrix} = -137.83$$

(Values used in these calculations were carried to several decimal places before rounding.) When these procedures are repeated for groups 2 and 3, the full set of classification equations is produced, as shown in Table 9.2.

TABLE 9.2   Classification Function Coefficients for Sample
Data of Table 9.1

	Group 1: MEMORY	Group 2: PERCEP	Group 3: COMMUN
PERF	1.92420	0.58704	1.36552
INFO	-17.56221	-8.69921	-10.58700
VERBEXP	5.54585	4.11679	2.97278
AGE	0.98723	5.01749	2.91135
(CONSTANT)	-137.82892	-71.28563	-71.24188

In its simplest form, classification proceeds as follows for the first case in group 1. Three classification scores, one for each group, are calculated for the case by applying Equation 9.3:

$$C_1 = -137.83 + (1.92)(87) + (-17.56)(5) + (5.55)(31) + (0.99)(6.4) = 119.80$$

$$C_2 = -71.29 + (0.59)(87) + (-8.70)(5) + (4.12)(31) + (5.02)(6.4) = 96.39$$

$$C_3 = -71.24 + (1.37)(87) + (-10.59)(5) + (2.97)(31) + (2.91)(6.4) = 105.69$$

Because this child has the highest classification score in group 1, the child is assigned to group 1, a correct classification in this case.

This simple classification scheme is most appropriate when equal group sizes are expected in the population. If unequal group sizes are expected, the classification procedure can be modified by setting a priori probabilities to group size. The classification equation for group j (C_j) then becomes

$$C_j = c_{j0} + \sum_{i=1}^{p} c_{ji} X_i + \ln(n_j/N) \tag{9.6}$$

where n_j = size of group j and N = total sample size.

It should be reemphasized that the classification procedures are highly sensitive to heterogeneity of variance-covariance matrices. Cases are more likely to be classified into the group with the greatest dispersion—that is, into the group for which the determinant of the within-group covariance matrix is greatest. Section 9.3.2.4 provides suggestions for dealing with this problem.

Uses of classification procedures are discussed more fully in Section 9.6.7.

9.4.3   Computer Analyses of Small-Sample Example

Syntax and selected output for computer analyses of the data in Table 9.1, using the simplest methods, are in Tables 9.3 and 9.4 for SPSS DISCRIMINANT and SAS DISCRIM, respectively.

SPSS DISCRIMINANT (Table 9.3) assigns equal prior probability for each group by default. The TABLE instruction requests a classification table. The output summarizing the Canonical Discriminant Functions appears in two tables. The first shows Eigenvalue, % of Variance, and Cumulative % of variance accounted for by each function, and Canonical Correlation for each discriminant function. Squared canonical correlations, the effect sizes for the discriminant functions, are

**TABLE 9.3   Syntax and Selected SPSS DISCRIMINANT Output
for Discriminant Analysis of Sample Data in Table 9.1**

```
DISCRIMINANT
  /GROUPS=GROUP(1 3)
  /VARIABLES=PERF INFO VERBEXP AGE
  /ANALYSIS ALL
  /PRIORS EQUAL
  /STATISTICS = TABLE
  /CLASSIFY=NONMISSING POOLED.
```

(continued)

TABLE 9.3   Continued

Summary of Canonical Discriminant Functions

Eigenvalues

Function	Eigenvalue	% of Variance	Cumulative %	Canonical Correlation
1	13.486[a]	70.7	70.7	.965
2	5.589[a]	29.3	100.0	.921

[a]First 2 canonical discriminant functions were used in the analysis.

Wilks' Lambda

Test of Function(s)	Wilks' Lambda	Chi-square	df	Sig.
1 through 2	.010	20.514	8	.009
2	.152	8.484	3	.037

Standardized Canonical Discriminant Function Coefficients

	Function	
	1	2
Performance IQ	−2.504	−1.474
Information	3.490	−.284
Verbal expression	−1.325	1.789
AGE	.503	.236

Structure Matrix

	Function	
	1	2
Information	.228*	.066
Verbal expression	−.022	.446*
Performance IQ	−.075	−.173*
AGE	−.028	−.149*

Pooled within-groups correlations between discriminating variables and standardized canonical discriminant functions
Variables ordered by absolute size of correlation within function

* Largest absolute correlation between each variable and any discriminant function

TABLE 9.3   Continued

Functions at Group Centroids

	Function	
Group	1	2
Memory	−4.102	.691
Perception	2.981	1.942
Communication	1.122	−2.633

Unstandardized canonical discriminant functions
evaluated at group means

Classification Statistics

Classification Results[a]

			Predicted Group Membership			
		Group	Memory	Perception	Communication	Total
Original	Count	Memory	3	0	0	3
		Perception	0	3	0	3
		Communication	0	0	3	3
	%	Memory	100.0	.0	.0	100.0
		Perception	.0	100.0	.0	100.0
		Communication	.0	.0	100.0	100.0

[a]100.0% of original grouped cases correctly classified.

$(.965)^2 = .93$ and $(.921)^2 = .85$, respectively. The first discriminant function accounts for 70.7% of the between-group (explained) variance in the solution while the second accounts for the remaining between-group variance. The Wilks' Lambda table shows the "peel off" significance tests of successive discriminant functions. For the combination of both discrimination functions, 1 through 2—all functions tested together, Chi square = 20.514. After the first function is removed, the test of function 2 shows that Chi square = 8.484 is still statistically significant at $\alpha = .05$ because Sig. = .0370. This means that the second discriminant function is significant as well as the first. If not, the second would not have been marked as one of the discriminated functions remaining in the analysis.

    Standardized Canonical Discriminant Function Coefficients (Equation 9.2) are given for deriving discriminant function scores from standardized predictors. Correlations (loadings) between predictors and discriminant functions are given in the Structure Matrix. These are ordered so that predictors loading on the first discriminant function are listed first, and those loading on the second discriminant function second. Then Functions at Group Centroids are shown, indicating the average discriminant score (aka centroid or multivariate mean) for each group on each function.

    In the Classification Results table, produced by the TABLE instruction in the STATISTICS paragraph, rows represent actual group membership and columns represent predicted group membership. Within each cell, the number and percent of cases correctly classified are shown. For this example, all of the diagonal cells show perfect classification (100.0%).

Syntax for SAS DISCRIM (Table 9.4) requests a `manova` table of the usual multivariate output. The request for `can` provides canonical correlations, loading matrices, and much more.

SAS DISCRIM output (Table 9.4) begins with a summary of input and degrees of freedom, followed by a summary of `GROUP`s, their `Frequency` (number of cases), `Weight` (sample sizes in this case), `Proportion` of cases in each group, and `Prior Probability` (set equal by default). This is followed by the multivariate results as per SAS GLM, produced by requesting `manova`. Per the request for `can`, information about canonical correlation and eigenvalues for each of the discriminant functions follows, with information matching that of SPSS. SAS, however, explicitly includes squared canonical correlations. Significance tests of successive discriminant functions are of the "peel off" variety, as per SPSS. SAS DISCRIM uses F tests rather than the χ^2 tests used by SPSS. Also, the column labeled `Likelihood Ratio` is the Wilks' Lambda of SPSS. Results are consistent with those of SPSS, however. The structure matrix then appears in a table labeled `Pooled within Canonical Structure`, also produced by requesting `can` (a great deal of additional output produced by that request is omitted here). The `Class Means on Canonical Variables` are the group centroids of SPSS.

Equations for classification functions and classification coefficients (Equation 9.6) are given in the following matrix, labeled `Linear Discriminant Function for GROUP`. Finally, results of classification are presented in the table labeled `Number of Observations and Percent Classified into GROUP` where, as usual, rows represent actual group and columns represent predicted group. Cell values show number of cases classified and percentage correct. Number of erroneous classifications for each group are presented, and prior probabilities are repeated at the bottom of this table.

TABLE 9.4   Syntax and Selected SAS DISCRIM Output for Discriminant Analysis of Small-Sample Data of Table 9.1

```
proc discrim data=SASUSER.SS_DISC manova can;
      class GROUP;
      var PERF INFO VERBEXP AGE;
run;
```

                    The DISCRIM Procedure

            Observations   9   DF Total   8
            Variables   4   DF Within Classes   6
            Classes   3   DF Between Classes   2

                    Class Level Information

              Variable   Prior
    GROUP   Name   Frequency   Weight   Proportion   Probability

        1   _1   3   3.0000   0.333333   0.333333
        2   _2   3   3.0000   0.333333   0.333333
        3   _3   3   3.0000   0.333333   0.333333

TABLE 9.4 Continued

Multivariate Statistics and F Approximations

S=2   M=0.5   N=0.5

Statistic	Value	F Value	Num DF	Den DF	Pr > F
Wilks' Lambda	0.01047659	6.58	8	6	0.0169
Pillai's Trace	1.77920446	8.06	8	8	0.0039
Hotelling-Lawley Trace	19.07512732	8.18	8	2.8235	0.0627
Roy's Greatest Root	13.48590176	13.49	4	4	0.0136

NOTE: F Statistic for Roy's Greatest Root is an upper bound.
NOTE: F Statistic for Wilks' Lambda is exact.

Canonical Discriminant Analysis

	Canonical Correlation	Adjusted Canonical Correlation	Approximate Standard Error	Squared Canonical Correlation
1	0.964867	0.944813	0.024407	0.930967
2	0.920998	.	0.053656	0.848237

Test of H0: The canonical correlations in
the current row and all that follow are zero

Eigenvalues of Inv(E)*H
= CanRsq/(1-CanRsq)

	Eigenvalue	Difference	Proportion	Cumulative	Likelihood Ratio	Approximate F Value	Num DF	Den DF	PR > F
1	13.4859	7.8967	0.7070	0.7070	0.01047659	6.58	8	6	0.0169
2	5.5892		0.2930	1.0000	0.15176290	7.45	3	4	0.0409

(continued)

393

TABLE 9.4 Continued

Pooled Within Canonical Structure

Variable	Can1	Can2
PERF	-0.075459	-0.173408
INFO	0.227965	0.066418
VERBEXP	-0.022334	0.446298
AGE	-0.027861	-0.148606

Class Means on Canonical Variables

GROUP	Can1	Can2
1	-4.102343810	0.690967850
2	2.980678873	1.941686409
3	1.121664938	-2.632654259

Linear Discriminant Function

$$\text{Constant} = -.5\ \bar{X}_j'\ \text{COV}^{-1}\ \bar{X}_j \qquad \text{Coefficient Vector} = \text{COV}^{-1}\ \bar{X}_j$$

Linear Discriminant Function for GROUP

Variable	1	2	3
Constant	-137.81247	-71.28575	-71.24170
PERF	1.92420	0.58704	1.36552
INFO	-17.56221	-8.69921	-10.58700
VERBEXP	5.54585	4.11679	2.97278
AGE	0.98723	5.01749	2.91135

Classification Summary for Calibration Data: SASUSER.SS_DISC
Resubstitution Summary using Linear Discriminant Function

Generalized Squared Distance Function

$$D_j^2(\bar{X}) = (X-\bar{X}_j)'\ \text{COV}^{-1}\ (X-\bar{X}_j)$$

Posterior Probability of Membership in Each GROUP

$$Pr(j|X) = \exp(-.5\ D_j^2(X))\ /\ \text{SUM}_k\ \exp(-.5\ D_k^2(X))$$

TABLE 9.4   Continued

```
    Number of Observations and Percent Classified into GROUP

    From GROUP            1            2            3        Total

        1                 3            0            0            3
                     100.00         0.00         0.00       100.00

        2                 0            3            0            3
                       0.00       100.00         0.00       100.00

        3                 0            0            3            3
                       0.00         0.00       100.00       100.00

    Total                 3            3            3            9
                      33.33        33.33        33.33       100.00

    Priors          0.33333      0.33333      0.33333


            Error Count Estimates for GROUP

                        1            2            3        Total

    Rate           0.0000       0.0000       0.0000       0.0000
    Priors         0.3333       0.3333       0.3333
```

9.5   Types of Discriminant Analyses

The three types of discriminant analyses—standard (direct), sequential, and statistical (stepwise)—are analogous to the three types of multiple regressions discussed in Section 5.5. Criteria for choosing among the three strategies are the same as those discussed in Section 5.5.4 for multiple regression.

9.5.1   Direct Discriminant Analysis

In standard (direct) DISCRIM, like standard multiple regression, all predictors enter the equations at once and each predictor is assigned only the unique association it has with groups. Variance shared among predictors contributes to the total relationship, but not to any one predictor.

The overall test of relationship between predictors and groups in direct DISCRIM is the same as the test of main effect in MANOVA where all discriminant functions are combined and DVs are considered simultaneously. Direct DISCRIM is the model demonstrated in Section 9.4.1. All the computer programs described later in Table 9.18 perform direct DISCRIM; the use of some of them for that purpose is shown in Tables 9.3 and 9.4.

9.5.2   Sequential Discriminant Analysis

Sequential (or, as some prefer to call it, hierarchical) DISCRIM is used to evaluate contributions to prediction of group membership by predictors as they enter the equations in an order determined by the researcher. The researcher assesses improvement in classification when a new predictor is added to a set of prior predictors. Does classification of cases into groups improve reliably when the new predictor or predictors are added (cf. Section 9.6.7.3)?

If predictors with early entry are viewed as covariates and an added predictor is viewed as a DV, DISCRIM is used for analysis of covariance. Indeed, sequential DISCRIM can be used to perform stepdown analysis following MANOVA (cf. Section 7.5.3.2) because stepdown analysis is a sequence of ANCOVAs.

Sequential DISCRIM also is useful when a reduced set of predictors is desired and there is some basis for establishing a priority order among them. If, for example, some predictors are easy or inexpensive to obtain and they are given early entry, a useful, cost-effective set of predictors may be found through the sequential procedure.

Neither SPSS DISCRIMINANT nor SAS DISCRIM[8] provides convenient methods for entering predictors in priority order. Instead, the sequence is set up by running a separate discriminant analysis for each step, the first with the highest priority variable, the second with the two highest priority variables entering simultaneously, and so on. One or more variables may be added at each step. However, the test for the significance of improvement in prediction is tedious in the absence of very large samples (Section 9.6.7.3). If you have only two groups and sample sizes are approximately equal, you might consider performing sequential discriminant analysis through SPSS REGRESSION or interactive SAS REGRESS where the DV is a dichotomous variable representing group membership, with groups coded 0 and 1. If classification is desired, preliminary multiple regression analysis with fully flexible entry of predictors could be followed by discriminant analysis to provide classification.

If you have more than two groups or your group sizes are very unequal, sequential logistic regression is the procedure of choice, and, as seen in Chapter 10, most programs classify cases.

9.5.3   Stepwise (Statistical) Discriminant Analysis

When the researcher has no reasons for assigning some predictors higher priority than others, statistical criteria can be used to determine order of entry in preliminary research. That is, if a researcher wants a reduced set of predictors but has no preferences among them, stepwise DISCRIM can be used to produce the reduced set. Entry of predictors is determined by user-specified statistical criteria, of which several are available as discussed in Section 9.6.1.2.

Stepwise DISCRIM has the same controversial aspects as stepwise procedures in general (see Section 5.5.3). Order of entry may be dependent on trivial differences in relationships among predictors in the sample that do not reflect population differences. However, this bias is reduced if cross-validation is used (cf. Sections 9.6.7.1 and 9.7.2). Costanza and Afifi (1979) recommend a probability to enter criterion more liberal than .05. They suggest a choice in the range of .15 to .20 to ensure entry of important variables.

---

[8]Sequential discriminant analysis is available in BMDP4M (shown in Tabachnick and Fidell, 1989) and, interactively, through SYSTAT DISCRIM (shown in Tabachnick and Fidell, 2007).

In SAS, stepwise discriminant analysis is provided through a separate program—STEPDISC. Three entry methods are available (cf. Section 9.6.1.2), as well as additional statistical criteria for two of them. SPSS DISCRIMINANT has several methods for statistical discriminant analysis, summarized in Table 9.18.

9.6   Some Important Issues

9.6.1   Statistical Inference

Section 9.6.1.1 contains a discussion of criteria for evaluating the overall statistical significance of a set of predictors for predicting group membership. Section 9.6.1.2 summarizes methods for directing the progression of stepwise discriminant analysis and statistical criteria for entry of predictors.

9.6.1.1   *Criteria for Overall Statistical Significance*

Criteria for evaluating overall statistical significance in DISCRIM are the same as those in MANOVA. The choice between Wilks' lambda, Roy's gcr, Hotelling's trace, and Pillai's criterion is based on the same considerations as discussed in Section 7.5.2. Different statistics are available in different programs, as noted in Section 9.8.

Two additional statistical criteria, Mahalanobis' D^2 and Rao's V, are especially relevant to stepwise DISCRIM. Mahalanobis' D^2 is based on distance between pairs of group centroids which is then generalizable to distances over multiple pairs of groups. Rao's V is another generalized distance measure that attains its largest value when there is greatest overall separation among groups.

These two criteria are available both to direct the progression of stepwise discriminant analysis and to evaluate the reliability of a set of predictors to predict group membership. Like Wilks' lambda, Mahalanobis' D^2 and Rao's V are based on all discriminant functions rather than one. Note that lambda, D^2, and V are descriptive statistics; they are not, themselves, inferential statistics, although inferential statistics are applied to them.

9.6.1.2   *Stepping Methods*

Related to criteria for statistical inference is the choice among methods to direct the progression of entry of predictors in stepwise discriminant analysis. Different methods of progression maximize group differences along different statistical criteria.

Selection of stepping method depends on the availability of programs and choice of statistical criterion. If, for example, the statistical criterion is Wilks' lambda, it is beneficial to choose the stepping method that minimizes Λ. (In SPSS DISCRIMINANT, Λ is the least expensive method, and is recommended in the absence of contrary reasons.) Or, if the statistical criterion is "change in Rao's V," the obvious choice of stepping method is RAO.

Statistical criteria also can be used to modify stepping. For example, the user can modify minimum F for a predictor to enter, minimum F to avoid removal, and so on. SAS allows forward, backward, and "stepwise" stepping (cf. Section 5.5.3). Either partial R^2 or significance level is chosen for variables or enter (forward stepping) or stay (backward stepping) in the model. Tolerance (the proportion of variance for a potential predictor that is not already accounted for by predictors in the equation) can be modified in SAS and SPSS stepwise programs. Comparison of programs with respect to these stepwise statistical criteria is provided in Table 9.18.

9.6.2   Number of Discriminant Functions

A number of discriminant functions are extracted in discriminant analysis with more than two groups. The maximum number of functions is the lesser of either degrees of freedom for groups or, as in canonical correlation, principal components analysis and factor analysis, equal to the number of predictors. As in these other analyses, some functions often carry no worthwhile information. It is frequently the case that the first one or two discriminant functions account for the lion's share of discriminating power, with no additional information forthcoming from the remaining functions.

Many of the programs evaluate successive discriminant functions. For the SPSS DISCRIMINANT example of Table 9.3, note that eigenvalues, percents of variance, and canonical correlations are given for each discriminant function for the small-sample data of Table 9.1. With both functions included, the $\chi^2(8)$ of 20.514 indicates a relationship between groups and predictors that is highly unlikely to be due to chance. With the first discriminant function removed, there is still a reliable relationship between groups and predictors as indicated by the $\chi^2(3) = 8.484$, $p = .037$. This finding indicates that the second discriminant function is also reliable.

How much between-group variability is accounted for by each discriminant function? The **% of Variance Values** (in the **Eigenvalues** table) associated with discriminant functions indicate the relative proportion of between-group variability accounted for by each function. In the small-sample example of Table 9.3, 70.70% of the between-group variability is accounted for by the first discriminant function and 29.30% by the second. These values appear as **Proportions** in the eigenvalue section of SAS DISCRIM output.

SPSS DISCRIMINANT offers the most flexibility with regard to number of discriminant functions (through syntax mode only). The user can choose the number of functions, the critical value for proportion of variance accounted for (with succeeding discriminant functions dropped once that value is exceeded), or the significance level of additional functions. SAS DISCRIM and CANDISC (but not STEPDISC) provide tests of successive functions.

9.6.3   Interpreting Discriminant Functions

If a primary goal of analysis is to discover and interpret the combinations of predictors (the discriminant functions) that separate groups in various ways, then the next two sections are relevant. Section 9.6.3.1 reveals how groups are spaced out along the various discriminant functions. Section 9.6.3.2 discusses correlations between predictors and the discriminant functions.

9.6.3.1   Discriminant Function Plots

Groups are spaced along the various discriminant functions according to their centroids. Recall from Section 9.4.1 that centroids are mean discriminant scores for each group on a function. Discriminant functions form axes and the centroids of the groups are plotted along the axes. If there is a big difference between the centroid of one group and the centroid of another along a discriminant function axis, the discriminant function separates the two groups. If there is not a big distance, the discriminant function does not separate the two groups. Many groups can be plotted along a single axis.

An example of a discriminant function plot is illustrated in Figure 9.2 for the data of Section 9.4. Centroids are obtained from the section called Functions at Group Centroids and Class

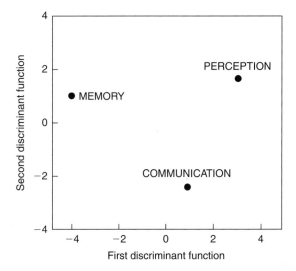

**FIGURE 9.2   Centroids of three learning disability
groups on the two discriminant functions derived
from sample data of Table 9.1.**

Means on Canonical Variables means in Table 9.3. They are also available in SAS DISCRIM with a request for `canonical` information (Table 9.4).

The plot emphasizes the utility of both discriminant functions in separating the three groups. On the first discriminant function (X axis), the MEMORY group is some distance from the other two groups, but the COMMUNICATION and PERCEPTION groups are close together. On the second function (Y axis) the COMMUNICATION group is far from the MEMORY and PERCEPTION groups. It takes both discriminant functions, then, to separate the three groups from each other.

If there are four or more groups and, therefore, more than two statistically significant discriminant functions, then pairwise plots of axes are used. One discriminant function is the X axis and another is the Y axis. Each group has a centroid for each discriminant function; paired centroids are plotted with respect to their values on the X and Y axes. Because centroids are only plotted pairwise, three significant discriminant functions require three plots (function 1 vs. function 2; function 1 vs. function 3; and function 2 vs. function 3), and so on.

SPSS DISCRIMINANT provides a plot of group centroids for the first pair of discriminant functions. Cases as well as means are plotted, making separations among groups harder to see than with simpler plots, but facilitating evaluation of classification.

Plots of centroids on additional pairs of statistically significant discriminant functions have to be prepared by hand, or discriminant scores can be passed to a "plotting" program such as SPSS PLOT. SAS passes the discriminant scores to plotting programs.

With factorial designs (Section 9.6.6), separate sets of plots are required for each significant main effect and interaction. Main effect plots have the same format as Figure 9.2, with one centroid per group per margin. Interaction plots have as many centroids as cells in the design.

9.6.3.2   *Structure Matrix of Loadings*

Plots of centroids tell you how groups are separated by a discriminant function, but they do not reveal the meaning of the discriminant function. A variety of matrices exist to reveal the nature of the combination of predictors that separate the groups. Matrices of standardized discriminant (canonical) functions are basically regression weights, the weights you would apply to the score of each case to find a standardized discriminant score for that case (Equation 9.2). These suffer from the same difficulties in interpretation as standardized regression coefficients, discussed in Section 5.6.1.

The structure matrix (a.k.a. loading matrix) contains correlations between predictors and discriminant functions. The meaning of the function is inferred by a researcher from this pattern of correlations (loadings). Correlations between predictors and functions are called loadings in discriminant function analysis, canonical correlation analysis (Chapter 12), and factor analysis (see Chapter 13). If predictors X_1, X_2, and X_3 load (correlate) highly with the function but predictors X_4 and X_5 do not, the researcher attempts to understand what X_1, X_2, and X_3 have in common with each other that is different from X_4 and X_5; the meaning of the function is determined by this understanding. (Read Section 13.6.5 for further insights into the art of interpreting loadings.)

Mathematically, the matrix of loadings is the pooled within-group correlation matrix multiplied by the matrix of standardized discriminant function coefficients.

$$\mathbf{A} = \mathbf{R}_w\mathbf{D} \qquad\qquad (9.7)$$

The structure matrix of correlations between predictors and discriminant functions, **A,** is found by multiplying the matrix of within-group correlations among predictors, $\mathbf{R}_w$, by a matrix of standardized discriminant function coefficients, **D** (standardized using pooled within-group standard deviations).

For the example of Table 9.1, the **Structure Matrix** appears as the middle matrix in Table 9.3. Structure matrices are read in columns; the column is the discriminant **Function** (1 or 2), the rows are predictors (**information to age**), and the entries in the column are correlations. For this example, the first discriminant function correlates most highly with **Information** (WISC Information scores, $r = .228$), while the second function correlates most highly with **Verbal expression** (ITPA Verbal Expression scale, $r = .446$). The structure matrix is available in SAS DISCRIM with the `canoni-cal` instruction and is labeled `Pooled Within Canonical Structure` as seen in Table 9.4.

These findings are related to discriminant function plots (e.g., Figure 9.2) for full interpretation. The first discriminant function is largely a measure of INFOrmation, and it separates the group with MEMORY problems from the groups with PERCEPTION and COMMUNICATION problems. The second discriminant function is largely a measure of VERBEXP (verbal expression) and it separates the group with COMMUNICATION problems from the groups with PERCEPTION and MEMORY problems. Interpretation in this example is reasonably straightforward because only one predictor is highly correlated with each discriminant function; interpretation is much more interesting when several predictors correlate with a discriminant function.

Consensus is lacking regarding how high correlations in a structure matrix must be to be interpreted. By convention, correlations in excess of .33 (10% of variance) may be considered eligible while lower ones are not. Guidelines suggested by Comrey and Lee (1992) are included in Section 13.6.5. However, the size of loadings depends both on the value of the correlation in the population and on the homogeneity of scores in the sample taken from it. If the sample is unusually homogeneous with respect to a predictor, the loadings for the predictor are lower and it may be wise to lower the criterion for determining whether or not to interpret the predictor as part of a discriminant function.

Caution is always necessary in interpreting loadings, however, because they are full, not partial or semipartial, correlations. The loading could be substantially lower if correlations with other predictors were partialed out. For a review of this material, read Section 5.6.1. Section 9.6.4 deals with methods for interpreting predictors after variance associated with other predictors is removed, if that is desired.

In some cases, rotation of the structure matrix may facilitate interpretation, as discussed in Chapter 13. SPSS DISCRIMINANT and MANOVA allow rotation of discriminant functions. But rotation of discriminant structure matrices is considered problematic and not recommended for the novice.

9.6.4 Evaluating Predictor Variables

Another tool for evaluating contribution of predictors to separation of groups is available through SAS and SPSS GLM in which means for predictors for each group are contrasted with means for other groups pooled. For instance, if there are three groups, means on predictors for group 1 are contrasted with pooled means from groups 2 and 3; then means for group 2 are contrasted with pooled means from groups 1 and 3; and finally means for group 3 are contrasted with pooled means from groups 1 and 2. This procedure is used to determine which predictors are important for isolating one group from the rest.

Twelve GLM runs are required in the example of Table 9.1: four for each of the three contrasts. Within each of the three contrasts, which isolates the means from each group and contrasts them with the means for the other groups, there are separate runs for each of the four predictors, in which each predictor is adjusted for the remaining predictors. In these runs, the predictor of interest is labeled the DV and remaining predictors are labeled CVs. The result is a series of tests of the significance of each predictor after adjusting for all other predictors in separating out each group from the remaining groups.

In order to avoid overinterpretation, it is probably best to consider only predictors with F ratios "significant" after adjusting error for the number of predictors in the set. The adjustment is made on the basis of Equation 7.12 of Section 7.5.3.1. This procedure is demonstrated in the complete example of Section 9.7. Even with this adjustment, there is danger of inflation of Type I error rate because multiple nonorthogonal contrasts are performed. If there are numerous groups, further adjustment might be considered such as multiplication of critical F by $k - 1$, where $k =$ number of groups. Or interpretation can proceed very cautiously, de-emphasizing statistical justification.

The procedures detailed in this section are most useful when the number of groups is small and the separations among groups are fairly uniform on the discriminant function plot for the first two functions. With numerous groups, some closely clustered, other kinds of contrasts might be suggested by the discriminant function plot (e.g., groups 1 and 2 might be pooled and contrasted with pooled groups 3, 4, and 5). Or, with a very large number of groups, the procedures of Section 9.6.3 may suffice.

If there is logical basis for assigning priorities to predictors, a sequential rather than standard approach to contrasts can be used. Instead of evaluating each predictor after adjustment for all other predictors, it is evaluated after adjustment by only higher priority predictors. This strategy is accomplished through a series of SPSS MANOVA runs, in which Roy-Bargmann stepdown F's (cf. Chapter 7) are evaluated for each contrast.

All the procedures for evaluation of DVs in MANOVA apply to evaluation of predictor variables in DISCRIM. Interpretation of stepdown analysis, univariate F, pooled within-group correlations among predictors, or standardized discriminant function coefficients is as appropriate (or inappropriate) for DISCRIM as for MANOVA. These procedures are summarized in Section 7.5.3.

9.6.5   Effect Size

Three types of effect sizes are of interest in discriminant analysis: those that describe variance associated with the entire analysis and two types that describe variance associated with individual predictors. The η^2 (or partial η^2) that can be found from Wilks' lambda or from associated F and df using Smithson's (2003) procedure provides the effect size for the entire analysis. Smithson's procedure also can be used to find the confidence intervals around these F values, as seen in complete examples of Chapters 6, 7, and 8. SAS DISCRIM provides the required values of F and df in the section labeled `Multivariate Statistics and F Approximations`.

SPSS DISCRIMINANT uses χ^2 rather than F to evaluate statistical significance but provides Λ for calculating effect size.

Calculating partial η^2 from Λ in Table 9.3,

$$\text{partial } \eta^2 = 1 - \Lambda^{1/3} = 1 - .01^{1/3} = 1 - .215 = .78$$

Steiger and Fouladi's (1992) software can be used to provide confidence intervals around the measure. The `Confidence Interval` is chosen as the `Option` and `Maximize Accuracy` is chosen as the `Algorithm`. Using the R^2 value of .78, Figure 9.3a shows setup values of 9 observations, 6 variables (including 4 predictors 2 variables for the group df, considered the criterion) and probability value of .95. As seen in Figure 9.3b, the R2 program provides 95% confidence limits from .00 to .88.

Separate effect sizes for each discriminant function are available as squared canonical correlations. SPSS DISCRIMINANT shows canonical correlations in the output section labeled `Eigenvalues` (see Table 9.3). Thus, the squared canonical correlations for the two functions are, respectively, $(.965)^2 = .93$ and $(.921)^2 = .85$. For SAS, the squared values are given directly in the first section of output labeled `Canonical Discriminant Analysis`. Steiger's program may also be used to find confidence limits for these values.

The structure matrix provides loadings in the form of correlations of each predictor with each discriminant function. These correlations, when squared, are effect sizes, indicating the proportion of variance shared between each predictor and each function. The structure matrix in SAS DISCRIM is labeled `Pooled Within Canonical Structure`. For the small-sample example then, the effect size (r^2) for INFO at discriminant function 1 is $(.228)^2 = .05$; 5% of variance is shared between the information subtest and the first discriminant function.

Another form of effect size is the η^2 that can be found when contrasts are run between each group and the remaining groups, with each predictor adjusted for all other predictors (Section 9.6.4). The contrasts conveniently are provided with F values and associated df, so that confidence limits also are available through Smithson's (2003) procedure. This is demonstrated in Section 9.7.

(a)

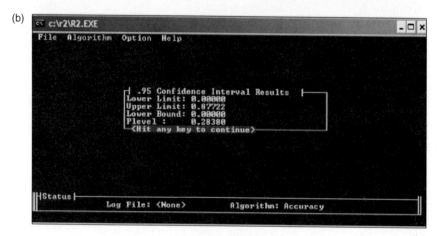

FIGURE 9.3   Confidence limits around R^2 using
Steiger and Fouladi's (1992) software:
(a) setup and (b) results.

9.6.6   Design Complexity: Factorial Designs

The notion of placing cases into groups is easily extended to situations where groups are formed by differences on more than one dimension. An illustration of factorial arrangement of groups is the complete example of Section 7.6.2, where women are classified by femininity (high or low) and also by masculinity (high or low) on the basis of scores on the Bem Sex Role Inventory (BSRI). Dimensions of femininity and masculinity (each with two levels) are factorially combined to form four groups: high-high, high-low, low-high, low-low. Unless you want to classify cases, factorial designs are best analyzed through MANOVA. If classification is your goal, however, some issues require attention.

A two-stage analysis is often best. First, questions about the statistical significance of separation of groups by predictors are answered through MANOVA. Second, if classification is desired after MANOVA, it is found through DISCRIM programs.

Formation of groups for DISCRIM depends on the outcome of MANOVA. If the interaction is statistically significant, groups are formed for the cells of the design. That is, in a two-by-two design, four groups are formed and used as the grouping variable in DISCRIM. Note that main effects as well as interactions influence group means (cell means) in this procedure, but for most purposes classification of cases into cells seems reasonable.

If an interaction is not statistically significant, classification is based on significant main effects. For example, interaction is not statistically significant in the data of Section 7.6.2, but both main effect of masculinity and main effect of femininity are statistically significant. One DISCRIM run is used to produce the classification equations for main effect of masculinity and a second run is used to produce the classification equations for main effect of femininity. That is, classification of main effects is based on marginal groups.

9.6.7   Use of Classification Procedures

The basic technique for classifying cases into groups is outlined in Section 9.4.2. Results of classification are presented in tables such as the Classification results of SPSS (Table 9.3), or Number of Observations and Percents Classified into GROUP of SAS (Table 9.4) where actual group membership is compared to predicted group membership. From these tables, one finds the percent of cases correctly classified and the number and nature of errors of classification.

But how good is the classification? When there are equal numbers of cases in every group, it is easy to determine the percent of cases that should be correctly classified by chance alone to compare to the percent correctly classified by the classification procedure. If there are two equally sized groups, 50% of the cases should be correctly classified by chance alone (cases are randomly assigned into two groups and half of the assignments in each group are correct), while three equally sized groups should produce 33% correct classification by chance, and so forth. However, when there are unequal numbers of cases in the groups, computation of the percent of cases that should be correctly classified by chance alone is a bit more complicated.

The easier way to find it[9] is to first compute the number of cases in each group that should be correct by chance alone and then add across the groups to find the overall expected percent correct. Consider an example where there are 60 cases, 10 in Group 1, 20 in Group 2, and 30 in Group 3. If prior probabilities are specified as .17, .33, and .50, respectively, the programs will assign 10, 20, and 30 cases to the groups. If 10 cases are assigned at random to Group 1, .17 of them (or 1.7) should be correct by chance alone. If 20 cases are randomly assigned to Group 2, .33 (or 6.6) of them should be correct by chance alone, and if 30 cases are assigned to Group 3, .50 of them (or 15) should be correct by chance alone. If 20 cases are randomly assigned to Group 2, .33 (or 6.6) of them should be correct by chance alone, and if 30 cases are assigned to Group 3, .50 of them (or 15) should be correct by chance alone. Adding together 1.7, 6.6, and 15 gives 23.3 cases correct by chance alone, 39% of the total. The percent correct using classification equations has to be substantially larger than the percent expected correct by chance alone if the equations are to be useful.

Some of the computer programs offer sophisticated additional features that are helpful in many classification situations.

---

[9]The harder way to find it is to expand the multinomial distribution, a procedure that is more technically correct but produces identical results to those of the simpler method presented here.

9.6.7.1 Cross-Validation and New Cases

Classification is based on classification coefficients derived from samples and they usually work too well for the sample from which they were derived. Because the coefficients are only estimates of population classification coefficients, it is often most desirable to know how well the coefficients generalize to a new sample of cases. Testing the utility of coefficients on a new sample is called cross-validation. One form of cross-validation involves dividing a single large sample randomly in two parts, deriving classification functions on one part, and testing them on the other. A second form of cross-validation involves deriving classification functions from a sample measured at one time, and testing them on a sample measured at a later time. In either case, cross-validation techniques are especially well developed in the SAS DISCRIM and SPSS DISCRIMINANT programs.

For a large sample randomly divided into parts, you simply omit information about actual group membership for some cases (hide it in the program) as shown in Section 9.7.2. SPSS DIS-CRIMINANT does not include these cases in the derivation of classification functions, but does include them in the classification phase. In SAS DISCRIM, the withheld cases are put in a separate data file. The accuracy with which the classification functions predict group membership for cases in this data file is then examined. This "calibration" procedure is demonstrated in Section 9.7.2. (Note that SAS refers to this as calibration, not cross-validation. The latter term is used to label what other programs call jackknifing; cf. Section 9.6.7.2.)

When the new cases are measured at a later time classifying them is somewhat more compli-cated unless you use SAS DISCRIM (in the same way that you would for cross-validation/calibra-tion). This is because other computer programs for DISCRIM do not allow classification of new cases without repeated entry of the original cases to derive the classification functions. You "hide" the new cases, derive the classification functions from the old cases, and test classification on all cases. Or, you can input the classification coefficients along with raw data for the new cases and run the data only through the classification phase. Or, it may be easiest to write your own program based on the classification coefficients to classify cases as shown in Section 9.4.2.

9.6.7.2 Jackknifed Classification

Bias enters classification if the coefficients used to assign a case to a group are derived, in part, from the case. In jackknifed classification, the data from the case are left out when the coefficients used to assign it to a group are computed. Each case has a set of coefficients that are developed from all other cases. Jackknifed classification gives a more realistic estimate of the ability of predictors to separate groups.

SAS DISCRIM and SPSS DISCRIMINANT provide for jackknifed classification (SAS calls it `crossvalidate`; SPSS calls it Leave-one-out classification). When the procedure is used with all predictors forced into the equation (i.e., direct or sequential with all predictors included), bias in classification is eliminated. When it is used with stepwise entry of predictors (where they may not all enter), bias is reduced. An application of jackknifed classification is shown in Section 9.7.

9.6.7.3 Evaluating Improvement in Classification

In sequential DISCRIM, it is useful to determine if classification improves as a new set of predictors is added to the analysis. McNemar's repeated-measures chi square provides a simple, straightforward (but tedious) test of improvement. Cases are tabulated one by one, by hand, as to whether they are cor-rectly or incorrectly classified before the step and after the step where the predictors are added.

Early Step Classification

		Correct	Incorrect
Later Step Classification	Correct	(A)	B
	Incorrect	C	(D)

Cases that have the same result at both steps (either correctly classified—cell A—or incorrectly classified—cell D) are ignored because they do not change. Therefore, χ^2 for change is

$$\chi^2 = \frac{(|B - C| - 1)^2}{B + C} \qquad df = 1 \tag{9.8}$$

Ordinarily, the researcher is only interested in improvement in χ^2, that is, in situations where $B > C$ because more cases are correctly classified after addition of predictors. When $B > C$ and χ^2 is greater than 3.84 (critical value of χ^2 with 1 df at $\alpha = .05$), the added predictors reliably improve classification.

With very large samples hand tabulation of cases is not reasonable. An alternative, but possibly less desirable, procedure is to test the significance of the difference between two lambdas, as suggested by Frane (1977). Wilks' lambda from the step with the larger number of predictors, Λ_2, is divided by lambda from the step with fewer predictors, Λ_1, to produce Λ_D

$$\Lambda_D = \frac{\Lambda_2}{\Lambda_1} \tag{9.9}$$

Wilks' lambda for testing the significance of the difference between two lambdas (Λ_D) is calculated by dividing the smaller lambda (Λ_2) by the larger lambda (Λ_1).

Λ_D is evaluated with three degree of freedom parameters: p, the number of predictors after addition of predictors; df_{bg}, the number of groups minus 1; and the df_{wg} at the step with the added predictors. Approximate F is found according to procedures in Section 9.4.1.

For example, suppose the small-sample data[10] were analyzed with only AGE as a predictor (not shown), yielding Wilks' lambda of .882. Using this, one can test whether addition of INFO, PERF, and VERBEXP for the full analysis ($\Lambda_2 = .010$) reliably improves classification of cases over that achieved with only AGE in the equation ($\Lambda_1 = .882$).

$$\Lambda_D = \frac{.010}{.882} = .0113$$

where

$$df_p = p = 4$$
$$df_{bg} = k - 1 = 2$$
$$df_{wg} = N - k = 6$$

[10]This procedure is inappropriate for such a small sample, but is shown here for illustrative purposes.

Finding approximate F from Section 9.4.1,

$$s = \min\,(p,\, k - 1) = 2$$

$$y = .0113^{1/2} = .1063$$

$$\mathrm{df}_1 = (4)(2) = 8$$

$$\mathrm{df}_2 = (2)\left[6 - \frac{4 - 2 + 1}{2}\right] - \left[\frac{4(2) - 2}{2}\right] = 6$$

$$\text{Approximate } F(8,\, 6) = \left(\frac{1 - .1063}{.1063}\right)\left(\frac{6}{8}\right) = 6.40$$

Because critical $F(8, 6)$ is 4.15 at $\alpha = .05$, there is statistically significant improvement in classification into the three groups when INFO, PERF, and VERBEXP scores are added to AGE scores.

9.7 Complete Example of Discriminant Analysis

The example of direct discriminant analysis in this section explores how role-dissatisfied housewives, role-satisfied housewives, and employed women differ in attitudes. The sample of 465 women is described in Appendix B, Section B.1. The grouping variable is role-dissatisfied housewives (UNHOUSE), role-satisfied housewives (HAPHOUSE), and working women (WORKING). Data are in DISCRIM.*.

Predictors are internal vs. external locus of control (CONTROL), satisfaction with current marital status (ATTMAR), attitude toward women's role (ATTROLE), and attitude toward housework (ATTHOUSE). Scores are scaled so that low values represent more positive or "desirable" attitudes. A fifth attitudinal variable, attitude toward paid work, was dropped from analysis because data were available only for women who had been employed within the past five years and use of this predictor would have involved nonrandom missing values (cf. Chapter 4). The example of DISCRIM, then, involves prediction of group membership from the four attitudinal variables.

The direct discriminant analysis allows us to evaluate the distinctions among the three groups on the basis of attitudes. We explore the dimensions on which the groups differ, the predictors contributing to differences among groups on these dimensions, and the degree to which we can accurately classify members into their own groups. We also evaluate efficiency of classification with a cross-validation sample.

9.7.1 Evaluation of Assumptions

The data are first evaluated with respect to practical limitations of DISCRIM.

9.7.1.1 Unequal Sample Sizes and Missing Data

In a screening run through SAS Interactive Data Analysis (cf. Section 4.2.2.1), seven cases had missing values among the four attitudinal predictors. Missing data were scattered over predictors and groups in apparently random fashion, so that deletion of the cases was deemed appropriate.[11] The full data set includes 458 cases once cases with missing values are deleted.

During classification, unequal sample sizes are used to modify the probabilities with which cases are classified into groups. Because the sample is randomly drawn from the population of interest, sample sizes in groups are believed to represent some real process in the population that should be reflected in classification. For example, knowledge that over half the women are employed implies that greater weight should be given the WORKING group.

9.7.1.2 Multivariate Normality

After deletion of cases with missing data, there are still over 80 cases per group. Although SAS MEANS run reveals skewness in ATTMAR, sample sizes are large enough to suggest normality of sampling distributions of means. Therefore there is no reason to expect distortion of results due to failure of multivariate normality.

9.7.1.3 Linearity

Although ATTMAR is skewed, there is no expectation of curvilinearity between this and the remaining predictors. At worst, ATTMAR in conjunction with the remaining continuous, well-behaved predictors may contribute to a mild reduction in association.

9.7.1.4 Outliers

To identify univariate outliers, z-scores associated with minimum and maximum values on each of the four predictors are investigated through SAS MEANS for each group separately, as per Section 4.2.2. There are some questionable values on ATTHOUSE, with a few exceptionally positive (low) scores. These values are about 4.5 standard deviations below their group means, making them candidates for deletion or alteration. However, the cases are retained for the search for multivariate outliers.

Multivariate outliers are sought through SAS REG by subsets (groups) and a request for an output table containing leverage statistics, as seen in Table 9.5. Data first are sorted by WORKSTAT, which then becomes the **by** variable in the **proc reg** run. Leverage values (**H**) are saved in a file labeled DISC_OUT. Table 9.5 shows a portion of the output data file for the working women (WORKSTAT = 1).

Outliers are identified as cases with too large a Mahalanobis D^2 for their own group, evaluated as χ^2 with degrees of freedom equal to the number of predictors. Critical χ^2 with 4 df at $\alpha = .001$ is 18.467; any case with $D^2 > 18.467$ is an outlier. Translating this critical value to leverage (h_{ii}) for the first group using the variation on Equation 4.3:

$$h_{ii} = \frac{\text{Mahalanobis distance}}{N-1} + \frac{1}{N} = \frac{18.467}{240} + \frac{1}{241} = .081$$

In Table 9.5, CASESEQ 346 (H = .0941) and CASESEQ 407 (H = .0898) are identified as outliers in the group of WORKING women. No additional outliers were found.

---

[11]Alternative strategies for dealing with missing data are discussed in Chapter 4.

TABLE 9.5   Identification of Multivariate Outliers (SAS SORT and REG Syntax and Selected Portion of Output File from SAS REG)

```
proc sort data = Sasuser.Discrim;
  by WORKSTAT;
run;

proc data=Sasuser.Discrim;
    by WORKSTAT;
  model CASESEQ= CONTROL ATTMAR ATTROLE ATTHOUSE/ selection=

    RSQUARE COLLIN;
output out=SASUSER.DISC_OUT H=H;
run;
```

SASUSER.DISC_OUT

	Int	Int	Int	Int	Int	Int	Int
465	CASESEQ	WORKSTAT	CONTROL	ATTMAR	ATTROLE	ATTHOUSE	H
136	345	1	8	19	38	19	0.0144
137	346	1	5	20	41	2	0.0941
138	347	1	6	28	34	26	0.0079
139	348	1	7	23	26	24	0.0103
140	349	1	5	25	31	30	0.0211
141	355	1	6	25	27	30	0.0148
142	357	1	6	17	38	22	0.0087
143	358	1	8	42	36	26	0.0276
144	359	1	7	21	22	30	0.0204
145	362	1	8	24	35	24	0.0088
146	365	1	5	14	36	18	0.0186
147	369	1	7	23	27	34	0.0249
148	372	1	7	18	45	13	0.0284
149	378	1	5	26	23	21	0.0284
150	380	1	7	.	29	28	.
151	381	1	6	35	30	20	0.0224
152	383	1	7	30	44	25	0.0176
153	384	1	7	25	41	27	0.0136
154	386	1	7	20	30	25	0.0067
155	387	1	7	23	29	22	0.0082
156	397	1	5	16	35	15	0.0245
157	398	1	6	30	23	33	0.0267
158	399	1	9	12	25	24	0.0361
159	400	1	7	25	23	29	0.0160
160	401	1	5	42	35	21	0.0393
161	403	1	6	35	27	29	0.0191
162	404	1	7	20	30	21	0.0087
163	406	1	7	39	35	18	0.0306
164	407	1	6	20	42	2	0.0898
165	425	1	6	30	33	22	0.0098

TABLE 9.6 Syntax and Selected Output from SAS DISCRIM to Check Homogeneity of Variance-Covariance Matrices

```
proc discrim data=Sasuser.Discrim short noclassify
      pool=test slpool=.001;
   class workstat;
   var CONTROL ATTMAR ATTROLE ATTHOUSE;
   priors proportional;
   where CASESEQ^=346 and CASESEQ^=407;
run;
```

```
              Test of Homogeneity of Within Covariance Matrices

Notation: K     = Number of Groups

          P     = Number of Variables

          N     = Total Number of Observations - Number of Groups

          N(i) = Number of Observations in the i'th Group - 1

                    __                            N(i)/2
                    ||  |Within SS Matrix(i)|
          V     = ----------------------------------
                                             N/2
                    |Pooled SS Matrix|

                    _                   _      2
                   |           1     1   | 2P + 3P - 1
          RHO  = 1.0 - | SUM ----- - --- | ------------
                   |         N(i)    N  _| 6(P+1)(K-1)
                   |_

          DF    = .5(K_1)P(P+1)

                                        _               _
                                       |    PN/2         |
                                       |  N       V      |
Under the null hypothesis:    -2 RHO ln | --------------- |
                                       |           PN(i)/2 |
                                       |  __             |
                                       |_ || N(i)       _|

is distributed approximately as Chi-Square(DF).

              Chi-Square        DF        Pr > ChiSq
              50.753826         20          0.0002

Since the Chi-Square value is significant at the 0.001 level,
the within covariance matrices will be used in the discriminant
function.
Reference: Morrison, D.F. (1976) Multivariate Statistical
Methods p252.
```

The multivariate outliers are the same cases that have extreme univariate scores on ATTHOUSE. Because transformation is questionable for ATTHOUSE (where it seems unreasonable to transform the predictor for only two cases) it is decided to delete multivariate outliers.

Therefore, of the original 465 cases, 7 are lost due to missing values and 2 are multivariate outliers, leaving a total of 456 cases for analysis.

9.7.1.5   Homogeneity of Variance-Covariance Matrices

A SAS DISCRIM run, Table 9.6, deletes the outliers in order to evaluate homogeneity of variance-covariance matrices. Most output has been omitted here. The instruction to produce the test of homogeneity of variance-covariance matrices is `pool=test`.

This test shows significant heterogeneity of variance-covariance matrices. The program uses separate matrices in the classification phase of discriminant analysis if `pool=test` is specified and the test shows significant heterogeneity.

9.7.1.6   Multicollinearity and Singularity

Because SAS DISCRIM, used for the major analysis, protects against multicollinearity through checks of tolerance, no formal evaluation is necessary (cf. Chapter 4). However, the SAS REG syntax of Table 9.6 that evaluates multivariate outliers also requests collinearity information, shown in Table 9.7. No problems with multicollinearity are noted.

TABLE 9.7   SAS REG Output Showing Collinearity Information for All Groups Combined (Syntax Is in Table 9.6)

Collinearity Diagnostics

Number	Eigenvalue	Condition Index
1	4.83193	1.00000
2	0.10975	6.63518
3	0.03169	12.34795
4	0.02018	15.47559
5	0.00645	27.37452

Collinearity Diagnostics

Number	Intercept	CONTROL	ATTMAR	ATTROLE	ATTHOUSE
1	0.00036897	0.00116	0.00508	0.00102	0.00091481
2	0.00379	0.00761	0.94924	0.02108	0.00531
3	0.00188	0.25092	0.04175	0.42438	0.10031
4	0.00266	0.61843	0.00227	0.01676	0.57008
5	0.99129	0.12189	0.00166	0.53676	0.32339

Proportion of Variation

9.7.2   Direct Discriminant Analysis

Direct DISCRIM is performed through SAS DISCRIM with the 4 attitudinal predictors all forced in the equation. The program instructions and some of the output appear in Table 9.8. Simple statistics are requested to provide predictor means, helpful in interpretation. The anova and manova instructions request univariate statistics on group differences separately for each of the variables and a multivariate test for the difference among groups. Pcorr requests the pooled within-groups correlation matrix, and crossvalidate requests jackknifed classification. The priors proportional instruction specifies prior probabilities for classification proportional to sample sizes.

When all 4 predictors are used, the F of 6.274 (with 8 and 900 df based on Wilks' lambda) is highly significant. That is, there is statistically significant separation of the three groups based on all four predictors combined, as discussed in Section 9.6.1.1. Partial η^2 and associated 95% confidence limits are found through Smithson's (2003) NoncF2.sas procedure (as in Table 8.16), yielding $\eta^2 = .05$ with limits from .02 to .08.

Canonical correlations (in the section of output following multivariate analysis) for each discriminant function (.267 and .184), although small, are relatively equal for the two discriminant functions. The adjusted values are not very much different with this relatively large sample. The "peel down" test shows that both functions significantly discriminate among the groups. That is, even after the first function is removed, there remains significant discrimination, Pr > F = 0.0014. Because there are only two possible discriminant functions, this is a test of the second one. Steiger's R2 program (demonstrated in Section 5.6.2.4) may be used to find confidence limits around the Squared Canonical Correlations of .07 and .02. With 6 variables (four predictors and two variables for the 2 df for groups) and 456 observations, the limits for the first discriminant function are .03 to .11, and for the second function they are .00 to .06.

The loading matrix (correlations between predictors and discriminant functions) appears in the section of output labeled Pooled Within Canonical Structure. Class means on canonical variables are centroids on the discriminant functions for the groups, discussed in Sections 9.4.1 and 9.6.3.1.

A plot of the placement of the centroids for the three groups on the two discriminant functions (canonical variables) as axes appears in Figure 9.4. The points that are plotted are given in Table 9.9 as Class means on canonical variables.

TABLE 9.8   Syntax And Partial Output From SAS DISCRIM Analysis of Four Attitudinal Variables

```
proc discrim data=Sasuser.Discrim simple anova manova pcorr can
                             crossvalidate pool=test;

    class workstat;
    var CONTROL ATTMAR ATTROLE ATTHOUSE;
    priors proportional;
    where CASESEQ^=346 and CASESEQ^=407;
run;
```

TABLE 9.8 Continued

Pooled Within-Class Correlation Coefficients / Pr > |r|

Variable		CONTROL	ATTMAR	ATTROLE	ATTHOUSE
CONTROL	Locus-of-control	1.00000	0.17169	0.00912	0.15500
			0.0002	0.8463	0.0009
ATTMAR	Attitude toward current marital status	0.17169	1.00000	-0.07010	0.28229
		0.0002		0.1359	<.0001
ATTROLE	Attitudes toward role of women	0.00912	-0.07010	1.00000	-0.29145
		0.8463	0.1359		<.0001
ATTHOUSE	Attitudes toward housework	0.15500	0.28229	-0.29145	1.00000
		0.0009	<.0001	<.0001	

Simple Statistics

Total-Sample

Variable	Label	N	Sum	Mean	Variance	Standard Deviation
CONTROL	Locus-of-control	456	3078	6.75000	1.60769	1.2679
ATTMAR	Attitude toward current marital Status	456	10469	22.95833	72.73892	8.5287
ATTROLE	Attitudes toward role of women	456	16040	35.17544	45.68344	6.7590
ATTHOUSE	Attitudes toward housework	456	10771	23.62061	18.30630	4.2786

WORKSTAT = 1

Variable	Label	N	Sum	Mean	Variance	Standard Deviation
CONTROL	Locus-of-control	239	1605	6.71548	1.53215	1.2378
ATTMAR	Attitude toward current marital status	239	5592	23.39749	72.76151	8.5300
ATTROLE	Attitudes toward role of women	239	8093	33.86192	48.38842	6.9562
ATTHOUSE	Attitudes toward housework	239	5691	23.81172	19.85095	4.4554

(continued)

413

TABLE 9.8 Continued

WORKSTAT = 2

Variable	Label	N	Sum	Mean	Variance	Standard Deviation
CONTROL	Locus-of-control	136	902.00000	6.63235	1.71569	1.3098
ATTMAR	Attitude toward current marital status	136	2802	20.60294	43.87081	6.6235
ATTROLE	Attitudes toward role of women	136	5058	37.19118	41.71133	6.4584
ATTHOUSE	Attitudes toward housework	136	3061	22.50735	15.08143	3.8835

Simple Statistics

WORKSTAT = 3

Variable	Label	N	Sum	Mean	Variance	Standard Deviation
CONTROL	Locus-of-control	81	571.00000	7.04938	1.57253	1.2540
ATTMAR	Attitude toward current marital status	81	2075	25.61728	106.03920	10.2975
ATTROLE	Attitudes toward role of women	81	2889	35.66667	33.17500	5.7598
ATTHOUSE	Attitudes toward housework	81	2019	24.92593	15.66944	3.9585

Univariate Test Statistics

F Statistics, Num DF=2, Den DF=453

Variable	Label	Total Standard Deviation	Pooled Standard Deviation	Between Standard Deviation	R-Square	R-Square / (1-RSq)	F Value	Pr > F
CONTROL	Locus-of-control	1.2679	1.2625	0.1761	0.0129	0.0131	2.96	0.0530
ATTMAR	Attitude toward current marital status	8.5287	8.3683	2.1254	0.0415	0.0433	9.81	<.0001
ATTROLE	Attitudes toward role of women	6.7590	6.6115	1.7996	0.0474	0.0497	11.26	<.0001
ATTHOUSE	Attitudes toward housework	4.2786	4.2061	1.0184	0.0379	0.0393	8.91	0.0002

TABLE 9.8 Continued

Average R-Square

Unweighted	0.0348993
Weighted by Variance	0.0426177

Multivariate Statistics and F Approximations

S=2 M=0.5 N=224

Statistic	Value	F Value	Num DF	Den DF	Pr > F
Wilks' Lambda	0.89715033	6.27	8	900	<.0001
Pillai's Trace	0.10527259	6.26	8	902	<.0001
Hotelling-Lawley Trace	0.11193972	6.29	8	640.54	<.0001
Roy's Greatest Root	0.07675307	8.65	4	451	<.0001

NOTE: F Statistic for Roy's Greatest Root is an upper bound.
NOTE: F Statistic for Wilks' Lambda is exact.

Canonical Discriminant Analysis

	Canonical Correlation	Adjusted Canonical Correlation	Approximate Standard Error	Squared Canonical Correlation
1	0.266987	0.245497	0.043539	0.071282
2	0.184365	0.182794	0.045287	0.033991

Test of H0: The canonical
correlations in the current
row and all that follow are zero

	Likelihood Ratio	Approximate F Value	Num DF	Den DF	Pr > F
1	0.89715033	6.27	8	900	<.0001
2	0.96600937	5.29	3	451	0.0014

Eigenvalues of Inv(E)*H
= CanRsq/(1-CanRsq)

	Eigenvalue	Difference	Proportion	Cumulative
1	0.0768	0.0416	0.6857	0.6857
2	0.0352		0.3143	1.0000

(continued)

415

TABLE 9.8 Continued

Canonical Discriminant Analysis

Pooled Within Canonical Structure

Variable	Label	Can1	Can2
CONTROL	Locus-of-control	0.281678	0.444939
ATTMAR	Attitude toward current marital status	0.718461	0.322992
ATTROLE	Attitudes toward role of women	-0.639249	0.722228
ATTHOUSE	Attitudes toward housework	0.679447	0.333315

Class Means on Canonical Variables

WORKSTAT	Can1	Can2
1	0.1407162321	-.1505321835
2	-.4160079128	0.0539321812
3	0.2832826750	0.3536100644

The DISCRIM Procedure

Classification Summary for Calibration Data: SASUSER.DISCRIM

Resubstitution Summary using Quadratic Discriminant Function

Generalized Squared Distance Function

$$D_j^2(X) = (X-\bar{X}_j)'\ COV_j^{-1}\ (X-\bar{X}_j) + \ln\ |COV_j|\ -\ 2\ \ln\ PRIOR_j$$

Posterior Probability of Membership in Each WORKSTAT

$$Pr(j|X) = \exp(-.5\ D_j^2(X))\ /\ SUM_k\ \exp(-.5\ D_k^2(X))$$

Number of Observations and Percent Classified into WORKSTAT

From WORKSTAT	1	2	3	Total
1	184	48	7	239
	76.99	20.08	2.93	100.00
2	73	59	4	136
	53.68	43.38	2.94	100.00
3	59	12	10	81
	72.84	14.81	12.35	100.00
Total	316	119	21	456
	69.30	26.10	4.61	100.00
Priors	0.52412	0.29825	0.17763	

TABLE 9.8   Continued

```
              Error Count Estimates for WORKSTAT
                     1          2          3       Total
         Rate     0.2301     0.5662     0.8765    0.4452
         Priors   0.5241     0.2982     0.1776
```

Classification Summary for Calibration Data: SASUSER.DISCRIM
Cross-validation Summary using Quadratic
Discriminant Function

Generalized Squared Distance Function

$$D_j^2(X) = (X - \bar{X}_{(X)j})' COV_{(X)j}^{-1} (X - \bar{X}_{(X)j}) + \ln |COV_{(X)j}| - 2 \ln PRIOR_j$$

Posterior Probability of Membership
in Each WORKSTAT

$$Pr(j|X) = \exp(-.5\, D_j^2(X)) / SUM_k \exp(-.5\, D_k^2(X))$$

Number of Observations and Percent
Classified into WORKSTAT

From WORKSTAT	1	2	3	Total
1	179	50	10	239
	74.90	20.92	4.18	100.00
2	78	53	5	136
	57.35	38.97	3.68	100.00
3	60	13	8	81
	74.07	16.05	9.88	100.00
Total	317	116	23	456
	69.52	25.44	5.04	100.00
Priors	0.52412	0.29825	0.17763	

```
              Error Count Estimates for WORKSTAT
                     1          2          3       Total
         Rate     0.2510     0.6103     0.9012    0.4737
         Priors   0.5241     0.2982     0.1776
```

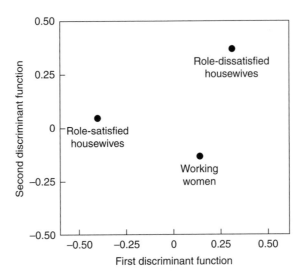

FIGURE 9.4 **Plots of three group centroids
on two discriminant functions derived
from four attitudinal variables.**

Table 9.8 shows the classification functions used to classify cases into the three groups (see Equation 9.3) and the results of that classification, with and without jackknifing (see Section 9.6.7). In this case, classification is made on the basis of a modified equation in which unequal prior probabilities are used to reflect unequal group sizes by the use of `prior proportional` in the syntax. Classification is based on the quadratic discriminant function to compensate for heterogeneity of various covariance matrices.

A total of 55% (`1 - Error Count Rate of 0.4452`) of cases are correctly classified by normal procedures, and 52% by jackknifed procedures. How do these compare with random assignment? Prior probabilities, specified as .52 (WORKING), .30 (HAPHOUSE), and .18 (UNHOUSE), put 237 cases (.52 × 456) in the WORKING group, 137 in the HAPHOUSE group, and 82 in the UNHOUSE group. Of those randomly assigned to the WORKING group, 123 (.52 × 237) should be correct, while 41.1 (.30 × 137) and 14.8 (.18 × 82) should be correct by chance in the HAPHOUSE and UNHOUSE groups, respectively. Over all three groups 178.9 out of the 456 cases or 39% should be correct by chance alone. Both classification procedures correctly classify substantially more than that.

An additional SAS DISCRIM run for cross-validation is shown in Table 9.9. SAS DISCRIM has no direct procedure of forming and using a cross-validation sample. Instead, other procedures must be used to split the file into the "training" cases, used to develop (calibrate) the classification equations, and the "testing" cases, used to validate the classification.

First a new data set is created: data SASUSER.DISCRIMX. The original data set is identified as set SASUSER.DISCRIM. Then outliers and cases with missing data are omitted. Finally, a variable is created on which to split the data set, here called TEST1, which is set to zero, and then changed to 1 for 25% of the cases. Then an additional two files are created on the basis of `TEST1` with set

TABLE 9.9　Cross-Validation of Classification of Cases by Four Attitudinal Variables (Syntax for SAS DATA: Syntax and Selected Output from SAS DISCRIM)

```
data Sasuser.Discrimx;
    set SASUSER.DISCRIM;
    if ATTHOUSE=2 or ATTHOUSE=. or ATTMAR=. or ATTROLE=.
        or CONTROL=. then delete;
    TEST1=0;
    if uniform(11738) <= .25 then TEST1=1;
run;

data Sasuser.Disctrng;
    set Sasuser.Discrimx;
    where TEST1=0;
data Sasuser.Disctest;
    set Sasuser.Discrimx;
    where TEST1=1;
run;

proc discrim data=SASUSER.Disctrng outstat=INFO pool=test;
    class WORKSTAT;
    var CONTROL ATTMAR ATTROLE ATTHOUSE;
    priors proportional;
run;

proc discrim data=INFO testdata=SASUSER.Disctest pool=test;
    class WORKSTAT;
    var CONTROL ATTMAR ATTROLE ATTHOUSE;
    priors proportional;
run;
```

```
                    The DISCRIM Procedure
            Classification Summary for Calibration
                    Data: SASUSER.DISCTRNG
            Resubstitution Summary using Quadratic
                    Discriminant Function

            Generalized Squared Distance Function
```

$$D^2_j(X) = (X-\bar{X}_j)'\,COV^{-1}_j\,(X-\bar{X}_j) + \ln|COV_j| - 2\ln\,PRIOR_j$$

```
        Posterior Probability of Membership in Each WORKSTAT
```

$$Pr(j|X) = \exp(-.5\,D^2_j(X))\,/\,SUM_k\,\exp(-.5\,D^2_k(X))$$

(continued)

TABLE 9.9   Continued

Number of Observations and Percent Classified into WORKSTAT

From WORKSTAT	1	2	3	Total
1	129	32	11	172
	75.00	18.60	6.40	100.00
2	49	46	7	102
	48.04	45.10	6.86	100.00
3	45	9	11	65
	69.23	13.85	16.92	100.00
Total	223	87	29	339
	65.78	25.66	8.55	100.00
Priors	0.50737	0.30088	0.19174	

Error Count Estimates for WORKSTAT

	1	2	3	Total
Rate	0.2500	0.5490	0.8308	0.4513
Priors	0.5074	0.3009	0.1917	

Classification Summary for Test Data: SASUSER.DISCTEST
Classification Summary using Quadratic Discriminant Function

Generalized Squared Distance Function

$$D_j^2(X) = (X-\bar{X}_j)'\ COV_j^{-1}\ (X-\bar{X}_j) + \ln |COV_j| - 2 \ln PRIOR_j$$

Posterior Probability of Membership in Each WORKSTAT

$$Pr(j|X) = \exp(-.5\ D_j^2(X))\ /\ SUM_k\ \exp(-.5\ D_k^2(X))$$

Number of Observations and Percent Classified into WORKSTAT

From WORKSTAT	1	2	3	Total
1	40	16	11	67
	59.70	23.88	16.42	100.00
2	17	15	2	34
	50.00	44.12	5.88	100.00
3	10	2	4	16
	62.50	12.50	25.00	100.00
Total	67	33	17	117
	57.26	28.21	14.53	100.00
Priors	0.50737	0.30088	0.19174	

Error Count Estimates for WORKSTAT

	1	2	3	Total
Rate	0.4030	0.5588	0.7500	0.5164
Priors	0.5074	0.3009	0.1917	

S A S U S E R . D I S C R I M X: a calibration (training) file, through data S A S U S E R . D I S C T R N G, and a cross-validation (test) file through data S A S U S E R . D I S C T E S T. Finally, a discriminant analysis on the training file (with 339 cases) is run which saves the calibration information in a file called I N F O, and then applies the calibration information to the test file (with 117 cases). Again, the quadratic classification procedure is used.

A summary of information appropriate for publication appears in Table 9.10. In the table are the loadings, univariate F for each predictor, and pooled within-group correlations among predictors.

SAS DISCRIM has no contrast procedure, nor does it provide F or t ratios for predictor variables adjusted for all other variables. However, the information is available using contrasts with separate analyses of covariance for each variable in GLM. In each analysis of covariance, the variable of interest is declared the DV and the remaining variables are declared covariates. The process is demonstrated for the twelve contrast runs needed in Tables 9.11 to 9.13; means on each predictor adjusted for all other predictors for each group are contrasted with the pooled means for the other two groups. WORKING women are contrasted with the pooled means for HAPHOUSE and UNHOUSE to determine which predictors distinguish WORKING women from others in Table 9.11. Table 9.12 has the HAPHOUSE group contrasted with the other two groups; Table 9.13 shows the UNHOUSE group contrasted with the other two groups. Note that df for error $= N - k - c - 1 = 450$.

Based on familywise $\alpha = .05$, $\alpha_i = .0125$, the predictor that most clearly distinguishes the WORKING group from the other two is ATTROLE after adjustment for the other predictors. The HAPHOUSE group differs from the other two groups on the basis of ATTMAR after adjustment for the remaining predictors. The UNHOUSE group does not differ from the other two when each predictor is adjusted for all others. Separate runs without covariates would be needed if there is interest in which predictors separate each group from the others *without* adjustment for the other predictors. Table 9.14 summarizes the results of Smithson's procedure for finding effect sizes and 98.75% confidence limits for all twelve runs.

A checklist for a direct discriminant function analysis appears in Table 9.15. It is followed by an example of a Results section, in journal format, for the analysis just described.

TABLE 9.10   Results of Discriminant Analysis of Attitudinal Variables

Predictor Variable	Correlations of Predictor Variables with Discriminant Functions		Univariate $F(2, 453)$	Pooled Within-Group Correlations among Predictors		
	1	*2*		*ATTMAR*	*ATTROLE*	*ATTHOUSE*
CONTROL	.28	.44	2.96	.17	.01	.16
ATTMAR	.72	.32	9.81		−.07	.28
ATTROLE	−.64	.72	11.26			−.29
ATTHOUSE	.68	.33	8.91			
Canonical R	.27	.18				
Eigenvalue	.08	.04				

TABLE 9.11 Syntax and Highly Abbreviated Output of SAS GLM Contrasting the WORKING Group with the Other Two Groups

```
proc glm data=Sasuser.Discrim;
    class WORKSTAT;
    model ATTHOUSE = WORKSTAT CONTROL ATTMAR ATTROLE ;
        where CASESEQ^=346 and CASESEQ^=407;
    contrast  WORKING VS. OTHERS' WORKSTAT -2 1 1 ;
run;
proc glm data=Sasuser.Discrim;
    class WORKSTAT;
    model ATTROLE = WORKSTAT ATTHOUSE CONTROL ATTMAR ;
        where CASESEQ^=346 and CASESEQ^=407;
    contrast  WORKING VS. OTHERS' WORKSTAT -2 1 1 ;
run;
proc glm data=Sasuser.Discrim;
    class WORKSTAT;
    model ATTMAR = WORKSTAT CONTROL ATTHOUSE ATTROLE ;
        where CASESEQ^=346 and CASESEQ^=407;
    contrast  WORKING VS. OTHERS' WORKSTAT -2 1 1 ;
run;
proc glm data=Sasuser.Discrim;
    class WORKSTAT;
    model CONTROL = WORKSTAT ATTROLE ATTHOUSE ATTMAR ;
        where CASESEQ^=346 and CASESEQ^=407;
    contrast 'WORKING VS. OTHERS' WORKSTAT -2 1 1 ;
run;
```

Dependent Variable: ATTHOUSE Attitudes toward housework

Contrast	DF	Contrast SS	Mean Square	F Value	Pr > F
WORKING VS. OTHERS	1	12.32545468	12.32545468	0.83	0.3626

Dependent Variable: ATTROLE Attitudes toward role of women

Contrast	DF	Contrast SS	Mean Square	F Value	Pr > F
WORKING VS. OTHERS	1	676.9471257	676.9471257	16.87	<.0001

Dependent Variable: ATTMAR Attitude toward current marital status

Contrast	DF	Contrast SS	Mean Square	F Value	Pr > F
WORKING VS. OTHERS	1	13.99801413	13.99801413	0.22	0.6394

Dependent Variable: CONTROL Locus-of-control

Contrast	DF	Contrast SS	Mean Square	F Value	Pr > F
WORKING VS. OTHERS	1	1.20936265	1.20936265	0.79	0.3749

TABLE 9.12   Syntax and Highly Abbreviated Output of SAS DISCRIM Contrasting the HAPHOUSE GROUP with the Other Two Groups

```
proc glm data=Sasuser.Discrim;
    class WORKSTAT;
    model ATTHOUSE = WORKSTAT CONTROL ATTMAR ATTROLE ;
        where CASESEQ^=346 and CASESEQ^=407;
    contrast 'HAPHOUSE VS. OTHERS' WORKSTAT 1 -2 1 ;
run;

proc glm data=Sasuser.Discrim;
    class WORKSTAT;
    model ATTROLE = WORKSTAT ATTHOUSE CONTROL ATTMAR ;
        where CASESEQ^=346 and CASESEQ^=407;
    contrast 'HAPHOUSE VS. OTHERS' WORKSTAT 1 -2 1 ;
run;

proc glm data=Sasuser.Discrim;
    class WORKSTAT;
    model ATTMAR = WORKSTAT CONTROL ATTHOUSE ATTROLE ;
        where CASESEQ^=346 and CASESEQ^=407;
    contrast 'HAPHOUSE VS. OTHERS' WORKSTAT 1 -2 1 ;
run;

proc glm data=Sasuser.Discrim;
    class WORKSTAT;
    model CONTROL = WORKSTAT ATTROLE ATTHOUSE ATTMAR ;
        where CASESEQ^=346 and CASESEQ^=407;
    contrast 'HAPHOUSE VS. OTHERS' WORKSTAT 1 -2 1 ;
run;
```

Dependent Variable: ATTHOUSE   Attitudes toward housework

Contrast	DF	Contrast SS	Mean Square	F Value	Pr > F
HAPHOUSE VS. OTHERS	1	60.74947570	60.74947570	4.09	0.0436

Dependent Variable: ATTROLE   Attitudes toward role of women

Contrast	DF	Contrast SS	Mean Square	F Value	Pr > F
HAPHOUSE VS. OTHERS	1	218.5434340	218.5434340	5.45	0.0201

Dependent Variable: ATTMAR   Attitude toward current marital status

Contrast	DF	Contrast SS	Mean Square	F Value	Pr > F
HAPHOUSE VS. OTHERS	1	615.1203307	615.1203307	9.66	0.0020

Dependent Variable: CONTROL Locus-of-control

Contrast	DF	Contrast SS	Mean Square	F Value	Pr > F
HAPHOUSE VS. OTHERS	1	1.18893484	1.18893484	0.78	0.3789

TABLE 9.13 **Syntax and Highly Abbreviated Output of SAS DISCRIM Contrasting the UNHOUSE GROUP with the Other Two Groups**

```
proc glm data=Sasuser.Discrim;
    class WORKSTAT;
    model ATTHOUSE = WORKSTAT CONTROL ATTMAR ATTROLE ;
        where CASESEQ^=346 and CASESEQ^=407;
    contrast 'UNHOUSE VS. OTHERS' WORKSTAT 1 1 -2 ;
run;
proc glm data=Sasuser.Discrim;
    class WORKSTAT;
    model ATTROLE = WORKSTAT ATTHOUSE CONTROL ATTMAR ;
        where CASESEQ^=346 and CASESEQ^=407;
    contrast 'UNHOUSE VS. OTHERS' WORKSTAT 1 1 -2 ;
run;
proc glm data=Sasuser.Discrim;
    class WORKSTAT;
    model ATTMAR = WORKSTAT CONTROL ATTHOUSE ATTROLE ;
        where CASESEQ^=346 and CASESEQ^=407;
    contrast 'UNHOUSE VS. OTHERS' WORKSTAT 1 1 -2 ;
run;
proc glm data=Sasuser.Discrim;
    class WORKSTAT;
    model CONTROL = WORKSTAT ATTROLE ATTHOUSE ATTMAR ;
        where CASESEQ^=346 and CASESEQ^=407;
    contrast 'UNHOUSE VS. OTHERS' WORKSTAT 1 1 -2 ;
run;
```

Dependent Variable: ATTHOUSE Attitudes toward housework

Contrast	DF	Contrast SS	Mean Square	F Value	Pr > F
UNHOUSE VS. OTHERS	1	92.00307841	92.00307841	6.20	0.0131

Dependent Variable: ATTROLE Attitudes toward role of women

Contrast	DF	Contrast SS	Mean Square	F Value	Pr > F
UNHOUSE VS. OTHERS	1	45.69837169	45.69837169	1.14	0.2865

Dependent Variable: ATTMAR Attitude toward current marital status

Contrast	DF	Contrast SS	Mean Square	F Value	Pr > F
UNHOUSE VS. OTHERS	1	354.1278220	354.1278220	5.56	0.0188

Dependent Variable: CONTROL Locus-of-control

Contrast	DF	Contrast SS	Mean Square	F Value	Pr > F
UNHOUSE VS. OTHERS	1	3.27205950	3.27205950	2.13	0.1447

TABLE 9.14   Effect Sizes and 98.75% Confidence Limits for Contrasts of Each Group with the Two Other Groups Pooled for Each Predictor Adjusted for the Three Other Predictors

		Predictor (adjusted for all others)			
Contrast		Attitude toward housework	Attitude toward role of women	Attitude toward marriage	Locus-of-control
Working women vs. others	Effect Size	.00	.04	.00	.00
	98.75% CL	.00–.03	.01–.09	.00–.02	.00–.02
Role-satisfied housewives vs. others	Effect Size	.01	.01	.02	.01
	98.75% CL	.00–.04	.00–.05	.00–.07	.00–.02
Role-dissatisfied housewives vs. others	Effect Size	.01	.01	.01	.01
	98.75% CL	.00–.05	.00–.03	.00–.05	.00–.03

TABLE 9.15   Checklist for Direct Discriminant Analysis

1. Issues
   a. Unequal sample sizes and missing data
   b. Normality of sampling distributions
   c. Outliers
   d. Linearity
   e. Homogeneity of variance-covariance matrices
   f. Multicollinearity and singularity
2. Major analysis
   a. Significance of discriminant functions. If significant:
      (1) Variance accounted for and confidence limits for each significant function
      (2) Plot(s) of discriminant functions
      (3) Structure matrix
   b. Effect size and confidence limits for solution
   c. Variables separating each group with effect sizes and confidence limits
3. Additional analyses
   a. Group means and standard deviations for high-loading variables
   b. Pooled within-group correlations among predictor variables
   c. Classification results
      (1) Jackknifed classification
      (2) Cross-validation
   d. Change in Rao's V (or stepdown F) plus univariate F for predictors

Results

A direct discriminant analysis was performed using four attitudinal variables as predictors of membership in three groups. Predictors were locus of control, attitude toward marital status, attitude toward role of women, and attitude toward homemaking. Groups were working women, role-satisfied housewives, and role-dissatisfied housewives.

Of the original 465 cases, seven were dropped from analysis because of missing data. Missing data appeared to be randomly scattered throughout groups and predictors. Two additional cases were identified as multivariate outliers with $p < .001$ and were also deleted. Both of the outlying cases were in the working group; they were women with extraordinarily favorable attitudes toward housework. For the remaining 456 cases (239 working women, 136 role-satisfied housewives, and 81 role-dissatisfied housewives), evaluation of assumptions of linearity, normality, multicollinearity or singularity were satisfactory. Statistically significant heterogeneity of variance-covariance matrices ($p < .10$) was observed, however, so a quadratic procedure was used by SAS PROC DISCRIM for analysis.

Two discriminant functions were calculated, with a combined $F(8, 900) = 6.27$, $p < .01$, $\eta^2 = .05$ with 95% confidence limits from .02 to .08. After removal of the first function, there was still strong association between groups and predictors, $F(3, 451) = 5.29$, $p < .01$. Canonical $R^2 = .07$ with 95% confidence limits from .03 to .11 for the first discriminant function and .03 with limits from .00 to .06 for the second discriminant function. Thus, the two functions accounted for about 7% and 3% of the total relationship between predictors and groups. The two discriminant functions account for 69% and 31%, respectively, of the between-group variability. [*F values, squared canonical correlations, and percents of variance are from Table 9.8; cf. Section 9.6.2.*] As shown in Figure 9.4, the first discriminant function maximally separates role-satisfied housewives from the other two groups. The second discriminant function discriminates role-dissatisfied housewives from working women, with role satisfied housewives falling between these two groups.

The structure (loading) matrix of correlations between predictors and discriminant functions, as seen in Table 9.10, suggests that the best predictors for distinguishing between role-satisfied housewives and the other two groups (first function) are attitudes toward current marital status, toward women's role, and toward homemaking. Role-satisfied housewives have more favorable attitudes toward marital status (mean = 20.60, SD = 6.62) than working women (mean = 23.40, SD = 8.53) or role-dissatisfied housewives (mean = 25.62, SD = 10.30), and more conservative attitudes toward women's role (mean = 37.19, SD = 6.46) than working women (mean = 33.86, SD = 6.96) or dissatisfied housewives (mean = 35.67, SD = 5.76). Role-satisfied women are more favorable toward homemaking (mean = 22.51, SD = 3.88) than either working women (mean = 23.81, SD = 4.55) or role-dissatisfied housewives (mean = 24.93, SD = 3.96). [*Group means and standard deviations are shown in Table 9.8.*] Loadings less than .50 are not interpreted.

One predictor, attitudes toward women's role, has a loading in excess of .50 on the second discriminant function, which separates role-dissatisfied housewives from working women. Role-dissatisfied housewives have more conservative attitudes toward the role of women than working women (means have already been cited).

Twelve contrasts were performed where each group, in turn, was contrasted with the other two groups, pooled, to determine which predictors reliably separate each group from the other two groups after adjustment for the other predictors. Table 9.14 shows effect sizes for the 12 contrasts and their 98.75% confidence limits (keeping overall confidence level at .95). When working women were contrasted with the pooled groups of housewives, after adjustment for all other predictors only attitude toward women's role significantly separates working women from the other two groups, $F(1, 450) = 16.87$, $p < .05$.

Role-satisfied housewives differ from the other two groups on attitudes toward marital status, $F(1, 450) = 9.66$, $p < .05$.

The group of role-dissatisfied housewives does not differ from the other two groups on any predictor after adjustment for all other predictors.

Thus, the three groups of women differ most notably on their attitudes toward the proper role of women in society. Working women have the most liberal attitudes, followed by role-dissatisfied housewives, with role-satisfied housewives showing the most conservative attitudes. Role-satisfied housewives also have more positive attitudes toward marriage than the combination of the other two groups.

Pooled within-group correlations among the four predictors are shown in Table 9.8. Of the six correlations, four would show statistical significance at $\alpha = .01$ if tested individually. There is a small positive relationship between locus of control and attitude toward marital status, with $r(454) = .17$ indicating that women who are more satisfied with their current marital status are less likely to attribute control of reinforcements to external sources. Attitude toward homemaking is positively correlated with locus of control, $r(454) = .16$, and attitude toward marital status, $r(454)$ = .28, and negatively correlated with attitude toward women's role, $r(454)$ = -.29. This indicates that women with negative attitudes toward homemaking are likely to attribute control to external sources, to be dissatisfied with their current marital status, and to have more liberal attitudes toward women's role.

With the use of a jackknifed (one case at a time deleted) quadratic classification procedure for the total usable sample of 456 women, 240 (53%) were classified correctly, compared with 178.9 (39%) who would be correctly classified by chance alone. The 53% classification rate was achieved by classifying a disproportionate number of cases as working women. Although 52% of the women actually were employed, the classification scheme, using sample proportions as prior probabilities, classified 70% of the women as employed [*317/456 from Cross-validation classification matrix in Table 9.8*]. This means that the working women were more likely to be correctly classified (75% correct classifications) than either the role-satisfied housewives (39% correct classifications) or the role-dissatisfied housewives (10% correct classifications).

The stability of the classification procedure was checked by a cross-validation run. Approximately 25% of the cases were withheld from calculation of the classification functions in this run. For the 75% of the cases from whom the functions were derived, there was a 54% correct classification rate. For the cross-validation cases, classification was 55%. This indicates a high degree of consistency in the classification scheme, although there is some gain in correct classification for working women at the expense of role-satisfied housewives.

9.8   Comparison of Programs

There are numerous programs for discriminant analysis in statistical packages, some general and some special purpose. SPSS has a general purpose discriminant analysis program that performs direct, sequential, or stepwise DISCRIM with classification. In addition, SPSS MANOVA performs DISCRIM, but not classification. SAS has two programs, with a separate one for stepwise analysis. SYSTAT has a single DISCRIM program. Finally, if the only question is reliability of predictors to separate groups, any of the MANOVA programs discussed in Chapter 7 is appropriate. Table 9.16 compares features of direct discriminant programs. Features for stepwise discriminant function are compared in Table 9.17.

9.8.1   SPSS Package

SPSS DISCRIMINANT, features of which are described in both Tables 9.16 and 9.17, is the basic program in this package for DISCRIM. The program provides direct (standard), sequential, or step-wise entry of predictors with numerous options, but some features are available only in syntax mode. Strong points include several types of plots and plenty of information about classification. Territorial maps are handy for classification using discriminant function scores if there are only a few cases to classify. In addition, a test of homogeneity of variance-covariance matrices is provided through plots and, should heterogeneity be found, classification may be based on separate matrices. Other useful features are evaluation of successive discriminant functions and default availability of structure matrices.

SPSS MANOVA can also be used for DISCRIM and has some features unobtainable in any of the other DISCRIM programs. SPSS MANOVA is described rather fully in Table 7.34 but some aspects especially pertinent to DISCRIM are featured in Table 9.16. MANOVA offers a variety of statistical criteria for testing the significance of the set of predictors (cf. Section 7.6.1). Many matrices can be printed out, and these, along with determinants, are useful for the more sophisticated researcher. Successive discriminant functions (roots) are evaluated, as in SPSS DISCRIMINANT.

SPSS MANOVA provides discriminant functions for more complex designs such as factorial arrangements with unequal sample sizes. The program is limited, however, in that it includes no classification phase. Further, only standard DISCRIM is available, with no provision for stepwise or sequential analysis other than Roy-Bargmann stepdown analysis as described in Chapter 7.

9.8.2   SAS System

In SAS, there are three separate programs to deal with different aspects of discriminant analysis, with surprisingly little overlap between the stepwise and direct programs. However, the older direct program, CANDISC, has been replaced by DISCRIM and is not reviewed here. Both of the SAS programs for discriminant analysis are especially rich in output of SSCP, correlation, and covariance matrices.

The most comprehensive program is DISCRIM, but it does not perform stepwise or sequential analysis. This program is especially handy for classifying new cases or performing cross-validation (Section 9.6.7.1) and in testing and dealing with violation of homogeneity of variance-covariance matrices. DISCRIM offers alternate inferential tests, dimension reduction analysis, and all of the standard matrices of discriminant results.

TABLE 9.16   Comparison of Programs for Direct Discriminant Analysis

Feature	SAS DISCRIM	SPSS DISCRIMINANT	SPSS MANOVA[a]	SYSTAT DISCRIM
Input				
Optional matrix input	Yes	Yes	Yes	No
Missing data options	No	Yes	No	No
Restrict number of discriminant functions	NCAN	Yes	No	No
Specify cumulative % of sum of eigenvalues	No	Yes	No	No
Specify significance level of functions to retain	No	Yes	ALPHA	No
Factorial arrangement of groups	No	No	Yes	CONTRASTS
Specify tolerance	SINGULAR	Yes	No	Yes
Rotation of discriminant functions	No	Yes	Yes	No
Quadratic discriminant analysis	POOL=NO	No	No	Yes
Optional prior probabilities	Yes	Yes	N.A.[b]	Yes
Specify separate covariance matrices for classification	POOL=NO	Yes	N.A.	No
Threshold for classification	Yes	No	N.A.	No
Nonparametric classification method	Yes	No	N.A.	No
Output				
Wilks' lambda with approx. F	Yes	Yes	Yes	Yes
χ^2	No	Yes	No	No
Generalized distance between groups (Mahalanobis D^2)	Yes	Yes	No	No
Hotelling's trace criterion	Yes	No	Yes	Yes
Roy's gcr (maximum root)	Yes	No	Yes	No
Pillai's criterion	Yes	No	Yes	Yes
Tests of successive dimensions (roots)	Yes	Yes	Yes	No[c]
Univariate F ratios	Yes	Yes	Yes	Yes
Group means	Yes	Yes	Yes	PRINT MEDIUM
Total and within-group standardized group means	Yes	No	No	No
Group standard deviations	Yes	Yes	Yes	No

(continued)

TABLE 9.16 Continued

Feature	SAS DISCRIM	SPSS DISCRIMINANT	SPSS MANOVA[a]	SYSTAT DISCRIM
Output (continued)				
Total, within-group and between-group standard deviations	Yes	No	No	No
Standardized discriminant function (canonical) coefficients	Yes	Yes	Yes	PRINT MEDIUM
Unstandardized (raw) discriminant function (canonical) coefficients	Yes	Yes	Yes	PRINT MEDIUM
Group centroids	Yes	Yes	No	Yes
Pooled within-groups (residual) SSCP matrix	Yes	No	Yes	No
Between-groups SSCP matrix	Yes	No	No	No
Hypothesis SSCP matrix	No	No	Yes	No
Total SSCP matrix	Yes	No	No	No
Group SSCP matrices	Yes	No	No	No
Pooled within-groups (residual) correlation matrix	Yes	Yes	Yes	PRINT LONG
Determinant of within-group correlation matrix	No	No	Yes	No
Between-groups correlation matrix	Yes	No	No	No
Group correlation matrices	Yes	No	No	PRINT LONG
Total correlation matrix	Yes	No	No	PRINT LONG
Total covariance matrix	Yes	Yes	No	PRINT LONG
Pooled within-groups (residual) covariance matrix	Yes	Yes	Yes	PRINT LONG
Group covariance matrices	Yes	Yes	No	No
Between-group covariance matrix	Yes	No	No	No
Determinants of group covariance matrices	Yes	No	Yes	No
Homogeneity of variance-covariance matrices	Yes	Yes	Yes	Yes
F matrix, pairwise group comparison	No	Yes	No[d]	Yes
Canonical correlations	Yes	Yes	Yes	Yes
Adjusted canonical correlations	Yes	No	No	No

TABLE 9.16   Continued

Feature	SAS DISCRIM	SPSS DISCRIMINANT	SPSS MANOVA[a]	SYSTAT DISCRIM
Output *(continued)*				
Eigenvalues	Yes	Yes	Yes	Yes
SMCs for each variable	R-Squared	No	No	No
SMC divided by tolerance for each variable	RSQ/ $(1 - RSQ)$	No	No	No
Structure (loading) matrix (pooled within-groups)	Yes	Yes	No	No
Total structure matrix	Yes	No	No	No
Between structure matrix	Yes	No	No	No
Individual discriminant (canonical variate) scores	Data file	Yes	No	Yes
Classification features				
Classification of cases	Yes	Yes	N.A.[b]	Yes
Classification function coefficients	Yes[e]	Yes	N.A.	PRINT MEDIUM
Classification matrix	Yes	Yes	N.A.	Yes
Posterior probabilities for classification	Data file	Yes	N.A.	PRINT LONG
Mahalanobis' D^2 or leverage for cases (outliers)	No	Yes[f]	N.A.	PRINT LONG
Jackknifed (leave-one-out) classification matrix	Yes	Yes	N.A.	Yes
Classification with a cross-validation sample	Yes	Yes	N.A.	Yes
Plots				
All groups scatterplot	No	Yes	N.A.	Yes
Centroid included in all groups scatterplot	No	Yes	N.A.	No
Separate scatterplots by group	No	Yes	N.A.	No
Territorial map	No	Yes	N.A.	No

[a]Additional features reviewed in Section 7.7.

[b]SPSS MANOVA does not classify cases.

[c]Available in GLM with PRINT-LONG. See Chapter 7 for additional features.

[d]Can be obtained through CONTRAST procedure.

[e]Labeled Linear Discriminant Function.

[f]Outliers in the solution.

TABLE 9.17   Comparison of Programs for Stepwise and Sequential Discriminant Analysis

Feature	SPSS DISCRIMINANT	SAS STEPDISC	SYSTAT DISCRIM
Input			
Optional matrix input	Yes	Yes	Yes
Missing data options	Yes	No	No
Specify contrast	No	No	Yes
Factorial arrangement of groups	No	No	CONTRAST
Suppress intermediate steps	No	No	No
Suppress all but summary table	NOSTEP	SHORT	No
Optional methods for order of entry/removal	3	5	2
Forced entry by level (sequential)	No	No	Yes
Force some variables into model	Yes	INCLUDE	FORCE
Specify tolerance	Yes	SINGULAR	Yes
Specify maximum number of steps	Yes	Yes	No
Specify number of variables in final stepwise model	No	STOP=	No
Specify F to enter/remove	FIN/FOUT	No	FEnter/Fremove
Specify significance of F to enter/remove	PIN/POUT	SLE/SLS	Enter/Remove
Specify partial R^2 to enter/remove	No	PR2E/PR2S	No
Restrict number of discriminant functions	Yes	No	No
Specify cumulative % of sum of eigenvalues	Yes	No	No
Specify significance level of functions to retain	Yes	No	No
Rotation of discriminant functions	Yes	No	No
Prior probabilities optional	Yes	N.A.[a]	Yes
Specify separate covariance matrices for classification	Yes	N.A.	No
Output			
Wilks' lambda with approximate F	Yes	Yes	PRINT MEDIUM
χ^2	Yes	No	No
Mahalanobis' D^2 (between groups)	Yes	No	No
Rao's V	Yes	No	No
Pillai's criterion	No	Yes	PRINT MEDIUM
Tests of successive dimensions (roots)	Yes	No	No
Univariate F ratios	Yes	STEP 1 F	Yes[b]
Group means	Yes	Yes	PRINT MEDIUM
Group standard deviations	Yes	Yes	No
Total and pooled within-group standard deviations	No	Yes	No

TABLE 9.17 Continued

Feature	SPSS DISCRIMINANT	SAS[a] STEPDISC	SYSTAT DISCRIM
Output *(continued)*			
Standardized discriminant function (canonical) coefficients	Yes	No	PRINT MEDIUM
Unstandardized discriminant function (canonical) coefficients	Yes	No	PRINT MEDIUM
Group centroids	Yes	No	Yes[c]
Pooled within-group correlation matrix	Yes	Yes	PRINT LONG
Total correlation matrix	No	Yes	PRINT LONG
Total covariance matrix	Yes	Yes	PRINT LONG
Total SSCP matrix	No	Yes	No
Pooled within-group covariance matrix	Yes	Yes	PRINT LONG
Pooled within-group SSCP matrix	No	Yes	No
Group covariance matrices	Yes	Yes	PRINT LONG
Group correlation matrices	No	Yes	PRINT LONG
Group SSCP matrices	No	Yes	No
Between-group correlation matrix	No	Yes	No
Between-group covariance matrix	No	Yes	No
Between-group SSCP matrix	No	Yes	No
Homogeneity of variance-covariance matrices	Yes	No	Yes
F matrix, pairwise group comparison	Yes	No	Yes
Canonical correlations, each discriminant function	Yes	No	Yes
Canonical correlations, average	No	Yes	No
Eigenvalues	Yes	No	Yes
Structure (loading) matrix	Yes	No	No
Partial R^2 (or tolerance) to enter/remove, each step	Yes	Yes	Yes
F to enter/remove, each step	Yes	Yes	Yes
Classification features			
Classification of cases	Yes	N.A.[a]	Yes
Classification function coefficients	Yes	N.A.	PRINT MEDIUM
Classification matrix	Yes	N.A.	Yes
Individual discriminant (canonical variate) scores	Yes	N.A.	PRINT LONG

(continued)

TABLE 9.17 Continued

Feature	SPSS DISCRIMINANT	SAS STEPDISC	SYSTAT DISCRIM
Output *(continued)*			
Posterior probabilities for classification	Yes	N.A.	PRINT LONG
Mahalanobis' D^2 for cases (outliers)	Yes[d]	N.A.	PRINT LONG
Jackknifed classification matrix	CROSS VALIDATE	N.A.	Yes
Classification with a cross-validation sample	Yes	N.A.	Yes
Classification information at each step	No	N.A.	No
Plots			
Plot of group centroids alone	No	N.A.	No
All groups scatterplot	Yes	N.A.	Yes
Separate scatterplots by group	Yes	N.A.	No
Territorial map	Yes	N.A.	No

[a]SAS STEPDISC does not classify cases (see SAS DISCRIM, Table 9.16).

[b]F-to-enter prior to first step.

[c]Canonical scores of group means.

[d]Outliers in the solution.

Stepwise (but not sequential) analysis is accomplished through STEPDISC. As seen in Table 9.17, very few additional amenities are available in this program. There is no classification, nor is there information about the discriminant functions. On the other hand, this program offers plenty of options for entry and removal of predictors.

9.8.3 SYSTAT System

SYSTAT DISCRIM is the discriminant analysis program. The program deals with all varieties of DISCRIM. Automatic (forward and backward) and interactive stepping are available, as well as a contrast procedure to control entry of variables. The contrast procedure also is useful for comparing means of one group with pooled means of the other groups. Jackknifed classification is produced by default, and cross-validation may be done as well. Dimension reduction analysis is no longer available, but can be obtained by rephrasing the problem as MANOVA and running it through GLM with PRINT=LONG. Such a strategy also is well suited to factorial arrangements of unequal-n groups.

Scatterplot matrices (SYSTAT SPLOM) may be used to evaluate homogeneity of variance-covariance matrices; quadratic discrimination analysis is available through DISCRIM should the assumption be violated. Several univariate and multivariate inferential tests also are available. SYSTAT DISCRIM can be used to assess outliers through Mahalanobis distance of each case to each group centroid.

10 Logistic Regression

10.1 General Purpose and Description

Logistic regression allows one to predict a discrete outcome such as group membership from a set of variables that may be continuous, discrete, dichotomous, or a mix. Because of its popularity in the health sciences, the discrete outcome in logistic regression is often disease/no disease. For example, can presence or absence of hay fever be diagnosed from geographic area, season, degree of nasal stuffiness, and body temperature?

Logistic regression is related to, and answers the same questions as, discriminant analysis, the logit form of multiway frequency analysis with a discrete DV, and multiple regression analysis with a dichotomous DV. However, logistic regression is more flexible than the other techniques. Unlike discriminant analysis, logistic regression has no assumptions about the distributions of the predictor variables; in logistic regression, the predictors do not have to be normally distributed, linearly related, or of equal variance within each group. Unlike multiway frequency analysis, the predictors do not need to be discrete; the predictors can be any mix of continuous, discrete and dichotomous variables. Unlike multiple regression analysis, which also has distributional requirements for predictors, logistic regression cannot produce negative predicted probabilities.

There may be two or more outcomes (groups) in logistic regression. If there are more than two outcomes, they may or may not have order (e.g., no hay fever, moderate hay fever, severe hay fever). Logistic regression emphasizes the probability of a particular outcome for each case. For example, it evaluates the probability that a given person has hay fever, given that person's pattern of responses to questions about geographic area, season, nasal stuffiness, and temperature.

Logistic regression analysis is especially useful when the distribution of responses on the DV is expected to be nonlinear with one or more of the IVs. For example, the probability of heart disease may be little affected (say 1%) by a 10-point difference among people with low blood pressure (e.g., 110 vs. 120) but may change quite a bit (say 5%) with an equivalent difference among people with high blood pressure (e.g., 180 vs. 190). Thus, the relationship between heart disease and blood pressure is not linear.

Bennett and colleagues (1991) compared non-ulcer dyspeptic (NUD) patients with controls in an application of logistic regression to a case-control study. Subjects were matched on age, sex, and social status, with one control for each patient. Predictor variables included a variety of life stress, personality, mood state, and coping measures. While univariate analyses showed that patients differed from controls on 17 psychological variables, logistic regression analysis showed that a single predictor, highly threatening chronic difficulties, alone provided a highly adequate model. While 98% of NUDs were exposed to at least one such stressor, only 2% of controls were so exposed. In its presence, no other predictor could improve the model.

Fidell and colleagues (1995) studied the probability of awakening from sleep at night associated with some event as a function of four noise characteristics, three personal characteristics, three time-related characteristics and three pre-sleep characteristics. Seven variables successfully predicted behavioral awakenings associated with noise events in a standard logistic regression analysis. Predictive noise characteristics were sound level (positive relationship) and ambient level (negative relationship). Personal characteristics predicting probability of awakening were number of spontaneous awakenings (negatively related) and age (also negatively related). Time since retiring was the strongest predictor (positive relationship), with duration of residence a positive, statistically significant, but trivial predictor. Rating of tiredness the night before strongly and positively predicted awakening. However, the model with all predictors included accounted for only 13% of the variance in probability of awakening. Correct prediction of non-awakening was 97%, but correct prediction of awakening, a rare event, was only 8%.

Kirkpatrick and Messias (2003) looked at predictors of substance abuse among schizophrenics in an epidemiological catchment area study. Three groups of schizophrenics were defined on the basis of substance abuse: alcohol and marijuana abuse, polysubstance abuse, and those with no substance abuse serving as a reference group. Covariates were gender, age, and race. Clinical features used as predictors were thought disorganization, prevalence of deficits (e.g., blunted affect, 2 weeks of depressive mood), and hallucinations/delusions. The presence of thought disorganization was associated with a 6-fold increase in risk of alcohol/marijuana abuse. The group identified with greater prevalence of deficits was associated with a more than 5-fold increase in risk of polysubstance abuse.

Because the model produced by logistic regression is nonlinear, the equations used to describe the outcomes are slightly more complex than those for multiple regression. The outcome variable, $\hat{Y}$, is the probability of having one outcome or another based on a nonlinear function of the best linear combination of predictors; with two outcomes:

$$\hat{Y}_i = \frac{e^u}{1 + e^u} \tag{10.1}$$

where $\hat{Y}_i$ is the estimated probability that the ith case ($i = 1,..., n$) is in one of the categories and u is the usual linear regression equation:

$$u = A + B_1 X_1 + B_2 X_2 + \cdots + B_k X_k \tag{10.2}$$

with constant A, coefficients B_j, and predictors, X_j for k predictors ($j = 1, 2,..., k$).

This linear regression equation creates the *logit* or log of the odds:

$$\ln\left(\frac{\hat{Y}}{1 - \hat{Y}}\right) = A + \sum B_j X_{ij} \tag{10.3}$$

That is, the linear regression equation is the natural log ($\log_e$) of the probability of being in one group divided by the probability of being in the other group. The procedure for estimating coefficients is maximum likelihood, and the goal is to find the best linear combination of predictors to maximize the likelihood of obtaining the observed outcome frequencies. Maximum likelihood estimation is an iterative procedure that starts with arbitrary values of coefficients for the set of predictors and determines the direction and size of change in the coefficients that will maximize the likelihood of obtaining the observed frequencies. Then residuals for the predictive model based on those coefficients are tested and another determination of direction and size of change in coefficients is made, and so on, until the coefficients change very little, i.e., convergence is reached. In effect, maximum likelihood estimates are those parameter estimates that maximize the probability of finding the sample data that actually have been found (Hox, 2002).

Logistic regression, like multiway frequency analysis, can be used to fit and compare models. The simplest (and worst-fitting) model includes only the constant and none of the predictors. The most complex (and "best"-fitting) model includes the constant, all predictors, and, perhaps, interactions among predictors. Often, however, not all predictors (and interactions) are related to the outcome. The researcher uses goodness-of-fit tests to choose the model that does the best job of prediction with the fewest predictors.

10.2 Kinds of Research Questions

The goal of analysis is to correctly predict the category of the outcome for individual cases. The first step is to establish that there is a relationship between the outcome and the set of predictors. If a relationship is found, one usually tries to simplify the model by eliminating some predictors while still maintaining strong prediction. Once a reduced set of predictors is found, the equation can be used to predict outcomes for new cases on a probabilistic basis.

10.2.1 Prediction of Group Membership or Outcome

Can outcome be predicted from the set of variables? For example, can hay fever be predicted from geographic area, season, degree of nasal stuffiness, and body temperature? Several tests of relationship are available in logistic regression. The most straightforward compares a model with the constant plus predictors with a model that has only the constant. A statistically significant difference between the models indicates a relationship between the predictors and the outcome. This procedure is demonstrated in Section 10.4.2.

An alternative is to test a model with only some predictors against the model with all predictors (called a full model). The goal is to find a nonsignificant χ^2, indicating no statistically significant difference between the model with only some predictors and the full model. The use of these and other goodness-of-fit tests is discussed in Section 10.6.1.1.

10.2.2 Importance of Predictors

Which variables predict the outcome? How do variables affect the outcome? Does a particular variable increase or decrease the probability of an outcome, or does it have no effect on outcome? Does inclusion of information about geographic area improve prediction of hay fever and is a particular area associated with an increase or decrease in the probability that a case has hay fever? Several methods of answering these questions are available in logistic regression. One may, for instance, ask how much the model is harmed by eliminating a predictor, or one may assess the statistical significance of the coefficients associated with each of the predictors, or one may ask how much the odds of observing an outcome are changed by a predictor. These procedures are discussed in Sections 10.4 and 10.6.8.

10.2.3 Interactions among Predictors

As in multiway frequency (logit) analysis, a model can also include interactions among the predictor variables: two-way interactions and, if there are many predictor variables, higher-order interactions. For example, knowledge of geographic area and season, in combination, might be useful in the prediction of hay fever; or knowledge of degree of nasal stuffiness combined with fever. Geographic area may be associated with hay fever only in some seasons; stuffiness might only matter with no fever. Other combinations such as between temperature and geographic areas may, however, not be helpful. If there are interactions among continuous variables (or powers of them), multicollinearity is avoided by centering the variables (Section 5.6.6).

Like individual predictors, interactions may complicate a model without significantly improving the prediction. Decisions about including interactions are made in the same way as decisions about including individual predictors. Section 10.6.7 discusses decisions about whether inclusion of interactions also presumes inclusion of their individual components.

10.2.4 Parameter Estimates

The parameter estimates in logistic regression are the coefficients of the predictors included in a model. They are related to the *A* and *B* values of Equation 10.2. Section 10.4.1 discusses methods for calculating parameter estimates. Section 10.6.3 shows how to use parameter estimates to calculate and interpret odds. For example, what are the odds that someone has hay fever in the spring, given residence in the Midwest, nasal stuffiness, and no fever?

10.2.5 Classification of Cases

How good is a statistically significant model at classifying cases for whom the outcome is known? For example, how many people with hay fever are diagnosed correctly? How many people without hay fever are diagnosed correctly? The researcher establishes a cutpoint (say, .5) and then asks, for instance: How many people with hay fever are correctly classified if everyone with a predicted probability of .5 or more is diagnosed as having hay fever? Classification of cases is discussed in Section 10.6.6.

10.2.6 Significance of Prediction with Covariates

The researcher may consider some of the predictors to be covariates and others to be independent variables. For example, the researcher may consider stuffiness and temperature covariates, and geographic area and season independent variables in an analysis that asks if knowledge of geographic area and season added to knowledge of physical symptoms reliably improves prediction over knowledge of physical symptoms alone. Section 10.5.2 discusses sequential logistic regression and a complete example of Section 10.7.3 demonstrates sequential logistic regression.

10.2.7 Effect Size

How strong is the relationship between outcome and the set of predictors in the chosen model? What proportion of variance in outcome is associated with the set of predictors? For example, what proportion of the variability in hay fever is accounted for by geographic area, season, stuffiness, and temperature?

The logic of assessing effect size is different in routine statistical hypothesis testing from situations where models are being evaluated. In routine statistical hypothesis testing, one might not report effect size for a nonsignificant effect. However, in model testing, the goal is often to find *non-significance*, to find effect size for a model that is not reliably different from a full model. However, when samples are large, there may be a statistically significant deviation from the full model, even when a model does a fine job of prediction. Therefore, effect size is also reported with a model that deviates significantly from chance. Measures of effect size are discussed in Section 10.6.2.

10.3 Limitations to Logistic Regression Analysis

Logistic regression is relatively free of restrictions and, with the capacity to analyze a mix of all types of predictors (continuous, discrete, and dichotomous), the variety and complexity of data sets that can be analyzed are almost unlimited. The outcome variable does have to be discrete, but a continuous variable can be converted to a discrete one when there is reason to do so.

10.3.1 Theoretical Issues

The usual cautions about causal inference apply, as in all analyses in which one variable is an outcome. To say that the probability of correctly diagnosing hay fever is related to geographic area, season, nasal stuffiness, and fever is not to imply that any of those variables cause hay fever.

As a flexible alternative to both discriminant analysis and the logit form of multiway frequency analysis, the popularity of logistic regression analysis is growing. The technique has the sometimes useful property of producing predicted values that are probabilities between 0 and 1. However, when assumptions regarding the distributions of predictors are met, discriminant analysis may be a more powerful and efficient analytic strategy. On the other hand, discriminant analysis sometimes overestimates the size of the association with dichotomous predictors (Hosmer & Lemeshow, 2000). Multiple regression is likely to be more powerful than logistic regression when the outcome is continuous and the assumptions regarding it and the predictors are met.

When all the predictors are discrete, multiway frequency analysis offers some convenient screening procedures that may make it the more desirable option.

As in all research, the importance of selecting predictors on the basis of a well-justified, theoretical model cannot be overemphasized. In logistic regression, as in other modeling strategies, it is tempting (and often common in the research community) to amass a large number of predictors and then, on the basis of a single data set, eliminate those that are not statistically significant. This a practice widely adopted in the research community, but is especially dangerous in logistic regression because the technique is often used to address life-and-death issues in medical policy and practice (Harlow, 2002).

10.3.2 Practical Issues

Although assumptions regarding the distributions of predictors are not required for logistic regression, multivariate normality and linearity among the predictors may enhance power, because a linear combination of predictors is used to form the exponent (see Equations 10.1 and 10.2). Also, it is assumed that continuous predictors are linear with the logit of the DV. Other limitations are mentioned below.

10.3.2.1 Ratio of Cases to Variables

A number of problems may occur when there are too few cases relative to the number of predictor variables. Logistic regression may produce extremely large parameter estimates and standard errors, and, possibly, failure of convergence when combinations of discrete variables result in too many cells with no cases. If this occurs, *collapse categories, delete the offending category, or delete the discrete variable if it is not important to the analysis.*

   A maximum likelihood solution also is impossible when outcome groups are perfectly separated. Complete separation of groups by a dichotomous predictor occurs when all cases in one outcome group have a particular value of a predictor (e.g., all those with hay fever have sniffles) while all those in another group have another value of a predictor (e.g., no hay fever and no sniffles.) This is likely to be a result of too small a sample rather than a fortuitous discovery of the perfect predictor that will generalize to the population. Complete separation of groups also can occur when there are too many variables relative to the few cases in one outcome (Hosmer & Lemeshow, 2000). This is essentially a problem of overfitting, as occurs with a small case-to-variable ratio in multiple regression (cf. Section 5.3.2.1).

   Extremely high parameter estimates and standard errors are indications that a problem exists. These estimates also increase with succeeding iterations (or the solution fails to converge). If this occurs, *increase the number of cases or eliminate one or more predictors.* Overfitting with small samples is more difficult to spot in logistic regression than in multiple regression because there is no form of "adjusted R^2" which, when very different from unadjusted R^2 in multiple regression, signals an inadequate sample size.

10.3.2.2 Adequacy of Expected Frequencies and Power

When a goodness-of-fit test is used that compares observed with expected frequencies in cells formed by combinations of discrete variables, the analysis may have little power if expected frequencies are too small. *If you plan to use such a goodness-of-fit test, evaluate expected cell frequencies for all pairs of discrete variables, including the outcome variable.* Recall from garden-variety χ^2 (Section 3.6) that expected frequencies = [(row total) × (column total)]/grand total. It is best if all expected frequencies are greater than one, and that no more than 20% are less than five. Should either of these conditions fail, the choices are: (1) accept lessened power for the analysis, (2) collapse categories for variables with more than two levels, (3) delete discrete variables to reduce the number of cells, or (4) use a goodness-of-fit criterion that is not based on observed versus expected frequencies of cells formed by categorical variables, as discussed in Section 10.6.1.1.

   As with all statistical techniques, power increases with sample size. Some statistical software available for determining sample size and power specifically for a logistic regression analysis includes NCSS PASS (Hintze, 2002) and nQuery Advisor (Elashoff, 2000). The Internet is a never-ending but ever-changing source of free power programs.

10.3.2.3 Linearity in the Logit

Logistic regression assumes a linear relationship between continuous predictors and the logit transform of the DV (see Equation 10.3), although there are no assumptions about linear relationships among predictors themselves.

There are several graphical and statistical methods for testing this assumption; the Box-Tidwell approach (Hosmer & Lemeshow, 2000) is among the simplest. In this approach, *terms are added to the logistic regression model which are composed of the interactions between each predictor and its natural logarithm.* The assumption is violated if one or more of the added interaction terms is statistically significant. Violation of the assumptions leads to transformation (Section 4.1.6) of the offending predictor(s). Tests of linearity of the logit are demonstrated in Section 10.7.

10.3.2.4 Absence of Multicollinearity

Logistic regression, like all varieties of multiple regression, is sensitive to extremely high correlations among predictor variables, signaled by exceedingly large standard errors for parameter estimates and/or failure of a tolerance test in the computer run. To find a source of multicollinearity among the discrete predictors, use multiway frequency analysis (cf. Chapter 16) to find very strong relationships among them. To find a source of multicollinearity among the continuous predictors, replace the discrete predictors with dichotomous dummy variables and then use the procedures of Section 4.1.7. *Delete one or more redundant variables from the model to eliminate multicollinearity.*

10.3.2.5 Absence of Outliers in the Solution

One or more of the cases may be very poorly predicted by the solution; a case that actually is in one category of outcome may show a high probability for being in another category. If there are enough cases like this, the model has poor fit. Outlying cases are found by examination of residuals, which can also aid in interpreting the results of the logistic regression analysis. Section 10.4.4 discusses how to examine residuals to evaluate outliers.

10.3.2.6 Independence of Errors

Logistic regression assumes that responses of different cases are independent of each other. That is, it is assumed that each response comes from a different, unrelated, case. Thus, logistic regression basically is a between-subjects strategy.

However, if the design is repeated measures, say the levels of the outcome variable are formed by the time period in which measurements are taken (before and after some treatment) or the levels of outcome represent experimental vs. control subjects who have been matched on a 1 to 1 basis (called a matched case-control study), the usual logistic regression procedures are inappropriate because of correlated errors.

The effect of non-independence in logistic regression is to produce overdispersion, a condition in which the variability in cell frequencies is greater than expected by the underlying model. This results in an inflated Type I error rate for tests of predictors. One remedy is to do multilevel modeling with a categorical DV in which such dependencies are considered part of the model (cf. Section 15.5.4).

There are two fixes which provide conservative tests of predictors to compensate for the increased Type I error rate due to non-independence. A simple remedy for overdispersion in a logistic regression model is to rescale the Wald standard errors for each parameter (Section 10.4.1) by a variance inflation factor. This is done by multiplying the calculated standard error by $(\chi^2/df)^{1/2}$, where χ^2 and df are from the deviance or Pearson goodness-of-fit statistics (available in SAS LOGISTIC and in SPSS NOMREG, which also can be instructed to do the scaling). Indeed, one indication that overdispersion is a problem is a large discrepancy between the Pearson and deviance test statistics. The larger of the two values is to be used to compute the variance inflation factor.

SAS LOGISTIC permits a more sophisticated remedy if all predictors are discrete, by scaling the standard errors through the `scale` instruction plus an `aggregate` instruction that specifies the variable indicating the matching identifier, e.g., individual or pair number. This provides the appropriate standard errors for tests of parameters, but the deviance and Pearson χ^2 test cannot be used to evaluate goodness of fit of the model.

Also, special procedures are available in both statistical packages for matched case-control studies, as described in Section 10.6.9. Within-subjects (repeated measures) analysis also is available through SAS CATMOD, but, again, predictors must be discrete. SPSS COMPLEX SAMPLES may be used for repeated-measures designs with a dichotomous DV when cases are defined as clusters.

10.4   Fundamental Equations for Logistic Regression

Table 10.1 shows a hypothetical data set in which falling down (0 = not falling, 1 = falling) on a ski run is tested against the difficulty of the run (on an ordered scale from 1 to 3, treated as if continuous) and the season (a categorical variable where 1 = autumn, 2 = winter, and 3 = spring). Data

TABLE 10.1   Small Sample of Hypothetical Data for Illustration of Logistic Regression Analysis

Fall	Difficulty	Season
1	3	1
1	1	1
0	1	3
1	2	3
1	3	2
0	2	2
0	1	2
1	3	1
1	2	3
1	2	1
0	2	2
0	2	3
1	3	2
1	2	2
0	3	1

from 15 skiers are presented. Logistic regression uses procedures similar to both multiple regression and multiway frequency analysis. Like multiple regression, the prediction equation includes a linear combination of the predictor variables. For example, with three predictors and no interactions:

$$\hat{Y}_i = \frac{e^{A+B_1X_1+B_2X_2+B_3X_3}}{1 + e^{A+B_1X_1+B_2X_2+B_3X_3}} \tag{10.4}$$

The difference between multiple regression and logistic regression is that the linear portion of the equation $(A + B_1X_1 + B_2X_2 + B_3X_3)$, the logit, is not the end in itself, but is used to find the odds of being in one of the categories of the DV given a particular combination of scores on the Xs. Similar to multiway frequency analysis, models are evaluated by assessing the (natural log) likelihood for each model. Models are then compared by calculating the difference between their log-likelihoods.

In this section the simpler calculations are illustrated in detail, while those involving calculus or matrix inversion are merely described and left to computer software for solution. At the end of the section, the most straightforward program in each package (SPSS LOGISTIC REGRESSION and SAS LOGISTIC) is demonstrated for the small data set. Additional programs for analysis of more complex data sets are described in Section 10.8.

Before analysis, discrete variables are recoded into a series of dichotomous (dummy) variables, one fewer than there are categories. Thus two dichotomous variables, called season(1) and season(2), are created to represent the three categories of season. Following the convention of most software, season(1) is coded 1 if the season is autumn, and 0 otherwise; season(2) is coded 1 if winter, and 0 otherwise. Spring is identified by codes of 0 on both dummy variables.

10.4.1   Testing and Interpreting Coefficients

Solving for logistic regression coefficients A and B and their standard errors involves calculus, in which values are found using maximum likelihood methods. These values, in turn, are used to evaluate the fit of one or more models (cf. Section 10.4.2). If an acceptable model is found, the statistical significance of each of the coefficients is evaluated[1] using the Wald test where the squared coefficient is divided by its squared standard error:

$$W_j = \frac{B_j^2}{SE_{B_j^2}} \tag{10.5}$$

A squared parameter estimate divided by its squared standard error is a χ^2 statistic. Table 10.2 shows these coefficients, their standard errors, and the Wald test obtained from statistical software. None of the predictors is statistically significant in this tiny data set.

The logistic regression equation is:

$$Prob(fall) = \hat{Y}_i = \frac{e^{-1.776+(1.010)(DIFF)+(0.927)(SEAS1)+(-0.418)(SEAS2)}}{1 + e^{-1.776+(1.010)(DIFF)+(0.927)(SEAS1)+(-0.418)(SEAS2)}}$$

---

[1]It is convenient for didactic purposes to first illustrate coefficients and then show how they are used to develop goodness-of-fit tests for models.

445

CHAPTER 10

TABLE 10.2 Coefficients, Standard Errors, and Wald Test Derived for Small-Sample Example

Term	Coefficient	Standard Wald Error	Test (χ^2)
(CONSTANT)	−1.776	1.89	0.88
DIFFICULTY	1.010	0.90	1.27
SEASON(1)	0.927	1.59	0.34
SEASON(2)	−0.418	1.39	0.91

Because the equation is solved for the outcome "falling," coded 1,[2] the derived probabilities are also for falling. Since none of the coefficients is significant, the equation would normally not be applied to any cases. However, for illustration, the equation is applied below to the first case, a skier who actually did fall on a difficult run in the autumn. The probability is:

$$Prob(fall) = \hat{Y}_i = \frac{e^{-1.776+(1.010)(3)+(0.927)(1)+(-0.418)(0)}}{1 + e^{-1.776+(1.010)(3)+(0.927)(1)+(-0.418)(0)}}$$

$$= \frac{e^{2.181}}{1 + e^{2.181}}$$

$$= \frac{8.855}{9.855} = .899$$

Prediction is quite good for this case, since the probability of falling is .899 (with a residual of $1 - .899 = .101$, where 1 represents the actual outcome: falling). Section 10.6.3 discusses further interpretation of coefficients.

10.4.2 Goodness-of-Fit

For a candidate model, a log-likelihood is calculated, based on summing the probabilities associated with the predicted and actual outcomes for each case:

$$\text{log-likelihood} = \sum_{i=1}^{N} [Y_i \ln(\hat{Y}_i) + (1 - Y_i) \ln(1 - \hat{Y}_i)] \tag{10.6}$$

Table 10.3 shows the actual outcome (Y) and predicted probability of falling for the 15 cases in the small-sample example, along with the values needed to calculate log-likelihood.

---

[2]Some texts and software solve the equation for the outcome coded 0 by default, in the example above "not falling."

TABLE 10.3   Calculation of Log-Likelihood for Small-Sample Example

Outcome	Predicted Probability				Log-Likelihood
Y	$\hat{Y}$	$1 - \hat{Y}$	$Y \ln \hat{Y}$	$(1-Y)\ln(1-\hat{Y})$	$\sum [Y \ln \hat{Y} + (1-Y)\ln(1-\hat{Y})]$
1	.899	.101	−0.106	0	−0.106
1	.540	.460	−0.616	0	−0.616
0	.317	.683	0	−0.381	−0.381
1	.516	.439	−0.578	0	−0.578
1	.698	.302	−3.360	0	−0.360
0	.457	.543	0	−0.611	−0.611
0	.234	.766	0	−0.267	−0.267
1	.899	.101	−0.106	0	−0.106
1	.451	.439	−0.578	0	−0.578
1	.764	.236	−0.267	0	−0.269
0	.457	.543	0	−0.611	−0.611
0	.561	.439	0	−0.823	−0.823
1	.698	.302	−0.360	0	−0.360
1	.457	.543	−0.783	0	−0.783
0	.899	.101	0	−2.293	−2.293
					SUM $= -8.74^{a}$

[a] −2 * log-likelihood = 17.48.

Two models are compared by computing the difference in their log-likelihoods (times −2) and using chi square. The bigger model is the one to which predictors have been added to the smaller model. Models must be nested to be compared; all the components of the smaller model must also be in the bigger model.

$$\chi^2 = [(-2 * [(\text{log-likelihood for smaller model}) - (-2 * \text{log-likelihood for the bigger model})] \quad (10.7)$$

When the bigger model contains all predictors and the smaller model contains only the intercept, a conventional notation for Equation 10.7 is:

$$\chi^2 = 2[\text{LL}(B) - \text{LL}(0)]$$

For this example, the log-likelihood for the smaller model that contains only the constant is −10.095. When all predictors are in the bigger model, the log-likelihood, as shown in Table 10.3, is −8.740. The difference between log-likelihoods is multiplied by two to create a statistic that is distributed as chi square. In the example, the difference (times two) is:

$$\chi^2 = 2[(-8.740) - (-10.095)] = 2.71$$

Degrees of freedom are the difference between degrees of freedom for the bigger and smaller models. The constant-only model has 1 df (for the constant) and the full model has 4 df (1 df for each individual effect and one for the constant); therefore χ^2 is evaluated with 3 df. Because χ^2 is not statistically significant at $\alpha = .05$, the model with all predictors is no better than one with no predictors, an expected result because of the failure to find any statistically significant predictors. Additional goodness-of-fit statistics are described in Section 10.6.1.1.

10.4.3   Comparing Models

This goodness-of-fit χ^2 process is also used to evaluate predictors that are eliminated from the full model, or predictors (and their interactions) that are added to a smaller model. In general, as predictors are added/deleted, log-likelihood decreases/increases. The question in comparing models is, Does the log-likelihood decrease/increase significantly with the addition/deletion of predictor(s)?

For example, the $-2 *$ log-likelihood for the small-sample example with difficulty removed is 18.87. Compared to the full model, χ^2 is, using Equation 10.7:

$$\chi^2 = (18.87 - 17.48) = 1.39$$

with 1 df, indicating no significant enhancement to prediction of falling by knowledge of difficulty of the ski run.

10.4.4   Interpretation and Analysis of Residuals

As shown in Section 10.4.2, the first case has a residual of .101; the predicted probability of falling, .899, was off by .101 from the actual outcome of falling for that case of 1.00. Residuals are calculated for each case and then standardized to assist in the evaluation of the fit of the model to each case.

There are several schemes for standardizing residuals. The one used here is common among software packages, and defines a standardized residual for a case as:

$$std\ residual_i = \frac{(Y_i - \hat{Y}_i)/\hat{Y}_i(1 - \hat{Y}_i)}{\sqrt{1 - h_i}} \tag{10.8}$$

where

$$h_i = \hat{Y}_i(1 - \hat{Y}_i)x_i'(X'VX)^{-1}x_i \tag{10.9}$$

and where x_i is the vector of predictors for the case, X is the data matrix for the whole sample including the constant, and V is a diagonal matrix with general element:

$$\hat{Y}_i(1 - \hat{Y}_i)$$

Table 10.4 shows residuals and standardized residuals for each case in the small-sample example. There is a very large residual for the last case, a skier who did not fall, but had a predicted prob-

TABLE 10.4 Residuals and Standardized Residuals for Small-Sample Example

Outcome	Residual	Standardized Residual
1	.101	0.375
1	.460	1.403
0	−.317	−0.833
1	.439	1.034
1	.302	0.770
0	−.457	−1.019
0	−.234	−0.678
1	.101	0.375
1	.439	1.034
1	.236	0.646
0	−.457	−1.019
0	−.561	−1.321
1	.302	0.770
1	.543	1.213
0	−.899	3.326

ability of falling of about .9. This is the case the model predicts most poorly. With a standardized residual $(z) = 3.326$ in a sample of this size, the case is an outlier in the solution.

10.4.5 Computer Analyses of Small-Sample Example

Syntax and selected output for computer analyses of the data in Table 10.1 appear in Tables 10.5 and 10.6: SAS LOGISTIC in Table 10.5 and SPSS LOGISTIC REGRESSION in Table 10.6.

As seen in Table 10.5, SAS LOGISTIC uses the CLASS instruction to designate categorical predictors; param=glm produces internal coding that matches that of hand calculations and default SPSS coding.

The **Response Profile** in the output shows the coding and numbers of cases for each outcome group. Three **Model Fit Statistics** are given for the **Intercept Only** model and the full model (**Intercept and Covariates**), including **−2 Log L**, which is −2 times the log-likelihood of Section 10.4.2. Under **Testing Global Null Hypotheses: BETA=0** are three χ^2 goodness-of-fit tests for the overall model. The **Likelihood Ratio** test is the test of the full model vs. the constant-only model (cf. Section 10.4.2).

Analysis of Type 3 Effects shows tests of significance for each of the predictors, combining the df for SEASON into a single test. The **Analysis of Maximum Likelihood Estimates** provides B weights (**Parameter Estimates**) for predicting the probability of *not* falling (code of 0 for FALL; SAS solves for the 0 code), the **Standard Error** of B, and **Wald Chi-Square** together with its **Pr**(obability). **Standardized** (parameter), **Estimates**, and **Odds Ratios** along with their 95% confidence limits are also provided. SAS LOGISTIC additionally provides several measures of effect size for the set of predictors: Somers' D, Gamma, Tau-a, and Tau-c (see Section 10.6.2).

TABLE 10.5   Syntax and Selected Output from SAS LOGISTIC Analysis of Small-Sample Example

```
proc logistic data=SASUSER.SSLOGREG;
    class season / param=glm;
    model FALL=DIFFCLTY SEASON;
run;
```

The LOGISTIC Procedure

Response Profile

Ordered Value	FALL	Total Frequency
1	0	6
2	1	9

Probability modeled is FALL=0.

Class Level Information

		Design Variables		
Class	Value	1	2	3
SEASON	1	1	0	0
	2	0	1	0
	3	0	0	1

Model Fit Statistics

Criterion	Intercept Only	Intercept and Covariates
AIC	22.190	25.481
SC	22.898	28.313
-2 Log L	20.190	17.481

Testing Global Null Hypothesis: BETA=0

Test	Chi-Square	DF	Pr > ChiSq
Likelihood Ratio	2.7096	3	0.4386
Score	2.4539	3	0.4837
Wald	2.0426	3	0.5636

Type 3 Analysis of Effects

Effect	DF	Wald Chi-Square	Pr > ChiSq
DIFFCLTY	1	1.2726	0.2593
SEASON	2	0.8322	0.6596

TABLE 10.5   Continued

```
            Analysis of Maximum Likelihood Estimates

                              Standard           Wald
Parameter       DF   Estimate   Error      Chi-Square   Pr > ChiSq

Intercept       1     1.7768   1.8898        0.8841       0.3471
DIFFCLTY        1    -1.0108   0.8960        1.2726       0.2593
SEASON    1     1    -0.9275   1.5894        0.3406       0.5595
SEASON    2     1     0.4185   1.3866        0.0911       0.7628
SEASON    3     0        0        .             .            .
```

Odds Ratio Estimates

| | Point | 95% Wald | |
Effect	Estimate	Confidence Limits	
DIFFCLTY	0.364	0.063	2.107
SEASON 1 vs 3	0.396	0.018	8.914
SEASON 2 vs 3	1.520	0.100	23.016

Association of Predicted Probabilities
and Observed Responses

Percent Concordant	72.2	Somers' D	0.556
Percent Discordant	16.7	Gamma	0.625
Percent Tied	11.1	Tau-a	0.286
Pairs	54	c	0.778

SPSS LOGISTIC REGRESSION (accessed in the menu system as Binary Logistic Regression) uses Indicator coding by default with the last category (falling in this example) as reference. The ENTER instruction assures that all of the predictors enter the logistic regression equation simultaneously on Step Number 1.

The first two tables of output show the coding for the outcome and predictor variables. After information about the constant-only model (not shown), overall χ^2 tests are given for the step, the block (only interesting for sequential logistic regression), and the model in the table labeled Omnibus Tests of Model Coefficients. Note that the test of the model matches the test for the difference between the constant-only and full model in SAS.

The Model Summary table provides −2 Log-Likelihood (cf. Table 10.3 footnote). Effect size measures (R Square) are discussed in Section 10.6.2. The Classification Table follows, showing the results of classifying all cases with predicted values below .5 as 0 (not falling) and all cases above .5 as 1 (falling). Of the skiers who did not fall, 66.67% are correctly classified by the model; of those who did fall, 88.89% are correctly classified. The Variables in the Equation table provides B coefficients, standard errors of B (S.E.), a Wald test ($\chi^2 = B^2/S.E.^2$) for each coefficient, and e^B.

TABLE 10.6   Syntax and Selected Output from SPSS LOGISTIC REGRESSION Analysis of Small-Sample Example

```
LOGISTIC REGRESSION VAR=FALL
 /METHOD=ENTER DIFFCLTY SEASON
 /CONTRAST (SEASON)=INDICATOR (1)
 /CRITERIA-PIN(.05) POUT(.10) ITERATE(20) CUT(.5).
```

Logistic Regression

Case Processing Summary

Unweighted Cases[a]		N	Percent
Selected Cases	Included in Analysis	15	100.0
	Missing Cases	0	.0
	Total	15	100.0
Unselected Cases		0	.0
Total		15	100.0

[a]If weight is in effect, see classification table for the total number of cases.

Dependent Variable Encoding

Original Value	Internal Value
0	0
1	1

Categorical Variables Coding

		Frequency	Parameter coding	
			(1)	(2)
SEASON	1	5	1.000	.000
	2	6	.000	1.000
	3	4	.000	.000

Block 1: Method = Enter

Omnibus Tests of Model Coefficients

		Chi-square	df	Sig.
Step 1	Step	2.710	3	.439
	Block	2.710	3	.439
	Model	2.710	3	.439

TABLE 10.6   Continued

Model Summary

Step	−2 Log likelihood	Cox & Snell R Square	Nagelkerke R Square
1	17.481[a]	.165	.223

[a]Estimation terminated at iteration number 4 because parameter estimates changed by less than .001.

Classification Table[a]

			Predicted		
			FALL		Percentage Correct
Observed			0	1	
Step 1 FALL	0		4	2	66.7
	1		1	8	88.9
Overall Percentage					80.0

[a]The cut value is .500

Variables in the Equation

	B	S.E.	Wald	df	Sig.	Exp(B)
Step 1[a] DIFFCLTY	1.011	.896	1.273	1	.259	2.748
SEASON			.832	2	.660	
SEASON(1)	−1.346	1.478	.829	1	.362	.260
SEASON(2)	−.928	1.589	.341	1	.560	.396
Constant	−.849	2.179	.152	1	.697	.428

[a]Variable(s) entered on step 1: DIFFCLTY, SEASON.

10.5   Types of Logistic Regression

As in multiple regression and discriminant analysis, there are three major types of logistic regression: direct (standard), sequential, and statistical. Logistic regression programs tend to have more options for controlling equation-building than discriminant programs, but fewer options than multiple regression programs.

10.5.1   Direct Logistic Regression

In direct logistic regression, all predictors enter the equation simultaneously (as long as tolerance is not violated, cf. Chapter 4). As with multiple regression and discriminant analyses, this is the method of choice if there are no specific hypotheses about the order or importance of predictor variables. The method allows evaluation of the contribution made by each predictor over and above that of the other predictors. In other words, each predictor is evaluated as if it entered the equation last.

This method has the usual difficulties with interpretation when predictors are correlated. A predictor that is highly correlated with the outcome by itself may show little predictive capability in the presence of the other predictors (cf. Section 5.5.1, Figure 5.2b).

SAS LOGISTIC and SPSS LOGISTIC REGRESSION produce direct logistic regression analysis by default (Tables 10.5 and 10.6).

10.5.2   Sequential Logistic Regression

Sequential logistic regression is similar to sequential multiple regression and sequential discriminant analysis in that the researcher specifies the order of entry of predictors into the model. SPSS LOGIS-TIC REGRESSION allows sequential entry of one or more predictors by the use of successive ENTER instructions. In SAS LOGISTIC you can specify sequential entry of predictors, to enter in the order listed in the model instruction, but only one predictor at a time. You also have to specify selection=forward and select large significance values (e.g., .9) for slentry and slstay to ensure that all of your variables enter the equation and stay there.

Another option with any logistic regression program is simply to do multiple runs, one for each step of the proposed sequence. For example, one might start with a run predicting hay fever from degree of stuffiness and temperature. Then, in a second run, geographic area and season are added to stuffiness and temperature. The difference between the two models is evaluated to determine if geographic area and season significantly add to prediction above that afforded by symptoms alone, using the technique described in Section 10.4.3 and illustrated in the large-sample example of Section 10.7.3.

Table 10.7 shows sequential logistic regression for the small-sample example through SPSS LOGISTIC REGRESSION. Difficulty is given highest priority because it is expected to be the strongest predictor of falling. The sequential process asks if season adds to prediction of falling beyond that of difficulty of the ski run.

Block $\chi^2(2, N = 15) = 0.906, p > .05$ at Block 2, indicating no significant improvement with addition of SEASON as a predictor. Note that this improvement in fit statistic is the difference between –2 Log Likelihood for Block 1 (18.387) and –2 Log Likelihood for the full model (17.481). The classification table, model summary, and the logistic regression equation at the end of the second block, with all predictors in the equation, are the same in direct and sequential logistic regression (not shown).

10.5.3   Statistical (Stepwise) Logistic Regression

In statistical logistic regression, inclusion and removal of predictors from the equation are based solely on statistical criteria. Thus, statistical logistic regression is best seen as a screening or hypothesis-generating technique, as it suffers from the same problems as statistical multiple regression and discriminant analysis (Sections 5.5.3 and 9.5.3). When statistical analyses are used, it is very easy to

TABLE 10.7   Syntax and Selected Sequential Logistic Regression Output from SPSS LOGISTIC REGRESSION Analysis of Small-Sample Example

```
LOGISTIC REGRESSION VAR=FALL
 /METHOD=ENTER DIFFCLTY /METHOD=ENTER SEASON
 /CONTRAST (SEASON)=INDICATOR
 /CRITERIA-PIN(.05) POUT(.10) ITERATE(20) CUT(.5).
```

Block 1: Method = Enter

Omnibus Tests of Model Coefficients

		Chi-square	df	Sig.
Step 1	Step	1.804	1	.179
	Block	1.804	1	.179
	Model	1.804	1	.179

Model Summary

Step	−2 Log likelihood	Cox & Snell R Square	Nagelkerke R Square
1	18.387[a]	.113	.153

[a]Estimation terminated at iteration number 4 because parameter estimates changed by less than .001.

Variables in the Equation

		B	S.E.	Wald	df	Sig.	Exp(B)
Step 1[a]	DIFFCLTY	1.043	.826	1.593	1	.207	2.837
	Constant	−1.771	1.783	.986	1	.321	.170

[a]Variable(s) entered on step 1: DIFFCLTY.

Block 2: Method = Enter

Omnibus Tests of Model Coefficients

		Chi-square	df	Sig.
Step 1	Step	.906	2	.636
	Block	.906	2	.636
	Model	2.710	3	.439

Model Summary

Step	−2 Log likelihood	Cox & Snell R Square	Nagelkerke R Square
1	17.481	.165	.223

misinterpret the exclusion of a predictor; the predictor may be very highly correlated with the outcome but not included in the equation because it was "bumped" out by another predictor or combination of predictors. The practice of basing decisions on data-driven rather than theory-driven models is especially hazardous in logistic regression, with its frequent application to life-and-death biomedical issues. At the very least, sequential logistic regression should be part of a cross-validation strategy to investigate the extent to which sample results may be more broadly generalized. Hosmer and Lemeshow (2000) recommend a criterion for inclusion of a variable that is less stringent than .05; they suggest that something in the range of .15 or .20 is more appropriate to ensure entry of variables with coefficients different from zero.

Both computer packages reviewed offer statistical logistic regression allow specification of alternative stepping methods and criteria. SPSS LOGISTIC REGRESSION offers forward or backward statistical regression, either of which can be based on either the Wald or maximum likelihood-ratio statistic, with user specified tail probabilities.

SAS LOGISTIC allows specification of forward, backward or "stepwise" stepping. (In forward selection, a variable once in the equation stays there; if stepwise is chosen, variables once in the equation may leave.) The researcher can specify the maximum number of steps in the process, the significance level for entry or for staying in the model, variables to be included in all models, and maximum number of variables to be included. The researcher can also specify the removal or entry of variables based on the residual chi square.

You may want to consider including interactions as potential predictors if you do statistical model building. Hosmer and Lemeshow (2000) discuss issues surrounding use of interactions and appropriate scaling of continuous variables for them (pp. 70–74).

10.5.4   Probit and Other Analyses

Probit analysis is highly related to logistic regression and is often used to analyze dose-response data in biomedical applications. For example, what is the median dosage of aspirin required to prevent future heart attacks in half the population of heart attack victims?

Both probit analysis and logistic regression focus on proportions of cases in two or more categories of the DV. Both are akin to multiple regression in that the DV (a proportion in both) is predicted from a set of variables that are continuous or coded to be dichotomous. Both produce an estimate of the probability that the DV is equal to 1 given a set of predictor variables.

The difference between logistic regression and probit analysis lies in the transformation applied to the proportions forming the DV that, in turn, reflects assumptions about the underlying distribution of the DV. Logistic regression uses a logit transform of the proportion, as seen in Equation 10.3 (where the proportion is expressed as $\hat{Y}$). Probit analysis uses the probit transform where each observed proportion is replaced by the value of the standard normal curve (z value) below which the observed proportion is found. Thus, logistic regression assumes an underlying qualitative DV (or ordered DV in some applications) and probit analysis assumes an underlying normally distributed DV.

Both transformations produce a value of zero when the proportion is .5, for probit because half of the cases in a normal distribution fall below $z = 0$. For the logit transform,

$$\ln\left(\frac{.5}{1 - .5}\right) = 0$$

It is at the extremes that the values differ; with a proportion of .95, for example, the probit $(z) = 1.65$ and the logit is 2.94. However, the shapes of the logit and probit distributions are quite similar and, as long as proportions are not extreme, the results of the two types of analyses are very similar. Nevertheless, the assumption that the underlying distribution is normal makes probit analysis a bit more restrictive than logistic regression. Thus, logistic regression is considered better than probit analysis if there are too many cases with very high or very low values so that an underlying normal distribution is untenable.

Probit coefficients represent how much difference a unit change in the predictor makes in the cumulative normal probability of the outcome (i.e., the effect of the predictor on the z value for the outcome). This probability of outcome depends on the levels of the predictors; a unit change at the mean of a predictor has a different effect on the probability of the outcome than a unit change at an extreme value of the predictor. Therefore, a reference point for the predictors is necessary and is usually set at the sample means of all predictors.

The two procedures also differ in their emphasis with respect to results. Logistic regression emphasizes odds ratios. Probit analysis often focuses on effective values of predictors for various rates of response, for example, median effective dose of a medication, lethal dose, and so on.

Both software packages have PROBIT modules that provide likelihood-ratio χ^2 tests of models and parameter estimates for predictors. SPSS PROBIT is the more complete, with confidence intervals for expected dosages (lethal or whatever), comparisons of effective doses for different groups, and expected doses for different agents. SAS PROBIT permits transforms other than probit, including logit and Gompertz (for a nonsymmetrical gombit model), and prints confidence intervals for effective doses.

SAS LOGISTIC permits Poisson regression, useful when the DV is in the form of counts that are separated either in time or space. For example, the number of books checked out from the university library might be predicted by major and semester.

10.6   Some Important Issues

10.6.1   Statistical Inference

Logistic regression has two types of inferential tests: tests of models and tests of individual predictors.

10.6.1.1   *Assessing Goodness-of-Fit of Models*

There are numerous models in logistic regression: a constant- (intercept-) only model that includes no predictors, an incomplete model that includes the constant plus some predictors, a full model that includes the constant plus all predictors (including, possibly, interactions and variables raised to a power), and a perfect (hypothetical) model that would provide an exact fit of expected frequencies to observed frequencies if only the right set of predictors were measured.

As a consequence, there are numerous comparisons: between the constant-only model and the full model, between the constant-only model and an incomplete model, between an incomplete model and the full model, between two incomplete models, between a chosen model and the perfect model, between . . . well, you get the picture.

Not only are there numerous possible comparisons among models but also there are numerous tests to evaluate goodness of fit. Because no single test is universally preferred, the computer programs report several tests for differences among the models. Worse, sometimes a good fit is indicated by a nonsignificant result (when, for example, an incomplete model is tested against a perfect model) whereas other times a good fit is indicated by a significant result (when, for example, the full model is tested against a constant-only model).

Sample size also is relevant because if sample size is very large, almost any difference between models is likely to be statistically significant even if the difference has no practical importance and classification is wonderful with either model. Therefore, the analyst needs to keep both the effects of sample size (big = more likely to find significance) and the way the test works (good fit = significant, or good fit = not significant) in mind while interpreting results.

10.6.1.1.1 Constant-Only vs. Full Model

A common first step in any analysis is to ask if the predictors, as a group, contribute to prediction of the outcome. In logistic regression, this is the comparison of the constant-only model with a model that has the constant plus all predictors. If no improvement is found when all predictors are added, the predictors are unrelated to outcome.

The log-likelihood technique for comparing the constant-only model with the full model is shown in Section 10.4.2, Equation 10.7. Both computer programs do a log-likelihood test, but use different terms to report it. Table 10.8 summarizes the test as it is presented in the two programs.

One hopes for a statistically significant difference between the full model and the constant-(intercept-) only model at a level of at least $p < .05$. SAS LOGISTIC provides a second statistic labeled $Score$ that is interpreted in the same way as the difference between log-likelihoods.

These same procedures are used to test the adequacy of an incomplete model (only some predictors) against the constant-only model by inclusion of some but not all predictors in the syntax.

10.6.1.1.2 Comparison with a Perfect (Hypothetical) Model

The perfect model contains exactly the right set of predictors to duplicate the observed frequencies. Either the full model (all predictors) or an incomplete model (some predictors) can be tested against the perfect model in several different ways. However, these statistics are based on differences between observed and expected frequencies and assume adequate expected cell frequencies between pairs of discrete predictors, as discussed in Section 10.3.2.2. In this context, the set of pre-

TABLE 10.8   Summary of Software Labels for the Test of Constant-Only vs. Full Model

Program	Label for χ^2 Test
SPSS LOGISTIC REGRESSION	Model Chi square in the table labeled Omnibus Test of Model Coefficients
SAS LOGISTIC	`Likelihood Ratio Chi Square` in the table labeled `Testing Global Null Hypotheses`

dictors is sometimes called the covariate pattern where covariate pattern refers to combinations of scores on all predictors, both continuous and discrete.

With these statistics a *nonsignificant* difference is desired. A nonsignificant difference indicates that the full or incomplete model being tested is not reliably different from the perfect model. Put another way, a nonsignificant difference indicates that the full or incomplete model adequately duplicates the observed frequencies at the various levels of outcome.

10.6.1.1.3 Deciles of Risk

Deciles-of-risk statistics evaluate goodness-of-fit by creating ordered groups of subjects and then comparing the number actually in each group with the number predicted into each group by the logistic regression model.

Subjects are first put in order by their estimated probability on the outcome variable. Then subjects are divided into 10 groups according to their estimated probability; those with estimated probability below .1 (in the lowest decile) form one group, and so on, up to those with estimated probability .9 or higher (in the highest decile).[3] The next step is to further divide the subjects into two groups on the outcome variable (e.g., didn't fall, did fall) to form a 2×10 matrix of observed frequencies. Expected frequencies for each of the 20 cells are obtained from the model. If the logistic regression model is good, then most of the subjects with outcome 1 are in the higher deciles of risk and most with outcome 0 in the lower deciles of risk. If the model is not good, then subjects are roughly evenly spread among the deciles of risk for both outcomes 1 and 0. Goodness-of-fit is formally evaluated using the Hosmer-Lemeshow statistic where a good model produces a nonsignificant chi-square. The Hosmer-Lemeshow statistic is available in SPSS LOGISTIC REGRESSION with a request for GOODFIT. The program also produces the observed vs. expected frequencies for each decile of risk, separately for each outcome group, reported as Contingency Table for Hosmer and Lemeshow Test.

10.6.1.2 Tests of Individual Variables

Three types of tests are available to evaluate the contribution of an individual predictor to a model: (1) the Wald test, (2) evaluation of the effect of omitting a predictor, and (3) the score (Lagrange multiplier) test. For all these tests, a significant result indicates a predictor that is reliably associated with outcome.

The Wald test is the simplest; it is the default option, called Wald in SPSS and WALD Chi Square in SAS. As seen in Section 10.4.1, this test is the squared logistic regression coefficient divided by its squared standard error.[4] However, several sources express doubt about use of the Wald statistic. For instance, Menard (2001) points out that when the absolute value of the regression coefficient is large, the estimated standard error tends to become too large, resulting in increased Type II error: making the test too conservative.

The test that compares models with and without each predictor (sometimes called the likelihood-ratio test) is considered superior to the Wald test, but is highly computer intensive. Each predictor is evaluated by testing the improvement in model fit when that predictor is added to the

---

[3]Sometimes subjects are divided into 10 groups by putting the first $N/10$ subjects in the first group, the second $N/10$ in the second group, and so on; however, Hosmer and Lemeshow (2000) report that the other procedure is preferable.

[4]SYSTAT LOGIC reports a *t*-ratio which is the parameter estimated divided by its standard error.

model or, conversely, the decrease in model fit when that predictor is removed (using Equation 10.7). Both basic programs require runs of models with and without each predictor to produce the likelihood ratio test to assess the statistical significance of improvement in fit when a predictor is included in the model. However, the test is available in SPSS NOMREG.

The score test is reported in SAS and may be advantageous in stepwise logistic regression.

10.6.2 Effect Size for a Model

A number of measures have been proposed in logistic regression as an analog to R^2 in multiple linear regression. None of these has the same variance interpretation as R^2 for linear regression, but all approximate it. One option (but only for a two-category outcome model) is to calculate R^2 directly from actual outcome scores (1 or 0) and predicted scores, which may be saved from any of the logistic regression programs. A bivariate regression run provides r. Or an ANOVA may be run with predicted scores as the DV and actual outcome as a grouping variable, with η^2 as the measure of effect size (Equation 3.25).

McFadden's ρ^2 (Maddala, 1983) is a transformation of the likelihood ratio statistic intended to mimic an R^2 with a range of 0 to 1.

$$\text{McFadden's } \rho^2 = 1 - \frac{\text{LL}(B)}{\text{LL}(0)} \tag{10.10}$$

where LL(B) is the log-likelihood of the full model and LL(0) is the log-likelihood of the constant-only model. SAS and SPSS LOGISTIC programs provide log-likelihoods in the form of -2 log-likelihood. For the small-sample example,

$$\text{McFadden's } \rho^2 = 1 - \frac{-8.74}{-10.095} = .134$$

However, McFadden's ρ^2 tends to be much lower than R^2 for multiple regression with values in the .2 to .4 range considered highly satisfactory (Hensher & Johnson, 1981). McFadden's ρ^2 is provided by SPSS NOMREG.

SPSS LOGISTIC REGRESSION and NOMREG also provide R^2 measures devised by Nagelkerke as well as Cox and Snell (Nagelkerke, 1991). The Cox and Snell measure is based on log-likelihoods and takes into account sample size.

$$R^2_{CS} = 1 - \exp\left[-\frac{2}{n}[\text{LL}(B) - \text{LL}(0)]\right] \tag{10.11}$$

For the small-sample example,

$$R^2_{CS} = 1 - \exp\left[-\frac{2}{15}[-8.74 - (-10.095)]\right] = 1 - .835 = .165$$

Cox and Snell R^2, however, cannot achieve a maximum value of 1.

The Nagelkerke measure adjusts Cox and Snell so that a value of 1 could be achieved.

$$R_N^2 = \frac{R_{CS}^2}{R_{MAX}^2} \tag{10.12}$$

where $R_{MAX}^2 = 1 - \exp[2(N^{-1})LL(0)]$.

For the small-sample example,

$$R_{MAX}^2 = 1 - \exp[2(15^{-1})(-10.095)] = 1 - .26 = .74$$

and

$$R_N^2 = \frac{.165}{.74} = .223$$

Steiger and Fouladi's (1992) software can be used to provide confidence intervals around the measures (cf. Figure 9.3), although they cannot be interpreted as explained variance. For example, the R_N^2 value used is .233, $N = 15$, number of variables (including predictors and criterion) = 4 and probability value is set to .95. The number of predictors is considered to be 3 rather than 2 to take into account the 2 df for the SEASON variable. Using these values, the 95% confidence interval for R^2 ranges from 0 to .52. Inclusion of zero indicates lack of statistical significance at $\alpha = .05$. Note that the probability level of .40819 approximates the chi-square significance level of .439 in the output of Tables 10.5 and 10.6.

SAS LOGISTIC also provides a number of measures of association: Somer's D, Gamma, Tau-a, and -c. These are various methods of dealing with concordant and discordant pairs of outcomes and are best understood in the context of a two-category outcome and a single two-category predictor. A pair of outcomes is concordant if the response with the larger value also has the higher probability of occurring. The four correlation measures (which need to be squared to be interpreted as effect size) differ in how they deal with the number of concordant and discordant pairs and how they deal with tied pairs. All are considered rank order correlations (cf. on-disk documentation).

The final SAS measure, c, is the area under the receiver operating characteristic (ROC) curve when the response is binary. Aficionados of the Theory of Signal Detectability will recognize this as a form of d'. This may be interpreted as the probability of a correct classification of a randomly selected pair of cases from each outcome category. It varies from .5 (indicating chance prediction) to 1.0 (indicating perfect prediction).

Another measure of effect size is the odds ratio (Section 10.6.3), appropriate for a 2 × 2 contingency table in which one dimension represents an outcome and the other dimension represents a predictor (such as treatment).

10.6.3   Interpretation of Coefficients Using Odds

The odds ratio is the change in odds of being in one of the categories of outcome when the value of a predictor increases by one unit. The coefficients, B, for the predictors are the natural logs of the odds ratios; odds ratio $= e^B$. Therefore, a change of one unit on the part of a predictor multiples the odds by e^B. For example, in Table 10.5 the odds of falling on a ski run increase by a multiplicative factor of 2.75 as the difficulty level of the run increases from 1 to 2 (or 2 to 3); the odds that a skier will fall are almost three times greater on a ski run rated 2 as on a ski run rated 1.

Odds ratios greater than 1 reflect the increase in odds of an outcome of 1 (the "response" category) with a one-unit increase in the predictor; odds ratios less than one reflect the decrease in odds of that outcome with a one-unit change. For example, an odds ratio of 1.5 means that odds of the outcome labeled 1 are 1.5 greater with a one-unit increase in a predictor. That is, the odds are increased by 50%. An odds ratio of 0.8 indicates that the odds of an outcome labeled 1 are 0.8 less with a one unit increase in the predictor; the odds are decreased by 20%.

As in linear regression, coefficients are interpreted in the context of the other predictor variables. That is, the odds of falling as a function of difficulty level are interpreted after adjusting for all other predictors. (Usually only statistically significant coefficients are interpreted; this example is for illustrative purposes only.)

The odds ratio has a clear, intuitive meaning for a 2×2 table; it is the odds of an outcome for cases in a particular category of a predictor divided by the odds of that outcome for the other category of the predictor. Suppose the outcome is hyperactivity in a child and the predictor is familial history of hyperactivity:

		Familial History of Hyperactivity	
		Yes	No
Hyperactivity	Yes	15	9
	No	5	150

$$odds\ ratio = \frac{15/5}{9/150} = 50$$

The odds of hyperactivity in children with a familial history are 50 times greater than the odds of hyperactivity among those without a familial history. Odds are 3:1 for hyperactivity in a child with familial history; odds are 9:150 for hyperactivity in a child without familial history. Therefore, the ratio of odds is $3/.6 = 50$. This also may be expressed as the reciprocal, $1/50 = 0.02$. The interpretation for this reciprocal odds ratio is that odds of hyperactivity in children without a familial history are 0.02 times as great as the odds of hyperactivity among those with familial history of hyperactivity (i.e., there is a reduction in the overall odds from .06 when there is no familial history of hyperactivity). Either interpretation is equally correct; a good choice is the one that is easiest to communicate. For example, if there is a treatment to reduce the occurrence of disease, it is the reduction in disease that may be of greatest interest. Further discussion of this issue is in Section 10.6.4.

Odds ratios are produced directly by SPSS LOGISTIC REGRESSION and SAS LOGISTIC. It is called **Odds Ratio** in SAS LOGISTIC, and it is called Exp(B) by SPSS LOGISTIC REGRESSION and NOMREG.

Keep in mind that the odds ratios are for the outcome coded 1 in SPSS and the outcome coded 0 in SAS[5] (unless you respecify the reference category). Using default SAS coding, the odds ratio for the example is 0.02. Interpretation of odds ratios therefore depends on how the outcome is coded. You need to take care that the outcome variable is coded in a direction that reflects your eventual desired interpretation.

The coding of categorical predictors also is important, as indicated in Section 10.6.4. For example, in SPSS you may want to specify that the reference category for a categorical predictor variable such as familial history is first (0, "no family history of hyperactivity"), rather than last (the default, "family history of hyperactivity"). With the default option in place, SPSS reports an odds ratio of 50. With the reference category on the predictor changed, the value is reported as 0.02. The reverse is true for SAS. Declaring the predictor to be a categorical (**class**) variable with default outcome coding results in an odds ratio of 50; default coding of both the predictor and the outcome results in an odds ratio of 0.02.

The odds ratio is interpretable as an effect size; the closer the odds ratio is to 1, the smaller the effect. As for any other effect size, it is desirable to report confidence limits around the estimated value. Both SPSS and SAS provide confidence limits around odds ratios—SAS by default and SPSS by request. Chinn (2000) shows how to convert an odds ratio to Cohen's d, which in turn can be converted into η^2 (Cohen, 1988). First, $d = \ln(\text{odds ratio})/1.81$. In this example, $\ln(50)/1.81 = 3.91/1.81 = 2.16$. The conversion from d to η^2 is:

$$\eta^2 = \frac{d^2}{d^2 + 4} = \frac{2.16^2}{2.16^2 + 4} + \frac{4.67}{8.67} = .54$$

Relative risk is a similar measure, used in biomedical research when the predictor is treatment and the outcome is disease or some other application when the predictor clearly precedes the outcome. The difference is that the ratios are formed on the basis of column totals. In the example (arguably inappropriate for this type of analysis), RR (the relative risk ratio) is

$$RR = \frac{15/(15 + 5)}{9/(9 + 150)} = \frac{0.75}{0.57} = 13.16$$

The risk (probability) of developing hyperactivity with a family history of it is 75%; the risk of developing hyperactivity without a family history of it is less than 6%; the relative risk of hyperactivity is 13 times greater for those with a family history than for those without a family history of hyperactivity. Typically, odds ratios are used for retrospective studies (e.g., observing a sample of hyperactive children and nonhyperactive children and reviewing their family history), whereas RR is used for prospective, experimental studies in which a sample at risk is assigned to treatment versus control and the outcome is then observed. For example, a sample of hyperactive children might be assigned to educational therapy versus waiting list control, and presence of reading disability assessed after a period of one year. If the first row is very rare (e.g., the presence of hyperactivity in the example), the difference between odds ratio and RR becomes small.

---

[5]Both SPSS and SAS convert codes for predictor and outcome variables to 0, 1. The output indicates conversions from your original coding, if any.

10.6.4   Coding Outcome and Predictor Categories

The way that outcome categories are coded determines the direction of the odds ratios as well as the sign of the B coefficient. The interpretation is simplified, therefore, if you pay close attention to coding your categories. Most software programs solve the logistic regression equation for the dichotomous outcome category coded 1 but a few solve for the category coded 0. If the odds ratio is 4 in a problem run through the first type of program, it will be 0.25 in the second type of program.

A convenient way of setting up the coding follows the jargon of SYSTAT LOGIT; for the outcome, the category coded 1 is the "response" category (e.g., illness) and the category coded 0 is the "reference" category (wellness).[6] It is often helpful, then, to think of the response group in comparison to the reference group, e.g., compare people who are ill to those who are well. The solution tells you the odds of being in the response group given some value on a predictor. If you also give higher codes to the category of a predictor most likely associated with "response," interpretation is facilitated because the parameter estimates are positive. For example, if people over 60 are more likely to be ill, code both "wellness" and "under 60" 0 and both "illness" and "over 60" 1.

This recommendation is extended to predictors with multiple discrete levels where dummy variables are formed for all but one level of the discrete predictor (e.g., Season1 and Season2 for the three seasons in the small-sample example). Each dummy variable is coded 1 for one level of a predictor and 0 for the other levels. If possible, code levels likely to be associated with the "reference" group 0 and code levels likely to be associated with the "response" group 1. Odds ratios are calculated in the standard manner, and the usual interpretation is made of each dummy variable (e.g., Season1).

Other coding schemes may be used, such as orthogonal polynomial coding (trend analysis) but interpretation via odds ratios is far more difficult. However, in some contexts, significance tests for trends may be more interesting than odds ratios.

The SPSS programs (LOGISTIC REGRESSION and NOMREG) solve for the outcome coded 1. However, SAS LOGISTIC routinely solves for the outcome coded 0. You might want to consider using `param = glm` with SAS (as per Table 10.5) if that eases interpretation.

There are other methods of coding discrete variables, each with its own impact on interpretation. For example, dichotomous $(1, -1)$ coding might be used. Or discrete categories might be coded for trend analysis, or whatever. Hosmer and Lemeshow (2000) discuss the desirability of various coding schemes and the effects of them on parameter estimates (pp. 48–56). Further discussion of coding schemes available in the computer packages is in Section 10.8.

10.6.5   Number and Type of Outcome Categories

Logistic regression analysis can be applied with two or more categories of outcome, and, when there are more than two categories of outcome, they may or may not have order. That is, outcomes with more than two categories can be either nominal (without order) or ordinal (with order). Logistic regression is more appropriate than multiple regression when the distribution of responses over a set of categories seriously departs from normality, making it difficult to justify using an ordered categorical variable as if it were continuous.

---

[6]Hosmer and Lemeshow (2000) recommend coding all dichotomous variables 0 and 1.

When there are more than two categories the analysis is called multinomial or polychotomous logistic regression or MLOGIT, and there is more than one logistic regression model/equation. In fact, like discriminant analysis, there are as many models (equations) as there are degrees of freedom for the outcome categories; the number of models is equal to the number of categories minus one.

When the outcome is ordered, the first equation finds the probability that a case is above the first (lowest) category. The second equation finds the probability that the case is above the second category, and so forth, as seen in Equation 10.13,

$$P(Y > j) = \frac{e^u}{1 + e^u} \tag{10.13}$$

where u is the linear regression equation as in Equation 10.1, Y is the outcome, and j indicates the category.

An equation is solved for each category except the last, since there are no cases above the last category. (SAS LOGISTIC bases the equations on the probability that a case is below rather than above a category, so it is the lowest category that is omitted.)

When there are more than two categories of outcome, but they are not ordered, each equation predicts the probability that a case is (or is not) in a particular category. Equations are built for all categories except the last. Logistic regression with this type of outcome is illustrated in the large sample example of Section 10.7.3.

With the exception of SPSS LOGISTIC REGRESSION, which analyzes only two-category outcomes, the programs handle multiple-category outcomes but have different ways of summarizing the results of multiple models. Multiple-category outcomes are handled by SPSS NOMREG, available since Version 9.0, and PLUM, available since Version 10.0.

SPSS NOMREG assumes unordered categories; SPSS PLUM assumes ordered categories. Classification and prediction success tables (cf. Section 10.6.6) are used to evaluate the success of the equations taken together.

SAS LOGISTIC treats categories in all multinomial models as ordered; there is no provision for unordered categories. The logistic regression coefficients for individual predictors are for the combined set of equations. Classification and prediction success tables and effect size measures are used to evaluate the set of equations as a whole. Parameter estimates for unordered models may be approximated by running analyses using two categories at a time. For example, if there are three groups, an analysis is done with groups 1 and 3, and another analysis with groups 2 and 3.

Of course, it is always possible to reduce a multinomial/polychotomous model to a two-category model if that is of research interest. Simply recode the data so that one category becomes the response category and all of the others are combined into the "reference" category. Section 10.7.3 demonstrates analysis by SPSS NOMREG with unordered outcome categories.

Table 10.9 shows the analysis of a data set through SAS LOGISTIC, which assumes that response categories are ordered. This is an analysis of the frequency with which psychotherapists reported that they had been sexually attracted to the therapists in their own psychotherapy.[7] Higher numbered categories represent greater frequency of sexual attraction (0 = not at all, etc.). Predictors

---

[7]Most psychotherapists undergo psychotherapy themselves as part of their training.

TABLE 10.9   Logistic Regression with Ordered Categories (Syntax and SAS LOGISTIC Output)

```
proc logistic data=SASUSER.LOGMULT;
    model ATTRACT = AGE SEX THEORET;
run;
```

Response Profile

Ordered Value	ATTRACT	Total Frequency
1	0	232
2	1	31
3	2	52
4	3	26
5	4	23

NOTE: 112 observation(s) were deleted due to missing values for the response or explanatory variables.

Model Convergence Status

Convergence criterion (GCONV=1E-8) satisfied.

Score Test for the Proportional Odds Assumption

Chi-Square	DF	Pr > ChiSq
12.9147	9	0.1665

Model Fit Statistics

Criterion	Intercept Only	Intercept and Covariates
AIC	836.352	805.239
SC	851.940	832.519
-2 Log L	828.352	791.239

The LOGISTIC Procedure

Testing Global Null Hypothesis: BETA=0

Test	Chi-Square	DF	Pr > ChiSq
Likelihood Ratio	37.1126	3	<.0001
Score	35.8860	3	<.0001
Wald	33.7303	3	<.0001

TABLE 10.9   Continued

Analysis of Maximum Likelihood Estimates

Parameter	DF	Estimate	Standard Error	Wald Chi-Square	Pr > ChiSq
Intercept0	1	0.0653	0.7298	0.0080	0.9287
Intercept1	1	0.4940	0.7302	0.4577	0.4987
Intercept2	1	1.4666	0.7352	3.9799	0.0460
Intercept3	1	2.3372	0.7496	9.7206	0.0018
AGE	1	-0.00112	0.0116	0.0094	0.9227
SEX	1	-1.1263	0.2307	23.8463	<.0001
THEORET	1	0.7555	0.2222	11.5619	0.0007

Odds Ratio Estimates

Effect	Point Estimate	95% Wald Confidence Limits	
AGE	0.999	0.977	1.022
SEX	0.324	0.206	0.510
THEORET	2.129	1.377	3.290

Association of Predicted Probabilities and Observed Responses

Percent Concordant	62.5	Somers' D	0.317
Percent Discordant	30.8	Gamma	0.340
Percent Tied	6.7	Tau-a	0.177
Pairs	36901	c	0.659

are age, sex and theoretical orientation (psychodynamic or not) of the therapist in their own psychotherapy

The Score Test for the Proportional Odds Assumption (cf. Chapter 11) shows that the odds ratios between adjacent outcome categories are not significantly different ($p = 0.1664$). The Likelihood Ratio test in the Testing Global Null... table shows a significant difference between the constant-only model and the full model, indicating a good model fit with the set of three predictors (covariates).

The following table in the output shows Wald tests for the three predictors, indicating that SEX and THEORETical orientation, but not AGE, significantly predict response category. Estimates for Parameters show direction of relationships between predictor and outcome variables (recall that SAS LOGISTIC solves for the probability that a response is below a particular category). Thus the negative value for SEX indicates that male therapists are associated with lower numbered categories (less frequently sexually attracted to their own therapists); the positive value for theoretical orientation indicates that psychodynamically oriented therapists are associated with higher numbered categories (more frequently sexually attracted to their own therapists). Note that AGE is not a significant predictor in this model in which categories are ordered and parameter estimates are combined over all categories. Again, measures of association between the set of predictors and attraction are small.

10.6.6 Classification of Cases

One method of assessing the success of a model is to evaluate its ability to predict correctly the outcome category for cases for whom outcome is known. If a case has hay fever, for instance, we can see if the case is correctly classified as diseased on the basis of degree of stuffiness, temperature, geographic area, and season. Classification is available for two-category outcomes only.

As in hypothesis testing, there are two types of errors: classifying a truly nondiseased individual as diseased (Type I error or false alarm) or classifying a truly diseased individual as nondiseased (Type II error or miss). For different research projects, the costs associated with the two types of errors may be different. A Type II error is very costly, for instance, when there is an effective treatment, but the case will not receive it if classified nondiseased. A Type I error is very costly, for instance, when there is considerable risk associated with treatment, particularly for a nondiseased individual. Some of the computer programs allow different cutoffs for asserting "diseased" or "nondiseased." Because extreme cutoffs could result in everyone or no one being classified as diseased, intermediate values for cutoffs are recommended, but these can be chosen to reflect the relative costs of Type I and Type II errors. However, the only way to improve the overall accuracy of classification is to find a better set of predictors.

Because the results of logistic regression analysis are in terms of probability of a particular outcome (e.g., having hay fever), the cutoff chosen for assignment to a category is critical in evaluating the success of the model.

Classification is available only for two-category outcomes through SAS LOGISTIC. However, the program does allow specification of the cutoff criterion, and prints results for many additional cutoff criteria. The classification procedure in SAS LOGISTIC includes jackknifing (cf. Section 9.6.7.2).

SPSS LOGISTIC REGRESSION, which analyzes data with two-outcome categories only, prints a classification table by default; assignment is based on a cutoff probability criterion of .5. Although the criterion cannot be changed in SPSS, the CLASSPLOT instruction produces a histogram of predicted probabilities, showing whether incorrectly classified cases had probabilities near the criterion. SPSS LOGISTIC REGRESSION also has a SELECT instruction, by which certain cases are selected for use in computing the equations. Classification is then performed on all of the cases in a form of cross validation. This offers a less biased estimate of the classification results. SPSS NOMREG classifies cases into the category with the highest predicted probability.

10.6.7 Hierarchical and Nonhierarchical Analysis

The distinction between hierarchical and nonhierarchical logistic regression is the same as in multiway frequency analysis. When interactions among predictors are included in a model,[8] the model is hierarchical if all main effects and lower order interactions of those predictors are also included in the model.

---

[8]Hosmer and Lemeshow (2000) discuss issues surrounding the use of interactions in a model, particularly the problem of adjusting for confounding variables (covariates) in the presence of interactions (pp. 70–74).

In most programs, interactions are specified in the MODEL instruction, using a convention in which interaction components are joined by asterisks (e.g., X1*X2). However, SAS LOGISTIC requires creation of a new variable in the DATA step to form an interaction, which is then used like any other variable in the `model` instruction.

All of the reviewed programs allow nonhierarchical models, although Steinberg and Colla (1991) advise against including interactions without their main effects.

10.6.8 Importance of Predictors

The usual problems of evaluating the importance of predictors in regression apply to logistic regression. Further, there is no comparable measure to sr_i^2 in logistic regression. One strategy is to evaluate odds ratios: The statistically significant predictors that change the odds of the outcome the most are interpreted as the most important. That is, the farther the odds ratio from 1, the more influential the predictor.

Another strategy is to calculate standardized regression coefficients comparable to the β weights in multiple regression. These are not available in most statistical packages; the standardized estimates offered by SAS LOGISTIC are only partially standardized and more likely to wander outside the bounds of -1 and 1 than fully standardized coefficients (Menard, 2001, p. 55). The simplest way to get standardized regression coefficients is to standardize the predictors before the analysis, and then interpret the coefficients that are produced as standardized.

10.6.9 Logistic Regression for Matched Groups

Although usually logistic regression is a between-subjects analysis, there is a form of it called conditional logistic regression for matched subjects or case-control analysis. Cases with disease are matched by cases without disease on variables such as age, gender, socioeconomic status, and so forth. There may be only one matched control subject for each disease subject, or more than one matched control for each disease subject. When there is only one matched control subject the outcome has two categories (disease and control), while multiple matched control subjects lead to multiple outcome categories (disease, control1, control2, etc.). The model is, as usual, based on the predictors that are included, but there is no constant (or intercept).

SAS LOGISTIC uses a conditional logistic regression procedure for case-control studies with a single control by specifying `noint` (no intercept) and requires that each matched pair be transformed into a single observation, where the response variable is the difference in scores between each case and its control. SPSS NOMREG also uses a procedure in which the outcome is the difference between each case and its control and the intercept is suppressed.

10.7 Complete Examples of Logistic Regression

Data for these analyses are from the data set described in Appendix B. Two complete examples are demonstrated. The first is a simple direct logistic regression analysis, in which work status (employed vs. unemployed) is the two-category outcome that is predicted from four attitudinal variables through SAS LOGISTIC.

The second analysis is more complex, involving three categories of outcome and sequential entry of predictors. The goal in the second analysis is to predict membership in one of three categories of outcome formed from work status and attitude toward that status. Variables are entered in two sets: demographic and then attitudinal. The major question is whether attitudinal variables significantly enhance prediction of outcome after prediction by demographic variables. Demographic variables include a variety of continuous, discrete and dichotomous variables: marital status (discrete), presence of children (dichotomous), religious affiliation (discrete), race (dichotomous), socioeconomic level (continuous), age (continuous), and attained educational level (continuous). The attitudinal variables used for both analyses are all continuous; they are locus of control, attitude toward current marital status, attitude toward role of women, and attitude toward housework. (Note that prediction of outcome in these three categories on the basis of attitudinal variables alone is addressed in the large sample discriminant analysis of Chapter 9.) This analysis is run through SPSS NOMREG. Data files for both analyses are LOGREG.*. Linearity in the logit is tested separately for each analysis.

10.7.1   Evaluation of Limitations

10.7.1.1   Ratio of Cases to Variables and Missing Data

Sections 10.7.2 and 10.7.3 show no inordinately large parameter estimates or standard errors. Therefore, there is no reason to suspect a problem with too many empty cells or with outcome groups perfectly predicted by any variable. Table 10.10 shows an SPSS MVA run to investigate the pattern missing data and evaluate its randomness after declaring that values of zero on SEL are to be considered missing. All variables to be used in either analysis are investigated, with categorical variables identified. The EM algorithm is chosen for imputing missing values, and a full data set is saved to a file labeled LOGREGC.SAV. Separate variance t tests, in which the grouping variable is missing vs. nonmissing, are requested for all quantitative (continuous) variables that are missing 1% or more of their values. Although there is some concern about bias associated with the SPSS MVA implementation of EM, the number of missing values here (about 5%) is low enough that parameter estimates

TABLE 10.10   Analysis of Missing Values Through SPSS MVA (Syntax and Selected Output)

```
RECODE AGE SEL (0 = SYSMIS).
MVA
    CONTROL ATTMAR ATTROLE SEL ATTHOUSE AGE EDUC WORKSTAT MARITAL CHILDREN
    RELIGION RACE
    /MAXCAT = 25
    /CATEGORICAL = WORKSTAT MARITAL CHILDREN RELIGION RACE
    /NOUNIVARIATE
    /TTEST PROB PERCENT=1
    /CROSSTAB PERCENT=1
    /MPATTERN DESCRIBE=CONTROL ATTMAR ATTROLE SEL ATTHOUSE AGE EDUC WORKSTAT
    MARITAL CHILDREN RELIGION
    /EM (TOLERANCE=0.001 CONVERGENCE=0.0001 ITERATIONS=25
    OUTFILE='C:\DATA\BOOK.5TH\LOGISTIC\LOGREGC.SAV').
```

TABLE 10.10 Continued

MVA

Separate Variance t Tests[a]

		CONTROL	ATTMAR	ATTROLE	SEL	ATTHOUSE	AGE	EDUC
ATTMAR	t	.6	.	3.3	1.8	−1.1	.0	−.1
	df	4.5	.	4.2	3.0	4.2	4.2	4.0
	P(2-tail)	.582	.	.027	.164	.336	.995	.914
	# Present	459	460	460	449	459	456	460
	# Missing	5	0	5	4	5	5	5
	Mean(Present)	6.7495	22.9804	35.2065	52.7016	23.5251	4.3947	13.2391
	Mean(Missing)	6.6000	.	28.6000	28.5000	25.0000	4.4000	13.4000
SEL	t	.0	−.6	−.4	.	−.3	−2.3	.8
	df	11.4	10.3	11.8	.	11.5	11.5	11.2
	P(2-tail)	.996	.532	.686	.	.786	.040	.463
	# Present	452	449	453	453	452	449	453
	# Missing	12	11	12	0	12	12	12
	Mean(Present)	6.7478	22.9287	35.1170	52.4879	23.5310	4.3541	13.2605
	Mean(Missing)	6.7500	25.0909	35.8333	.	23.9167	5.9167	12.5000

For each quantitative variable, pairs of groups are formed by indicator variables (present, missing).

[a]Indicator variables with less than 1% missing are not displayed.

Crosstabulations of Categorical Versus Indicator Variables

WORKSTAT

			Total	Working	Role-satisfied housewives	Role-dissatisfied housewives
ATTMAR	Present	Count	460	242	137	81
		Percent	98.9	98.4	100.0	98.8
	Missing	% SysMis	1.1	1.6	.0	1.2
SEL	Present	Count	453	243	132	78
		Percent	97.4	98.8	96.4	95.1
	Missing	% SysMis	2.6	1.2	3.6	4.9

Indicator variables with less than 1% missing are not displayed.

(continued)

TABLE 10.10 Continued

Missing Patterns (cases with missing values)

Case	# Missing	% Missing	\<—Missing and Extreme Value Patterns[a]—\> SEL	ATTMAR	AGE	RELIGION	ATTHOUSE	EDUC	\<—Variable Values—\> CONTROL	ATTMAR	ATTROLE	SEL	ATTHOUSE	AGE	EDUC	WORKSTAT	MARITAL	CHILDREN	RELIGION
37	1	8.3	S					−	7.00	11.00	39.00	.	21.00	4.00	9.00	Role-satisfied housewives	Broken	Yes	Catholic
95	1	8.3	S						5.00	14.00	32.00	.	16.00	8.00	16.00	Role-satisfied housewives	Married	No	Protestant
118	1	8.3	S					−	7.00	16.00	45.00	.	28.00	8.00	10.00	Role-satisfied housewives	Single	No	Catholic
159	1	8.3	S						7.00	11.00	45.00	.	22.00	5.00	11.00	Working	Married	Yes	Protestant
196	1	8.3	S						8.00	35.00	38.00	.	29.00	8.00	6.00	Role-dissatisfied housewives	Broken	Yes	Catholic
219	1	8.3	S						6.00	39.00	37.00	.	25.00	5.00	16.00	Role-dissatisfied housewives	Broken	No	Protestant
265	1	8.3	S						6.00	25.00	27.00	.	30.00	2.00	16.00	Working	Married	No	Jewish
314	1	8.3	S						5.00	35.00	32.00	.	23.00	8.00	13.00	Role-satisfied housewives	Broken	Yes	Protestant
341	1	8.3	S					−	10.00	35.00	40.00	.	27.00	3.00	9.00	Role-dissatisfied housewives	Broken	Yes	Protestant
448	1	8.3	S						5.00	35.00	35.00	.	22.00	8.00	13.00	Role-satisfied housewives	Broken	No	Protestant
457	1	8.3	S						8.00	20.00	29.00	.	16.00	8.00	16.00	Working	Married	Yes	Catholic
300	2	16.7	S	A					7.00	.00	31.00	7.00	28.00	4.00	15.00	Role-dissatisfied housewives	Broken	Yes	Protestant
135	1	8.3		A					6.00	.00	30.00	7.00	21.00	4.00	12.00	Working	Single	No	Protestant
280	1	8.3		A					7.00	.00	29.00	61.00	28.00	6.00	12.00	Working	Broken	Yes	Jewish
317	1	8.3		A				+	7.00	.00	21.00	7.00	24.00	2.00	18.00	Working	Broken	No	None-or-other
113	1	8.3		A				−	6.00	.00	32.00	39.00	24.00	6.00	10.00	Working	Broken	Yes	Catholic
83	1	8.3				A			5.00	21.00	50.00	81.00	16.00	8.00	8.00	Role-satisfied housewives	Married	Yes	.00
80	1	8.3				A			6.00	29.00	26.00	62.00	23.00	4.00	15.00	Role-satisfied housewives	Married	Yes	.00
437	1	8.3				A			6.00	16.00	36.00	52.00	27.00	7.00	12.00	Working	Married	Yes	.00
208	1	8.3			S				8.00	13.00	39.00	13.00	20.00	.	13.00	Role-dissatisfied housewives	Married	Yes	Protestant
209	1	8.3			S				8.00	42.00	29.00	25.00	25.00	.	13.00	Working	Married	Yes	Protestant
206	1	8.3			S				5.00	29.00	28.00	53.00	24.00	.	15.00	Role-dissatisfied housewives	Married	No	Protestant
207	1	8.3			S				8.00	31.00	39.00	87.00	29.00	1.00	14.00	Working	Married	Yes	Catholic
253	1	8.3					A		9.00	44.00	32.00	45.00	1.00	1.00	13.00	Working	Single	No	Catholic
303	1	8.3							.00	18.00	40.00	62.00	23.00	4.00	13.00	Role-satisfied housewives	Married	Yes	Protestant

− indicates an extreme low value, while + indicates an extreme high value. The range used is (Q1 − 1.5*IQR, Q3 + 1.5*IQR).
aCases and variables are sorted on missing patterns.

472

TABLE 10.10   Continued

EM Estimated Statistics

EM Correlations[a]

	CONTROL	ATTMAR	ATTROLE	SEL	ATTHOUSE	AGE	EDUC
CONTROL	1.000						
ATTMAR	.195	1.000					
ATTROLE	.001	−.096	1.000				
SEL	−.132	−.027	−.196	1.000			
ATTHOUSE	.185	.304	−.307	−.022	1.000		
AGE	−.126	−.061	.246	.121	−.082	1.000	
EDUC	−.092	−.056	−.379	.341	.100	−.009	1.000

[a]Little's MCAR test: Chisquare = 38.070, df = 35, Prob = .331

are expected to be appropriate even though standard errors are inflated. Results of inferential statistics will be interpreted with caution.

The Separate Variance t Tests show for the two quantitative variables with 1% or more values missing (ATTMAR and SEL), the relationship between missingness and other quantitative variables. For example, there is a suggestion that whether data are missing on ATTMAR might be related to ATTROLE, $t(4.2) = 3.3, p = .027$. However, an adjustment for familywise Type I error rate for the 6 t tests for each variable places criterion $\alpha = .008$ for each test. Using this criterion, there is no worrisome relationship between missing data on ATTMAR or SEL and any of the other quantitative variables.

The relationship between missingness on ATTMAR and SEL and the categorical variables is in the section labeled Crosstabulations of Categorical Versus Indicator Variables. Only the table for WORKSTAT is shown, and there is no great difference among groups in percentages missing for the two variables.

The Missing Patterns table shows, for each case with at least one missing value, the variable(s) on which data are missing, variables on which the case has an extreme value as indicated by a quartile criterion, and the values for that case on all other variables. Note that 3 of the cases are missing values on RELIGION, a categorical variable. These missing values are not imputed.

The most critical part of the output is Little's MCAR test, which appears at the bottom of the EM Correlations table. This shows that there is no significant deviation from a pattern of values that are "missing completely at random," $\chi^2(35) = 38.07, p = .331$. Thus, there is support for imputation of missing values using the EM algorithm. Remaining analyses use the data set with imputed values.

10.7.1.2   *Multicollinearity*

Analyses in Sections 10.7.2 and 10.7.3 show no problem with convergence, nor are the standard errors for parameters exceedingly large. Therefore, no multicollinearity is evident.

10.7.1.3 Outliers in the Solution

Sections 10.7.2 and 10.7.3 show adequate model fits. Therefore, there is no need to search for outliers in the solution.

10.7.2 Direct Logistic Regression with Two-Category Outcome and Continuous Predictors

This analysis uses only two of the three WORKSTAT categories, therefore WORKSTAT needs to be recoded. Table 10.11 shows the syntax to create a new file, LOGREG from the original file, LOGREGCC.

10.7.2.1 Limitation: Linearity in the Logit

The main analysis has four continuous attitudinal variables. Interactions between each predictor and its natural log are added to test the assumption. SAS Interactive Data Analysis is used to create interactions between continuous variables and their natural logarithms (not shown) and add them to the data set, which is saved as LOGREGIN. In Table 10.12, a two-category direct logistic regression analysis is performed with the four original continuous variables and four interactions as predictors, using the new data set which also has recoded values for WORKSTAT.

The only hint of violation is for ATTROLE, with Pr > ChiSq = .0125. However, a reasonable criterion for determining significance for this test with nine terms is $\alpha = .05/9 = .006$. Therefore, the model is run as originally proposed.

10.7.2.2 Direct Logistic Regression with Two-Category Outcome

Table 10.13 shows the results of the main direct logistic regression analysis with two outcomes. The instructions for the SAS LOGISTIC run include a request for 95% confidence intervals around odds ratios (CLODDS=WALD) and tables showing success of prediction (CTABLE).

The sample is split into 245 working women (coded 0) and 217 housewives (coded 1). The comparison of the constant-only model with the full model (Likelihood Ratio) shows a highly significant probability value, $\chi^2(4, N = 440) = 23.24, p < .0001$, indicating that the predictors, as a set, reliably predict work status.

The table of parameters shows that the only successful predictor is attitude toward role of women; working women and housewives differ significantly only in how they view the proper role

TABLE 10.11   SAS DATA Syntax for Recoding WORKSTAT
into Two Categories

```
data Sasuser.Logreg;
    set Sasuser.Logregcc;
    if WORKSTAT=1 then WORKSTAT=0;
    if WORKSTAT=3 or WORKSTAT=2 then WORKSTAT=1;
run;
```

TABLE 10.12   Direct Logistic Regression to Test Linearity in the Logit (SAS Data and Logistic Syntax and Selected Logistic Output)

```
proc logistic data=Sasuser.Logreg;
model WORKSTAT = CONTROL ATTMAR ATTROLE ATTHOUSE LIN_CTRL
                 LIN_ATMR LIN_ATRL LIN_ATHS;
run;
```

The LOGISTIC Procedure

Analysis of Maximum Likelihood Estimates

Parameter	DF	Estimate	Error	Chi-Square	Pr > ChiSq
Intercept	1	12.8762	7.4039	3.0245	0.0820
CONTROL	1	1.9706	2.1771	0.8193	0.3654
ATTMAR	1	0.2215	0.2344	0.8923	0.3449
ATTROLE	1	-1.5562	0.6231	6.2375	0.0125
ATTHOUSE	1	-0.8193	0.6314	1.6836	0.1944
LIN_CTRL	1	-0.6863	0.7396	0.8610	0.3534
LIN_ATMR	1	-0.0475	0.0549	0.7481	0.3871
LIN_ATRL	1	0.3263	0.1362	5.7391	0.0166
LIN_ATHS	1	0.1905	0.1529	1.5531	0.2127

of women. Nonsignificant coefficients are produced for locus of control, attitude toward marital status, and attitude toward housework. The negative coefficient for attitude toward the role of women means that working women (coded 0: see **Probability modeled is WORKSTAT=0**) have lower scores on the variable, indicating more liberal attitudes. **Adjusted Odds Ratios** are omitted here because they duplicate the nonadjusted ones. (Note that these findings are consistent with the results of the contrast of working women vs. the other groups in Table 9.12 of the discriminant analysis.) Somers' D indicates about 7% ($.263^2 = .07$) of shared variance between work status and the set of predictors. Using Steiger and Fouladi's (1992) software (cf. Figure 9.3), the 95% confidence interval ranges from .03 to .12. Thus, the gain in prediction is unimpressive.

SAS LOGISTIC uses jackknife classification in the **Classification Table**. At a **Prob Level** of 0.500, the correct classification rate is 57.8%. Sensitivity is the proportion of cases in the "response" category (working women coded 0) correctly predicted. Specificity is the proportion of cases in the "reference" category (housewives) correctly predicted.

Because of difficulties associated with the Wald test (cf. Section 10.6.1.2), an additional run is prudent to evaluate the predictors in the model. Another SAS LOGISTIC run (Table 10.14) evaluates a model without attitude toward women's role. Applying Equation 10.7, the difference between that model and the model that includes ATTROLE is:

$$\chi^2 = 636.002 - 615.534 = 20.468$$

with df = 1, $p < .01$, reinforcing the finding of the Wald test that attitude toward women's role significantly enhances prediction of work status. Note also that Table 10.14 shows no statistically

TABLE 10.13   Syntax and Output of SAS LOGISTIC for Logistic Regression Analysis of Work Status with Attitudinal Variables

```
proc logistic data=Sasuser.Logreg;
model WORKSTAT = CONTROL ATTMAR ATTROLE ATTHOUSE / CTABLE
     CLODDS=WALD;
run;
```

Response Profile

Ordered Value	WORKSTAT	Total Frequency
1	0	245
2	1	217

Probability modeled is WORKSTAT=0.

Model Fit Statistics

Criterion	Intercept Only	Intercept and Covariates
AIC	640.770	625.534
SC	644.906	646.212
−2 Log L	638.770	615.534

Testing Global Null Hypothesis: BETA=0

Test	Chi-Square	DF	Pr > ChiSq
Likelihood Ratio	23.2362	4	0.0001
Score	22.7391	4	0.0001
Wald	21.7498	4	0.0002

Analysis of Maximum Likelihood Estimates

Parameter	DF	Estimate	Standard Error	Wald Chi-Square	Pr > ChiSq
Intercept	1	3.1964	0.9580	11.1319	0.0008
CONTROL	1	−0.0574	0.0781	0.5410	0.4620
ATTMAR	1	0.0162	0.0120	1.8191	0.1774
ATTROLE	1	−0.0681	0.0155	19.2977	<.0001
ATTHOUSE	1	−0.0282	0.0238	1.4002	0.2367

Odds Ratio Estimates

Effect	Point Estimate	95% Wald Confidence Limits	
CONTROL	0.944	0.810	1.100
ATTMAR	1.016	0.993	1.041
ATTROLE	0.934	0.906	0.963
ATTHOUSE	0.972	0.928	1.019

TABLE 10.13   Continued

```
Association of Predicted Probabilities and Observed Responses
```

Percent Concordant	62.9	Somers' D 0.263
Percent Discordant	36.6	Gamma 0.265
Percent Tied	0.5	Tau-a 0.131
Pairs	53165	c 0.632

```
                            Classification Table
```

		Correct		Incorrect			Percentages			
Prob		Non-			Non-		Sensi-	Speci-	False	False
Level	Event	Event	Event	Event	Correct	tivity	ficity	POS	NEG
0.200	245	0	217	0	53.0	100.0	0.0	47.0	.
0.220	244	0	217	1	52.8	99.6	0.0	47.1	100.0
0.240	244	0	217	1	52.8	99.6	0.0	47.1	100.0
0.260	244	2	215	1	53.2	99.6	0.9	46.8	33.3
0.280	243	3	214	2	53.2	99.2	1.4	46.8	40.0
0.300	241	5	212	4	53.2	98.4	2.3	46.8	44.4
0.320	237	9	208	8	53.2	96.7	4.1	46.7	47.1
0.340	232	10	207	13	52.4	94.7	4.6	47.2	56.5
0.360	231	15	202	14	53.2	94.3	6.9	46.7	48.3
0.380	226	25	192	19	54.3	92.2	11.5	45.9	43.2
0.400	220	34	183	25	55.0	89.8	15.7	45.4	42.4
0.420	207	48	169	38	55.2	84.5	22.1	44.9	44.2
0.440	197	67	150	48	57.1	80.4	30.9	43.2	41.7
0.460	189	74	143	56	56.9	77.1	34.1	43.1	43.1
0.480	180	90	127	65	58.4	73.5	41.5	41.4	41.9
0.500	164	103	114	81	57.8	66.9	47.5	41.0	44.0
0.520	151	117	100	94	58.0	61.6	53.9	39.8	44.5
0.540	142	129	88	103	58.7	58.0	59.4	38.3	44.4
0.560	127	138	79	118	57.4	51.8	63.6	38.3	46.1
0.580	114	153	64	131	57.8	46.5	70.5	36.0	46.1
0.600	90	164	53	155	55.0	36.7	75.6	37.1	48.6
0.620	71	185	32	174	55.4	29.0	85.3	31.1	48.5
0.640	53	196	21	192	53.9	21.6	90.3	28.4	49.5
0.660	39	201	16	206	51.9	15.9	92.6	29.1	50.6
0.680	28	208	9	217	51.1	11.4	95.9	24.3	51.1
0.700	16	214	3	229	49.8	6.5	98.6	15.8	51.7
0.720	9	215	2	236	48.5	3.7	99.1	18.2	52.3
0.740	7	216	1	238	48.3	2.9	99.5	12.5	52.4
0.760	2	216	1	243	47.2	0.8	99.5	33.3	52.9
0.780	1	216	1	244	47.0	0.4	99.5	50.0	53.0
0.800	1	216	1	244	47.0	0.4	99.5	50.0	53.0
0.820	0	216	1	245	46.8	0.0	99.5	100.0	53.1
0.840	0	217	0	245	47.0	0.0	100.0	.	53.0

TABLE 10.14   Syntax and Selected SAS LOGISTIC Output for Model That Excludes ATTROLE

```
proc logistic data=Sasuser.Logreg;
    model WORKSTAT = CONTROL ATTMAR ATTHOUSE;
run;
```

```
                  The LOGISTIC Procedure

                  Model Fit Statistics

                                              Intercept
                               Intercept         and
            Criterion            Only        Covariates

               AIC             640.770         644.002
               SC              644.906         660.544
               -2 Log L        638.770         636.002
```

significant difference between the model with the three remaining predictors and the constant-only model, confirming that these predictors are unrelated to work status.

Table 10.15 summarizes the statistics for the predictors. Table 10.16 contains a checklist for direct logistic regression with a two-category outcome. A Results section follows that might be appropriate for submission to a journal.

TABLE 10.15   Logistic Regression Analysis of Work Status as a Function of Attitudinal Variables

Variables	B	Wald Chi-Square	Odds Ratio	95% Confidence Interval for Odds Ratio	
				Lower	Upper
Locus of control	−0.06	0.54	0.94	0.81	1.10
Attitude toward marital status	0.02	1.82	1.02	0.99	1.04
Attitude toward role of women	−0.07	19.30	0.93	0.91	0.96
Attitude toward housework	−0.03	1.40	0.97	0.93	1.02
(Constant)	3.20	11.13			

TABLE 10.13   Continued

Association of Predicted Probabilities and Observed Responses

Percent Concordant	62.9	Somers' D	0.263
Percent Discordant	36.6	Gamma	0.265
Percent Tied	0.5	Tau-a	0.131
Pairs	53165	c	0.632

Classification Table

Prob Level	Correct Event	Correct Non-Event	Incorrect Event	Incorrect Non-Event	Correct	Percentages Sensi-tivity	Speci-ficity	False POS	False NEG
0.200	245	0	217	0	53.0	100.0	0.0	47.0	.
0.220	244	0	217	1	52.8	99.6	0.0	47.1	100.0
0.240	244	0	217	1	52.8	99.6	0.0	47.1	100.0
0.260	244	2	215	1	53.2	99.6	0.9	46.8	33.3
0.280	243	3	214	2	53.2	99.2	1.4	46.8	40.0
0.300	241	5	212	4	53.2	98.4	2.3	46.8	44.4
0.320	237	9	208	8	53.2	96.7	4.1	46.7	47.1
0.340	232	10	207	13	52.4	94.7	4.6	47.2	56.5
0.360	231	15	202	14	53.2	94.3	6.9	46.7	48.3
0.380	226	25	192	19	54.3	92.2	11.5	45.9	43.2
0.400	220	34	183	25	55.0	89.8	15.7	45.4	42.4
0.420	207	48	169	38	55.2	84.5	22.1	44.9	44.2
0.440	197	67	150	48	57.1	80.4	30.9	43.2	41.7
0.460	189	74	143	56	56.9	77.1	34.1	43.1	43.1
0.480	180	90	127	65	58.4	73.5	41.5	41.4	41.9
0.500	164	103	114	81	57.8	66.9	47.5	41.0	44.0
0.520	151	117	100	94	58.0	61.6	53.9	39.8	44.5
0.540	142	129	88	103	58.7	58.0	59.4	38.3	44.4
0.560	127	138	79	118	57.4	51.8	63.6	38.3	46.1
0.580	114	153	64	131	57.8	46.5	70.5	36.0	46.1
0.600	90	164	53	155	55.0	36.7	75.6	37.1	48.6
0.620	71	185	32	174	55.4	29.0	85.3	31.1	48.5
0.640	53	196	21	192	53.9	21.6	90.3	28.4	49.5
0.660	39	201	16	206	51.9	15.9	92.6	29.1	50.6
0.680	28	208	9	217	51.1	11.4	95.9	24.3	51.1
0.700	16	214	3	229	49.8	6.5	98.6	15.8	51.7
0.720	9	215	2	236	48.5	3.7	99.1	18.2	52.3
0.740	7	216	1	238	48.3	2.9	99.5	12.5	52.4
0.760	2	216	1	243	47.2	0.8	99.5	33.3	52.9
0.780	1	216	1	244	47.0	0.4	99.5	50.0	53.0
0.800	1	216	1	244	47.0	0.4	99.5	50.0	53.0
0.820	0	216	1	245	46.8	0.0	99.5	100.0	53.1
0.840	0	217	0	245	47.0	0.0	100.0	.	53.0

TABLE 10.14   Syntax and Selected SAS LOGISTIC Output for Model That Excludes ATTROLE

```
proc logistic data=Sasuser.Logreg;
    model WORKSTAT = CONTROL ATTMAR ATTHOUSE;
run;
```

                    The LOGISTIC Procedure

                    Model Fit Statistics

                                                Intercept
                                   Intercept   and
          Criterion   Only   Covariates

          AIC   640.770   644.002
          SC   644.906   660.544
          -2 Log L   638.770   636.002

significant difference between the model with the three remaining predictors and the constant-only model, confirming that these predictors are unrelated to work status.

Table 10.15 summarizes the statistics for the predictors. Table 10.16 contains a checklist for direct logistic regression with a two-category outcome. A Results section follows that might be appropriate for submission to a journal.

TABLE 10.15   Logistic Regression Analysis of Work Status as a Function of Attitudinal Variables

Variables	*B*	Wald Chi-Square	Odds Ratio	95% Confidence Interval for Odds Ratio	
				Lower	*Upper*
Locus of control	−0.06	0.54	0.94	0.81	1.10
Attitude toward marital status	0.02	1.82	1.02	0.99	1.04
Attitude toward role of women	−0.07	19.30	0.93	0.91	0.96
Attitude toward housework	−0.03	1.40	0.97	0.93	1.02
(Constant)	3.20	11.13			

TABLE 10.16   Checklist for Standard Logistic Regression with Dichotomous Outcome

1. Issues
   a. Ratio of cases to variables and missing data
   b. Adequacy of expected frequencies (if necessary)
   c. Outliers in the solution (if fit inadequate)
   d. Multicollinearity
   e. Linearity in the logit
2. Major analysis
   a. Evaluation of overall fit. If adequate:
      (1) Significance tests for each predictor
      (2) Parameter estimates
   b. Effect size for model
   c. Evaluation of models without predictors
3. Additional analyses
   a. Odds ratios
   b. Classification or prediction success table
   c. Interpretation in terms of means and/or percentages

Results

A direct logistic regression analysis was performed on work status as outcome and four attitudinal predictors: locus of control, attitude toward current marital status, attitude toward women's role, and attitude toward housework. Analysis was performed using SAS LOGISTIC. Twenty-two cases with missing values on continuous predictors were imputed using the EM algorithm through SPSS MVA after finding no statistically significant deviation from randomness using Little's MCAR test, $p = .331$. After deletion of 3 cases with missing values on religious affiliation, data from

462 women were available for analysis: 217 housewives and 245 women who work outside the home more than 20 hours a week for pay.

A test of the full model with all four predictors against a constant-only model was statistically significant, $\chi^2(4, N = 440)$ = 23.24, $p < .001$, indicating that the predictors, as a set, reliably distinguished between working women and housewives. The variance in work status accounted for is small, however, with McFadden's D = .263, with a 95% confidence interval for the effect size of .07 ranging from .02 to .12 using Steiger and Fouladi's (1992) R2 software. Classification was unimpressive, with 67% of the working women and 48% of the housewives correctly predicted, for an overall success rate of 58%.

Table 10.15 shows regression coefficients, Wald statistics, odds ratios, and 95% confidence intervals for odds ratios for each of the four predictors. According to the Wald criterion, only attitude toward role of women reliably predicted work status, $\chi^2(1, N = 440) = 19.30$, $p < .0001$. A model run with attitude toward role of women omitted was not reliably different from a constant-only model, however this model was reliably different from the full model, $\chi^2(1, N = 440) = 20.47$, $p < .001$. This confirms the finding that attitude toward women's role is the only statistically significant predictor of work status among the four attitudinal variables. However, the odds ratio of .94 shows little change in the likelihood of working on the basis of a one-unit change in attitude toward women's role.

Thus, attitude towards the proper role of women in society distinguishes between women who do and do not work outside the home at least 20 hours per week, but the distinction is not a very strong one.

10.7.3   Sequential Logistic Regression with Three Categories of Outcome

The sequential analysis is done through two major runs, one with and one without demographic predictors.

10.7.3.1   *Limitations of Multinomial Logistic Regression*

10.7.3.1.1   *Adequacy of Expected Frequencies*

Only if a goodness-of-fit criterion is to be used to compare observed and expected frequencies is there a limitation to logistic regression. As discussed in Section 10.3.2.2, the expected frequencies for all pairs of discrete predictors must meet the usual "chi-square" requirements. (This requirement is only for this section on sequential analysis because the predictors in the direct analysis are all continuous.)

After filtering out cases with missing data on RELIGION, Table 10.17 shows the results of an SPSS CROSSTABS run to check the adequacy of expected frequencies for all pairs of discrete predictors. (Only the first 7 tables are shown, the remaining 3 are omitted.) Observed (COUNT) and EXPECTED frequencies are requested.

The last crosstabs in the table show one expected cell frequency under 5: 2.4 for single, non-white women. This is the only expected frequency that is less than 5, so that in no two-way table do more than 20% of the cells have frequencies less than 5, nor are any expected frequencies less than 1. Therefore, there is no restriction on the goodness-of-fit criteria used to evaluate the model.

10.7.3.1.2   *Linearity in the Logit*

An SPSS NOMREG run to test for linearity of the logit is shown in Table 10.18. Interactions between continuous variables and their natural logarithms are formed, for example, as LIN_SEL=SEL*LN(SEL). The NOMREG instruction identifies workstat as the DV; marital, children, religion, and race as categorical "factors" (discrete predictors, following the BY instruction); and the remaining variables as covariates (continuous predictors, following the WITH instruction). The added interactions are included as covariates. The MODEL includes main effects by default; interactions among original predictors may be included by request.

Table 10.18 shows no serious violation of the assumption of linearity of the logit.

10.7.3.2   *Sequential Multinomial Logistic Regression*

Table 10.19 shows the results of logistic regression analysis through SPSS NOMREG predicting the three categories of outcome (working, role-satisfied housewives, and role-dissatisfied housewives) from the set of seven demographic variables. This is a baseline model, used to evaluate improvement in the model when attitudinal predictors are added. That is, we are interested in evaluating the predictive ability of attitudinal variables after adjusting for demographic differences. Only minimal output—goodness-of-fit and classification—is requested for this baseline model.

**TABLE 10.17   Syntax and Partial Output of SPSS CROSSTABS for Screening
All Two-Way Tables for Adequacy of Expected Frequencies**

```
USE ALL.
COMPUTE FILTER_$=(RELIGION < 9).
VARIABLE LABEL FILTER_$ 'RELIGION < 9 (FILTER)'.
VALUE LABELS FILTER_$ 0 'NOT SELECTED' 1 'SELECTED'.
FORMAT FILTER_$ (F1.0).
FILTER BY FILTER_$.
EXECUTE.
CROSSTABS
   /TABLES=MARITAL CHILDREN RELIGION RACE BY WORKSTAT
   /FORMAT= AVALUE TABLES
   /CELLS= COUNT EXPECTED.
CROSSTABS
   /TABLES=CHILDREN RELIGION RACE BY MARITAL
   /FORMAT= AVALUE TABLES
   /CELLS= COUNT EXPECTED.
```

Crosstabs

Current marital status * Current work status Crosstabulation

			Current work status			Total
			Working	Role-satisfied housewives	Role-dissatisfied housewives	
Current marital status	Single	Count	24	3	4	31
		Expected Count	16.4	9.1	5.5	31.0
	Married	Count	168	127	64	359
		Expected Count	190.4	104.9	63.7	359.0
	Broken	Count	53	5	14	72
		Expected Count	38.2	21.0	12.8	72.0
Total		Count	245	135	82	462
		Expected Count	245.0	135.0	82.0	462.0

Presence of children * Current work status Crosstabulation

			Current work status			Total
			Working	Role-satisfied housewives	Role-dissatisfied housewives	
Presence of children	No	Count	57	1.3	12	82
		Expected Count	43.5	24.0	14.6	82.0
	Yes	Count	188	122	70	380
		Expected Count	201.5	111.0	67.4	380.0
Total		Count	245	135	82	462
		Expected Count	245.0	135.0	82.0	462.0

TABLE 10.17 Continued

Religious affiliation * Current work status Crosstabulation

			Current work status			Total
			Working	Role-satisfied housewives	Role-dissatisfied housewives	
Religious affiliation	None-or-other	Count	46	21	9	76
		Expected Count	40.3	22.2	13.5	76.0
	Catholic	Count	63	29	27	119
		Expected Count	63.1	34.8	21.1	119.0
	Protestant	Count	92	52	31	175
		Expected Count	92.8	51.1	31.1	175.0
	Jewish	Count	44	33	15	92
		Expected Count	48.8	26.9	16.3	92.0
Total		Count	245	135	82	462
		Expected Count	245.0	135.0	82.0	462.0

RACE * Current work status Crosstabulation

			Current work status			Total
			Working	Role-satisfied housewives	Role-dissatisfied housewives	
RACE	White	Count	218	131	73	422
		Expected Count	223.8	123.3	74.9	422.0
	Non-white	Count	27	4	9	40
		Expected Count	21.2	11.7	7.1	40.0
Total		Count	245	135	82	462
		Expected Count	245.0	135.0	82.0	462.0

(continued)

TABLE 10.17 Continued

Presence of children * Current marital status Crosstabulation

			Current marital status			
			Single	Married	Broken	Total
Presence of children	No	Count	29	38	15	82
		Expected Count	5.5	63.7	12.8	82.0
	Yes	Count	2	321	57	380
		Expected Count	25.5	295.3	59.2	380.0
Total		Count	31	359	72	462
		Expected Count	31.0	359.0	72.0	462.0

Religious affiliation * Current marital status Crosstabulation

			Current marital status			
			Single	Married	Broken	Total
Religious affiliation	None-or-other	Count	9	48	19	76
		Expected Count	5.1	59.1	11.8	76.0
	Catholic	Count	4	98	17	119
		Expected Count	8.0	92.5	18.5	119.0
	Protestant	Count	9	136	30	175
		Expected Count	11.7	136.0	27.3	175.0
	Jewish	Count	9	77	6	92
		Expected Count	6.2	71.5	14.3	92.0
Total		Count	31	359	72	462
		Expected Count	31.0	359.0	72.0	462.0

RACE * Current marital status Crosstabulation

			Current marital status			
			Single	Married	Broken	Total
RACE	White	Count	28	330	64	422
		Expected Count	28.3	327.9	65.8	422.0
	Non-white	Count	3	29	8	40
		Expected Count	2.7	31.1	6.2	40.0
Total		Count	31	359	72	462
		Expected Count	31.0	359.0	72.0	462.0

TABLE 10.18   Syntax and Selected Output of SPSS NOMREG for Test of Linearity of the Logit for a Logistic Regression Analysis of Work Status with Demographic and Attitudinal Variables

```
NOMREG
 WORKSTAT BY MARITAL CHILDREN RELIGION RACE WITH
 CONTROL ATTMAR ATTROLE SEL ATTHOUSE AGE EDUC
 LIN_CTRL LIN_ATMR LIN_ATRL LIN_ATHS LIN_SEL LIN_AGE LIN_EDUC
 /CRITERIA = CIN(95) DELTA(0) MXITER(100) MXSTEP(5) LCONVERGE(0)
 PCONVERGE(1.0E-6) SINGULAR(1.0E-8)
 /MODEL
 /PRINT = LRT.
```

Likelihood Ratio Tests

Effect	−2 Log Likelihood of Reduced Model	Chi-Square	df	Sig.
Intercept	781.627[a]	.000	0	.
CONTROL	783.778	2.151	2	.341
ATTMAR	785.478	3.851	2	.146
ATTROLE	489.738	8.111	2	.017
SEL	786.633	5.006	2	.082
ATTHOUSE	784.800	3.173	2	.205
AGE	784.032	2.405	2	.300
EDUC	782.310	.683	2	.711
LIN_CTRL	783.700	2.073	2	.355
LIN_ATMR	785.861	4.234	2	.120
LIN_ATRL	789.239	7.612	2	.022
LIN_ATHS	784.925	3.298	2	.192
LIN_SEL	787.330	5.703	2	.058
LIN_AGE	783.332	1.705	2	.426
LIN_EDUC	782.214	.587	2	.746
MARITAL	794.476	12.849	4	.012
CHILDREN	785.607	3.980	2	.137
RELIGION	789.446	7.819	6	.252
RACE	494.095	12.468	2	.002

The chi-square statistic is the difference in −2 log-likelihoods between the final model and a reduced model. The reduced model is formed by omitting an effect from the final model. The null hypothesis is that all parameters of that effect are 0.

[a]This reduced model is equivalent to the final model because omitting the effect does not increase the degrees of freedom.

TABLE 10.19   Syntax and Selected Output of SPSS NOMREG for Logistic Regression Analysis of Work Status with Demographic Variables Only

```
NOMREG
  WORKSTAT BY CHILDREN RELIGION RACE MARITAL WITH
  SEL AGE EDUC
  /CRITERIA = CIN(95) DELTA(0) MXITER(100) MXSTEP(5) LCONVERGE(0)
  PCONVERGE(1.0E-6) SINGULAR(1.0E-8)
  /MODEL
  /PRINT = CLASSTABLE FIT STEP MFI.
```

Model Fitting Information

Model	−2 Log Likelihood	Chi-Square	df	Sig.
Intercept Only	910.459			
Final	832.404	78.055	20	.000

Goodness-of-Fit

	Chi-Square	df	Sig.
Pearson	892.292	864	.245
Deviance	816.920	864	.872

Classification

		Predicted		
Observed	Working	Role-satisfied housewives	Role-dissatisfied housewives	Percent Correct
Working	200	43	2	81.6%
Role-satisfied housewives	86	48	1	35.6%
Role-dissatisfied housewives	59	23	0	.0%
Overall Percentage	74.7%	24.7%	.6%	53.7%

The model provides an acceptable fit to the data. Goodness-of-Fit statistics (comparing observed with expected frequencies) with all predictors in the model show good fit with $p = .872$ by the Deviance criterion and with $p = .245$ for the Pearson criterion. Correct classification on the basis of demographic variables alone is 54% overall: with 82% for working women (the largest group) but no correct classifications for role-dissatisfied housewives.

Table 10.20 shows the results of logistic regression analysis through SPSS NOMREG predicting the three categories of outcome (working, role-satisfied housewives, and role-dissatisfied housewives) from the set of seven demographic and four attitudinal variables. For purposes of single-df tests (comparisons among groups) the reference (**BASE**) group is set to the first category, working

TABLE 10.20 Syntax and Selected Output of SPSS NOMREG for Logistic Regression Analysis of Work Status with Demographic and Attitudinal Variables

```
NOMREG
   WORKSTAT (BASE=FIRST ORDER=ASCENDING) BY MARITAL CHILDREN RELIGION
      RACE WITH CONTROL ATTMAR ATTROLE SEL ATTHOUSE AGE EDUC
   /CRITERIA = CIN(95) DELTA(0) MXITER(100) MXSTEP(5) LCONVERGE(0)
   PCONVERGE(1.0E-6) SINGULAR(1.0E-8)
   /MODEL
   /PRINT = CLASSTABLE FIT PARAMETER SUMMARY LRT CPS MFI .
```

Nominal Regression

Case Processing Summary

		N	Marginal Percentage
Work status	Working	245	53.0%
	Role-satisfied housewives	135	29.2%
	Role-dissatisfied housewives	82	17.7%
Current marital status	Single	31	6.7%
	Married	359	77.7%
	Broken	72	15.6%
Presence of children	No	82	17.7%
	Yes	380	82.3%
Religious affiliation	None-or-other	76	16.5%
	Catholic	119	25.8%
	Protestant	175	37.9%
	Jewish	92	19.9%
RACE	White	422	91.3%
	Non-white	40	8.7%
Valid		462	100.0%
Missing		0	
Total		462	
Subpopulation		462[a]	

[a]The dependent variable has only one value observed in 462 (100.0%) subpopulations.

(continued)

TABLE 10.20 Continued

Model Fitting Information

Model	−2 Log Likelihood	Chi-Square	df	Sig.
Intercept Only	926.519			
Final	806.185	120.334	28	.000

Goodness-of-Fit

	Chi-Square	df	Sig.
Pearson	930.126	894	.195
Deviance	806.246	894	.983

Pseudo R-Square

Cox and Snell	.229
Nagelkerke	.265
McFadden	.130

Likelihood Ratio Tests

Effect	−2 Log Likelihood of Reduced Model	Chi-Square	df	Sig.
Intercept	806.246[a]	.000	0	.
CONTROL	807.098	.851	2	.653
ATTMAR	809.404	3.157	2	.206
ATTROLE	824.412	18.166	2	.000
SEL	813.061	6.814	2	.033
ATTHOUSE	814.700	8.453	2	.015
AGE	810.898	4.651	2	.098
EDUC	813.287	7.040	2	.030
MARITAL	821.474	15.228	4	.004
CHILDREN	810.283	4.036	2	.133
RELIGION	813.149	6.903	6	.330
RACE	817.969	11.723	2	.003

The chi-square statistic is the difference in −2 log-likelihoods between the final model and a reduced model. The reduced model is formed by omitting an effect from the final model. The null hypothesis is that all parameters of that effect are 0.

[a]This reduced model is equivalent to the final model because omitting the effect does not increase the degrees of freedom.

TABLE 10.20 Continued

Parameter Estimates

Work Status[a]		B	Std. Error	Wald	df	Sig.	Exp(B)	95% Confidence Interval for Exp(B)	
								Lower Bound	Upper Bound
Role-satisfied housewives	Intercept	-4.120	1.853	4.945	1	.026			
	CONTROL	-.016	.101	.025	1	.874	.984	.808	1.199
	ATTMAR	-.012	.019	.384	1	.536	.988	.953	1.025
	ATTROLE	.088	.022	16.479	1	.000	1.092	1.047	1.140
	SEL	.015	.006	6.612	1	.010	1.015	1.004	1.027
	ATTHOUSE	-.029	.031	.911	1	.340	.971	.915	1.031
	AGE	-.091	.063	2.118	1	.146	.913	.808	1.032
	EDUC	-.126	.067	3.526	1	.060	.882	.773	1.006
	[MARITAL=1.00]	.551	.884	.388	1	.533	1.735	.307	9.821
	[MARITAL=2.00]	1.758	.536	10.761	1	.001	5.803	2.029	16.591
	[MARITAL=3.00]	0[b]	.	.	0	.	.	.	.
	[CHILDREN=.00]	-.811	.419	3.744	1	.053	.445	.196	1.011
	[CHILDREN=1.00]	0[b]	.	.	0	.	.	.	.
	[RELIGION=1.00]	.226	.399	.321	1	.571	.798	.365	1.743
	[RELIGION=2.00]	-.620	.371	2.786	1	.095	.538	.260	1.114
	[RELIGION=3.00]	-.494	.328	2.272	1	.132	.610	.321	1.160
	[RELIGION=4.00]	0[b]	.	.	0	.	.	.	.
	[RACE=l.00]	1.789	.598	8.938	1	.003	5.983	1.852	19.330
	[RACE=2.00]	0[b]	.	.	0	.	.	.	.

aThis reference category is: Working.
bThis parameter is set to zero because it is redundant.

(continued)

TABLE 10.20 Continued

Parameter Estimates

Work Status[a]		B	Std. Error	Wald	df	Sig.	Exp(B)	95% Confidence Interval for Exp(B)	
								Lower Bound	Upper Bound
Role-dissatisfied housewives	Intercept	-4.432	1.994	4.940	1	.026			
	CONTROL	.088	.108	.657	1	.418	1.092	.883	1.350
	ATTMAR	.025	.018	1.958	1	.162	1.025	.990	1.061
	ATTROLE	.049	.024	4.143	1	.042	1.050	1.002	1.101
	SEL	.005	.006	.709	1	.400	1.005	.993	1.018
	ATTHOUSE	.084	.036	5.381	1	.020	1.087	1.013	1.167
	AGE	-.135	.070	3.712	1	.054	.874	.761	1.002
	EDUC	-.164	.074	4.864	1	.027	.849	.734	.982
	[MARITAL=1.00]	-.280	.746	.141	1	.707	.756	.175	3.258
	[MARITAL=2.00]	.439	.400	1.202	1	.273	1.551	.708	3.399
	[MARITAL=3.00]	0[b]	.	.	0	.	.	.	.
	[CHILDREN=.00]	-.352	.456	.595	1	.441	.704	.288	1.719
	[CHILDREN=1.00]	0[b]	.	.	0	.	.	.	.
	[RELIGION=1.00]	-.445	.501	.787	1	.375	.641	.240	1.712
	[RELIGION=2.00]	-.218	.422	.268	1	.604	1.244	.544	2.844
	[RELIGION=3.00]	-.023	.393	.003	1	.954	1.023	.473	2.211
	[RELIGION=4.00]	0[b]	.	.	0	.	.	.	.
	[RACE=I.00]	.658	.481	1.875	1	.171	1.932	.753	4.957
	[RACE=2.00]	0[b]	.	.	0	.	.	.	.

aThe reference category is: Working
bThis parameter is set to zero because it is redundant.

TABLE 10.20   Continued

Observed	Predicted			
	Working	Role-satisfied housewives	Role-dissatisfied housewives	Percent Correct
Working	199	39	7	81.2%
Role-satisfied housewives	65	68	2	50.4%
Role dissatisfied housewives	57	16	9	11.0%
Overall Percentage	69.5%	1626.6%	3.9%	59.7%

women. Parameter estimates, effect-size statistics (SUMMARY), likelihood ratio tests (LRT), and case processing summary (CPS) are requested in this final model, in addition to statistics requested in the base model.

Table 10.20 shows the number of respondents in each of the three outcomes and in each category of discrete predictors. Goodness-of-Fit statistics (comparing observed with expected frequencies) with all predictors in the model show an excellent fit with $p = .983$ by the Deviance criterion and with $p = .195$ for the Pearson criterion. Using the Nagelkerke measure with the Steiger and Fouladi (1992) software, $R^2 = .265$ with a 95 confidence interval ranging from .15 to .29. Note that the number of variables is considered to be 28 (i.e., the df for the Final Model Chi-Square test).

Likelihood ratio tests show three of the predictors to significantly add to prediction of work status using a critical value for each test that sets $\alpha = .0045$ to compensate for inflation in familywise error rate associated with the 11 predictors, as well as possible bias introduced by use of EM imputation. Critical values for the predictors depend on their df: 10.81 for 2 df, 15.089 for 4df, and 18.81 for 6 df. Thus, attitudes toward role of women, marital status, and race significantly distinguish among the three groups of women.

Two tables of Parameter Estimates are shown, one for each degree of freedom for outcome. The first table compares working women with role-satisfied housewives, while the second compares working with role-dissatisfied women. Using a criterion $\alpha = .0045$ (to compensate for inflated Type I error rate with 11 predictors and bias associated with EM imputation), the critical value for χ^2 with 1 df = 8.07. By this criterion, attitudes toward role of women, marital status, and race reliably separate working women from role-satisfied housewives. Role-satisfied women score higher (i.e., more conservatively) than working women on their attitudes towards the role of women in society. The odds of being married are almost 6 times greater for role-satisfied women than for working women, as are the odds for being Caucasian. No predictor reliably separates working women from role-dissatisfied housewives. Parameter estimates and odds ratios with their 95% confidence limits are in Tables 10.21 and 10.22 in a form suitable for reporting.

The classification table shows that 60% of the cases now are correctly classified: ranging from 81% of the working women (about the same number as in the base model) but now 11% of the role-dissatisfied housewives (increasing from no correct classifications for that group).

**TABLE 10.21   Logistic Regression Analysis of Work Status as a Function
of Attitudinal Variables: Working Women vs. Role-Satisfied Housewives**

Variables	B	Wald χ^2-test	Odds Ratio	95% Confidence Interval for Odds Ratio	
				Lower	*Upper*
Locus of control	−0.02	0.03	0.98	0.81	1.20
Attitude toward marital status	−0.01	0.38	0.99	0.95	1.03
Attitude toward role of women	0.09	16.48	1.09	1.05	1.14
Socioeconomic level	0.02	6.61	1.02	1.00	1.03
Attitude toward housework	−0.03	0.91	0.97	0.92	1.03
Age	−0.09	2.12	0.91	0.81	1.03
Educational level	−0.13	3.53	0.88	0.77	1.01
Single vs. broken marriage	0.55	0.39	1.74	0.32	9.82
Married vs. broken marriage	1.76	10.73	5.80	2.03	16.59
Presence vs. absence of children	−0.81	3.74	0.45	0.20	1.01
Protestant vs. no or other religion	−0.23	0.32	0.80	0.37	4.17
Catholic vs. no or other religion	−0.62	2.79	0.54	0.26	1.11
Jewish vs. no or other religion	−0.49	2.27	0.61	0.32	1.16
Caucasian vs. non-white	1.79	8.94	5.98	1.85	19.33
(Constant)	−4.12	4.95			

Evaluation of the addition of attitudinal variables as a set is most easily accomplished by calculating the difference between the two models. When both demographic and attitudinal predictors are included, $\chi^2 = 120.273$ with 28 df; when demographic predictors alone are included, $\chi^2 = 78.055$ with 20 df. Applying Equation 10.7 to evaluate improvement in fit,

$$\chi^2 = 120.273 - 78.055 = 42.22$$

with df $= 8, p < .05$. This indicates statistically significant improvement in the model with the addition of attitudinal predictors.

The remaining issue, interpretation of the statistically significant effects, is difficult through SPSS NOMREG because of the separation of effects for both outcome and predictors into single degree of freedom dummy variables. For categorical predictors, group differences are observed in proportions of cases in each category of predictor for each category of outcome. This information is available from the screening run of Table 10.17. For example, we see that 69% (168/245) of the

TABLE 10.22   Logistic Regression Analysis of Work Status as a Function of Attitudinal and Demographic Variables: Working Women vs. Role-Dissatisfied Housewives

Variables	B	Wald χ^2-test	Odds Ratio	95% Confidence Interval for Odds Ratio	
				Lower	*Upper*
Locus of control	0.09	0.66	1.09	0.87	1.35
Attitude toward marital status	0.03	1.96	1.03	0.99	1.06
Attitude toward role of women	0.05	4.14	1.05	1.00	1.10
Attitude toward housework	0.08	5.38	1.09	1.01	1.17
Socioeconomic level	0.01	0.71	1.01	0.99	1.02
Age	−0.14	3.71	0.87	0.76	1.00
Educational level	−0.16	4.86	0.85	0.73	0.98
Single vs. broken marriage	−0.28	0.14	0.76	0.18	3.26
Married vs. broken marriage	0.44	1.20	1.55	0.71	3.40
Presence vs. absence of children	−0.35	0.60	0.70	0.29	1.72
Protestant vs. no or other religion	−0.45	0.79	0.64	0.24	1.71
Catholic vs. no or other religion	0.22	0.27	1.24	0.54	2.84
Jewish vs. no or other religion	0.02	0.00	1.02	0.47	2.21
Caucasian vs. non-white	0.66	1.88	1.93	0.75	4.96
(Constant)	−4.32	4.94			

working women are currently married, while 94% and 78% of the role-satisfied and role-dissatisfied housewives, respectively, are currently married.

For continuous predictors, interpretation is based on mean differences for significant predictors for each category of outcome. Table 10.23 shows SPSS DESCRIPTIVES output giving means

TABLE 10.23   Syntax and Partial Output of SPSS DESCRIPTIVES Showing Group Means for Atthouse

```
SORT CASES BY WORKSTAT.
SPLIT FILE
 SEPARATE BY WORKSTAT.
DESCRIPTIVES
 VARIABLES=ATTROLE
 /STATISTICS=MEAN STDDEV.
```

TABLE 10.23   Continued

Descriptives

Current work status = Working

Descriptive Statistics[a]

	N	Mean	Std. Deviation
Attitudes toward role of women	245	33.8122	6.96577
Valid N (listwise)	245		

[a]Current work status = Working.

Current work status = Role-satisfied housewives

Descriptive Statistics[a]

	N	Mean	Std. Deviation
Attitudes toward role of women	135	37.2000	6.31842
Valid N (listwise)	135		

[a]Current work status = Role-satisfied housewives.

Current work status = Role-dissatisfied housewives

Descriptive Statistics[a]

	N	Mean	Std. Deviation
Attitudes toward role of women	82	35.6098	5.74726
Valid N (listwise)	82		

[a]Current work status = Role-dissatisfied housewives.

for each category of outcome on the statistically significant predictor, ATTROLE, after splitting the file into WORKSTAT groups.

Table 10.24 summarizes results of the sequential analysis. Table 10.25 shows contingency tables for statistically significant discrete predictors. Table 10.26 provides a checklist for sequential logistic regression with more than two outcomes. A Results section in journal format follows.

**TABLE 10.24   Logistic Regression Analysis of Work Status
as a Function of Demographic and Attitudinal Variables**

Variables	χ^2 to Remove	df	Model χ^2
Demographic			
Marital status	15.23*	4	
Presence of children	4.04	2	
Religion	6.90	6	
Race	11.72*	2	
Socioeconomic level	6.81	2	
Age	4.65	2	
Educational level	7.04	2	
All demographic variables			78.06
Attitudinal			
Locus of control	0.85	2	
Attitude toward marital status	3.16	2	
Attitude toward role of women	18.17*	2	
Attitude toward housework	8.45	2	
All variables			120.27

*$p < .0045$.

**TABLE 10.25   Marital Status and Race
as a Function of Work Status**

	Work Status[a]			
	1	*2*	*3*	*Total*
Marital Status				
Single	24	3	4	31
Married	168	127	64	359
Broken	53	5	14	72
Total	245	135	82	462
Race				
White	218	131	73	422
Nonwhite	27	4	9	40
Total	245	135	82	462

[a]1 = Working; 2 = Role-satisfied housewives;
3 = Role-dissatisfied housewives.

TABLE 10.26   Checklist for Sequential Logistic Regression with Multiple Outcomes

1. Issues
   a. Ratio of cases to variables and missing data
   b. Adequacy of expected frequencies (if necessary)
   c. Outliers in the solution (if fit inadequate)
   d. Multicollinearity
   e. Linearity in the logit
2. Major analyses
   a. Evaluation of overall fit at each step.
      (1) Significance tests for each predictor at each step of interest
      (2) Parameter estimates at each step of interest
      (3) Effect size at each step of interest
   b. Evaluation of improvement in model at each step
3. Additional analyses
   a. Odds ratios
   b. Classification and/or prediction success table
   c. Interpretation in terms of means and/or percentages
   d. Evaluation of models without individual predictors

```
                            Results
     A sequential logistic regression analysis was performed
through SPSS NOMREG to assess prediction of membership in one of
three categories of outcome (working women, role-satisfied house-
wives, and role-dissatisfied housewives), first on the basis of
seven demographic predictors and then after addition of four atti-
tudinal predictors. Demographic predictors were children (presence
or absence), race (Caucasian or other), socioeconomic level, age,
religious affiliation (Protestant, Catholic, Jewish, none/other),
and marital status (single, married, broken). Attitudinal predic-
tors were locus of control, attitude toward marital status, atti-
tude toward role of women, and attitude toward housework.
```

Values for 22 cases with missing data on continuous predictors were imputed using the EM algorithm through SPSS MVA after finding no statistically significant deviation from randomness using Little's MCAR test, p = .331. After deletion of 3 cases with missing values on religious affiliation, data from 462 women were available for analysis: 217 housewives and 245 women who work outside the home more than 20 hours a week for pay. Evaluation of adequacy of expected frequencies for categorical demographic predictors revealed no need to restrict model goodness-of-fit tests. No serious violation of linearity in the logit was observed.

There was a good model fit (discrimination among groups) on the basis of the seven demographic predictors alone, χ^2(864, N = 462) = 816.92, p = .87, using a deviance criterion. After addition of the four attitudinal predictors, χ^2(894, N = 462) = 806.25, p = .98, Nagelkerke R^2 = .27 with a 95% confidence interval ranging from .15 to .29 (Steiger and Fouladi, 1992). Comparison of log-likelihood ratios (see Table 10.24) for models with and without attitudinal variables showed statistically significant improvement with the addition of attitudinal predictors, χ^2(8, N = 462) = 42.22, p < .05.

Overall classification was unimpressive. On the basis of 7 demographic variables alone, correction classification rates were 82% for working women, 36% for role-satisfied, and 0% for role-dissatisfied women; the overall correct classification rate was 54%. The improvement to 60% with the addition of four attitudinal predictors reflected success rates of 81%, 50%, and 11% for the three groups, respectively. Clearly, cases were overclassified into the largest group: working women.

Table 10.24 shows the contribution of the individual predictors to the model by comparing models with and without each predictor. Two predictors from the demographic and one from the attitudinal set statistically significant enhanced prediction, $p < .0045$. Outcome was predictable from marital status, race, and attitude toward role of women.

Tables 10.20 and 10.21 show regression coefficients and chi-square tests of them as well as odds ratios and the 95% confidence intervals around them. Role-satisfied housewives were more likely than working women to have positive attitudes toward housework. The odds of being currently married and of being Cauasian are almost 6 times as great for role-satisfied women as for working women.

Table 10.25 shows the relationship between work status and the two categorical demographic predictors. Working women are less likely to be currently married (69%) than are role-satisfied house-wives (94%) or role-dissatisfied housewives (78%). Role-satisfied housewives are more likely to be Caucasian (97%) than are working women or role-dissatisfied housewives (89% for both groups).

Mean group differences in attitudes toward role of women and homemaking were not large. However, role-satisfied housewives had more conservative attitudes (mean = 37.2) than role-dissatisfied housewives (mean = 35.6) or working women (mean = 33.8).

Thus, the three groups of women are distinguished on the basis of three predictors. Intact marriage is most common among role-satisfied housewives and least common among working women. Role-satisfied women are also more likely to be Caucasian than the other two groups and have more conservative attitudes toward the proper role of women in society. The most liberal attitudes toward role of women are held by women who work for pay outside the home for at least 20 hours per week, however these attitudinal differences are not large.

10.8 Comparison of Programs

Only the programs in the three reviewed packages that are specifically designed for logistic regression are discussed here. All of the major packages also have programs for nonlinear regression, which perform logistic regression if the researcher specifies the basic logistic regression equation.

 The logistic regression programs differ in how they code the outcome: some base probabilities and other statistics on outcome coded "1" (i.e., success, disease) and some base statistics on outcomes coded "0." The manuals are also sometimes inconsistent and confusing in their labeling of predictors where the terms independent variable, predictor, and covariate are used interchangeably. Table 10.27 compares five programs from the three major packages.

10.8.1 SPSS Package

SPSS LOGISTIC REGRESSION[9] handles only dichotomous outcomes and bases statistics on the outcome coded "1." SPSS NOMREG (nominal regression) and PLUM are the programs that handle multiple outcome categories.

 SPSS LOGISTIC REGRESSION offers the most flexible procedures for controlling the entry of predictors. All can be entered in one step, or sequential steps can be specified where one or more predictors enter at each step. For statistical logistic regression both forward and backward stepping are available based on either the Wald or likelihood-ratio statistics at the user's option. The user also has a choice among several criteria for terminating the iterative procedure used to find the optimal solution: change in parameter estimates, maximum number of iterations, percent change in log-likelihood, probability of score statistics for predictor entry, probability of Wald or likelihood ratio statistic to remove a variable, and the epsilon value used for redundancy checking.

 SPSS LOGISTIC REGRESSION has a simple SELECT instruction to select a subset of cases on which to compute the logistic regression equation; classification is then performed on all cases as a test of the generalizability of the equation.

 This program provides a comprehensive set of residuals statistics which can be displayed or saved, including predicted probability, predicted group, difference between observed and predicted values (residuals), deviance, the logit of the residual, studentized residual, normalized residual, leverage value, Cook's influence, and difference in betas as a result of leaving that case out of the logistic regression equation. The listing can be restricted to outliers with user-specified criteria for determining an outlier.

 SPSS NOMREG handles multinomial models (multiple outcome categories) and, since Version 12, does statistical analysis and offers a variety of diagnostics. Extensive diagnostic values are saved to the data set, such as estimated response probabilities for each case in each category, predicted category for each case, and predicted and actual category probabilities for each case. Discrete predictors are specified as "factors" and continuous predictors as "continuous." The default model includes main effects of all predictors, but full factorial models may be specified as well as just about any desired customized model of main effects, interactions, and forced or statistical entry of them. The program assumes that the multiple outcome categories are unordered, and gives regression coefficients for each outcome category (except the one designated BASE, by default the last category) vs.

---

[9]The PROBIT program can also do logistic regression.

TABLE 10.27 Comparison of Programs for Logistic Regression

Feature	SPSS LOGISTIC REGRESSION	SPSS NOMREG	SPSS PLUM	SAS LOGISTIC	SYSTAT LOGIT
Input					
Accepts discrete predictors without recoding	Yes	Yes	Yes	CLASS	Yes
Alternative coding schemes for discrete predictors	8	No	No	Yes	2
Accepts tabulated data	No	No	No	Yes	Yes
Specify reference category and order for parameter estimates	Yes	Yes	NA	Yes	No
Specify inclusion of intercept in model	Yes	Yes	Yes	Yes	Yes
Specify how covariate patterns are defined	No	No	No	No	No
Specify exact logical regression	Yes	Yes	No	Yes	No
Specify stepping methods and criteria	Yes	No	No	Yes	Yes
Specify sequential order of entry and test of predictors	No	No	No	Yes	Interactive
Specify a case-control design (conditional)	No	No	No	STRATA	Yes
Can specify size of confidence limits for odds ratio (e^b)	Yes	Yes	N.A.	No	No
Specify cutoff probability for classification table	No	No	N.A.	Yes	No
Accepts multiple unordered outcome categories	No	Yes	No	No	Yes
Deals with multiple ordered outcome categories	No	No	Yes	Yes	No
Can specify equal odds model	N.A.	No	No	No	No
Can specify discrete choice models	No	No	No	No	Yes
Can specify repeated-measures outcome variable	No[d]	No	No	Yes[c]	No
Can specify Poisson regression	No	No	No	Yes	No
Syntax to select a subset of cases for classification only	Yes	No	No	No	SP
Score new data sets without refitting model	No	No	No	Yes	No
Specify quasi-maximum likelihood covariance matrix	No	No	No	No	Yes
Specify case weights	No	No	No	Yes	Yes
Specify start values	No	No	No	No	Yes
Specify link function for response probabilities	No	No	LINK	No	No
Can restrict printing of diagnostics to outliers	Yes	No	No	No	No

TABLE 10.27 Continued

Feature	SPSS LOGISTIC REGRESSION	SPSS NOMREG	SPSS PLUM	SAS LOGISTIC	SYSTAT LOGIT
Input (continued)					
Add delta to observed cell frequencies	No	DELTA	Yes	No	No
Specify log-likelihood convergence criterion	LCON	LCONVERGE	LCONVERGE	No	CONVERG
Specify maximum number of iterations	ITERATE	MXITER	MXITER	MAXITER	No
Specify maximum step-halving allowed	No	MXSTEP	MXSTEP	MAXSTEP	No
Parameter estimates convergence criterion	BCON	PCONVERGE	PCONVERGE	CONVERGE	Yes
Additional convergence criteria	No	Yes	Yes	Yes	No
Specify tolerance	No	SINGULAR	SINGULAR	SINGULAR	TOL
Epsilon value used for redundancy checking	EPS	No	No	No	No
Specify scale component	No	No	Yes	No	No
Specify correction for overdispersion	No	Yes	No	Yes	No
Regression output					
Log-likelihood (or -2 log-likelihood) for full model	Yes	Yes	Yes	Yes	Yes
Log-likelihood (or -2 log-likelihood) for constant-only model	Yes	Yes	Yes	Yes	Yes
Deviance and Pearson goodness-of-fit statistics	No	Yes	Yes	SCALE	No
Hosmer-Lemeshow goodness-of-fit χ^2	Yes	Yes	Yes	Yes	Yes
Goodness-of-fit χ^2: constant-only vs. full model	Yes	Yes	Yes	Yes	Yes
Goodness-of-fit χ^2: based on observed vs. expected frequencies	Yes	Pearson	Pearson	No	Yes
Akaike information index (AIC)	No	No	No	Yes	No
Schwartz criterion	No	No	No	Yes	No
Score statistic	No	No	No	Yes	Yes
Improvement in goodness-of-fit since last step	Yes	N.A.	No	Yes	No
Goodness-of-fit χ^2 tests for individual predictors in specified model	Yes(LR)	Yes	Yes	Yes	No
Wald tests for predictors combined over multiple categories	Yes	No	No	No	Yes[c]
Regression coefficient	B	B	Estimate	Parameter estimate	ESTIMATE
Standard error of the regression coefficient	S.E.	Std. Error	Std. Error	Yes	Yes

(continued)

TABLE 10.27 Continued

Feature	SPSS LOGISTIC REGRESSION	SPSS NOMREG	SPSS PLUM	SAS LOGISTIC	SYSTAT LOGIT
Regression output (*continued*)					
Regression coefficient divided by standard error	No	No	No	No	t-ratio
Squared regression coefficient divided by squared standard error	Wald	Wald	Wald	Wald Chi-Square	No
Probability value for coefficient divided by standard error	Sig	Sig	Sig	Pr > ChiSq	p-value
Partially standardized regression coefficient	No	No	No	Yes	No
e^B (odds ratio)	Exp(B)	Exp(B)	No	Odds ratio	Odds Ratio
McFadden's rho squared for model	No	Yes	No	No	Yes
Cox and Snell R^2 for model	Yes	Yes	Yes	No	No
Nagelkerke R^2 for model	Yes	Yes	Yes	No	No
Association measures between observed responses and predicted probabilities	No	No	No	Yes	No
Partial correlations between outcome and each predictor variable (R)	Yes	No	No	No	No
Correlations among regressions coefficients	Yes	Yes	Yes	Yes	No
Covariances among regression coefficients	No	Yes	Yes	Yes	No
Classification table	Yes	Yes	No	Yes[b]	Yes
Prediction success table	No	No	No	Yes[b]	Yes
Histograms of predicted probabilities for each group	CLASSPLOT	No	No	No	No
Quantile table	No	No	No	No	QNTL
Derivative tables	No[e]	No	No	No	Yes
Plot of predicted probability as a function of the logit	Yes	No	No	No	No
Diagnostics saved to file					
Predicted probability of success for each case	Yes	No	Yes	Yes	Yes
Options for predicted probabilities	No	No	No	Yes	No
Raw residual for each case	Yes	No	No[a]	No	No
Standardized (Pearson) residual for each case	Yes	Yes[a]	Yes[a]	Yes[b]	Yes

TABLE 10.27 Continued

Diagnostics saved to file (*continued*)

Feature	SPSS LOGISTIC REGRESSION	SPSS NOMREG	SPSS PLUM	SAS LOGISTIC	SYSTAT LOGIT
Variance of Standardized (Pearson) residual for each case	No	No	No	No	Yes
Standardized (normed) residual for each case	Yes	No	No	No	Yes
Studentized residual for each case	Yes	No	No	No	No
Logit residual for each case	Yes	No	No	No	No
Predicted log odds for each pattern of predictors	No	No	No	No	No
Deviance for each case	Yes	No	No	Yes[b]	Yes
Diagonal of the hat matrix (leverage)	Yes	No	No	Yes[b]	Yes
Cook's distance for each case	Yes	No	No	Yes[b]	No
Cumulative residuals	No	No	No	No	No
Total χ^2 for pattern of predictors (covariates)	No	No	No	No	Yes
Deviance residual for each case	Yes	No	No	Yes[b]	Yes
Change in Pearson χ^2 for each case	Yes	No	No	Yes[b]	Yes
Change in betas for each case	Yes	No	Yes	Yes[b]	Yes
Confidence interval displacement diagnostics for each case	No	No	No	Yes[b]	No
Special diagnostics for ordered response variables	No	No	No	Yes	No
Predicted category for each case	Yes	Yes	Yes	No	Yes
Estimated response probability for each case in each category	No	Yes	Yes	No	Yes
Predicted category probability for each case	No	Yes	Yes	No	Yes
Actual category probability for each case	No	Yes	Yes	No	No

[a] Available for each cell (covariate pattern).

[b] For two-category outcome analysis only.

[c] Discrete predictors only, also available in SAS CATMOD.

[d] May be done through SPSS COMPLEX SAMPLES LOGISTIC REGRESSION.

the base category. Three R^2 measures are routinely printed; Pearson and deviance goodness-of-fit statistics are available. SPSS NOMREG also has a scaling instruction in the event of overdispersion.

SPSS PLUM is the newest logistic regression program and analyzes models with ordered multi-category outcomes. PLUM (accessed in the menu system as Ordinal Regression) has most of the features of NOMREG as well as a few others. Classification tables are not produced but they may be constructed by cross-tabulating predicted with actual categories for each case.

PLUM offers several alternative link functions, including Cauchit (for outcomes with many extreme values), complementary log-log (makes higher categories more probable), negative log-log (makes lower categories more probable, and probit (assumes a normally distributed latent variable). The user also is given an opportunity to scale the results to one or more of the predictors to adjust for differences in variability over predictor categories. PLUM has a test for parallel lines, to evaluate whether the parameters are the same for all categories.

10.8.2   SAS System

SAS LOGISTIC handles multiple as well as dichotomous response categories, but assumes multiple categories are ordered. There is no default coding of categorical predictors; the coding is user-specified before invoking PROC LOGISTIC. Statistics for dichotomous outcomes are based on the category coded "0" unless otherwise specified.

This is the most flexible of the logistic regression programs in that alternative (to logit) link functions can be specified, including `normit` (inverse standardized normal probability integral function) and complementary log-log functions. The program also does Poisson regression. SAS LOGISTIC also provides correction for overdispersion when predictors are discrete, in which the variance in cell frequencies is greater than that assumed by the underlying model, a common condition when there are repeated measures.

SAS LOGISTIC has the basic goodness-of-fit statistics, as well as exclusive ones such as Akaike's Information Criterion and the Schwartz Criterion. Strength of association between the set of predictors and the outcome is assessed using Somers' D, Gamma, Tau-a, or Tau-c, a more extensive set of strength of association statistics than the other programs.

Classification is done with jackknifing. A cutoff criterion may be specified, but results are shown for a variety of cutoff criteria as well. Results include number correct for each category as well as percentages for sensitivity, specificity, false positives, and false negatives. Classification is available only for analyses with a two-category outcome. Exact logistic regression is available for SAS LOGISTIC since Version 8. This permits analysis of smaller data sets than can be legitimately analyzed with the usual asymptotic procedures.

A full set of regression diagnostics is available, accessible through saved files or through an "influence plot" to find outliers in the solution. The plot consists of a case-by-case listing of the values of regression statistics along with a plot of deviations from those statistics. Through IPLOT an additional set of plots can be produced in which the value of each statistic is plotted as a function of case numbers.

10.8.3   SYSTAT System

LOGIT is the major program for logistic regression analysis in SYSTAT. While logistic regression can also be performed through the NONLIN (nonlinear estimation) module, it is much more painful to do so.

SYSTAT LOGIT is highly flexible in the types of models permitted. For dichotomous outcomes, the statistics are based on the category coded "1." With more than two categories of outcome, the categories are considered unordered. Two types of coding (including the usual dummy variable 0, 1 coding) are permitted for discrete predictor variables. A case-control model can be specified. This is the only program that allows specification of quasi-maximum likelihood covariance, which corrects problems created by misspecified models. And only this program provides prediction success tables in addition to classification tables.

The basic output is rather sparse, but includes the test of the full model against the constant-only model, with McFadden's ρ^2 as a measure of strength of association, and the odds ratio and its confidence interval for each predictor component. Hosmer-Lemeshow goodness-of-fit tests are available through options. Stepping options are plentiful. Although improvement in fit at each step is not given, it can be hand-calculated easily from the log-likelihood ratio that is available at each step. Plenty of diagnostics are available for each case. However, they must be saved into a file for viewing; they are not available as part of printout of results.

11

Survival/Failure Analysis

11.1 General Purpose and Description

Survival/failure analysis is a family of techniques dealing with the time it takes for something to happen: a cure, a failure, an employee leaving, a relapse, a death, and so on. The term *survival analysis* is based on medical applications in which time to relapse or death is studied between groups who have received different medical treatments. Is survival longer for a group receiving one chemotherapy rather than another? Does survival time also depend on the age, gender, or marital status of the patient? In manufacturing, the less propitious term *failure analysis* is used in which time until a manufactured component fails is recorded. Does time to failure depend on the material used? Does it depend on the temperature of the room where the component is fabricated? Does it still depend on material used if room temperature is controlled? We generally use the more optimistic survival analysis in this chapter.

One interesting feature of the analysis is that survival time (the DV) often is unknown for a number of cases at the conclusion of the study. Some of the cases are still in the study but have not yet failed: some employees have not yet left, some components are still functioning, some patients are still apparently well, or some patients are still living. For other cases, the outcome is simply unknown because they have withdrawn from the study or are for some reason lost to follow-up. Cases whose DV values—survival time—are unknown for whatever reason are referred to as *censored.*[1]

Within the family of survival-analysis techniques, different procedures are used depending on the nature of the data and the kinds of questions that are of greatest interest. Life tables describe the survival (or failure) times for cases, and often are accompanied by a graphical representation of the survival rate as a function of time, called a *survivor function.* Survivor functions are frequently plotted side-by-side for two or more groups (e.g., treated and untreated patients) and statistical tests are used to test the differences between groups in survival time.

Another set of procedures is used when the goal is to determine if survival time is influenced by some other variables (e.g., is longevity within a company influenced by age or gender of an employee?). These are basically regression procedures in which survival time is predicted from a set of variables, where the set may include one or more treatment variables. However, the analysis accommodates censored cases and, like logistic regression and multiway frequency analysis, uses a

---

[1]It is interesting that the living cases are the censored ones.

506

log-linear rather than a linear model, which tends to be more forgiving in terms of assumptions about censoring. As in logistic regression, analysis can be direct or sequential in which differences between models that do and do not include a treatment variable are studied.

A potential source of confusion is that all of the predictors, including the treatment IV, if there is one, are called *covariates*. In most previous analyses, the word "covariates" was used for variables that enter a sequential equation early and the analysis adjusts for their relationship with the DV before variables of greater interest enter. There is sequential survival analysis in which the term covariates is used in the traditional way, but covariates is also used for the IVs even when the analysis is not sequential.

This chapter emphasizes those techniques within the survival-analysis repertoire that address group differences in survival, whether those differences arise from experimental or naturally occurring treatments.

Mayo and colleagues (1991) modeled time from admission to achievement of independent function for stroke patients monitored daily while undergoing physical rehabilitation. Four variables were found to influence recovery time: (1) age influenced the rate of recovery walking and stair climbing; (2) perceptual impairment influenced the rate of achieving independent sitting and stair climbing; and (3) depression and (4) comprehension influenced walking.

Nolan and co-authors (1991) studied the impact of behavior modification with and without laxatives for children with primary fecal incontinence. Children with the multimodal treatment achieved remission significantly sooner than children given behavioral modification alone; the difference in remission curves was most striking in the first 30 weeks of follow-up.

Van der Pol, Ooms, van't Hof, and Kuper (1998) determined conditions under which burn-in of integrated circuits can be eliminated. Their review of prior failure analyses showed that failures were dominated by defects, with no wear-out observed. On that basis they developed a general model of product failure rate as a function of batch yield. They found that high-yield batches generated fewer failures than low-yield batches, despite burn-in of the latter. They also found that for many applications, the use of screens is more effective in finding latent defects than a standard burn-in.

A comprehensive treatment of survival/failure analysis is provided by Singer and Willett (2003) in their highly readable text on this and other techniques for modeling longitudinal data. Allison (1995) provides a lucid description of the many faces of survival analysis as well as practical guidelines for using SAS to do survival analysis.

11.2   Kinds of Research Questions

The primary goal of one type of survival analysis is to describe the proportion of cases surviving at various times, within a single group or separately for different groups. The analysis extends to statistical tests of group differences. The primary goal of the other type of survival analysis is to assess the relationship between survival time and a set of covariates (predictors), with treatment considered one of the covariates, to determine whether treatment differences are present after statistically controlling for the other covariates.

11.2.1   Proportions Surviving at Various Times

What is the survival rate at various points in time? For example, what proportion of employees last three months? What proportion of components fail within the first year? *Life tables* describe the proportion of cases surviving (and failing) at various times. For example, it may be found that 19% of employees have left by the end of three months, 35% by the end of six months, and so on. A survivor function displays this information graphically. Section 11.4 demonstrates survivor functions and several statistics associated with them.

11.2.2   Group Differences in Survival

If there are different groups, are their survival rates different? Do employees who are in a program to lessen attrition stay longer than those who are not in such a program? Several tests are available to evaluate group differences; one is demonstrated in Section 11.4.5. If statistically significant group differences are found, separate life tables and survivor functions are developed for each group.

11.2.3   Survival Time with Covariates

11.2.3.1   Treatment Effects

Do survival times differ among treatment groups after controlling for other variable(s)? For example, does average employee longevity differ for treated and control groups after adjusting for differences in age at employment and starting salary? Several tests of the relationship between survival and level of treatment are available in the regression forms of survival analysis. Test statistics for treatment effects are discussed in Section 11.6.4.1, and tests of treatment differences after controlling for other covariates are described in Section 11.5.2.

11.2.3.2   Importance of Covariates

Which covariates are associated with survival time? If covariates affect survival time, do they increase it or decrease it? Does longevity on a job depend on age when employed? Do employees who start when they are older stay longer or shorter than those who start when they are younger? Does beginning salary matter? Do higher starting salaries tend to keep employees on the job longer? Tests of these questions are essentially tests of regression coefficients, as demonstrated in Section 11.6.4.2.

11.2.3.3   Parameter Estimates

The parameter estimates in survival analysis are the regression coefficients for the covariates. Because the regression is log-linear, the coefficients often are expressed as odds. For example, what are the odds that someone will stay on the job four years, given that the job was started at age 30 with an annual salary of $40,000? Section 11.6.5 shows how to interpret regression coefficients in terms of odds.

11.2.3.4   Contingencies among Covariates

Some covariates may be related to differences among treatment groups when considered alone, but not after adjustment for other covariates. For example, salary level may modify treatment effects on job longevity when considered by itself, but not after adjusting for differences in age at entry into the job. These contingencies are examined through sequential survival analysis as demonstrated in Section 11.5.2.2.

11.2.3.5 *Effect Size and Power*

How strong is the association between failure/survival and the set of covariates? On a scale of 0 to 1, how well does the combination of age and salary predict longevity on the job? None of the reviewed statistical packages provides this information directly, but Section 11.6.3 shows how to calculate a form of R^2 from output provided by several programs and discusses issues related to power.

11.3 Limitations to Survival Analysis

11.3.1 Theoretical Issues

One problem with survival analysis is the nature of the outcome variable, time itself. Events must occur before survival or failure time can be analyzed: Components must fail, employees must leave, patients must succumb to the illness. However, the purpose of treatment often is to delay this occurrence or prevent it altogether. The more successful the treatment, then, the less able the researcher is to collect data in a timely fashion.

Survival analysis is subject to the usual cautions about causal inference. For example, a difference in survival rates among groups cannot be attributed to treatment unless assignment to levels of treatment and implementation of those levels, with control, are properly experimental.

11.3.2 Practical Issues

In the descriptive use of survival analysis, assumptions regarding the distributions of covariates and survival times are not required. However, in the regression forms of survival analysis in which covariates are assessed, multivariate normality, linearity, and homoscedasticity among covariates, although not required, often enhance the power of the analysis to form a useful linear equation of predictors.

11.3.2.1 *Sample Size and Missing Data*

Some statistical tests in survival analyses are based on maximum likelihood methods. Typically these tests are trustworthy only with larger samples. Eliason (1993) suggests a sample size of 60 if 5 or fewer parameters for covariates (including treatment) are to be estimated. Larger sample sizes are needed with more covariates. Different sample sizes among treatment groups pose no special difficulty.

Missing data can occur in a variety of ways in survival analysis. The most common is that the case survives to the end of the study so time to failure is not yet known. Or a case may withdraw or be lost to follow-up before the end of the study, although it was intact when last seen. These are called right-censored cases. Alternatively, the critical event may have occurred at an uncertain time before monitoring began. For example, you know that a disease process began before the first observation time, but you don't know exactly when. These are called left-censored cases and are much less common. Section 11.6.2 discusses various forms of censoring.

Missing data can also occur in the usual fashion if some covariate scores are missing for some of the cases. Section 4.1.3 discusses issues associated with missing data: randomness, amount, and solutions.

11.3.2.2   Normality of Sampling Distributions, Linearity, and Homoscedasticity

Although the assumptions of multivariate normality, linearity, and homoscedasticity (Chapter 4) are not necessary for survival analysis, meeting them often results in greater power, better prediction, and less difficulty in dealing with outliers. It is therefore useful and relatively easy to assess the distribution of each covariate by statistical or graphical methods, as described in Section 4.1.5, prior to analysis.

11.3.2.3   Absence of Outliers

Those few cases that are very discrepant from others in their group have undue influence on results and must be dealt with. Outliers can occur among covariates singly or in combination. Outliers affect the inferential tests of the relationships between survival time and the set of covariates (including covariates representing treatment groups). Methods of detecting outliers and remedies for them are discussed in Section 4.1.4.

11.3.2.4   Differences between Withdrawn and Remaining Cases

It is assumed in survival analysis that censored cases, ones lost to study, do not differ systematically from those whose fate is known at the conclusion of the study. If the assumption is violated, it is essentially a missing data problem with nonrandom loss of cases. If the study started as an experiment, it is no longer an experiment if cases with missing data are systematically different from cases with complete data because cases available for analysis at the end of the study are no longer the product of random assignment to treatment groups.

11.3.2.5   Change in Survival Conditions over Time

It is assumed that the same things that affect survival at the beginning of the study affect survival at the end of the study and that other conditions have not changed. For example, in the experiment to lessen employee attrition, it is assumed that the factors that influence attrition at the beginning of observation also influence it at the end. This assumption is violated if other working conditions change during the study and they affect survival.

11.3.2.6   Proportionality of Hazards

One of the most popular models for evaluating effects of predictors on survival, the Cox proportional-hazards model, assumes that the shape of the survival function over time is the same for all cases and, as an extension, for all groups. Otherwise there is an interaction between groups and time in survival rates, or between other covariates and time. Section 11.6.1 shows how to test this assumption and discusses evaluation of differences between groups when the assumption is violated.

11.3.2.7   Absence of Multicollinearity

Survival analysis with covariates is sensitive to extremely high correlations among covariates. As in multiple regression, multicollinearity is signaled by extremely high standard errors for parameter estimates and/or failure of a tolerance test in the computer analysis. The source of multicollinearity may be found through multiple regression procedures in which each covariate, in turn, is treated as a DV with the remaining covariates treated as IVs. *Any covariate with a squared multiple correlation (SMC) in excess of .90 is redundant and deleted from further analysis.* Section 11.7.1.6 demonstrates evaluation of multicollinearity through SPSS FACTOR; Section 4.1.7 provides further discussion of this issue.

11.4 Fundamental Equations for Survival Analysis

Table 11.1 shows a hypothetical data set for evaluating how long a belly dancer continues to take classes (survives) as a function of treatment and, later, age. In this example, the DV, months, is the number of months until a dancer dropped out of class during a 12-month follow-up period. Dancing is the censoring variable that indicates whether she has dropped out at the end of the study (1 = dropped out, 0 = still dancing). No cases were withdrawn from the study. Therefore, only the last case is censored because her total survival time remains unknown at the end of the year-long follow-up period. The 12 cases belong to one of two groups (0 = control, 1 = treatment). Treatment consists of a preinstructional night out with dinner at a mid-Eastern restaurant, live music, and belly dancers: 7 cases are in the control group and 5 in the treatment group. Age when beginning to belly dance is included as a covariate for later analyses.

11.4.1 Life Tables

Life tables are built around time intervals; in this example (Table 11.2), the intervals are of width 1.2 months. The *survivor function, P*, is the *cumulative proportion of cases surviving* to the beginning of the $i + 1^{st}$ interval, estimated as:

$$P_{i+1} = p_i P_i \qquad (11.1)$$

where:

$$p_i = 1 - q_i$$

$$q_i = \frac{d_i}{r_i}$$

TABLE 11.1   Small Sample of Hypothetical Data for Illustration of Survival Analysis

Case	Months	Dancing	Treatment	Age
1	1	1	0	16
2	2	1	0	24
3	2	1	0	18
4	3	1	0	27
5	4	1	0	25
6	5	1	0	21
7	7	1	1	26
8	8	1	1	36
9	10	1	1	38
10	10	1	1	45
11	11	1	0	55
12	12	0	1	47

TABLE 11.2   Survivor Functions for Dancers with and without Preinstructional Night Out

Interval (month)	Entered	Censored[a]	Dropped	Proportion Dropped	Proportion Surviving	Cumulative Proportion Surviving
Control Group with No Preinstructional Night Out						
0.0–1.2	7	0	1	.1429	.8571	1.000
1.2–2.4	6	0	2	.3333	.6667	.8571
2.4–3.6	4	0	1	.2500	.7500	.5714
3.6–4.8	3	0	1	.3333	.6667	.4286
4.8–6.0	2	0	1	.5000	.5000	.2857
6.0–7.2	1	0	0	.0000	1.0000	.1429
7.2–8.4	1	0	0	.0000	1.0000	.1429
8.4–9.6	1	0	0	.0000	1.0000	.1429
9.6–10.8	1	0	0	.0000	1.0000	.1429
10.8–12.00	1	0	1	1.0000	.0000	.0000
Treatment Group with Preinstructional Night Out						
0.0–1.2	5	0	0	.0000	1.0000	1.000
1.2–2.4	5	0	0	.0000	1.0000	1.000
2.4–3.6	5	0	0	.0000	1.0000	1.000
3.6–4.8	5	0	0	.0000	1.0000	1.000
4.8–6.0	5	0	0	.0000	1.0000	1.000
6.0–7.2	5	0	1	.2000	.8000	1.000
7.2–8.4	4	0	1	.2500	.7500	.8000
8.4–9.6	3	0	0	.0000	1.0000	.6000
9.6–10.8	3	0	2	.6667	.3333	.6000
10.8–12.00	1	1	0	.0000	1.0000	.2000

[a]Note that in Table 11.1, codes for dancing are 1 = No, 0 = Yes. Codes in this table for censored are the reverse.

and

where d_i = number responding (dropping out) in the interval and

$$r_i = n_i - \frac{1}{2}c_i$$

where n_i = number entering the interval, and c_i = number censored in the interval (lost to follow up for reasons other than dropping out).

In words, the proportion of cases surviving to the $i + 1^{st}$ interval is the proportion who survived to the start of ith interval times the probability of surviving to the end of the ith interval (by not dropping out or being censored during that interval).

For the first interval (0 to 1.2), all 7 dancers in the control group enter the interval; thus the cumulative proportion surviving to the beginning of the first interval is 1. During the first interval, no cases are censored but one case in the control group drops out. Therefore,

$$r_1 = 7 - \frac{1}{2}(0) = 7$$

$$q_1 = \frac{1}{7} = .1429$$

$$p_1 = 1 - .1429 = .8571$$

That is, 85.71% $(.8571)(7) = 6$ of the cases have survived the first interval.

For the second interval (1.2 to 2.4), the cumulative proportion surviving to the beginning of the interval is:

$$P_2 = p_1 P_1 = (.8571)(1) = .8571$$

The six remaining dancers enter the interval, none is censored but two drop out, so that:

$$r_2 = 6 - \frac{1}{2}(0) = 6$$

$$q_2 = \frac{2}{6} = .3333$$

$$p_2 = 1 - .3333 = .6667$$

For the third interval (2.4 to 3.6), the cumulative proportion surviving to the beginning of the interval is:

$$P_3 = p_2 P_2 = (.6667)(.8571) = .5714$$

and so on.

In the control group, then, over half the cases have dropped out by the middle of the third month (beginning of fourth interval), and only 14% survive to the start of the sixth month. All have stopped taking classes by the end of the study. In the treatment group, over half the cases survive to the middle of the ninth month (beginning of the ninth interval), and one is still taking classes at the end of the study.

Various statistics and standard errors for them are developed to facilitate inferential tests of survival functions, as described below.

11.4.2 Standard Error of Cumulative Proportion Surviving

The standard error of a cumulative proportion of cases surviving an interval is approximately:

$$s.e.(P_i) \cong P_i \sqrt{\sum_{j=1}^{i-1} \frac{q_j}{r_j p_j}} \qquad (11.2)$$

For the control group in the second interval (month 1.20–2.40):

$$s.e.(P_2) \cong .8571 \sqrt{\frac{.1429}{7(.8571)}} \cong .1323$$

11.4.3   Hazard and Density Functions

The *hazard* (also sometimes called the failure rate) is the rate of not surviving to the midpoint of an interval, given survival to the start of the interval.

$$\lambda_i = \frac{2q_i}{h_i(1 + p_i)} \tag{11.3}$$

where h_i = the width of the ith interval.

For the dancers in the control group in the second interval, the hazard is:

$$\lambda_2 = \frac{2(.3333)}{1.2(1 + .6667)} = .3333$$

That is, the drop-out rate is one-third by the middle of the second interval for cases surviving to the beginning of the second interval. Of the 6 cases entering the second interval, the expected rate of drop-out is $(.3333)(6) = 2$ of them by the middle of that interval.

The approximate standard error of the hazard is:

$$s.e.(\lambda_i) \cong \lambda_i \sqrt{\frac{1 - (h_i\lambda_i/2)^2}{r_iq_i}} \tag{11.4}$$

The standard error of the hazard for the control group in the second interval, then, is:

$$s.e.(\lambda_2) \cong .3333 \sqrt{\frac{1 - (1.2(.3333/2))^2}{6(.3333)}} \cong .2309$$

The *probability density* is the probability of not surviving to the midpoint of an interval, given survival to the start of the interval:

$$f_i = \frac{P_iq_i}{h_i} \tag{11.5}$$

For the control group in the second interval:

$$f_2 = \frac{.8571(.3333)}{1.2} = .2381$$

For any of one of the six dancers in the control group who are still dancing at 1.2 months, the probability of dropping out by 1.8 months is .2381.

The approximate standard error of density is:

$$s.e.(f_i) \cong \frac{P_i q_i}{h_i} \sqrt{\frac{p_i}{r_{iq_i}} + \sum_{j=1}^{i-1} \frac{q_j}{r_j P_j}} \tag{11.6}$$

For the control group in the second interval the approximate standard error of density is:

$$s.e.(f_2) \cong \frac{.8571(.3333)}{1.2} \sqrt{\frac{.6667}{6(.3333)} + \frac{.1429}{7(.8571)}} \cong .1423$$

Note the distinction between the hazard function and the probability density function. The hazard function is the instantaneous *rate* of dropping out at a specific time (e.g., 1.8 months which is the midpoint of the second interval), among cases who survived to at least the beginning of that time interval (e.g., 1.2 months). The probability density function is *probability* of a given case dropping out at a specified time point.

11.4.4 Plot of Life Tables

Life table plots are simply the cumulative proportion surviving (P_i) plotted as a function of each time interval. Referring to Table 11.2, for example, the first point plotted is the interval beginning at 0 months, and the cumulative proportion surviving is 1.0 for both of the groups. At the beginning of the second interval, 1.2 months, the cumulative proportion surviving is still 1.0 for those dancers in the treatment group but is .8571 for dancers in the control group, and so on. Figure 11.1 shows the resulting survival function.

11.4.5 Test for Group Differences

Group differences in survival are tested through χ^2 with degrees of freedom equal to the number of groups minus 1. Of the several tests that are available, the one demonstrated here is labeled Log-Rank in SAS LIFETEST and SPSS KM. When there are only two groups, the overall test is:

$$\chi^2 = \frac{v_j^2}{V_j} \tag{11.7}$$

χ^2 equals the squared value of the observed minus expected frequencies of number of survivors summed over all intervals for one of the groups (v_j^2) under the null hypothesis of no group differences, divided by V_j, the variance of the group.

The degrees of freedom for this test is (number of groups − 1). When there are only two groups, the value in the numerator is the same for both groups, but opposite in sign. The value in the denominator is also the same for both groups. Therefore, computations for either group produce the same χ^2.

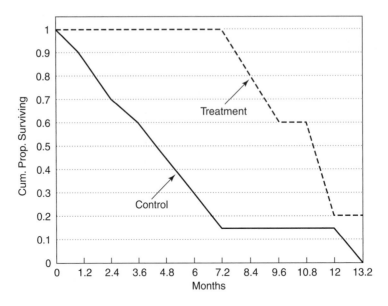

**FIGURE 11.1   Cumulative proportion of treatment
and control dancers surviving.**

The value, v_j, of observed minus expected frequencies is calculated separately for each interval for one of the groups. For the control group of dancers (the group coded 0):

$$v_0 = \sum_{i=1}^{k} (d_{0i} - n_{0i} d_{Ti}/n_{Ti}) \tag{11.8}$$

The difference between observed and expected frequencies for a group, v_0, is the summed differences over the intervals between the number of survivors in each interval, d_{0i}, minus the ratio of the number of cases at risk in the interval (n_{0i}) times the total number of survivors in that interval summed over all groups (d_{Ti}), divided by the total number of cases at risk in that interval summed over all groups (n_{Ti}).

For example using the control group, in the first interval there are 6 survivors (out of a possible 7 at risk), and a total of 11 survivors (out of a possible 12 at risk) for the two groups combined. Therefore,

$$v_{01} = 6 - 7(11)/12 = -0.417$$

For the second interval, there are 4 survivors (out of a possible 6 at risk) in the control group, and a total of 9 survivors (out of a possible 11 at risk) for the two groups combined. Therefore,

$$v_{02} = 4 - 6(9)/11 = -0.909$$

and so on.

The sum over all 10 intervals for the control group is -2.854. (For the treated group it is 2.854.) The variance, V, for a group is:

$$V_0 = \sum_{i=1}^{k} [(n_{Ti}n_{0i} - n_{0i}^2)d_{Ti}s_{Ti}]/[n_{Ti}^2(n_{Ti} - 1)] \tag{11.9}$$

The variance for the control group, V_0, is the sum over all intervals of the difference between the total number of survivors in an interval (n_{Ti}) times the number of survivors in the control group in the interval (n_{0i}) minus the squared number of survivors in the control group in the interval, this difference multiplied by the product of the total number of survivors in the interval (d_{Ti}) times $s_{Ti}(= n_{Ti} - d_{Ti})$; all of this is divided by the squared total number of survivors in the interval (n_{Ti}^2) times the total number of survivors in the interval minus one.

In jargon, the total number of cases that have survived to an interval (n_{Ti}) is called the *risk set*. The variance for the control group, V_0, for the first interval is:

$$V_{01} = [(12 \cdot 5 - 25)11(1)]/[144(12 - 1)] = 385/1584 = 0.24306$$

and for the second interval:

$$V_{02} = [(11 \cdot 5 - 25)9(2)]/[121(11 - 1)] = 540/1210 = 0.4462$$

and so on. The value, summed over all 10 intervals is 2.1736. (This also is the value for the second group when there are only two groups.)

Using these values in Equation 11.7:

$$\chi^2 = \frac{(-2.854)^2}{2.1736} = 3.747$$

Table C.4 shows that the critical value of χ^2 with 1 df at $\alpha = .05$ is 3.84. Therefore, the groups are not significantly different by the log-rank test. Matrix equations are more convenient to use if there are more than two groups. (The matrix procedure is not at all convenient to use if there are only two groups because the procedure requires inversion of a singular variance-covariance matrix.)

11.4.6   Computer Analyses of Small-Sample Example

Tables 11.3 and 11.4 show syntax and selected output for computer analyses of the data in Table 11.1 for SPSS SURVIVAL and SAS LIFETEST, respectively. Syntax and output from SPSS SURVIVAL are in Table 11.3. The DV (MONTHS), and the grouping variable (TREATMNT) and its levels, are shown in the TABLE instruction. The dropout variable and the level indicating dropout are indicated by the STATUS=DANCING(1) instruction as per Table 11.1. SPSS requires explicit instruction as to the setup of time intervals and a request for a survival (or hazard) plot. The COMPARE instruction requests a test of equality of survival functions for the two groups.

Life tables are presented for each group, control group first, showing the Number of cases Entering the Interval, the number Withdrawing from the study during Interval (censored), and the number of cases who quit classes in each interval in a column labeled Number of Terminal Events. The remaining columns of proportions and their standard errors are as described in Sections 11.4.1 through 11.4.3, with some change in notation. Number Exposed to Risk is the Number Entering Interval minus the Number Withdrawing during Interval. Terminal events are drop-outs. Note that Cumulative Proportion Surviving at End of Interval is the cumulative proportion surviving at the *end* of an interval rather than at the *beginning* of an interval as in Table 11.2.

Median survival times are given for each group. The survival function comparing the two groups is then shown. The two groups are compared using the Wilcoxon (Gehan) test. With $\chi^2(1, N = 12) = 4.840$, $p = 0.28$; this test shows a significant difference in survival between groups. Finally, a summary table shows the overall censoring rates for the two groups, as well as a mean score for each group.

Table 11.4 shows syntax and output for SAS LIFETEST. This program requires an explicit request for actuarial tables (method=life), as well as specification of the time intervals and a request for survival plot(s). The strata TREATMNT instruction identifies the grouping variable (IV). The time MONTHS*DANCING(0) instruction identifies MONTHS as the time variable and DANCING as the response variable, with 0 indicating the data that are censored (the DV value is not known).

Output is somewhat different from that of SPSS, partly because some statistics are evaluated at the median rather than at the beginning of each interval. Number Failed corresponds to Number of Terminal Events in SPSS and Number Censored to Number Withdrawing during Interval. Effective Sample Size corresponds to Number Exposed to Risk (or Number Entering this Interval). Conditional Probability of failure is the proportion not surviving within an interval (Proportion Terminating in SPSS); proportion surviving is one minus that value. The column labeled Survival is the cumulative proportion surviving to the beginning of an interval; the column labeled Failure is one minus that value.

TABLE 11.3   Syntax and Output for Small-Sample Example through SPSS SURVIVAL

```
SURVIVAL
 TABLE=MONTHS BY TREATMNT (0 1)
 /INTERVAL=THRU 12 BY 1.2
 /STATUS=DANCING(1)
 /PRINT=TABLE
 /PLOTS (SURVIVAL)=MONTHS BY TREATMNT
 /COMPARE=MONTHS BY TREATMNT.
```

TABLE 11.3 Continued

Life Table

First-order Controls	Interval Start Time	Number Entering Interval	Number Withdrawing during Interval	Number Exposed to Risk	Number of Terminal Events	Proportion Terminating	Proportion Surviving	Cumulative Proportion Surviving at End of Interval	Std. Error of Cumulative Proportion Surviving at End of Interval	Probability Density	Std. Error of Probability Density	Hazard Rate	Std. Error of Hazard Rate
TREATMENT 0	0	7	0	7.000	1	.14	.86	.86	.13	.119	.110	.13	.13
	1.2	6	0	6.000	2	.33	.67	.57	.19	.238	.142	.33	.23
	2.4	4	0	4.000	1	.25	.75	.43	.19	.119	.110	.24	.24
	3.6	3	0	3.000	1	.33	.67	.29	.17	.119	.110	.33	.33
	4.8	2	0	2.000	1	.50	.50	.14	.13	.119	.110	.56	.52
	6	1	0	1.000	0	.00	1.00	.14	.13	.000	.000	.00	.00
	7.2	1	0	1.000	0	.00	1.00	.14	.13	.000	.000	.00	.00
	8.4	1	0	1.000	0	.00	1.00	.14	.13	.000	.000	.00	.00
	9.6	1	0	1.000	0	.00	1.00	.14	.13	.000	.000	.00	.00
	10.8	1	0	1.000	1	1.00	.00	.00	.00	.119	.110	1.67	.00
1	0	5	0	5.000	0	.00	1.00	1.00	.00	.000	.000	.00	.00
	1.2	5	0	5.000	0	.00	1.00	1.00	.00	.000	.000	.00	.00
	2.4	5	0	5.000	0	.00	1.00	1.00	.00	.000	.000	.00	.00
	3.6	5	0	5.000	0	.00	1.00	1.00	.00	.000	.000	.00	.00
	4.8	5	0	5.000	0	.00	1.00	1.00	.00	.000	.000	.00	.00
	6	5	0	5.000	1	.20	.80	.80	.18	.167	.149	.19	.18
	7.2	4	0	4.000	1	.25	.75	.60	.22	.167	.149	.24	.24
	8.4	3	0	3.000	0	.00	1.00	.60	.22	.000	.000	.00	.00
	9.6	3	0	3.000	2	.67	.33	.20	.18	.333	.183	.83	.51
	10.8	1	0	1.000	0	.00	1.00	.20	.18	.000	.000	.00	.00

(continued)

519

TABLE 11.3   Continued

Median Survival Times

First-order Controls		Med Time
TREATMNT	0	3.000
	1	9.900

First-order Control: TREATMNT

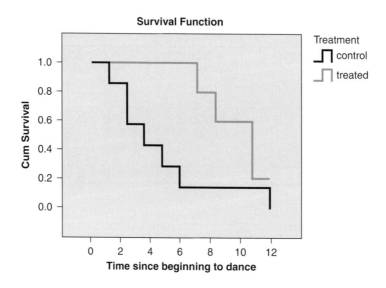

Comparisons for Control Variable: TREATMNT

Overall Comparisons[a]

Wilcoxon (Gehan) Statistics	df	Sig.
4.480	1	.028

[a]Comparisons are exact.

TABLE 11.4 Syntax and Output for Small-Sample Example through SAS LIFETEST

```
proc lifetest data=SASUSER.SURVIVAL
     plots=(s) method=life interval=0 to 12 BY 1.2;
     time MONTHS*DANCING(0);
     strata TREATMNT;
run;
```

The LIFETEST Procedure

Stratum 1: TREATMNT = 0

Life Table Survival Estimates

Interval [Lower, Upper)	Number Failed	Number Censored	Effective Sample Size	Conditional Probability of Failure	Conditional Probability Standard Error	Survival	Failure
0 1.2	1	0	7.0	0.1429	0.1323	1.0000	0
1.2 2.4	2	0	6.0	0.3333	0.1925	0.8571	0.1429
2.4 3.6	1	0	4.0	0.2500	0.2165	0.5714	0.4286
3.6 4.8	1	0	3.0	0.3333	0.2722	0.4286	0.5714
4.8 6	1	0	2.0	0.5000	0.3536	0.2857	0.7143
6 7.2	0	0	1.0	0	0	0.1429	0.8571
7.2 8.4	0	0	1.0	0	0	0.1429	0.8571
8.4 9.6	0	0	1.0	0	0	0.1429	0.8571
9.6 10.8	0	0	1.0	0	0	0.1429	0.8571
10.8 12	1	0	1.0	1.0000	0	0.1429	0.8571

Interval [Lower, Upper)	Survival Standard Error	Median Residual Lifetime	Median Standard Error	PDF	PDF Standard Error	Hazard	Hazard Standard Error
0 1.2	0	3.0000	1.5875	0.1190	0.1102	0.128205	0.127825
1.2 2.4	0.1323	2.4000	1.4697	0.2381	0.1423	0.333333	0.23094
2.4 3.6	0.1870	2.4000	1.2000	0.1190	0.1102	0.238095	0.235653
3.6 4.8	0.1870	1.8000	1.0392	0.1190	0.1102	0.333333	0.326599
4.8 6	0.1707	1.2000	0.8485	0.1190	0.1102	0.555556	0.523783
6 7.2	0.1323	.	.	0	.	0	.
7.2 8.4	0.1323	.	.	0	.	0	.
8.4 9.6	0.1323	.	.	0	.	0	.
9.6 10.8	0.1323	.	.	0	.	0	.
10.8 12	0.1323	.	.	0.1190	0.1102	1.666667	0

Evaluated at the Midpoint of the Interval

(continued)

521

TABLE 11.4 Continued

```
                    Stratum 2: TREATMNT = 1
                Life Table Survival Estimates
```

Interval [Lower, Upper)	Number Failed	Number Censored	Effective Sample Size	Conditional Probability of Failure	Conditional Probability Standard Error	Survival	Failure
0 – 1.2	0	0	5.0	0	0	1.0000	0
1.2 – 2.4	0	0	5.0	0	0	1.0000	0
2.4 – 3.6	0	0	5.0	0	0	1.0000	0
3.6 – 4.8	0	0	5.0	0	0	1.0000	0
4.8 – 6	0	0	5.0	0	0	1.0000	0
6 – 7.2	1	0	5.0	0.2000	0.1789	1.0000	0
7.2 – 8.4	1	0	4.0	0.2500	0.2165	0.8000	0.2000
8.4 – 9.6	0	0	3.0	0	0	0.6000	0.4000
9.6 – 10.8	2	0	3.0	0.6667	0.2722	0.6000	0.4000
10.8 – 12	0	0	1.0	0	0	0.2000	0.8000
12 – .	0	1	0.5	0	0	0.2000	0.8000

```
                  Evaluated at the Midpoint of the Interval
```

Interval [Lower, Upper)	Survival Standard Error	Median Residual Lifetime	Median Standard Error	PDF	PDF Standard Error	Hazard	Hazard Standard Error
0 – 1.2	0	9.9000	0.6708	0	.	0	.
1.2 – 2.4	0	8.7000	0.6708	0	.	0	.
2.4 – 3.6	0	7.5000	0.6708	0	.	0	.
3.6 – 4.8	0	6.3000	0.6708	0	.	0	.
4.8 – 6	0	5.1000	0.6708	0	.	0	.
6 – 7.2	0	3.9000	0.6708	0.1667	0.1491	0.185185	0.184039
7.2 – 8.4	0.1789	3.0000	0.6000	0.1667	0.1491	0.238095	0.235653
8.4 – 9.6	0.2191	2.1000	0.5196	0	.	0	.
9.6 – 10.8	0.2191	0.9000	0.5196	0.3333	0.1826	0.833333	0.51031
10.8 – 12	0.1789	.	.	0	.	0	.
12 – .	0.1789	.	.	.	.	.	.

```
        Summary of the Number of Censored and Uncensored Values
```

Stratum	TREATMNT	Total	Failed	Censored	Percent Censored
1	0	7	7	0	0.00
2	1	5	4	1	20.00
Total		12	11	1	8.3333

Testing Homogeneity of Survival Curves for MONTHS over Strata

TABLE 11.4   Continued

```
                    Rank Statistics

       TREATMNT    Log-Rank    Wilcoxon

          0         2.8539      27.000
          1        -2.8539     -27.000

   Covariance Matrix for the Log-Rank Statistics

       TREATMNT           0              1

          0         2.17360      -2.17360
          1        -2.17360       2.17360

   Covariance Matrix for the Wilcoxon Statistics

       TREATMNT           0              1

          0         148.000      -148.000
          1        -148.000       148.000

           Test of Equality over Strata

                                         Pr >
       Test       Chi-Square    DF    Chi-Square

    Log-Rank        3.7472       1       0.0529
    Wilcoxon        4.9257       1       0.0265
    -2Log(LR)       3.1121       1       0.0777
```

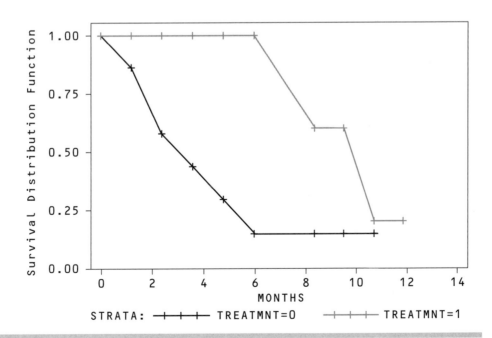

SAS LIFETEST shows `Median Residual Lifetime` (and its standard error), which is the amount of time elapsed before the number of at-risk cases is reduced to half. `PDF` refers to the probability density function, Probability Density in SPSS. These statistics are followed by the usual summary table of censored and failed values. Then the matrices used in calculating group differences are shown (see Equations 11.7 through 11.9), and finally a table shows the results of three chi-square tests. The `Wilcoxon test` corresponds to a `GENERALIZED WILCOXON (BRESLOW)` test, as discussed in Section 11.6.4.1. The high resolution survival function graph is then shown, produced in a separate window from the printed output in SAS.

11.5   Types of Survival Analyses

There are two major types of survival analyses: life tables (including proportions of survivors at various times and survivor functions, with tests of group differences) and prediction of survival time from one or more covariates (some of which may represent group differences). Life tables are estimated by either the actuarial or the product-limit (Kaplan-Meier) method. Prediction of survival time from covariates most often involves the Cox proportional-hazards model (Cox regression).

11.5.1   Actuarial and Product-Limit Life Tables and Survivor Functions

The actuarial method for calculating life tables and testing differences among groups is illustrated in Section 11.4. An alternative for forming life tables and testing differences among groups is the product-limit method. The product-limit method does not use a specified interval size but rather calculates survival statistics each time an event is observed. The two methods produce identical results when there is no censoring and intervals contain no more than one time unit. The product-limit method (also known as the Kaplan-Meier method) is the most widely used, particularly in biomedicine (Allison, 1995). It has the advantage of producing a single statistic, such as mean or median, that summarizes survival time.

SAS LIFETEST offers a choice of either the actuarial or product-limit method. SPSS has SURVIVAL for actuarial tables and KM for product-limit. Table 11.5 shows SPSS KM syntax and output for a product-limit analysis of the small-sample data.

The output is organized by time at which events occur rather than by time interval. For example, there are two lines of output for the two control dancers who dropped out during the second month and for the two treated dancers who dropped out in the tenth month. Both mean and median survival time are given, along with their standard errors and 95% confidence intervals. The survival-function chart differs slightly from that of the actuarial method in Table 11.3 by including information about cases that are censored. Group differences are tested through the Log Rank test, among others that can be requested, rather than the Wilcoxon test produced by SPSS SURVIVAL.

11.5.2   Prediction of Group Survival Times from Covariates

Prediction of survival (or failure) time from covariates is similar to logistic regression (Chapter 10) but with provision for censored data. This method also differs in analyzing the time between events rather than predicting the occurrence of events. Cox proportional hazards (Cox regression) is the most popular method. Accelerated failure-time models are also available for the more sophisticated user.

As in other forms of regression (cf. Chapter 5), analysis of survival can be direct, sequential, or statistical. A treatment IV, if present, is analyzed the same as any other discrete covariate. When there are only two levels of treatment, the treated group is usually coded 1 and the control group 0. If there are more than two levels of treatment, dummy variable coding is used to represent group

TABLE 11.5   Syntax and Output for SPSS Kaplan-Meier Analysis of Small-Sample Data

```
KM
 MONTHS BY TREATMNT
 /STATUS=DANCING(1)
 /PRINT TABLE MEAN
 /PLOT SURVIVAL
 /TEST LOGRANK
 /COMPARE OVERALL POOLED.
```

Kaplan-Meier

Case Processing Summary

| Treatment | Total N | N of Events | Censored | |
			N	Percent
control	7	7	0	.0%
treated	5	4	1	20.0%
Overall	12	11	1	8.3%

Survival Table

| Treatment | | Time | Status | Cumulative Proportion Surviving at the Time | | N of Cumulative Events | N of Cumulative Events |
				Estimate	Std. Error		
control	1	1.000	dropped out	.857	.132	1	6
	2	2.000	dropped out	.	.	2	5
	3	2.000	dropped out	.571	.187	3	4
	4	3.000	dropped out	.429	.187	4	3
	5	4.000	dropped out	.286	.171	5	2
	6	5.000	dropped out	.143	.132	6	1
	7	11.000	dropped out	.000	.000	7	0
treated	1	7.000	dropped out	.800	.179	1	4
	2	8.000	dropped out	.600	.219	2	3
	3	10.000	dropped out	.	.	3	2
	4	10.000	dropped out	.200	.179	4	1
	5	12.000	still dancing	.	.	4	0

(continued)

TABLE 11.5   Continued

Means and Medians for Survival Time

	Mean[a]				Median			
			95% Confidence Interval				95% Confidence Interval	
Treatment	Estimate	Std. Error	Lower Bound	Upper Bound	Estimate	Std. Error	Lower Bound	Upper Bound
control	4.000	1.272	1.506	6.494	3.000	1.309	.434	5.566
treated	9.400	.780	7.872	10.928	10.000	.894	8.247	11.753
Overall	6.250	1.081	4.131	8.369	5.000	2.598	.000	10.092

a Estimation is limited to the largest survival time if it is censored.

Overall Comparisons

	Chi-Square	df	Sig.
Log Rank (Mantel-Cox)	3.747	1	.053

Test of equality of survival distributions for the different levels of Treatment.

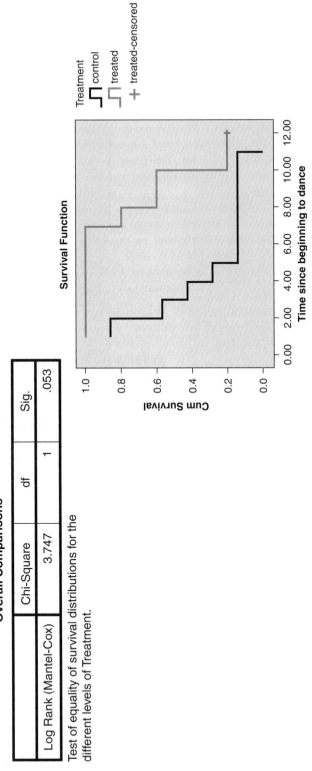

membership, as described in Section 5.2. Successful prediction on the basis of this IV indicates significant treatment effects.

11.5.2.1 *Direct, Sequential, and Statistical Analysis*

The three major analytic strategies in survival analysis with covariates are direct (standard), sequential (hierarchical), and statistical (stepwise or setwise). Differences among the strategies involve what happens to overlapping variability due to correlated covariates (including treatment groups) and who determines the order of entry of covariates into the equation.

In the direct, or simultaneous, model all covariates enter the regression equation at one time and each is assessed as if it entered last. Therefore, each covariate is evaluated as to what it adds to prediction of survival time that is different from the prediction afforded by all the other covariates.

In the sequential (sometimes called hierarchical) model, covariates enter the equation in an order specified by the researcher. Each covariate is assessed by what it adds to the equation at its own point of entry. Covariates are entered one at a time or in blocks. The analysis proceeds in steps, with information about the covariates in the equation given at each step. A typical strategy in survival analysis with an experimental IV is to enter all the nontreatment covariates at the first step, and then enter the covariate(s) representing the treatment variable at the second step. Output after the second step indicates the importance of the treatment variable to prediction of survival after statistical adjustment for the effects of other covariates.

Statistical regression (sometimes generically called stepwise regression) is a controversial procedure in which order of entry of variables is based solely on statistical criteria. The meaning of the variables is not relevant. Decisions about which variables are included in the equation are based solely on statistics computed from the particular sample drawn; minor differences in these statistics can have a profound effect on the apparent importance of a covariate, including the one representing treatment groups. The procedure is typically used during early stages of research, when nontreatment covariates are being assessed for their relationship with survival. Covariates which contribute little to prediction are then dropped from subsequent research into the effects of treatment. As with logistic regression, data-driven strategies are especially dangerous when important decisions are based on results that may not generalize beyond the sample chosen. Cross-validation is crucial if statistical/stepwise techniques are used for any but the most preliminary investigations.

Both of the reviewed programs provide direct analysis. SPSS COXREG also provides both sequential and stepwise analysis. SAS LIFEREG provides only direct analysis, but SAS PHREG provides direct, sequential, stepwise, and setwise analysis (in which models including all possible combinations of covariates are evaluated).

11.5.2.2 *Cox Proportional-Hazards Model*

This method models event (failure, death) rates as a log-linear function of predictors, called covariates. Regression coefficients give the relative effect of each covariate on the survivor function. Cox modeling is available through SPSS COXREG and SAS PHREG.

Table 11.6 shows the results of a direct Cox regression analysis through SAS PHREG using the small-sample data of Table 11.1. Treatment (a dichotomous variable) and age are considered covariates for purposes of this analysis. This analysis assumes proportionality of hazard (that

TABLE 11.6   Syntax and Output for Direct Cox Regression Analysis through SAS PHREG

```
proc phreg data=SASUSER.SURVIVE;
    model months*dancing(0) = age treatmnt;
    run;
```

The PHREG Procedure

Model Information

Data Set	SASUSER.SURVIVE
Dependent Variable	MONTHS
Censoring Variable	DANCING
Censoring Value(s)	0
Ties Handling	BRESLOW

Summary of the Number of Event and Censored Values

Total	Event	Censored	Percent Censored
12	11	1	8.33

Model Fit Statistics

Criterion	Without Covariates	With Covariates
_2 LOG L	40.740	21.417
AIC	40.740	25.417
SBC	40.740	26.213

Testing Global Null Hypothesis: BETA=0

Test	Chi-Square	DF	Pr > ChiSq
Likelihood Ratio	19.3233	2	<.0001
Score	14.7061	2	0.0006
Wald	6.6154	2	0.0366

Analysis of Maximum Likelihood Estimates

Variable	DF	Parameter Estimate	Standard Error	Chi-Square	Pr > ChiSq	Hazard Ratio
AGE	1	-0.22989	0.08945	6.6047	0.0102	0.795
TREATMNT	1	-2.54137	1.54632	2.7011	0.1003	0.079

the shapes of survivor functions are the same for all levels of treatment). Section 11.6.1 shows how to test the assumption and evaluate group differences if it is violated. The time variable (months) and the response variable (dancing, with 0 indicating censored data) are identified in a model instruction and are predicted by two covariates: age and treatment.

Model Fit Statistics are useful for comparing models, as in logistic regression analysis (Chapter 10). Three tests are available to evaluate the hypothesis that all regression coefficients are zero. For example, the Likelihood Ratio test indicates that the combination of age and treatment significantly predicts survival time, $\chi^2(2, N = 12) = 19.32, p < .0001$. (Note that this is a large-sample test and cannot be taken too seriously with only 12 cases.) Significance tests for the individual predictors also are shown as Chi-Square tests. Thus, age significantly predicts survival time, after adjusting for differences in treatment, $\chi^2(1, N = 12) = 6.60, p = .01$. However, treatment does not predict survival time, after adjusting for differences in age, $\chi^2(1, N = 12) = 2.70, p = .10$. Thus this analysis shows no significant treatment effect. Regression coefficients Parameter Estimate) for significant effects and odds ratios (Hazard Ratio) may be interpreted as per Section 11.6.5.

Sequential Cox regression analysis through SPSS COXREG is shown in Table 11.7. Age enters the prediction equation first, followed by treatment. Block 0 shows the model fit, -2 Log Likelihood, corresponding to -2 Log L in SAS, useful for comparing models (cf. Section 10.6.1.1).

Step one (Block1) shows a significant effect of age alone, by both the Wald test (the squared z test with the coefficient divided by its standard error, $p = .006$) and the likelihood ratio test [$\chi^2(1, N = 12) = 15.345, p < .001$]. However, the results for treatment differ for the two criteria. The Wald test gives the same result as reported for the direct analysis above in which treatment is adjusted for differences in age and is not statistically significant. The likelihood ratio test, on the other hand, shows a significant change with the addition of treatment as a predictor, $\chi^2(1, N = 12) = 3.98, p < .05$. With a sample size this small, it is probably safer to rely on the Wald test indicating no statistically significant treatment effect.

11.5.2.3 *Accelerated Failure-Time Models*

These models replace the general hazard function of the Cox model with a specific distribution (exponential, normal, or some other). However, greater user sophistication is required to choose the distribution. Accelerated failure-time models are handled by SAS LIFEREG. SPSS has no program for accelerated failure-time modeling.

Table 11.8 shows an accelerated failure-time analysis corresponding to the Cox model of Section 11.5.2.2 through SAS LIFEREG, using the default Weibull distribution.

It is clear in this analysis that both age and treatment significantly predict survival (Pr>Chi is less than .05), leading to the conclusion that treatment significantly affects survival in belly dance classes, after adjusting for differences in age at which instruction begins. The Type III Analysis of Effects is useful only when a categorical predictor has more than 1 df.

Choice of distributions in accelerated failure-time models has implications for hazard functions, so that modeling based on different distributions may lead to different interpretations. Distributions available in SAS LIFEREG are Weibull, normal, logistic, gamma, exponential, log-normal, and log-logistic. Table 11.9 summarizes the various distributions available in SAS LIFEREG for modeling accelerated failure time.

TABLE 11.7   Syntax and Output for Sequential Cox Regression Analysis through SPSS COXREG

```
COXREG
  MONTHS /STATUS=DANCING(1)
  /CONTRAST (TREATMNT)=indicator
  /METHOD=ENTER AGE /METHOD=ENTER TREATMNT
  /CRITERIA=PIN(.05) POUT(.10) ITERATE(20).
```

Cox Regression

Case Processing Summary

		N	Percent
Cases available in analysis	Event[a]	11	91.7%
	Censored	1	8.3%
	Total	12	100.0%
Cases dropped	Cases with missing values	0	.0%
	Cases with non-positive time	0	.0%
	Censored cases before the earliest event in a stratum	0	.0%
	Total	0	.0%
Total		12	100.0%

[a]Dependent Variable: Time since beginning to dance

Categorical Variable Codings[a,b]

		Frequency	(1)
TREATMNT	.00=control	7	1.000
	1.00=treated	5	.000

[a]Indicator Parameter Coding
[b]Category variable: TREATMNT (Treatment)

Block 0: Beginning Block

Omnibus Tests of Model Coefficients

−2 Log Likelihood
40.740

TABLE 11.7 Continued

Omnibus Tests of Model Coefficients[a,b]

−2 Log Likelihood	Overall (score)			Change From Previous Step			Change From Previous Block		
	Chi-square	df	Sig.	Chi-square	df	Sig.	Chi-square	df	Sig.
25.395	11.185	1	.001	15.345	1	.000	15.345	1	.000

[a] Beginning Block Number 0, initial Log Likelihood function: −2 Log likelihood: −40.740
[b] Beginning Block Number 1. Method: Enter

Block1: Method = Enter

Variables in the Equation

	B	SE	Wald	df	Sig.	Exp(B)
AGE	−.199	.072	7.640	1	.006	.819

Variables not in the Equation[a]

	Score	df	Sig.
TREATMNT	3.477	1	.062

[a] Residual Chi Square = 3.477 with 1 df Sig. = .062

Block2: Method = Enter

Omnibus Tests of Model Coefficients[a,b]

−2 Log Likelihood	Overall (score)			Change From Previous Step			Change From Previous Block		
	Chi-square	df	Sig.	Chi-square	df	Sig.	Chi-square	df	Sig.
21.417	14.706	2	.001	3.978	1	.046	3.978	1	.046

[a] Beginning Block Number 0, initial Log Likelihood function: −2 Log likelihood: −40.740
[b] Beginning Block Number 2. Method: Enter

(continued)

TABLE 11.7   Continued

Variables in the Equation

	B	SE	Wald	df	Sig.	Exp(B)
AGE	-.230	.089	6.605	1	.010	.795
TREATMNT	2.542	1.546	2.701	1	.100	12.699

Covariate Means

	Mean
AGE	31.500
TREATMNT	.583

Use of the exponential distribution assumes that the percentage for the hazard rate remains the same within a particular group of cases over time. For example, a hazard rate of .1 per month indicates that 10% of the remaining cases fail in each succeeding month. Thus, the hazard rate does not depend on time.

The Weibull distribution, on the other hand, permits the hazard rate for a particular group of cases to change over time. For example, the probability of failure of a hard disk is much higher in the first few months than in later months. Thus, the failure rate depends on time, either increasing or decreasing. The exponential model is a special case of the Weibull model. Because the exponential model is nested within the Weibull model, you can test the difference between results based on the two of them (as shown by Allison, 1995, p. 89).

The log-normal distribution is basically an inverted U-shaped distribution in which the hazard function rises to a peak and then declines as time goes by. This function is often associated with repeatable events, such as marriage or residential moves. (SAS LIFEREG also permits specification of a normal distribution, in which there is no log-transform of the response variable, time.)

The log-logistic distribution also is an inverted U-shaped distribution when the scale parameter (σ) is less than 1, but behaves like the Weibull distribution when $\sigma \geq 1$. It is a proportional-odds model, meaning that the *change* in log-odds is constant over time. A logistic distribution (without log-transform of time) may be specified in SAS LIFEREG.

The gamma model (the one available in SAS LIFEREG is the generalized gamma model) is the most general model. Exponential, Weibull, and log-normal models are all special cases of it. Because of this relationship, differences between gamma and these other three models can be evaluated through likelihood ratio chi-square statistics (Allison, 1995, p. 89).

The gamma model can also have shapes that other models cannot, such as a U shape in which the hazard decreases over time to a minimum and then increases. Human mortality (and perhaps hard disks) over the whole life span follow this distribution (although hard disks are likely to become obsolete in both size and speed before the increase in hazard occurs). Because there is no more general model than the gamma, there is no test of its adequacy as an underlying distribution. And because the model is so general, it will always provide at least as good a fit as any other model.

TABLE 11.8   Syntax and Output for Accelerated Failure-Time Model through SAS LIFEREG

```
proc      lifereg data=SASUSER.SURVIVE;
          model MONTHS*DANCING(0)= AGE TREATMNT;
run;
```

The LIFEREG Procedure

Model Information

Data Set	SASUSER.SURVIVE
Dependent Variable	Log(MONTHS)
Censoring Variable	DANCING
Censoring Value(s)	0
Number of Observations	12
Noncensored Values	11
Right Censored Values	1
Left Censored Values	0
Interval Censored Values	0
Name of Distribution	Weibull
Log Likelihood	−4.960832864
Number of Observations Read	12
Number of Observations Used	12

Algorithm converged.

Type III Analysis of Effects

Effect	DF	Wald Chi-Square	Pr > ChiSq
AGE	1	15.3318	<.0001
TREATMNT	1	5.6307	0.0176

Analysis of Parameter Estimates

Parameter	DF	Estimate	Standard Error	95% Confidence Limits		Chi-Square	Pr > ChiSq
Intercept	1	0.4355	0.2849	−0.1229	0.9939	2.34	0.1264
AGE	1	0.0358	0.0091	0.0179	0.0537	15.33	<.0001
TREATMNT	1	0.5034	0.2121	0.0876	0.9192	5.63	0.0176
Scale	1	0.3053	0.0744	0.1894	0.4923		
Weibull Shape	1	3.2751	0.7984	2.0311	5.2811		

TABLE 11.9 **Distributions Available for Accelerated Failure-Time Models in SAS LIFEREG**

Model Distribution	Shape of Hazard Function	Parameters	Syntax	Comments
Exponential	Constant over time	1 (location); scale constrained to 1.	D=EXPONENTIAL	Simplest model, hazard rate does not depend on time
Weibull	Increases or decreases over time	2 (location and scale)	D=WIEBULL	Most commonly used model
Log-normal	Single-peaked	2 (location and scale)	D=LNORMAL	Often appropriate for repeatable events
Normal	Single-peaked	2 (location and scale)	D=NORMAL	As per log-normal, but without log transform of response
Log-logistic	Decreases over time or single-peaked	2 (location and scale)	D=LLOGISTIC	Has properties of Weibull or log-normal depending on scale parameter
Logistic	Decreases over time or single peaked	2 (location and scale)	D=LOGISTIC	As per log-logistic, but without log transform of response
Gamma	Decreases over time, increases over time, or constant over time	3 (location, scale, and shape)	D=GAMMA	Nonparsimonious, computer intensive

However, considerations of parsimony limit its use, as well as the more practical considerations of greater computation time and failure to converge on a solution.

Choice of models is made on the basis of logic, graphical fit, or, in the case of nested models, goodness-of-fit tests. For example, the expected failure rate of mechanical equipment should logically increase over time, indicating that a Weibull distribution is most appropriate. Exponential, log-normal, or log-logistic models could easily be ruled out on the basis of logic alone.

Allison (1995) provides guidance for producing graphs of appropriate transformations to Kaplan-Meier estimates. A resulting linear plot indicates that the distribution providing the transformation is the appropriate one. Allison also illustrates procedures for applying goodness-of-fit statistics to statistically evaluate competing models. As seen in Table 11.8, accelerated failure-time analyses produce χ^2 log-likelihood values (in which negative values closer to zero indicate better fits of data to models). Thus, twice the difference between nested models provides a likelihood-ratio χ^2 statistic. Of the nested models, gamma is the most general, followed by Weibull, then exponential, and finally log-normal. That is, log-normal is nested within exponential, which in turn is nested within Weibull, etc.

For example, Table 11.8 shows a log-likelihood $= -4.96$ with a Weibull distribution (default for SAS LIFEREG). A run with gamma specified (not shown) produces a log-likelihood $= -3.88$. The likelihood ratio is:

$$2[(-3.88) - (-4.96)] = 2.16$$

with df $= 1$ (because there is only one nesting "step" between Weibull and gamma). With critical $\chi^2 (\text{df} = 1) = 3.84$ at $\alpha = .05$, there is no significant difference between the Weibull and gamma models. Therefore, the Weibull model is preferred because it is the more parsimonious.

On the other hand, the log-likelihood for the exponential model (not shown) is -12.43. Comparing that with the Weibull model,

$$2[(-4.96) - (-12.43)] = 14.94, \qquad \text{df} = 1,$$

clearly a significant difference at $\alpha = .05$. The Weibull model remains the one of choice, because the exponential model is significantly worse.

11.5.2.4   *Choosing a Method*

The most straightforward way to analyze survival data with covariates is Cox regression. It is more robust than accelerated failure-time methods (Allison, 1995) and requires no choice among the distributions in Table 11.9 on the part of the researcher. However, the Cox model does have the assumption of proportionality of hazards over time, as described in the following section.

11.6   Some Important Issues

Issues in survival analysis include testing the assumption of proportionality of hazards, dealing with censored data, assessing effect size of models and individual covariates, choosing among the variety of statistical tests for differences among treatment groups and contributions of covariates, and interpreting odds ratios.

11.6.1   Proportionality of Hazards

When Cox regression is used to analyze differences between levels of a discrete covariate such as treatment, it is assumed that the shapes of the survival functions are the same for all groups over time. That is, the time until failures begin to appear may be longer for one group than another, but once failures start, they proceed at the same rate for all groups. When the assumption is met, the lines for the survival functions for different groups are roughly parallel, as seen in Figure 11.1 and Tables 11.3 and 11.4. Although inspection of the plots is helpful, a formal test of the assumption is also required.

The assumption is similar to homogeneity of regression in ANCOVA, which requires that the relationship between the DV and the covariate(s) is the same for all levels of treatment. The proportionality of hazards assumption is that the relationship between survival rate and time is the same for all levels of treatment (or any other covariate). In ANCOVA, violation of homogeneity of regression signals interaction between the covariate(s) and levels of treatment. In survival analysis, violation of proportionality signals interaction between time and levels of treatment (or any other covariate). To test the assumption, a time variable is constructed, and its interaction with levels of treatment (and other covariates) tested.

Both treatment and age are covariates in the examples of Cox regression in Tables 11.6 and 11.7. The question is whether either (or both) of these covariates interact with time; if so, there is violation of the proportionality of hazards. To test the assumption, a time variable, either continuous or discrete, is created and tested for interaction with each covariate. Table 11.10 shows a test for proportionality of hazards through SAS PHREG. The model includes two new predictors: `MONTHTRT` and `MONTHAGE`. These are defined in the following instructions as interactions between each covariate and the natural logarithm of the time variable, `MONTHS`, e.g., `MONTHTRT=TREATMNT*LOG(MONTHS)`. The logarithmic transform is recommended to compensate for numeric problems if the time variable takes on large values (Cantor, 1997).

The proportionality assumption is met in this example where neither `MONTHAGE` nor `MONTHTRT` is significant. Therefore, the Cox regression analyses of Tables 11.6 and 11.7 provide appropriate tests of treatment effects. If there is violation of the assumption, the test for treatment effects requires inclusion of the interaction(s) along with the other covariate(s) in either direct or sequential regression. That is, the test of differences due to treatment is conducted after adjustment for the interaction(s) along with other covariates.

Another remedy is to use a covariate that interacts with time as a stratification variable if it is discrete (or transformed into discrete) and *not* of direct interest. For example, suppose there is a significant interaction between age and time and age is not of research interest. Age can be divided into at least two levels and then used as a stratification variable in a Cox regression.

SPSS COXREG has a built-in procedure for creating and testing time-dependent covariates. The procedure is demonstrated in Section 11.7.1.5.

TABLE 11.10   Syntax and Partial Output for Proportionality Test through SAS PHREG

```
Proc phreg data=SASUSER.SURVIVAL2;
        Model MONTHS*DANCING(0)=AGE TREATMNT MONTHAGE MONTHTRT;
        MONTHTRT=TREATMNT*LOG(MONTHS); MONTHAGE=AGE*LOG(MONTHS);
run;
```

The PHREG Procedure

Analysis of Maximum Likelihood Estimates

Variable	DF	Parameter Estimate	Standard Error	Wald Chi-Square	Pr > ChiSq	Hazard Ratio
AGE	1	−0.22563	0.21790	1.0722	0.3004	0.798
TREATMNT	1	−3.54911	8.57639	0.1713	0.6790	0.029
MONTHAGE	1	−0.0001975	0.12658	0.0000	0.9988	1.000
MONTHTRT	1	0.47902	3.97015	0.0146	0.9040	1.614

11.6.2   Censored Data

Censored cases are those for whom the time of the event being studied (dropout, death, failure, graduation) is unknown or only vaguely known. If failure has not occurred by the end of the study (e.g., the part is still functioning, the dancer is still in classes, and so on), the case is considered right-censored. Or, if you know that failure occurred before a particular time (e.g., a tumor is already developed when the case enters your study), the case is considered left-censored. Or, if you know only that failure occurred sometime within a wide time interval, the case is considered interval-censored. A data set can contain a mix of cases with several forms of censoring.

11.6.2.1   *Right-Censored Data*

Right-censoring is the most common form of censoring and occurs when the event being studied has not occurred by the time data collection ceases. When the term "censoring" is used generically in some texts and computer programs, it refers to right-censoring.

Sometimes right-censoring is under the control of the researcher. For example, the researcher decides to monitor cases until some predetermined number has failed, or until every case has been followed for three years. Cases are censored, then, because the researcher terminates data collection before the event occurs for some cases. Other times the researcher has no control over right-censoring. For example, a case might be lost because a participant refuses to continue to the end of the study or dies for some reason other than the disease under study. Or, survival time may be unknown because the entry time of a case is not under the control of the researcher. For example, cases are monitored until a predetermined time, but the time of entry into the study (e.g., the time of surgery) varies randomly among cases so that total survival time is unknown. That is, all you know about the time of occurrence of an event (failure, recovery) is that it occurred after some particular time, that is, it is greater than some value (Allison, 1995).

Most methods of survival analysis do not distinguish among types of right-censoring, but cases that are lost from the study may pose problems because it is assumed that there are no systematic differences between them and the cases that remain (Section 11.3.2.4). This assumption is likely to be violated when cases voluntarily leave the study. For example, students who drop out of a graduate program are unlikely to have graduated (had they stayed) as soon as students who continued. Instead, those who drop out are probably among those who would have taken longer to graduate. About the only solution to the problem is to try to include covariates that are related to this form of censoring.

All of the programs reviewed here deal with right-censored data, but none distinguishes among the various types of right-censoring. Therefore, results are misleading if assumptions about censoring are violated.

11.6.2.2   *Other Forms of Censoring*

A case is left-censored if the event of interest occurred before an observed time, so that you know only that survival time is less than the total observation time. Left-censoring is unlikely to occur in an experiment, because random assignment to conditions is normally made only for intact cases. However, left-censoring can occur in a nonexperimental study. For example, if you are studying the failure time of a component, some components may have failed before you start your observations, so you don't know their total survival time.

With interval censoring, you know the interval within which the event occurred, but not the exact time within the interval. Interval censoring is likely to occur when events are monitored infrequently. Allison (1995) provides as an example annual testing for HIV infection in which a person who tested negative at year 2, but tested positive at year 3, is interval-censored between 2 and 3.

SAS LIFEREG handles right-, left-, and interval-censoring by requiring two time variables, upper time and lower time. For right-censored cases, the upper time value is missing; for left-censored cases, the lower time value is missing. Interval-censoring is indicated by different values for the two variables.

11.6.3 Effect Size and Power

Cox and Snell (1989) provide a measure of effect size for logistic regression that is demonstrated for survival analysis by Allison (1995). It is based on G^2, a likelihood-ratio chi-square statistic (Section 11.6.4.2), that can be calculated from SAS PHREG and LIFEREG and SPSS COXREG.

Models are fit both with and without covariates, and a difference G^2 is found by:

$$G^2 = [(-2 \text{ log-likelihood for smaller model}) - (-2 \text{ log-likelihood for larger model})] \quad (11.10)$$

Then, R^2 is found by

$$R^2 = 1 - e^{(-G^2/n)} \quad (11.11)$$

When applied to experiments, the R^2 of greatest interested is the association between survival and treatment, after adjustment for other covariates. Therefore the smaller model is the one that includes covariates but not treatment, and the larger model is the one that includes covariates *and* treatment.

For the example of Table 11.7 (the sequential analysis in which treatment is adjusted for age), -2 log-likelihood with age alone is 25.395 and -2 log-likelihood with age and treatment is 21.417, so that

$$G^2 = 25.395 - 21.417 = 3.978$$

for treatment. (Note that this value is also provided by SPSS COXREG, as Change from Previous Block.)

Applying Equation 11.11:

$$R^2 = 1 - e^{(-3.978/12)} = 1 - .7178 = .28$$

Steiger and Fouladi's (1992) software may be used to find confidence limits around this value (cf. Figure 9.3). The number of variables (including the criterion but not the covariate) is 2, with $N = 12$. The software provides a 95% confidence limit ranging from 0 to .69.

Allison (1995) points out that this R^2 is not the proportion of variance in survival that is explained by the covariates, but merely represents relative association between survival and the covariates tested, in this case treatment after adjustment for age.

Power in survival analysis is, as usual, enhanced by larger sample sizes and covariates with stronger effects. Amount of censoring and patterns of entry of cases into the study also affect power, as does the relative size of treatment groups. Unequal sample sizes reduce power while equal sample sizes increase it. Estimating sample sizes and power for survival analysis is not included in the software discussed in this book except for NCSS (2002) PASS which provides power and sample size estimates for a survival test, based on Lachin and Foulkes (1986). Another, stand-alone, program provides power analysis for several types of survival analyses: nQuery Advisor 4.0 (Elashoff, 2000).

11.6.4 Statistical Criteria

Numerous statistical tests are available for evaluating group differences due to treatment effects from an actuarial life table or product-limit analysis, as discussed in Section 11.6.4.1. Tests for evaluating the relationships among survival time and various covariates (including treatment) are discussed in Section 11.6.4.2.

11.6.4.1 Test Statistics for Group Differences in Survival Functions

Several statistical tests are available for evaluating group differences, and there is inconsistent labeling among programs. The tests differ primarily in how cases are weighted, with weighting based on the time that groups begin to diverge during the course of survival. For example, if the groups begin to diverge right away (untreated cases fail quickly but treated cases do not), statistics based on heavier weighting of cases that fail quickly show greater group differences than statistics for which all cases are weighted equally. Table 11.11 summarizes statistics for differences among groups that are available in the programs.

SAS LIFETEST provides three tests: The `Log-Rank` and `Wilcoxon` statistics and the likelihood-ratio test, labeled `-2Log(LR)`, which assumes an exponential distribution of failures in each of the groups. SPSS KM offers three statistics as part of the Kaplan-Meier analysis: the `Log`

TABLE 11.11 Tests for Differences among Groups in Actuarial and Product-Limits Methods

	Nomenclature			
Test	*SAS[a]* *LIFETEST*	*SPSS* *SURVIVAL*	*SPSS* *KM*	**Comments**
1	Log-Rank	N.A.	Log Rank	Equal weight to all observations
2	Tarone	N.A.	Tarone- Ware	Slightly greater weight to early observations, between test 1 and test 3
3	Wilcoxon	N.A.	Breslow	Greater weight to early observations
4	N.A.	Wilcoxon (Gehan)	N.A.	Differs slightly from test 3
5	-2Log(LR)	N.A.	N.A.	Assumes an exponential distribution of failures in each group

[a]Additional SAS tests are listed in Table 11.24.

Rank test, the `Tarone-Ware` statistic, and the `Breslow` statistic, which is equivalent to the `Wilcoxon` statistic of SAS. SPSS SURVIVAL provides an alternative form of the `Wilcoxon` test, the `Gehan` statistic, which appears to use weights intermediate between Breslow (Wilcoxon) and Tarone-Ware.

11.6.4.2   Test Statistics for Prediction from Covariates

Log-likelihood chi-square tests (G^2 as described in Section 11.6.3) are used both to test the hypothesis that all regression coefficients for covariates are zero in a Cox proportional hazards model and to evaluate differences in models with and without a particular set of covariates, as illustrated in Section 11.6.3. The latter application, using Equation 11.10, most often evaluates the effects of treatment after adjustment for other covariates. All of these likelihood-ratio statistics are large sample tests and are not to be taken seriously with small samples such as the example of Section 11.4.

Statistics are also available to test regression coefficients separately for each covariate. These Wald tests are z tests where the coefficient is divided by its standard error. When the test is applied to the treatment covariate, it is another test of the effect of treatment after adjustment for all other covariates.

SPSS COXREG provides all of the required information in a sequential run, as illustrated in Table 11.7. The last step (in which treatment is included) shows `Chi-Square` for `Change` `(-2 Log Likelihood) from Previous Block` as the likelihood ratio test of treatment as well as Wald tests for both treatment and age, the covariates.

SAS PHREG provides `Model Chi-Square` which is overall G^2, with and without all covariates. A likelihood-ratio test for models with and without treatment (in which other covariates are included in both models) requires a sequential run followed by application of Equation 11.10 to the models with and without treatment. SAS LIFEREG, on the other hand, provides no overall chi-square likelihood-ratio test but does provide chi-square tests for each covariate, adjusted for all others, based on the squares of coefficients divided by their standard errors. A log-likelihood value for the whole model is also provided, so that two runs, one with and the other without treatment, provide the statistics necessary for Equation 11.10.

11.6.5   Predicting Survival Rate

11.6.5.1   Regression Coefficients (Parameter Estimates)

Statistics for predicting survival from covariates require calculating regression coefficients for each covariate where one or more of the "covariates" may represent treatment. The regression coefficients give the relative effect of each covariate on the survival function, but the size depends on the scale of the covariate. These coefficients may be used to develop a regression equation for risk as a DV. An example of this is in Section 11.7.2.2.

11.6.5.2   Odds Ratios

Because survival analysis is based on a linear combination of effects in the exponent (like logistic regression, Chapter 10) rather than a simple linear combination of effects (like multiple regression, Chapter 5), effects are most often interpreted as odds. How does a covariate change the odds of surviving? For example, how does a one-year increase in age change the odds of surviving in dance classes?

Odds are found from a regression coefficient (B) as e^B. However, for correct interpretation, you also have to consider the direction of coding for survival. In the small-sample example (Table 11.1), dropping out is coded 1 and "surviving" (still dancing) is coded 0. Therefore, a positive regression coefficient means that an increase in age increases the likelihood of dropping out while a negative regression coefficient means that an increase in age decreases the likelihood of dropping out. Treatment is also coded 1, 0 where 1 is used for the group that had a preinstruction night out on the town and 0 for the control group. For this variable, a change in the value of the treatment covariate from 0 to 1 means that the dancer is more likely to drop out following a night out if the regression coefficient is positive, and less likely to drop out following a night out if the regression coefficient is negative. This is because a positive regression coefficient leads to an odds ratio greater than one while a negative coefficient leads to an odds ratio less than one.

Programs for Cox proportional-hazards models show both the regression coefficients and odds ratios (see Tables 11.6 and 11.7). Regression coefficients are labeled B or Estimate. Odds ratios are labeled Exp (B) or Hazard Ratio.

Table 11.7 shows that age is significantly related to survival as a belly dancer. The negative regression coefficient (and odds ratio less than 1) indicates that older dancers are less likely to drop out. Recall that $e^B = 0.79$; this indicates that the odds of dropping out are decreased by about 21% $[(1 - .79)100]$ with each year of increasing age. The hazard of dropping out for a 25-year-old, for instance, is only 79% of that for a 24-year-old. (If the odds [hazard] ratio were .5 for age, it would indicate that the risk of dropping out is halved with each year of increasing age.)

In some tests the treatment covariate fails to reach statistical significance, but if we attribute this to lack of power with such a small sample, rather than a lack of treatment effectiveness, we can interpret the odds ratio for illustrative purposes. The odds ratio of .08 ($e^{-2.542}$) for treatment indicates that treatment decreases the odds of dropping out by 92%.

11.6.5.3 *Expected Survival Rates*

More complex methods are required for predicting expected survival rates at various time periods for particular values of covariates, as described using SAS procedures by Allison (1995, pp. 171–172). For example, what is the survivor function for 25-year-olds in the control group? This requires creating a data set with the particular covariate values of interest, e.g., 0 for treatment and 25 for age. The model is run with the original data set, and then a print procedure applied to the newly created data set. Table 11.12 shows syntax and partial output for prediction runs for two cases: a 25-year-old dancer in the control group and 30-year-old dancer in the treated group.

The likelihood of survival, column **s**, for a 25-year-old in the control condition drops quickly after the first month and is very low by the fifth month; on the other hand, the likelihood of survival for a 30-year-old in the treated condition stays pretty steady through the fifth month.

11.7   Complete Example of Survival Analysis

These experimental data are from a clinical trial of a new drug (D-penicillamine) versus a placebo for treatment of primary biliary cirrhosis (PBC) conducted at the Mayo Clinic between 1974 and 1984. The data were copied to the Internet from Appendix D of Fleming and Harrington (1991), who describe the data set as follows:

A total of 424 PBC patients, referred to Mayo Clinic during that ten-year interval, met eligibility criteria for the randomized placebo controlled trial of the drug D-penicillamine. The first 312 cases in the data set participated in the randomized trial and contain largely complete data (p. 359).

Thus, differences in survival time following treatment with either the experimental drug or the placebo are examined in the 312 cases with nearly complete data who participated in the trial. Coding for drug is 1 = D-penicillamine and 2 = placebo. Additional covariates are those in the Mayo model for "assessing survival in relation to the natural history of primary biliary cirrhosis" (Markus et al., 1989, p. 1710). These include age (in days), serum bilirubin in mg/dl, serum albumin in gm/dl, prothrombin time in seconds, and presence of edema. Edema has three levels treated as continuous:

TABLE 11.12   Predicted Survivor Functions for 25-Year-Old Control Dancers and 30-Year-Old Treated Dancers (Syntax and Partial Output Using SAS PHREG)

```
data      surv;
  set SASUSER.SURVIVE;
data      covals;
          input TREATMNT AGE;
          datalines;
0 25
1 30
run;
proc      phreg data=SASUSER.SURVIVE;
          model MONTHS*DANCING(0)= AGE TREATMNT;
          baseline out=predict covariates=covals survival=s
          lower=lcl upper=ucl / nomean;
run;
proc      print data=predict;
run;
```

                          Model Fit Statistics

Criterion	Without Covariates	With Covariates
-2 LOG L	40.740	21.417
AIC	40.740	25.417
SBC	40.740	26.213

                Testing Global Null Hypothesis: BETA=0

Test	Chi-Square	DF	Pr > ChiSq
Likelihood Ratio	19.3233	2	<.0001
Score	14.7061	2	0.0006
Wald	6.6154	2	0.0366

                Analysis of Maximum Likelihood Estimates

Variable	DF	Parameter Estimate	Standard Error	Chi-Square	Pr > ChiSq	Hazard Ratio
AGE	1	-0.22989	0.08945	6.6047	0.0102	0.795
TREATMNT	1	-2.54137	1.54632	2.7011	0.1003	0.079

TABLE 11.12   Continued

Obs	AGE	TREATMNT	MONTHS	s	lcl	ucl
1	25	0	0	1.00000	.	.
2	25	0	1	0.93135	0.82423	1.00000
3	25	0	2	0.69114	0.48766	0.97953
4	25	0	3	0.53449	0.28983	0.98567
5	25	0	4	0.38533	0.16137	0.92013
6	25	0	5	0.09404	0.02674	0.33066
7	25	0	7	0.00000	0.00000	1.00000
8	25	0	8	0.00000	0.00000	1.00000
9	25	0	10	0.00000	0.00000	1.00000
10	25	0	11	0.00000	0.00000	1.00000
11	30	1	0	1.00000	.	.
12	30	1	1	0.99823	0.99159	1.00000
13	30	1	2	0.99083	0.96435	1.00000
14	30	1	3	0.98449	0.93576	1.00000
15	30	1	4	0.97649	0.90254	1.00000
16	30	1	5	0.94272	0.83716	1.00000
17	30	1	7	0.46191	0.23216	0.91903
18	30	1	8	0.02934	0.00024	1.00000
19	30	1	10	0.00001	0.00000	1.00000
20	30	1	11	0.00000	0.00000	1.00000

(1) no edema and no diuretic therapy for edema, coded 0.00; (2) edema present without diuretics or edema resolved by diuretics, coded 0.50; and (3) edema despite diuretic therapy, coded 1.00. Codes for STATUS are 0 = censored, 1 = liver transplant, and 2 = event (nonsurvival).

Remaining variables in the data set are sex, presence versus absence of ascites, presence or absence of hepatomegaly, presence or absence of spiders, serum cholesterol in mg/dl, urine copper in ug/day, alkaline phosphatase in U/liter, SGOT in U/ml, triglyceride in mg/dl, platelets per cubic ml/100, and histologic stage of disease. These variables were not used in the present analysis.

The primary goal of the clinical trial is to assess the effect of the experimental drug on survival time after statistically adjusting for the other covariates. A secondary goal is to assess the effects of the other covariates on survival time. Data files are SURVIVAL.*.

11.7.1   Evaluation of Assumptions

11.7.1.1   Accuracy of Input, Adequacy of Sample Size, Missing Data, and Distributions

SPSS DESCRIPTIVES is used for a preliminary look at the data, as seen in Table 11.13. The SAVE request produces standard scores for each covariate for each case used to assess univariate outliers.

The values for most of the covariates appear reasonable; for example, the average age is about 50. The sample size of 312 is adequate for survival analysis, and cases are evenly split between experimental and placebo groups (mean = 1.49 with coding of 1 and 2 for the groups).

TABLE 11.13 Description of Covariates through SPSS DESCRIPTIVES (Syntax and Output)

DESCRIPTIVES
VARIABLES=AGE ALBUMIN BILIRUBI DRUG EDEMA PROTHOM
/SAVE
/STATISTICS=MEAN STDDEV MIN MAX KURTOSIS SKEWNESS.

Descriptives

Descriptive Statistics

	N	Minimum	Maximum	Mean	Std.	Skewness		Kurtosis	
	Statistic	Statistic	Statistic	Statistic	Statistic	Statistic	Std. Error	Statistic	Std. Error
Age in days	312	9598.00	28650.00	18269.44	3864.805	.168	.138	-.534	.275
Albumin in gm/dl	312	1.96	4.64	3.5200	.41989	-.582	.138	.946	.275
Serum bilirubin in mg/dl	312	.30	28.00	3.2561	4.53032	2.848	.138	8.890	.275
Experimental drug	312	1.00	2.00	1.4936	.50076	.026	.138	-2.012	.275
Edema presence	312	.00	1.00	.1106	.27451	2.414	.138	4.604	.275
Prothrombin time in seconds	312	9.00	17.10	10.7256	1.00432	1.730	.138	6.022	.275
Valid N (listwise)	312								

None of the covariates has missing data. However, except for age and drug (the treatment), all of the covariates are seriously skewed, with z-scores for skewness ranging from $(-.582)/0.138 = -4.22$ for serum albumin to $(2.85)/0.138 = 20.64$ for bilirubin. Kurtosis values listed in Table 11.13 pose no problem in this large sample (cf. Section 11.3.2.2). Decisions about transformation are postponed until outliers are assessed.

11.7.1.2 Outliers

Univariate outliers are assessed by finding $z = (Y - \overline{Y})/S$ for each covariate's lowest and highest scores. The /SAVE instruction in the SPSS DESCRIPTIVES run of Table 11.13 adds a column to the data file of z-scores for each case on each covariate. An SPSS DESCRIPTIVES run on these standard scores shows minimum and maximum values (Table 11.14).

Using $|z| = 3.3$ as the criterion (cf. Section 11.3.2.3), the lowest albumin score is a univariate outlier, as are the highest scores on bilirubin and prothrombin time. Considering the skewness in these distributions, the decision is made to transform them to deal with both outliers and the possibility of diminished predictability of survival time as a result of nonnormality of covariates. Tests of multivariate outliers are performed on the transformed variables.

A logarithmic transform of bilirubin [LBILIRUB = LG10(BILIRUBI)] diminishes its skewness (although $z > 4.6$) and kurtosis and brings outlying cases to within acceptable limits. However, various transformations (log, inverse, square root) of prothrombin time and albumin do not remove the outliers, so the decision is made to retain the original scales of these variables. An additional transform is performed on age (in days) into years of age: Y_AGE = (AGE/365.25) to facilitate interpretation. Table 11.15 shows descriptive statistics for transformed age and log of bilirubin.

Mahalanobis distance to assess multivariate outliers is computed through SPSS REGRESSION and examined through SPSS SUMMARIZE. Table 11.16 first shows the SPSS REGRESSION syntax

TABLE 11.14 Description of Standard Scores through SPSS DESCRIPTIVES (Syntax and Selected Output)

```
DESCRIPTIVES
 VARIABLES=ZAGE ZALBUMIN ZBILIRUB ZDRUG ZEDEMA ZPROTHOM
 /STATISTICS=MIN MAX.
```

Descriptives

Descriptive Statistics

	N	Minimum	Maximum
Zscore: Age in days	312	−2.24369	2.68592
Zscore: Albumin in gm/dl)	312	−3.71524	2.66735
Zscore: Serum bilirubin in mg/dl	312	−.65251	5.46185
Zscore: Experimental drug	312	−.98568	1.01128
Zscore: Edema presence	312	−.40282	3.24008
Zscore: Prothrombin time in seconds	312	−1.71821	6.34692
Valid N (listwise)	312		

TABLE 11.15 Description of Transformed Covariates through SPSS DESCRIPTIVES (Syntax and Output)

```
DESCRIPTIVES
VARIABLES=LBILIRUBI Y_AGE
/SAVE
/STATISTICS=MEAN STDDEV MIN MAX KURTOSIS SKEWNESS.
```

Descriptives

Descriptive Statistics

	N	Minimum	Maximum	Mean	Std.	Skewness		Kurtosis	
	Statistic	Statistic	Statistic	Statistic	Statistic	Statistic	Std. Error	Statistic	Std. Error
LBILIRUB	312	−.52	1.45	.2500	.44827	.637	.138	−.376	.275
Y_AGE	312	26.28	78.44	50.0190	10.58126	.168	.138	−.534	.275
Valid N (listwise)	312								

TABLE 11.16 Mahalanobis Distances and Covariate Scores for Multivariate Outliers (Syntax and Selected Output for SPSS REGRESSION and SUMMARIZE)

```
REGRESSION
 /MISSING LISTWISE
 /STATISTICS COEFF OUTS R ANOVA
 /CRITERIA=PIN(.05) POUT(.10)
 /NOORIGIN
 /DEPENDENT ID
 /METHOD=ENTER ALBUMIN DRUG EDEMA PROTHOM LBILIRUB Y_AGE
 /SAVE MAHAL.
USE ALL.
COMPUTE filter_$=(MAH_1>22.458).
VARIABLE LABEL filter_$ 'MAH_1>22.458 (FILTER)'.
VALUE LABELS filter_$ 0 'NOT SELECTED' 1 'SELECTED'.
FORMAT filter_$ (f1.0).
FILTER BY filter_$.
EXECUTE
SUMMARIZE
 /TABLES=ALBUMIN DRUG EDEMA PROTHOM LBILIRUB Y_AGE MAH_1 ID
 /FORMAT=VALIDLIST NOCASENUM TOTAL LIMIT=100
 /TITLE='Case Summaries' /FOOTNOTE"
 /MISSING=VARIABLE
 /CELLS=COUNT.
```

Summarize

Case Summaries[a]

	Albumin in gm/dl	Experimental drug	Edema presence	Prothrombin time in seconds	LBILIRUB	Y_AGE	Mahalanobis Distance	ID
1	2.27	Placebo	Edema despite therapy	11.00	-.10	56.22	23.28204	14.00
2	4.03	Placebo	No edema	17.10	-.22	62.52	58.81172	107.00
3	3.35	Placebo	No edema	15.20	1.39	52.69	28.79138	191.00
Total N	3	3	3	3	3	3	3	3

a.Limited to first 100 cases.

that saves the Mahalanobis distance for each case into a column of the data file as a variable labeled mah_1. The critical value of χ^2 with 6 df at $\alpha = .001$ is 22.458. Cases with mah_1 greater than 22.458 are selected for printing through SPSS SUMMARIZE with their case ID number, scores for the four continuous covariates for those cases, and Mahalanobis distance. Syntax for the selection of multivariate outliers is shown in Table 11.16 along with the output of SPSS SUMMARIZE.

Three cases are multivariate outliers. Table 11.17 shows the results of a regression analysis on case number 14 to determine which covariates distinguish it from the remaining 311 cases. A dichotomous DV, labeled dummy, is created based on case identification number and then SPSS REGRESSION is run to determine which variables significantly predict that dummy DV. Note that the selection based on Mahalanobis distance must be altered for each run so the file again includes all cases.

Covariates with levels of Sig. less than .05 contribute to the extremity of the multivariate out-lier where a positive coefficient indicates a higher score on the variable for the case (because the out-lier has a higher code (1) than the remaining cases (0) on the dummy DV). Case 14 differs from the

**TABLE 11.17   Identification of Covariates Causing Multivariate Outliers
(SPSS REGRESSION Syntax and Selected Output)**

```
USE ALL.
COMPUTE          DUMMY = 0.
IF          (id EQ 14) DUMMY=1.
REGRESSION
 /MISSING LISTWISE
 /STATISTICS COEFF OUTS R ANOVA
 /NOORIGIN
 /DEPENDENT dummy
 /METHOD=ENTER albumin drug edema prothom lbilirub y_age.
```

Regression

Coefficients[a]

Model		Unstandardized Coefficients B	Unstandardized Coefficients Std. Error	Standardized Coefficients Beta	t	Sig.
1	(Constant)	.106	.051		2.079	.038
	Albumin in gm/dl	−.023	.009	−.168	−2.661	.008
	Experimental drug	.008	.006	.068	1.216	.225
	Edema presence	.041	.013	.200	3.069	.002
	Prothrombin time in seconds	−.003	.004	−.046	−.740	.460
	LBILIRUB	−.021	.008	−.167	−2.630	.009
	Y_AGE	.000	.000	−.023	−.387	.699

[a]Dependent Variable: DUMMY

remaining cases in the combination of low scores on albumin and the logarithm of bilirubin along with a high score on edema. Table 11.17 shows the values of those scores: 2.27 on albumin as compared with a mean of 3.52 (seen in Table 11.13); 0.26 on the logarithm of bilirubin as compared with a mean of 0.49; and a score of 1 on edema as compared with a mean of 0.11. Similar regression analyses for the two remaining outliers (not shown) indicate that case 107 is an outlier because of a high score on prothrombin time (17.10), and case 191 is an outlier because of a high score on prothrombin time (15.20) and a low score on edema (0). The decision is made to eliminate these multivariate outlying cases from subsequent analyses and report details about them in the results section. A rerun of syntax in Tables 11.13 and 11.14 with multivariate outliers removed (not shown) indicates that only one of the univariate outliers from Table 11.14 remains; the case with a z-score of -3.715 on albumin. It is decided to retain this case in subsequent analyses because it did not appear as a multivariate outliers and is not inordinately extreme considering the sample size.

11.7.1.3   *Differences between Withdrawn and Remaining Cases*

Several cases were censored because they were withdrawn from this clinical trial for liver transplantation. It is assumed that the remaining censored cases were alive at the end of the study. Table 11.18 shows a regression analysis where status is used to form a dichotomous DV (labeled xplant) where cases who were withdrawn for liver transplant have a value of 1 and the other cases have a value of 0. The six covariates serve as the IVs for the regression analysis. Note that the multivariate outliers are omitted from this and all subsequent analyses.

There is a significant difference between those undergoing liver transplantation and the remaining cases, however the difference is limited to age with $\alpha = .008$ using a Bonferroni-type correction for inflated Type I error associated with the six covariates. The negative coefficient indicates that liver transplants were done on younger cases, on average, and not surprisingly. Because age is the only variable distinguishing these cases from the remaining ones, the decision is made to leave them in the analysis, grouped with the other censored cases at the end of the test period.

11.7.1.4   *Change in Survival Experience over Time*

There is no indication in the data set or supporting documentation of a change in procedures over the ten-year period of the study. Other factors, such as pollution and economic climate, of course, remain unknown and uncontrolled potential sources of variability. Because of the random assignment of cases to drug conditions, however, there is no reason to expect that these environmental sources of variance differ for the two drug conditions.

11.7.1.5   *Proportionality of Hazards*

Proportionality of hazards is checked prior to Cox regression analysis, to determine if the assumption is violated. Table 11.19 shows the test for proportionality of hazards through SPSS COXREG. The TIME PROGRAM instruction sets up the internal time variable, T_ (a reserved name for the transformed time variable). Then COMPUTE is used to create T_COV_ as the natural logarithm (LN) of time. All of the covariate*T_COV_ interactions are included in the COXREG instruction.

Only the terms representing interaction of T_COV_ with covariates are used to evaluate proportionality of hazards. (Ignore the rest of the output for now, especially the drug result.) If $\alpha = .008$ is used because of the number of time-covariate interactions being evaluated, none of the covariates significantly interacts with time. Therefore we consider the assumption met.

TABLE 11.18   SPSS REGRESSION for Differences between Liver Transplant and Remaining Cases (Syntax and Selected Output)

```
USE ALL.
COMPUTE filter_$=(MAH_1 LE 22.458).
VARIABLE LABEL filter_$ 'MAH_1 LE 22.458 (FILTER)'.
VALUE LABELS filter_$ 0 'Not Selected' 1 'Selected'.
FORMAT filter_$ (f1.0).
FILTER BY filter_$.
EXECUTE.

COMPUTE XPLANT = 0.
IF        (STATUS EQ 1) XPLANT = 1.
REGRESSION
 /MISSING LISTWISE
 /STATISTICS COEFF OUTS R ANOVA
 /CRITERIA=PIN(.05) POUT(.10)
 /NOORIGIN
 /DEPENDENT XPLANT
 /METHOD=ENTER ALBUMIN DRUG EDEMA LBILIRUB PROTHOM Y_AGE.
```

Regression

ANOVA[b]

Model		Sum of Squares	df	Mean Square	F	Sig.
1	Regression	1.144	6	.191	3.449	.003[a]
	Residual	16.688	302	.055		
	Total	17.832	308			

[a]Predictors: (Constant), Y_AGE, LBILIRUB, Experimental drug, Prothrombin time in seconds, Albumin in gm/dl, Edema presence
[b]Dependent Variable: XPLANT

Coefficients[a]

Model		Unstandardized Coefficients		Standardized Coefficients	t	Sig.
		B	Std. Error	Beta		
1	(Constant)	.536	.243		2.208	.028
	Albumin in gm/dl	.006	.037	.010	.164	.870
	Experimental drug	−.018	.027	−.037	−.662	.509
	Edema presence	−.015	.060	−.017	−.255	.799
	LBILIRUB	.075	.035	.139	2.143	.033
	Prothrombin time in seconds	−.025	.017	−.092	−1.422	.156
	Y_AGE	−.004	.001	−.197	−3.351	.001

[a]Dependent Variable: XPLANT

TABLE 11.19   Test for Proportionality of Hazards through SPSS COXREG (Syntax and Selected Output)

```
TIME PROGRAM.
COMPUTE T_COV_ = LN(T_).
COXREG
 DAYS /STATUS=STATUS(2)
 /METHOD=ENTER ALBUMIN T_COV_*ALBUMIN DRUG T_COV_*DRUG EDEMA
          T_COV_*EDEMA
 PROTHOM T_COV_*PROTHOM LBILIRUB T_COV_*LBILIRUB Y_AGE T_COV_*Y_AGE
 /CRITERIA=PIN(.05) POUT(.10)ITERATE(20).
```

Variables in the Equation

	B	SE	Wald	df	Sig.	Exp(B)
ALBUMIN	−1.821	1.892	.927	1	.336	.162
DRUG	2.379	1.382	2.963	1	.085	10.798
EDEMA	5.685	2.390	5.657	1	.017	294.353
PROTHOM	1.449	.734	3.904	1	.048	4.261
LBILIRUB	−.371	1.815	.042	1	.838	.690
Y_AGE	.087	.075	1.349	1	.245	.1091
T_COV_*ALBUMIN	.129	.272	.226	1	.635	1.138
T_COV_*DRUG	−.319	.197	2.637	1	.104	.727
T_COV_*EDEMA	−.778	.364	4.569	1	.033	.459
T_COV_*PROTHOM	−.164	.106	2.387	1	.122	.849
T_COV_*LBILIRUB	.339	.261	1.690	1	.194	1.404
T_COV_*Y_AGE	−.008	.011	.542	1	.461	.992

11.7.1.6   Multicollinearity

Survival-analysis programs protect against statistical problems associated with multicollinearity. However, the analysis is best served by a set of covariates that are not too highly related. Thus, it is worthwhile to investigate how highly each of the covariates is related to the remaining ones.

Squared multiple correlations (SMCs) are available through SPSS FACTOR by specifying principal axis factoring because this type of factor analysis begins with SMCs as initial communalities (Section 13.6.1). Table 11.20 shows the syntax and selected output for SPSS FACTOR for the set of covariates used in the survival analysis.

Redundant covariates are those with Initial Communalities (SMCs) in excess of .90. As seen in Table 11.20, there is no danger of either conceptual or statistical multicollinearity among this set, with the highest SMC = .314 for presence of edema.

11.7.2   Cox Regression Survival Analysis

SPSS COXREG is used to evaluate the effects of drug and other covariates on survival time of patients with primary biliary cirrhosis of the liver. Table 11.21 shows the syntax and output for the

TABLE 11.20   SMCs (Communalities) Produced by SPSS FACTOR (Syntax and Selected Output)

```
SELECT IF mah_1 LE 22.458
FACTOR
 /VARIABLES ALBUMIN DRUG EDEMA PROTHOM LBILIRUB Y_AGE /MISSING LISTWISE
 /ANALYSIS ALBUMIN DRUG EDEMA PROTHOM LBILIRUB Y_AGE
 /PRINT INITIAL EXTRACTION
 /CRITERIA MINEIGEN(1) ITERATE(25)
 /EXTRACTION PAF
 /ROTATION NOROTATE
 /METHOD=CORRELATION.
```

Factor Analysis

Communalities

	Initial	Extraction
Albumin in gm/dl	.239	.319
Experimental drug	.026	.048
Edema presence	.314	.472
Prothrombin time in seconds	.266	.356
LBILIRUB	.264	.455
Y_AGE	.105	.466

Extraction Method: Principal Axis Factoring.

sequential Cox regression analysis in which covariates other than drug are entered first, as a set, followed by drug treatment. This permits a likelihood-ratio test of the effect of drug treatment, after statistical adjustment for the other covariates.

11.7.2.1   Effect of Drug Treatment

The effect of drug treatment with D-penicillamine versus the placebo is evaluated as Change from Previous Block at Block 2. A value of Sig for Chi-Square less than .05 is required for drug treatment to successfully predict survival time after adjusting for the other covariates. Here, $\chi^2(1) = 0.553$, $p = .457$, revealing that drug treatment has no statistically significant effect on survival time of PBC patients after taking into account their age, serum albumin level, condition of edema, prothrombin time, and the logarithm of the level of serum bilirubin. That is, length of survival is unaffected by the D-penicillamine drug. Survival curves for the two groups are not shown because there is no statistically significant difference between groups.

11.7.2.2   Evaluation of Other Covariates

The output of Block1 of Table 11.21 reveals the relationship between survival time and the other covariates. None of these variables is experimentally manipulated in this study; however, as a group, they form the Mayo model for predicting survival of PBC patients. Change from Previous Step $\chi^2(5)$ of 192.857, $p < .0005$, shows that, as a set, the covariates reliably predict survival time. Applying equation 11.11, the effect size of the set of covariates and survival time is:

$$R^2 = 1 - e^{(-192.867/309)} = .46$$

with a 95% confidence interval from .37 to .53 using Steiger and Fouladi's (1992) R2 software (see Figure 9.3 for demonstration of the use of the software).

TABLE 11.21   Cox Regression Analysis for PBC Patients through SPSS COXREG (Syntax and Output)

```
SELECT IF mah_1 LE 22.458.
COXREG
 DAYS /STATUS=STATUS(2)
 /METHOD=ENTER ALBUMIN EDEMA PROTHOM LBILIRUB Y_AGE
 /METHOD=ENTER DRUG
 /CRITERIA=PIN(.05) POUT(.10) ITERATE(20).
```

Cox Regression

Case Processing Summary

		N	Percent
Cases available in analysis	Event[a]	123	39.8%
	Censored	186	60.2%
	Total	309	100.0%
Cases dropped	Cases with missing values	0	.0%
	Cases with negative time	0	.0%
	Censored cases before the earliest event in a stratum	0	.0%
	Total	0	.0%
Total		309	100.0%

[a]Dependent Variable: DAYS

Block 0: Beginning Block

Omnibus Tests of Model Coefficients

−2 Log Likelihood
1255.756

(continued)

TABLE 11.21   Continued

Block1: Method = Enter

Omnibus Tests of Model Coefficients[a,b]

-2 Log Likelihood	Overall (score)			Change From Previous Step			Change From Previous Block		
	Chi-square	df	Sig.	Chi-square	df	Sig.	Chi-square	df	Sig.
1062.899	261.098	5	.000	192.857	5	.000	192.857	5	.000

[a]Beginning Block Number 0, initial Log Likelihood function: -2 Log likelihood: -1255.756
[b]Beginning Block Number 1. Method: Enter

Variables in the Equation

	B	SE	Wald	df	Sig.	Exp(B)
ALBUMIN	-.884	.242	13.381	1	.000	.413
EDEMA	.743	.311	5.712	1	.017	2.101
PROTHOM	.307	.104	8.668	1	.003	1.359
LBILIRUB	1.988	.235	71.799	1	.000	7.298
Y_AGE	.034	.009	15.592	1	.000	1.035

Variables not in the Equation[a]

	Score	df	Sig.
DRUG	.555	1	.456

[a]Residual Chi Square = .555 with 1 df Sig. = .456

Block2: Method = Enter

Omnibus Tests of Model Coefficients[a,b]

-2 Log Likelihood	Overall (score)			Change From Previous Step			Change From Previous Block		
	Chi-square	df	Sig.	Chi-square	df	Sig.	Chi-square	df	Sig.
1062.346	261.200	6	.000	.553	1	.457	.553	1	.457

[a]Beginning Block Number 0, initial Log Likelihood function: -2 Log likelihood: -1255.756
[b]Beginning Block Number 2. Method: Enter

TABLE 11.21 Continued

Variables in the Equation

	B	SE	Wald	df	Sig.	Exp(B)
ALBUMIN	−.894	.241	13.735	1	.000	.409
EDEMA	.742	.308	5.795	1	.016	2.100
PROTHOM	.306	.104	8.736	1	.003	1.358
LBILIRUB	1.994	.234	72.305	1	.000	7.342
Y_AGE	.036	.009	16.005	1	.000	1.036
DRUG	.139	.187	.555	1	.456	1.150

Covariate Means

	Mean
ALBUMIN	3.523
EDEMA	.108
PROTHOM	10.690
LBILIRUB	.249
Y_AGE	49.950
DRUG	1.489

The contribution of each covariate, adjusted for all others, is evaluated in the section labeled Variables in the Equation for the first block. If $\alpha = .01$ is used to adjust for inflated familywise error rate with five covariates, there are statistically significant differences due to age, serum albumin level, prothrombin time, and the logarithm of the level of serum bilirubin. (If $\alpha = .05$ is used, instead, edema is also statistically significant.) Because STATUS is coded 2 for death and 0 or 1 for survival, negative coefficients are associated with longer survival time. Thus, higher serum albumin predicts longer survival, but shorter survival is associated with greater age (no surprise), greater prothrombin time, and higher levels of the logarithm of serum bilirubin. An overall risk score for survival time is:

$$\text{Risk} = -.88 \text{ (albumin in g/dl)} + .31 \text{ (prothrombin time in sec.)} + 1.99 \log_{10} \text{ (bilirubin in mg/dl)} + .03 \text{ (age in years)}$$

Exp(B) is the hazard ratio for each covariate (cf. Section 11.6.5) where a negative sign for the associated B value implies an increase in survival and a positive sign implies an increase in the probability of death. For each one-point increase in serum albumin level, the risk of death decreases by about 60%: $(1 - 0.4132)100$. For each one-unit change in the edema measure, the risk of death more than doubles. For each one-second increase in prothrombin time, the risk of death increases by about 36%. For each one-point increase in the logarithm of serum bilirubin level, the risk of death increases more than seven times. Finally, for each year of age, the risk of death increases by 3.5% (odds ratio = 1.0347). Table 11.22 summarizes the results of the analysis of nondrug covariates.

TABLE 11.22 Cox Regression Analysis of Non-Drug Variables on Survival Time of PBC Patients

Covariate	B	df	Prob.	Odds Ratio
Serum albumin	−0.884	1	.0003	0.413
Edema	0.743	1	.0168	2.101
Prothrombin time	0.307	1	.0032	1.359
Logarithm (serum bilirubin)	1.988	1	.0000	7.298
Age in years	0.034	1	.0001	1.035

Figure 11.2 shows that the expected five-year survival rate of a patient at the mean of all covariates (see end of Table 11.21) is a bit under 80% (1,826.25 days). The ten-year survival rate is about 40%.

Table 11.23 is a checklist for predicting survival from covariates. An example of a Results section, in journal format, follows for the study just described.

```
COXREG
  DAYS /STATUS=STATUS(2)
  /METHOD=ENTER ALBUMIN EDEMA PROTHOM LBILIRUB Y_AGE
  /PLOT=SURVIVAL
  /CRITERIA=PIN(.05) POUT(.10) ITERATE(20).
```

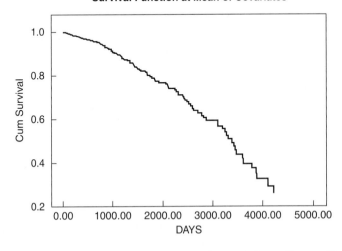

Survival Function at Mean of Covariates

FIGURE 11.2   Survival function at mean of five covariates: Serum albumin level, edema score, prothrombin time, logarithm of bilirubin level, and age in years.

TABLE 11.23   Checklist for Predicting Survival from Covariates, Including Treatment

1. Issues
   a. Adequacy of sample sizes and missing data
   b. Normality of distributions
   c. Absence of outliers
   d. Differences between withdrawn and remaining cases
   e. Changes in survival experience over time
   f. Proportionality of hazards
   g. Multicollinearity
2. Major analyses
   a. Test of treatment effect, if significant:
      (1) Treatment differences in survival
      (2) Parameter estimates, including odds ratios
      (3) Effect size and confidence limits
      (4) Survival function showing groups separately
   b. Effects of covariates, for significant ones:
      (1) Direction of effect(s)
      (2) Parameter estimates, including odds ratios
      (3) Effect size and confidence limits
3. Additional analyses
   a. Contingencies among covariates
   b. Survival function based on covariates alone

```
                        Results

     A Cox regression survival analysis was performed to assess the

effectiveness of the drug D-penicillamine for primary biliary cir-

rhosis in a random clinical trial after adjusting for the effects of

the five covariates found to be predictive of survival in the Mayo

clinic model: age, degree of edema (mild, moderate, or severe),

serum bilirubin in mg/dl, prothrombin time, and serum albumin in

gm/dl. A logarithmic transform reduced skewness and the influence of

outliers for bilirubin level. However, three multivariate outliers

remained. One case had an unusual combination of low scores on serum

albumin and logarithm of bilirubin with severe edema, the second had
```

an extremely high prothrombin time, and the third combined an extremely high prothrombin time with a low edema score. Three hundred nine cases remained after deletion of the three outliers, 186 censored either because they were alive at the end of the 10-year trial or had withdrawn from the trial for liver transplant. (Withdrawn cases differed from those who remained in the study only in that they were younger.) The cases were about evenly split between those who were given the drug and those given a placebo.

There was no statistically significant effect of drug treatment after adjusting for the five covariates, $G^2(1) = 0.553$, $p = .46$. Survival time, however, was fairly well predicted by the set of covariates, $R^2 = .46$ with a 95% confidence interval from .37 to .53 using Steiger and Fouladi's (1992) R2 software. All of the covariates except edema reliably predicted survival time at $\alpha = .01$: Risk = -.88 (albumin in g/dl) + .31 (prothrombin time in sec.) + 1.99 $\log_{10}$ (bilirubin in mg/dl) + .03 (age in years). Table 11.22 shows regression coefficients, degrees of freedom, p values, and hazard ratios for each covariate. The greatest contribution was by the logarithm of serum bilirubin level; each increase of one point increases the risk of death about seven times. Risk of death is increased by 3.5% with each year of age, and by about 36% with each one-point increase in prothrombin time. On the other hand, a one-point increase in serum albumin level decreases the risk of death by about 60%.

At the mean of the covariates, the five-year survival rate is just under 80% and the ten-year survival rate is about 40%, as seen in Figure 11.2.

Thus, survival time is predicted by several covariates but not by drug treatment. Increases in risk are associated with high serum bilirubin level, age, and prothrombin level but risk decreases with high serum albumin level.

11.8   Comparison of Programs

SPSS and SAS have two or more programs that do different types of analysis. SAS has one program for survival functions and two for regression-type problems: one for proportional-hazards models and the other for various non–proportional-hazards models. SPSS has three programs as well: one for proportional-hazards models and two for survival functions (one for actuarial and one for product-limit methods). SYSTAT has a single program for survival analysis. Table 11.24 compares programs for survival curves; Table 11.25 compares programs for prediction of survival from covariates.

11.8.1   SAS System

SAS has LIFETEST for life tables and survivor functions, and LIFEREG and PHREG for predicting survival from covariates. SAS LIFETEST offers both actuarial and product-limit methods for survivor functions; however, median survival for each group is only available for the product-limit method. LIFETEST is the only survivor function program that lets you specify the α level for survival confidence limits. A summary table is provided for each group.

SAS LIFEREG and PHREG are quite different programs. LIFEREG offers a variety of models; PHREG is limited to Cox proportional-hazards models. LIFEREG does direct analyses only; PHREG does direct, sequential, and stepwise modeling and is the only program reviewed that does best-subsets modeling. LIFEREG allows you to analyze discrete covariates with more than two levels, but PHREG does not. Instead, you need to dummy-code discrete variables. However, the Test procedure in PHREG allows a simultaneous test of a hypothesis about a set of regression coefficients, so you can do a test of the null hypothesis that all dummy-coded variables for a single covariate are zero.

The LIFEREG program permits separate analyses by groups, but no stratification variables. PHREG, on the other hand, allows you to specify a stratification variable which does the analysis without making the proportional-hazards assumption (cf. Section 11.6.1). PHREG also permits you to specify time-dependent covariates and has several options for dealing with tied data. PHREG provides the initial log-likelihood estimate, without any covariates, as well as score and Wald chi-square statistics for the full set of covariates in the model. PHREG provides odds ratios (risk ratios) and their standard errors for each covariate; LIFEREG provides neither. Both programs save predicted scores and their standard errors on request, but only PHREG also provides residuals, change in regression coefficients if a case is omitted from the analysis, and log(time) of the response. LIFEREG also shows a Type III analysis of effects, useful when categorical predictors have more than two levels.

11.8.2   SPSS Package

SPSS has an unusual group of programs for survival analysis. There are separate programs for actuarial (SURVIVAL, Life Tables in the Survival menu) and product-limit (KM, Kaplan-Meier in the Survival menu) methods for survivor functions, but only one program for predicting survival from covariates (COXREG, Cox Regression in the Survival menu). Other (nonproportional) modeling methods are not implemented within the SPSS package.

Both SURVIVAL and KM permit tests of group differences, as well as pairwise comparisons if there are more than two groups. KM also provides comparisons of groups when they are ordered. Only KM also allows testing of strata pooled over groups, with separate plots provided for each

TABLE 11.24   Comparison of Programs for Life Tables and Survivor Functions

Feature	SAS LIFETEST	SPSS SURVIVAL	SPSS KM	SYSTAT SURVIVAL
Input				
Actuarial method	Yes	Yes	No	Yes
Product-limit (Kaplan-Meier) method	Yes	No	Yes	Yes
Missing data options	Yes	Yes	No	No
Group comparisons	STRATA	COMPARE	COMPARE	STRATA
Ordered group comparisons	No	No	Yes	No
Pairwise comparisons among groups	No	Yes	Yes	No
Specify exact or approximate comparisons	N.A.	Yes	N.A.	N.A.
Test strata pooled over groups	No	No	STRATA	No
Use tabular data as input	No	Yes	No	No
Specify a frequency variable to indicate number of cases	Yes	No	No	No
Specify tolerance	SINGULAR	No	No	TOLERANCE
Specify α for survival confidence limits	Yes	No	No	No
Specify percentiles for combinations of groups and strata	No	No	Yes	No
Specify confidence bands for the survivor function	SURVIVAL	No	No	No
Specify interval-censoring	Yes	No	No	Yes
Specify left-censoring	Yes	No	No	No
Output				
Mantel-Cox log rank test	LOGRANK	No	Yes	KM only
Breslow test (Generalized Wilcoxon)	Yes	No	Yes	No
Peto-Prentice test (Generalized Wilcoxon)	PETO	No	No	No
Modified Peto-Prentice Test	MODPETO	No	No	No
Tarone-Ware test	TARONE	No	Yes	KM only
Gehan (Wilcoxon) – Log Rank	WILCOXON	Yes	No	KM only
Fleming-Harrington G^2 family of tests	FLEMING	No	No	No
Likelihood ratio test statistic	Yes	No	No	No
Number entering each interval	Yes	Yes	N.A.	KM only
Number lost (failed, dead, terminating) each interval	Yes	Yes	N.A.	KM only
Number censored each interval	Yes	Yes	No	No
Number remaining /effective sample size	Yes	No	Yes	No
Proportion failures/cond. probability of failure	Yes	Yes	No	No

TABLE 11.24   Continued

Feature	SAS LIFETEST	SPSS SURVIVAL	SPSS KM	SYSTAT SURVIVAL
Output *(continued)*				
Proportion surviving	No	Yes	No	KM only
Survival standard error	Yes	No	No	KM only
Cumulative proportion surviving	Yes	Yes	Yes	No
Standard error of cumulative proportion surviving	No	Yes	Yes	No
Cumulative proportion failure	Yes	No	No	No
Standard error of cumulative proportion failing	Yes	No	No	No
Cumulative events	No	N.A.	Yes	No
Hazard and standard error	Yes	Yes	No	ACT only
Density (PDF) and standard error	Yes	Yes	No	ACT only
Median survival, each group	KM only	Yes	Yes	No
Standard error of median survival, each group	No	No	Yes	No
Confidence interval for median survival	No	No	Yes	No
75th quantile surviving and standard error, each group	KM only	No	No	No
25th quantile surviving and standard error, each group	KM only	No	No	No
Other survival quantiles	No	No	No	KM only
Mean survival time	Yes	No	Yes	KM only
Standard error of mean survival time	KM only	No	Yes	No
Confidence interval for mean survival	No	No	Yes	No
Median Residual Lifetime, each interval	ACT only	No	No	No
Median standard error, each interval	ACT only	No	No	No
Summary table	Yes	No	Yes	No
Rank statistics and matrices for tests of groups	Yes	No	No	No
Plots				
Cumulative survival function	SURVIVAL	SURV	SURVIVAL	Yes (default)
Cumulative survival function on log scale	LOGSURV	LOGSURV	LOGSURV	TLOG
Cumulative survival function on a log-log scale	LOGLOGS	No	No	No
Cumulative hazard function	HAZARD	HAZARD	HAZARD	CHAZ
Cumulative hazard function on log scale	No	No	No	LHAZ
Cumulative density function	PDF	DENSITY	No	No

TABLE 11.25   Comparison of Programs for Prediction of Survival Time from Covariates

Feature	SAS LIFEREG	SAS PHREG	SPSS COXREG	SYSTAT SURVIVAL
Input				
Specify a frequency variable to indicate number of cases	No	Yes	No	Yes
Missing data options	No	No	Yes	No
Differential case weighting	Yes	Yes	No	No
Specify strata in addition to covariates	No	STRATA	STRATA	STRATA
Specify categorical covariates	Yes	No	Yes	No
Choice among contrasts for categorical covariates	No	No	Yes	No
Test linear hypotheses about regression coefficients	No	TEST	No	No
Specify time-dependent covariates	No	Yes	Yes	Yes
Specify interval-censoring	Yes	Yes	No	Yes
Specify left-censoring	Yes	Yes	No	No
Options for finding solution:				
Maximum number of iterations	MAXITER	MAXITER	ITERATE	MAXIT
One or more convergence criteria	CONVERGE	Several	LCON	CONVERGE
Change in parameter estimates	N.A.	Yes	BCON	No
Tolerance	SINGULAR	SINGULAR	No	TOLERANCE
Specify start values	Yes	No	No	Yes
Hold scale and shape parameters fixed	Yes	No	No	No
Direct analysis	Yes	Yes	ENTER	Yes
Sequential analysis	No	Yes	Yes	No
Best-subsets analysis	No	SCORE	No	No
Types of stepwise analyses:				
Forward stepping	N.A.	Yes	FSTEP	Yes
Backward stepping	N.A.	Yes	BSTEP	Yes
Interactive stepping	N.A.	Yes	No	Yes
Test statistics for removal in stepwise analysis				
Conditional statistic	N.A.	No	COND	No
Wald statistic	N.A.	No	WALD	No
Likelihood ratio	N.A.	STOPRES	LR	No

TABLE 11.25   Continued

Feature	SAS LIFEREG	SAS PHREG	SPSS COXREG	SYSTAT SURVIVAL
Input *(continued)*				
Criteria for stepwise analysis				
Maximum number of steps	N.A.	MAXSTEP	No	MAXSTEP
Probability of score statistic for entry	N.A.	SLENTRY	PIN	ENTER
Probability of statistic for removal	N.A.	SLSTAY	POUT	REMOVE
Force a number of covariates into model	N.A.	INCLUDE	No	FORCE
Specify pattern of covariate values for plots and tables	No	No	PATTERN	No
Types of models (distribution functions, cf. Table 11.9):				
Cox proportional hazards	No	Yes	Yes	Yes
Weibull	Yes	No	No	Yes
Nonaccelerated Weibull	No	No	No	Yes
Logistic	Yes	No	No	No
Log-logistic	Yes	No	No	Yes
Exponential	Yes	No	No	Yes
Nonaccelerated exponential	No	No	No	Yes
Normal	Yes	No	No	No
Log-normal	Yes	No	No	Yes
Gamma	Yes	No	No	Yes
Request no log transform of response	NOLOG	No	No	No
Request no intercept	NOINT	No	No	No
Specify survival analyses by groups	Yes	No	No	Yes
Special features to deal with tied data	No	Yes	No	No
Output				
Number of observations and number censored	Yes	Yes	Yes	Yes
Percent of events censored	No	Yes	Yes	No
Descriptive statistics for each covariate	No	Yes	No	No
Initial log-likelihood	No	Yes (-2)	Yes (-2)	No
Log-likelihood after each step	N.A.	Yes (-2)	Yes (-2)	Yes
Final log-likelihood	Yes	Yes (-2)	Yes (-2)	Yes
Overall (score) chi-square	No	Yes	Yes	No
Overall Wald chi-square	No	Yes	No	No

(continued)

TABLE 11.25 Continued

Feature	SAS LIFEREG	SAS PHREG	SPSS COXREG	SYSTAT SURVIVAL
Output *(continued)*				
Chi-square for change in likelihood from previous block	N.A.	No	Yes	No
Chi-square for change in likelihood from previous step	N.A.	No	Yes	No
Residual chi-square at each step	N.A.	Yes	Yes	No
For each covariate in the equation:				
Regression coefficient, *B*	Estimate	Parameter Estimate	B	Estimate
Standard error of regression coefficient	Std Err	Standard Error	S.E.	S.E.
Confidence interval for regression coefficient	No	No	No	Yes
Wald statistic (or $B/S.E.$ or χ^2) with df and significance level	Chi Square	Chi-Square	Wald	t-ratio
Odds ratio, e^b	No	Risk Ratio	Exp(B)	No
Confidence interval for odds ratio	No	Yes	Yes	No
Type III SS analysis (combining multiple df effects)	Yes	No	No	No
Summary table for stepwise results	N.A.	Yes	No	No
Covariate means	No	No	Yes	Yes
For covariates not in the equation				
Score statistic with df and significance level	N.A.	Yes	Score	No
Estimate of partial correlation with response variable	N.A.	No	R	No
t-ratio (or chi-square to enter) and significance	N.A.	No	No	Yes
Chi-square (with df and significance level) for model if last entered term removed (or chi-square to remove)	N.A.	No	Loss Chi-Square	No
Correlation/covariance matrix of parameter estimates	Yes	Yes	Yes	Yes
Baseline cumulative hazard table for each stratum	No	No	Yes	No
Survival function(s)	No	No	No	Yes
Printed residuals	No	No	No	No
Print iteration history	ITPRINT	ITPRINT	No	Yes

TABLE 11.25   Continued

Feature	SAS LIFEREG	SAS PHREG	SPSS COXREG	SYSTAT SURVIVAL
Plots				
Cumulative survival distribution	No	No	SURVIVAL	Yes (default)
Cumulative survival function on a log scale	No	No	No	TLOG
Cumulative hazard function	No	No	HAZARD	CHAZ
Cumulative hazard function on a log scale	No	No	No	LHAZ
Log-minus-log-of-survival function	No	No	LML	No
Saved on request				
Coefficients from final model	No	No	Yes	No
Survival table	No	No	Yes	No
For each case:				
Survival function	Yes	Yes	Yes	No
Change in coefficient for each covariate if current case is removed	No	DFBETA	Yes	No
Residuals and/or partial residuals for each covariate	No	Yes	Yes	No
Estimates of linear predictors	XBETA	XBETA	XBETA	No
Standard errors of estimated linear predictors	STD	STDXBETA	No	No
Linear combination of mean-corrected covariate times regression coefficients	No	No	Yes	No
Case weight	No	No	No	Yes
Time or log(time) of response	No	Yes	No	LOWER
Quantile estimates and standard errors	QUANTILE	No	No	No

stratum. SURVIVAL, but not KM, can use tabular data as input. KM provides median and mean survival times with standard errors and confidence intervals; SURVIVAL only provides median survival time. Both provide a variety of plots.

COXREG permits specification of strata as well as discrete covariates and is the only program reviewed that provides a choice among contrasts for discrete covariates. Direct, sequential, and stepwise analyses are available. Model parameters and a survival table can be saved to a file, and an additional file can be requested for residuals, predicted scores, and other statistics.

11.8.3   SYSTAT System

SYSTAT has a single program, SURVIVAL, for all types of survival analyses, including life tables and survivor functions as well as proportional- and non–proportional-hazards models for predicting survival from covariates. Group differences in survivor functions can be specified, however, they are only tested if the product-limit method is chosen. The program also does not allow much flexibility in defining intervals for survivor functions based on the actuarial method. Mean survival times and their standard errors are provided for the product-limit method only.[2]

Prediction of survival from covariates can be done using the widest variety of possible distribution functions of any single program reviewed here, and time-dependent covariates can be specified. Direct and stepwise analyses are available, and this is the only program reviewed that implements interactive stepping. Covariate means are provided, but odds ratios for each covariate and their confidence intervals are not. (Confidence intervals are given for regression coefficients, however.) The combining of life tables and prediction functions into a single program provides you with a variety of survivor plots in a modeling run. Information saved to file is rather sparse and does not include predicted scores or residuals.

---

[2]Is there perhaps a bias toward the product-limit method here?

CHAPTER 12

Canonical Correlation

12.1 General Purpose and Description

The goal of canonical correlation is to analyze the relationships between two sets of variables. It may be useful to think of one set of variables as IVs and the other set as DVs, or it may not. In any event, canonical correlation provides a statistical analysis for research in which each subject is measured on two sets of variables and the researcher wants to know if and how the two sets relate to each other.

Suppose, for instance, a researcher is interested in the relationship between a set of variables measuring medical compliance (willingness to buy drugs, to make return office visits, to use drugs, to restrict activity) and a set of demographic characteristics (educational level, religious affiliation, income, medical insurance). Canonical analysis might reveal that there are two statistically significant ways that the two sets of variables are related. The first way is between income and insurance on the demographic side and purchase of drugs and willingness to make return office visits on the medical-compliance side. Together, these results indicate a relationship between compliance and demography based on ability to pay for medical services. The second way is between willingness to use drugs and restrict activity on the compliance side and religious affiliation and educational level on the demographic side, interpreted, perhaps, as a tendency to accede to authority (or not).

The easiest way to understand canonical correlation is to think of multiple regression. In regression, there are several variables on one side of the equation and a single variable on the other side. The several variables are combined into a predicted value to produce, across all subjects, the highest correlation between the predicted value and the single variable. The combination of variables can be thought of as a dimension among the many variables that predicts the single variable.

In canonical correlation, the same thing happens except that there are several variables on both sides of the equation. Sets of variables on each side are combined to produce, for each side, a predicted value that has the highest correlation with the predicted value on the other side. The combination of variables on each side can be thought of as a dimension that relates the variables on one side to the variables on the other.

There is a complication, however. In multiple regression, there is only one combination of variables because there is only a single variable to predict on the other side of the equation. In canonical correlation, there are several variables on both sides and there may be several ways to recombine the variables on both sides to relate them to each other. In the example, the first way of combining the variables had to do with economic issues and the second way had to do with authority. Although there are potentially as many ways to recombine the variables as there are variables in the smaller set, usually only the first two or three combinations are statistically significant and need to be interpreted.

A good deal of the difficulty with canonical correlation is due to jargon. First, there are variables, then there are canonical variates, and, finally, there are pairs of canonical variates. Variables refers to the variables measured in research (e.g., income). Canonical variates are linear combinations of variables, one combination on the IV side (e.g., income and medical insurance) and a second combination on the DV side (e.g., purchase of drugs and willingness to make return office visits). These two combinations form a pair of canonical variates. However, there may be more than one significant pair of canonical variates (e.g., a pair associated with economics and a pair associated with authority).

Canonical analysis is one of the most general of the multivariate techniques. In fact, many other procedures—multiple regression, discriminant analysis, MANOVA—are special cases of it. But it is also the least used and most impoverished of the techniques, for reasons that are discussed in what follows.

Although not the most popular of multivariate techniques, examples of canonical correlation are found across disciplines. Mann (2004) examined the relationship between variables associated with college student adjustment and certain personality variables. Variables on one side of the equation were academic adjustment, social adjustment, personal-emotional adjustment, and institutional attachment; variables on the other side were shame-proneness, narcissistic injury, self-oriented perfectionism, other-oriented perfectionism, and socially prescribed perfectionism. A single canonical composite was found, accounting for 33% of overlapping variance. Those low in institutional attachment were high in narcissistic injury, low in self-oriented perfectionism, and high in other-oriented and social prescribed perfectionism.

Gebers and Peck (2003) examined the relationship between traffic citations and accidents in a subsequent 3-year period from those variables plus a variety of demographic variables in the prior 3-year period. Two canonical variable pairs were identified that, taken together, predicted subsequent traffic incidents (accidents and citations) better than prediction afforded by prior citations alone. Increasing traffic incidents were associated with more prior citations, more prior accidents, young age, and male gender. A cross-validation sample confirmed the efficacy of the equations.

12.2   Kinds of Research Questions

Although a large number of research questions are answered by canonical analysis in one of its specialized forms (such as discriminant analysis), relatively few intricate research questions are readily answered through direct application of computer programs currently available for canonical correlation. In part, this has to do with the programs themselves, and in part, it has to do with the kinds of questions researchers consider appropriate in a canonical correlation.

In its present stage of development, canonical correlation is best considered a descriptive technique or a screening procedure rather than a hypothesis-testing procedure. The following sections, however, contain questions that can be addressed with the aid of SPSS and SAS programs.

12.2.1   Number of Canonical Variate Pairs

How many significant canonical variate pairs are there in the data set? Along how many dimensions are the variables in one set related to the variables in the other? In the example: Is the pair associated with economic issues significant? And if so, Is the pair associated with authority also significant? Because canonical variate pairs are computed in descending order of magnitude, the first one or two pairs are often significant and remaining ones are not. Significance tests for canonical variate pairs are described in Sections 12.4 and 12.5.1.

12.2.2   Interpretation of Canonical Variates

How are the dimensions that relate two sets of variables to be interpreted? What is the meaning in the combination of variables that compose one variate in conjunction with the combination composing the other in the same pair? In the example, all the variables that are important to the first pair of canonical variates have to do with money, so the combination is interpreted as an economic dimension. Interpretation of pairs of canonical variates usually proceeds from matrices of correlations between variables and canonical variates, as described in Sections 12.4 and 12.5.2.

12.2.3   Importance of Canonical Variates

There are several ways to assess the importance of canonical variates. The first is to ask how strongly the variate on one side of the equation relates to the variate on the other side of the equation; that is, how strong is the correlation between variates in a pair? The second is to ask how strongly the variate on one side of the equation relates to the variables on its own side of the equation. The third is to ask how strongly the variate on one side of the equation relates to the variables on the other side of the equation.

For the example, what is the correlation between the economic variate on the compliance side and the economic variate on the demographic side? Then, how much variance does the economic variate on the demographic side extract from the demographic variables? Finally, how much variance does the economic variate on the demographic side extract from the compliance variables? These questions are answered by the procedures described in Sections 12.4 and 12.5.1. Confidence limits for canonical correlations are not readily available.

12.2.4   Canonical Variate Scores

Had it been possible to measure directly the canonical variates from both sets of variables, what scores would subjects have received on them? For instance, if directly measurable, what scores would the first subject have received on the economic variate from the compliance side and the economic variate from the demographic side? Examination of canonical variate scores reveals deviant cases, the shape of the relationship between two canonical variates, and the shape of the relationships between canonical variates and the original variables, as discussed briefly in Sections 12.3 and 12.4.

If canonical variates are interpretable, scores on them might be useful as IVs or DVs in other analyses. For instance, the researcher might use scores on the economic variate from the compliance side to examine the effects of publicly supported medical facilities. Canonical scores also may be useful for comparing canonical correlations in a manner that is generalized from the comparison of two sets of predictors (Section 5.6.2.5). Steiger (1980) and Steiger and Browne (1984) provide the basic rationale and examples of various procedures for comparing correlations.

12.3   Limitations

12.3.1   Theoretical Limitations[1]

Canonical correlation has several important theoretical limitations that help explain its scarcity in the literature. Perhaps the most critical limitation is interpretability; procedures that maximize correlation

---

[1]The authors are indebted to James Fleming for many of the insights of this section.

do not necessarily maximize interpretation of pairs of canonical variates. Therefore, canonical solutions are often mathematically elegant but uninterpretable. And, although it is common practice in factor analysis and principal components analysis (Chapter 13) to rotate a solution to improve interpretation, rotation of canonical variates is not common practice or even available in some computer programs.

The algorithm used for canonical correlation maximizes the linear relationship between two sets of variables. If the relationship is nonlinear, the analysis misses some or most of it. If a nonlinear relationship between dimensions in a pair is suspected, use of canonical correlation may be inappropriate unless variables are transformed or combined to capture the nonlinear component.

The algorithm also computes pairs of canonical variates that are independent of all other pairs. In factor analysis (Chapter 13), one has a choice between an orthogonal (uncorrelated) and an oblique (correlated) solution, but in canonical analysis, only the orthogonal solution is routinely available. In the example, if there was a possible relationship between economic issues and authority, canonical correlation might be inappropriate.

An important concern is the sensitivity of the solution in one set of variables to the variables included *in the other set*. In canonical analysis, the solution depends both on correlations among variables in each set and on correlations among variables between sets. Changing the variables in one set may markedly alter the composition of canonical variates in the other set. To some extent, this is expected given the goals of analysis, yet the sensitivity of the procedure to apparently minor changes is a cause for concern.

It is especially important in canonical analysis to emphasize that the use of terms IV and DV does not imply a causal relationship. Both sets of measures are manipulated by nature rather than by an experimenter's design, and there is nothing in the statistics that changes that arrangement—this is truly a correlational technique. Canonical analysis conceivably could be used when one set of measures is indeed experimentally manipulated, but it is difficult to imagine that MANOVA could not do a better, more interpretable job with such data.

Much of the benefit in studying canonical correlation analysis is in its introduction to the notion of dimensionality, and in providing a broad framework with which to understand other techniques in which there are multiple variables on both sides of a linear equation.

12.3.2 Practical Issues

12.3.2.1 Ratio of Cases to IVs

The number of cases needed for analysis depends on the reliability of the variables. For variables in the social sciences where reliability is often around .80, about 10 cases are needed for every variable. However, if reliability is very high, as, for instance, in political science where the variables are measures of the economic performance of countries, then a much lower ratio of cases to variables is acceptable.

Power considerations are as important in canonical correlation as in other techniques, but software is less likely to be available to provide aid in determining sample sizes for expected effect sizes and desired power.

12.3.2.2 Normality, Linearity, and Homoscedasticity

Although there is no requirement that the variables be normally distributed when canonical correlation is used descriptively, the analysis is enhanced if they are. However, inference regarding number of significant canonical variate pairs proceeds on the assumption of multivariate normality. Multivariate normality is the assumption that all variables and all linear combinations of variables are normally distributed. It is not itself an easily testable hypothesis (most tests available are too strict), but the likelihood of multivariate normality is increased if the variables are all normally distributed.

Linearity is important to canonical analysis in at least two ways. The first is that the analysis is performed on correlation or variance-covariance matrices that reflect only linear relationships. If the relationship between two variables is nonlinear, it is not "captured" by these statistics. The second is that canonical correlation maximizes the linear relationship between a variate from one set of variables and a variate from the other set. Canonical analysis misses potential nonlinear components of relationships between canonical variate pairs.

Finally, canonical analysis is best when relationships among pairs of variables are homoscedastic, that is, when the variance of one variable is about the same at all levels of the other variable.

Normality, linearity, and homoscedasticity can be assessed through normal screening procedures or through the distributions of canonical variate scores produced by a preliminary canonical analysis. If routine screening is undertaken, variables are examined individually for normality through one of the descriptive programs such as SPSS FREQUENCIES or SAS Interactive Data Analysis. Pairs of variables, both within sets and across sets, are examined for nonlinearity or heteroscedasticity through programs such as SAS PLOT or SPSS GRAPH. If one or more of the variables is in violation of the assumptions, transformation is considered, as discussed in Chapter 4 and illustrated in Section 12.6.1.2.

Alternatively, distributions of canonical variate scores produced by a preliminary canonical analysis are examined for normality, linearity, and homoscedasticity, and, if found, screening of the original variables is not necessary. Scatterplots, where pairs of canonical variates are plotted against each other, are available through SAS CANCORR if canonical variates scores are written to a file for processing through a scatterplot program. The SPSS CANCORR macro automatically adds canonical variate scores to the original data set. If, in the scatterplots, there is evidence of failure of normality, linearity, and/or homoscedasticity, screening of the variables is undertaken. This procedure is illustrated in Section 12.6.1.2.

In the event of persistent heteroscedasticity, you might consider weighting cases based on variables producing unequal variance or adding a variable that accounts for unequal variance (cf. Section 5.3.2.4).

12.3.2.3 Missing Data

Levine (1977) gives an example of a dramatic change in a canonical solution with a change in procedures for handling missing data. Because canonical correlation is quite sensitive to minor changes in a data set, consider carefully the methods of Chapter 4 for estimating values or eliminating cases with missing data.

12.3.2.4 Absence of Outliers

Cases that are unusual often have undue impact on canonical analysis. The search for univariate and multivariate outliers is conducted separately within each set of variables. Consult Chapter 4 and Section 12.6.1.3 for methods of detecting and reducing the effects of both univariate and multivariate outliers.

12.3.2.5 Absence of Multicollinearity and Singularity

For both logical and computational reasons, it is important that the variables in each set and across sets are not too highly correlated with each other. This restriction applies to values in $\mathbf{R}_{xx}$, $\mathbf{R}_{yy}$, and

$\mathbf{R}_{xy}$ (see Equation 12.1). Consult Chapter 4 for methods of identifying and eliminating multi-collinearity and singularity in correlation matrices.

12.4   Fundamental Equations for Canonical Correlation

A data set that is appropriately analyzed through canonical correlation has several subjects, each measured on four or more variables. The variables form two sets with at least two variables in the smaller set. A hypothetical data set, appropriate for canonical correlation, is presented in Table 12.1. Eight intermediate- and advanced-level belly dancers are rated on two sets of variables, the quality of their "top" shimmies (TS), "top" circles (TC), and the quality of their "bottom" shimmies (BS), and "bottom" circles (BC). Each characteristic of the dance is rated by two judges on a 7-point scale (with larger numbers indicating higher quality) and the ratings averaged. The goal of analysis is to discover patterns, if any, between the quality of the movements on top and the quality of movements on bottom.

You are cordially invited to follow (or dance along with) this example by hand and by computer. Examples of syntax and output for this analysis using several popular computer programs appear at the end of this section.

The first step in a canonical analysis is generation of a correlation matrix. In this case, however, the correlation matrix is subdivided into four parts: the correlations between the DVs ($\mathbf{R}_{yy}$), the correlations between the IVs ($\mathbf{R}_{xx}$), and the two matrices of correlations between DVs and IVs ($\mathbf{R}_{xy}$ and $\mathbf{R}_{yx}$).[2] Table 12.2 contains the correlation matrices for the data in the example.

There are several ways to write the fundamental equation for canonical correlation, some more intuitively appealing than others. The equations are all variants on the following equation:

$$\mathbf{R} = \mathbf{R}_{yy}^{-1}\mathbf{R}_{yx}\mathbf{R}_{xx}^{-1}\mathbf{R}_{xy} \tag{12.1}$$

The canonical correlation matrix is a product of four correlation matrices, between DVs (inverted), between IVs (inverted), and between DVs and IVs.

TABLE 12.1   Small Sample of Hypothetical Data for Illustration of Canonical Correlation Analysis

ID	TS	TC	BS	BC
1	1.0	1.0	1.0	1.0
2	7.0	1.0	7.0	1.0
3	4.6	5.6	7.0	7.0
4	1.0	6.6	1.0	5.9
5	7.0	4.9	7.0	2.9
6	7.0	7.0	6.4	3.8
7	7.0	1.0	7.0	1.0
8	7.0	1.0	2.4	1.0

[2]Although in this example the sets of variables are neither IVs nor DVs, it is useful to use the terms when explaining the procedure.

TABLE 12.2   Correlation Matrices for the Data Set in Table 12.1

	$\mathbf{R}_{xx}$		$\mathbf{R}_{xy}$	
	$\mathbf{R}_{yx}$		$\mathbf{R}_{yy}$	
	TS	**TC**	**BS**	**BC**
TS	1.000	−.161	.758	−.341
TC	−.161	1.000	.110	.857
BS	.758	.110	1.000	.051
BC	−.341	.857	.051	1.000

It is helpful conceptually to compare Equation 12.1 with Equation 5.6 for regression. Equation 5.6 indicates that regression coefficients for predicting Y from a set of Xs are a product of (the inverse of) the matrix of correlations among the Xs and the matrix of correlations between the Xs and Y. Equation 12.1 can be thought of as a product of regression coefficients for predicting Xs from Ys $(\mathbf{R}_{yy}^{-1}\mathbf{R}_{yx})$ and regression coefficients for predicting Ys from Xs $(\mathbf{R}_{xx}^{-1}\mathbf{R}_{xy})$.

12.4.1   Eigenvalues and Eigenvectors

Canonical analysis proceeds by solving for the eigenvalues and eigenvectors of the matrix $\mathbf{R}$ of Equation 12.1. Eigenvalues are obtained by analyzing the matrix in Equation 12.1. Eigenvectors are obtained for the Y variables first and then calculated for the Xs using Equation 12.6 in Section 12.4.2. As discussed in Chapter 13 and in Appendix A, solving for the eigenvalues of a matrix is a process that redistributes the variance in the matrix, consolidating it into a few composite variates rather than many individual variables. The eigenvector that corresponds to each eigenvalue is transformed into the coefficients (e.g., regression coefficients, canonical coefficients) used to combine the original variables into the composite variate.

Calculation of eigenvalues and corresponding eigenvectors is demonstrated in Appendix A but is difficult and not particularly enlightening. For this example, the task is accomplished with assistance from SAS CANCORR (see Table 12.4 where the eigenvalues are called canonical correlations). The goal is to redistribute the variance in the original variables into a very few pairs of canonical variates, each pair capturing a large share of variance and defined by linear combinations of IVs on one side and DVs on the other. Linear combinations are chosen to maximize the canonical correlation for each pair of canonical variates.

Although computing eigenvalues and eigenvectors is best left to the computer, the relationship between a canonical correlation and an eigenvalue[3] is simple, namely,

$$\lambda_i = r_{ci}^2 \tag{12.2}$$

[3]SPSS and SAS use the terms Sq. Cor and Squared Canonical Correlation, respectively, in place of eigenvalue and use the term *eigenvalue* in a different way.

Each eigenvalue, λ_i, is equal to the squared canonical correlation, r_{ci}^2, for the pair of canonical variates.

Once the eigenvalue is calculated for each pair of canonical variates, canonical correlation is found by taking the square root of the eigenvalue. Canonical correlation, r_{ci}, is interpreted as an ordinary Pearson product-moment correlation coefficient. When r_{ci}, is squared, it represents, as usual, overlapping variance between two variables, or, in this case, variates. Because $r_{ci}^2 = \lambda_i$, the eigenvalues themselves represent overlapping variance between pairs of canonical variates.

For the data set of Table 12.1, two eigenvalues are calculated, one for each variable in the smaller set (both sets in this case). The first eigenvalue is .83566, which corresponds to a canonical correlation of .91414. The second eigenvalue is .58137, so canonical correlation is .76247. That is, the first pair of canonical variates correlate .91414 and overlap 83.57% in variance, and the second pair correlate .76247 and overlap 58.14% in variance.

Note, however, that the variance in the original variables accounted for by the solution cannot exceed 100%. Rather, the squared canonical correlation from the second pair of canonical variates is the proportion of variance extracted from the residual after the first pair has been extracted.

Significance tests (Bartlett, 1941) are available to test whether one or a set of r_cs differs from zero.[4]

$$\chi^2 = -\left[N - 1 - \left(\frac{k_x + k_y + 1}{2} \right) \right] \ln \Lambda_m \qquad (12.3)$$

The significance of one or more canonical correlations is evaluated as a chi-square variable, where N is the number of cases, k_x is the number of variables in the IV set, k_y is the number in the DV set, and the natural logarithm of lambda, Λ, is defined in Equation 12.4. This chi square has $(k_x)(k_y)$ df.

$$\Lambda_m = \prod_{i=1}^{m} (1 - \lambda_i) \qquad (12.4)$$

Lambda, Λ, is the product of differences between eigenvalues and unity, generated across m canonical correlations.

For the example, to test if the canonical correlations as a set differ from zero:

$$\Lambda_2 = (1 - \lambda_1)(1 - \lambda_2) = (1 - .84)(1 - .58) = .07$$

$$\chi^2 = -\left[8 - 1 - \left(\frac{2 + 2 + 1}{2} \right) \right] \ln .07$$

$$= -(4.5)(-2.68)$$

$$= 12.04$$

---

[4]Some researchers (e.g., Harris, 2001) prefer a strategy that concentrates only on the first eigenvalue. See Section 7.5.2 for a discussion of this issue.

This χ^2 is evaluated with $(k_x)(k_y) = 4$ df. The two canonical correlations differ from zero: $\chi^2(4) = 12.04, p < .02$. The results of this test are interpreted to mean that there is significant overlap in variability between the variables in the IV set and the variables in the DV set, that is, that there is some relationship between quality of top movements and of bottom movements. This result is often taken as evidence that at least the first canonical correlation is significant.

With the first canonical correlate removed, is there still a significant relationship between the two sets of variables?

$$\Lambda_1 = (1 - .58) = .42$$

$$\chi^2 = -\left[8 - 1 - \left(\frac{2+2+1}{2}\right)\right]\ln .42$$

$$= -(4.5)(-.87)$$

$$= 3.92$$

This chi square has $(k_x - 1)(k_y - 1) = 1$ df and also differs significantly from zero: $\chi^2(1) = 3.92$, $p < .05$. This result indicates that there is still significant overlap between the two sets of variables after the first pair of canonical variates is removed. It is taken as evidence that the second canonical correlation is also significant.

Significance of canonical correlations is also evaluated using the F distribution as, for example, in SAS CANCORR and SPSS MANOVA.

12.4.2   Matrix Equations

Two sets of canonical coefficients (analogous to regression coefficients) are required for each canonical correlation, one set to combine the DVs and the other to combine the IVs. The canonical coefficients for the DVs are found as follows:

$$\mathbf{B}_y = (\mathbf{R}_{yy}^{-1/2})'\hat{\mathbf{B}}_y \tag{12.5}$$

Canonical coefficients for the DVs are a product of (the transpose of the inverse of the square root of) the matrix of correlations between DVs and the normalized matrix of eigenvectors, $\hat{\mathbf{B}}_y$, for the DVs.

For the example:[5]

$$\mathbf{B}_y = \begin{bmatrix} 1.00 & -0.03 \\ -0.03 & 1.00 \end{bmatrix}\begin{bmatrix} -0.45 & 0.89 \\ 0.89 & 0.47 \end{bmatrix} = \begin{bmatrix} -0.48 & 0.88 \\ 0.90 & 0.44 \end{bmatrix}$$

Once the canonical coefficients are computed, coefficients for the IVs are found using the following equation:

$$\mathbf{B}_x = \mathbf{R}_{xx}^{-1}\mathbf{R}_{xy}\mathbf{B}_y^* \tag{12.6}$$

[5]These calculations, like others in this section, were carried to several decimal places and then rounded back. The results agree with computer analyses of the same data but the rounded-off figures presented here do not always check out to both decimals.

Coefficients for the IVs are a product of (the inverse of the square root of) the matrix of correlations between the IVs, the matrix of correlations between the IVs and DVs, and the matrix formed by the coefficients for the DVs, each divided by their corresponding canonical correlations.

For the example:

$$\mathbf{B}_x = \begin{bmatrix} 1.03 & 0.17 \\ 0.17 & 1.03 \end{bmatrix} \begin{bmatrix} 0.76 & -0.34 \\ 0.11 & 0.86 \end{bmatrix} \begin{bmatrix} -0.48/.91 & 0.89/.76 \\ 0.90/.91 & 0.44/.76 \end{bmatrix} = \begin{bmatrix} -0.63 & 0.80 \\ 0.69 & 0.75 \end{bmatrix}$$

The two matrices of canonical coefficients are used to estimate scores on canonical variates:

$$\mathbf{X} = \mathbf{Z}_x \mathbf{B}_x \tag{12.7}$$

and

$$\mathbf{Y} = \mathbf{Z}_y \mathbf{B}_y \tag{12.8}$$

Scores on canonical variates are estimated as the product of the standardized scores on the original variates, $\mathbf{Z}_x$ and $\mathbf{Z}_y$, and the canonical coefficients used to weight them, $\mathbf{B}_x$ and $\mathbf{B}_y$.

For the example:

$$\mathbf{X} = \begin{bmatrix} -1.54 & -0.91 \\ 0.66 & -0.91 \\ -0.22 & 0.76 \\ -1.54 & 1.12 \\ 0.66 & 0.50 \\ 0.66 & 1.26 \\ 0.66 & -0.91 \\ 0.66 & -0.91 \end{bmatrix} \begin{bmatrix} -0.63 & 0.80 \\ 0.69 & 0.75 \end{bmatrix} = \begin{bmatrix} 0.34 & -1.91 \\ -1.04 & -0.15 \\ 0.66 & 0.39 \\ 1.73 & -0.40 \\ -0.07 & 0.90 \\ 0.45 & 1.47 \\ -1.04 & -0.15 \\ -1.04 & -0.15 \end{bmatrix}$$

$$\mathbf{Y} = \begin{bmatrix} -1.36 & -0.81 \\ 0.76 & -0.81 \\ 0.76 & 1.67 \\ -1.36 & 1.22 \\ 0.76 & -0.02 \\ 0.55 & 0.35 \\ 0.76 & -0.81 \\ -0.86 & -0.81 \end{bmatrix} \begin{bmatrix} -0.48 & 0.88 \\ 0.90 & 0.44 \end{bmatrix} = \begin{bmatrix} -0.07 & -1.54 \\ -1.09 & 0.31 \\ 1.14 & 1.39 \\ 1.75 & -0.66 \\ -0.38 & 0.66 \\ 0.05 & 0.63 \\ -1.09 & 0.31 \\ -0.31 & -1.11 \end{bmatrix}$$

The first belly dancer, in standardized scores (and appropriate costume), has a z-score of -1.35 on TS, -0.81 on TC, -1.54 on BS, and -0.91 on BC. When these z-scores are weighted by canonical coefficients, this dancer is estimated to have a score of 0.34 on the first canonical variate and a score of -1.91 on the second canonical variate for the IVs (the Xs), and scores of -0.07 and -1.54 on the first and second canonical variates, respectively, for the DVs (the Ys).

The sum of canonical scores for all belly dancers on each canonical variate is zero, within rounding error. These scores, like factor scores (Chapter 13), are estimates of scores the dancers would receive if they were judged directly on the canonical variates.

Matrices of correlations between the variables and the canonical coefficients, called *loading matrices,* are used to interpret the canonical variates.

$$A_x = R_{xx} B_x \tag{12.9}$$

and

$$A_y = R_{yy} B_y \tag{12.10}$$

Correlations between variables and canonical variates are found by multiplying the matrix of correlations between variables by the matrix of canonical coefficients.

For the example:

$$A_x = \begin{bmatrix} 1.00 & -0.16 \\ -0.16 & 1.00 \end{bmatrix} \begin{bmatrix} -0.63 & 0.80 \\ 0.69 & 0.75 \end{bmatrix} = \begin{bmatrix} -0.74 & 0.68 \\ 0.79 & 0.62 \end{bmatrix}$$

$$A_y = \begin{bmatrix} 1.00 & 0.05 \\ 0.05 & 1.00 \end{bmatrix} \begin{bmatrix} -0.48 & 0.88 \\ 0.90 & 0.44 \end{bmatrix} = \begin{bmatrix} -0.44 & 0.90 \\ 0.88 & 0.48 \end{bmatrix}$$

The loading matrices for these data are summarized in Table 12.3. Results are interpreted down columns across sets of variables. For the first canonical variate pair (the first column), TS correlates $-.74$, TC .79, BS $-.44$, and BC .88. The first pair of canonical variates links low scores on TS and high scores on TC (in the first set of variables) with high scores on BC (in the second set), indicating that poor-quality top shimmies and high-quality top circles are associated with high-quality bottom circles.

For the second canonical variate pair (the second column), TS correlates .68, TC .62, BS .90, and BC .48; the second canonical variate pair indicates that high scores on bottom shimmies are associated

TABLE 12.3   Loading Matrix for the Data Set in Table 12.1

	Variable Sets	Canonical Variate Pairs	
		First	Second
First	TS	−0.74	0.68
	TC	0.79	0.62
Second	BS	−0.44	0.90
	BC	0.88	0.48

with high scores on both top circles and top shimmies. Taken together, these results suggest that ability to do bottom circles is related to ability to do top circles but inability to do top shimmies, whereas ability to do bottom shimmies is associated with ability to do both top movements well.

Figure 12.1 shows, in general, the relationships among variables, canonical variates, and the first pair of canonical variates.

Figure 12.2 shows the path diagrams for the two pairs of canonical variates in the small-sample example.

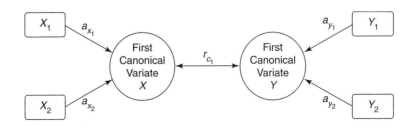

X_i = Variable in X set
Y_i = Variable in Y set
a_{x_i} = Loading of (correlation with) ith X variable on canonical variate X
a_{y_i} = Loading of (correlation with) ith Y variable on canonical variate Y
r_{c_1} = Canonical correlation for the first pair of canonical variates

FIGURE 12.1   Relationships among variables, canonical variates, and the first pair of canonical variates.

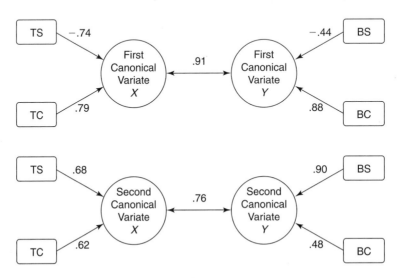

FIGURE 12.2   Loadings and canonical correlations for both canonical variate pairs for the data in Table 12.1.

12.4.3   Proportions of Variance Extracted

How much variance does each of the canonical variates extract from the variables on its own side of the equation? The proportion of variance extracted from the IVs by the canonical variates of the IVs is

$$pv_{xc} = \sum_{i=1}^{k_x} \frac{a_{ixc}^2}{k_x} \tag{12.11}$$

and

$$pv_{yc} = \sum_{i=1}^{k_y} \frac{a_{iyc}^2}{k_y} \tag{12.12}$$

The proportion of variance extracted from a set of variables by a canonical variate of the set is the sum of the squared correlations divided by the number of variables in the set.

Thus, for the first canonical variate in the set of IVs,

$$pv_{x_1} = \frac{(-0.74)^2 + 0.79^2}{2} = .58$$

and for the second canonical variate of the IVs,

$$pv_{x_2} = \frac{0.68^2 + 0.62^2}{2} = .42$$

The first canonical variate extracts 58% of the variance in judgments of top movements, whereas the second canonical variate extracts 42% of the variance in judgments of top movements. In summing for the two variates, almost 100% of the variance in the IVs is extracted by the two canonical variates. As expected in summing the two variables, 100% of the variance in IVs is extracted by the two canonical variates. This happens when the number of variables on one side of the equation is equal to the number of canonical variates. The sum of the pv scores usually is less than 1.00 if there are more variables than canonical variates.

For the DVs and the first canonical variate,

$$pv_{y_1} = \frac{(-0.44)^2 + 0.88^2}{2} = .48$$

and for the second canonical variate,

$$pv_{y_2} = \frac{0.90^2 + 0.48^2}{2} = .52$$

That is, the first canonical variate extracts 48% of the variance in judgments of bottom movements, and the second canonical variate extracts 52% of variance in judgments of bottom movements. Together, the two canonical variates extract almost 100% of the variance in the DVs.

Often, however, one is interested in knowing how much variance the canonical variates from the IVs extract from the DVs, and vice versa. In canonical analysis, this variance is called *redundancy*.

$$rd = (pv)(r_c^2)$$

The redundancy in a canonical variate is the percentage of variance it extracts from its own set of variables times the squared canonical correlation for the pair.

Thus, for the example:

$$rd_{x_1 \to y} = \left[\frac{(-0.44)^2 + 0.88^2}{2} \right](.84) = 0.40$$

$$rd_{x_2 \to y} = \left[\frac{0.90^2 + 0.48^2}{2} \right](.58) = .30$$

$$rd_{y_1 \to x} = \left[\frac{(-0.74)^2 + 0.79^2}{2} \right](.84) = .48$$

and

$$rd_{y_2 \to x} = \left[\frac{0.68^2 + 0.62^2}{2} \right](0.58) = 0.24$$

So, the first canonical variate from the IVs extracts 40% of the variance in judgments of quality of bottom movements. The second canonical variate of the IVs extracts 30% of the variance in judgments of quality of bottom movements. Together the two variates extract 70% of the variance in the DVs.

The first and second canonical variates for the DVs extract 48% and 24% of the variance in judgments of quality of top movements, respectively. Together they extract 72% of the variance in judgments of quality of top movements.

12.4.4   Computer Analyses of Small-Sample Example

Tables 12.4 and 12.5 show analyses of this data set by SAS CANCORR and SPSS CANCORR (Macro), respectively.

In SAS CANCORR (Table 12.4), the one set of variables (DVs) is listed in the input statement that begins **var**, the other set (IVs) in the statement that begins **with**. Redundancy analysis also is available.

The first segment of output contains the canonical correlations for each of the canonical variates (labeled 1 and 2), including adjusted and squared correlations as well as standard errors for the correlations. The next part of the table shows the eigenvalues, the difference between eigenvalues, the proportion and the cumulative proportion of variance in the solution accounted for by each canonical variate pair. The **Test of H0: . . .** table shows "peel off" significance tests for canonical variate pairs evaluated through F followed in the next table by several multivariate significance

TABLE 12.4 Syntax and Selected SAS CANCORR Output for Canonical Correlation Analysis of Sample Data of Table 12.1

```
proc cancorr data=SASUSER.SSCANON
    var TS TC;
    with BS BC;
run;
```

The CANCORR Procedure

Canonical Correlation Analysis

	Canonical Correlation	Adjusted Canonical Correlation	Approximate Standard Error	Squared Canonical Correlation
1	0.914142	0.889541	0.062116	0.835656
2	0.762475	.	0.158228	0.581368

Test of H0: The canonical correlations in
the current row and all that follow are zero

Eigenvalues of Inv(E)*H
= CanRsq/(1−CanRsq)

	Eigenvalue	Difference	Proportion	Cumulative	Likelihood Ratio	Approximate F Value	Num DF	Den DF	Pr > F
1	5.0848	3.6961	0.7855	0.7855	0.06879947	5.62	4	8	0.0187
2	1.3887		0.2145	1.0000	0.41863210	6.94	1	5	0.0462

Multivariate Statistics and F Approximations

S=2   M=−0.5   N=1

Statistic	Value	F Value	Num DF	Den DF	Pr > F
Wilks' Lambda	0.06879947	5.62	4	8	0.0187
Pillai's Trace	1.41702438	6.08	4	10	0.0096
Hotelling-Lawley Trace	6.47354785	4.86	4	6	0.0433
Roy's Greatest Root	5.08481559	12.71	2	5	0.0109

NOTE: F Statistic for Roy's Greatest Root is an upper bound.
NOTE: F Statistic for Wilks' Lambda is exact.

(continued)

581

TABLE 12.4 Continued

Canonical Correlation Analysis

Raw Canonical Coefficients for the VAR Variables

	V1	V2
TS	-0.2297894988	0.2929561178
TC	0.2488132088	0.2703732438

Raw Canonical Coefficients for the WITH Variables

	W1	W2
BS	-0.169694016	0.3087617628
BC	0.3721067975	0.1804101724

Canonical Correlation Analysis

Standardized Canonical Coefficients for the VAR Variables

	V1	V2
TS	-0.6253	0.7972
TC	0.6861	0.7456

Standardized Canonical Coefficients for the WITH Variables

	W1	W2
BS	-0.4823	0.8775
BC	0.9010	0.4368

TABLE 12.4 Continued

Canonical Structure

Correlations Between the VAR Variables and Their Canonical Variables

	V1	V2
TS	-0.7358	0.6772
TC	0.7868	0.6172

Correlations Between the WITH Variables and Their Canonical Variables

	W1	W2
BS	-0.4363	0.8998
BC	0.8764	0.4816

Correlations Between the VAR Variables and the Canonical Variables of the WITH Variables

	W1	W2
TS	-0.6727	0.5163
TC	0.7193	0.4706

Correlations Between the WITH Variables
and the Canonical Variables of the VAR Variables

	V1	V2
BS	-0.3988	0.6861
BC	0.8011	0.3672

tests. Matrices of raw and standardized canonical coefficients for each canonical variate labeled `'VAR'` and `'WITH'` in the syntax follow; loading matrices are labeled `Correlations Between the ... Variables and Their Canonical Variables`. The portion labeled Canonical Structure is part of the redundancy analysis and shows another type of loading matrices: the correlations between each set of variables and the canonical variates of the other set.

Table 12.5 shows the canonical correlation analysis as run through SPSS CANCORR, a macro available through syntax. (SPSS MANOVA also may be used through syntax for a canonical analysis, but the output is much more difficult to interpret.) The INCLUDE instruction invokes the SPSS CANCORR macro by running the syntax file: canonical correlation.sps.[6]

The rather compact output begins with correlation matrices for both sets of variables individually and together. `Canonical Correlations` are then given, followed by their peel down χ^2 tests. Standardized and raw canonical coefficients and loadings are then shown, in the same format as SAS. Correlations between one set of variables and the canonical variates of the other set are labeled `Cross Loadings`. A redundancy analysis is produced by default, showing for each set the proportion of variance associated with its own and the other set. Compare these values with results of Equations 12.11 through 12.13. The program writes canonical scores to the data file and writes a scoring program to another file.

TABLE 12.5   Syntax and Selected SPSS CANCORR Output for Canonical Correlation Analysis on Sample Data in Table 12.1

```
INCLUDE 'Canonical correlation.sps'.
CANCORR SET1 = ts, tc /
         SET2 = bs, bc /.

Run MATRIX procedure:

Correlations for Set-1
          TS         TC
TS    1.0000    -.1611
TC    -.1611    1.0000

Correlations for Set-2
          BS         BC
BS    1.0000     .0511
BC     .0511    1.0000

Correlations Between Set-1 and Set-2
          BS         BC
TS     .7580    -.3408
TC     .1096     .8570

Canonical Correlations
1        .914
2        .762
```

[6]A copy of this syntax file is included with the SPSS data files for this book online.

TABLE 12.5   Continued

```
Test that remaining correlations are zero:

    Wilk's   Chi-SQ     DF    Sig.
1    .069   12.045   4.000    .017
2    .419    3.918   1.000    .048


Standardized Canonical Coefficients for Set-1
          1      2
TS    -.625   .797
TC     .686   .746


Raw Canonical Coefficients for Set-1
          1      2
TS    -.230   .293
TC     .249   .270


Standardized Canonical Coefficients for Set-2
          1      2
BS    -.482   .878
BC     .901   .437


Raw Canonical Coefficients for Set-2
          1      2
BS    -.170   .309
BC     .372   .180


Canonical Loadings for Set-1
          1      2
TS    -.736   .677
TC     .787   .617


Cross Loadings for Set-1
          1      2
TS    -.673   .516
TC     .719   .471


Canonical Loadings for Set-2
          1      2
BS    -.436   .900
BC     .876   .482


Cross Loadings for Set-2
          1      2
BS    -.399   .686
BC     .801   .367
```

(continued)

TABLE 12.5   Continued

Redundancy Analysis:

Proportion of Variance of Set-1 Explained by Its Own Can. Var.
```
        Prop Var
CV1-1     .580
CV1-2     .420
```

Proportion of Variance of Set-1 Explained by Opposite Can.Var.
```
        Prop Var
CV2-1     .485
CV2-2     .244
```

Proportion of Variance of Set-2 Explained by Its Own Can. Var.
```
        Prop Var
CV2-1     .479
CV2-2     .521
```

Proportion of Variance of Set-2 Explained by Opposite Can. Var.
```
        Prop Var
CV1-1     .400
CV1-2     .303
```
—END MATRIX—

12.5   Some Important Issues

12.5.1   Importance of Canonical Variates

As in most statistical procedures, establishing significance is usually the first step in evaluating a solution. Conventional statistical procedures apply to significance tests for the number of canonical variate pairs. The results of Equations 12.3 and 12.4, or a corresponding F test, are available in all programs reviewed in Section 12.7. But the number of statistically significant pairs of canonical variates is often larger than the number of interpretable pairs if N is at all sizable.

The only potential source of confusion is the meaning of the chain of significance tests. The first test is for all pairs taken together and is a test of independence between the two sets of variables. The second test is for all pairs of variates with the first and most important pair of canonical variates removed; the third is done with the first two pairs removed, and so forth. If the first test, but not the second, reaches significance, then only the first pair of canonical variates is interpreted.[7] If the first and second tests are significant but the third is not, then the first two pairs of variates are interpreted, and so on. Because canonical correlations are reported out in descending order of importance, usually only the first few pairs of variates are interpreted.

Once significance is established, amount of variance accounted for is of critical importance. Because there are two sets of variables, several assessments of variance are relevant. First, there is

---

[7]It is possible that the first canonical variate pair is not, by itself, significant, but rather achieves significance only in combination with the remaining canonical variate pairs. To date, there is no significance test for each pair by itself.

variance overlap between variates in a pair. Second is variance overlap between a variate and its own set of variables. Third is variance overlap between a variate and the other set of variables.

The first, and easiest, is the variance overlap between each significant set of canonical variate pairs. As indicated in Equation 12.2, the squared canonical correlation is the overlapping variance between a pair of canonical variates. Most researchers do not interpret pairs with a canonical correlation lower than .30, even if interpreted,[8] because r_c values of .30 or lower represent, squared, less than a 10% overlap in variance.

The next consideration is the variance a canonical variate extracts from its own set of variables. A pair of canonical variates may extract very different amounts of variance from their respective sets of variables. Equations 12.11 and 12.12 indicate that the variance extracted, pv, is the sum of squared loadings on a variate divided by the number of variables in the sets.[9] Because canonical variates are independent of one another (orthogonal), pvs are summed across all significant variates to arrive at the total variance extracted from the variables by all the variates of the set.

The last consideration is the variance a variate from one set extracts from the variables in the other set, called *redundancy* (Stewart & Love, 1968; Miller & Farr, 1971). Equation 12.12 shows that redundancy is the percent of variance extracted by a canonical variate times the canonical correlation for the pair. A canonical variate from the IVs may be strongly correlated with the IVs, but weakly correlated with the DVs (and vice versa). Therefore, the redundancies for a pair of canonical variates are usually not equal. Because canonical variates are orthogonal, redundancies for a set of variables are also added across canonical variates to get a total for the DVs relative to the IVs, and vice versa.

12.5.2   Interpretation of Canonical Variates

Canonical correlation creates linear combinations of variables, canonical variates, that represent mathematically viable combinations of variables. However, although mathematically viable, they are not necessarily interpretable. A major task for the researcher is to discern, if possible, the meaning of pairs of canonical variates.

Interpretation of significant pairs of canonical variates is based on the loading matrices, $\mathbf{A}_x$ and $\mathbf{A}_y$ (Equations 12.9 and 12.10, respectively). Each pair of canonical variates is interpreted as a pair, with a variate from one set of variables interpreted vis-à-vis the variate from the other set. A variate is interpreted by considering the pattern of variables highly correlated (loaded) with it. Because the loading matrices contain correlations, and because squared correlations measure overlapping variance, variables with correlations of .30 (9% of variance) and above are usually interpreted as part of the variate, and variables with loadings below .30 are not. Deciding on a cutoff for interpreting loadings is, however, somewhat a matter of taste, although guidelines are presented in Section 13.6.5.

12.6   Complete Example of Canonical Correlation

For an example of canonical correlation, variables are selected from among those made available by research described in Appendix B, Section B.1. The goal of analysis is to discover the dimensions, if any, along which certain attitudinal variables are related to certain health characteristics. Files are CANON.*.

---

[8]Significance depends, to a large extent, on N.

[9]This calculation is identical to the one used in factor analysis for the same purpose, as shown in Table 13.4.

Selected attitudinal variables (Set 1) include attitudes toward the role of women (ATTROLE), toward locus of control (CONTROL), toward current marital status (ATTMAR), and toward self (ESTEEM). Larger numbers indicate increasingly conservative attitudes about the proper role of women, increasing feelings of powerlessness to control one's fate (external as opposed to internal locus of control), increasing dissatisfaction with current marital status, and increasingly poor self-esteem.

Selected health variables (Set 2) include mental health (MENHEAL), physical health (PHYHEAL), number of visits to health professionals (TIMEDRS), attitude toward use of medication (ATTDRUG), and a frequency-duration measure of use of psychotropic drugs (DRUGUSE). Larger numbers reflect poorer mental and physical health, more visits, greater willingness to use drugs, and more use of them.

12.6.1   Evaluation of Assumptions

12.6.1.1   Missing Data

A screening run through SAS MEANS, illustrated in Table 12.6, finds missing data for 6 of the 465 cases. One woman lacks a score on CONTROL, and five lack scores on ATTMAR. With deletion of these cases (less than 2%), remaining $N = 459$.

12.6.1.2   Normality, Linearity, and Homoscedasticity

SAS provides a particularly flexible scheme for assessing normality, linearity, and homoscedasticity between pairs of canonical variates. Canonical variate scores are saved to a data file, and then PROC PLOT permits a scatterplot of them.

Figure 12.3 shows two scatterplots produced by PROC PLOT for the example using default size values for the plots. The CANCORR syntax runs a preliminary canonical correlation analysis and saves the canonical variate scores (as well as the original data) to a file labeled LSSCORES. The four canonical variates for the first set are labeled V1 through V4; the canonical variates for the second set are labeled W1 through W4. Thus, the Plot syntax requests scatterplots that are between the first and second pairs of canonical variates, respectively. V1 is canonical variate scores, first set, first variate; W1 is canonical variate scores, second set, first variate. V2 is canonical variate scores, first set, second variate; W2 is canonical variate scores, second set, second variate.

The shapes of the scatterplots reflect the low canonical correlations for the solution (see Section 12.6.2), particularly for the second pair of variates where the overall shape is nearly circular except for a few extreme values in the lower third of the plot. There are no obvious departures from linearity or homoscedasticity because the overall shapes do not curve and they are of about the same width throughout.

Deviation from normality is evident, however, for both pairs of canonical variates: on both plots, the 0–0 point departs from the center of the vertical and horizontal axes. If the points are projected as a frequency distribution to the vertical or horizontal axes of the plots, there is further evidence of skewness. For the first plot, there is a pileup of cases at low scores and a smattering of cases at high scores on both axes, indicating positive skewness. In plot 2, there are widely scattered cases with extremely low scores on W2, with no corresponding high scores, indicating negative skewness.

Departure from normality is confirmed by the output of SAS MEANS. By using Equation 4.7 to compute the standard error for skewness,

$$s_s = \sqrt{\frac{6}{N}} = \sqrt{\frac{6}{465}} = 0.1136$$

TABLE 12.6   Syntax and Selected SAS MEANS Output for Initial Screening of Canonical Correlation Data Set

```
proc means data=SASUSER.CANON   vardef=DF
    N NMISS MIN MAX MEAN VAR STD SKEWNESS KURTOSIS;
  var TIMEDRS ATTDRUG PHYHEAL MENHEAL ESTEEM CONTROL ATTMAR
      DRUGUSE ATTROLE;
run;
```

The MEANS Procedure

Variable	Label	N	N Miss	Minimum	Maximum	Std Dev
TIMEDRS	Visits to health professionals	465	0	0	81.0000000	10.9484932
ATTDRUG	Attitude toward use of medication	465	0	5.0000000	10.0000000	1.1560925
PHYHEAL	Physical health symptoms	465	0	2.0000000	15.0000000	2.3882961
MENHEAL	Mental health symptoms	465	0	0	18.0000000	4.1935945
ESTEEM	Self-esteem	465	0	8.0000000	29.0000000	3.9425425
CONTROL	Locus of control	464	1	5.0000000	10.0000000	1.2655067
ATTMAR	Attitude toward current marital status	460	5	11.0000000	58.0000000	8.5534108
DRUGUSE	Use of psychotropic drugs	465	0	0	66.0000000	10.1130461
ATTROLE	Attitudes toward role of women	465	0	18.0000000	55.0000000	6.7582109

Variable	Label	Mean	Variance
TIMEDRS	Visits to health professionals	7.9010753	119.8695032
ATTDRUG	Attitude toward use of medication	7.6860215	1.3365499
PHYHEAL	Physical health symptoms	4.9720430	5.7039581
MENHEAL	Mental health symptoms	6.1225806	17.5862347
ESTEEM	Self-esteem	15.8344086	15.5436411
CONTROL	Locus of control	6.7478448	1.6015072
ATTMAR	Attitude toward current marital status	22.9804348	73.1608364
DRUGUSE	Use of psychotropic drugs	9.0021505	102.2737023
ATTROLE	Attitudes toward role of women	35.1354839	45.6734149

Variable	Label	Skewness	Kurtosis
TIMEDRS	Visits to health professionals	3.2481170	13.1005155
ATTDRUG	Attitude toward use of medication	-0.1225099	-0.4470689
PHYHEAL	Physical health symptoms	1.0313360	1.1235075
MENHEAL	Mental health symptoms	0.6024595	-0.2921355
ESTEEM	Self-esteem	0.4812032	-0.2916191
CONTROL	Locus of control	0.4895045	-0.3992646
ATTMAR	Attitude toward current marital status	1.0035327	0.8119797
DRUGUSE	Use of psychotropic drugs	1.7610005	4.2601383
ATTROLE	Attitudes toward role of women	0.0498862	-0.4009358

```
proc cancorr data=SASUSER.CANON
   out=WORK.LSSCORES
   sing=1E-8;
   var ESTEEM CONTROL ATTMAR ATTROLE;
   with MENHEAL PHYHEAL TIMEDRS ATTDRUG DRUGUSE;
run;
proc plot data=WORK.LSSCORES;
   plot w1*v1;
   plot w2*v2;
run;
```

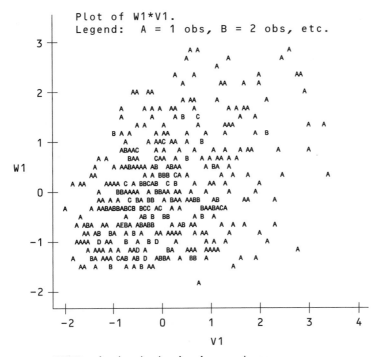

NOTE: 6 obs had missing values.

FIGURE 12.3 SAS CANCORR and PLOT syntax and output showing scatterplots between first and second pairs of canonical variates.

and Equation 4.8 to compute z for skewness, Table 12.7 shows extreme positive skewness for TIMEDRS ($z = 3.248/0.1136 = 28.59$) as well as strong skewness for PHYHEAL, ATTMAR, and DRUGUSE. Logarithmic transformation of these variables results in variables that are far less skewed. The transformed variables are named LATTMAR, LDRUGUSE, LTIMEDRS, and LPHY-HEAL. Moderate skewness also is noted for MENHEAL, ESTEEM, and CONTROL. However, histograms and normal probability plots through SAS UNIVARIATE (not shown) indicated no serious departure from normality.

A second SAS MEANS run provides univariate statistics for both transformed and untrans-formed variables. Table 12.7 shows the SAS DATA step to delete the cases with missing data on

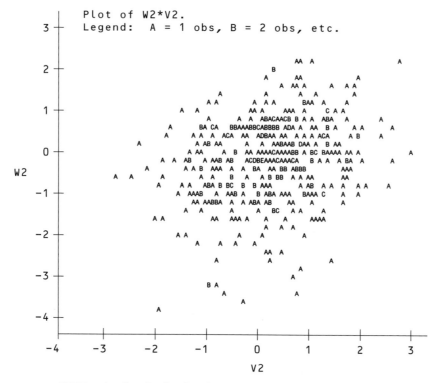

FIGURE 12.3   Continued.

CONTROL and ATTMAR and accomplish the logarithmic transforms for four of the variables. New values (along with old ones) are saved in a data set called CANONT. Only skewness and kurtosis are requested for the new variables.

Compare the skewness and kurtosis of ATTMAR and DRUGUSE in Table 12.6 with that of LATTMAR and LDRUGUSE in Table 12.7. SAS PLOT scatterplots based on transformed variables (not shown) confirm improvement in normality with transformed variables, particularly for the second pair of canonical variates.

12.6.1.3   Outliers

Standard scores are created by SAS STANDARD (Table 12.8) specifying MEAN=0 and STD=1. These scores are saved into a new file labeled CANONS, which is then used by SAS MEANS to print minimum and maximum values for all of the variables to be used in the analysis.

Minimum and maximum standard scores are within a range of ± 3.29 with the exception of a large score on ESTEEM ($z = 3.34$), not disturbing in a sample of over 400 cases.

SAS REGRESSION is used to screen multivariate outliers by requesting that leverage values be saved in a new data file. Table 12.9 shows syntax to run the regression analysis on the first set of variables and save the H (leverage) values into a data file labeled CANLEV. Leverage values for the first few cases are shown in the table.

TABLE 12.7 SAS DATA and MEANS Syntax and Output Showing Skewness and Kurtosis

```
data SASUSER.CANONT;
  set SASUSER.CANON;
  if CONTROL =. or ATTMAR =. then delete;
  LTIMEDRS = log10(TIMEDRS+1);
  LPHYHEAL = log10(PHYHEAL);
  LATTMAR = log10(ATTMAR);
  LDRUGUSE = log10(DRUGUSE + 1);
run;
proc means data=SASUSER.CANONT    vardef=DF
  N NMISS SKEWNESS KURTOSIS MEAN;
  var LTIMEDRS LPHYHEAL LATTMAR LDRUGUSE;
run;
```

The MEANS Procedure

Variable	Label	N	N Miss	Skewness	Kurtosis	Mean
LTIMEDRS	log(TIMEDRS + 1)	459	0	0.2296331	-0.1861264	0.7413859
LPHYHEAL	log(PHYHEAL)	459	0	-0.0061454	-0.6984603	0.6476627
LATTMAR	log(ATTMAR)	459	0	0.2291448	-0.5893927	1.3337398
LDRUGUSE	log(DRUGUSE + 1)	459	0	-0.1527641	-1.0922599	0.7637207

TABLE 12.8 SAS STANDARD and MEANS Syntax and Output to Evaluate Univariate Outliers

```
proc standard data=SASUSER.CANONT out=SASUSER.CANONS
     vardef=DF MEAN=0 STD=1;
     var LTIMEDRS LPHYHEAL LATTMAR LDRUGUSE
         ATTDRUG MENHEAL ESTEEM CONTROL ATTROLE;
run;
proc means data=SASUSER. CANONS vardef=DF
     N MIN MAX;
     var LTIMEDRS LPHYHEAL LATTMAR LDRUGUSE
         ATTDRUG MENHEAL ESTEEM CONTROL ATTROLE;
run;
```

Variable	Label	N	Minimum	Maximum
LTIMEDRS	log(TIMEDRS + 1)	459	-1.7788901	2.8131374
LPHYHEAL	log(PHYHEAL)	459	-1.6765455	2.5558303
LATTMAR	log(ATTMAR)	459	-1.8974189	2.7888035
LDRUGUSE	log(DRUGUSE + 1)	459	-1.5628634	2.1739812
ATTDRUG	Attitude toward use of medication	459	-2.3195715	2.0119310
MENHEAL	Mental health symptoms	459	-1.4747464	2.8675041
ESTEEM	Self-esteem	459	-1.9799997	3.3391801
CONTROL	Locus of control	459	-1.3761468	2.5569253
ATTROLE	Attitudes toward role of women	459	-2.5470346	2.9333009

TABLE 12.9 SAS REG Syntax and Selected Portion of Data File for Identification of Multivariate Outliers for the First Set of Variables

```
proc reg data=SASUSER.CANONT;
    model SUBNO= ESTEEM CONTROL LATTMAR ATTROLE;
    output out=SASUSER.CANONLEV H=H;
run;
```

SASUSER.CANONLEV

	LTIMEDRS	LPHYHEAL	LATTMAR	LDRUGUSE	H
1	0.3010	0.6990	1.5563	0.6021	0.0163
2	0.6021	0.6021	1.3222	0.0000	0.0080
3	0.0000	0.4771	1.3010	0.6021	0.0111
4	1.1461	0.3010	1.3802	0.7782	0.0049
5	1.2041	0.4771	1.1761	1.3979	0.0121
6	0.6021	0.6990	1.4472	0.6021	0.0080
7	0.4771	0.6990	1.4314	0.0000	0.0100
8	0.0000	0.6021	1.2553	0.0000	0.0128
9	0.9031	0.6990	1.0792	0.4771	0.0114
10	0.6990	0.4771	1.7243	1.0414	0.0208
11	1.2041	0.7782	1.0414	1.2788	0.0150
12	0.0000	0.4771	1.2041	0.3010	0.0120
13	0.4771	0.4771	1.2304	0.6021	0.0184
14	1.1461	0.7782	1.3010	1.4771	0.0076
15	0.4771	0.4771	1.1761	0.0000	0.0102
16	0.4771	0.4771	1.0792	0.3010	0.0092
17	0.3010	0.4771	1.2553	0.0000	0.0090
18	0.4771	0.8451	1.0414	1.1761	0.0149
19	0.7782	0.6021	1.3222	1.3010	0.0066
20	0.7782	0.8451	1.4161	0.9031	0.0147
21	0.6021	0.6021	1.4150	0.3010	0.0096
22	0.6990	0.3010	1.3979	0.6021	0.0133
23	0.4771	0.4771	1.3222	0.6021	0.0037
24	0.0000	0.6990	1.5798	0.9031	0.0219
25	1.1461	0.8541	1.5051	1.4314	0.0123
26	0.9031	0.9031	1.3617	1.1139	0.0109
27	0.4771	0.7782	1.3222	0.4771	0.0098
28	1.1139	0.9542	1.3979	1.5563	0.0049
29	0.4771	0.4771	1.5441	0.4471	0.0094
30	0.7782	0.8451	1.2788	1.3424	0.0229
31	0.6990	0.9031	1.2788	0.9031	0.0039
32	0.8451	0.9031	1.2041	0.4471	0.0126
33	0.4771	0.7782	1.3222	0.4771	0.0241
34	0.6021	0.6021	1.2041	0.0000	0.0050

Critical value of Mahalanobis distance with four variables at $\alpha = .001$ is 18.467. Using Equation 4.3 to convert this to a critical leverage value:

$$h_{ii} = \frac{18.467}{459 - 1} + \frac{1}{459} = 0.0425$$

There are no outliers in the segment of the data set shown in Table 12.9 or any other in either set of variables.

12.6.1.4 *Multicollinearity and Singularity*

SAS CANCORR protects against multicollinearity and singularity by setting a value for tolerance (`sing`) in the main analysis. It is not necessary to further check multicollinearity unless there is reason to expect large SMCs among variables in either set and there is a desire to eliminate logically redundant variables.

12.6.2   Canonical Correlation

The number and importance of canonical variates are determined using procedures from Section 12.5.1 (Table 12.10). `RED` requests redundancy statistics.

Significance of the relationships between the sets of variables is reported directly by SAS CANCORR, as shown in Table 12.10. With all four canonical correlations included, $F(20, 1,493.4) = 5.58, p < .001$. With the first and second canonical correlations removed, F values are not significant; $F(6, 904) = 0.60, p = .66$. Therefore, only significant relationships are in the first two pairs of canonical variates and these are interpreted.

Canonical correlations (r_c) and eigenvalues (r_c^2) are also in Table 12.10. The first canonical correlation is .38 (.36 adjusted), representing 14% overlapping variance for the first pair of canonical variates (see Equation 12.2). The second canonical correlation is .27 (.26 adjusted), representing 7% overlapping variance for the second pair of canonical variates. Although highly significant, neither of these two canonical correlations represents a substantial relationship between pairs of canonical variates. Interpretation of the second canonical correlation and its corresponding pair of canonical variates is marginal.

Loading matrices between canonical variates and original variables are in Table 12.11. Interpretation of the two significant pairs of canonical variates from loadings follows procedures mentioned in Section 12.5.2. Correlations between variables and variates (loadings) in excess of .3 are interpreted. Both the direction of correlations in the loading matrices and the direction of scales of measurement are considered when interpreting the canonical variates.

The first pair of canonical variates has high loadings on ESTEEM, CONTROL, and LATTMAR (.596, .784, and .730, respectively) on the attitudinal set and on LPHYHEAL and MENHEAL (.408 and .968) on the health side. Thus, low self-esteem, external locus of control, and dissatisfaction with marital status are related to poor physical and mental health.

The second pair of canonical variates has high loadings on ESTEEM, LATTMAR, and ATTROLE (.601, −.317, and .783) on the attitudinal side and on LTIMEDRS, ATTDRUG, and LDRUGUSE (−.359, .559, and −.548) on the health side. Big numbers on ESTEEM, little numbers on LATTMAR, and big numbers on ATTROLE go with little numbers on LTIMEDRS, big numbers on ATTDRUG, and little numbers on LDRUGUSE. That is, low self-esteem, satisfaction with marital status, and conservative attitudes toward the proper role of women in society go with few visits to physicians, favorable attitudes toward use of drugs, and little actual use of them. (Figure that one out!)

Loadings are converted to *pv* values by application of Equations 12.11 and 12.12. These values are shown in the output in sections `Standardized Variance of the ... Variables Explained by Their Own Canonical Variables` (Table 12.11). The values for the first pair of canonical variates are .38 for the first set of variables and .24 for the second set of variables. That is, the first canonical variate pair extracts 38% of variance from the attitudinal variables and 24% of variance from the health variables. The values for the second pair of canonical variates are .27 for the first set of variables and .15 for the second set; the second canonical variate pair

TABLE 12.10 Syntax and Selected Portion of SAS CANCORR Output Showing Canonical Correlations and Significance Levels for Sets of Canonical Correlations

```
proc cancorr data=SASUSER.CANONT RED
out=SASUSER.LSSCORNW
sing=1E-8;
var ESTEEM CONTROL LATTMAR ATTROLE;
with LTIMEDRS ATTDRUG LPHYHEAL MENHEAL LDRUGUSE;
run;
```

The CANCORR Procedure

Canonical Correlation Analysis

	Canonical Correlation	Adjusted Canonical Correlation	Approximate Standard Error	Squared Canonical Correlation
1	0.378924	0.357472	0.040018	0.143583
2	0.268386	0.255318	0.043361	0.072031
3	0.088734	.	0.046359	0.007874
4	0.034287	.	0.046672	0.001176

Eigenvalues of Inv(E)*H = CanRsq/(1-CanRsq)

	Eigenvalue	Difference	Proportion	Cumulative
1	0.1677	0.0900	0.6590	0.6590
2	0.0776	0.0697	0.3051	0.9642
3	0.0079	0.0068	0.0312	0.9954
4	0.0012		0.0046	1.0000

Test of H0: The canonical correlations in the current row and all that follow are zero

	Likelihood Ratio	Approximate F Value	Num DF	Den DF	Pr > F
1	0.78754370	5.58	20	1493.4	<.0001
2	0.91957989	3.20	12	1193.5	0.0002
3	0.99095995	0.69	6	904	0.6613
4	0.99882441	0.27	2	453	0.7661

TABLE 12.11 Selected SAS CANCORR Output of Loading Matrices for the Two Sets of Variables in the Example. Syntax Is in Table 12.10

Canonical Structure

Correlations Between the VAR Variables and Their Canonical Variables

		V1	V2	V3	V4
ESTEEM	Self-esteem	0.5958	0.6005	-0.2862	-0.4500
CONTROL	Locus of control	0.7836	0.1478	-0.1771	0.5769
LATTMAR	log(ATTMAR)	0.7302	-0.3166	0.4341	-0.4221
ATTROLE	Attitudes toward role of women	-0.0937	0.7829	0.6045	0.1133

Correlations Between the WITH Variables and Their Canonical Variables

		W1	W2	W3	W4
LTIMEDRS	log(TIMEDRS + 1)	0.1229	-0.3589	-0.8601	0.2490
ATTDRUG	Attitude toward use of medication	0.0765	0.5593	-0.0332	0.4050
LPHYHEAL	log(PHYHEAL)	0.4082	-0.0479	-0.6397	-0.5047
MENHEAL	Mental health symptoms	0.9677	-0.1434	-0.1887	0.0655
LDRUGUSE	log(DRUGUSE + 1)	0.2764	-0.5479	0.0165	-0.0051

TABLE 12.12   Selected SAS CANCORR Output Showing Percents of Variance and Redundancy for First and Second Set of Canonical Variates. Syntax Is in Table 12.10

Canonical Redundancy Analysis

Standardized Variance of the VAR Variables Explained by

	Their Own Canonical Variables		The Opposite Canonical Variables	
Canonical Variable Number	Proportion	Cumulative Proportion	Canonical R–Square	Proportion	Cumulative Proportion
1	0.3777	0.3777	0.1436	0.0542	0.0542
2	0.2739	0.6517	0.0720	0.0197	0.0740
3	0.1668	0.8184	0.0079	0.0013	0.0753
4	0.1816	1.0000	0.0012	0.0002	0.0755

Standardized Variance of the WITH Variables Explained by

	Their Own Canonical Variables		The Opposite Canonical Variables	
Canonical Variable Number	Proportion	Cumulative Proportion	Canonical R–Square	Proportion	Cumulative Proportion
1	0.2401	0.2401	0.1436	0.0345	0.0345
2	0.1529	0.3930	0.0720	0.0110	0.0455
3	0.2372	0.6302	0.0079	0.0019	0.0474
4	0.0970	0.7272	0.0012	0.0001	0.0475

extracts 27% of variance from the attitudinal variables and 15% of variance from the health variables. Together, the two canonical variates account for 65% of variance (38% plus 27%) in the attitudinal set, and 39% of variance (24% and 15%) in the health set.

Redundancies for the canonical variates are found in SAS CANCORR in the sections labeled **Variance of the ... Variables Explained by The Opposite Canonical Variables** (Table 12.12). That is, the first health variate accounts for 5% of the variance in the attitudinal variables, and the second health variate accounts for 2% of the variance. Together, two health variates "explain" 7% of the variance in attitudinal variables. The first attitudinal variate accounts for 3% and the second 1% of the variance in the health set. Together, the two attitudinal variates overlap the variance in the health set, 4%.

If a goal of analysis is production of scores on canonical variates, coefficients for them are readily available. Table 12.13 shows both standardized and unstandardized coefficients for production of canonical variates. Scores on the variates themselves for each case are also produced by SAS CANCORR if an output file is requested (see syntax in Table 12.9). A summary table of information appropriate for inclusion in a journal article appears in Table 12.14.

A checklist for canonical correlation appears in Table 12.15. An example of a Results section, in journal format, follows for the complete analysis described in Section 12.6.

TABLE 12.13 Selected SAS CANCORR Output of Unstandardized and Standardized Canonical Variate Coefficients. Syntax Is in Table 12.10

```
               Canonical Correlation Analysis
```

Raw Canonical Coefficients for the VAR Variables

		V1	V2	W1	W2
ESTEEM	Self-esteem	0.0619490461	0.1551478815	-0.1589070745	-0.172696778
CONTROL	Locus of control	0.465185442	0.0213951395	-0.086711685	0.6995969512
LATTMAR	log(ATTMAR)	3.4017147794	-2.916036329	4.756204374	-2.4133550645
ATTROLE	Attitudes toward role of women	-0.012933183	0.0919237888	0.1182827337	0.0303123963

Raw Canonical Coefficients for the WITH Variables

		W1	W2
LTIMEDRS	log(TIMEDRS + 1)	-0.64318212	-0.925366063
ATTDRUG	Attitude toward use of medication	0.0396639542	0.6733109169
LPHYHEAL	log(PHYHEAL)	0.2081929915	2.1591824353
MENHEAL	Mental health symptoms	0.2563584017	0.0085859621
LDRUGUSE	log(DRUGUSE + 1)	-0.12004514	-1.693255583

Raw Canonical Coefficients for the WITH Variables

		W3	W4
LTIMEDRS	log(TIMEDRS + 1)	-2.0510523	1.8736861531
ATTDRUG	Attitude toward use of medication	-0.03925235	0.3880909448
LPHYHEAL	log(PHYHEAL)	-2.144723348	-5.73998122
MENHEAL	Mental health symptoms	0.0368750419	0.0917261729
LDRUGUSE	log(DRUGUSE + 1)	1.0486138709	-0.102427113

(continued)

TABLE 12.13 Continued

Standardized Canonical Coefficients for the VAR Variables

		V1	V2	V3	V4
ESTEEM	Self-esteem	0.2446	0.6125	-0.6276	-0.6818
CONTROL	Locus of control	0.5914	0.0272	-0.1102	0.8894
LATTMAR	log(ATTMAR)	0.5241	-0.4493	0.7328	-0.3718
ATTROLE	Attitudes toward role of women	-0.0873	0.6206	0.7986	0.2047

Standardized Canonical Coefficients for the WITH Variables

		W1	W2	W3	W4
LTIMEDRS	log(TIMEDRS + 1)	-0.2681	-0.3857	-0.8548	0.7809
ATTDRUG	Attitude toward use of medication	0.0458	0.7772	-0.0453	0.4480
LPHYHEAL	log(PHYHEAL)	0.0430	0.4464	-0.4434	-1.1868
MENHEAL	Mental health symptoms	1.0627	0.0356	0.1529	0.3802
LDRUGUSE	log(DRUGUSE + 1)	-0.0596	-0.8274	0.5124	-0.0501

TABLE 12.14   Correlations, Standardized Canonical Coefficients, Canonical Correlations, Proportions of Variance, and Redundancies between Attitudinal and Health Variables and Their Corresponding Canonical Variates

	First Canonical Variate		Second Canonical Variate	
	Correlation	*Coefficient*	*Correlation*	*Coefficient*
Attitudinal set				
Locus of control	.78	.59	.15	.03
Attitude toward current marital status		.52	−.32	−.45
(logarithm)	.73			
Self-esteem	.60	.25	.60	.61
Attitude toward role of women	−.09	−.09	.78	.62
Percent of variance	.38		.27	Total = .65
Redundancy	.05		.02	Total = .07
Health set				
Mental health	.97	1.06	−.14	.04
Physical health (logarithm)	.41	.04	−.05	.45
Visits to health professionals (logarithm)	.12	−.27	−.36	−.39
Attitude toward use of medication	.08	.05	.56	.78
Use of psychotropic drugs (logarithm)	.28	−.06	−.55	−.83
Percent of variance	.24		.15	Total = .39
Redundancy	.03		.01	Total = .04
Canonical correlation	.38		.27	

TABLE 12.15   Checklist for Canonical Correlation

1. Issues
    a. Missing data
    b. Normality, linearity, homoscedasticity
    c. Outliers
    d. Multicollinearity and singularity
2. Major analyses
    a. Significance of canonical correlations
    b. Correlations of variables and variates
    c. Variance accounted for
        (1) By canonical correlations
        (2) By same-set canonical variates
        (3) By other-set canonical variates (redundancy)
3. Additional analyses
    a. Canonical coefficients
    b. Canonical variates scores

Results

Canonical correlation was performed between a set of attitudinal variables and a set of health variables using SAS CANCORR. The attitudinal set included attitudes toward the role of women, toward locus of control, toward current marital status, and toward self-worth. The health set measured mental health, physical health, visits to health professionals, attitude toward use of medication, and use of psychotropic drugs. Increasingly large numbers reflected more conservative attitudes toward women's role, external locus of control, dissatisfaction with marital status, low self-esteem, poor mental health, poor physical health, more numerous health visits, favorable attitudes toward drug use, and more drug use.

To improve linearity of relationship between variables and normality of their distributions, logarithmic transformations were applied to attitude toward marital status, visits to health professionals, physical health, and drug use. No within-set multivariate outliers were identified at $p < .001$, although six cases were found to be missing data on locus of control or attitude toward marital status and were deleted, leaving $N = 459$. Assumptions regarding within-set multicollinearity were met.

The first canonical correlation was .38 (14% overlapping variance); the second was .27 (7% overlapping variance). The remaining two canonical correlations were effectively zero. With all four canonical correlations included, $\chi^2(20) = 108.19$, $p < .001$, and with the first canonical correlation removed, $\chi^2(12) = 37.98$, $p < .001$. Subsequent χ^2 tests were not statistically significant. The first two pairs of canonical variates, therefore, accounted for the significant relationships between the two sets of variables.

Data on the first two pairs of canonical variates appear in Table 12.14. Shown in the table are correlations between the variables and the canonical variates, standardized canonical variate coefficients, within-set variance accounted for by the canonical variates (proportion of variance), redundancies, and canonical correlations. Total proportion of variance and total redundancy indicate that the first pair of canonical variates was moderately related, but the second pair was only minimally related; interpretation of the second pair is questionable.

With a cutoff correlation of .3, the variables in the attitudinal set that were correlated with the first canonical variate were locus of control, (log of) attitude toward marital status, and self-esteem. Among the health variables, mental health and (log of) physical health correlated with the first canonical variate. The first pair of canonical variates indicate that those with external locus of control (.78), feelings of dissatisfaction toward marital status (.73), and lower self-esteem (.60) are associated with more numerous mental health symptoms (.97) and more numerous physical health symptoms (.41).

The second canonical variate in the attitudinal set was composed of attitude toward role of women, self-esteem, and negative of (log of) attitude toward marital status, and the corresponding canonical variate from the health set was composed of negative of (log of) drug use, attitude toward drugs, and negative of (log of) visits to health professionals. Taken as a pair, these variates suggest that a combination of more conservative

attitudes toward the role of women (.78), lower self-esteem
(.60), but relative satisfaction with marital status (-.32) is
associated with a combination of more favorable attitudes toward
use of drugs (.56), but lower psychotropic drug use (-.55),
and fewer visits to health professionals (-.36).

That is, women who have conservative attitudes toward role of
women and are happy with their marital status but have lower self-
esteem are likely to have more favorable attitudes toward drug use
but fewer visits to health professionals and lower use of psychotropic drugs.

12.7   Comparison of Programs

One program is available in the SAS package for canonical analyses. SPSS has two programs that
may be used for canonical analysis. Table 12.16 provides a comparison of important features of the
programs. If available, the program of choice is SAS CANCORR and, with limitations, the SPSS
CANCORR macro.

12.7.1   SAS System

SAS CANCORR is a flexible program with abundant features and ease of interpretation. Along with
the basics, you can specify easily interpretable labels for canonical variates and the program accepts
several types of input matrices.

Multivariate output is quite detailed, with several test criteria and voluminous redundancy
analyses. Univariate output is minimal, however, and if plots are desired, case statistics such as
canonical scores are written to a file to be analyzed by the SAS PLOT procedure. If requested, the
program does separate multiple regressions with each variable predicted from the other set. You can
also do separate canonical correlation analyses for different groups.

12.7.2   SPSS Package

SPSS has two programs for canonical analysis, both available only through syntax: SPSS MANOVA
and a CANCORR macro (see Table 12.5). A complete canonical analysis is available through SPSS
MANOVA, which provides loadings, percents of variance, redundancy, and much more. But problems arise with reading the results, because MANOVA is not designed specifically for canonical
analysis and some of the labels are confusing.

TABLE 12.16   Comparison of SPSS, SAS, and SYSTAT Programs for Canonical Correlation

Feature	SPSS MANOVA[a]	SPSS CANCORR	SAS CANCORR	SYSTAT SETCOR
Input				
Correlation matrix	Yes	Yes	Yes	Yes
Covariance matrix	No	No	Yes	No
SSCP matrix	No	No	Yes	No
Number of canonical variates	No	Yes	Yes	No
Tolerance	No	No	Yes	No
Minimum canonical correlation	Specify alpha	No	No	No
Labels for canonical variates	No	No	Yes	No
Error df if residuals input	No	No	Yes	No
Specify partialing covariates	No	No	Yes	Yes
Output				
Univariate:				
Means	Yes	No	Yes	No
Standard deviations	Yes	No	Yes	No
Confidence intervals	Yes	No	No	No
Normal plots	Yes	No	No	No
Multivariate:				
Canonical correlations	Yes	Yes	Yes	Yes
Eigenvalues (r_c^2)	Yes	No	Yes	No
Significance test	F	χ^2	F	χ^2
Lambda	Yes	No	Yes	RAO
Additional test criteria	Yes	No	Yes	Yes
Correlation matrix	Yes	Yes	Yes	Yes
Covariance matrix	Yes	No	No	No
Loading matrix	Yes	Yes	Yes	Yes
Loading matrix for opposite set	No	Yes	Yes	No
Raw canonical coefficients	Yes	Yes	Yes	Yes
Standardized canonical coefficients	Yes	Yes	Yes	Yes
Canonical variate scores	No	Data file	Data file	No
Proportion of variance	Yes	Yes	Yes	No
Redundancies	Yes	Yes	Yes	Yes
Stewart-Love Redundancy Index	No	No	No	Yes
Between-sets SMCs	No	No	Yes	No
Multiple-regression analyses	DVs only	No	Yes	DVs only
Separate analyses by groups	No	No	Yes	No

[a]Additional features are listed in Section 6.7 and 7.7.

Canonical analysis is requested through MANOVA by calling one set of variables DVs and the other set covariates; no IVs are listed. Although SPSS MANOVA provides a rather complete canonical analysis, it does not calculate canonical variate scores, nor does it offer multivariate plots. Tabachnick and Fidell (1996) show an SPSS MANOVA analysis of the small-sample example of Table 12.1.

Syntax and output for the SPSS CANCORR macro is much simpler and easier to interpret. All of the critical information is available, however, with the peel-down tests, and a full set of correlations, canonical coefficients, and loadings. A redundancy analysis is included by default, and canonical variate scores are written to the original data set for plotting.

12.7.3   SYSTAT System

Canonical analysis currently is most readily done through SETCOR (SYSTAT Software Inc., 2002). The program provides all of the basics of canonical correlation and several others. There is a test of overall association between the two sets of variables, as well as tests of prediction of each DV from the set of IVs. The program also provides analyses in which one set is partialed from the other set, useful for statistical adjustment of irrelevant sources of variance (as per covariates in ANCOVA) as well as representation of curvilinear relationships and interactions. These features are well-explained in the manual. The Stewart-Love canonical redundancy index also is provided. Canonical factors may be rotated.

Canonical analysis also may be done through the multivariate general linear model GLM program in SYSTAT. But to get all the output, the analysis must be done twice, once with the first set of variables defined as the DVs, and a second time with the other set of variables defined as DVs. The advantages over SETCOR are that canonical variate scores may be saved in a data file, and that standardized canonical coefficients are provided.

CHAPTER

13 Principal Components and Factor Analysis

13.1 General Purpose and Description

Principal components analysis (PCA) and factor analysis (FA) are statistical techniques applied to a single set of variables when the researcher is interested in discovering which variables in the set form coherent subsets that are relatively independent of one another. Variables that are correlated with one another but largely independent of other subsets of variables are combined into factors.[1] Factors are thought to reflect underlying processes that have created the correlations among variables.

Suppose, for instance, a researcher is interested in studying characteristics of graduate students. The researcher measures a large sample of graduate students on personality characteristics, motivation, intellectual ability, scholastic history, familial history, health and physical characteristics, etc. Each of these areas is assessed by numerous variables; the variables all enter the analysis individually at one time and correlations among them are studied. The analysis reveals patterns of correlation among the variables that are thought to reflect underlying processes affecting the behavior of graduate students. For instance, several individual variables from the personality measures combine with some variables from the motivation and scholastic history measures to form a factor measuring the degree to which a person prefers to work independently, an independence factor. Several variables from the intellectual ability measures combine with some others from scholastic history to suggest an intelligence factor.

A major use of PCA and FA in psychology is in development of objective tests for measurement of personality and intelligence and the like. The researcher starts out with a very large number of items reflecting a first guess about the items that may eventually prove useful. The items are given to randomly selected subjects and factors are derived. As a result of the first factor analysis, items are added and deleted, a second test is devised, and that test is given to other randomly selected subjects. The process continues until the researcher has a test with numerous items forming several factors that represent the area to be measured. The validity of the factors is tested in research where predictions are made regarding differences in the behavior of persons who score high or low on a factor.

---

[1]PCA produces components while FA produces factors, but it is less confusing in this section to call the results of both analyses factors.

The specific goals of PCA or FA are to summarize patterns of correlations among observed variables, to reduce a large number of observed variables to a smaller number of factors, to provide an operational definition (a regression equation) for an underlying process by using observed variables, or to test a theory about the nature of underlying processes. Some or all of these goals may be the focus of a particular research project.

PCA and FA have considerable utility in reducing numerous variables down to a few factors. Mathematically, PCA and FA produce several linear combinations of observed variables, each linear combination a factor. The factors summarize the patterns of correlations in the observed correlation matrix and can, with varying degrees of success, be used to reproduce the observed correlation matrix. But since the number of factors is usually far fewer than the number of observed variables, there is considerable parsimony in using the factor analysis. Further, when scores on factors are estimated for each subject, they are often more reliable than scores on individual observed variables.

Steps in PCA or FA include selecting and measuring a set of variables, preparing the correlation matrix (to perform either PCA or FA), extracting a set of factors from the correlation matrix, determining the number of factors, (probably) rotating the factors to increase interpretability, and, finally, interpreting the results. Although there are relevant statistical considerations to most of these steps, an important test of the analysis is its interpretability.

A good PCA or FA "makes sense"; a bad one does not. Interpretation and naming of factors depend on the meaning of the particular combination of observed variables that correlate highly with each factor. A factor is more easily interpreted when several observed variables correlate highly with it and those variables do not correlate with other factors.

Once interpretability is adequate, the last, and very large, step is to verify the factor structure by establishing the construct validity of the factors. The researcher seeks to demonstrate that scores on the latent variables (factors) covary with scores on other variables, or that scores on latent variables change with experimental conditions as predicted by theory.

One of the problems with PCA and FA is that there are no readily available criteria against which to test the solution. In regression analysis, for instance, the DV is a criterion and the correlation between observed and predicted DV scores serves as a test of the solution—similarly for the two sets of variables in canonical correlation. In discriminant function analysis, logistic regression, profile analysis, and multivariate analysis of variance, the solution is judged by how well it predicts group membership. But in PCA and FA there is no external criterion such as group membership against which to test the solution.

A second problem with FA or PCA is that, after extraction, there is an infinite number of rotations available, all accounting for the same amount of variance in the original data, but with the factors defined slightly differently. The final choice among alternatives depends on the researcher's assessment of its interpretability and scientific utility. In the presence of an infinite number of mathematically identical solutions, researchers are bound to differ regarding which is best. Because the differences cannot be resolved by appeal to objective criteria, arguments over the best solution sometimes become vociferous. However, those who expect a certain amount of ambiguity with respect to choice of the best FA solution will not be surprised when other researchers choose a different one. Nor will they be surprised when results are not replicated exactly, if different decisions are made at one, or more, of the steps in performing FA.

A third problem is that FA is frequently used in an attempt to "save" poorly conceived research. If no other statistical procedure is applicable, at least data can usually be factor analyzed.

Thus, in the minds of many, the various forms of FA are associated with sloppy research. The very power of PCA and FA to create apparent order from real chaos contributes to their somewhat tarnished reputations as scientific tools.

There are two major types of FA: exploratory and confirmatory. In exploratory FA, one seeks to describe and summarize data by grouping together variables that are correlated. The variables themselves may or may not have been chosen with potential underlying processes in mind. Exploratory FA is usually performed in the early stages of research, when it provides a tool for consolidating variables and for generating hypotheses about underlying processes. Confirmatory FA is a much more sophisticated technique used in the advanced stages of the research process to test a theory about latent processes. Variables are carefully and specifically chosen to reveal underlying processes. Confirmatory FA is often performed through structural equation modeling (Chapter 14).

Before we go on, it is helpful to define a few terms. The first terms involve correlation matrices. The correlation matrix produced by the observed variables is called the *observed correlation matrix*. The correlation matrix produced from factors is called the *reproduced correlation matrix*. The difference between observed and reproduced correlation matrices is the *residual correlation matrix*. In a good FA, correlations in the residual matrix are small, indicating a close fit between the observed and reproduced matrices.

A second set of terms refers to matrices produced and interpreted as part of the solution. Rotation of factors is a process by which the solution is made more interpretable without changing its underlying mathematical properties. There are two general classes of rotation: orthogonal and oblique. If rotation is *orthogonal* (so that all the factors are uncorrelated with each other), a *loading* matrix is produced. The loading matrix is a matrix of correlations between observed variables and factors. The sizes of the loadings reflect the extent of relationship between each observed variable and each factor. Orthogonal FA is interpreted from the loading matrix by looking at which observed variables correlate with each factor.

If rotation is *oblique* (so that the factors themselves are correlated), several additional matrices are produced. The *factor correlation* matrix contains the correlations among the factors. The loading matrix from orthogonal rotation splits into two matrices for oblique rotation: a *structure* matrix of correlations between factors and variables and a *pattern* matrix of unique relationships (uncontaminated by overlap among factors) between each factor and each observed variable. Following oblique rotation, the meaning of factors is ascertained from the pattern matrix.

Lastly, for both types of rotations, there is a *factor-score* coefficients matrix, a matrix of coefficients used in several regression-like equations to predict scores on factors from scores on observed variables for each individual.

FA produces *factors*, while PCA produces *components*. However, the processes are similar except in preparation of the observed correlation matrix for extraction and in the underlying theory. Mathematically, the difference between PCA and FA is in the variance that is analyzed. In PCA, all the variance in the observed variables is analyzed. In FA, only shared variance is analyzed; attempts are made to estimate and eliminate variance due to error and variance that is unique to each variable. The term *factor* is used here to refer to both components and factors unless the distinction is critical, in which case the appropriate term is used.

Theoretically, the difference between FA and PCA lies in the reason that variables are associated with a factor or component. Factors are thought to "cause" variables—the underlying construct (the factor) is what produces scores on the variables. Thus, exploratory FA is associated with theory development and confirmatory FA is associated with theory testing. The question in exploratory FA is: What are the underlying processes that could have produced correlations among these variables? The question in confirmatory FA is: Are the correlations among variables consistent with a hypothesized factor structure? Components are simply aggregates of correlated variables. In that sense, the variables "cause"—or produce—the component. There is no underlying theory about which variables should be associated with which factors; they are simply empirically associated. It is understood that any labels applied to derived components are merely convenient descriptions of the combination of variables associated with them, and do not necessarily reflect some underlying process.

Parinet, Lhote, and Legube (2004) used principal components analysis without rotation to provide an empirical summary of 18 variables measured repeatedly from nine lakes on the Ivory Coast. Two components were identified that accounted for 62% of the total variance in feedback effects. The first component summarized measures of the mineral content of the lakes while the second summarized colonization by phytoplankton. Using these results and the pattern of correlations among the variables, it was possible to further reduce the number of variables to four which were easily and cheaply measured but provided good prediction.

Mudrack (2004) used principal components analysis followed by orthogonal rotation to examine the structure of the "Belief in a Just World" inventory originally devised by Rubin and Peplau (1975). The original 20 items were empirically summarized by 11 components that accounted for 83.8% of the variance. The first two components were characterized as "deserved misfortune" (because of bad behavior) and "deserved good fortune" (because of good behavior), respectively. Mudrack attributes the clarity of the structure to retention of the correct number of components.

Collins, Litman, and Spielberger (2004) used principal factors extraction with oblique rotation to investigate the nature of perceptual curiosity. Two factors of six items each were identified for men and for women. The first factor included items related to ". . . exploring new places and seeking a broad range of perceptual stimulation . . ." while the second included items related to "engaging in a closer inspection of a specific stimulus" (p. 1137). The two factors were, however, correlated 0.52 for women and 0.48 for men, providing support for generalized perceptual curiosity. The utility of the factors was verified by their pattern of correlations with scales of sensation seeking and novelty experiencing.

13.2   Kinds of Research Questions

The goal of research using PCA or FA is to reduce a large number of variables to a smaller number of factors, to concisely describe (and perhaps understand) the relationships among observed variables, or to test theory about underlying processes. Some of the specific questions that are frequently asked are presented in Sections 13.2.1 through 13.2.5.

13.2.1   Number of Factors

How many reliable and interpretable factors are there in the data set? How many factors are needed to summarize the pattern of correlations in the correlation matrix? In the graduate student example, two factors are discussed; are these both reliable? Are there any more factors that are reliable? Strategies for choosing an appropriate number of factors and for assessing the correspondence between observed and reproduced correlation matrices are discussed in Section 13.6.2.

13.2.2   Nature of Factors

What is the meaning of the factors? How are the factors to be interpreted? Factors are interpreted by the variables that correlate with them. Rotation to improve interpretability is discussed in Section 13.6.3; interpretation itself is discussed in Section 13.6.5.

13.2.3   Importance of Solutions and Factors

How much variance in a data set is accounted for by the factors? Which factors account for the most variance? In the graduate student example, does the independence or intellectual ability factor account for more of the variance in the measured variables? How much variance does each account for? In a good factor analysis, a high percentage of the variance in the observed variables is accounted for by the first few factors. And, because factors are computed in descending order of magnitude, the first factor accounts for the most variance, with later factors accounting for less and less of the variance until they are no longer reliable. Methods for assessing the importance of solutions and factors are in Section 13.6.4.

13.2.4   Testing Theory in FA

How well does the obtained factor solution fit an expected factor solution? If the researcher had generated hypotheses regarding both the number and the nature of the factors expected of graduate students, comparisons between the hypothesized factors and the factor solution provide a test of the hypotheses. Tests of theory in FA are addressed, in preliminary form, in Sections 13.6.2 and 13.6.7.

More highly developed techniques are available for testing theory in complex data sets in the form of structural equation modeling, which can also be used to test theory regarding factor structure. These techniques are sometimes known by the names of the most popular programs for doing them, EQS and LISREL. Structural equation modeling is the focus of Chapter 14. Confirmatory factor analysis is demonstrated in Section 14.7.

13.2.5   Estimating Scores on Factors

Had factors been measured directly, what scores would subjects have received on each of them? For instance, if each graduate student were measured directly on independence and intelligence, what scores would each student receive for each of them? Estimation of factor scores is the topic of Section 13.6.6.

13.3   Limitations

13.3.1   Theoretical Issues

Most applications of PCA or FA are exploratory in nature; FA is used primarily as a tool for reducing the number of variables or examining patterns of correlations among variables. Under these circumstances, both the theoretical and the practical limitations to FA are relaxed in favor of a frank exploration of the data. Decisions about number of factors and rotational scheme are based on pragmatic rather than theoretical criteria.

The research project that is designed specifically to be factor analyzed, however, differs from other projects in several important respects. Among the best detailed discussions of the differences is the one found in Comrey and Lee (1992), from which some of the following discussion is taken.

The first task of the researcher is to generate hypotheses about factors believed to underlie the domain of interest. Statistically, it is important to make the research inquiry broad enough to include five or six hypothesized factors so that the solution is stable. Logically, in order to reveal the processes underlying a research area, all relevant factors have to be included. Failure to measure some important factor may distort the apparent relationships among measured factors. Inclusion of all relevant factors poses a logical, but not statistical, problem to the researcher.

Next, one selects variables to observe. For each hypothesized factor, five or six variables, each thought to be a relatively pure measure of the factor, are included. Pure measures are called *marker variables*. Marker variables are highly correlated with one and only one factor and load on it regardless of extraction or rotation technique. Marker variables are useful because they define clearly the nature of a factor; adding potential variables to a factor to round it out is much more meaningful if the factor is unambiguously defined by marker variables to begin with.

The complexity of the variables is also considered. Complexity is indicated by the number of factors with which a variable correlates. A pure variable, which is preferred, is correlated with only one factor, whereas a complex variable is correlated with several. If variables differing in complexity are all included in an analysis, those with similar complexity levels may "catch" each other in factors that have little to do with underlying processes. Variables with similar complexity may correlate with each other because of their complexity and not because they relate to the same factor. Estimating (or avoiding) the complexity of variables is part of generating hypotheses about factors and selecting variables to measure them.

Several other considerations are required of the researcher planning a factor analytic study. It is important, for instance, that the sample chosen exhibits spread in scores with respect to the variables and the factors they measure. If all subjects achieve about the same score on some factor, correlations among the observed variables are low and the factor may not emerge in analysis. Selection of subjects expected to differ on the observed variables and underlying factors is an important design consideration.

One should also be wary of pooling the results of several samples, or the same sample with measures repeated in time, for factor analytic purposes. First, samples that are known to be different with respect to some criterion (e.g., socioeconomic status) may also have different factors. Examination of group differences is often quite revealing. Second, underlying factor structure may shift in time for the same subjects with learning or with experience in an experimental setting and these differences may also be quite revealing. Pooling results from diverse groups in FA may obscure differences rather than illuminate them. On the other hand, if different samples do produce the same factors, pooling them is desirable because of increase in sample size. For example, if men and women produce the same factors, the samples should be combined and the results of the single FA reported.

13.3.2   Practical Issues

Because FA and PCA are exquisitely sensitive to the sizes of correlations, it is critical that honest correlations be employed. Sensitivity to outlying cases, problems created by missing data, and degradation of correlations between poorly distributed variables all plague FA and PCA. A review of these issues in Chapter 4 is important to FA and PCA. Thoughtful solutions to some of the problems, including variable transformations, may markedly enhance FA, whether performed for exploratory or confirmatory purposes. However, the limitations apply with greater force to confirmatory FA.

13.3.2.1 Sample Size and Missing Data

Correlation coefficients tend to be less reliable when estimated from small samples. Therefore, it is important that sample size be large enough that correlations are reliably estimated. The required sample size also depends on magnitude of population correlations and number of factors: if there are strong correlations and a few, distinct factors, a smaller sample size is adequate.

Comrey and Lee (1992) give as a guide sample sizes of 50 as very poor, 100 as poor, 200 as fair, 300 as good, 500 as very good, and 1000 as excellent. *As a general rule of thumb, it is comforting to have at least 300 cases for factor analysis.* Solutions that have several high loading marker variables (> .80) do not require such large sample sizes (about 150 cases should be sufficient) as solutions with lower loadings and/or fewer marker variables (Guadagnoli & Velicer, 1988). Under some circumstances 100—or even 50—cases are sufficient (Sapnas & Zeller, 2002; Zeller, 2005).

If cases have missing data, either the missing values are estimated, the cases deleted, or a missing data (pairwise) correlation matrix is analyzed. Consult Chapter 4 for methods of finding and estimating missing values and cautions about pairwise deletion of cases. Consider the distribution of missing values (is it random?) and remaining sample size when deciding between estimation and deletion. If cases are missing values in a nonrandom pattern or if sample size becomes too small, estimation is in order. However, beware of using estimation procedures (such as regression) that are likely to overfit the data and cause correlations to be too high. These procedures may "create" factors.

13.3.2.2 Normality

As long as PCA and FA are used descriptively as convenient ways to summarize the relationships in a large set of observed variables, assumptions regarding the distributions of variables are not in force. If variables are normally distributed, the solution is enhanced. To the extent that normality fails, the solution is degraded but may still be worthwhile.

However, multivariate normality is assumed when statistical inference is used to determine the number of factors. Multivariate normality is the assumption that all variables, and all linear combinations of variables, are normally distributed. Although tests of multivariate normality are overly sensitive, *normality among single variables is assessed by skewness and kurtosis* (see Chapter 4 and Section 13.7.1.2). If a variable has substantial skewness and kurtosis, variable transformation is considered.

13.3.2.3 Linearity

Multivariate normality also implies that relationships among pairs of variables are linear. The analysis is degraded when linearity fails, because correlation measures linear relationship and does not reflect nonlinear relationship. *Linearity among pairs of variables is assessed through inspection of scatterplots.* Consult Chapter 4 and Section 13.7.1.3 for methods of screening for linearity. If nonlinearity is found, transformation of variables is considered.

13.3.2.4 Absence of Outliers among Cases

As in all multivariate techniques, cases may be outliers either on individual variables (univariate) or on combinations of variables (multivariate). Such cases have more influence on the factor solution than other cases. Consult Chapter 4 and Section 13.7.1.4 for methods of detecting and reducing the influence of both univariate and multivariate outliers.

13.3.2.5 Absence of Multicollinearity and Singularity

In PCA, multicollinearity is not a problem because there is no need to invert a matrix. For most forms of FA and for estimation of factor scores in any form of FA, singularity or extreme multicollinearity is a problem. For FA, if the determinant of **R** and eigenvalues associated with some factors approach 0, multicollinearity or singularity may be present.

To investigate further, look at the SMCs for each variable where it serves as DV with all other variables as IVs. If any of the SMCs is one, singularity is present; if any of the SMCs is very large (near one), multicollinearity is present. Delete the variable with multicollinearity or singularity. Chapter 4 and Section 13.7.1.5 provide examples of screening for and dealing with multicollinearity and singularity.

13.3.2.6 Factorability of R

A matrix that is factorable should include several sizable correlations. The expected size depends, to some extent, on *N* (larger sample sizes tend to produce smaller correlations), but if no correlation exceeds .30, use of FA is questionable because there is probably nothing to factor analyze. *Inspect **R** for correlations in excess of .30, and, if none is found, reconsider use of FA.*

High bivariate correlations, however, are not ironclad proof that the correlation matrix contains factors. It is possible that the correlations are between only two variables and do not reflect underlying processes that are simultaneously affecting several variables. For this reason, it is helpful to examine matrices of partial correlations where pairwise correlations are adjusted for effects of all other variables. If there are factors present, then high bivariate correlations become very low partial correlations. SPSS and SAS produce partial correlation matrices.

Bartlett's (1954) test of sphericity is a notoriously sensitive test of the hypothesis that the correlations in a correlation matrix are zero. The test is available in SPSS FACTOR but because of its sensitivity and its dependence on *N*, the test is likely to be significant with samples of substantial size even if correlations are very low. Therefore, use of the test is recommended only if there are fewer than, say, five cases per variable.

Several more sophisticated tests of the factorability of **R** are available through SPSS and SAS. Both programs give significance tests of correlations, the anti-image correlation matrix, and Kaiser's (1970, 1974) measure of sampling adequacy. Significance tests of correlations in the correlation matrix provide an indication of the reliability of the relationships between pairs of variables. If **R** is factorable, numerous pairs are significant. The anti-image correlation matrix contains the negatives of partial correlations between pairs of variables with effects of other variables removed. If **R** is factorable, there are mostly small values among the off-diagonal elements of the anti-image matrix. Finally, Kaiser's measure of sampling adequacy is a ratio of the sum of squared correlations to the sum of squared correlations plus sum of squared partial correlations. The value approaches 1 if partial correlations are small. Values of .6 and above are required for good FA.

13.3.2.7 Absence of Outliers among Variables

After FA, in both exploratory and confirmatory FA, variables that are unrelated to others in the set are identified. These variables are usually not correlated with the first few factors although they often correlate with factors extracted later. These factors are usually unreliable, both because they account for very little variance and because factors that are defined by just one or two variables are not stable. Therefore, one never knows whether these factors are "real." Suggestions for determining reliability of factors defined by one or two variables are in Section 13.6.2.

If the variance accounted for by a factor defined by only one or two variables is high enough, the factor is interpreted with great caution or ignored, as pragmatic considerations dictate. In confirmatory FA, the factor represents either a promising lead for future work or (probably) error variance, but its interpretation awaits clarification by more research.

A variable with a low squared multiple correlation with all other variables and low correlations with all important factors is an outlier among the variables. The variable is usually ignored in the current FA and either deleted or given friends in future research. Screening for outliers among variables is illustrated in Section 13.7.1.6.

13.4 Fundamental Equations for Factor Analysis

Because of the variety and complexity of the calculations involved in preparing the correlation matrix, extracting factors, and rotating them, and because, in our judgment, little insight is produced by demonstrations of some of these procedures, this section does not show them all. Instead, the relationships between some of the more important matrices are shown, with an assist from SPSS FACTOR for underlying calculations.

Table 13.1 lists many of the important matrices in FA and PCA. Although the list is lengthy, it is composed mostly of *matrices of correlations* (between variables, between factors, and between variables and factors), *matrices of standard scores* (on variables and on factors), *matrices of regression weights* (for producing scores on factors from scores on variables), and the *pattern matrix* of unique relationships between factors and variables after oblique rotation.

Also in the table are the matrix of eigenvalues and the matrix of their corresponding eigenvectors. Eigenvalues and eigenvectors are discussed here and in Appendix A, albeit scantily, because of their importance in factor extraction, the frequency with which one encounters the terminology, and the close association between eigenvalues and variance in statistical applications.

A data set appropriate for FA consists of numerous subjects each measured on several variables. A grossly inadequate data set appropriate for FA is in Table 13.2. Five subjects who were trying on ski boots late on a Friday night in January were asked about the importance of each of four variables to their selection of a ski resort. The variables were cost of ski ticket (COST), speed of ski lift (LIFT), depth of snow (DEPTH), and moisture of snow (POWDER). Larger numbers indicate greater importance. The researcher wanted to investigate the pattern of relationships among the variables in an effort to understand better the dimensions underlying choice of ski area.

Notice the pattern of correlations in the correlation matrix as set off by the vertical and horizontal lines. The strong correlations in the upper left and lower right quadrants show that scores on COST and LIFT are related, as are scores on DEPTH and POWDER. The other two quadrants show that scores on DEPTH and LIFT are unrelated, as are scores on POWDER and LIFT, and so on. With luck, FA will find this pattern of correlations, easy to see in a small correlation matrix but not in a very large one.

TABLE 13.1   **Commonly Encountered Matrices in Factor Analyses**

Label	Name	Rotation	Size[a]	Description
R	Correlation matrix	Both orthogonal and oblique	$p \times p$	Matrix of correlations between variables
Z	Variable matrix	Both orthogonal and oblique	$N \times p$	Matrix of standardized observed variable scores
F	Factor-score matrix	Both orthogonal and oblique	$N \times m$	Matrix of standardized scores on factors or components
A	Factor loading matrix / Pattern matrix	Orthogonal / Oblique	$p \times m$	Matrix of regression-like weights used to estimate the unique contribution of each factor to the variance in a variable. If orthogonal, also correlations between variables and factors
B	Factor-score coefficients matrix	Both orthogonal and oblique	$p \times m$	Matrix of regression-like weights used to generate factor scores from variables
C	Structure matrix[b]	Oblique	$p \times m$	Matrix of correlations between variables and (correlated) factors
Φ	Factor correlation matrix	Oblique	$m \times m$	Matrix of correlations among factors
L	Eigenvalue matrix[c]	Both orthogonal and oblique	$m \times m$	Diagonal matrix of eigenvalues, one per factor[e]
V	Eigenvector matrix[d]	Both orthogonal and oblique	$p \times m$	Matrix of eigenvectors, one vector per eigenvalue

[a]Row by column dimensions where

  p = number of variables

  N = number of subjects

  m = number of factors or components.

[b]In most textbooks, the structure matrix is labeled **S**. However, we have used **S** to represent the sum-of-squares and cross-products matrix elsewhere and will use **C** for the structure matrix here.

[c]Also called characteristic roots or latent roots.

[d]Also called characteristic vectors.

[e]If the matrix is of full rank, there are actually p rather than m eigenvalues and eigenvectors. Only m are of interest, however, so the remaining $p - m$ are not displayed.

13.4.1   Extraction

An important theorem from matrix algebra indicates that, under certain conditions, matrices can be diagonalized. Correlation and covariance matrices are among those that often can be diagonalized. When a matrix is diagonalized, it is transformed into a matrix with numbers in the positive diagonal[2]

[2]The positive diagonal runs from upper left to lower right in a matrix.

**TABLE 13.2   Small Sample of Hypothetical Data
for Illustration of Factor Analysis**

	Variables			
Skiers	COST	LIFT	DEPTH	POWDER
S_1	32	64	65	67
S_2	61	37	62	65
S_3	59	40	45	43
S_4	36	62	34	35
S_5	62	46	43	40

Correlation Matrix

	COST	LIFT	DEPTH	POWDER
COST	1.000	−.953	−.055	−.130
LIFT	−.953	1.000	−.091	−.036
DEPTH	−.055	−.091	1.000	.990
POWDER	−.130	−.036	.990	1.000

and zeros everywhere else. In this application, the numbers in the positive diagonal represent variance from the correlation matrix that has been repackaged as follows:

$$\mathbf{L} = \mathbf{V'RV} \qquad (13.1)$$

Diagonalization of **R** is accomplished by post- and pre-multiplying it by the matrix **V** and its transpose.

The columns in **V** are called eigenvectors, and the values in the main diagonal of **L** are called eigenvalues. The first eigenvector corresponds to the first eigenvalue, and so forth.

Because there are four variables in the example, there are four eigenvalues with their corresponding eigenvectors. However, because the goal of FA is to summarize a pattern of correlations with as few factors as possible, and because each eigenvalue corresponds to a different potential factor, usually only factors with large eigenvalues are retained. In a good FA, these few factors almost duplicate the correlation matrix.

In this example, when no limit is placed on the number of factors, eigenvalues of 2.02, 1.94, .04, and .00 are computed for each of the four possible factors. Only the first two factors, with values over 1.00, are large enough to be retained in subsequent analyses. FA is rerun specifying extraction of just the first two factors; they have eigenvalues of 2.00 and 1.91, respectively, as indicated in Table 13.3.

TABLE 13.3 Eigenvectors and Corresponding Eigenvalues for the Example

Eigenvector 1	Eigenvector 2
−.283	.651
.177	−.685
.658	.252
.675	.207

Eigenvalue 1	Eigenvalue 2
2.00	1.91

Using Equation 13.1 and inserting the values from the example, we obtain

$$
\mathbf{L} = \begin{bmatrix} -.283 & .177 & .658 & .675 \\ .651 & -.685 & .252 & .207 \end{bmatrix} \begin{bmatrix} 1.000 & -.953 & -.055 & -.130 \\ -.953 & 1.000 & -.091 & -.036 \\ -.055 & -.091 & 1.000 & .990 \\ -.130 & -.036 & .990 & 1.000 \end{bmatrix} \begin{bmatrix} -.283 & .651 \\ .177 & -.685 \\ -.658 & .252 \\ .675 & .207 \end{bmatrix}
$$

$$
= \begin{bmatrix} 2.00 & .00 \\ .00 & 1.91 \end{bmatrix}
$$

(All values agree with computer output. Hand calculation may produce discrepancies due to rounding error.)

The matrix of eigenvectors pre-multiplied by its transpose produces the identity matrix with ones in the positive diagonal and zeros elsewhere. Therefore, pre- and post-multiplying the correlation matrix by eigenvectors does not change it so much as repackage it.

$$
\mathbf{V'V} = \mathbf{I} \tag{13.2}
$$

For the example:

$$
\begin{bmatrix} -.283 & .177 & .658 & .675 \\ .651 & -.685 & .252 & .207 \end{bmatrix} \begin{bmatrix} -.283 & .651 \\ .177 & -.685 \\ -.658 & .252 \\ .675 & .207 \end{bmatrix} = \begin{bmatrix} 1.000 & .000 \\ .000 & 1.000 \end{bmatrix}
$$

The important point is that because correlation matrices often meet requirements for diagonalizability, it is possible to use on them the matrix algebra of eigenvectors and eigenvalues with FA as the result. When a matrix is diagonalized, the information contained in it is repackaged. In FA, the variance in the correlation matrix is condensed into eigenvalues. The factor with the largest eigenvalue has the most variance and so on, down to factors with small or negative eigenvalues that are usually omitted from solutions.

Calculations for eigenvectors and eigenvalues are extremely laborious and not particularly enlightening (although they are illustrated in Appendix A for a small matrix). They require solving p equations in p unknowns with additional side constraints and are rarely performed by hand. Once the eigenvalues and eigenvectors are known, however, the rest of FA (or PCA) more or less "falls out," as is seen from Equations 13.3 to 13.6.

Equation 13.1 can be reorganized as follows:

$$\mathbf{R} = \mathbf{VLV}' \tag{13.3}$$

The correlation matrix can be considered a product of three matrices—the matrices of eigenvalues and corresponding eigenvectors.

After reorganization, the square root is taken of the matrix of eigenvalues.

$$\mathbf{R} = \mathbf{V}\sqrt{\mathbf{L}}\sqrt{\mathbf{L}}\mathbf{V}'$$

or

$$\mathbf{R} = (\mathbf{V}\sqrt{\mathbf{L}})(\sqrt{\mathbf{L}}\mathbf{V}') \tag{13.4}$$

If $\mathbf{V}\sqrt{\mathbf{L}}$ is called $\mathbf{A}$, and $\sqrt{\mathbf{L}}\mathbf{V}'$ is $\mathbf{A}'$, then

$$\mathbf{R} = \mathbf{AA}' \tag{13.5}$$

The correlation matrix can also be considered a product of two matrices, each a combination of eigenvectors and the square root of eigenvalues.

Equation 13.5 is frequently called the fundamental equation for FA.[3] It represents the assertion that the correlation matrix is a product of the factor loading matrix, $\mathbf{A}$, and its transpose.

Equations 13.4 and 13.5 also reveal that the major work of FA (and PCA) is calculation of eigenvalues and eigenvectors. Once they are known, the (unrotated) factor loading matrix is found by straightforward matrix multiplication, as follows.

$$\mathbf{A} = \mathbf{V}\sqrt{\mathbf{L}} \tag{13.6}$$

For the example:

$$\mathbf{A} = \begin{bmatrix} -.283 & .651 \\ .177 & -.685 \\ .658 & .252 \\ .675 & .207 \end{bmatrix} \begin{bmatrix} \sqrt{2.00} & 0 \\ 0 & \sqrt{2.00} \end{bmatrix} = \begin{bmatrix} -.400 & .900 \\ .251 & -.947 \\ .932 & .348 \\ .956 & .286 \end{bmatrix}$$

The factor loading matrix is a matrix of correlations between factors and variables. The first column is correlations between the first factor and each variable in turn, COST($-.400$), LIFT (.251), DEPTH (.932), and POWDER (.956). The second column is correlations between the second factor

---

[3]In order to reproduce the correlation matrix exactly, as indicated in Equations 13.4 and 13.5, all eigenvalues and eigenvectors are necessary, not just the first few of them.

and each variable in turn, COST (.900), LIFT (−.947), DEPTH (.348), and POWDER (.286). A factor is interpreted from the variables that are highly correlated with it—that have high loadings on it. Thus, the first factor is primarily a snow conditions factor (DEPTH and POWDER), while the second reflects resort conditions (COST and LIFT). Subjects who score high on the resort conditions factor (Equation 13.11) tend to assign high value to COST and low value to LIFT (the negative correlation); subjects who score low on the resort conditions factor value LIFT more than COST.

Notice, however, that all the variables are correlated with both factors to a considerable extent. Interpretation is fairly clear for this hypothetical example, but most likely would not be for real data. Usually a factor is most interpretable when a few variables are highly correlated with it and the rest are not.

13.4.2   Orthogonal Rotation

Rotation is ordinarily used after extraction to maximize high correlations between factors and variables and minimize low ones. Numerous methods of rotation are available (see Section 13.5.2) but the most commonly used, and the one illustrated here, is *varimax*. Varimax is a variance maximizing procedure. The goal of varimax rotation is to maximize the variance of factor loadings by making high loadings higher and low ones lower for each factor.

This goal is accomplished by means of a transformation matrix Λ (as defined in Equation 13.8), where

$$\mathbf{A}_{\text{unrotated}} \Lambda = \mathbf{A}_{\text{rotated}} \tag{13.7}$$

The unrotated factor loading matrix is multiplied by the transformation matrix to produce the rotated loading matrix.

For the example:

$$\mathbf{A}_{\text{rotated}} = \begin{bmatrix} -.400 & .900 \\ .251 & -.947 \\ .932 & .348 \\ .956 & .286 \end{bmatrix} \begin{bmatrix} .946 & -.325 \\ .325 & .946 \end{bmatrix} = \begin{bmatrix} -.086 & .981 \\ -.071 & -.977 \\ .994 & .026 \\ .997 & -.040 \end{bmatrix}$$

Compare the rotated and unrotated loading matrices. Notice that in the rotated matrix the low correlations are lower and the high ones higher than in the unrotated loading matrix. Emphasizing differences in loadings facilitates interpretation of a factor by making unambiguous the variables that correlate with it.

The numbers in the transformation matrix have a spatial interpretation.

$$\Lambda = \begin{bmatrix} \cos \Psi & -\sin \Psi \\ \sin \Psi & \cos \Psi \end{bmatrix} \tag{13.8}$$

The transformation matrix is a matrix of sines and cosines of an angle Ψ.

TABLE 13.4   Relationships among Loadings, Communalities, SSLs, Variance, and Covariance of Orthogonally Rotated Factors

	Factor 1	Factor 2	Communalities (h^2)
COST	−.086	.981	$\sum a^2 = .970$
LIFT	−.071	−.977	$\sum a^2 = .960$
DEPTH	.994	.026	$\sum a^2 = .989$
POWDER	.997	−.040	$\sum a^2 = .996$
SSLs	$\sum a^2 = 1.994$	$\sum a^2 = 1.919$	3.915
Proportion of variance	.50	.48	.98
Proportion of covariance	.51	.49	

For the example, the angle is approximately 19°. That is, cos 19 < .946 and sin 19 < .325. Geometrically, this corresponds to a 19° swivel of the factor axes about the origin. Greater detail regarding the geometric meaning of rotation is in Section 13.5.2.3.

13.4.3   Communalities, Variance, and Covariance

Once the rotated loading matrix is available, other relationships are found, as in Table 13.4. The communality for a variable is the variance accounted for by the factors. It is the squared multiple correlation of the variable as predicted from the factors. Communality is the sum of squared loadings (SSL) for a variable across factors. In Table 13.4, the communality for COST is $(-.086)^2 + .981^2 = .970$. That is, 97% of the variance in COST is accounted for by Factor 1 plus Factor 2.

The proportion of variance *in the set of variables* accounted for by a factor is the SSL for the factor divided by the number of variables (if rotation is orthogonal).[4] For the first factor, the proportion of variance is $[(-.086)^2 + (-.071)^2 + .994^2 + .997^2]/4 = 1.994/4 = .50$. Fifty percent of the variance in the variables is accounted for by the first factor. The second factor accounts for 48% of the variance in the variables and, because rotation is orthogonal, the two factors together account for 98% of the variance in the variables.

The proportion of variance *in the solution* accounted for by a factor—the proportion of covariance—is the SSL for the factor divided by the sum of communalities (or, equivalently, the sum of the SSLs). The first factor accounts for 51% of the variance in the solution (1.994/3.915) while the second factor accounts for 49% of the variance in the solution (1.919/3.915). The two factors together account for all of the covariance.

---

[4]For unrotated factors only, the sum of the squared loadings for a factor is equal to the eigenvalue. Once loadings are rotated, the sum of squared loadings is called SSL and is no longer equal to the eigenvalue.

The reproduced correlation matrix for the example is generated using Equation 13.5:

$$\overline{\mathbf{R}} = \begin{bmatrix} -.086 & .981 \\ -.071 & -.977 \\ .994 & .026 \\ .997 & -.040 \end{bmatrix} \begin{bmatrix} -.086 & -.071 & .994 & .997 \\ .981 & -.977 & .026 & -.040 \end{bmatrix}$$

$$= \begin{bmatrix} .970 & -.953 & -.059 & -.125 \\ -.953 & .962 & -.098 & -.033 \\ -.059 & -.098 & .989 & .990 \\ -.125 & -.033 & .990 & .996 \end{bmatrix}$$

Notice that the reproduced correlation matrix differs slightly from the original correlation matrix. The difference between the original and reproduced correlation matrices is the residual correlation matrix:

$$\mathbf{R}_{res} = \mathbf{R} - \overline{\mathbf{R}} \qquad (13.9)$$

The residual correlation matrix is the difference between the observed correlation matrix and the reproduced correlation matrix.

For the example, with communalities inserted in the positive diagonal of $\mathbf{R}$:

$$\mathbf{R}_{res} = \begin{bmatrix} .970 & -.953 & -.055 & -.130 \\ -.953 & .960 & -.091 & -.036 \\ -.055 & -.091 & .989 & .990 \\ -.130 & -.036 & .990 & .996 \end{bmatrix} - \begin{bmatrix} .970 & -.953 & -.059 & -.125 \\ -.953 & .960 & -.098 & -.033 \\ -.059 & -.098 & .989 & .990 \\ -.125 & -.033 & .990 & .996 \end{bmatrix}$$

$$= \begin{bmatrix} .000 & .000 & .004 & -.005 \\ .000 & .000 & .007 & -.003 \\ .004 & .007 & .000 & .000 \\ -.005 & -.003 & .000 & .000 \end{bmatrix}$$

In a "good" FA, the numbers in the residual correlation matrix are small because there is little difference between the original correlation matrix and the correlation matrix generated from factor loadings.

13.4.4 Factor Scores

Scores on factors can be predicted for each case once the loading matrix is available. Regression-like coefficients are computed for weighting variable scores to produce factor scores. Because $\mathbf{R}^{-1}$ is the inverse of the matrix of correlations among variables and $\mathbf{A}$ is the matrix of correlations between factors and variables, Equation 13.10 for factor score coefficients is similar to Equation 5.6 for regression coefficients in multiple regression.

$$\mathbf{B} = \mathbf{R}^{-1}\mathbf{A} \tag{13.10}$$

Factor score coefficients for estimating factor scores from variable scores are a product of the inverse of the correlation matrix and the factor loading matrix.

For the example:[5]

$$\mathbf{B} = \begin{bmatrix} 25.485 & 22.689 & -31.655 & 35.479 \\ 22.689 & 21.386 & -24.831 & 28.312 \\ -31.655 & -24.831 & 99.917 & -103.950 \\ 35.479 & 28.312 & -103.950 & 109.567 \end{bmatrix} \begin{bmatrix} -.087 & .981 \\ -.072 & -.978 \\ .994 & .027 \\ .997 & -.040 \end{bmatrix}$$

$$= \begin{bmatrix} 0.082 & 0.537 \\ 0.054 & -0.461 \\ 0.190 & 0.087 \\ 0.822 & -0.074 \end{bmatrix}$$

To estimate a subject's score for the first factor, then, all of the subject's scores on variables are standardized and then the standardized score on COST is weighted by 0.082, LIFT by 0.054, DEPTH by 0.190, and POWDER by 0.822 and the results are added. In matrix form,

$$\mathbf{F} = \mathbf{ZB} \tag{13.11}$$

Factor scores are a product of standardized scores on variables and factor score coefficients.

For the example:

$$\mathbf{F} = \begin{bmatrix} -1.22 & 1.14 & 1.15 & 1.14 \\ 0.75 & -1.02 & 0.92 & 1.01 \\ 0.61 & -0.78 & -0.36 & -0.47 \\ -0.95 & 0.98 & -1.20 & -1.01 \\ 0.82 & -0.30 & -0.51 & -0.67 \end{bmatrix} \begin{bmatrix} 0.082 & 0.537 \\ 0.054 & -0.461 \\ 0.190 & 0.087 \\ 0.822 & -0.074 \end{bmatrix}$$

$$= \begin{bmatrix} 1.12 & -1.16 \\ 1.01 & 0.88 \\ -0.45 & 0.69 \\ -1.08 & -0.99 \\ -0.60 & 0.58 \end{bmatrix}$$

[5]The numbers in **B** are different from the factor score coefficients generated by computer for the small data set. The difference is due to rounding error following inversion of a multicollinear correlation matrix. Note also that the **A** matrix contains considerable rounding error.

The first subject has an estimated standard score of 1.12 on the first factor and -1.16 on the second factor, and so on for the other four subjects. The first subject strongly values both the snow factor and the resort factor, one positive and the other negative (indicating primary value assigned to speed of LIFT). The second subject values both the snow factor and the resort factor (with more value placed on COST than LIFT); the third subject places more value on resort conditions (particularly COST) and less value on snow conditions, and so forth. The sum of standardized factor scores across subjects for a single factor is zero.

Predicting scores on variables from scores on factors is also possible. The equation for doing so is

$$\mathbf{Z} = \mathbf{FA}'$$ (13.12)

Predicted standardized scores on variables are a product of scores on factors weighted by factor loadings.

For example:

$$\mathbf{Z} = \begin{bmatrix} 1.12 & -1.16 \\ 1.01 & 0.88 \\ -0.45 & 0.69 \\ -1.08 & -0.99 \\ -0.60 & 0.58 \end{bmatrix} \begin{bmatrix} -.086 & -.072 & .994 & .997 \\ .981 & -.978 & .027 & -.040 \end{bmatrix}$$

$$= \begin{bmatrix} -1.23 & 1.05 & 1.08 & 1.16 \\ 0.78 & -0.93 & 1.03 & 0.97 \\ 0.72 & -0.64 & -0.43 & -0.48 \\ -0.88 & 1.05 & -1.10 & -1.04 \\ 0.62 & -0.52 & -0.58 & -0.62 \end{bmatrix}$$

That is, the first subject (the first row of $\mathbf{Z}$) is predicted to have a standardized score of -1.23 on COST, 1.05 on LIFT, 1.08 on DEPTH, and 1.16 on POWDER. Like the reproduced correlation matrix, these values are similar to the observed values if the FA captures the relationship among the variables.

It is helpful to see these values written out because they provide an insight into how scores on variables are conceptualized in factor analysis. For example, for the first subject,

$$-1.23 = -.086(1.12) + .981(-1.16)$$

$$1.05 = -.072(1.12) - .978(-1.16)$$

$$1.08 = .994(1.12) + .027(-1.16)$$

$$1.16 = .997(1.12) - .040(-1.16)$$

Or, in algebraic form,

$$z_{COST} = a_{11}F_1 + a_{12}F_2$$
$$z_{LIFT} = a_{21}F_1 + a_{22}F_2$$
$$z_{DEPTH} = a_{31}F_1 + a_{32}F_2$$
$$z_{POWDER} = a_{41}F_1 + a_{42}F_2$$

A score on an observed variable is conceptualized as a properly weighted and summed combination of the scores on factors that underlie it. The researcher believes that each subject has the same latent factor structure, but different scores on the factors themselves. A particular subject's score on an observed variable is produced as a weighted combination of that subject's scores on the underlying factors.

13.4.5 Oblique Rotation

All the relationships mentioned thus far are for orthogonal rotation. Most of the complexities of orthogonal rotation remain and several others are added when oblique (correlated) rotation is used. Consult Table 13.1 for a listing of additional matrices and a hint of the discussion to follow.

SPSS FACTOR is run on the data from Table 13.2 using the default option for oblique rotation (cf. Section 13.5.2.2) to get values for the pattern matrix, **A**, and factor-score coefficients, **B**.

In oblique rotation, the loading matrix becomes the pattern matrix. Values in the pattern matrix, when squared, represent the unique contribution of each factor to the variance of each variable but do not include segments of variance that come from overlap between correlated factors. For the example, the pattern matrix following oblique rotation is

$$\mathbf{A} = \begin{bmatrix} -.079 & .981 \\ -.078 & -.978 \\ .994 & .033 \\ .977 & -.033 \end{bmatrix}$$

The first factor makes a unique contribution of $-.079^2$ to the variance in COST, $-.078^2$ to LIFT, $.994^2$ to DEPTH, and $.997^2$ to POWDER.

Factor-score coefficients following oblique rotation are also found:

$$\mathbf{B} = \begin{bmatrix} 0.104 & 0.584 \\ 0.081 & -0.421 \\ 0.159 & -0.020 \\ 0.856 & 0.034 \end{bmatrix}$$

Applying Equation 13.11 to produce factor scores results in the following values:

$$\mathbf{F} = \begin{bmatrix} -1.22 & 1.14 & 1.15 & 1.14 \\ 0.75 & -1.02 & 0.92 & 1.01 \\ 0.61 & -0.78 & -0.36 & -0.47 \\ -0.95 & 0.98 & -1.20 & -1.01 \\ 0.82 & -0.30 & -0.51 & -0.67 \end{bmatrix} \begin{bmatrix} 0.104 & 0.584 \\ 0.081 & -0.421 \\ 0.159 & -0.020 \\ 0.856 & 0.034 \end{bmatrix}$$

$$= \begin{bmatrix} 1.12 & -1.18 \\ 1.01 & 0.88 \\ -0.46 & 0.68 \\ -1.07 & -0.98 \\ -0.59 & 0.59 \end{bmatrix}$$

Once the factor scores are determined, correlations among factors can be obtained. Among the equations used for this purpose is

$$\mathbf{\Phi} = \left(\frac{1}{N-1} \right) \mathbf{F'F} \tag{13.13}$$

One way to compute correlations among factors is from cross-products of standardized factor scores divided by the number of cases minus one.

The factor correlation matrix is a standard part of computer output following oblique rotation. For the example:

$$\mathbf{\Phi} = \frac{1}{4} \begin{bmatrix} 1.12 & 1.01 & -0.46 & -1.07 & -0.59 \\ -1.18 & 0.88 & 0.68 & -0.98 & 0.59 \end{bmatrix} \begin{bmatrix} 1.12 & -1.16 \\ 1.01 & 0.88 \\ -0.45 & 0.69 \\ -1.08 & -0.99 \\ -0.60 & 0.58 \end{bmatrix}$$

$$= \begin{bmatrix} 1.00 & -0.01 \\ -0.01 & 1.00 \end{bmatrix}$$

The correlation between the first and second factor is quite low, $-.01$. For this example, there is almost no relationship between the two factors, although considerable correlation could have been produced had it been warranted. Ordinarily one uses orthogonal rotation in a case like this because complexities introduced by oblique rotation are not warranted by such a low correlation among factors.

However, if oblique rotation is used, the structure matrix, **C**, is the correlations between variables and factors. These correlations assess the unique relationship between the variable and the factor (in the pattern matrix) plus the relationship between the variable and the overlapping variance among the factors. The equation for the structure matrix is

$$\mathbf{C} = \mathbf{A}\boldsymbol{\Phi} \tag{13.14}$$

The structure matrix is a product of the pattern matrix and the factor correlation matrix.

For example:

$$\mathbf{C} = \begin{bmatrix} -.079 & .981 \\ -.078 & -.978 \\ .994 & .033 \\ .997 & -.033 \end{bmatrix} \begin{bmatrix} 1.00 & -.01 \\ -.01 & 1.00 \end{bmatrix} = \begin{bmatrix} -.069 & .982 \\ -.088 & -.977 \\ .994 & .023 \\ .997 & -.043 \end{bmatrix}$$

COST, LIFT, DEPTH, and POWDER correlate $-.069$, $-.088$, .994, and .997 with the first factor and .982, $-.977$, .023, and $-.043$ with the second factor, respectively.

There is some debate as to whether one should interpret the pattern matrix or the structure matrix following oblique rotation. The structure matrix is appealing because it is readily understood. However, the correlations between variables and factors are inflated by any overlap between factors. The problem becomes more severe as the correlations among factors increase and it may be hard to determine which variables are related to a factor. On the other hand, the pattern matrix contains values representing the unique contributions of each factor to the variance in the variables. Shared variance is omitted (as it is with standard multiple regression), but the set of variables that composes a factor is usually easier to see. If factors are very highly correlated, it may appear that no variables are related to them because there is almost no unique variance once overlap is omitted.

Most researchers interpret and report the pattern matrix rather than the structure matrix. However, if the researcher reports either the structure or the pattern matrix and also $\boldsymbol{\Phi}$, then the interested reader can generate the other using Equation 13.14 as desired.

In oblique rotation, $\overline{\mathbf{R}}$ is produced as follows:

$$\overline{\mathbf{R}} = \mathbf{C}\mathbf{A}' \tag{13.15}$$

The reproduced correlation matrix is a product of the structure matrix and the transpose of the pattern matrix.

Once the reproduced correlation matrix is available, Equation 13.9 is used to generate the residual correlation matrix to diagnose adequacy of fit in FA.

13.4.6 Computer Analyses of Small-Sample Example

A two-factor principal factor analysis with varimax rotation using the example is shown for SPSS FACTOR and SAS FACTOR in Tables 13.5 and 13.6.

For a principal factor analysis with varimax rotation, SPSS FACTOR requires that you specify EXTRACTION PAF and ROTATION VARIMAX.[6] SPSS FACTOR (Table 13.5) begins by printing out SMCs for each variable, labeled Initial in the Communalities portion of the output. In the same table are final (Extraction) communalities. These show the portion of variance in each variable accounted for by the solution (h^2 in Table 13.4).

The next table shows a great deal of information about variance accounted for by the factors. Initial Eigenvalues, % of Variance, and percent of variance cumulated over the four factors (Cumulative %) are printed out for the four initial factors. (Be careful not to confuse factors with variables.) The remainder of the table shows the percent of variance (sums of squared loadings—see Table 13.3) accounted for by the two factors extracted with eigenvalues greater than 1 (the default value), after extraction and after rotation.

For the two extracted factors, an unrotated Factor (loading) Matrix is then printed. The Rotated Factor Matrix, which matches loadings in Table 13.4, is given along with the Factor Transformation Matrix (Equation 13.8) for orthogonal varimax rotation with Kaiser normalization.

SAS FACTOR (Table 13.6) requires a bit more instruction to produce a principal factor analysis with orthogonal rotation for two factors. You specify the type (method=prinit), initial communalities (priors=smc), number of factors to be extracted (nfactors=2), and the type of rotation (rotate=v). Prior Communality Estimates: SMCs are given, followed by Preliminary Eigenvalues for all four factors; also given is the Total of the eigenvalues and their Average. The next row shows Differences between successive eigenvalues. For example, there is a small difference between the first and second eigenvalues (0.099606) and between the third and fourth eigenvalues (0.020622), but a large difference between the second and third eigenvalues (1.897534). Proportion and Cumulative proportion of variance are then printed for each factor. This is followed by corresponding information for the Reduced Correlation Matrix (after factoring). Information on the iterative process is not shown.

The Factor Pattern matrix contains unrotated factor loadings for the first two factors. (Note that the signs of the FACTOR2 loadings are the reverse of those of SPSS.) SSLs for each factor are in the table labeled Variance explained by each factor. Both Final Communality Estimates (h^2) and the Total h^2 are then given. The Orthogonal Transformation Matrix for rotation (Equation 13.8) is followed by the rotated factor loadings in the Rotated Factor Pattern matrix. SSLs for rotated factors—Variance explained by each factor—appear below the loadings. Final Communality Estimates are then repeated.

---

[6]The defaults for SPSS FACTOR are principal components analysis with no rotation.

TABLE 13.5 Syntax and SPSS FACTOR Output for Factor Analysis on Sample Data of Table 13.2

```
FACTOR
/VARIABLES COST LIFT DEPTH POWDER /MISSING LISTWISE
/ANALYSIS COST LIFT DEPTH POWDER
/PRINT INITIAL EXTRACTION ROTATION
/CRITERIA MINEIGEN(1) ITERATE(25)
/EXTRACTION PAF
/CRITERIA ITERATE(25)
/ROTATION VARIMAX
/METHOD=CORRELATION.
```

Communalities

	Initial	Extraction
COST	.961	.970
LIFT	.953	.960
DEPTH	.990	.989
POWDER	.991	.996

Extraction Method: Principal Axis Factoring.

Total Variance Explained

Factor	Initial Eigenvalues			Extraction Sums of Squared Loadings			Rotation Sums of Squared Loadings		
	Total	% of Variance	Cumulative %	Total	% of Variance	Cumulative %	Total	% of Variance	Cumulative %
1	2.016	50.408	50.408	2.005	50.118	50.118	1.995	49.866	49.866
2	1.942	48.538	98.945	1.909	47.733	97.852	1.919	47.986	97.852
3	.038	.945	99.891						
4	.004	.109	100.000						

Extraction Method: Principal Axis Factoring.

(continued)

TABLE 13.5 **Continued**

Factor Matrix[a]

	Factor	
	1	2
COST	−.400	.900
LIFT	.251	−.947
DEPTH	.932	.348
POWDER	.956	.286

Extraction Method: Principal Axis
Factoring.

[a]2 factors extracted. 4 iterations
  required.

Rotated Factor Matrix[a]

	Factor	
	1	2
COST	−.086	−.981
LIFT	−.071	.977
DEPTH	.994	−.026
POWDER	.997	.040

Extraction Method: Principal Axis
Factoring.
Rotation Method: Varimax with Kaiser
Normalization.

[a]Rotation converged in 3 iterations.

Factor Transformation Matrix

Factor	1	2
1	.946	.325
2	.325	−.946

Extraction Method: Principal Axis
Factoring.
Rotation Method: Varimax with Kaiser
Normalization.

TABLE 13.6 Syntax and Selected SAS FACTOR Output for Factor Analysis of Sample Data of Table 13.2

```
proc factor data=SASUSER.SSFACTOR
method=prinit priors=smc nfactors=2 rotate=v;
  var cost lift depth powder;
run;
```

The FACTOR Procedure

Initial Factor Method: Iterated Principal Factor Analysis

Prior Communality Estimates: SMC

COST	LIFT	DEPTH	POWDER
0.96076070	0.95324069	0.98999165	0.99087317

Preliminary Eigenvalues: Total = 3.89486621 Average = 0.97371655

	Eigenvalue	Difference	Proportion	Cumulative
1	2.00234317	0.09960565	0.5141	0.5141
2	1.90273753	1.89753384	0.4885	1.0026
3	0.00520369	0.02062186	0.0013	1.0040
4	-.01541817		-0.0040	1.0000

2 factors will be retained by the NFACTOR criterion.

WARNING: Too many factors for a unique solution.

Eigenvalues of the Reduced Correlation Matrix: Total = 3.91277649 Average = 0.97819412

	Eigenvalue	Difference	Proportion	Cumulative
1	2.00473399	0.09539900	0.5124	0.5124
2	1.90933499	1.90037259	0.4880	1.0003
3	0.00896240	0.01921730	0.0023	1.0026
4	-.01025490		-0.0026	1.0000

(continued)

TABLE 13.6   Continued

Initial Factor Method: Iterated Principal Factor Analysis

Factor Pattern

	Factor1	Factor2
COST	−0.40027	0.89978
LIFT	0.25060	−0.94706
DEPTH	0.93159	0.34773
POWDER	0.95596	0.28615

Variance Explained by Each Factor

Factor1	Factor2
2.0047340	1.9093350

Final Communality Estimates: Total = 3.914069

COST	LIFT	DEPTH	POWDER
0.96982841	0.95972502	0.98877384	0.99574170

Rotation Method: Varimax

Orthogonal Transformation Matrix

	1	2
1	0.94565	−0.32519
2	0.32519	0.94565

Rotated Factor Pattern

	Factor1	Factor2
COST	−0.08591	0.98104
LIFT	−0.07100	−0.97708
DEPTH	0.99403	0.02588
POWDER	0.99706	−0.04028

Variance Explained by Each Factor

Factor1	Factor2
1.9946455	1.9194235

Final Communality Estimates: Total = 3.914069

COST	LIFT	DEPTH	POWDER
0.96982841	0.95972502	0.98877384	0.99574170

13.5   Major Types of Factor Analyses

Numerous procedures for factor extraction and rotation are available. However, only those procedures available in SPSS and SAS packages are summarized here. Other extraction and rotational techniques are described in Mulaik (1972), Harman (1976), Rummel (1970), Comrey and Lee (1992), and Gorsuch (1983), among others.

13.5.1   Factor Extraction Techniques

Among the extraction techniques available in the packages are principal components (PCA), principal factors, maximum likelihood factoring, image factoring, alpha factoring, and unweighted and generalized (weighted) least squares factoring (see Table 13.7). Of these, PCA and principal factors are the most commonly used.

All the extraction techniques calculate a set of orthogonal components or factors that, in combination, reproduce **R**. Criteria used to establish the solution, such as maximizing variance or mini-

TABLE 13.7   Summary of Extraction Procedures

Extraction Technique	Program	Goal of Analysis	Special Features
Principal components	SPSS SAS	Maximize variance extracted by orthogonal components	Mathematically determined, empirical solution with common, unique, and error variance mixed into components
Principal factors	SPSS SAS	Maximize variance extracted by orthogonal factors	Estimates communalities to attempt to eliminate unique and error variance from variables
Image factoring	SPSS SAS (Image and Harris)	Provides an empirical factor analysis	Uses variances based on multiple regression of a variable with all other variables as communalities to generate a mathematically determined solution with error variance and unique variance eliminated
Maximum likelihood factoring	SAS SPSS	Estimate factor loadings for population that maximize the likelihood of sampling the observed correlation matrix	Has significance test for factors; especially useful for confirmatory factor analysis
Alpha factoring	SPSS SAS	Maximize the generalizability of orthogonal factors	
Unweighted least squares	SPSS SAS	Minimize squared residual correlations	
Generalized least squares	SPSS SAS	Weights variables by shared variance before minimizing squared residual correlations	

TABLE 13.8   Results of Different Extraction Methods on Same Data Set

Variables	Factor 1				Factor 2			
	PCA	PFA	Rao	Alpha	PCA	PFA	Rao	Alpha
			Unrotated factor loadings					
1	.58	.63	.70	.54	.68	.68	−.54	.76
2	.51	.48	.56	.42	.66	.53	−.47	.60
3	.40	.38	.48	.29	.71	.55	−.50	.59
4	.69	.63	.55	.69	−.44	−.43	.54	−.33
5	.64	.54	.48	.59	−.37	−.31	.40	−.24
6	.72	.71	.63	.74	−.47	−.49	.59	−.40
7	.63	.51	.50	.53	−.14	−.12	.17	−.07
8	.61	.49	.47	.50	−.09	−.09	.15	−.03
			Rotated factor loadings (varimax)					
1	.15	.15	.15	.16	.89	.91	.87	.92
2	.11	.11	.10	.12	.83	.71	.72	.73
3	−.02	.01	.02	.00	.81	.67	.69	.66
4	.82	.76	.78	.76	−.02	−.01	−.03	.01
5	.74	.62	.62	.63	.01	.04	.03	.04
6	.86	.86	.87	.84	.04	−.02	−.01	−.03
7	.61	.49	.48	.50	.20	.18	.21	.17
8	.57	.46	.45	.46	.23	.20	.20	.19

Note: The largest difference in communality estimates for a single variable between extraction techniques was 0.08.

mizing residual correlations, differ from technique to technique. But differences in solutions are small for a data set with a large sample, numerous variables and similar communality estimates. In fact, one test of the stability of a FA solution is that it appears regardless of which extraction technique is employed. Table 13.8 shows solutions for the same data set after extraction with several different techniques, followed by varimax rotation. Similarities among the solutions are obvious.

None of the extraction techniques routinely provides an interpretable solution without rotation. All types of extractions may be rotated by any of the procedures described in Section 13.5.2.

Lastly, when using FA the researcher should hold in abeyance well-learned proscriptions against data snooping. It is quite common to use PCA or PFA as a preliminary extraction technique, followed by one or more of the other procedures, perhaps varying number of factors, communality estimates, and rotational methods with each run. Analysis terminates when the researcher decides on the preferred solution.

13.5.1.1   PCA vs. FA

One of the most important decisions is the choice between PCA and FA. Mathematically, the difference involves the contents of the positive diagonal in the correlation matrix (the diagonal that

contains the correlation between a variable and itself). In either PCA or FA, the variance that is analyzed is the sum of the values in the positive diagonal. In PCA ones are in the diagonal and there is as much variance to be analyzed as there are observed variables; each variable contributes a unit of variance by contributing a 1 to the positive diagonal of the correlation matrix. All the variance is distributed to components, including error and unique variance for each observed variable. So if all components are retained, PCA duplicates exactly the observed correlation matrix and the standard scores of the observed variables.

In FA, on the other hand, only the variance that each observed variable shares with other observed variables is available for analysis. Exclusion of error and unique variance from FA is based on the belief that such variance only confuses the picture of underlying processes. Shared variance is estimated by *communalities*, values between 0 and 1 that are inserted in the positive diagonal of the correlation matrix.[7] The solution in FA concentrates on variables with high communality values. The sum of the communalities (sum of the SSLs) is the variance that is distributed among factors and is less than the total variance in the set of observed variables. Because unique and error variances are omitted, a linear combination of factors approximates, but does not duplicate, the observed correlation matrix and scores on observed variables.

PCA analyzes variance; FA analyzes covariance (communality). The goal of PCA is to extract maximum variance from a data set with a few orthogonal components. The goal of FA is to reproduce the correlation matrix with a few orthogonal factors. PCA is a unique mathematical solution, whereas most forms of FA are not unique.

The choice between PCA and FA depends on your assessment of the fit between the models, the data set, and the goals of the research. If you are interested in a theoretical solution uncontaminated by unique and error variability and have designed your study on the basis of underlying constructs that are expected to produce scores on your observed variables, FA is your choice. If, on the other hand, you simply want an empirical summary of the data set, PCA is the better choice.

13.5.1.2   *Principal Components*

The goal of PCA is to extract maximum variance from the data set with each component. The first principal component is the linear combination of observed variables that maximally separates subjects by maximizing the variance of their component scores. The second component is formed from residual correlations; it is the linear combination of observed variables that extracts maximum variability uncorrelated with the first component. Subsequent components also extract maximum variability from residual correlations and are orthogonal to all previously extracted components.

The principal components are ordered, with the first component extracting the most variance and the last component least variance. The solution is mathematically unique and, if all components are retained, exactly reproduces the observed correlation matrix. Further, since the components are orthogonal, their use in other analyses (e.g., as DVs in MANOVA) may greatly facilitate interpretation of results.

PCA is the solution of choice for the researcher who is primarily interested in reducing a large number of variables down to a smaller number of components. PCA is also useful as an initial step in FA where it reveals a great deal about maximum number and nature of factors.

---

[7]Maximum likelihood extraction manipulates off-diagonal elements rather than values in the diagonal.

13.5.1.3   Principal Factors

Principal factors extraction differs from PCA in that estimates of communality, instead of ones, are in the positive diagonal of the observed correlation matrix. These estimates are derived through an iterative procedure, with SMCs (squared multiple correlations of each variable with all other variables) used as the starting values in the iteration. The goal of analysis, like that for PCA, is to extract maximum orthogonal variance from the data set with each succeeding factor. Advantages to principal factors extraction are that it is widely used (and understood) and that it conforms to the factor analytic model in which common variance is analyzed with unique and error variance removed. Because the goal is to maximize variance extracted, however, principal factors is sometimes not as good as other extraction techniques in reproducing the correlation matrix. Also, communalities must be estimated and the solution is, to some extent, determined by those estimates. Principal factor analysis is available through both SPSS and SAS.

13.5.1.4   Image Factor Extraction

The technique is called image factoring because the analysis distributes among factors the variance of an observed variable that is *reflected* by the other variables, in a manner similar to the SMC. Image factor extraction provides an interesting compromise between PCA and principal factors. Like PCA, image extraction provides a mathematically unique solution because there are fixed values in the positive diagonal of **R**. Like principal factors, the values in the diagonal are communalities with unique and error variability excluded.

Image scores for each variable are produced by multiple regression, with each variable, in turn, serving as a DV. A covariance matrix is calculated from these image (predicted) scores. The variances from the image score covariance matrix are the communalities for factor extraction. Care is necessary in interpreting the results of image analysis, because loadings represent covariances between variables and factors rather than correlations.

Image factoring is available through SPSS, and SAS FACTOR (with two types—"image" and Harris component analysis).

13.5.1.5   Maximum Likelihood Factor Extraction

The maximum likelihood method of factor extraction was developed originally by Lawley in the 1940s (see Lawley & Maxwell, 1963). Maximum likelihood extraction estimates population values for factor loadings by calculating loadings that maximize the probability of sampling the observed correlation matrix from a population. Within constraints imposed by the correlations among variables, population estimates for factor loadings are calculated that have the greatest probability of yielding a sample with the observed correlation matrix. This method of extraction also maximizes the canonical correlations between the variables and the factors (see Chapter 12).

Maximum likelihood extraction is available through SPSS FACTOR and SAS FACTOR.

13.5.1.6   Unweighted Least Squares Factoring

The goal of unweighted least squares factor extraction is to minimize squared differences between the observed and reproduced correlation matrices. Only off-diagonal differences are considered; communalities are derived from the solution rather than estimated as part of the solution. Thus,

unweighted least squares factoring can be seen as a special case of principal factors analysis in which communalities are estimated after the solution.

The procedure, originally called minimum residual, was developed by Comrey (1962) and later modified by Harman and Jones (1966). The latter procedure is available through SPSS FACTOR and SAS FACTOR.

13.5.1.7 *Generalized (Weighted) Least Squares Factoring*

Generalized least squares extraction also seeks to minimize (off-diagonal) squared differences between observed and reproduced correlation matrices but in this case weights are applied to the variables. Variables that have substantial shared variance with other variables are weighted more heavily than variables that have substantial unique variance. In other words, variables that are not as strongly related to other variables in the set are not as important to the solution. This method of extraction is available through SPSS FACTOR and SAS FACTOR.

13.5.1.8 *Alpha Factoring*

Alpha factor extraction, available through SPSS FACTOR and SAS FACTOR, grew out of psychometric research where the interest is in discovering which common factors are found consistently when repeated samples of variables are taken from a population of *variables*. The problem is the same as identifying mean differences that are found consistently among samples of subjects taken from a population of subjects—a question at the heart of most univariate and multivariate statistics.

In alpha factoring, however, the concern is with the reliability of the common factors rather than with the reliability of group differences. Coefficient alpha is a measure derived in psychometrics for the reliability (also called generalizability) of a score taken in a variety of situations. In alpha factoring, communalities that maximize coefficient alpha for the factors are estimated using iterative procedures (and sometimes exceed 1.0).

Probably the greatest advantage to the procedure is that it focuses the researcher's attention squarely on the problem of sampling variables from the domain of variables of interest. Disadvantages stem from the relative unfamiliarity of most researchers with the procedure and the reason for it.

13.5.2 Rotation

The results of factor extraction, unaccompanied by rotation, are likely to be hard to interpret regardless of which method of extraction is used. After extraction, rotation is used to improve the interpretability and scientific utility of the solution. It is *not* used to improve the quality of the mathematical fit between the observed and reproduced correlation matrices because all orthogonally rotated solutions are mathematically equivalent to one another and to the solution before rotation.

Just as different methods of extraction tend to give similar results with a good data set, so also different methods of rotation tend to give similar results if the pattern of correlations in the data is fairly clear. In other words, a stable solution tends to appear regardless of the method of rotation used.

A decision is required between orthogonal and oblique rotation. In orthogonal rotation, the factors are uncorrelated. Orthogonal solutions offer ease of interpreting, describing, and reporting results; yet they strain "reality" unless the researcher is convinced that underlying processes are

almost independent. The researcher who believes that underlying processes are correlated uses an oblique rotation. In oblique rotation the factors may be correlated, with conceptual advantages but practical disadvantages in interpreting, describing, and reporting results.

Among the dozens of rotational techniques that have been proposed, only those available in both reviewed packages are included in this discussion (see Table 13.9). The reader who wishes to know more about these or other techniques is referred to Gorsuch (1983), Harman (1976), or Mulaik (1972). For the industrious, a presentation of rotation by hand is in Comrey and Lee (1992).

13.5.2.1 Orthogonal Rotation

Varimax, quartimax, and equamax—three orthogonal techniques—are available in both packages. Varimax is easily the most commonly used of all the rotations available.

Just as the extraction procedures have slightly different statistical goals, so also the rotational procedures maximize or minimize different statistics. The goal of varimax rotation is to simplify factors by maximizing the variance of the loadings within factors, across variables. The spread in loadings is maximized—loadings that are high after extraction become higher after rotation and loadings that are low become lower. Interpreting a factor is easier because it is obvious which variables correlate with it. Varimax also tends to reapportion variance among factors so that they become relatively equal in importance; variance is taken from the first factors extracted and distributed among the later ones.

Quartimax does for variables what varimax does for factors. It simplifies variables by increasing the dispersion of the loadings within variables, across factors. Varimax operates on the columns of the loading matrix, quartimax operates on the rows. Quartimax is not nearly as popular as varimax because one is usually more interested in simple factors than in simple variables.

Equamax is a hybrid between varimax and quartimax that tries simultaneously to simplify the factors and the variables. Mulaik (1972) reports that equamax tends to behave erratically unless the researcher can specify the number of factors with confidence.

Thus, varimax rotation simplifies the factors, quartimax the variables, and equamax both. They do so by setting levels on a simplicity criterion—such as Γ (gamma)—of 1, 0, and $1/2$, respectively. Gamma can also be continuously varied between 0 (variables simplified) and 1 (factors simplified) by using the orthogonal rotation that allows the user to specify Γ level. In SAS FACTOR, this is done through orthomax with Γ. Parsimax in SAS uses a formula incorporating numbers of factors and variables to determine Γ (see Table 13.9).

Varimax is the orthogonal rotation of choice for many applications; it is the default option of packages that have defaults.

13.5.2.2 Oblique Rotation

An *embarrasse de richesse* awaits the researcher who uses oblique rotation (see Table 13.9). Oblique rotations offer a continuous range of correlations between factors. The amount of correlation permitted between factors is determined by a variable called delta (δ) by SPSS FACTOR. The values of delta and gamma determine the maximum amount of correlation permitted among factors. When the value is less than zero, solutions are increasingly orthogonal; at about -4 the solution is orthogonal. When the value is zero, solutions can be fairly highly correlated. Values near 1 can produce factors that are very highly correlated. Although there is a relationship between values of delta or gamma

TABLE 13.9   Summary of Rotational Techniques

Rotational Technique	Program	Type	Goals of Analysis	Comments
Varimax	SAS SPSS	Orthogonal	Minimize complexity of factors (simplify columns of loading matrix) by maximizing variance of loadings on each factor.	Most commonly used rotation; recommended as default option
Quartimax	SAS SPSS	Orthogonal	Minimize complexity of variables (simplify rows of loading matrix) by maximizing variance of loadings on each variable.	First factor tends to be general, with others subclusters of variables.
Equamax	SAS SPSS	Orthogonal	Simplify both variables and factors (rows and columns); compromise between quartimax and varimax.	May behave erratically
Orthogonal with gamma (orthomax)	SAS	Orthogonal	Simplify either factors or variables, depending on the value of gamma (Γ).	Gamma (Γ) continuously variable
Parsimax	SAS	Orthogonal	Simplifies both variables and factors: $\Gamma = (p^*(m-1))/p + m - 2.$	
Direct oblimin	SPSS	Oblique	Simplify factors by minimizing cross-products of loadings.	Continuous values of gamma, or delta, δ (SPSS), available; allows wide range of factor intercorrelations
(Direct) quartimin	SPSS	Oblique	Simplify factors by minimizing sum of cross-products of squared loadings in pattern matrix.	Permits fairly high correlations among factors. Achieved in SPSS by setting $\delta = 0$ with direct oblimin.
Orthoblique	SAS (HK) SPSS	Both orthogonal and oblique	Rescale factor loadings to yield orthogonal solution; non-rescaled loadings may be correlated.	
Promax	SAS	Oblique	Orthogonal factors rotated to oblique positions.	Fast and inexpensive
Procrustes	SAS	Oblique	Rotate to target matrix.	Useful in confirmatory FA

and size of correlation, the maximum correlation at a given size of gamma or delta depends on the data set.

It should be stressed that factors do not necessarily correlate when an oblique rotation is used. Often, in fact, they do not correlate and the researcher reports the simpler orthogonal rotation.

The family of procedures used for oblique rotation with varying degrees of correlation in SPSS is direct oblimin. In the special case where Γ or $\delta = 0$ (the default option for the programs), the procedure is called direct quartimin. Values of gamma or delta greater than zero permit high correlations among factors, and the researcher should take care that the correct number of factors is chosen. Otherwise, highly correlated factors may be indistinguishable one from the other. Some trial and error, coupled with inspection of the scatterplots of relationships between pairs of factors, may be required to determine the most useful size of gamma or delta. Or, one might simply trust to the default value.

Orthoblique rotation uses the quartimax algorithm to produce an orthogonal solution *on rescaled factor loadings*; therefore the solution may be oblique with respect to the original factor loadings.

In promax rotation, available through SAS and SPSS, an orthogonally rotated solution (usually varimax) is rotated again to allow correlations among factors. The orthogonal loadings are raised to powers (usually powers of 2, 4, or 6) to drive small and moderate loadings to zero while larger loadings are reduced, but not to zero. Even though factors correlate, simple structure is maximized by clarifying which variables do and do not correlate with each factor. Promax has the additional advantage of being fast and inexpensive.

In Procrustean rotation, available in SAS, a target matrix of loadings (usually zeros and ones) is specified by the researcher and a transformation matrix is sought to rotate extracted factors to the target, if possible. If the solution can be rotated to the target, then the hypothesized factor structure is said to be confirmed. Unfortunately, as Gorsuch (1983) reports, with Procrustean rotation, factors are often extremely highly correlated and sometimes a correlation matrix generated by random processes is rotated to a target with apparent ease.

13.5.2.3 *Geometric Interpretation*

A geometric interpretation of rotation is in Figure 13.1 where 13.1(a) is the unrotated and 13.1(b) the rotated solution to the example in Table 13.2. Points are represented in two-dimensional space by listing their coordinates with respect to X and Y axes. With the first two unrotated factors as axes, unrotated loadings are COST ($-.400$, $.900$), LIFT ($.251$, $-.947$), DEPTH ($.932$, $.348$), and POWDER ($.956$, $.286$).

The points for these variables are also located with respect to the first two rotated factors as axes in Figure 13.1(b). The position of points does not change, but their coordinates change in the new axis system. COST is now (-0.86, $.981$), LIFT ($-.071$, $-.977$), DEPTH ($.994$, $.026$), and POWDER ($.997$, $-.040$). Statistically, the effect of rotation is to amplify high loadings and reduce low ones. Spatially, the effect is to rotate the axes so that they "shoot through" the variable clusters more closely.

Factor extraction yields a solution in which observed variables are vectors that run from the origin to the points indicated by the coordinate system. The factors serve as axes for the system. The coordinates of each point are the entries from the loading matrix for the variable. If there are three factors, then the space has three axes and three dimensions, and each observed variable is positioned by three coordinates. The length of the vector for each variable is the communality of the variable.

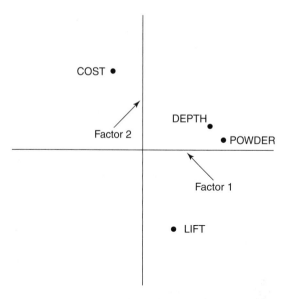

(a) Location of COST, LIFT, DEPTH, and POWDER after extraction, before rotation

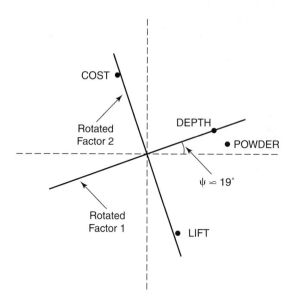

(b) Location of COST, LIFT, DEPTH, and POWDER vis-à-vis rotated axes

FIGURE 13.1   Illustration of rotation of axes to provide a better definition of factors vis-à-vis the variables with which they correlate.

If the factors are orthogonal, the factor axes are all at right angles to one another and the coordinates of the variable points are correlations between the common factors and the observed variables. Correlations (factor loadings) are read directly from these graphs by projecting perpendicular lines from each point to each of the factor axes.

One of the primary goals of PCA or FA, and the motivation behind extraction, is to discover the minimum number of factor axes needed to reliably position variables. A second major goal, and the motivation behind rotation, is to discover the meaning of the factors that underlie responses to observed variables. This goal is met by interpreting the factor axes that are used to define the space. Factor rotation repositions factor axes so as to make them maximally interpretable. Repositioning the axes changes the coordinates of the variable points but not the positions of the points with respect to each other.

Factors are usually interpretable when some observed variables load highly on them and the rest do not. And, ideally, each variable loads on one, and only one, factor. In graphic terms this means that the point representing each variable lies far out along one axis but near the origin on the other axes, that is, that coordinates of the point are large for one axis and near zero for the other axes.

If you have only one observed variable, it is trivial to position the factor axis—variable point and axis overlap in a space of one dimension. However, with many variables and several factor axes, compromises are required in positioning the axes. The variables form a "swarm" in which variables that are correlated with one another form a cluster of points. The goal is to shoot an axis to the swarm of points. With luck, the swarms are about 90° away from one another so that an orthogonal solution is indicated. And with lots of luck, the variables cluster in just a few swarms with empty spaces between them so that the factor axes are nicely defined.

In oblique rotation the situation is slightly more complicated. Because factors may correlate with one another, factor axes are not necessarily at right angles. And, although it is easier to position each axis near a cluster of points, axes may be very near each other (highly correlated), making the solution harder to interpret. See Section 13.6.3 for practical suggestions of ways to use graphic techniques to judge the adequacy of rotation.

13.5.3 Some Practical Recommendations

Although an almost overwhelmingly large number of combinations of extraction and rotation techniques is available, in practice differences among them are often slight (Velicer and Jackson, 1990; Fava and Velicer, 1992). The results of extraction are similar regardless of which method is used when there is a large number of variables with some strong correlations among them, with the same, well-chosen number of factors, and with similar values for communality. Further, differences that are apparent after extraction tend to disappear after rotation.

Most researchers begin their FA by using principal components extraction and varimax rotation. From the results, one estimates the factorability of the correlation matrix (Section 13.3.2.6), the rank of the observed correlation matrix (Sections 13.3.2.5 and 13.7.1.5), the likely number of factors (Section 13.6.2), and variables that might be excluded from subsequent analyses (Sections 13.3.2.7 and 13.7.1.6).

During the next few runs, researchers experiment with different numbers of factors, different extraction techniques, and both orthogonal and oblique rotations. Some number of factors with some combination of extraction and rotation produces the solution with the greatest scientific utility, consistency, and meaning; this is the solution that is interpreted.

13.6   Some Important Issues

Some of the issues raised in this section can be resolved through several different methods. Usually different methods lead to the same conclusion; occasionally they do not. When they do not, results are judged by the interpretability and scientific utility of the solutions.

13.6.1   Estimates of Communalities

FA differs from PCA in that communality values (numbers between 0 and 1) replace ones in the positive diagonal of **R** before factor extraction. Communality values are used instead of ones to remove the unique and error variance of each observed variable; only the variance a variable shares with the factors is used in the solution. But communality values are estimated, and there is some dispute regarding how that should be done.

The SMC of each variable as DV with the others in the sample as IVs is usually the starting estimate of communality. As the solution develops, communality estimates are adjusted by iterative procedures (which can be directed by the researcher) to fit the reproduced to the observed correlation matrix with the smallest number of factors. Iteration stops when successive communality estimates are very similar.

Final estimates of communality are also SMCs, but now between each variable as DV and the factors as IVs. Final communality values represent the proportion of variance in a variable that is predictable from the factors underlying it. Communality estimates do not change with orthogonal rotation.

Image extraction and maximum likelihood extraction are slightly different. In image extraction, variances from the image covariance matrix are used as the communality values throughout. Image extraction produces a mathematically unique solution because communality values are not changed. In maximum likelihood extraction, number of factors instead of communality values are estimated and off-diagonal correlations are "rigged" to produce the best fit between observed and reproduced matrices.

SPSS and SAS provide several different starting statistics for communality estimation. SPSS FACTOR permits user supplied values for principal factor extraction only, but otherwise uses SMCs. SAS FACTOR offers, for each variable, a choice of SMC, SMC adjusted so that the sum of the communalities is equal to the sum of the maximum absolute correlations, maximum absolute correlation with any other variable, user-specified values, or random numbers between 0 and 1.

The seriousness with which estimates of communality should be regarded depends on the number of observed variables. If the number of variables exceeds, say, 20, sample SMCs probably provide reasonable estimates of communality. Furthermore, with 20 or more variables, the elements in the positive diagonal are few compared with the total number of elements in R, and their sizes do not influence the solution very much. Actually, if the communality values for all variables in FA are of approximately the same magnitude, results of PCA and FA are very similar (Velicer & Jackson, 1990; Fava & Velicer, 1992).

If communality values equal or exceed 1, problems with the solution are indicated. There is too little data, or starting communality values are wrong, or the number of factors extracted is wrong; addition or deletion of factors may reduce the communality below 1. Very low communality values, on the other hand, indicate that the variables with them are unrelated to other variables in the set (Sections 13.3.2.7 and 13.7.1.6). SAS FACTOR has two alternatives for dealing with communalities > 1:

HEYWOOD sets them to 1, and ULTRAHEYWOOD allows them to exceed 1, but warns that doing so can cause convergence problems.

13.6.2 Adequacy of Extraction and Number of Factors

Because inclusion of more factors in a solution improves the fit between observed and reproduced correlation matrices, adequacy of extraction is tied to number of factors. The more factors extracted, the better the fit and the greater the percent of variance in the data "explained" by the factor solution. However, the more factors extracted, the less parsimonious the solution. To account for all the variance (PCA) or covariance (FA) in a data set, one would normally have to have as many factors as observed variables. It is clear, then, that a trade-off is required: One wants to retain enough factors for an adequate fit, but not so many that parsimony is lost.

Selection of the number of factors is probably more critical than selection of extraction and rotational techniques or communality values. In confirmatory FA, selection of the number of factors is really selection of the number of theoretical processes underlying a research area. You can partially confirm a hypothesized factor structure by asking if the theoretical number of factors adequately fits the data.

There are several ways to assess adequacy of extraction and number of factors. For a highly readable summary of these methods, not all currently available through the statistical packages, see Gorsuch (1983) and Zwick and Velicer (1986). Reviewed below are methods available through SPSS and SAS.

A first quick estimate of the number of factors is obtained from the sizes of the eigenvalues reported as part of an initial run with principal components extraction. Eigenvalues represent variance. Because the variance that each standardized variable contributes to a principal components extraction is 1, a component with an eigenvalue less than 1 is not as important, from a variance perspective, as an observed variable. The number of components with eigenvalues greater than 1 is usually somewhere between the number of variables divided by 3 and the number of variables divided by 5 (e.g., 20 variables should produce between 7 and 4 components with eigenvalues greater than 1). If this is a reasonable number of factors for the data, if the number of variables is 40 or fewer, and if sample size is large, the number of factors indicated by this criterion is probably about right. In other situations, this criterion is likely to overestimate the number of factors in the data set.

A second criterion is the scree test (Cattell, 1966) of eigenvalues plotted against factors. Factors, in descending order, are arranged along the abscissa with eigenvalue as the ordinate. The plot is appropriately used with principal components or factor analysis at initial and later runs to find the number of factors. The scree plot is available through SPSS and SAS FACTOR.

Usually the scree plot is negatively decreasing—the eigenvalue is highest for the first factor and moderate but decreasing for the next few factors before reaching small values for the last several factors, as illustrated for real data through SPSS in Figure 13.2. You look for the point where a line drawn through the points changes slope. In the example, a single straight line can comfortably fit the first four eigenvalues. After that, another line, with a noticeably different slope, best fits the remaining eight points. Therefore, there appear to be about four factors in the data of Figure 13.2.

Unfortunately, the scree test is not exact; it involves judgment of where the discontinuity in eigenvalues occurs and researchers are not perfectly reliable judges. As Gorsuch (1983) reports, results of the scree test are more obvious (and reliable) when sample size is large, communality val-

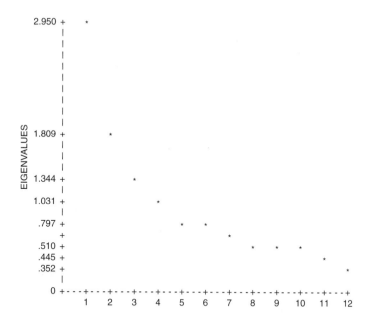

FIGURE 13.2   Scree output produced by SPSS FACTOR. Note break in size of eigenvalues between the fourth and fifth factors.

ues are high, and each factor has several variables with high loadings. Zoski and Jurs (1996) recommend a refinement to the visual scree test that involves computing the standard error of the eigenvalues for the last few components.

Horn (1965) proposed parallel analysis as an alternative to retaining all principal components with eigenvalues larger than 1. This is a three step process. First, a randomly generated data set with the same number of cases and variables is generated. Next, principal components analysis is repeatedly performed on the randomly generated data set and all eigenvalues noted for each analysis. Those eigenvalues are then averaged for each component and compared to the results from the real data set. Only components from the real data set whose eigenvalues exceed the averaged eigenvalue from the randomly generated data set are retained. A major advantage to this procedure is to remind the user that even randomly generated data can have relationships based on chance that produce components with eigenvalues larger than 1, sometimes substantially so.

As an alternative, Velicer (1976) proposed the minimum average partial correlation (MAP) test. The first step is to perform PCA with one component. Partial correlation is used to take the variance of the first component from the variable intercorrelations before the mean squared coefficient of all partial correlations (the values off of the main diagonal) is computed. Then PCA is performed with two components, and the procedure is repeated. Mean squared partial correlations are computed for all solutions until the minimum squared partial correlation is identified. The number of components that produces the minimum mean squared partial correlation is the number of components to retain. Gorsuch (1976) points out that this procedure does not work well when some components have only a few variables that load on them.

Zwick and Velicer (1986) tested the scree test, Horn's parallel test, and Velicer's MAP test (among others) in simulation studies using a data set with a clear factor structure. Both the parallel test and the minimum average partial test seemed to work well. These procedures have been extended successfully to principal factor analysis. O'Connor (2000) provides programs for conducting the parallel test and the minimum average partial test through both SPSS and SAS.

Once you have determined the number of factors, it is important to look at the rotated loading matrix to determine the number of variables that load on each factor (see Section 13.6.5). If only one variable loads highly on a factor, the factor is poorly defined. If two variables load on a factor, then whether or not it is reliable depends on the pattern of correlations of these two variables with each other and with other variables in **R**. If the two variables are highly correlated with each other (say, $r > .70$) and relatively uncorrelated with other variables, the factor may be reliable. Interpretation of factors defined by only one or two variables is hazardous, however, under even the most exploratory factor analysis.

For principal components extraction and maximum likelihood extraction in confirmatory factor analysis there are significance tests for number of factors. Bartlett's test evaluates all factors together and each factor separately against the hypothesis that there are no factors. However, there is some dispute regarding use of these tests. The interested reader is referred to Gorsuch (1983) or one of the other newer factor analysis texts for discussion of significance testing in FA.

There is debate about whether it is better to retain too many or too few factors if the number is ambiguous. Sometimes a researcher wants to rotate, but not interpret, marginal factors for statistical purposes (e.g., to keep some factors with communality values < 1). Other times the last few factors represent the most interesting and unexpected findings in a research area. These are good reasons for retaining factors of marginal reliability. However, if the researcher is interested in using only demonstrably reliable factors, the fewest possible factors are retained.

13.6.3 Adequacy of Rotation and Simple Structure

The decision between orthogonal and oblique rotation is made as soon as the number of reliable factors is apparent. In many factor analytic situations, oblique rotation seems more reasonable on the face of it than orthogonal rotation because it seems more likely that factors are correlated than that they are not. However, reporting the results of oblique rotation requires reporting the elements of the pattern matrix (**A**) and the factor correlation matrix (**Φ**), whereas reporting orthogonal rotation requires only the loading matrix (**A**). Thus, simplicity of reporting results favors orthogonal rotation. Further, if factor scores or factorlike scores (Section 13.6.6) are to be used as IVs or DVs in other analyses, or if a goal of analysis is comparison of factor structure in groups, then orthogonal rotation has distinct advantages.

Perhaps the best way to decide between orthogonal and oblique rotation is to request oblique rotation with the desired number of factors and look at the correlations among factors. The oblique rotations available by default in SPSS and SAS calculate factors that are fairly highly correlated if necessary to fit the data. However, if factor correlations are not driven by the data, the solution remains nearly orthogonal.

Look at the factor correlation matrix for correlations around .32 and above. If correlations exceed .32, then there is 10% (or more) overlap in variance among factors, enough variance to warrant oblique rotation unless there are compelling reasons for orthogonal rotation. Compelling reasons include a desire to compare structure in groups, a need for orthogonal factors in other analyses, or a theoretical need for orthogonal rotation.

Once the decision is made between orthogonal and oblique rotation, the adequacy of rotation is assessed several ways. Perhaps the simplest way is to compare the pattern of correlations in the correlation matrix with the factors. Are the patterns represented in the rotated solution? Do highly correlated variables tend to load on the same factor? If you included marker variables, do they load on the predicted factors?

Another criterion is simple structure (Thurstone, 1947). If simple structure is present (and factors are not too highly correlated), several variables correlate highly with each factor and only one factor correlates highly with each variable. In other words, the columns of **A**, which define factors vis-à-vis variables, have several high and many low values while the rows of **A**, which define variables vis-à-vis factors, have only one high value. Rows with more than one high correlation correspond to variables that are said to be complex because they reflect the influence of more than one factor. It is usually best to avoid complex variables because they make interpretation of factors more ambiguous.

Adequacy of rotation is also ascertained through the PLOT instructions of the four programs. In the figures, factors are considered two at a time with a different pair of factors as axes for each plot. Look at the *distance*, *clustering*, and *direction* of the points representing variables relative to the factor axes in the figures.

The *distance* of a variable point from the origin reflects the size of factor loadings; variables highly correlated with a factor are far out on that factor's axis. Ideally, each variable point is far out on one axis and near the origin on all others. *Clustering* of variable points reveals how clearly defined a factor is. One likes to see a cluster of several points near the end of each axis and all other points near the origin. A smattering of points at various distances along the axis indicates a factor that is not clearly defined, while a cluster of points midway between two axes reflects the presence of another factor or the need for oblique rotation. The *direction* of clusters after orthogonal rotation may also indicate the need for oblique rotation. If clusters of points fall between factor axes after orthogonal rotation, if the angle between clusters with the respect to the origin is not 90°, then a better fit to the clusters is provided by axes that are not orthogonal. Oblique rotation may reveal substantial correlations among factors. Several of these relationships are depicted in Figure 13.3.

13.6.4   Importance and Internal Consistency of Factors

The importance of a factor (or a set of factors) is evaluated by the proportion of variance or covariance accounted for by the factor after rotation. The proportion of variance attributable to individual factors differs before and after rotation because rotation tends to redistribute variance among factors somewhat. Ease of ascertaining proportions of variance for factors depends on whether rotation was orthogonal or oblique.

After orthogonal rotation, the importance of a factor is related to the size of its SSLs (Sum of Squared Loadings from **A** after rotation). SSLs are converted to proportion of variance for a factor by dividing by p, the number of variables. SSLs are converted to proportion of covariance for a factor by dividing its SSL by the sum of SSLs or, equivalently, sum of communalities. These computations are illustrated in Table 13.4 and Section 13.4 for the example.

The proportion of variance accounted for by a factor is the amount of variance in the original variables (where each has contributed one unit of variance) that has been condensed into the factor.

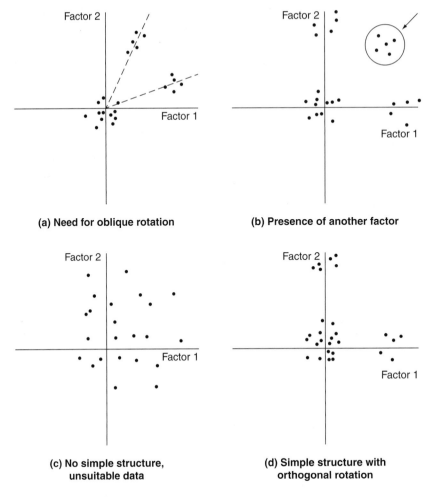

(a) Need for oblique rotation

(b) Presence of another factor

(c) No simple structure, unsuitable data

(d) Simple structure with orthogonal rotation

FIGURE 13.3   Pairwise plots of factor loadings following orthogonal rotation and indicating: (a) need for oblique rotation; (b) presence of another factor; (c) unsuitable data; and (d) simple structure.

Proportion of *variance* is the variance of a factor relative to the variance in the variables. The proportion of covariance accounted for by a factor indicates the relative importance of the factor to the total covariance accounted for by all factors. Proportion of *covariance* is the variance of a factor relative to the variance in the solution. The variance in the solution is likely to account for only a fraction of the variance in the original variables.

In oblique rotation, proportions of variance and covariance can be obtained from **A** *before* rotation by the methods just described, but they are only rough indicators of the proportions of variance and covariance of factors after rotation. Because factors are correlated, they share overlapping variability, and assignment of variance to individual factors is ambiguous. After oblique rotation the size of the SSL associated with a factor is a rough approximation of its importance—factors with bigger SSLs are more important—but proportions of variance and covariance cannot be specified.

An estimate of the internal consistency of the solution—the certainty with which factor axes are fixed in the variable space—is given by the squared multiple correlations of factor scores predicted from scores on observed variables. In a good solution, SMCs range between 0 and 1; the larger the SMCs, the more stable the factors. A high SMC (say, .70 or better) means that the observed variables account for substantial variance in the factor scores. A low SMC means the factors are poorly defined by the observed variables. If an SMC is negative, too many factors have been retained. If an SMC is above 1, the entire solution needs to be reevaluated.

SPSS FACTOR prints these SMCs as the diagonal of the covariance matrix for estimated regression factor scores. In SAS FACTOR, SMCs are printed along with factor score coefficients by the SCORE option.

13.6.5   Interpretation of Factors

To interpret a factor, one tries to understand the underlying dimension that unifies the group of variables loading on it. In both orthogonal and oblique rotations, loadings are obtained from the loading matrix, $\mathbf{A}$, but the meaning of the loadings is different for the two rotations.

After orthogonal rotation, the values in the loading matrix are correlations between variables and factors. The researcher decides on a criterion for meaningful correlation (usually .32 or larger), collects together the variables with loadings in excess of the criterion, and searches for a concept that unifies them.

After oblique rotation, the process is the same, but the interpretation of the values in $\mathbf{A}$, the pattern matrix, is no longer straightforward. The loading is not a correlation but is a measure of the unique relationship between the factor and the variable. Because factors correlate, the correlations between variables and factors (available in the structure matrix, $\mathbf{C}$) are inflated by overlap between factors. A variable may correlate with one factor through its correlation with another factor rather than directly. The elements in the pattern matrix have overlapping variance among factors "partialed out," but at the expense of conceptual simplicity.

Actually, the reason for interpretation of the pattern matrix rather than the structure matrix is pragmatic—it's easier. The difference between high and low loadings is more apparent in the pattern matrix than in the structure matrix.

As a rule of thumb, only variables with loadings of .32 and above are interpreted. The greater the loading, the more the variable is a pure measure of the factor. Comrey and Lee (1992) suggest that loadings in excess of .71 (50% overlapping variance) are considered excellent, .63 (40% overlapping variance) very good, .55 (30% overlapping variance) good, .45 (20% overlapping variance) fair, and .32 (10% overlapping variance) poor. Choice of the cutoff for size of loading to be interpreted is a matter of researcher preference. Sometimes there is a gap in loadings across the factors and, if the cutoff is in the gap, it is easy to specify which variables load and which do not. Other times, the cutoff is selected because one can interpret factors with that cutoff but not with a lower cutoff.

The size of loadings is influenced by the homogeneity of scores in the sample. If homogeneity is suspected, interpretation of lower loadings is warranted. That is, if the sample produces similar scores on observed variables, a lower cutoff is used for interpretation of factors.

At some point, a researcher usually tries to characterize a factor by assigning it a name or a label, a process that involves art as well as science. Rummel (1970) provides numerous helpful hints on interpreting and naming factors. Interpretation of factors is facilitated by output of the matrix of sorted loadings where variables are grouped by their correlations with factors. Sorted loadings are produced routinely by REORDER in SAS FACTOR, and SORT in SPSS.

The replicability, utility, and complexity of factors are also considered in interpretation. Is the solution replicable in time and/or with different groups? Is it trivial or is it a useful addition to scientific thinking in a research area? Where do the factors fit in the hierarchy of "explanations" about a phenomenon? Are they complex enough to be intriguing without being so complex that they are uninterpretable?

13.6.6   Factor Scores

Among the potentially more useful outcomes of PCA or FA are factor scores. Factor scores are estimates of the scores subjects would have received on each of the factors had they been measured directly.

Because there are normally fewer factors than observed variables, and because factor scores are nearly uncorrelated if factors are orthogonal, use of factor scores in other analyses may be very helpful. Multicollinear matrices can be reduced to orthogonal components using PCA, for instance. Or, one could use PCA to reduce a large number of DVs to a smaller number of components for use as DVs in MANOVA. Alternatively, one could reduce a large number of IVs to a small number of factors for purposes of predicting a DV in multiple regression or group membership in discriminant analysis or logistic regression. If factors are few in number, stable, and interpretable, their use enhances subsequent analyses. In the context of a theoretical FA, factor scores are estimates of the values that would be produced if the underlying constructs could be measured directly.

Procedures for estimating factor scores range between simple-minded (but frequently adequate) and sophisticated. Comrey and Lee (1992) describe several rather simple-minded techniques for estimating factor scores. Perhaps the simplest is to sum scores on variables that load highly on each factor. Variables with bigger standard deviations contribute more heavily to the factor scores produced by this procedure, a problem that is alleviated if variable scores are standardized first or if the variables have roughly equal standard deviations to begin with. For many research purposes, this "quick and dirty" estimate of factor scores is entirely adequate.

There are several sophisticated statistical approaches to estimating factors. All produce factor scores that are correlated, but not perfectly, with the factors. The correlations between factors and factor scores are higher when communalities are higher and when the ratio of variables to factors is higher. But as long as communalities are estimated, factor scores suffer from indeterminacy because there is an infinite number of possible factor scores that all have the same mathematical characteristics. As long as factor scores are considered only estimates, however, the researcher is not overly beguiled by them.

The method described in Section 13.4 (especially Equations 13.10 and 13.11) is the regression approach to estimating factor scores. This approach results in the highest correlations between factors and factor scores. The distribution of each factor's scores has a mean of zero and a standard deviation of 1 (after PCA) or equal to the SMC between factors and variables (after FA). However, this regression method, like all others (see Chapter 5), capitalizes on chance relationships among variables so that factor-score estimates are biased (too close to "true" factor scores). Further, there are often correlations among scores for factors even if factors are orthogonal and factor scores sometimes correlate with other factors (in addition to the one they are estimating).

The regression approach to estimating factor scores is available through SAS and SPSS. Both packages write component/factor scores to files for use in other analyses. SAS and SPSS print standardized component/factor score coefficients.

SPSS FACTOR provides two additional methods of estimating factor scores. In the Bartlett method, factor scores correlate only with their own factors and the factor scores are unbiased (that is, neither systematically too close nor too far away from "true" factor scores). The factor scores correlate with the factors almost as well as in the regression approach and have the same mean and standard deviation as in the regression approach. However, factor scores may still be correlated with each other.

The Anderson-Rubin approach (discussed by Gorsuch, 1983) produces factor scores that are uncorrelated with each other even if factors are correlated. Factor scores have mean zero, standard deviation 1. Factor scores correlate with their own factors almost as well as in the regression approach, but they sometimes also correlate with other factors (in addition to the one they are estimating) and they are somewhat biased. If you need uncorrelated scores, the Anderson-Rubin approach is best; otherwise the regression approach is probably best simply because it is best understood and most widely available.

13.6.7   Comparisons among Solutions and Groups

Frequently, a researcher is interested in deciding whether or not two groups that differ in experience or characteristics have the same factors. Comparisons among factor solutions involve the *pattern* of the correlations between variables and factors, or both the *pattern and magnitude* of the correlations between them. Rummel (1970), Levine (1977), and Gorsuch (1983) have excellent summaries of several comparisons that might be of interest. Some of the simpler of these techniques are described in an earlier version of this book (Tabachnick & Fidell, 1989).

Tests of theory (in which theoretical factor loadings are compared with those derived from a sample) and comparisons among groups are currently the province of structural equation modeling. These techniques are discussed in Chapter 14.

13.7   Complete Example of FA

During the second year of the panel study described in Appendix B Section B.1, participants completed the Bem Sex Role Inventory (BSRI; Bem, 1974). The sample included 369 middle-class, English-speaking women between the ages of 21 and 60 who were interviewed in person.

Forty-five items from the BSRI were selected for this research, where 20 items measure femininity, 20 masculinity,[8] and 5 social desirability. Respondents attribute traits (e.g., "gentle," "shy," "dominant") to themselves by assigning numbers between 1 ("never or almost never true of me") and 7 ("always or almost always true of me") to each of the items. Responses are summed to produce separate masculine and feminine scores. Masculinity and femininity are conceived as orthogonal

---

[8]Due to clerical error, one of the masculine items, "aggression," was omitted from the questionnaires.

dimensions of personality with both, one, or neither descriptive of any given individual. Files are FACTOR.*.

Previous factor analytic work had indicated the presence of between three and five factors underlying the items of the BSRI. Investigation of the factor structure for this sample of women is a goal of this analysis.

13.7.1 Evaluation of Limitations

Because the BSRI was neither developed through nor designed for factor analytic work, it meets only marginally the requirements listed in Section 13.3.1. For instance, marker variables are not included and variables from the feminine scale differ in social desirability as well as in meaning (e.g., "tender" and "gullible"), so some of these variables are likely to be complex.

13.7.1.1 Sample Size and Missing Data

Data are available initially from 369 women with no missing values. With outlying cases deleted (see below), the FA is conducted on responses of 344 women. Using the guidelines of Section 13.3.2.1, over 300 cases provide a good sample size for factor analysis.

13.7.1.2 Normality

Distributions of the 44 variables are examined for skewness through SAS MEANS (cf. Chapter 12). Many of the variables are negatively skewed and a few are positively skewed. However, because the BSRI is already published and in use, no deletion of variables or transformations of them is performed.

Because the variables fail in normality, significance tests are inappropriate. And because the direction of skewness is different for different variables, we also anticipate a weakened analysis due to lowering of correlations in **R**.

13.7.1.3 Linearity

The differences in skewness for variables suggest the possibility of curvilinearity for some pairs of variables. With 44 variables, however, examination of all pairwise scatterplots (about 1,000 plots) is impractical. Therefore, a spot check on a few plots is run through SAS PLOT. Figure 13.4 shows the plot expected to be among the worst—between LOYAL (with strong negative skewness) and MASCULIN (with strong positive skewness). Although the plot is far from pleasing and shows departure from linearity as well as the possibility of outliers, there is no evidence of true curvilinearity (Section 4.1.5.2). And again, transformations are viewed with disfavor considering the variable set and the goals of analysis.

13.7.1.4 Outliers

Multivariate outliers are identified using SAS REG (cf. Chapter 12) which adds a leverage variable to the data set, now labeled FACTLEV. Using a criterion of $\alpha = .001$ with 44 df, critical $\chi^2 = 78.75$, and using Equation 4.3, $h_{ii} = 0.2167$. With this criterion, 25 women are identified as outliers,

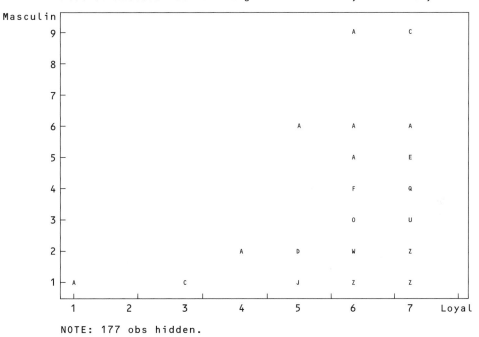

FIGURE 13.4 Spot check for linearity among variables.
Syntax and output from SAS PLOT.

leaving 344 nonoutlying cases. Outliers are then sought among the reduced data set, with critical $h_{ii} = 0.2325$. Eleven more cases are identified as potential outliers, however, only one of these, with $h_{ii} = 0.2578$, exceeds the criterion suggested by Lunneborg (1994) of critical $h_{ii} = 2(k/N) = 0.2558$. Therefore, the decision is made not to delete any more cases, and to run remaining analyses on the data set with 344 cases.

Because of the large number of outliers and variables, a case-by-case analysis (cf. Chapter 4) is not feasible. Instead, a stepwise discriminant analysis is used to identify variables that significantly discriminate between outliers and nonoutliers. First, a variable labeled DUMMY is added to the data set, in which each outlier is coded 1 and the remaining cases are labeled 0. Then DUMMY is declared the class (grouping) variable in the stepwise regression run through SAS STEPDISC, as seen in Table 13.10. Means in each group are requested for all variables. On the last step of the discriminant analysis, two variables (RELIANT and FLATTER) discriminate outliers as a group with $p < .001$.

A reduced data set that includes only the 344 nonoutlying cases is created, called FACTORR to be used for all subsequent analyses.

TABLE 13.10 Description of Variables Causing Multivariate Outliers Using SAS REG (Syntax and Selected Output)

```
proc stepdisc data=SASUSER.FACTLEV simple;
  class DUMMY;
  var HELPFUL RELIANT DEFBEL YIELDING CHEERFUL INDPT ATHLET SHY ASSERT
  STRPERS FORCEFUL AFFECT FLATTER LOYAL ANALYT FEMININE SYMPATHY MOODY SENSITIV UNDSTAND
  COMPASS LEADERAB SOOTHE RISK DECIDE SELFSUFF CONSCIEN DOMINANT MASCULIN STAND HAPPY
  SOFTSPOK WARM TRUTHFUL TENDER GULLIBLE LEADACT CHILDLIK INDIV FOULLANG LOVECHIL
  COMPETE AMBITIOU GENTLE;
run;
```

The STEPDISC Procedure
Simple Statistics

DUMMY = 0

Variable	N	Sum	Mean	Variance	Standard Deviation
HELPFUL	344	2089	6.07267	0.91307	0.9555
RELIANT	344	2068	6.01163	1.06109	1.0301
DEFBEL	344	2056	5.97674	1.46301	1.2096
YIELDING	344	1569	4.56105	1.57061	1.2532
CHEERFUL	344	2005	5.82849	1.00548	1.0027
INDPT	344	2033	5.90988	1.59244	1.2619
ATHLET	344	1258	3.65698	3.65459	1.9117
SHY	344	1020	2.96512	2.41860	1.5552
ASSERT	344	1605	4.66570	2.02494	1.4230
STRPERS	344	1757	5.10756	2.16624	1.4718
FORCEFUL	344	1365	3.96802	2.74241	1.6560
AFFECT	344	2058	5.98256	1.18337	1.0878
FLATTER	344	1553	4.51453	2.75197	1.6589
.					
.					
.					

TABLE 13.10 Continued

DUMMY = 1

Variable	N	Sum	Mean	Variance	Standard Deviation
HELPFUL	25	143.00000	5.72000	2.12667	1.4583
RELIANT	25	120.00000	4.80000	5.91667	2.4324
DEFBEL	25	123.00000	4.92000	5.49333	2.3438
YIELDING	25	103.00000	4.12000	4.36000	2.0881
CHEERFUL	25	140.00000	5.60000	3.00000	1.7321
INDPT	25	127.00000	5.08000	4.91000	2.2159
ATHLET	25	83.00000	3.32000	4.89333	2.2121
SHY	25	87.00000	3.48000	3.92667	1.9816
ASSERT	25	111.00000	4.44000	5.59000	2.3643
STRPERS	25	116.00000	4.64000	5.24000	2.2891
FORCEFUL	25	93.00000	3.72000	5.71000	2.3896
AFFECT	25	134.00000	5.36000	3.24000	1.8000
FLATTER	25	82.00000	3.28000	4.79333	2.1894

. . .

(continued)

TABLE 13.10 Continued

Stepwise Selection Summary

Step	Number In	Entered	Removed	Partial R-Square	F Value	Pr > F	Wilks' Lambda	Pr > Lambda	Average Squared Canonical Correlation	Pr > ASCC
1	1	RELIANT		0.0633	24.82	<.0001	0.93665946	<.0001	0.06334054	<.0001
2	2	FLATTER		0.0330	12.50	0.0005	0.90573479	<.0001	0.09426521	<.0001
3	3	TRUTHFUL		0.0275	10.32	0.0014	0.88082728	<.0001	0.11917272	<.0001
4	4	LEADACT		0.0144	5.33	0.0215	0.86811036	<.0001	0.13188964	<.0001
5	5	LEADERAB		0.0170	6.28	0.0127	0.85334921	<.0001	0.14665079	<.0001
6	6	FEMININE		0.0139	5.11	0.0244	0.84147148	<.0001	0.15852852	<.0001
7	7	MASCULIN		0.0187	6.88	0.0091	0.82573101	<.0001	0.17426899	<.0001
8	8	FOULLANG		0.0111	4.04	0.0452	0.81656744	<.0001	0.18343256	<.0001
9	9	SELFSUFF		0.0089	3.23	0.0732	0.80928660	<.0001	0.19071340	<.0001
10	10	CHILDLIK		0.0128	4.64	0.0319	0.79893591	<.0001	0.20106409	<.0001
11	11	DEFBEL		0.0099	3.57	0.0597	0.79103028	<.0001	0.20896972	<.0001
12	12	HAPPY		0.0079	2.83	0.0933	0.78478880	<.0001	0.21521120	<.0001
13	13	CHEERFUL		0.0209	7.57	0.0062	0.76839511	<.0001	0.23160489	<.0001
14	14	YIELDING		0.0083	2.96	0.0861	0.76201822	<.0001	0.23798178	<.0001

13.7.1.5   Multicollinearity and Singularity

Nonrotated PCA runs through SAS FACTOR reveal that the smallest eigenvalue is 0.126, not dangerously close to 0. The largest SMC between variables where each, in turn, serves as DV for the others is .76, not dangerously close to 1 (Table 13.11). Multicollinearity is not a threat in this data set.

The SPSS FACTOR correlation matrix (not shown) reveals numerous correlations among the 44 items, well in excess of .30, therefore patterns in responses to variables are anticipated. Table 13.11 syntax produces Kaiser's measures of sampling adequacy (**msa**), which are acceptable because all are greater than .6 (not shown). Most of the values in the negative anti-image correlation matrix (also not shown) are small, another requirement for good FA.

13.7.1.6   Outliers among Variables

SMCs among variables (Table 13.11) are also used to screen for outliers among variables, as discussed in Section 13.3.2.7. The lowest SMC among variables is .11. It is decided to retain all 44 variables although many are largely unrelated to others in the set. (In fact, 45% of the 44 variables in the analysis have loadings too low on all the factors to assist interpretation in the final solution.)

13.7.2   Principal Factors Extraction with Varimax Rotation

Principal components extraction with varimax rotation through SAS FACTOR is used in an initial run to estimate the likely number of factors from eigenvalues.[9] The first 13 eigenvalues are shown in Table 13.12. The maximum number of factors (eigenvalues larger than 1) is 11. However, retention of 11 factors seems unreasonable so sharp breaks in size of eigenvalues are sought using the scree test (Section 13.6.2).

Eigenvalues for the first four factors are all larger than two, and, after the sixth factor, changes in successive eigenvalues are small. This is taken as evidence that there are probably between 4 and 6 factors. The scree plot visually suggests breaks between 4 and 6 factors. These results are consistent with earlier research suggesting 3 to 5 factors on the BSRI.

A common factor extraction model that removes unique and error variability from each variable is used for the next several runs and the final solution. Principal factors is chosen from among methods for common factor extraction. Several PFA runs specifying 4 to 6 factors are planned to find the optimal number of factors.

The trial PFA run with 5 factors has 5 eigenvalues larger than 1 among unrotated factors. But after rotation, the eigenvalue for the fifth factor is below 1 and it has no loadings larger than .45, the criterion for interpretation chosen for this research. The solution with four factors, on the other hand, meets the goals of interpretability, so four factors are chosen for follow-up runs. The first six eigenvalues from the four-factor solution are shown in Table 13.13.

As another test of adequacy of extraction and number of factors, it is noted (but not shown) that most values in the residual correlation matrix for the four-factor orthogonal solution are near zero. This is further confirmation that a reasonable number of factors is 4.

The decision between oblique and orthogonal rotation is made by requesting principal factor extraction with oblique rotation of four factors. Promax is the oblique method employed; power = 2

---

[9]Principal components extraction is chosen to estimate the maximum number of factors that might be interesting. Principal factor analysis, which produces fewer eigenvalues greater than 1, is a reasonable alternative for estimation.

TABLE 13.11   Syntax and Selected SAS Factor Output to Assess Multicollinearity

```
proc factor data=SASUSER.FACTORR prior=smc msa;
    var HELPFUL RELIANT DEFBEL YIELDING CHEERFUL INDPT ATHLET SHY ASSERT
STRPERS FORCEFUL AFFECT FLATTER LOYAL ANALYT FEMININE SYMPATHY MOODY SENSITIV UNDSTAND
COMPASS LEADERAB SOOTHE RISK DECIDE SELFSUFF CONSCIEN DOMINANT MASCULIN STAND HAPPY
SOFTSPOK WARM TRUTHFUL TENDER GULLIBLE LEADACT CHILDLIK INDIV FOULLANG LOVECHIL
COMPETE AMBITIOU GENTLE;
run;
```

Prior Communality Estimates: SMC

HELPFUL	RELIANT	DEFBEL	YIELDING	CHEERFUL	INDPT	ATHLET	SHY
0.37427495	0.46116308	0.41691063	0.22995039	0.49202401	0.53847238	0.25760550	0.32497651

Prior Communality Estimates: SMC

ASSERT	STRPERS	FORCEFUL	AFFECT	FLATTER	LOYAL	ANALYT	FEMININE
0.53767397	0.59340397	0.56564996	0.55263932	0.29585781	0.39072689	0.24184436	0.35791153

Prior Communality Estimates: SMC

SYMPATHY	MOODY	SENSITIV	UNDSTAND	COMPASS	LEADERAB	SOOTHE	RISK
0.45290025	0.38090682	0.48603768	0.61662633	0.64933699	0.76269041	0.43513523	0.42237094

Prior Communality Estimates: SMC

DECIDE	SELFSUFF	CONSCIEN	DOMINANT	MASCULIN	STAND	HAPPY	SOFTSPOK
0.48929870	0.63267630	0.39916244	0.56213138	0.31595289	0.57283829	0.53576835	0.40179017

Prior Communality Estimates: SMC

WARM	TRUTHFUL	TENDER	GULLIBLE	LEADACT	CHILDLIK	INDIV	FOULLANG
0.61522559	0.35627431	0.60454474	0.29683146	0.76136184	0.29603932	0.37905349	0.11346293

Prior Communality Estimates: SMC

LOVECHIL	COMPETE	AMBITIOU	GENTLE
0.28419396	0.46467594	0.45870303	0.57953547

TABLE 13.12 Eigenvalues and Proportions of Variance for First 13 Components (SAS FACTOR PCA Syntax and Selected Output)

```
proc factor data=SASUSER.FACTORR simple corr scree;
    var HELPFUL RELIANT DEFBEL YIELDING CHEERFUL INDPT ATHLET SHY ASSERT
STRPERS FORCEFUL AFFECT FLATTER LOYAL ANALYT FEMININE SYMPATHY MOODY SENSITIV UNDSTAND
COMPASS LEADERAB SOOTHE RISK DECIDE SELFSUFF CONSCIEN DOMINANT MASCULIN STAND HAPPY
SOFTSPOK WARM TRUTHFUL TENDER GULLIBLE LEADACT CHILDLIK INDIV FOULLANG LOVECHIL
COMPETE AMBITIOU GENTLE;
run;
```

Eigenvalues of the Correlation Matrix: Total = 44 Average = 1

	Eigenvalue	Difference	Proportion	Cumulative
1	8.19403261	3.04053048	0.1862	0.1862
2	5.15350213	2.56303643	0.1171	0.3034
3	2.59046570	0.51750786	0.0589	0.3622
4	2.07295785	0.42538555	0.0471	0.4093
5	1.64757230	0.23237531	0.0374	0.4468
6	1.41519699	0.12450020	0.0322	0.4789
7	1.29069678	0.06948058	0.0293	0.5083
8	1.22121620	0.11167570	0.0278	0.5360
9	1.10954050	0.03190183	0.0252	0.5613
10	1.07763867	0.04595329	0.0245	0.5857
11	1.03168538	0.08043037	0.0234	0.6092
12	0.95125501	0.00960040	0.0216	0.6308
13	0.94165462	0.05995335	0.0214	0.6522

.
.

11 factors will be retained by the MINEIGEN criterion.

(continued)

659

TABLE 13.12   Continued

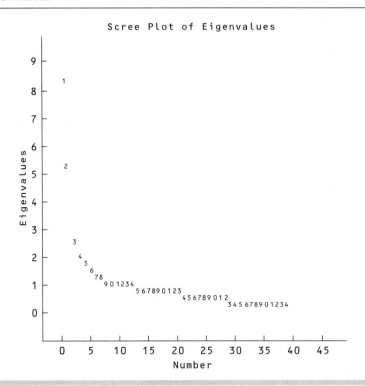

sets the degree of allowable correlation among factors. The highest correlation (.299) is between factors 2 and 3 (see Table 13.14).

The request for an output data set (`outfile=SASUSER.FACSCORE`) in the syntax produces factor scores, which are plotted in Figure 13.5. The generally oblong shape of the scatterplot of factor scores between these two factors confirms the correlation. This level of correlation can be considered borderline between accepting an orthogonal solution versus dealing with the complexities of interpreting an oblique solution. The simpler, orthogonal, solution is chosen.

The solution that is evaluated, interpreted, and reported is the run with principal factors extraction, varimax rotation, and 4 factors. In other words, after "trying out" oblique rotation, the decision is made to interpret the earlier run with orthogonal rotation. Syntax for this run is in Table 13.13.

Communalities are inspected to see if the variables are well defined by the solution. Communalities indicate the percent of variance in a variable that overlaps variance in the factors. As seen in Table 13.15, communality values for a number of variables are quite low (e.g., FOULLANG). Ten of the variables have communality values lower than .2 indicating considerable heterogeneity among the variables. It should be recalled, however, that factorial purity was not a consideration in development of the BSRI.

TABLE 13.13 Eigenvalues and Proportions of Variance for First Six Factors.
Principal Factors Extraction and Varimax Rotation (SAS FACTOR Syntax and Selected Output)

```
proc factor data=SASUSER.FACTORR prior=smc nfact=4 method=prinit rotate=varimax plot
reorder residuals out=SASUSER.FACSCPFA;
  var HELPFUL RELIANT DEFBEL YIELDING CHEERFUL INDPT ATHLET SHY ASSERT STRPERS FORCEFUL
AFFECT FLATTER LOYAL ANALYT FEMININE SYMPATHY MOODY SENSITIV UNDSTAND COMPASS LEADERAB
SOOTHE RISK DECIDE SELFSUFF CONSCIEN DOMINANT MASCULIN STAND HAPPY SOFTSPOK WARM
TRUTHFUL TENDER GULLIBLE LEADACT CHILDLIK INDIV FOULLANG LOVECHIL COMPETE AMBITIOU
GENTLE;
run;
```

         Initial Factor Method: Iterated Principal Factor Analysis

Eigenvalues of the Reduced Correlation Matrix: Total = 15.6425418 Average = 0.35551231

	Eigenvalue	Difference	Proportion	Cumulative
1	7.61934168	3.01821307	0.4871	0.4871
2	4.60112861	2.64348628	0.2941	0.7812
3	1.95764234	0.49300719	0.1251	0.9064
4	1.46463515	0.52864417	0.0936	1.0000
5	0.93599097	0.16446358	0.0598	1.0598
6	0.77152739	0.21427798	0.0493	1.1092

TABLE 13.14 Syntax and Selected SAS FACTOR PFA Output of Correlations among Factors Following Promax Rotation

```
proc factor data=SASUSER.FACTORR prior=smc nfact=4 method=prinit rotate=promax power=2
reorder out=SASUSER.FACSCORE;
   var HELPFUL RELIANT DEFBEL YIELDING CHEERFUL INDPT ATHLET SHY ASSERT STRPERS FORCEFUL
AFFECT FLATTER LOYAL ANALYT FEMININE SYMPATHY MOODY SENSITIV UNDSTAND COMPASS LEADERAB
SOOTHE RISK DECIDE SELFSUFF CONSCIEN DOMINANT MASCULIN STAND HAPPY SOFTSPOK WARM
TRUTHFUL TENDER GULLIBLE LEADACT CHILDLIK INDIV FOULLANG LOVECHIL COMPETE AMBITIOU
GENTLE;
run;
```

Inter-Factor Correlations

	Factor1	Factor2	Factor3	Factor4
Factor1	1.00000	0.13743	0.10612	0.15100
Factor2	0.13743	1.00000	0.29925	0.01143
Factor3	0.10612	0.29925	1.00000	0.03143
Factor4	0.15100	0.01143	0.03143	1.00000

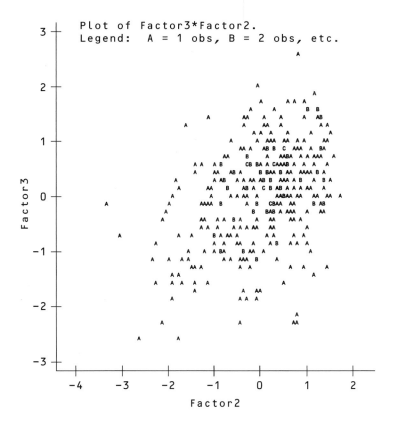

FIGURE 13.5 Scatterplot of factor scores with pairs of factors (2 and 3) as axes following oblique rotation.

Adequacy of rotation (Section 13.6.3) is assessed, in part, by scatterplots with pairs of rotated factors as axes and variables as points, as partially shown in Figure 13.6. Ideally, variable points are at the origin (the unmarked middle of figures) or in clusters at the ends of factor axes. Scatterplots between factor 1 and factor 2 (the only one shown), between factor 2 and factor 4, and between factor 3 and factor 4 seem reasonably clear. The scatterplots between other pairs of factors show evidence of correlation among factors as found during oblique rotation. Otherwise, the scatterplots are disappointing but consistent with the other evidence of heterogeneity among the variables in the BSRI.

Simplicity of structure (Section 13.6.3) in factor loadings following orthogonal rotation is assessed from **Rotated Factor Pattern** table (see Table 13.16). In each column there are a few high and many low correlations between variables and factors. There are also numerous moderate loadings so several variables will be complex (load on more than one factor) unless a fairly high cutoff for interpreting loadings is established. Complexity of variables (Section 13.6.5) is assessed by

TABLE 13.15 Communality Values (Four Factors), Selected Output from SAS FACTOR PFA (See Table 13.13 for Syntax)

Final Communality Estimates: Total = 15.642748

Variable	Value	Variable	Value
HELPFUL	0.28247773	SYMPATHY	0.44050701
RELIANT	0.39792405	MOODY	0.27127119
DEFBEL	0.24898804	SENSITIV	0.44399028
YIELDING	0.15114135	UNDSTAND	0.58130105
CHEERFUL	0.35981928	COMPASS	0.68459538
INDPT	0.45412974	LEADERAB	0.57710696
ATHLET	0.18372573	SOOTHE	0.38766356
SHY	0.15681146	RISK	0.27628362
ASSERT	0.44027299	DECIDE	0.37646744
STRPERS	0.50741884	SELFSUFF	0.63646751
FORCEFUL	0.46350807	CONSCIEN	0.35022727
AFFECT	0.47961520	DOMINANT	0.54004032
FLATTER	0.20018392	MASCULIN	0.18936093
LOYAL	0.29379728	STAND	0.43845627
ANALYT	0.15138438	HAPPY	0.44207474
FEMININE	0.15620355	SOFTSPOK	0.27748592
WARM	0.63155117	TENDER	0.53456942
TRUTHFUL	0.16826304	GULLIBLE	0.22140178
LEADACT	0.54070079	CHILDLIK	0.19201877
INDIV	0.23621643	FOULLANG	0.02475343
LOVECHIL	0.13659523	COMPETE	0.33755360
AMBITIOU	0.26450504	GENTLE	0.51391804

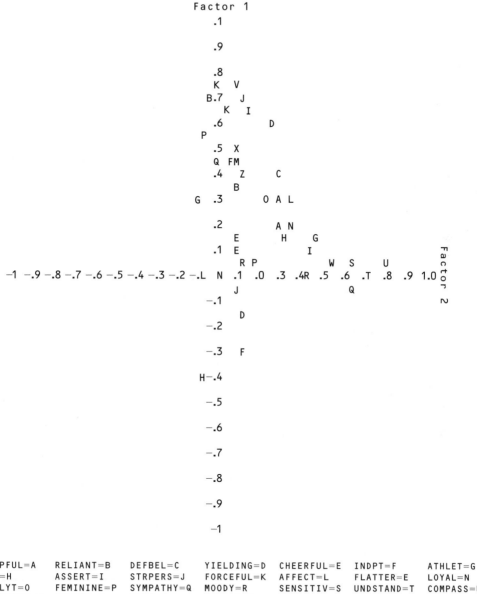

Plot of Factor Pattern for Factor1 and Factor2

FIGURE 13.6   Selected SAS FACTOR PFA output showing scatterplot of variable loadings with factors 1 and 2 as axes. (Syntax in Table 13.13.)

TABLE 13.16   Factor Loadings for Principal Factors Extraction and Varimax Rotation of Four Factors. Selected SAS FACTOR Output (Syntax Appears in Table 13.13)

Rotated Factor Pattern

	Factor1	Factor2	Factor3	Factor4
LEADERAB	0.73903	0.08613	0.04604	0.14629
LEADACT	0.72702	-0.02385	0.02039	0.10565
STRPERS	0.70096	0.10223	-0.05164	-0.05439
DOMINANT	0.67517	-0.06437	-0.28063	0.03583
FORCEFUL	0.64508	0.05473	-0.21028	-0.01292
ASSERT	0.64259	0.14184	-0.08400	0.01351
STAND	0.59253	0.24355	0.04322	0.16179
COMPETE	0.54071	-0.08335	0.16245	-0.10883
RISK	0.49569	0.08158	0.15386	0.01561
DECIDE	0.48301	0.08503	0.12984	0.34508
AMBITIOU	0.46606	0.00019	0.19928	0.08705
INDIV	0.43516	0.09442	0.06890	0.18217
DEFBEL	0.41270	0.27996	0.00103	0.01695
ATHLET	0.32396	-0.12167	0.24707	-0.05413
HELPFUL	0.31087	0.26951	0.29585	0.16024
MASCULIN	0.30796	-0.10533	-0.28704	-0.03220
ANALYT	0.27719	0.23310	-0.05488	0.13117
SHY	-0.38348	-0.07433	-0.04977	-0.04188
COMPASS	0.05230	0.81101	0.15098	0.03640
UNDSTAND	0.02375	0.73071	0.17725	0.12402
SENSITIV	0.05619	0.65980	0.06857	0.02820
SYMPATHY	-0.04187	0.64934	0.12865	-0.02375
SOOTHE	0.06957	0.53975	0.29652	-0.05973
AFFECT	0.29979	0.39154	0.39101	-0.28905
LOYAL	0.20039	0.38769	0.31916	-0.03842
TRUTHFUL	0.13882	0.32001	0.13851	0.16553
HAPPY	0.12217	0.06907	0.64118	0.10615
WARM	0.14939	0.48347	0.59415	-0.14995
CHEERFUL	0.16664	0.08795	0.55905	0.10853
GENTLE	0.02278	0.44682	0.55386	-0.08360
TENDER	0.10734	0.44629	0.55110	-0.14200
SOFTSPOK	-0.29038	0.12946	0.38838	0.15991
YIELDING	-0.13886	0.11282	0.34504	-0.00906
FEMININE	0.05666	0.18883	0.32403	0.11109
LOVECHIL	0.02370	0.20065	0.28216	-0.12712
FOULLANG	-0.01697	0.03248	0.14994	0.03046
MOODY	0.03005	0.10334	-0.37409	-0.34605
SELFSUFF	0.41835	0.10969	0.13556	0.65654
INDPT	0.46602	0.04291	0.03648	0.48351
RELIANT	0.36502	0.08295	0.16676	0.47958
CONSCIEN	0.20263	0.28468	0.23462	0.41603
FLATTER	0.16489	0.09539	0.21485	-0.34313
CHILDLIK	0.00494	-0.06847	-0.11126	-0.41824
GULLIBLE	-0.04076	0.08513	0.11042	-0.44755

TABLE 13.16   Continued

Variance Explained by Each Factor			
Factor1	Factor2	Factor3	Factor4
6.0140024	4.0025570	3.3995972	2.2265911

examining loadings for a variable across factors. With a loading cut of .45 only two variables, WARM and INDPT load on more than one factor.

The importance of each factor (Sections 13.4 and 13.6.4) is assessed by the percent of variance and covariance it represents. SSLs, called **Variance Explained by Each Factor** below the loadings in Table 13.16, are used in the calculations. It is important to use SSLs from rotated factors, because the variance is redistributed during rotation. Proportion of variance for a factor is SSL for the factor divided by number of variables. Proportion of covariance is SSL divided by sum of SSLs. Results, converted to percent, are shown in Table 13.17. Each of the factors accounts for between 4 and 16% of the variance in the set of variables, not an outstanding performance. Only the first factor accounts for substantial covariance.

Internal consistency of the factors (Section 13.6.4) is assessed through SMCs, available in SAS FACTOR when factor scores are requested (**out=SASUSER.FACSCPFA** in the syntax of Table 13.13). These are found in the **Squared Multiple Correlations of the Variables with Each Factor** table, in which factors serve as DVs with variables as IVs. Factors that are well defined by the variables have high SMCs, whereas poorly defined factors have low SMCs. As can be seen in Table 13.18, all factors are internally consistent. (The off-diagonal elements in these matrices are correlations among factor scores. Although uniformly low, the values are not zero. As discussed in Section 13.6.6, low correlations among scores on factors are often obtained even with orthogonal rotation.)

TABLE 13.17   Percents of Variance and Covariance Explained by Each of the Rotated Orthogonal Factors

	Factors			
	1	*2*	*3*	*4*
SSL	6.01	4.00	3.40	2.23
Percent of variance	13.66	9.09	7.73	5.07
Percent of covariance	38.42	25.57	21.74	14.26

TABLE 13.18   SMCs for Factors with Variables as IVs. Selected Output from SAS FACTOR PFA with Orthogonal (Varimax) Rotation (Syntax in Table 13.13)

```
Squared Multiple Correlations of the Variables with Each Factor

        Factor1         Factor2         Factor3         Factor4

     0.90317097      0.84911248      0.80546360      0.77855397
```

TABLE 13.19   Order (by Size of Loadings) in Which Variables Contribute to Factors

Factor 1: *Dominance*	Factor 2: *Empathy*	Factor 3: *Positive Affect*	Factor 4: *Independence*
Has leadership abilities	Compassionate	Happy	Self-sufficient
Acts as a leader	Understanding	Warm	Independent
Strong personality	Sensitive to needs of others	Cheerful	Self-reliant
Dominant	Sympathetic	Gentle	
Forceful	Eager to soothe hurt feelings	Tender	
Assertive	Warm		
Willing to take a stand			
Competitive			
Willing to take risks			
Makes decisions easily			
Independent			
Ambitious			

Note: Variables with higher loadings on the factor are nearer the top of the columns. Proposed labels are in italics.

Factors are interpreted through their factor loadings (Section 13.6.5) from Table 13.16. It is decided to use a loading of .45 (20% variance overlap between variable and factor). With the use of the .45 cut, Table 13.19 is generated to further assist interpretation. In more informal presentations of factor analytic results, this table might be reported instead of Table 13.16. Factors are put in columns and variables with the largest loadings are put on top. In interpreting a factor, items near the top of the columns are given somewhat greater weight. Variable names are written out in full detail and labels for the factors (e.g., Dominance) are suggested at the top of each column. Table 13.20 shows a more formal summary table of factor loadings, including communalities as well as percents of variance and covariance.

Table 13.21 provides a checklist for FA. A Results section in journal format follows for the data analyzed in this section.

**TABLE 13.20 Factor Loadings, Communalities (h^2), and
Percents of Variance and Covariance for Principal Factors
Extraction and Varimax Rotation on BSRI Items**

Item	F_1^a	F_2	F_3	F_4	h^2
Leadership ability	.74	.00	.00	.00	.58
Acts as leader	.73	.00	.00	.00	.54
Strong personality	.70	.00	.00	.00	.51
Dominant	.68	.00	.00	.00	.54
Forceful	.65	.00	.00	.00	.46
Assertive	.64	.00	.00	.00	.44
Takes stand	.59	.00	.00	.00	.44
Competitive	.54	.00	.00	.00	.34
Takes risks	.50	.00	.00	.00	.28
Makes decisions	.48	.00	.00	.00	.38
Independent	.47	.00	.00	.48	.25
Ambitious	.47	.00	.00	.00	.26
Compassionate	.00	.81	.00	.00	.68
Understanding	.00	.73	.00	.00	.58
Sensitive	.00	.66	.00	.00	.44
Sympathetic	.00	.65	.00	.00	.44
Eager to soothe hurt feelings	.00	.54	.00	.00	.39
Warm	.00	.48	.72	.00	.63
Happy	.00	.00	.64	.00	.44
Cheerful	.00	.00	.56	.00	.36
Gentle	.00	.00	.55	.00	.51
Tender	.00	.00	.55	.00	.53
Self-sufficient	.00	.00	.00	.66	.64
Self-reliant	.00	.00	.00	.48	.40
Affectionate	.00	.00	.00	.00	.48
Conscientious	.00	.00	.00	.00	.35
Defends beliefs	.00	.00	.00	.00	.25
Masculine	.00	.00	.00	.00	.19
Truthful	.00	.00	.00	.00	.17
Feminine	.00	.00	.00	.00	.16
Helpful	.00	.00	.00	.00	.28
Individualistic	.00	.00	.00	.00	.24
Shy	.00	.00	.00	.00	.16
Moody	.00	.00	.00	.00	.27
Percent of variance	13.66	9.09	7.73	5.06	
Percent of covariance	38.42	25.57	21.74	14.26	

[a]Factor labels:

F_1 Dominance

F_2 Empathy

F_3 Positive Affect

F_4 Independence

TABLE 13.21   Checklist for Factor Analysis

1. Limitations
    a. Outliers among cases
    b. Sample size and missing data
    c. Factorability of **R**
    d. Normality and linearity of variables
    e. Multicollinearity and singularity
    f. Outliers among variables
2. Major analyses
    a. Number of factors
    b. Nature of factors
    c. Type of rotation
    d. Importance of factors
3. Additional analyses
    a. Factor scores
    b. Distinguishability and simplicity of factors
    c. Complexity of variables
    d. Internal consistency of factors
    e. Outlying cases among the factors

Results

Principal factors extraction with varimax rotation was performed through SAS FACTOR on 44 items from the BSRI for a sample of 344 women. Principal components extraction was used prior to principal factors extraction to estimate number of factors, presence of outliers, absence of multicollinearity, and factorability of the correlation matrices. With an α = .001 cutoff level, 25 of 369 women produced scores that identified them as outliers; these cases were deleted from principal factors extraction.[10]

---

[10]Outliers were compared as a group to nonoutliers through discriminant analysis. As a group, at $p < .01$, the 25 women were less self-reliant and more easily flattered than women who were not outliers.

Four factors were extracted. As indicated by SMCs, all factors were internally consistent and well defined by the variables; the lowest of the SMCs for factors from variables was .78. [*Information on SMCs is from Table 13.18.*] The reverse was not true, however; variables were, by and large, not well defined by this factor solution. Communality values, as seen in Table 13.15, tended to be low. With a cutoff of .45 for inclusion of a variable in interpretation of a factor, 20 of 44 variables did not load on any factor. Failure of numerous variables to load on a factor reflects heterogeneity of items on the BSRI. However, only two of the variables in the solution, "warm" and "independent," were complex.

When oblique rotation was requested, factors interpreted as Empathy and Positive Affect correlated .30. However, because the correlation was modest and limited to one pair of factors, and because remaining correlations were low, orthogonal rotation was chosen.

Loadings of variables on factors, communalities, and percents of variance and covariance are shown in Table 13.20. Variables are ordered and grouped by size of loading to facilitate interpretation. Loadings under .45 (20% of variance) are replaced by zeros. Interpretive labels are suggested for each factor in a footnote.

In sum, the four factors on the BSRI for this group of women are dominance (e.g., leadership abilities and strong personality), empathy (e.g., compassion and understanding), positive affect (e.g., happy and warm), and independence (e.g., self-sufficient and self-reliant).

13.8   Comparison of Programs

SPSS, SAS, and SYSTAT each have a single program to handle both FA and PCA. The first two programs have numerous options for extraction and rotation and give the user considerable latitude in directing the progress of the analysis. Features of three programs are described in Table 13.22.

TABLE 13.22 Comparison of Factor Analysis Programs

Feature	SPSS FACTOR	SAS FACTOR	SYSTAT FACTOR
Input			
Correlation matrix	Yes	Yes	Yes
About origin	No	Yes	No
Covariance matrix	Yes	Yes	Yes
About origin	No	Yes	No
SSCP matrix	No	No	Yes
Factor loadings (unrotated pattern)	Yes	Yes	Yes
Factor-score coefficients	No	Yes	Data file
Factor loadings (rotated pattern) and factor correlations	No	Yes	Yes
Options for missing data	Yes	No	Yes
Analyze partial correlation or covariance matrix	No	Yes	No
Specify maximum number of factors	FACTORS	NFACT	NUMBER
Extraction method (see Table 13.7)			
PCA	PC	PRIN	PCA
PFA	PAF	PRINIT	IPA
Image (Little Jiffy, Harris)	IMAGE	Yes[a]	No
Maximum likelihood	ML	ML	MLA
Alpha	ALPHA	ALPHA	No
Unweighted least squares	ULS	ULS	No
Generalized least squares	GLS	Yes	No
Specify communalities	Yes	Yes	No
Specify minimum eigenvalues	MINEIGEN	MIN	EIGEN
Specify proportion of variance to be accounted for	No	PROPORTION	No
Specify maximum number of iterations	ITERATE	MAXITER	ITER
Option to allow communalities >1	No	HEYWOOD	No
Specify tolerance	No	SING	No
Specify convergence criterion for extraction	ECONVERGE	CONV	CONV
Specify convergence criterion for rotation	RCONVERGE	No	No
Rotation method (see Table 13.9)			
Varimax	Yes	Yes	Yes
Quartimax	Yes	Yes	Yes
Equamax	Yes	Yes	Yes
Orthogonal with gamma	No	ORTHOMAX	ORTHOMAX
Parsimax	No	Yes	No
Direct oblimin	Yes	No	Yes

TABLE 13.22   Continued

Feature	SPSS FACTOR	SAS FACTOR	SYSTAT FACTOR
Input *(continued)*			
Rotation method (see Table 13.9) *(continued)*			
Direct quartimin	DELTA = 0	No	No
Orthoblique	No	HK	No
Promax	No	Yes	No
Procrustes	No	Yes	No
Prerotation criteria	No	Yes	No
Optional Kaiser's normalization	Yes	Yes	Normalized only
Optional weighting by Cureton-Mulaik technique	No	Yes	No
Optional rescaling of pattern matrix to covariances	No	Yes	No
Weighted correlation matrix	No	WEIGHT	No
Alternate methods for computing factor scores	Yes	No	No
Output			
Means and standard deviations	Yes	Yes	No
Number of cases per variable (missing data)	Yes	No	No
Significance of correlations	Yes	No	No
Covariance matrix	Yes	Yes	Yes
Initial communalities	Yes	Yes	Yes
Final communalities	Yes	Yes	Yes
Eigenvalues	Yes	Yes	Yes
Difference between successive eigenvalues	No	Yes	No
Standard error for each eigenvector element	No	No	Yes
Percent of variance total variance explained by factors	Yes	No	Yes
Cumulative percent of variance	Yes	No	No
Percent of covariance	No	No	Yes
Unrotated factor loadings	Yes	Yes	Yes
Variance explained by factors for all loading matrices	No	Yes	Yes
Simplicity criterion, each rotation iteration	δ^{b}	No	No
Rotated factor loadings (pattern)	Yes	Yes	Yes
Rotated factor loadings (structure)	Yes	Yes	Yes
Eigenvectors	No	Yes	Yes
Standard error for each eigenvector element	No	No	Yes
Transformation matrix	Yes	Yes	No
Factor-score coefficients	Yes	Yes	Data file[c]

(continued)

TABLE 13.22 Continued

Feature	SPSS FACTOR	SAS FACTOR	SYSTAT FACTOR
Output *(continued)*			
Standardized factor scores	Data file	Data file	Data file[c]
Residual component scores	No	No	Data file[c]
Sum of squared residuals (Q)	No	No	Data file[c]
Probability for Q	No	No	Data file[c]
Scree plot	Yes	Yes	Yes
Plots of unrotated factor loadings	No	Yes	No
Plots of rotated factor loadings	Yes	Yes	Yes
Sorted rotated factor loadings	Yes	Yes	Yes
χ^2 test for number of factors (with maximum likelihood estimation)	No	Yes	No
χ^2 test that all eigenvalues are equal	No	No	Yes
χ^2 test that last n eigenvalues are equal	No	No	Yes
Standard errors of factor loadings (with maximum likelihood estimation and promax solutions)	No	Yes	No
Inverse of correlation matrix	Yes	Yes	No
Determinant of correlation matrix	Yes	No	No
Partial correlations (anti-image matrix)	AIC	MSA	No
Measure of sampling adequacy	AIC, KMO	MSA	No
Anti-image covariance matrix	AIC	No	No
Bartlett's test of sphericity	KMO	No	No
Residual correlation matrix	Yes	Yes	Yes
Reproduced correlation matrix	Yes	No	No
Correlations among factors	Yes	Yes	Yes

[a]Two types.
[b]Oblique only.
[c]PCA only.

13.8.1 SPSS Package

SPSS FACTOR does a PCA or FA on a correlation matrix or a factor loading matrix, helpful to the researcher who is interested in higher-order factoring (extracting factors from previous FAs). Several extraction methods and a variety of orthogonal rotation methods are available. Oblique rotation is done using direct oblimin, one of the best methods currently available (see Section 13.5.2.2).

Univariate output is limited to means, standard deviations, and number of cases per variable, so that the search for univariate outliers must be conducted through other programs. Similarly, there is no provision for screening for multivariate outliers among cases. But the program is very helpful in assessing factorability of **R**, as discussed in Section 13.3.2.6.

Output of extraction and rotation information is extensive. The residual and reproduced correlation matrices are provided as an aid to diagnosing adequacy of extraction and rotation. SPSS FACTOR is the only program reviewed that, under conditions requiring matrix inversion, prints out the determinant of the correlation matrix, helpful in signaling the need to check for multicollinearity and singularity (Sections 13.3.2.5 and 4.1.7). Determination of number of factors is aided by an optional printout of a scree plot (Section 13.6.2). Several estimation procedures for factor scores (Section 13.6.6) are available as output to a file.

13.8.2   SAS System

SAS FACTOR is another highly flexible, full-featured program for FA and PCA. About the only weakness is in screening for outliers. SAS FACTOR accepts rotated loading matrices, as long as factor correlations are provided, and can analyze a partial correlation or covariance matrix (with specification of variables to partial out). There are several options for extraction, as well as orthogonal and oblique rotation. Maximum-likelihood estimation provides a χ^2 test for number of factors. Standard errors may be requested for factor loadings with maximum-likelihood estimation and promax rotation. A target pattern matrix can be specified as a criterion for oblique rotation in confirmatory FA. Additional options include specification of proportion of variance to be accounted for in determining the number of factors to retain and the option to allow communalities to be greater than 1.0. The correlation matrix can be weighted to allow the generalized least squares method of extraction.

Factor scores can be written to a data file. SMCs of factors as DVs with variables as IVs are given, to evaluate the reliability of factors.

13.8.3   SYSTAT System

The current SYSTAT FACTOR program is less limited than earlier versions. Wilkinson (1990) advocated the use of PCA rather than FA because of the indeterminacy problem (Section 13.6.6). However, the program now does PFA (called IPA) as well as PCA and maximum likelihood (MLA) extraction. Four common methods of orthogonal rotation are provided, as well as provision for oblique rotation. SYSTAT FACTOR can accept correlation or covariance matrices as well as raw data.

The SYSTAT FACTOR program provides scree plots and plots of factor loadings and will optionally sort the loading matrix by size of loading to aid interpretation. Additional information is available by requesting that standardized component scores, their coefficients, and loadings be sent to a data file. Factor scores (from PFA or MLA) cannot be saved. Residual scores (actual minus predicted z-scores) also can be saved, as well as the sum of the squared residuals and a probability value for it.

14

Structural Equation Modeling

JODIE B. ULLMAN
California State University, San Bernardino

14.1 General Purpose and Description

Structural equation modeling (SEM) is a collection of statistical techniques that allow a set of relationships between one or more IVs, either continuous or discrete, and one or more DVs, either continuous or discrete, to be examined. Both IVs and DVs can be either factors or measured variables. Structural equation modeling is also referred to as causal modeling, causal analysis, simultaneous equation modeling, analysis of covariance structures, path analysis, or confirmatory factor analysis. The latter two are actually special types of SEM.

SEM allows questions to be answered that involve multiple regression analyses of factors. When exploratory factor analysis (EFA, Chapter 13) is combined with multiple regression analyses (Chapter 5), you have SEM. At the simplest level, a researcher posits a relationship between a single measured variable (say, success in graduate school) and other measured variables (say, undergraduate GPA, gender, and average daily caffeine consumption). This simple model is a multiple regression presented in diagram form in Figure 14.1. All four of the measured variables appear in boxes connected by lines with arrows indicating that GPA, gender, and caffeine (the IVs) predict graduate school success (the DV). A line with two arrows indicates a correlation among the IVs. The presence of a residual indicates imperfect prediction.

A more complicated model of success in graduate school appears in Figure 14.2. In this model, Graduate School Success is a latent variable (a factor) that is not directly measured but rather assessed indirectly using number of publications, grades, and faculty evaluations, three measured variables. Graduate School Success is, in turn, predicted by gender (a measured variable) and by Undergraduate Success, a second factor which is assessed through undergraduate GPA, faculty recommendations, and GRE scores (three additional measured variables). For clarity in the text, initial capitals are used for names of factors and lowercase letters for names of measured variables.

I would like to thank Barbara Tabachnick and Linda Fidell for the opportunity to write this chapter and also for their helpful comments on an earlier draft. I would also like to thank Peter Bentler who not only performed a detailed review of an earlier version of this chapter but who has also been responsible for shaping my thinking on SEM. I would also like to thank Lisa, Harlow, Jim Sidanius and an anonymous reviewer for their helpful suggestions. This chapter was supported in part by NIDA grant P0IDA01070-32.

Figures 14.1 and 14.2 are examples of *path diagrams.* These diagrams are fundamental to SEM because they allow the researcher to diagram the hypothesized set of relationships—the model. The diagrams are helpful in clarifying a researcher's ideas about the relationships among variables and they can be directly translated into the equations needed for the analysis.

Several conventions are used in developing SEM diagrams. Measured variables, also called *observed variables*, *indicators*, or *manifest variables* are represented by squares or rectangles. Factors have two or more indicators and are also called *latent variables*, *constructs*, or *unobserved variables.* Factors are represented by circles or ovals in path diagrams. Relationships between variables are indicated by lines; lack of a line connecting variables implies that no direct relationship has been

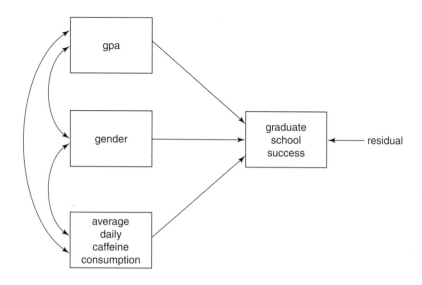

FIGURE 14.1   Path diagram of multiple regression.

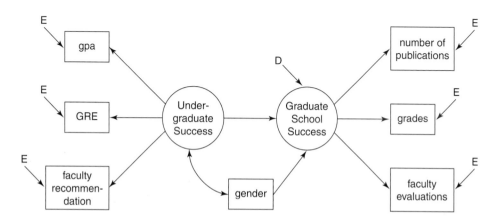

FIGURE 14.2   Path diagram of a structural model.

hypothesized. Lines have either one or two arrows. A line with one arrow represents a hypothesized direct relationship between two variables, and the variable with the arrow pointing to it is the DV. A line with an arrow at both ends indicates an unanalyzed relationship, simply a covariance between the two variables with no implied direction of effect.

In the model of Figure 14.2, Success in Graduate School is a latent variable (factor) that is predicted by gender (a measured variable) and Undergraduate Success (a factor). Notice the line with the arrow at both ends connecting Undergraduate Success and gender. This line with an arrow at both ends implies that there is a relationship between the variables but makes no prediction regarding the direction of effect. Also notice the direction of the arrows connecting the Graduate School Success construct (factor) to its indicators: The construct *predicts* the measured variables. The implication is that Graduate School Success drives, or creates, the number of publications, grades, and faculty evaluations of graduate students. It is impossible to measure this construct directly, so we do the next best thing and measure several indicators of success. We hope that we are able to tap into graduate students' true level of success by measuring a lot of observable indicators. This is the same logic as in factor analysis (Chapter 13).[1]

In Figure 14.2, GPA, GRE, faculty recommendations, Graduate School Success, number of publications, grades, and faculty evaluations are all DVs. They all have one-way arrows pointing to them. Gender and Undergraduate Success are IVs in the model. They have no one-way arrows pointing to them. Notice that all the DVs, both observed and unobserved, have arrows labeled "E" or "D" pointing toward them. Es (errors) point to measured variables; Ds (disturbances) point to latent variables (factors). As in multiple regression, nothing is predicted perfectly; there is always residual or error. In SEM, the residual not predicted by the IV(s) is included in the diagram with these paths.

The part of the model that relates the measured variables to the factors is sometimes called the *measurement model.* In this example, the two constructs (factors), Undergraduate Success and Graduate School Success, and the indicators of these constructs (factors) form the *measurement model.* The hypothesized relationships among the constructs, in this example, the one path between Undergraduate Success and Graduate School Success, is called the *structural model.*

Note, both models presented so far include hypotheses about relationships among variables (covariances) but not about means or mean differences. Mean differences associated with group membership can also be tested within the SEM framework.

When experiments are analyzed, with proper data collection, the adequacy of the manipulation can also be accounted for within the analysis (Feldman, Ullman, & Dunkel-Schetter, 1998). Experiments with or without a mean structure can be analyzed through SEM. For an example from the literature of an experiment analyzed through SEM, consider Feldman, Ullman, and Dunkel-Schetter (1998). Feldman and colleagues used SEM to analyze an experiment that examined the effects of perceived similarity and perceived vulnerability on attributions of victim blame. Aiken, Stein, and Bentler (1994), who employed SEM techniques to evaluate the effectiveness of a mammography screening program, provide an example of a treatment program evaluation. Even in a simple experiment, researchers are often interested in processes that are more complex than a standard analysis suggests. Consider the diagram in Figure 14.3.

---

[1]Now, thinking back to the chapters on factor analysis, MANOVA, discriminant analysis, and canonical correlation, what would the implication be if the arrows between the Graduate School Success factor and the measured indicators pointed the opposite way from the three indicators *toward* Graduate School Success? It would imply a principal component or linear combination.

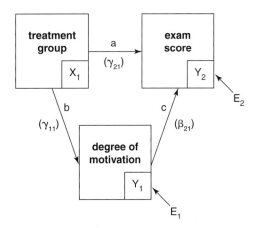

FIGURE 14.3   Path diagram of an experiment.

At the start of a semester, students are randomly assigned to one of two treatment conditions, a study skills training group or a waiting-list control. X_1 is a dummy-coded variable (cf. Section 1.2.1) that indicates the assigned group, where $0 = $ control, $1 = $ treatment. Final exam scores are recorded at the end of the semester. ANOVA essentially tests path *a*. But is it reasonable to suggest that mere assignment to a group creates the change? Perhaps not. Maybe, instead, study skills training increases a student's motivational level and higher motivation leads to a higher grade. Motivational level serves as an intervening variable between the treatment and the exam score (i.e., the treatment is associated with increased motivation and increased motivation is associated with increased exam scores). This is a different question than is posed in ANOVA or even ANCOVA or sequential regression. ANOVA asks simply "Is there a difference between the treatment and control group on exam score?" ANCOVA asks "Is there a difference between groups after the DV has been adjusted by a covariate (e.g., degree of motivation)?" These questions are distinct from the hypotheses illustrated in Figure 14.3 that involve a process or an *indirect effect*. The indirect effect can be tested by testing the product of paths *b* and *c*. This example uses only measured variables and is called *path analysis*; however, indirect effect hypotheses can be tested using both latent and observed variables.

The first step in a SEM analysis is specification of a model, so this is a *confirmatory* rather than an exploratory technique. The model is estimated, evaluated, and perhaps modified. The goal of the analysis might be to test a model, to test specific hypotheses about a model, to modify an existing model, or to test a set of related models.

There are a number of advantages to the use of SEM. When relationships among factors are examined, the relationships are free of measurement error because the error has been estimated and removed, leaving only common variance. Reliability of measurement can be accounted for explicitly within the analysis by estimating and removing the measurement error. Additionally, as was seen in Figure 14.2, complex relationships can be examined. When the phenomena of interest are complex and multidimensional, SEM is the only analysis that allows complete and simultaneous tests of all the relationships.

Unfortunately, there is a small price to pay for the flexibility that SEM offers. With the ability to analyze complex relationships among combinations of discrete and continuous variables, both observed and latent, comes more complexity and more ambiguity. Indeed, there is quite a bit of jargon and many choices of analytic techniques. But if you love to wallow in data, you'll adore SEM!

14.2   Kinds of Research Questions

The data set is an empirical covariance matrix and the model produces an estimated population covariance matrix. The major question asked by SEM is, "Does the model produce an estimated population covariance matrix that is consistent with the sample (observed) covariance matrix?" After the adequacy of the model is assessed, various other questions about specific aspects of the model are addressed.

14.2.1   Adequacy of the Model

Parameters (path coefficients, variances, and covariances of IVs) are estimated to create an estimated population covariance matrix. If the model is good the parameter estimates will produce an estimated matrix that is close to the sample covariance matrix. "Closeness" is evaluated primarily with the chi-square test statistic and fit indices. For the Graduate School Success model of Figure 14.2, is the estimated population covariance matrix generated by the model consistent with the sample covariance matrix of the data? This is discussed in Sections 14.4.5 and 14.5.3.

14.2.2   Testing Theory

Each theory (model) generates its own covariance matrix. Which theory produces an estimated population covariance matrix that is most consistent with the sample covariance matrix? Models representing competing theories in a specific research area are estimated, pitted against each other, and evaluated as demonstrated in Section 14.5.4.1.

14.2.3   Amount of Variance in the Variables Accounted for by the Factors

How much of the variance in the DVs, both latent and observed, is accounted for by the IVs? For example, how much variance in Graduate School Success is accounted for by gender and Undergraduate Success? Which of the variables included in the analysis account for the most variance? This question is answered through R^2-type statistics discussed in Section 14.5.5.

14.2.4   Reliability of the Indicators

How reliable is each of the measured variables? For the example, is the measure of faculty evaluations reliable? Reliability of measured variables and internal consistency measures of reliability are derived from SEM analyses and are discussed in Section 14.5.5.

14.2.5   Parameter Estimates

Estimates of parameters are fundamental to SEM analyses because they are used to generate the estimated population covariance matrix for the model. What is the path coefficient for a specific path? For example, what is the path coefficient for predicting Graduate School Success from Undergraduate Success? Does the coefficient differ significantly from 0? Within the model, what is the relative importance of various paths? For instance, is the path from Undergraduate Success more or less important to the prediction of Graduate School Success than the path from gender? Parameter estimates can also be compared across SEM models. When a single path is tested, it is called a test of a *direct effect*. Assessment of parameters is demonstrated in Sections 14.4.5, 14.6.1.3, and 14.6.2.4.

14.2.6   Intervening Variables

Does an IV directly affect a specific DV, or does the IV affect the DV through an intermediary, or mediating, variable? In the example of Figure 14.3, is the relationship between treatment group and exam score mediated by degree of motivation? Because motivation is an intervening variable, this is a test of *indirect effects*. Tests of indirect effects are demonstrated in Section 14.6.2.

14.2.7   Group Differences

Do two or more groups differ in their covariance matrices, regression coefficients, or means? For example, if the experiment described above (see Figure 14.3) is performed for both grade school and high school youngsters, does the same model fit both age groups? This analysis could be performed with or without means (c.f., Section 14.5.8). Multiple group modeling is briefly discussed in Section 14.5.7. Stein, Newcomb, and Bentler (1993) examined the effects of grandparent and parent drug use on behavior problems in boys and girls aged 2 to 8. Separate structural equations models were developed for boys and girls and then statistically compared.

14.2.8   Longitudinal Differences

Differences within and across people across time can also be examined. This time interval can be years, days, or microseconds. For the example of the experiment: How, if at all, does treatment change performance and motivation at several different time points in the semester? Longitudinal modeling is not illustrated in this chapter. Although there are several different approaches, one exciting new approach to analyzing longitudinal data with three or more time points is called Latent Growth Curve Modeling. This approach is innovative because it allows tests of *individual* growth patterns. Several hypotheses are tested with this analysis. How does a dependent variable (latent or observed), say, adolescent drug use, change across multiple time points, say, the teenage and young adult years? Is the change linear? quadratic? Do participants (teenagers) vary in their initial level of drug use? Do adolescents' drug use patterns change at the same rate?

14.2.9   Multilevel Modeling

Independent variables collected at different nested levels of measurement (e.g., students nested within classrooms nested within schools) are used to predict dependent variables at the same level or other levels of measurement. For example, using a multiple group model we could examine the effectiveness of an intervention given to classrooms of children from characteristics of the children, the classroom, and the school. In this example children are nested within classrooms and classrooms are nested within schools. This is briefly discussed in Section 14.5.7 and is the topic of Chapter 15.

14.3 Limitations to Structural Equation Modeling

14.3.1 Theoretical Issues

SEM is a *confirmatory technique* in contrast to exploratory factor analysis. It is used most often to test a theory—maybe just a personal theory—but a theory nonetheless. Indeed, one cannot do SEM without prior knowledge of, or hypotheses about, potential relationships among variables. This is perhaps the largest difference between SEM and other techniques in this book and one of its greatest strengths. Planning, driven by theory, is essential to any SEM analysis. The guidelines for planning an exploratory factor analysis, as outlined in Section 13.3.1, are also applicable to SEM analyses.

Although SEM is a confirmatory technique, there are ways to test a variety of different models (models that test specific hypotheses, or perhaps provide better fit) after a model has been estimated. However, if numerous modifications of a model are tested in hopes of finding the best-fitting model, the researcher has moved to exploratory data analysis and appropriate steps need to be taken to protect against inflated Type I error levels. Searching for the best model is appropriate provided significance levels are viewed cautiously and cross-validation with another sample is performed whenever possible.

SEM has developed a bad reputation in some circles, in part because of the use of SEM for exploratory work without the necessary controls. It may also be due, in part, to the use of the term *causal modeling* to refer to structural equation modeling. There is nothing causal, in the sense of inferring causality, about the use of SEM. Attributing causality is a design issue, not a statistical issue.

Unfortunately, SEM is often thought of as a technique strictly for nonexperimental or correlational designs. This is overly limiting. SEM, like regression, can be applied to both experimental and nonexperimental designs. In fact, there are some advantages to using SEM in the analysis of experiments: Mediational processes can be tested and information regarding the adequacy of the manipulations can be included in the analysis (Feldman, Ullman, & Dunkel-Schetter, 1998).

The same caveats regarding generalizing results apply to SEM as they do to the other techniques in this book. Results can only be generalized to the type of sample that was used to estimate and test the SEM model.

14.3.2 Practical Issues

14.3.2.1 Sample Size and Missing Data

Covariances, like correlations, are less stable when estimated from small samples. SEM is based on covariances. Parameter estimates and chi-square tests of fit are also very sensitive to sample size. SEM, then, like factor analysis, is a large sample technique. Velicer and Fava (1998) found that in exploratory factor analysis size of the factor loadings, the number of variables, and the size of the sample were important elements in obtaining a good factor model. This can be generalized to SEM models. Models with strong expected parameter estimates and reliable variables may require fewer participants. Although SEM is a large sample technique new test statistics have been developed that allow for estimation of models with as few as 60 participants (Bentler & Yuan, 1999). For estimating adequate sample size for power calculations, MacCallum, Browne, and Sugawara (1996) present tables of minimum sample sizes needed for tests of goodness of fit. These tables base sample size estimates on model degrees of freedom and effect size.

The Chapter 4 guidelines for the treatment of missing data apply to SEM analyses. However, as discussed in Chapter 4, problems are associated with either deleting or estimating missing data. An advantage of structural modeling is that the missing data mechanism can be included in the model. Some of the software packages now include procedures for estimating missing data, including the EM algorithm. Treatment of missing data patterns through SEM is not demonstrated in this chapter but the interested reader is referred to Allison (1987), Muthén, Kaplan, and Hollis (1987), and Bentler (1995).

14.3.2.2 *Multivariate Normality and Outliers*

Most of the estimation techniques used in SEM assume multivariate normality. To determine the extent and shape of nonnormally distributed data, *screen the measured variables for outliers, both univariate and multivariate, and the skewness and kurtosis of the measured variables examined in the manner described in Chapter 4.* All measured variables, regardless of their status as DVs or IVs, are screened together for outliers. (Some SEM packages test for the presence of multivariate outliers, skewness, and kurtosis.) If significant skewness is found, transformations can be attempted; however, often variables are still highly skewed or highly kurtotic even after transformation. Some variables, such as drug use variables, are not expected to be normally distributed in the population, anyway. If transformations do not restore normality, or a variable is not expected to be normally distributed in the population, an estimation method can be selected that addresses the nonnormality (Sections 14.5.2, 14.6.1, and 14.6.2).

14.3.2.3 *Linearity*

SEM techniques examine only linear relationships among variables. Linearity among latent variables is difficult to assess; however, *linear relationships among pairs of measured variables can be assessed through inspection of scatterplots.* If nonlinear relationships among measured variables are hypothesized, these relationships are included by raising the measured variables to powers, as in multiple regression. For example, if the relationship between graduate school success and average daily caffeine consumption is quadratic (a little caffeine is not enough, a few cups is good, but more than a few is detrimental), the square of average daily caffeine consumption is used.

14.3.2.4 *Absence of Multicollinearity and Singularity*

As with the other techniques discussed in the book, matrices need to be inverted in SEM. Therefore, if variables are perfect linear combinations of one another or are *extremely* highly correlated, the necessary matrices cannot be inverted. If possible *inspect the determinant of the covariance matrix. An extremely small determinant may indicate a problem with multicollinearity or singularity.* Generally, SEM programs abort and provide warning messages if the covariance matrix is singular. If you get such a message, check your data set. It often is the case that linear combinations of variables have been inadvertently included. Simply delete the variable causing the singularity. If true singularity is found, create composite variables and use them in the analysis.

14.3.2.5   Residuals

After model estimation, *the residuals should be small and centered around zero. The frequency distribution of the residual covariances should be symmetrical.* Residuals in the context of SEM are residual *covariances* not residual *scores* as discussed in other chapters. SEM programs provide diagnostics of residuals. Nonsymmetrically distributed residuals in the frequency distribution may signal a poor-fitting model; the model is estimating some of the covariances well and others poorly. It sometimes happens that one or two residuals remain quite large although the model fits reasonably well and the residuals appear to be symmetrically distributed and centered around zero. When large residuals are found it is often helpful to examine the Lagrange Multiplier (LM) test, discussed in Section 14.5.4.2, and consider adding paths to the model.

14.4   Fundamental Equations for Structural Equations Modeling

14.4.1   Covariance Algebra

The idea behind SEM is that the hypothesized model has a set of underlying parameters which correspond to (1) the regression coefficients, and (2) the variances and covariances of the independent variables in the model (Bentler, 1995). These parameters are estimated from the sample data to be a "best guess" about population values. The estimated parameters are then combined by means of covariance algebra to produce an estimated population covariance matrix. This estimated population covariance matrix is compared with the sample covariance matrix and, ideally, the difference is very small and not statistically significant.

Covariance algebra is a helpful tool in calculating variances and covariances in SEM models; however, matrix methods are generally employed because covariance algebra becomes extremely tedious as models become increasingly complex. Covariance algebra is useful to demonstrate how parameter estimates are combined to produce an estimated population covariance matrix for a small example.

The three basic rules in covariance algebra appear below where c is a constant and X_i is a random variable:

$$
\begin{aligned}
&1.\ COV(c, X_1) = 0 \\
&2.\ COV(cX_1, X_2) = cCOV(X_1, X_2) \\
&3.\ COV(X_1 + X_2, X_3) = COV(X_1, X_3) + COV(X_2, X_3)
\end{aligned}
\tag{14.1}
$$

By the first rule, the covariance between a variable and a constant is zero. By the second rule, the covariance between two variables where one is multiplied by a constant is the same as the constant multiplied by the covariance between the two variables. By the third rule, the covariance between the sum (or difference) of two variables and a third variable is the sum of the covariance of the first variable and the third and the covariance of the second variable and the third.

Figure 14.3 is used to illustrate some of the principles of covariance algebra. (Ignore for now the difference between γ and β; the difference is explained in Section 14.4.3.) In SEM, as in multiple regression, we assume that the residuals do not correlate with each other or with other variables in the models. In this model, both degree of motivation (Y_1) and exam score (Y_2) are DVs. Recall that a DV in SEM is any variable with a single-headed arrow pointing toward it. Treatment group (X_1) with no single-headed arrows pointing to it is an IV. To specify the model, a separate equation is written for each DV. For motivation, Y_1,

$$Y_1 = \gamma_{11}X_1 + \varepsilon_1 \tag{14.2}$$

Degree of motivation is a weighted function of treatment group plus error. Note that ε_1 in the equation corresponds to E_1 in Figure 14.3 and for exam score, Y_2,

$$Y_2 = \beta_{21}Y_1 + \gamma_{21}X_1 + \varepsilon_2 \tag{14.3}$$

Exam score is a weighted function of treatment group plus a weighted function of degree of motivation plus error.

To calculate the covariance between X_1 (treatment group) and Y_1 (degree of motivation) the first step is substituting in the equation for Y_1:

$$COV(X_1, Y_1) = COV(X_1, \gamma_{11}X_1 + \varepsilon_1) \tag{14.4}$$

The second step is distributing the first term, in this case X_1:

$$COV(X_1, Y_1) = COV(X_1\gamma_{11}X_1) + COV(X_1\varepsilon_1) \tag{14.5}$$

The last term in this equation, $COV(X_1\varepsilon_1)$, is equal to zero by assumption because it is assumed that there are no covariances between errors and other variables. Now,

$$COV(X_1, Y_1) = \gamma_{11}COV(X_1X_1) \tag{14.6}$$

by rule 2, and because the covariance of a variable with itself is just a variance,

$$COV(X_1Y_1) = \gamma_{11}\sigma_{x_1x_1} \tag{14.7}$$

The estimated population covariance between X_1 and Y_1 is equal to the path coefficient times the variance of X_1.

This is the population covariance between X_1 and Y_1 as estimated from the model. If the model is good, the product of $\gamma_{11}\sigma_{x_1x_1}$ produces a covariance that is very close to the sample covariance.
Following the same procedures, the covariance between Y_1 and Y_2 is:

$$COV(Y_1, Y_2) = COV(\gamma_{11}X_1 + \varepsilon_1, \beta_{21}Y_1 + \gamma_{21}X_1 + \varepsilon_2)$$
$$= COV(\gamma_{11}X_1\beta_{21}Y_1) + COV(\gamma_{11}X_1\gamma_{21}X_1) + COV(\gamma_{11}X_1\varepsilon_2)$$
$$+ COV(\varepsilon_1\beta_{21}Y_1) + COV(\varepsilon_1\gamma_{21}X_1) + COV(\varepsilon_1\varepsilon_2) \tag{14.8}$$
$$= COV(\gamma_{11}\beta_{21}\sigma_{x_1y_1}) + COV(\gamma_{11}\gamma_{21}\sigma_{x_1x_1})$$

because, as can be seen in the diagram, the error terms ε_1 and ε_2 do not correlate with any other variables.

All of the estimated covariances in the model could be derived in the same manner; but as is apparent even in this small example, covariance algebra rapidly becomes somewhat tedious. The "take home point" of this example is that covariance algebra can be used to estimate parameters and then estimate a population covariance matrix from them. Estimated parameters give us the estimated population covariance matrix.

14.4.2   Model Hypotheses

A truncated raw data set and corresponding covariance matrix appropriate for SEM analysis are presented in Table 14.1. This very small data set contains five continuous measured variables: (1) NUMYRS, the number of years a participant has skied, (2) DAYSKI, the total number of days a person has skied, (3) SNOWSAT, a Likert scale measure of overall satisfaction with the snow conditions, (4) FOODSAT, a Likert scale measure of overall satisfaction with the quality of the food at the resort, and (5) SENSEEK, a Likert scale measure of degree of sensation seeking. Note that hypothetical data are included for only 5 skiers although the analysis is performed with hypothetical data from 100 skiers. Matrix computations in SEM are tedious, at best, by hand. Therefore, MATLAB, a matrix manipulation program, is used to perform the calculations. Grab MATLAB or SYSTAT or SAS IML to perform matrix manipulations yourself as the example develops. Note also that the calculations presented here are rounded to two decimal places.

The hypothesized model for these data is diagrammed in Figure 14.4. Latent variables are represented with circles and measured variables are represented with squares. A line with an arrow

TABLE 14.1   Small Sample of Hypothetical Data for Structural Equation Modeling

	Covariance Matrix				
	NUMYRS	DAYSKI	SNOWSAT	FOODSAT	SENSEEK
NUMYRS	1.00				
DAYSKI	.70	11.47			
SNOWSAT	.62	.62	1.87		
FOODSAT	.44	.44	.95	1.17	
SENSEEK	.30	.21	.54	.38	1.00

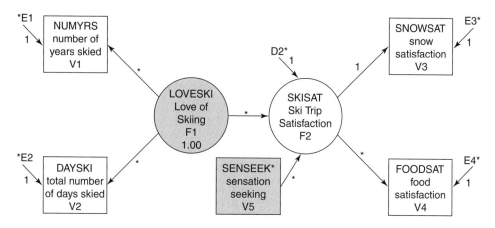

FIGURE 14.4   Hypothesized model for small-sample example.

indicates a hypothesized direct relationship between the variables. Absence of a line implies no hypothesized direct relationship. The asterisks indicate parameters to be estimated. Shading indicates that the variable is an IV. The variances of IVs are parameters of the model and are estimated or fixed to a particular value. The number 1 indicates that a parameter, either a path coefficient or a variance, has been set (fixed) to the value of 1. (At this point, don't worry about why we "fix" paths and variances to certain values like 1. This will be discussed in Section 14.5.1.)

This example contains two hypothesized latent variables (factors): Love of Skiing (LOVESKI), and Ski Trip Satisfaction (SKISAT). The Love of Skiing (LOVESKI) factor is hypothesized to have two indicators, number of years skied (NUMYRS) and number of days skied (DAYSKI). Greater Love of Skiing predicts more numerous years skied and days skied. Note that the direction of the prediction matches the direction of the arrows. The Ski Trip Satisfaction (SKISAT) factor also has two indicators; snow satisfaction (SNOWSAT) and food satisfaction (FOODSAT). Higher Ski Trip Satisfaction predicts a higher degree of satisfaction with both the snow and the food. This model also hypothesizes that both Love of Skiing and degree of sensation seeking (SENSEEK) predict level of Ski Trip Satisfaction; greater levels of Love of Skiing and sensation seeking predict higher levels of Ski Trip Satisfaction. Also notice that no arrow directly connects Love of Skiing with degree of sensation seeking. There is no hypothesized relationship, either predictive or correlational, between these variables. However, we can, and we will, test the hypothesis that there is a correlation between Love of Skiing and degree of sensation seeking.

As in the discussion of covariance algebra, these relationships are directly translated into equations and the model is then estimated. The analysis proceeds by specifying a model as in the diagram and then translating the model into a series of equations or matrices. Population parameters are then estimated that imply a covariance matrix. This estimated population covariance matrix is compared to the sample covariance matrix. The goal, as you might have guessed, is to estimate parameters that produce an estimated population covariance matrix that is not significantly different from the sample covariance matrix. This is similar to factor analysis (Chapter 13) where the reproduced correlation matrix is compared to the observed correlation matrix. One distinction between SEM and EFA is that in SEM the difference between the sample covariance matrix and the estimated population covariance matrix is evaluated with a chi-square test statistic.[2]

14.4.3   Model Specification

One method of model specification is the Bentler-Weeks method (Bentler & Weeks, 1980). In this method every variable in the model, latent or measured, is either an IV or a DV. The parameters to be estimated are the (1) regression coefficients, and (2) the variances and the covariances of the independent variables in the model (Bentler, 1995). In Figure 14.4 the regression coefficients and covariances to be estimated are indicated with an asterisk (*). The variances to be estimated are indicated by shading the independent variable.

In the example, SKISAT, SNOWSAT, FOODSAT, NUMYRS, and DAYSKI are all DVs because they all have at least one line with a single-headed arrow pointing to them. Notice that SKISAT is a latent variable and also a dependent variable. Whether or not a variable is observed makes no difference as to its status as a DV or IV. Although SKISAT is a factor, it is also a DV because it has arrows from both LOVESKI and SENSEEK. The seven IVs in this example are SENSEEK, LOVESKI, D2, E1, E2, E3, and E4.

Residual variables (errors) of measured variables are labeled E and errors of latent variables (called disturbances) are labeled D. It may seem odd that a residual variable is considered an IV but remember the familiar regression equation:

$$Y = X\beta + \varepsilon \tag{14.9}$$

where Y is the DV and X and ε are both IVs.

In fact the Bentler-Weeks model *is* a regression model, expressed in matrix algebra:

$$\boldsymbol{\eta} = \mathbf{B}\boldsymbol{\eta} + \boldsymbol{\gamma}\boldsymbol{\xi} \tag{14.10}$$

where, if q is the number of DVs and r is the number of IVs, then $\boldsymbol{\eta}$ (eta) is a $q \times 1$ vector of DVs, $\mathbf{B}$ (beta) is a $q \times q$ matrix of regression coefficients between DVs, $\boldsymbol{\gamma}$ (gamma) is a $q \times r$ matrix of regression coefficients between DVs and IVs, and $\boldsymbol{\xi}$ (xi) is an $r \times 1$ vector of IVs.

In the Bentler-Weeks model only independent variables have covariances and these covariances are in $\boldsymbol{\Phi}$ (phi), an $r \times r$ matrix. Therefore, the parameter matrices of the model are $\mathbf{B}$, $\boldsymbol{\gamma}$, and $\boldsymbol{\Phi}$. Unknown parameters in these matrices need to be estimated. The vectors of dependent variables, $\boldsymbol{\eta}$, and independent variables, $\boldsymbol{\xi}$, are not estimated.

---

[2]A chi-square test statistic can be used in EFA when maximum likelihood factor extraction is employed.

The diagram for the example is translated into the Bentler-Weeks model, with $r = 7$ and $q = 5$, as below.

$$
\boldsymbol{\eta} = \mathbf{B} \quad \boldsymbol{\eta} + \gamma \quad \boldsymbol{\xi}
$$

$$
\begin{bmatrix} V1 \text{ or } \eta_1 \\ V2 \text{ or } \eta_2 \\ V3 \text{ or } \eta_3 \\ V4 \text{ or } \eta_4 \\ F2 \text{ or } \eta_5 \end{bmatrix} =
\begin{bmatrix} 0 & 0 & 0 & 0 & 0 \\ 0 & 0 & 0 & 0 & 0 \\ 0 & 0 & 0 & 0 & 1 \\ 0 & 0 & 0 & 0 & * \\ 0 & 0 & 0 & 0 & 0 \end{bmatrix}
\begin{bmatrix} V1 \text{ or } \eta_1 \\ V2 \text{ or } \eta_2 \\ V3 \text{ or } \eta_3 \\ V4 \text{ or } \eta_4 \\ F2 \text{ or } \eta_5 \end{bmatrix} +
\begin{bmatrix} 0 & * & 1 & 0 & 0 & 0 & 0 \\ 0 & * & 0 & 1 & 0 & 0 & 0 \\ 0 & 0 & 0 & 0 & 1 & 0 & 0 \\ 0 & 0 & 0 & 0 & 0 & 1 & 0 \\ * & * & 0 & 0 & 0 & 0 & 1 \end{bmatrix}
\begin{bmatrix} V5 \text{ or } \xi_1 \\ F1 \text{ or } \xi_2 \\ E1 \text{ or } \xi_3 \\ E2 \text{ or } \xi_3 \\ E3 \text{ or } \xi_4 \\ E4 \text{ or } \xi_5 \\ D2 \text{ or } \xi_6 \end{bmatrix}
$$

Notice that $\boldsymbol{\eta}$ is on both sides of the equation. This is because DVs can predict one another in SEM. The diagram and matrix equations are identical. Notice that the asterisks in Figure 14.4 directly correspond to the asterisks in the matrices and these matrix equations directly correspond to simple regression equations. In the matrix equations the number 1 indicates that we have "fixed" the parameter, either a variance or a path coefficient, to the specific value of 1. Parameters are generally fixed for identification purposes. Identification will be discussed in more detail in Section 14.5.1. Parameters can be fixed to any number, most often, however, parameters are fixed to 1 or 0. The parameters that are fixed to 0 are also included in the path diagram but are easily overlooked because the 0 parameters are represented by the *absence* of a line in the diagram.

Carefully compare the model in Figure 14.4 with this matrix equation. The 5×1 vector of values to the left of the equal sign, the eta ($\boldsymbol{\eta}$) vector, is a vector of DVs listed in the order indicated, NUMYRS (V1), DAYSKI (V2), SNOWSAT (V3), FOODSAT (V4), and SKISAT (F2). The next matrix, just to the right of the equal sign, is a 5×5 matrix of regression coefficients among the DVs. The DVs are in the same order as above. The matrix contains 23 zeros, one 1, and one *. Remember that matrix multiplication involves cross multiplying and then summing the elements in the first row of the beta ($\mathbf{B}$) matrix with the first column in the eta ($\boldsymbol{\eta}$) matrix, and so forth (consult Appendix A as necessary). The zeros in the first, second, and fifth rows of the beta matrix indicate that no regression coefficients are to be estimated between DVs for V1, V2, and F2. The 1 at the end of the third row is the regression coefficient between F2 and SNOWSAT that was fixed to 1. The * at the end of the fourth row is the regression coefficient between F2 and V4 that is to be estimated.

Now look to the right of the plus sign. The 5×7 gamma matrix contains the regression coefficients that are used to predict the DVs from the IVs. The five DVs that are associated with the rows of this matrix are in the same order as above. The seven IVs that identify the columns are, in the order indicated, SENSEEK (V5), LOVESKI (F1), the four E (errors) for V1 to V4, and the D (disturbance) of F2. The 7×1 vector of IVs is in the same order. The first row of the γ (gamma) matrix times the ξ (Xi) vector produces the equation for NUMYRS. The * is the regression coefficient for predicting NUMYRS from LOVESKI (F1) and the 1 is the fixed regression coefficient for the relationship between NUMYRS and its E1. For example, consider the equation for NUMYRS (V1) reading from the first row in the matrices:

$$
\eta_1 = 0 \cdot \eta_1 + 0 \cdot \eta_2 + 0 \cdot \eta_3 + 0 \cdot \eta_4 + 0 \cdot \eta_5 + 0 \cdot \xi_1 + {} * \cdot \xi_2 + 1 \cdot \xi_3 + 0 \cdot \xi_4 + 0 \cdot \xi_5 + 0 \cdot \xi_6 + 0 \cdot \xi_7
$$

or, by dropping the zero-weighted products, and using the diagram's notation,

$$V1 = *F1 + E1$$

Continue in this fashion for the next four rows to be sure you understand their relationship to the diagrammed model.

In the Bentler-Weeks model only IVs have variances and covariances and these are in $\mathbf{\Phi}$ (phi), an $r \times r$ matrix. For the example, with seven IVs:

	V5 or ξ_1	F1 or ξ_2	E1 or ξ_3	E2 or ξ_4	E3 or ξ_5	E4 or ξ_6	D2 or ξ_7
$\mathbf{\Phi} = $ V5 or ξ_1	*	0	0	0	0	0	0
F1 or ξ_2	0	1	0	0	0	0	0
E1 or ξ_3	0	0	*	0	0	0	0
E2 or ξ_4	0	0	0	*	0	0	0
E3 or ξ_5	0	0	0	0	*	0	0
E4 or ξ_6	0	0	0	0	0	*	0
D2 or ξ_7	0	0	0	0	0	0	*

This 7×7 phi matrix contains the variances and covariances that are to be estimated for the IVs. The *s on the diagonal indicate the variances to be estimated for SENSEEK (V5), LOVESKI (F1), E1, E2, E3, E4, and D2. The 1 in the second row corresponds to the variance of LOVESKI (F1) that was set to 1. There are no covariances among IVs to be estimated, as indicated by the zeros in all the off-diagonal positions.

14.4.4 Model Estimation

Initial guesses (start values) for the parameters are needed to begin the modeling process. The more similar the guess and the start value, the fewer iterations needed to find a solution. There are many options available for start values (Bollen, 1989b). However, in most cases it is perfectly reasonable to allow the SEM computer program to supply initial start values. Computer program–generated start values are indicated with asterisks in the diagrams and in each of the three parameter matrices in the Bentler-Weeks model, $\hat{\mathbf{B}}$, $\hat{\mathbf{\gamma}}$, and $\hat{\mathbf{\Phi}}$, that follow below. The ^ (hat) over the matrices indicates that these are matrices of *estimated* parameters. The $\hat{\mathbf{B}}$ (beta hat) matrix is the matrix of regression coefficients between DVs where start values have been substituted for * (the parameters to be estimated). For the example:

$$\hat{\mathbf{B}} = \begin{bmatrix} 0 & 0 & 0 & 0 & 0 \\ 0 & 0 & 0 & 0 & 0 \\ 0 & 0 & 0 & 0 & 1 \\ 0 & 0 & 0 & 0 & .83 \\ 0 & 0 & 0 & 0 & 0 \end{bmatrix}$$

The matrix containing the start values for regression coefficients between DVs and IVs is $\hat{\gamma}$ (gamma hat). For the example:

$$\hat{\gamma} = \begin{bmatrix} 0 & .80 & 1 & 0 & 0 & 0 & 0 \\ 0 & .89 & 0 & 1 & 0 & 0 & 0 \\ 0 & 0 & 0 & 0 & 1 & 0 & 0 \\ 0 & 0 & 0 & 0 & 0 & 1 & 0 \\ .39 & .51 & 0 & 0 & 0 & 0 & 1 \end{bmatrix}$$

Finally, the matrix containing the start values for the variances and covariances of the IVs is $\hat{\Phi}$ (phi hat). For the example:

$$\hat{\Phi} = \begin{bmatrix} 1.00 & 0 & 0 & 0 & 0 & 0 & 0 \\ 0 & 1.00 & 0 & 0 & 0 & 0 & 0 \\ 0 & 0 & .39 & 0 & 0 & 0 & 0 \\ 0 & 0 & 0 & 10.68 & 0 & 0 & 0 \\ 0 & 0 & 0 & 0 & .42 & 0 & 0 \\ 0 & 0 & 0 & 0 & 0 & .73 & 0 \\ 0 & 0 & 0 & 0 & 0 & 0 & 1.35 \end{bmatrix}$$

To calculate the estimated population covariance matrix implied by the parameter estimates, selection matrices are first used to pull the measured variables out of the full parameter matrices. (Remember, the parameter matrices have both measured *and* latent variables as components.) The selection matrix is simply labeled **G** and has elements that are either 1s or 0s (Refer to Ullman, 2001, for a more detailed treatment of selection matrices.) The resulting vector is labeled **Y**,

$$\mathbf{Y} = \mathbf{G_y} * \mathbf{\eta} = \begin{bmatrix} V_1 \\ V_2 \\ V_3 \\ V_4 \end{bmatrix} \tag{14.11}$$

where **Y** is our name for the those measured variables that are dependent.

The independent measured variables are selected in a similar manner,

$$\mathbf{X} = \mathbf{G_x} * \mathbf{\xi} = V5 \tag{14.12}$$

where **X** is our name for the independent measured variables.

Computation of the estimated population covariance matrix proceeds by rewriting the basic structural modeling equation (14.10) as[3]:

$$\boldsymbol{\eta} = (\mathbf{I} - \mathbf{B})^{-1}\boldsymbol{\gamma}\boldsymbol{\xi} \tag{14.13}$$

where $\mathbf{I}$ is simply an identity matrix the same size as $\mathbf{B}$. This equation expresses the DVs as a linear combination of the IVs.

At this point the estimated population covariance matrix for the DVs, $\hat{\boldsymbol{\Sigma}}_{yy}$, is estimated using:

$$\hat{\boldsymbol{\Sigma}}_{yy} = \mathbf{G}_y(\mathbf{I} - \hat{\mathbf{B}})^{-1}\hat{\boldsymbol{\gamma}}\hat{\boldsymbol{\Phi}}\hat{\boldsymbol{\gamma}}'(\mathbf{I} - \hat{\mathbf{B}})^{-1}\mathbf{G}_y' \tag{14.14}$$

For the example:

$$\hat{\boldsymbol{\Sigma}}_{yy} = \begin{bmatrix} 1.04 & .72 & .41 & .34 \\ .72 & 11.48 & .45 & .38 \\ .41 & .45 & 2.18 & 1.46 \\ .34 & .38 & 1.46 & 1.95 \end{bmatrix}$$

The estimated population covariance matrix between IVs and DVs is obtained similarly by:

$$\hat{\boldsymbol{\Sigma}}_{yx} = \mathbf{G}_y(\mathbf{I} - \hat{\mathbf{B}})^{-1}\hat{\boldsymbol{\gamma}}\hat{\boldsymbol{\Phi}}\mathbf{G}_x' \tag{14.15}$$

For the example:

$$\hat{\boldsymbol{\Sigma}}_{yx} = \begin{bmatrix} 0 \\ 0 \\ .39 \\ .38 \end{bmatrix}$$

Finally, (phew!) the estimated population covariance matrix between IVs is estimated:

$$\hat{\boldsymbol{\Sigma}}_{xx} = \mathbf{G}_x\hat{\boldsymbol{\Phi}}\mathbf{G}_x' \tag{14.16}$$

For the example:

$$\hat{\boldsymbol{\Sigma}}_{xx} = 1.00$$

In practice a "super $\mathbf{G}$" matrix is used so that all the covariances are estimated in one step. The components of $\hat{\boldsymbol{\Sigma}}$ are then combined to produce the estimated population covariance matrix after one iteration.

---

[3]This rewritten equation is often called the "reduced form."

For the example, using initial start values supplied by EQS:

$$
\hat{\Sigma} =
\begin{bmatrix}
1.04 & .72 & .41 & .34 & 0 \\
.72 & 11.48 & .45 & .38 & 0 \\
.41 & .45 & 2.18 & 1.46 & .39 \\
.34 & .38 & 1.46 & 1.95 & .33 \\
0 & 0 & .39 & .33 & 1.00
\end{bmatrix}
$$

After initial start values are calculated, parameter estimates are changed incrementally (iterations continue) until the prespecified (in this case maximum likelihood) function (Section 14.5.2) is minimized (converges). After six iterations, the maximum likelihood function is at a minimum and the solution converges. The final estimated parameters are presented for comparison purposes in the **B**, **γ**, and **Φ** matrices; these unstandardized parameters are also presented in Figure 14.5.

$$
\hat{B} =
\begin{bmatrix}
0 & 0 & 0 & 0 & 0 \\
0 & 0 & 0 & 0 & 0 \\
0 & 0 & 0 & 0 & 1.00 \\
0 & 0 & 0 & 0 & .70 \\
0 & 0 & 0 & 0 & 0
\end{bmatrix}
$$

$$
\hat{\gamma} =
\begin{bmatrix}
0 & .81 & 1.00 & 0 & 0 & 0 & 0 \\
0 & .86 & 0 & 1.00 & 0 & 0 & 0 \\
0 & 0 & 0 & 0 & 1.00 & 0 & 0 \\
0 & 0 & 0 & 0 & 0 & 1.00 & 0 \\
.39 & .62 & 0 & 0 & 0 & 0 & 1.00
\end{bmatrix}
$$

$$
\hat{\Phi} =
\begin{bmatrix}
1.00 & 0 & 0 & 0 & 0 & 0 & 0 \\
0 & 1.00 & 0 & 0 & 0 & 0 & 0 \\
0 & 0 & .34 & 0 & 0 & 0 & 0 \\
0 & 0 & 0 & 10.72 & 0 & 0 & 0 \\
0 & 0 & 0 & 0 & .52 & 0 & 0 \\
0 & 0 & 0 & 0 & 0 & .51 & 0 \\
0 & 0 & 0 & 0 & 0 & 0 & .69
\end{bmatrix}
$$

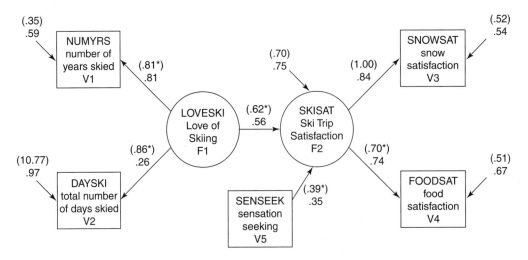

FIGURE 14.5 **Final model for small-sample example with standardized (and unstandardized) coefficients.**

The final estimated population covariance matrix is given by $\hat{\Sigma}$. For the example:

$$\hat{\Sigma} = \begin{bmatrix} 1.00 & .70 & .51 & .35 & 0 \\ .70 & 11.47 & .54 & .38 & 0 \\ .51 & .54 & 1.76 & .87 & .39 \\ .35 & .38 & .87 & 1.15 & .27 \\ 0 & 0 & .39 & .27 & 1.00 \end{bmatrix}$$

The final residual matrix is:

$$\mathbf{S} - \hat{\Sigma} = \begin{bmatrix} 0 & 0 & .12 & .08 & .30 \\ 0 & 0 & .08 & .06 & .21 \\ .12 & .08 & .12 & .08 & .15 \\ .08 & .06 & .08 & .06 & .11 \\ .30 & .21 & .15 & .11 & 0 \end{bmatrix}$$

14.4.5 Model Evaluation

A χ^2 statistic is computed based upon the function minimum when the solution has converged. The minimum of the function was .09432 in this example. This value is multiplied by $N - 1$ (N = number of participants) to yield the χ^2 value

$$(.09432)(99) = 9.337$$

This χ^2 is evaluated with degrees of freedom equal to the difference between the total number of degrees of freedom and the number of parameters estimated. The degrees of freedom in SEM are equal to the amount of unique information in the sample variance/covariance matrix (variances and covariances) minus the number of parameters in the model to be estimated (regression coefficients and variances and covariances of independent variables). In a model with a few variables it is easy to count the number of variances and covariances; however, in larger models, the number of data points is calculated as,

$$\text{number of data points} = \frac{p(p+1)}{2} \qquad (14.17)$$

where p equals the number of measured variables.

In this example with 5 measured variables, there are $(5(6)/2 =)$ 15 data points (5 variances and 10 covariances). The estimated model includes 11 parameters (5 regression coefficients and 6 variances) so χ^2 is evaluated with 4 dfs, χ^2 (99, df = 4) = 9.337, $p = .053$.

Because the goal is to develop a model that fits the data, a *nonsignificant* chi square is desired. This χ^2 is nonsignificant so we conclude that the model fits the data. However, chi-square values depend on sample sizes; in models with large samples, trivial differences often cause the χ^2 to be significant solely because of sample size. For this reason many fit indices have been developed that look at model fit while eliminating or minimizing the effect of sample size. All fit indices for this model indicate an adequate, but not spectacular, fit. Fit indices are discussed fully in Section 14.5.3.

The model fits, but what does it mean? The hypothesis is that the observed covariances among the measured variables arose because of the relationships between variables specified in the model; because the chi square is not significant, we conclude that we should retain our hypothesized model.

Next, researchers usually examine the statistically significant relationships within the model. If the unstandardized coefficients in the three parameter matrices are divided by their respective standard errors, a z-score is obtained for each parameter that is evaluated in the usual manner,[4]

$$z = \frac{\text{parameter estimate}}{\text{std error for estimate}} \qquad (14.18)$$

For NUMYRS predicted from LOVESKI $\quad \dfrac{.809}{.104} = 7.76, p < .05$

DAYSKI predicted from LOVESKI $\quad \dfrac{.865}{.106} = 8.25, p = .054$

FOODSAT predicted from SKISAT $\quad \dfrac{.701}{.127} = 5.51, p < .05$

SKISAT predicted from SENSEEK $\quad \dfrac{.389}{.108} = 3.59, p < .05$

SKISAT predicted from LOVESKI $\quad \dfrac{.625}{.128} = 4.89, p < .05$

---

[4]The standard errors are derived from the inverse of the information matrix.

Because of differences in scales, it is sometimes difficult to interpret unstandardized regression coefficients, therefore, researchers often examine standardized coefficients. Both the standardized and unstandardized regression coefficients for the final model are in Figure 14.5. The unstandardized coefficients are in parentheses. The paths from the factors to the variables are just standardized factor loadings. It could be concluded that number of years skied (NUMYRS) is a significant indicator of Love of Skiing (LOVESKI); the greater the Love of Skiing, the higher the number of years skied. Number of total days skied (DAYSKI) is a significant indicator of Love of Skiing, (i.e., greater Love of Skiing predicts more total days skied) because there was an *a priori* hypothesis that stated a positive relationship. Degree of food satisfaction (FOODSAT) is a significant indicator of ski trip satisfaction (SKISAT), higher Ski Trip Satisfaction predicts greater satisfaction with the food. Because the path from SKISAT to SNOWSAT is fixed to 1 for identification, a standard error is not calculated. If this standard error is desired, a second run is performed with the FOODSAT path fixed instead. Higher SENSEEK predicts higher SKISAT. Lastly, greater Love of Skiing (LOVESKI) significantly predicts Ski Trip Satisfaction (SKISAT) because this relationship is also tested as an *a priori*, unidirectional hypothesis.

14.4.6   Computer Analysis of Small-Sample Example

Tables 14.2, 14.4, and 14.5 show syntax and minimal selected output for computer analyses of the data in Table 14.1 using EQS, LISREL, and AMOS, respectively. The syntax and output for the programs are all quite different. Each of these programs offers the option of using a Windows "point and click" method in addition to the syntax approach. Additionally, EQS, AMOS, and LISREL allow for analyses based on a diagram. The sample example is shown only using the syntax approach. The "point and click" method and the diagram specification methods are just special cases of the syntax.

As seen in Table 14.2, and described in Section 14.4.3, the model is specified in EQS using a series of regression equations. In the /EQUATIONS section, as in ordinary regression, the DV appears on the left side of the equation, the IVs on the right side. Measured variables are referred to by the letter V and the number corresponding to the variable given in the /LABELS section. Errors associated with measured variables are indicated by the letter E and the number of the variable. Factors are referred to with the letter F and a number given in the /LABELS section. The errors, or disturbances, associated with factors are referred to by the letter D and the number corresponding to the factor. An asterisk indicates a parameter to be estimated. Variables included in the equation without asterisks are considered parameters fixed to the value 1. In this example start values are not specified and are estimated automatically by the program. The variances of IVs are parameters of the model and are indicated in the /VAR paragraph. The data appear as a covariance matrix in the paragraph labeled /MATRIX. In the /PRINT paragraph, FIT=ALL requests all goodness-of-fit indices available.

The output is heavily edited. After much diagnostic information (not included here), goodness-of-fit indices are given in the section labeled GOODNESS OF FIT SUMMARY. The independence model chi square is labeled INDEPENDENCE CHI-SQUARE. The independence chi square tests the hypothesis that there is no relationship among the variables. This chi square should always be significant, indicating that there is *some* relationship among the variables. CHI SQUARE is the model chi square that ideally should be nonsignificant. Several different goodness-of-fit indices are given (cf. Section 14.7.1) beginning with BENTLER–BONETT NORMED FIT INDEX. Significance tests for each parameter of the measurement portion of the model are found in the section labeled MEASUREMENT EQUATIONS WITH STANDARD ERRORS AND TEST STATISTICS. The

TABLE 14.2   Structural Equation Model of Small-Sample Example through EQS 6.1 (Syntax and Selected Output)

```
/TITLE
 EQS model created by EQS 6 for Windows—C:\JODIE\Papers\smallsample example
/SPECIFICATIONS
 DATA='C:\smallsample example 04.ESS';
 VARIABLES=5; CASES=100; GROUPS=1;
 METHODS=ML;
 MATRIX=covariance;
 ANALYSIS=COVARIANCE;
/LABELS
 V1=NUMYRS; V2=DAYSKI; V3=SNOWSAT; V4=FOODSAT; V5=SENSEEK;
 F1 = LOVESKI; F2=SKISAT;
/EQUATIONS
 !Love of Skiing Construct
 V1 = *F1 + E1;
 V2 = *F1 + E2;

 !Ski Trip Satisfaction Construct
 V3 = 1F2 + E3;
 V4 = *F2 + E4;

 F2 = *F1 + *V5 + D2;
/VARIANCES
 V5 = *;
 F1 = 1.00;
 E1 to E4 = *;
 D2 = *;
/PRINT
 EFFECT = YES;
 FIT=ALL;
 TABLE=EQUATION;
/LMTEST
/WTEST
/END


 MAXIMUM LIKELIHOOD SOLUTION (NORMAL DISTRIBUTION THEORY)
 GOODNESS OF FIT SUMMARY FOR METHOD = ML


 INDEPENDENCE MODEL CHI-SQUARE =          170.851 ON     10 DEGREES OF FREEDOM

    INDEPENDENCE AIC =    150.85057    INDEPENDENCE CAIC =    114.79887
          MODEL AIC =      1.33724          MODEL CAIC =     -13.08344


    CHI-SQUARE =         9.337 BASED ON     4 DEGREES OF FREEDOM
    PROBABILITY VALUE FOR THE CHI-SQUARE STATISTIC IS        .05320


    THE NORMAL THEORY RLS CHI-SQUARE FOR THIS ML SOLUTION IS         8.910.
```

(continued)

TABLE 14.2   Continued

```
FIT INDICES
----------
BENTLER-BONETT      NORMED FIT INDEX =              .945
BENTLER-BONETT NON-NORMED FIT INDEX =              .917
COMPARATIVE FIT INDEX (CFI)         =              .967
BOLLEN    (IFI)         FIT INDEX =                .968
MCDONALD (MFI)          FIT INDEX =                .974
LISREL    GFI           FIT INDEX =                .965
LISREL    AGFI          FIT INDEX =                .870
ROOT MEAN-SQUARE RESIDUAL (RMR)     =              .122
STANDARDIZED RMR                    =              .111
ROOT MEAN-SQUARE ERROR OF APPROXIMATION(RMSEA) = .116
90% CONFIDENCE INTERVAL OF RMSEA (        .000,        .214)


MEASUREMENT EQUATIONS WITH STANDARD ERRORS AND TEST STATISTICS
STATISTICS SIGNIFICANT AT THE 5% LEVEL ARE MARKED WITH @.



NUMYRS  =V1  =      .809*F1    + 1.000 E1
                   .104
                   7.755@


DAYSKI  =V2  =      .865*F1    + 1.000 E2
                   .105
                   8.250@


SNOWSAT =V3  =     1.000 F2    + 1.000 E3


FOODSAT =V4  =      .701*F2    + 1.000 E4
                   .127
                   5.511@


CONSTRUCT EQUATIONS WITH STANDARD ERRORS AND TEST STATISTICS
STATISTICS SIGNIFICANT AT THE 5% LEVEL ARE MARKED WITH @.



SKISAT  =F2  =      .389*V5    + .625*F1     + 1.000 D2
                   .108           .128
                   3.591@         4.888@



STANDARDIZED SOLUTION:                                    R-SQUARED

NUMYRS  =V1  =      .809*F1    + .588 E1                        .655
DAYSKI  =V2  =      .865*F1    + .502 E2                        .748
SNOWSAT =V3  =      .839 F2    + .544 E3                        .704
FOODSAT =V4  =      .738*F2    + .674 E4                        .545
SKISAT  =F2  =      .350*V5    + .562*F1     + .749 D2          .438
```

unstandardized coefficient appears on the first line, immediately below it is the standard error for that parameter. The z-score associated with the parameter (the unstandardized coefficient divided by the standard error) is given on the third line. The section labeled CONSTRUCT EQUATIONS WITH STANDARD ERRORS AND TEST STATISTICS contains the unstandardized regression coefficients, standard errors, and z-score significance tests for predicting factors from other factors and measured variables. The standardized parameter estimates appear in the section labeled STANDARDIZED SOLUTION.

LISREL offers two very different methods of specifying models. SIMPLIS uses equations and LISREL employs matrices. Neither program allows the exact model specified in Figure 14.4 to be tested. Underlying both these programs is the LISREL model, which, although similar to the Bentler-Weeks model, employs eight matrices instead of three. The matrices of the LISREL model that correspond to the Bentler-Weeks model are given in Table 14.3. Within the LISREL model there

TABLE 14.3 Equivalence of Matrices in Bentler-Weeks and LISREL Model Specifications

Bentler-Weeks Model			LISREL Model			
Symbol	Name	Contents	Symbol	Name	*LISREL* *Two Letter* *Specification*	Contents
B	Beta	matrix of regression coefficients of DVs predicting other DVs	1. **B**	1. Beta	1. BE	1. matrix of regression coefficients of latent DVs predicting other latent DVs
			2. Λ_y	2. Lambda y	2. LY	2. matrix of regression coefficients of measured DVs predicted by latent DVs
γ	Gamma	matrix of regression coefficients of DVs predicted by IVs	1. Γ	1. Gamma	1. GA	1. matrix of regression coefficient of latent DVs predicted by latent IVs
			2. Λ_x	2. Lambda x	2. LX	2. matrix of regression coefficients of measured DVs predicted by latent IVs
Φ	Phi	matrix of covariances among the IVs	1. Φ	1. Phi	1. PI	1. matrix of covariances among the latent IVs
			2. Ψ	2. Psi	2. PS	2. matrix of covariances of errors associated with latent DVs
			3. Θ_δ	3. Theta-Delta	3. TD	3. matrix of covariances among errors associated with measured DVs predicted from latent IVs
			4. Θ_ε	4. Theta-Epsilon	4. TE	4. matrix of covariances among errors associated with measured DVs predicted from latent DVs.

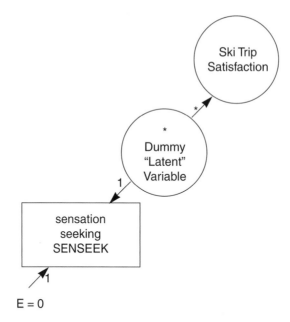

FIGURE 14.6   LISREL adaptation for small-sample example.

is no matrix of regression coefficients for predicting latent DVs from measured IVs. To estimate these parameters, a little trick, illustrated in Figure 14.6, is employed. A dummy "latent" variable with one indicator[5] is specified, in this example SENSEEK. The dummy latent variable then predicts SKISAT. The regression coefficient from the dummy "latent" variable to SENSEEK is fixed to one and the error variance of SENSEEK is fixed at zero. With this modification, the solutions are identical, because SENSEEK = (dummy latent variable) + 0.

LISREL uses matrices, rather than equations, to specify the model. Syntax and edited output are presented in Table 14.4. Matrices and commands are given with two-letter specifications defined in Table 14.3. CM with an asterisk indicates analysis of a covariance matrix. Following LA (for label) the measured variable names are given in the same order as the data. LISREL requires that the DVs appear before the IVs, so the specification SE (select) reorders the variables. The model specification begins with MO. The number of measured DVs is indicated after the key letters NY (number of Ys). The number of measured IVs is specified after the key letters NX (number of Xs). The latent DVs are specified after NE and the latent IVs are specified after NK. Labels are optional, but helpful. The labels for the latent DVs follow the key letters LE and labels for the latent IVs follow the key letters LK.

By default, elements of the matrices are either fixed at zero or are free. Additionally, matrices are one of four possible shapes: full nonsymmetrical, symmetrical, diagonal, or zero. Matrices are referred to by their two-letter designation, for example, LX (lambda x) is a full nonsymmetrical and fixed matrix of the regression coefficients predicting the measured DVs from latent IVs.

The model is specified by a combination of freeing (FR) or fixing (FI) elements of the relevant matrices. *Freeing* a parameter means estimating the parameter. When an element of a matrix is

---

[5]Note this dummy variable is not a true latent variable. A one indicator latent variable is simply a measured variable.

TABLE 14.4   Structural Equation Model for Small-Sample Example through LISREL 8.5.4 (Syntax and Edited Output)

```
TI Small Sample Example — LISREL
DA NI=5 NO=100 NG=1 MA=CM
CM
*
1.00
 .70 11.47
 .623  .623 1.874
 .436  .436  .95  1.173
 .3    .21   .54   .38  1.00
LA
NUMYRS DAYSKI SNOWSAT FOODSAT SENSEEK
SE
SNOWSAT FOODSAT NUMYRS DAYSKI SENSEEK
MO NY =2 NX = 3 NE =1 NK = 2
LE
SKISAT
LK
LOVESKI DUMMY
FR LX(1,1) LX(2,1) LY(2,1)
FI PH(2,1) TD(3,3)
VA 1 LX(3,2) LY(1,1) PH(1,1)
OU SC SE TV RS SS MI ND=3
```

LISREL Estimates (Maximum Likelihood)

LAMBDA-Y

	SKISAT
SNOWSAT	1.000
FOODSAT	0.701
	(0.137)
	5.120

LAMBDA-X

	LOVESKI	DUMMY
NUMYRS	0.809	- -
	(0.291)	
	2.782	
DAYSKI	0.865	- -
	(0.448)	
	1.930	
SENSEEK	- -	1.000

GAMMA

	LOVESKI	DUMMY
SKISAT	0.625	0.389
	(0.185)	(0.112)
	2.540	3.480

(continued)

TABLE 14.4 Continued

```
        Covariance Matrix of ETA and KSI

              SKISAT    LOVESKI     DUMMY
            --------   --------   --------
  SKISAT       1.236
 LOVESKI       0.625      1.000
   DUMMY       0.389      - -        1.000

        PHI
        Note: This matrix is diagonal.
              LOVESKI     DUMMY
            --------   --------
               1.000      1.000
                         (0.142)
                          7.036

        PSI

              SKISAT
            --------
               0.694
             (0.346)
               2.007

        Squared Multiple Correlations for Structural Equations
              SKISAT
            --------
               0.438

        THETA-EPS

              SNOWSAT    FOODSAT
            --------   --------
               0.520      0.507
             (0.223)    (0.126)
               2.327      4.015

        Squared Multiple Correlations for Y - Variables
              SNOWSAT    FOODSAT
            --------   --------
               0.704      0.546

        THETA-DELTA

              NUMYRS     DAYSKI     SENSEEK
            --------   --------   --------
               0.345     10.722      - -
             (0.454)    (1.609)
               0.760      6.664

        Squared Multiple Correlations for X - Variables
              NUMYRS     DAYSKI     SENSEEK
            --------   --------   --------
               0.655      0.065      1.000
```

TABLE 14.4   Continued

<pre>
                    Goodness of Fit Statistics
                  Degrees of Freedom = 4
         Minimum Fit Function Chi-Square = 9.337 (P = 0.0532)
  Normal Theory Weighted Least Squares Chi-Square = 8.910 (P = 0.0634)
           Estimated Non-centrality Parameter (NCP) = 4.910
        90 Percent Confidence Interval for NCP = (0.0 ; 17.657)

                  Minimum Fit Function Value = 0.0943
         Population Discrepancy Function Value (F0) = 0.0496
          90 Percent Confidence Interval for F0 = (0.0 ; 0.178)
        Root Mean Square Error of Approximation (RMSEA) = 0.111
       90 Percent Confidence Interval for RMSEA = (0.0 ; 0.211)
          P-Value for Test of Close Fit (RMSEA < 0.05) = 0.127

            Expected Cross-Validation Index (ECVI) = 0.312
       90 Percent Confidence Interval for ECVI = (0.263 ; 0.441)
                 ECVI for Saturated Model = 0.303
                ECVI for Independence Model = 1.328

 Chi-Square for Independence Model with 10 Degrees of Freedom = 121.492
                  Independence AIC = 131.492
                      Model AIC = 30.910
                    Saturated AIC = 30.000
                 Independence CAIC = 149.518
                     Model CAIC = 70.567
                   Saturated CAIC = 84.078

                 Normed Fit Index (NFI) = 0.923
               Non-Normed Fit Index (NNFI) = 0.880
           Parsimony Normed Fit Index (PNFI) = 0.369
              Comparative Fit Index (CFI) = 0.952
              Incremental Fit Index (IFI) = 0.955
               Relative Fit Index (RFI) = 0.808

                  Critical N (CN) = 141.770

           Root Mean Square Residual (RMR) = 0.122
                  Standardized RMR = 0.0974
             Goodness of Fit Index (GFI) = 0.965
        Adjusted Goodness of Fit Index (AGFI) = 0.870
        Parsimony Goodness of Fit Index (PGFI) = 0.257
</pre>

Completely Standardized Solution

<pre>
        LAMBDA-Y

              SKISAT

  SNOWSAT   0.839
  FOODSAT   0.769
</pre>

(continued)

TABLE 14.4 Continued

```
        LAMBDA-X

              LOVESKI      DUMMY
             --------    --------
  NUMYRS       0.809       - -
  DAYSKI       0.255       - -
  SENSEEK      - -         1.000

        GAMMA

              LOVESKI      DUMMY
             --------    --------
  SKISAT       0.562       0.350
```

Correlation Matrix of ETA and KSI

```
              SKISAT      LOVESKI      DUMMY
             --------    --------    --------
  SKISAT       1.000
  LOVESKI      0.562       1.000
  DUMMY        0.350       - -         1.000

        PSI

              SKISAT
             --------
               0.562

        THETA-EPS

              SNOWSAT     FOODSAT
             --------    --------
               0.296       0.454

        THETA-DELTA

              NUMYRS      DAYSKI      SENSEEK
             --------    --------    --------
               0.345       0.935       - -
```

Regression Matrix ETA on KSI (Standardized)

```
              LOVESKI      DUMMY
             --------    --------
  SKISAT       0.562       0.350
```

fixed with the key letters FI, it is fixed at zero. A command line begins with either FI (for fix) or FR (for free). Following this FI or FR specification, the particular matrix and specific element (row, column) that is to be freed or fixed is indicated. For example, from Table 14.4, FR LX(1,1) means free (FR) the element of the lambda x matrix (LX) that is in the first row and the first column (1,1), that is, the factor loading of NUMYRS on LOVESKI. Similarly FI PH(2,1) indicates that the covariance that is in the 2nd row, 1st column (2,1) of the phi matrix (PH) is fixed to zero (FI) (i.e., there is no relationship between LOVESKI and DUMMY).

In this example LX(LAMBDA-X) is a 3 × 2 full and fixed matrix of regression coefficients of measured variables predicted by latent IVs. The rows are the three measured variables that are the indicators of latent IVs: NUMYRS, DAYSKI, SENSEEK, and the columns are the latent IVs: LOVESKI and DUMMY. `LY` (`LAMBDA-Y`) is a full and fixed matrix of the regression coefficients predicting measured DVs from the latent DV. In this example `LY` is a 2×1 vector. The rows are the measured variables SNOWSAT and FOODSAT and the column is SKISAT. The `PH` (phi matrix) of covariances among latent IVs is by default symmetrical and free. In this example phi is a 2×2 matrix. No covariance is specified between the dummy latent variable and LOVESKI therefore `PH(2,1)` is fixed, `FI`. To estimate this model the error variance associated with SENSEEK must be fixed to zero. This is done by specifying `FI TD(3,3)`. `TD` refers to the theta delta matrix (errors associated with measured IVs serving as indicators of latent IVs); by default this matrix is diagonal and free. A diagonal matrix has zeros everywhere but the main diagonal. In the small-sample example it is a 3×3 matrix.

Only four of the eight LISREL matrices (`LX`, `LY`, `PH`, and `TD`) are included on the model (`MO`) line. LISREL matrices have particular shapes and elements specified by default. If these defaults are appropriate for the model there is no need to mention the unmodified matrices on the `MO` line. In this example the default specifications for `TE`, `GA`, `PS`, and `BE` are all appropriate. `TE` (theta epsilon) is diagonal and free by default. `TE` contains the covariances associated with the measured DVs associated with the latent DVs. In this example it is a 2×2 matrix. Gamma (`GA`) contains the regression coefficients of latent IVs predicting latent DVs. By default this matrix is full and free. In this example `GA` is a 1×2 vector. `PS` contains the covariances among errors associated with latent DVs, by default it is diagonal and free. In the small-sample example there is only 1 latent DV, therefore, `PS` is simply a scalar (a number). `BE` contains the regression coefficients among the latent DVs, by default a matrix of zeros. The small-sample example contains no relationships among latent DVs—there is only 1 latent DV—so there is no need to mention `BE`.

Finally, for identification, a path is fixed to 1 on each factor and the variance of LOVESKI is fixed at the value 1. (See Section 14.5.1 for a discussion of identification.) This is accomplished with the key letters `VA 1` and the relevant matrices and corresponding elements. The `OU` line specifies output options (`SC` completely standardized solution, `SE` standard errors, `TV` *t* values, `RS` residual information, `SS` standardized solution, and `ND` number of decimal places), not all of which are included in the edited output.

The highly edited output provides the unstandardized regression coefficients, standard errors for the regression coefficients, and *t* tests (unstandardized regression coefficient divided by standard error) by matrix in the section labeled `LISREL Estimates (Maximum Likelihood)`. The statistical significance of parameter estimates is determined with a table of *t* distributions (Table C.2). A *t* statistic greater than 1.96 is needed for significance at $p < .05$ and 2.56 for significance at $p < .01$. These are two-tailed tests. If the direction of the effect has been hypothesized a priori a one-tailed test can be employed, $t = 1.65$, $p < .05$, one-tailed. The goodness-of-fit summary is labeled `Goodness Of Fit Statistics`. A partially standardized solution appears, by matrix, in the section labeled `Completely Standardized Solution`. The regression coefficients, and variances and covariances, are completely standardized (latent variable mean of 0, sd = 1, observed variable mean = 0, sd = 1) and are identical to the standardized solution in EQS. The error variances given in `Completely Standardized Solution` for both measured variables and latent variables are not actually completely standardized and are different from EQS (Chou & Bentler, 1993). An option in LISREL (not shown) is the Standardized Solution, a second type of partially standardized solution in which the latent variables are standardized to a mean = 1 and sd = 0 but the observed variables remain in their original scale.

AMOS syntax uses equations to specify the model. Syntax and edited output are presented in Table 14.5. After an SEM new model is specified with `Dim Sem As New AmosEngine`, general

TABLE 14.5   Structural Equation Model of Small-Sample Example through AMOS (Syntax and Selected Output)

```
Sub Main
      Dim Sem As New AmosEngine
      Sem.TableOutput
      Sem.Standardized
      Sem.Mods 0

Sem.BeginGroup "UserGuide.xls", "smsample"
      Sem.Structure "numyrs<---loveski"
      Sem.Structure "dayski <---loveski"
      Sem.Structure "snowsat<---skisat (1)"
      Sem.Structure "foodsat<---skisat"
      Sem.Structure "loveski (1)"

      Sem.Structure "numyrs<---error1 (1)"
      Sem.Structure "dayski<---error2 (1)"
      Sem.Structure "snowsat <---error3 (1)"
      Sem.Structure "foodsat <---error4 (1)"

      Sem.Structure "skisat <---loveski"
      Sem.Structure "skisat <---senseek"
      Sem.Structure "skisat <---error5 (1)"
      Sem.Structure "loveski<--->senseek (0)"

End Sub
```

Computation of degrees of freedom (Model 1)

Number of distinct sample moments:	15
Number of distinct parameters to be estimated:	11
Degrees of freedom (15 - 11):	4

Result (Model 1)

Minimum was achieved
Chi-square = 9.337
Degrees of freedom = 4
Probability level = .053

Group number 1 (Group number 1 - Model 1)
Estimates (Group number 1 - Model 1)
Scalar Estimates (Group number 1 - Model 1)
Maximum Likelihood Estimates
Regression Weights: (Group number 1 - Model 1)

	Estimate	S.E.	C.R.	P	Label
skisat <-- loveski	.622	.245	2.540	.011	
skisat <-- senseek	.389	.112	3.480	***	
numyrs <-- loveski	.805	.289	2.782	.005	
dayski <-- loveski	.861	.446	1.930	.054	
snowsat <-- skisat	1.000				
foodsat <-- skisat	.701	.137	5.120	***	

TABLE 14.5   Continued

Standardized Regression Weights: (Group number 1 - Model 1)

	Estimate
skisat <-- loveski	.562
skisat <-- senseek	.350
numyrs <-- loveski	.809
dayski <-- loveski	.255
snowsat <-- skisat	.839
foodsat <-- skisat	.739

Covariances: (Group number 1 - Model 1)

	Estimate	S.E.	C.R.	P	Label
loveski <--> senseek	.000				

Model Fit Summary
CMIN

Model	NPAR	CMIN	DF	P	CMIN/DF
Default model	11	9.337	4	.053	2.334
Saturated model	15	.000	0		
Independence model	5	102.841	10	.000	10.284

RMR, GFI

Model	RMR	GFI	AGFI	PGFI
Default model	.121	.965	.870	.257
Saturated model	.000	1.000		
Independence model	.451	.671	.506	.447

Baseline Comparisons

Model	NFI Delta1	RFI rho1	IFI Delta2	TLI rho2	CFI
Default model	.909	.773	.946	.856	.943
Saturated model	1.000		1.000		1.000
Independence model	.000	.000	.000	.000	.000

Parsimony-Adjusted Measures

Model	PRATIO	PNFI	PCFI
Default model	.400	.364	.377
Saturated model	.000	.000	.000
Independence model	1.000	.000	.000

NCP

Model	NCP	LO 90	HI 90
Default model	5.337	.000	18.334
Saturated model	.000	.000	.000
Independence model	92.841	63.951	129.194

(continued)

TABLE 14.5 Continued

FMIN

Model	FMIN	F0	LO 90	HI 90
Default model	.094	.054	.000	.185
Saturated model	.000	.000	.000	.000
Independence model	1.039	.938	.646	1.305

RMSEA

Model	RMSEA	LO 90	HI 90	PCLOSE
Default model	.116	.000	.215	.110
Independence model	.306	.254	.361	.000

AIC

Model	AIC	BCC	BIC	CAIC
Default model	31.337	32.757	59.994	70.994
Saturated model	30.000	31.935	69.078	84.078
Independence model	112.841	113.486	125.867	130.867

ECVI

Model	ECVI	LO 90	HI 90	MECVI
Default model	.317	.263	.448	.331
Saturated model	.303	.303	.303	.323
Independence model	1.140	.848	1.507	1.146

HOELTER

	HOELTER	HOELTER
Model	.05	.01
Default model	101	141
Independence model	18	23

commands regarding output options are given. Each command begins with the letters Sem. Sem.TableOutput indicates that output be presented in table form similar to SPSS for Windows style. Other options are available. Sem.Standardized requests a completely standardized solution. Sem.Mods 0 requests all modification indices.

Heavily edited table output follows. The first section after Computation of degrees of freedom contains the model chi-square information. The model chi-square information is contained in the section labeled Chi-square. Detailed goodness-of-fit information follows in the section labeled Fit Measures. Significance tests for each parameter are given in the sections labeled Regression Weights. The first column in this table is parameter estimate, labeled Estimate. The next column, labeled S.E. contains the standard errors. The third column, labeled C.R., the critical ratio is the estimate divided by the S.E. The C.R. is the same as the z test in EQS. The final column, labeled P, contains the P value for the critical ratio. The completely standardized solution is given in the table labeled, Standardized Regression Weights.

The model specification begins by specifying the location of the data with the command, Sem-BeginGroup "UserGuide.xls", "smsample". In this example the data are in an Excel workbook "UserGuide.xls" in a worksheet called "smsample." Following specification of the data separate equations are written for each dependent variable in the model. The specific equation is stated double quotes (") after the command Sem.Structure. As with EQS paths that are fixed to 1 have (1) following the equation. AMOS automatically correlates the IV, in this example Love of Skiing (LOVESKI) and degree of sensation seeking (SENSEEK). To specify no relationship between these variables the command Sem.Structure "loveski <—> senseek(0)" is given. AMOS makes use of colors in the syntax specification method. When each line is correctly imputed the keywords, for example, Structure, on the line change color. If the line of syntax is incorrect there is no color change.

14.5   Some Important Issues

14.5.1   Model Identification

In SEM, a model is specified, parameters for the model are estimated using sample data, and the parameters are used to produce the estimated population covariance matrix. But only models that are identified can be estimated. A model is said to be identified if there is a unique numerical solution for each of the parameters in the model. For example, say both that the variance of $Y = 10$ and that the variance of $Y = \alpha + \beta$. Any two values can be substituted for α and β as long as they sum to 10. There is no unique numerical solution for either α or β; that is, there are an infinite number of combinations of two numbers that would sum to 10. Therefore, this single equation model is not identified. However, if we fix α to 0 then there is a unique solution for β, 10, and the equation is identified. It is possible to use covariance algebra to calculate equations and assess identification in very simple models; however, in large models this procedure quickly becomes unwieldy. For a detailed, technical discussion of identification, see Bollen (1989b). The following guidelines are rough, but may suffice for many models.

The first step is to count the numbers of data points and the number of parameters that are to be estimated. *The data in SEM are the variances and covariances in the sample covariance matrix.* The number of data points is the number of sample variances and covariances (found through Equation 14.17). The number of parameters is found by adding together the number of regression coefficients, variances, and covariances that are to be estimated (i.e., the number of asterisks in a diagram).

If there are more data points than parameters to be estimated, the model is said to be overidentified, a necessary condition for proceeding with the analysis. If there are the same number of data points as parameters to be estimated, the model is said to be just identified. In this case, the estimated parameters perfectly reproduce the sample covariance matrix, chi square and degrees of freedom are equal to zero, and the analysis is uninteresting because hypotheses about adequacy of the model cannot be tested. However, hypotheses about specific paths in the model can be tested. If there are fewer data points than parameters to be estimated, the model is said to be underidentified and parameters cannot be estimated. The number of parameters needs to be reduced by fixing, constraining, or deleting some of them. A parameter may be fixed by setting it to a specific value or constrained by setting the parameter equal to another parameter.

In the small-sample example of Figure 14.4, there are 5 measured variables so there are 15 data points: $5(5 + 1)/2 = 15$ (5 variances and 10 covariances). There are 11 parameters to be estimated in the hypothesized model: 5 regression coefficients and 6 variances. The hypothesized model has 4 fewer parameters than data points, so the model may be identified.

The second step in determining model identifiability is to examine the measurement portion of the model. The measurement part of the model deals with the relationship between the measured indicators and the factors. It is necessary both to establish the scale of each factor and to assess the identifiability of this portion of the model.

To establish the scale of a factor, you either fix the variance for the factor to 1, or fix to 1 the regression coefficient from the factor to one of the measured variables (perhaps a marker variable cf. Section 13.3.1). Fixing the regression coefficient to 1 gives the factor the same variance as the measured variable. If the factor is an IV, either alternative is acceptable. If the factor is a DV, most researchers fix the regression coefficient to 1. In the small-sample example, the variance of the Love of Skiing factor was set to 1 (normalized) and the scale of the Ski Trip Satisfaction factor was set equal to the scale of the snow satisfaction variable.

To establish the identifiability of the measurement portion of the model, look at the number of factors and the number of measured variables (indicators) loading on each factor. If there is only one factor, the model may be identified if the factor has at least three indicators with nonzero loading and the errors (residuals) are uncorrelated with one another. If there are two or more factors, again consider the number of indicators for each factor. If each factor has three or more indicators, the model may be identified if errors associated with the indicators are not correlated, each indicator loads on only one factor and the factors are allowed to covary. If there are only two indicators for a factor, the model may be identified if there are no correlated errors, each indicator loads on only one factor, and none of the variances or covariances among factors is equal to zero.

In the small-sample example, there are two indicators for each factor. The errors are uncorrelated and each indicator loads on only one factor. Additionally, the covariance between the factors is not zero. Therefore, this part of the model may be identified. Please note that identification may still be possible if errors are correlated or variables load on more than one factor, but it is more complicated.

The third step in establishing model identifiability is to examine the structural portion of the model, looking only at the relationships among the latent variables (factors). Ignore the measured variables for a moment; consider only the structural portion of the model that deals with the regression coefficients relating latent variables to one another. If none of the latent DVs predicts each other (the beta matrix is all zeros) the structural part of the model may be identified. The small-sample example has only one latent DV so this part of the model may be identified. If the latent DVs do predict one another, look at the latent DVs in the model and ask if they are recursive or nonrecursive. If the latent DVs are recursive there are no feedback loops among them, and there are no correlated disturbances (errors) among them. (In a feedback loop, DV1 predicts DV2 and DV2 predicts DV1. That is, there are two lines linking the factors, one with an arrow in one direction and the other line with an arrow in the other direction. Correlated disturbances are linked by single curved lines with double-headed arrows.) If the structural part of the model is recursive, it may be identifiable. These rules also apply to path analysis models with only measured variables. The small-sample example is a recursive model and therefore may be identified.

If a model is nonrecursive either there are feedback loops among the DVs or there are correlated disturbances among the DVs, or both, see Bollen 1989a.

Identification is often difficult to establish and frequently, despite the best laid plans, problems emerge. One extremely common error that leads to identification problems is failure to set the scale of a factor. In the small-sample example, if we had forgotten to set the scale of the Ski Trip Satisfaction factor, each of the programs would have indicated a problem. The error messages for each program are given in Table 14.6. Note that SIMPLIS fixes the problem automatically without printing out a warning message. Potentially, this could lead to some confusion.

Part (a) of Table 14.6 illustrates how EQS signals this type of identification problem. This message usually indicates an identification problem either with the particular variables mentioned, as in this case, or in the general neighborhood of the variables mentioned. Part (b) illustrates the LISREL message given the same identification problem. TE 2,2 refers to an element in the theta epsilon matrix (cf. Table 14.3). This indicates that the error variance for SNOWSAT may not be identified.

Part (c) of Table 14.6 shows the error message provided by AMOS. The first section indicates a possible identification problem and the second section indicates equations where the identification problem may have occurred.

When these messages occur, and despite the best of intentions they will, it is often helpful to compare the diagram of the model with the program input and be absolutely certain that everything on the diagram matches the input and that every factor has a scale. Application of these few basic principles will solve many identification problems.

Another common error is to fix both the factor variance to 1 and a path from the factor to an indicator to 1. This does not lead to an identification problem but does imply a very restricted model that almost certainly will not fit your data.

TABLE 14.6 Condition Codes When Latent Variable Variance Is Not Fixed

(a) EQS

```
PARAMETER              CONDITION CODE
   V3,F2                  LINEARLY DEPENDENT ON OTHER PARAMETERS
```

(b) LISREL

```
W_A_R_N_I_N_G: TE 2,2 may not be identified.
     Standard Errors, T-Values, Modification Indices,
     and Standardized Residuals cannot be computed.
```

(c) AMOS

Regression Weights

	Estimate	S.E.	C.R.	P	Label
skisat <-- loveski	Unidentified				
skisat <-- senseek	Unidentified				
skisat <-- error					
numyrs <-- loveski					
dayski <-- loveski					
snowsat <-- skisat	Unidentified				
foodsat <-- skisat	Unidentified				

Covariances

	Estimate	S.E.	C.R.	P	Label
loveski <-->senseek					

(continued)

TABLE 14.6   Continued

Variances

	Estimate	S.E.	C.R.	P	Label
loveski					
senseek					
error	Unidentified				
error1					
error2					
error3					
error4					

Computation of degrees of freedom

Number of distinct sample moments = 15
Number of distinct parameters to be estimated = 12
Degrees of freedom = 15-12 = 3

The model is probably unidentified. In order to achieve identifiability, it will probably be necessery to impose 1 additional constraint.

The (probably) unidentified parameters are marked.

TABLE 14.7   Summary of Estimation Techniques and Corresponding Function Minimized

Estimation Method	Function Minimized	Interpretation of W, the Weight Matrix
Unweighted Least Squares[a] (ULS)	$F_{\text{ULS}} = \frac{1}{2}\text{tr}[(\mathbf{S} - \mathbf{\Sigma}(\mathbf{\Theta}))^2]$	$\mathbf{W} = \mathbf{I}$, the identity matrix
Generalized Least Squares (GLS)	$F_{\text{GLS}} = \frac{1}{2}\text{tr}\{[(\mathbf{S} - \mathbf{\Sigma}(\mathbf{\Theta}))]\mathbf{W}^{-1}\}^2$	$\mathbf{W} = \mathbf{S}$. $\mathbf{W}$ is any consistent estimator of $\mathbf{\Sigma}$. Often the sample covariance matrix, $\mathbf{S}$, is used.
Maximum Likelihood (ML)	$F_{\text{ML}} = \log|\mathbf{\Sigma}| - \log|\mathbf{S}| + \text{tr}(\mathbf{S}\mathbf{\Sigma}^{-1}) - p$	$\mathbf{W} = \mathbf{\Sigma}^{-1}$, the inverse of the estimated population covariance matrix. The number of measured variables is p.
Elliptical Distribution Theory (EDT)	$F_{\text{EDT}} = \frac{1}{2}(\kappa + 1)^{-1}\text{tr}\{[\mathbf{S} - \mathbf{\Sigma}(\mathbf{\Theta})]\mathbf{W}^{-1}\}^2$ $- \delta\{\text{tr}[\mathbf{S} - \mathbf{\Sigma}(\mathbf{\Theta})]\mathbf{W}^{-1}\}^2$	$\mathbf{W} = $ any consistent estimator of $\mathbf{\Sigma}$. κ and δ are measures of kurtosis.
Asymptotically Distribution Free (ADF)	$F_{\text{ADF}} = [\mathbf{s} - \mathbf{\sigma}(\mathbf{\Theta})]'\mathbf{W}^{-1}[\mathbf{s} - \mathbf{\sigma}(\mathbf{\Theta})]$	$\mathbf{W}$ has elements, $\mathbf{w}_{ijkl} = \sigma_{ijkl} - \sigma_{ij}\sigma_{kl}$ (σ_{ijkl} is the kurtosis, σ_{ij} is the covariance).

[a]No χ^2 statistics or standard errors are available by the usual formulae, but some programs give these using more general computations.

14.5.2   Estimation Techniques

After a model is specified, population parameters are estimated with the goal of minimizing the difference between the observed and estimated population covariance matrices. To accomplish this goal a function, Q, is minimized where

$$Q = (\mathbf{s} - \boldsymbol{\sigma}(\boldsymbol{\Theta}))'\mathbf{W}(\mathbf{s} - \boldsymbol{\sigma}(\boldsymbol{\Theta})) \qquad (14.19)$$

> $\mathbf{s}$ is the vector of data (the observed sample covariance matrix stacked into a vector); $\boldsymbol{\sigma}$ is the vector of the estimated population covariance matrix (again, stacked into a vector) and $\boldsymbol{\Theta}$ indicates that $\boldsymbol{\sigma}$ is derived from the parameters (the regression coefficients, variances and covariances) of the model. $\mathbf{W}$ is the matrix that weights the squared differences between the sample and estimated population covariance matrix.

Recall that in factor analysis (Chapter 13) the observed and reproduced correlation matrices are compared. This notion is extended in SEM to include a statistical test of the difference. If the weight matrix, $\mathbf{W}$, is chosen correctly to minimize Q, Q multiplied by $(N - 1)$ yields a chi-square test statistic.

The trick is to select $\mathbf{W}$ to minimize the squared differences between observed and estimated population covariance matrices. In an ordinary chi square (Chapter 3), the weights are the set of expected frequencies in the denominators of the cells. If we use some other numbers instead of the expected frequencies, the result might be some sort of test statistic, but it would not be a χ^2 statistic; i.e., the weight matrix would be wrong.

In SEM, estimation techniques vary by the choice of $\mathbf{W}$. A summary of the most popular estimation techniques and the corresponding functions minimized is presented in Table 14.7.[6] Unweighted least squares estimation (ULS) does not usually yield a χ^2 statistic or standard errors. Because researchers are usually interested in the test statistic, ULS estimation is not discussed further (see Bollen, 1989b, for a further discussion of ULS).

Other estimation procedures are GLS (Generalized Least Squares), ML (Maximum Likelihood), EDT (Elliptical Distribution Theory), and ADF (Asymptotically Distribution Free). Satorra and Bentler (1988) have also developed an adjustment for nonnormality that can be applied to the chi-square test statistic following any estimation procedure. Briefly, the Satorra-Bentler scaled χ^2 is a correction to the χ^2 test statistic.[7] EQS also corrects the standard errors associated with the parameter estimates for the extent of the nonnormality (Bentler & Dijkstra, 1985). These adjustments to the standard errors and the Satorra-Bentler scaled chi square so far have been implemented only in the ML estimation procedure in EQS.

The performance of the χ^2 test statistic derived from these different estimation procedures is affected by several factors, among them (1) sample size, (2) nonnormality of the distribution of errors, of factors, and of errors and factors, and (3) violation of the assumption of independence of

---

[6]Really, it's not *that* technical! See Appendix A for additional guidance in deciphering the equations.

[7]The Satorra-Bentler Scaled χ^2 is the maximum likelihood test statistic (T_{ML}) adjusted using the following formula:

$$\text{Satorra-Bentler Scaled } \chi^2 = \frac{\text{dfs in the model}}{\text{tr}(\hat{\mathbf{U}}\mathbf{S}_y)} T_{ML}$$

where $\hat{\mathbf{U}}$ is the weight matrix and residual weight matrix under the model and $\mathbf{S}_y$ is the asymptotic covariance matrix.

factors and errors. The goal is to select an estimation procedure that, in Monte Carlo studies, produces a test statistic that neither rejects nor accepts the true model too many times as defined by a prespecified alpha level, commonly $p < .05$. Two studies provide guidelines for selection of an appropriate estimation method and test statistics. The following sections summarize the performance of estimation procedures examined in Monte Carlo studies by Hu, Bentler, and Kano (1992) and Bentler and Yuan (1999). Hu et al. (1992) varied sample size from 150 to 5,000 and Bentler and Yuan (1999) examined samples sizes ranging from 60 to 120. Both studies examined the performance of test statistics derived from several estimation methods when the assumptions of normality and independence of factors were violated.

14.5.2.1   Estimation Methods and Sample Size

Hu and colleagues (1992) found that when the normality assumption was reasonable, both the ML and the Scaled ML performed well with sample sizes over 500. When the sample size was less than 500, GLS performed slightly better. Interestingly, the EDT test statistic performed a little better than ML at small sample sizes. It should be noted that the elliptical distribution theory estimator (EDT) considers the kurtosis of the variables and assumes that all variables have the same kurtosis although the variables need not be normally distributed. (If the distribution is normal, there is no excess kurtosis.) Finally, the ADF estimator was poor with sample sizes under 2,500. Bentler and Yuan (1999) found that a test statistic similar to Hotelling's T, based on an adjustment to the ADF estimator, performed very well in models with small sample sizes ($N = 60$ to 120) and more subjects than the number of nonredundant variances and covariances in the sample covariance matrix (i.e., $[p(p + 1)]/2$, where p is the number of measured variables). This test statistic (Yuan-Bentler) adjusts the chi-square test statistic derived from the ADF estimator as,

$$T = \frac{[N - (p^* - q)]T_{ADF}}{[(N - 1)(p^* - q]} \qquad (14.20)$$

Where N is the number of subjects, $p^* = [p(p + 1)]/2$, where p is the number of measured variables, q is the number of parameters to be estimated, and T_{ADF} is the test statistic based on the ADF estimator.

14.5.2.2   Estimation Methods and Nonnormality

When the normality assumption was violated, Hu et al. (1992) found that the ML and GLS estimators worked well with sample sizes of 2,500 and greater. The GLS estimator was a little better with smaller sample sizes but led to acceptance of too many models. The EDT estimator accepted far too many models. The ADF estimator was poor with sample sizes under 2,500. Finally, the Scaled ML performed about the same as the ML and GLS estimators and better than the ADF estimator at all but the largest sample sizes.[8] With small sample sizes the Yuan-Bentler test statistic performed best.

---

[8]This is interesting in that the ADF estimator has no distributional assumptions and, theoretically, should perform quite well under conditions of nonnormality.

14.5.2.3 *Estimation Methods and Dependence*

The assumption that errors are independent underlies SEM and other multivariate techniques. Hu et al. (1992) also investigated estimation methods and test statistic performance when the errors and factors were dependent but uncorrelated.[9] ML and GLS performed poorly, always rejecting the true model. ADF was poor unless the sample size was greater than 2,500. EDT was better than ML, GLS, and ADF, but still rejected too many true models. The Scaled ML was better than the ADF at all but the largest sample sizes. The Scaled ML χ^2 performed best overall with medium to larger sample sizes, the Yuan-Bentler test statistic performed best with small samples.

14.5.2.4 *Some Recommendations for Choice of Estimation Method*

Sample size and plausibility of the normality and independence assumptions need to be considered in selection of the appropriate estimation technique and test statistic. ML, the Scaled ML, or GLS estimators may be good choices with medium to large samples and evidence of the plausibility of the normality and independence assumptions. The Scaled ML is fairly computer intensive. Therefore, if time or cost are an issue, ML and GLS are better choices when the assumptions seem plausible. ML estimation is currently the most frequently used estimation method in SEM. In medium to large samples the Scaled ML test statistic is a good choice with nonnormality or suspected dependence among factors and errors. Because scaled ML χ^2 is computer intensive and many model estimations may be required, it is often reasonable to use the ML χ^2 during model estimation and then scaled ML χ^2 for the final estimation. In small samples the Yuan-Bentler test statistic seems best. The test statistic based on ADF estimator (without adjustment) seems like a poor choice under all conditions unless the sample size is very large (>2,500).

14.5.3 Assessing the Fit of the Model

After the model has been specified and then estimated, the major question is, "Is it a good model?" One component of a "good" model is the fit between the sample covariance matrix and the estimated population covariance matrix. Like multiway frequency analysis (Chapter 16) and logistic regression (Chapter 10), a good fit is sometimes indicated by a nonsignificant χ^2. Unfortunately, assessment of fit is not always as straightforward as assessment of χ^2. With large samples, trivial differences between sample and estimated population covariance matrices are often significant because the minimum of the function is multiplied by $N - 1$. With small samples, the computed χ^2, may not be distributed as χ^2, leading to inaccurate probability levels. Finally, when assumptions underlying the χ^2 test statistic are violated, the probability levels are inaccurate (Bentler, 1995).

Because of these problems, numerous measures of model fit have been proposed. In fact, this is a lively area of research with new indices seemingly developed daily. One very rough "rule of thumb," however, directly related to the χ^2 value is that a good-fitting model may be indicated when the ratio of the χ^2 to the degrees of freedom is less than 2. The following discussion presents only

---

[9]Factors were dependent but uncorrelated by creating a curvilinear relationship between the factors and the errors. Correlation coefficients examine only linear relationships; therefore, although the correlation is zero between factors and errors, they are dependent.

some examples of each type of fit index. The interested reader is referred to Tanaka (1993), Browne and Cudeck (1993), and Williams and Holahan (1994) for excellent discussions of fit indices.

14.5.3.1 *Comparative Fit Indices*

One method of conceptualizing goodness of fit is by thinking of a series of models all nested within one another. Nested models are like the hierarchical models in log-linear modeling discussed in Chapter 16. Nested models are models that are subsets of one another. At one end of the continuum is the independence model: the model that corresponds to completely unrelated variables. This model would have degrees of freedom equal to the number of data points minus the variances that are estimated. At the other end of the continuum is the saturated (full or perfect) model with zero degrees of freedom. Fit indices that employ a comparative fit approach place the estimated model somewhere along this continuum. The Bentler-Bonett (1980) normed fit index (NFI) evaluates the estimated model by comparing the χ^2 value of the model to the χ^2 value of the independence model,

$$\text{NFI} = \frac{\chi^2_{indep} - \chi^2_{model}}{\chi^2_{indep}} \tag{14.21}$$

This yields a descriptive fit index that lies in the 0 to 1 range. For the small-sample example,

$$\text{NFI} = \frac{170.851 - 9.337}{170.851} = .945$$

High values (greater than .95) are indicative of a good-fitting model. Therefore, the NFI for the small-sample example indicates only a marginal fit as compared to a model with completely uncorrelated variables. Unfortunately, the NFI may underestimate the fit of the model in good-fitting models with small samples (Bearden, Sharma, & Teel, 1982). An adjustment to the NFI incorporating the degrees of freedom in the model yields the non-normed fit index (NNFI),

$$\text{NNFI} = \frac{\chi^2_{indep} - \dfrac{\text{df}_{indep}}{\text{df}_{model}} \chi^2_{model}}{\chi^2_{indep} - \text{df}_{indep}} \tag{14.22}$$

The adjustment improves on the problem of underestimating the fit in extremely good-fitting models but can sometimes yield numbers outside of the $0 - 1$ range. The NNFI can also be much too small in small samples, indicating a poor fit when other indices indicate an adequate fit (Anderson & Gerbing, 1984).

The problem of the large variability in the NNFI is addressed by the incremental fit index (IFI) (Bollen, 1989b),

$$\text{IFI} = \frac{\chi^2_{indep} - \chi^2_{model}}{\chi^2_{indep} - \text{df}_{model}} \tag{14.23}$$

The comparative fit index (CFI: Bentler, 1988) also assesses fit relative to other models as the name implies, but uses a different approach. The CFI employs the noncentral χ^2 distribution with noncentrality parameters, τ_i. The larger the value of τ_i, the greater the model misspecification: i.e., if the estimated model is perfect, $\tau_i = 0$. The CFI is defined as,

$$\text{CFI} = 1 - \frac{\tau_{\text{est. model}}}{\tau_{\text{indep. model}}} \tag{14.24}$$

So, clearly, the smaller the noncentrality parameter, τ_i, for the estimated model relative to the τ_i, for the independence model, the larger the CFI and the better the fit. The τ value for a model can be estimated by,

$$\tau_{\text{indep. model}} = \chi^2_{\text{indep. model}} - \text{df}_{\text{indep. model}}$$
$$\tau_{\text{est. model}} = \chi^2_{\text{est. model}} - \text{df}_{\text{est. model}} \tag{14.25}$$

For the small-sample example,

$$\tau_{\text{independence model}} = 170.851 - 10 = 160.851$$
$$\tau_{\text{estimated model}} = 9.337 - 4 = 5.337$$
$$\text{CFI} = 1 - \frac{5.337}{160.851} = .967$$

CFI values greater than .95 are often indicative of good-fitting models (Hu & Bentler, 1999). The CFI is normed to the $0 - 1$ range and does a good job of estimating model fit even in small samples (Bentler, 1989). It should be noted the values of all of these indices depend on the estimation method used.

The root mean square error of approximation (RMSEA; Browne & Cudeck, 1993) estimates the lack of fit in a model compared to a perfect (saturated) model. The equation for the estimated RMSEA is given by

$$\text{estimated RMSEA} = \sqrt{\frac{\hat{F}_o}{df_{\text{model}}}} \tag{14.26}$$

where $\hat{F}_o = \dfrac{\chi^2_{\text{model}} - df_{\text{model}}}{N}$ or 0 whichever is smaller but positive.

When the model is perfect, $\hat{F}_o = 0$. The greater the model misspecification the larger $\hat{F}_o$. Values of .06 or less indicate a good-fitting model relative to the model degrees of freedom (Hu & Bentler, 1999). Values larger than .10 are indicative of poor-fitting models (Browne & Cudeck, 1993). Hu and Bentler (1999) found that in small samples the RMSEA overrejected the true model, i.e., the value was too large. Because of this problem, this index may be less preferable with small samples. As with the CFI the choice of estimation method affects the size of the RMSEA.

For the small-sample example,

$$\hat{F} = \frac{9.337 - 4}{100} = .05337$$

therefore,

$$\text{RMSEA} = \sqrt{\frac{.05337}{4}} = .116$$

14.5.3.2 Absolute Fit Index

McDonald and Marsh (1990) have proposed an index that is absolute in that it does not depend on a comparison with another model such as the independence or saturated models (CFI) or the observed data (GFI). This index is illustrated with the small-sample example,

$$\text{MFI} = \exp\left[-.5\frac{(\chi^2_{\text{model}} - \text{df}_{\text{model}})}{N}\right]$$

$$\text{MFI} = \exp\left[-.5\frac{(9.337 - 4)}{100}\right] = .974$$

(14.27)

14.5.3.3 Indices of Proportion of Variance Accounted

Two widely available fit indices calculate a weighted proportion of variance in the sample covariance accounted for by the estimated population covariance matrix (Bentler, 1983; Tanaka & Huba, 1989). The goodness-of-fit index, GFI, can be defined by,

$$\text{GFI} = \frac{tr(\hat{\sigma}'\mathbf{W}\hat{\sigma})}{tr(s'\mathbf{W}s)}$$

(14.28)

where the numerator is the sum of the weighted variances from the estimated model covariance matrix and the denominator is the sum of the squared weighted variances from the sample covariance. $\mathbf{W}$ is the weight matrix that is selected by the choice of estimation method (Table 14.7).

Tanaka and Huba (1989) suggest that GFI is analogous to R^2 in multiple regression. This fit index can also be adjusted for the number of parameters estimated in the model. The adjusted fit index, labeled AGFI, is estimated by

$$\text{AGFI} = 1 - \frac{1 - \text{GFI}}{1 - \dfrac{\text{Number of est. parameters}}{\text{Number of data points}}}$$

(14.29)

For the small-sample example,

$$AGFI = 1 - \frac{1 - .965}{1 - \frac{11}{15}} = .87$$

The fewer the number of estimated parameters relative to the number of data points, the closer the AGFI is to the GFI. In this way the AGFI adjusts the GFI for the number of parameters estimated. The fit improves by estimating lots of parameters in SEM. However, a second goal of modeling is to develop a parsimonious model with as few parameters as possible.

14.5.3.4   Degree of Parsimony Fit Indices

Several indices have been developed that take into account the degree of parsimony in the model. Most simply, an adjustment can be made to the GFI (Mulaik et al., 1989), to produce PGFI

$$PGFI = \left[1 - \left(\frac{\text{Number of est. parameters}}{\text{Number of data points}} \right) \right] GFI \tag{14.30}$$

For the small-sample example,

$$PGFI = \left[1 - \left(\frac{11}{15} \right) \right].965 = .257$$

The larger the fit index the better (values closer to 1.00). Clearly, there is a heavy penalty for estimating a lot of parameters with this index. This index will always be substantially smaller than other indices unless the number of parameters estimated is much smaller than the number of data points.

Completely different methods of assessing fit that include a parsimony adjustment are the Akaike Information Criterion (AIC) and the Consistent Akaike Information Criterion (CAIC) (Akaike, 1987; Bozdogan, 1987). These indices are also functions of χ^2 and df,

$$\text{Model AIC} = \chi^2_{\text{model}} - 2df_{\text{model}} \tag{14.31}$$

$$\text{Model CAIC} = \chi^2_{\text{model}} - (\ln N + 1)df_{\text{model}} \tag{14.32}$$

For the small-sample example,

$$\text{Model AIC} = 9.337 - 2(4) = 1.337$$

$$\text{Model CAIC} = 9.337 - (\ln 100 + 1)4 = -13.08$$

Small values indicate a good-fitting, parsimonious model. How small is small enough? There is no clear answer because these indices are not normed to a $0 - 1$ scale. "Small enough" is small as compared to other competing models. This index is applicable to models estimated with maximum likelihood methods. It is useful for cross-validation because it is not dependent on sample data (Tanaka, 1993). EQS uses Equations 14.31 and 14.32 to calculate the AIC and CAIC. LISREL and AMOS, however, use

$$\text{AIC} = \chi^2_{\text{model}} + 2(\text{df}_{\text{number of est. parameters}}) \tag{14.33}$$

$$\text{CAIC} = \chi^2_{\text{model}} + (1 + \ln N)\text{df}_{\text{number of est. parameters}} \tag{14.34}$$

Both sets of equations are correct. LISREL and AMOS compute the AIC and CAIC with a constant included; EQS computes the AIC and CAIC without the constant. Therefore, although both sets of computations are correct, the AIC and CAIC computed in EQS are always smaller than the same values in LISREL and AMOS.

14.5.3.5   Residual-Based Fit Indices

Finally there are indices based on the residuals. The root mean square residual (RMR) and the standardized root mean square residual (SRMR) are the average differences between the sample variances and covariances and the estimated population variances and covariances. The root mean square residual is given by

$$\text{RMR} = \left[2 \sum_{i=1}^{q} \sum_{j=1}^{i} \frac{(s_{ij} - \hat{\sigma}_{ij})^2}{q(q+1)} \right]^{1/2} \tag{14.35}$$

The RMR is the square root (indicated by the power of $1/2$) of two times the sum, over all of the variables in the covariance matrix, of the average squared differences between each of the sample covariances (or variances) and the estimated covariances (or variances).

Good-fitting models have small RMR. It is sometimes difficult to interpret an unstandardized residual because the scale of the variables affect the size of the residual; therefore, a standardized root mean square residual (SRMR) is also available. Again, small values indicate good-fitting models. The SRMR has a range of 0 to 1, values of .08 or less are desired (Hu & Bentler, 1999).

14.5.3.6   Choosing among Fit Indices

Good-fitting models produce consistent results on many different indices in many, if not most, cases. If all the indices lead to similar conclusions, the issue of which indices to report is a matter of personal preference and, perhaps, the preference of the journal editor. The CFI and RMSEA are perhaps the most frequently reported fit indices. The RMSEA is particularly helpful if power calculations are to be performed. The AIC and CAIC are helpful indices to use when comparing models that are not nested. Often multiple indices are reported. If the results of the fit indices are inconsistent, the model should probably be re-examined; if the inconsistency cannot be resolved, consider reporting multiple indices. Hu and Bentler (1999) suggest reporting two types of fit indices; the SRMR and then a comparative fit index.

14.5.4   Model Modification

There are at least two reasons for modifying a SEM model: to improve fit (especially in exploratory work) and to test hypotheses (in theoretical work). The three basic methods of model modification are chi-square difference tests, Lagrange multiplier tests (LM), and Wald tests. All are asymptotically equivalent under the null hypothesis (act the same as the sample size approaches infinity) but approach model modification differently.

14.5.4.1   Chi-Square Difference Test

If models are nested (one model is a subset of another), the χ^2 value for the larger model is subtracted from the χ^2 value for the smaller nested model and the difference, also a χ^2, is evaluated with degrees of freedom equal to the difference between the degrees of freedom in the two models. When the data are normally distributed the chi squares can simply be subtracted. However, when the data are non-normal and the Satorra-Bentler scaled chi square is employed an adjustment is required so that the Satorra-Bentler chi square is distributed as a chi square (Satorra & Bentler, 2002). This will be demonstrated in Sections 14.6.2.3 and 14.6.2.4.

Recall that the residual between LOVESKI and SENSEEK is very high. We might allow these IVs to correlate and ask, "Does adding (estimating) this covariance improve the fit of the model?" Although our "theory" is that these variables are uncorrelated, is this aspect of theory supported by the data? To examine these questions, a second model is estimated in which LOVESKI and SENSEEK are allowed to correlate. The resulting model produces $\chi^2 = 0.084$, df = 3. In the small-sample example solution in Section 14.4.5, $\chi^2 = 9.337$, df = 4. The χ^2 difference test, (or likelihood ratio for maximum likelihood) is $9.337 - .084 = 9.253$, df $= 4 - 3 = -1, p < .05$. The model is significantly improved with the addition of this covariance; in fact, one of the fit indices (CFI) increases to 1 and the RMSEA drops to zero. Although the theory specifies independence between Sensation Seeking and Love of Skiing, the data support the notion that, indeed, these variables are correlated.

There are some disadvantages to the χ^2 difference test. Two models need to be estimated to get the χ^2 difference value and estimating two models for each parameter is time consuming with very large models and/or a slow computer. A second problem relates to χ^2 itself. Because of the relationship between sample size and χ^2, it is hard to detect a difference between models when sample sizes are small.

14.5.4.2   Lagrange Multiplier (LM) Test

The LM test also compares nested models but requires estimation of only one model. The LM test asks if the model is improved if one or more of the parameters in the model that are currently fixed are estimated. Or, equivalently, What parameters should be added to the model to improve the fit? This method of model modification is analogous to forward stepwise regression.

The LM test applied to the small-sample example indicates that if we add a covariance between LOVESKI and SENSEEK the approximate drop in χ^2 value is 8.801. This is one path, so the χ^2 value of 8.801 is evaluated with 1 df. The p level of this difference is .003. The model is then re-estimated if the decision is made to add the path. When the path is added, the drop is slightly larger, 9.253, but yields the same result.

The LM test can be examined either univariately or multivariately. There is a danger in examining only the results of univariate LM tests because overlapping variance between parameter estimates may make several parameters appear as if their addition would significantly improve the model. All of these parameters are candidates for inclusion by the results of univariate LM tests but the multivariate LM test identifies the single parameter that would lead to the largest drop in model χ^2 and calculates the expected change in χ^2. After this variance is removed, the parameter that accounts for the next largest drop in model χ^2 is assessed in a manner analogous to Roy-Bargmann stepdown analysis in MANOVA (Chapter 7).

EQS provides both univariate and multivariate LM tests. Additionally, several options are available for LM tests on specific sets of matrices and in specific orders of testing. The default LM test was requested for the small-sample example. Portions of the LM test output are presented in Table 14.8.

LM univariate output is presented first. The parameter that the LM test suggests adding is listed under the column labeled PARAMETER. The convention used in EQS is DV, IV, or IV,IV. Because both F1 and V5 are IVs, this refers to a covariance between LOVESKI and SENSEEK. The CHI-SQUARE column indicates the approximate chi square associated with this path, 8.801. The

TABLE 14.8   Edited Output from EQS for Lagrange Multiplier Tests (Syntax Appears in Table 14.2)

```
MAXIMUM LIKELIHOOD SOLUTION (NORMAL DISTRIBUTION THEORY)

LAGRANGE MULTIPLIER TEST (FOR ADDING PARAMETERS)

              ORDERED UNIVARIATE TEST STATISTICS:
```

NO	CODE		PARAMETER	CHI-SQUARE	PROBABILITY	PARAMETER CHANGE	STANDARDIZED CHANGE
1	2	2	F1,V5	8.801	.003	.366	.366
2	2	11	V1,V5	8.529	.003	.287	.287
3	2	20	V1,F2	8.529	.003	.738	.664
4	2	12	V3,F1	.000	.985	.005	.003
5	2	11	V3,V5	.000	.985	-.003	-.002
6	2	11	V4,V5	.000	.985	.002	.002
7	2	12	V4,F1	.000	.985	-.003	-.003
8	2	20	V2,F2	.000	1.000	.000	.000
9	2	11	V2,V5	.000	1.000	.000	.000
10	2	0	F1,F1	.000	1.000	.000	.000
11	2	0	V3,F2	.000	1.000	.000	.000

```
MULTIVARIATE LAGRANGE MULTIPLIER TEST BY SIMULTANEOUS PROCESS IN STAGE   1

PARAMETER SETS (SUBMATRICES) ACTIVE AT THIS STAGE ARE:

   PVV PFV PFF PDD GVV GVF GFV GFF BVF BFF
```

		CUMULATIVE MULTIVARIATE STATISTICS			UNIVARIATE INCREMENT	
STEP	PARAMETER	CHI-SQUARE	D.F.	PROBABILITY	CHI-SQUARE	PROBABILITY
1	F1,V5	8.801	1	.003	8.081	.003

probability level of this χ^2 is in the PROBABILITY column, $p = .003$. Significant χ^2 values are sought if improvement of the model is the goal. If hypotheses testing guides the use of the LM test, then the desired significance (or lack of significance) depends on the specific hypothesis. The PARAMETER CHANGE column indicates the approximate coefficient for the added parameter. But this series of univariate tests may include overlapping variance between parameters, so multivariate LM is examined.

Before presenting the multivariate test, EQS presents the list of default parameter matrices active at this stage of analysis. The matrices beginning with the letter P are phi matrices of covariances between IVs; the last two letters indicate what type of variables are included, where, for instance, VV indicates covariances between measured variables and FV covariances between factors and measured variables. The matrices beginning with the letter G refer to gamma matrices of regression coefficients between DVs and IVs. The matrices beginning with the letter B refer to matrices of regression coefficients of DVs with other DVs. At times only very particular parameters will be of interest for potential inclusion. For example, maybe the only parameters of interest are regression paths among the latent dependent variables, in these cases it is helpful to check that only the appropriate matrix is active (in this example, BFF).

The multivariate test also suggests adding the F1,V5 parameter. Indeed, after adding this one parameter, none of the other parameters significantly improves the model χ^2, therefore no other multivariate LM tests are shown.

LISREL presents only univariate LM tests, called Modification Indices. Edited LISREL printout for the small-sample example is given in Table 14.9. Modification indices are presented on a matrix-by-matrix basis. For each matrix in the LISREL model, four matrices of model modifications are included. The first, labeled Modification Indices For..., contains χ^2 values for the specific parameter. The second matrix, labeled Expected Change For..., contains unstandardized changes in parameter values. The third matrix, labeled Standardized Expected Change For..., contains the parameter changes where the latent variable has been standardized with a standard deviation of 1 but the original scale of the measured variables has been retained. The last matrix, labeled Completely Standardized Expected Change For..., contains the approximate parameter change when both the measured and latent variables (with the exception of the latent and observed errors) have been standardized with a standard deviation of 1. After all the LISREL matrices have been given, the parameter with the largest chi-square value is reported under Maximum Modification Index....

Table 14.10 shows edited output for modification indices from AMOS. Modification indices are grouped: covariances, variances, and regression weights. The modification index, the approximate chi square value, is given in the column labeled M.I. The approximate drop in parameter change, given this addition is in the column labeled Par Change. The probability associated with each modification index is not included.

14.5.4.3 Wald Test

While the LM test asks which parameters, if any, should be added to a model, the Wald test asks which, if any, could be deleted. Are there any parameters that are currently being estimated that could, instead, be fixed to zero? Or, equivalently, Which parameters are not necessary in the model? The Wald test is analogous to backward deletion of variables in stepwise regression where one seeks a nonsignificant change in the equation when variables are left out.

TABLE 14.9   Syntax and Edited Output from LISREL for Modification Indices

```
Modification Indices and Expected Change

No Non-Zero Modification Indices for LAMBDA-Y

       Modification Indices for LAMBDA-X

              LOVESKI       DUMMY
              --------     --------
    NUMYRS       - -         8.529
    DAYSKI       - -          - -
   SENSEEK      8.801         - -

       Expected Change for LAMBDA-X

              LOVESKI       DUMMY
              --------     --------
    NUMYRS       - -         0.287
    DAYSKI       - -          - -
   SENSEEK      0.366         - -

    Standardized Expected Change for LAMBDA-X

              LOVESKI       DUMMY
              --------     --------
    NUMYRS       - -         0.287
    DAYSKI       - -          - -
   SENSEEK     -0.366         - -

   Completely Standardized Expected Change for LAMBDA-X

              LOVESKI       DUMMY
              --------     --------
    NUMYRS       - -         0.287
    DAYSKI       - -          - -
   SENSEEK      0.366         - -

No Non-Zero Modification Indices for GAMMA

       Modification Indices for PHI

              LOVESKI       DUMMY
              --------     --------
   LOVESKI       - -
   DUMMY        8.801         - -

       Expected Change for PHI

              LOVESKI       DUMMY
              --------     --------
   LOVESKI       - -
   DUMMY        0.366         - -

    Standardized Expected Change for PHI

              LOVESKI       DUMMY
              --------     --------
   LOVESKI       - -
   DUMMY        0.366         - -

No Non-Zero Modification Indices for PSI
```

TABLE 14.9   Continued

```
            Modification Indices for THETA-DELTA-EPS

                  SNOWSAT    FOODSAT
                  --------   --------
    NUMYRS         0.000      0.000
    DAYSKI         0.000      0.000
   SENSEEK         0.000      0.000

            Expected Change for THETA-DELTA-EPS

                  SNOWSAT    FOODSAT
                  --------   --------
    NUMYRS         0.003     -0.002
    DAYSKI         0.000      0.000
   SENSEEK         0.003      0.002

    Completely Standardized Expected Change for THETA-DELTA-EPS

                  SNOWSAT    FOODSAT
                  --------   --------
    NUMYRS         0.002     -0.002
    DAYSKI         0.000      0.000
   SENSEEK        -0.002      0.002

            Modification Indices for THETA-DELTA
                  NUMYRS     DAYSKI     SENSEEK
                  --------   --------   --------
    NUMYRS          - -
    DAYSKI          - -        - -
   SENSEEK         8.529       - -        - -

            Expected Change for THETA-DELTA
                  NUMYRS     DAYSKI     SENSEEK
                  --------   --------   --------
    NUMYRS          - -
    DAYSKI          - -        - -
   SENSEEK         0.287       - -        - -

    Completely Standardized Expected Change for THETA-DELTA
                  NUMYRS     DAYSKI     SENSEEK
                  --------   --------   --------
    NUMYRS          - -
    DAYSKI          - -        - -
   SENSEEK         0.287       - -        - -
```

Maximum Modification Index is 8.80 for Element (3, 1) of LAMBDA-X

When the Wald test is applied to the small-sample example, the first candidate for deletion is error variance associated with NUMYRS. If this parameter is dropped, the χ^2 value increases by .578, a nonsignificant change ($p = .447$). The model is not significantly degraded by deletion of this parameter. However, because it is generally not reasonable to drop an error variance from a model, the decision is to retain the error variance associated with NUMYRS. Notice that, unlike the LM test, *nonsignificance* is desired when using the Wald test.

This illustrates an important point. Both the LM and Wald tests are based on statistical, not substantive, criteria. If there is conflict between these two criteria, substantive criteria are more important.

Table 14.11 presents edited output of both the univariate and multivariate Wald test from EQS. The specific candidate for deletion is indicated and the approximate multivariate χ^2 with its probability value is provided. The univariate tests are shown in the last two columns. In this example only the one parameter is suggested. By default, parameters are considered for deletion only if deletion does not cause the multivariate χ^2 associated with the Wald test to become significant. Remember,

TABLE 14.10   Edited Output from AMOS of Modification Indices (Syntax in Table 14.5)

Modification Indices (Group number 1 - Model 1)
Covariances: (Group number 1 - Model 1)

	M.I.	Par Change
loveski <-- senseek	8.324	.345
error5 <-- senseek	.000	.000
error5 <-- loveski	.000	.000
error4 <-- senseek	.000	.000
error4 <-- loveski	.000	−.000
error4 <-- error5	.000	.000
error3 <-- senseek	.000	−.000
error3 <-- loveski	.000	.000
error3 <-- error5	.000	.000
error3 <-- error4	.000	.000
error2 <-- senseek	.000	.000
error2 <-- loveski	.000	.000
error2 <-- error5	.000	.000
error2 <-- error4	.000	.000
error2 <-- error3	.000	.000
error1 <-- senseek	7.111	.237
error1 <-- loveski	.000	.000
error1 <-- error5	.000	.000
error1 <-- error4	.000	−.000
error1 <-- error3	.000	.000
error1 <-- error2	.000	.000

Variances: (Group number 1 - Model 1)

	M.I.	Par Change
senseek	.000	.000
loveski	.000	.000
error5	.000	.000
error4	.000	.000
error3	.000	.000
error2	.000	.000
error1	.000	.000

TABLE 14.10   Continued

Modification Indices (Group number 1 - Model 1)
Covariances: (Group number 1 - Model 1)
Regression Weights: (Group number 1 - Model 1)

	M.I.	Par Change
skisat <-- senseek	.000	.000
skisat <-- loveski	.000	.000
foodsat <-- senseek	.000	.001
foodsat <-- loveski	.000	−.001
foodsat <-- skisat	.000	.000
foodsat <-- snowsat	.000	.000
foodsat <-- dayski	.000	.000
foodsat <-- numyrs	.000	−.001
snowsat <-- senseek	.000	−.001
snowsat <-- loveski	.000	.001
snowsat <-- skisat	.000	.000
snowsat <-- foodsat	.000	.000
snowsat <-- dayski	.000	.000
snowsat <-- numyrs	.000	.001
dayski <-- senseek	.000	.000
dayski <-- loveski	.000	.000
dayski <-- skisat	.000	.000
dayski <-- foodsat	.000	.000
dayski <-- snowsat	.000	.000
dayski <-- numyrs	.000	.000
numyrs <-- senseek	7.111	.239
numyrs <-- loveski	.000	.000
numyrs <-- skisat	1.065	.092
numyrs <-- foodsat	.467	.058
numyrs <-- snowsat	.617	.053
numyrs <-- dayski	.000	.000

TABLE 14.11   Edited Output from EQS for Wald Test (Syntax Appears in Table 14.2)

```
WALD TEST (FOR DROPPING PARAMETERS)
MULTIVARIATE WALD TEST BY SIMULTANEOUS PROCESS
```

		CUMULATIVE MULTIVARIATE STATISTICS			UNIVARIATE INCREMENT	
STEP	PARAMETER	CHI-SQUARE	D.F.	PROBABILITY	CHI-SQUARE	PROBABILITY
1	E1,E1	.578	1	.447	.578	.447
2	V2,F1	3.737	2	.154	3.159	.076

the goal is to drop parameters that do contribute significantly to the model. LISREL and AMOS do not provide the Wald test.

14.5.4.4   *Some Caveats and Hints on Model Modification*

Because both the LM test and Wald test are stepwise procedures, Type I error rates are inflated but there are, as yet, no available adjustments as in ANOVA. A simple approach is to use a conservative probability value (say, $p < .01$) for adding parameters with the LM test. Cross validation with another sample is also highly recommended if modifications are made. If numerous modifications are made and new data are not available for cross-validation, compute the correlation between the estimated parameters from the original, hypothesized model and the estimated parameters from the final model using only parameters common to both models. If this correlation is high ($>.90$), relationships within the model have been retained despite the modifications (Tanaka & Huba, 1984).

Unfortunately, the order that parameters are freed or estimated can affect the significance of the remaining parameters. MacCallum (1986) suggests adding all necessary parameters before deleting unnecessary parameters. In other words, do the LM test before the Wald test. Because model modification easily gets very confusing, it is often wise to add, or delete, parameters one at a time.

A more subtle limitation is that tests leading to model modification examine overall changes in χ^2, not changes in individual parameter estimates. Large changes in χ^2 are sometimes associated with very small changes in parameter estimates. A missing parameter may be statistically needed but the estimated coefficient may have an uninterpretable sign. If this happens it is best not to add the parameter. Finally, if the hypothesized model is wrong, tests of model modification, by themselves, may be insufficient to reveal the true model. In fact, the "trueness" of any model is never tested directly, although cross validation does add evidence that the model is correct. Like other statistics, these tests must be used thoughtfully.

If model modifications are done in hopes of developing a good-fitting model, the fewer modifications the better, especially if a cross-validation sample is not available. If the LM test and Wald test are used to test specific hypotheses, the hypothesis will dictate the number of necessary tests.

14.5.5   Reliability and Proportion of Variance

Reliability is defined in the classic sense as the proportion of true variance relative to total variance (true plus error variance). Both the reliability and the proportion of variance of a measured variable are assessed through squared multiple correlation (SMC) where the measured variable is the DV and the factor is the IV. Each SMC is interpreted as the reliability of the measured variable in the analysis and as the proportion of variance in the variable that is accounted for by the factor, conceptually the same as a communality estimate in factor analysis.

To calculate a SMC:

$$\text{SMC}_{\text{var } i} = \frac{\lambda_i^2}{\lambda_i^2 + \Theta_{ii}} \tag{14.36}$$

The factor loading for variable i is squared and divided by that value plus the residual variance associated with the variable i.

This equation is applicable only for when there are no complex factor loadings or correlated errors.[10]

The proportion of variance in the variables accounted for by the factor is assessed as:

$$R_j^2 = 1 - D_j^2 \tag{14.37}$$

The disturbance (residual) for the DV factor j is squared and subtracted from 1.

14.5.6   Discrete and Ordinal Data

SEM assumes that measured variables are continuous and measured on an interval scale. Often, however, a researcher desires to include discrete and/or ordinally measured, categorical variables in an analysis. Because the data points in SEM are variances and covariances, the trick is to produce reasonable values for these types of variables for analysis.

Discrete (nominal level) measured variables such as favorite baseball team are included as IVs in a model by either dummy-coding the variable (i.e., Dodger fan or other) or by using a multiple group model where a model is tested for each team preference, as discussed in the next section.

Ordinal, categorical variables require special handling in SEM. Imagine that there is a normally distributed, continuous variable underlying each ordinal variable. To convert an ordinal variable to a continuous variable, the categories of the ordinal variables are converted to thresholds of the underlying (latent), normally distributed, continuous variable.

For example, say we ask people lingering outside a candy store if they: (1) hate, (2) like, or (3) love chocolate. As in Figure 14.7, underlying the ordinal variable is a normally distributed, latent, continuous construct representing love of chocolate. It is assumed that people who hate chocolate or absolutely loathe chocolate fall at, or below, the first threshold. Those who like chocolate fall between the two thresholds, and people who love chocolate, or who are enraptured by chocolate, fall at or above the second threshold. The proportion of people falling into each category is calculated and this proportion is used to calculate a z-score from a standardized normal table. The z-score is the threshold.

SEM proceeds by using polychoric correlations (between two ordinal variables), or polyserial correlations (between an ordinal and an interval variable) rather than covariance as the basis of the analysis.

Both EQS and LISREL (in PRELIS) compute thresholds and appropriate correlations. To incorporate categorical dependent variables in EQS, the statement `CATEGORY =` in the `SPECIFI-CATIONS` section is followed by the discrete variable labels, e.g., V1, V3. *All* measured variables must be DVs when models with categorical variables are estimated in EQS (Lee, Poon, & Bentler, 1994). If a model contains measured IVs, these IVs are first converted to factors in a method similar to LISREL. For example, if V1 was a measured IV, it is converted to a measured DV in the `\EQUA-TIONS` section with V1 = F1. F1 is then used in equations in place of V1.

Using PRELIS, the procedure file specifies both the continuous and ordinal variables and requests matrix output of polyserial and polychoric correlations with the statement `OUTPUT MATRIX = PMATRIX`. This matrix is then used in LISREL as the sample correlation matrix: PM is

---

[10]Note that factor loadings are denoted by λ_i in this chapter to stay consistent with general SEM terminology. In Chapter 13, factor loadings are denoted by a_{ij}. Both mean the same thing.

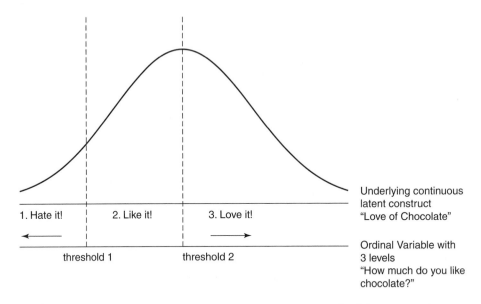

FIGURE 14.7 **Representation of thresholds underlying ordinal data categories.**

substituted for CM in the LISREL procedure file. AMOS does not accommodate categorical data. Although not illustrated in this chapter a particularly helpful SEM program for models with categorical data is Mplus (Muthén & Muthén, 2004).

14.5.7 Multiple Group Models

Although each of the models estimated in this chapter uses data from a single sample, it is also possible to estimate and compare models that come from two or more samples, called multiple group models. The general null hypothesis tested in multiple group models is that the data from each group are from the same population. For example, if data from a sample of men and a sample of women are drawn for the small-sample example, the general null hypothesis tested is that the two groups are drawn from the same population.

The analysis begins by developing good-fitting models in separate runs for each group. The models are then tested in one run with none of the parameters across models constrained to be equal. This unconstrained multiple group model serves as the baseline against which to judge more restricted models. Following baseline model estimation, progressively more stringent constraints are specified by constraining various parameters across all groups. When parameters are constrained they are forced to be equal to one another. After each set of constraints is added, a chi-square difference test is performed for each group between the less restrictive and more restrictive model. The goal is to not degrade the models by constraining parameters across the groups; therefore, you want a *nonsignificant* χ^2. If a significant difference in χ^2 is found between the models at any stage, the LM test is examined to locate the specific parameters that are different in the groups and these parameters are estimated separately in each group, i.e., the specific *across-group* parameter constraints are released.

Various hypotheses are tested in a specific order. The first step is usually to constrain the factor loadings (regression coefficients) between factors and their indices to equality across groups.

This step tests the hypothesis that the factor structure is the same in the different groups. Using the small-sample example with one model for men and another for women, we ask if men and women have the same underlying structure for the Love of Skiing and Ski Trip Satisfaction factors. If these constraints are reasonable, the χ^2 difference test between the restricted model and the baseline model is nonsignificant for both groups. If the difference between the restricted and nonrestricted models is significant, we need not throw in the towel immediately; rather, results of the LM test are examined and some equality constraints across the groups can be released. Naturally, the more parameters that differ across groups, the less alike the groups are. Consult Byrne, Shavelson, and Muthén (1989) for a technical discussion of these issues.

If the equality of factor structure is established, the second step is to ask if the factor variances and covariances are equal. In the small-sample example this is equivalent to asking if the variance of Ski Trip Satisfaction is equal for men and women. (Recall that the variance of Love of Skiing was set to 1 for identification.) If these constraints are feasible, the third step examines equality of the regression coefficients. In the small-sample example this is equivalent to testing the equality of the regression coefficient predicting Ski Trip Satisfaction from Love of Skiing. We could also test the equality of the regression path predicting Ski Trip Satisfaction from Sensation Seeking for men and women. If all of these constraints are reasonable, the last step is to examine the equality of residual variances across groups, an extremely stringent hypothesis not often tested. If all the regression coefficients, variances, and covariances are the same across groups, it is concluded that men and women represent the same population.

The groups are often similar in some respects but not others. For example, men and women could have identical factor structure except for one indicator on one factor. In this case, that loading is estimated separately for the two groups before further, more stringent, constraints are considered. Or men and women could have the same factor structure for Love of Skiing and Ski Trip Satisfaction but differ in the size of the regression coefficient linking Ski Trip Satisfaction to Love of Skiing.

Some cautions about multiple group modeling are in order before you dive head-first into this type of analysis. Multiple group models are often quite difficult to estimate. Critical to estimating a multiple group model with a good fit are single group models that fit well. It is extremely unlikely that the multiple group model will fit better than the individual group models. Good user-specified start values also seem to be more critical when estimating multiple group models. A demonstration of multiple group modeling is outside of the scope of this chapter, but see Bentler (1995) and Byrne, Shavelson, and Muthén (1989) for examples and detailed discussion of the process.

A different type of multiple group model is called a multilevel model. In this situation, separate models are developed from different levels of a nested hierarchy. For example, you might be interested in evaluating an intervention given to several classrooms of students in several different schools. One model is estimated for the schools, another for the classrooms that are nested within the schools, and a third for the children nested within the classrooms and schools. Predictors at each level are employed to test various within-level and cross-level hypotheses. An example of a multilevel model is Duncan, Alpert, and Duncan (1998).

14.5.8   Mean and Covariance Structure Models

The discussion so far has centered around modeling regression coefficients, variances, and covariances; however, means, both latent and observed, can also be modeled (Sörbom, 1974, 1982). Means of latent variables can be particularly interesting. Means are estimated in SEM models by adding a special intercept variable to the model.

Typically latent means are estimated in the context of a multiple group model. To ask, in the small-sample example, if men and women have the same mean Love of Skiing or if gender makes a difference in the mean of Ski Trip Satisfaction, the data are estimated as a two group model, constraints on the factor structure are made, the measured variable means estimated, and the latent means for both Love of Skiing and Ski Trip Satisfaction estimated. The most interpretable latent mean models are those in which the factor structure is identical or highly similar in both groups. For identification, the latent mean for one group is fixed at zero and the other estimated. The difference between the means is then estimated and evaluated with a z test like any other parameter, where the estimated parameter is divided by standard error. Demonstration of latent mean models is outside the scope of this chapter; however, Bentler (1995) and Byrne, Shavelson, and Muthén (1989) provide examples and detailed discussion of these models.

14.6   Complete Examples of Structural Equation Modeling Analysis

The first example is a confirmatory factor analysis (CFA) model performed through LISREL. The data used in this example are described in Appendix B. The factor structure underlying the subscales of the WISC in a sample of learning disabled children is examined. The model assesses the relationship between the indicators of IQ and two potential underlying constructs representing IQ. This type of model is sometimes referred to as a *measurement model*.

The second example is performed through EQS and has both measurement and structural components. In this example, mediators of the relationship between age, a life change measure, and latent variables representing Poor Sense of Self, Perceived Ill Health, and Health Care Utilization are examined. Data for the second example are from the women's health and drug study, described in Appendix B.1.

14.6.1   Confirmatory Factor Analysis of the WISC

The first example demonstrates confirmatory factor analysis (CFA) of 11 subtests of the Wechsler Intelligence Scale for Children (WISC) in a sample of learning-disabled children.

14.6.1.1   Model Specification for CFA

The hypothesized model is presented in Figure 14.8. In this model, a two-factor model is hypothesized: a Verbal factor (with the information, comprehension, arithmetic, similarities, vocabulary, and digit span subscales of the WISC as indicators) and a Performance factor (with the picture completion, picture arrangement, block design, object assembly, and coding subscales of the WISC serving as indicators). For clarity within the text, the labels of latent variables have initial capital letters and the measured variables do not. Two main hypotheses are of interest: (1) Does a two-factor model with simple structure (each variable loading only on one factor) fit the data? (2) Is there a significant covariance between the Verbal and Performance factors?

After the hypotheses are formulated and the model diagrammed, the first step of the modeling process is complete. At this point it is a good idea to do a preliminary check of the indentifiability of the model. Count the number of data points and the number of parameters to be estimated in the

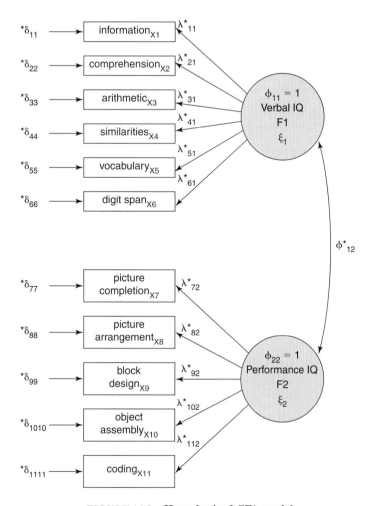

FIGURE 14.8   Hypothesized CFA model.

model. With 11 variables there are $(11(11 + 1))/2 = 66$ data points. The hypothesized model indicates that 23 parameters are to be estimated (11 regression coefficients, 1 covariance, and 11 variances have asterisks); therefore, the model is overidentified and is tested with 43 dfs ($66 - 23$).

14.6.1.2   Evaluation of Assumptions for CFA

Computer evaluation of assumptions is shown only when the procedure or output differs from that of other chapters in this book.

14.6.1.2.1   Sample Size and Missing Data

For this example there are 177 participants and 11 observed variables. The ratio of cases to observed variables is 16:1. The ratio of cases to estimated parameters is 8:1. This ratio is adequate given that the reliability of the subtests of the WISC-R is high. There are no missing data.

14.6.1.2.2 Normality and Linearity

Normality of the observed variables was assessed through examination of histograms using SPSS FREQUENCIES. None of the observed variables was significantly skewed or highly kurtotic. No variables had a standardized skewness greater than 3.75. It was not feasible to examine all pairwise scatterplots to assess linearity; therefore, randomly selected pairs of scatterplots were examined using SPSS GRAPHS SCATTER. All observed variables appeared to be linearly related, if at all.

14.6.1.2.3 Outliers

Using SPSS DESCRIPTIVES, one participant was found to have an extremely high score on the arithmetic subtest (19, $z = 4.11$) and was deleted. Using SPSS REGRESSION and Mahalanobis distance, 1 multivariate outlier was also detected and deleted ($p < .001$). This child had an extremely low comprehension subtest score and an extremely high arithmetic subtest score. The analysis was performed on 175 participants. SPSS was used to create a new file without the two outliers. (PRELIS could also have been used.) A new file was necessary because within the LISREL program itself outliers cannot be deleted (nor transformations made).

14.6.1.2.4 Multicollinearity and Singularity

The determinant of the covariance matrix is not given in LISREL output but the program converged so the covariance matrix was assumed to be nonsingular.

14.6.1.2.5 Residuals

Evaluation of the residuals is performed as part of evaluating the model.

14.6.1.3 CFA Model Estimation and Preliminary Evaluation

The syntax and edited output for the CFA analysis are presented in Table 14.12. As a first step it is helpful to check that the parameters that are indicated as free are those that were really intended to be estimated. It is also a good idea to check that the covariance matrix is correct, i.e., that it matches the covariance matrix from preliminary analyses. The output labeled `Parameter Specifications` lists each matrix specified in the model section and numbers each free parameter. `Lambda-X` is the matrix of regression coefficients to be estimated between indicators and factors. `PHI` is the matrix of covariances among factors. `THETA-DELTA` is the diagonal matrix of errors to be estimated for each measured variable. Only the diagonal of matrix is shown as all other entries in this matrix are zero, i.e., no correlated errors. In the other matrices, zeros indicate parameters that are fixed, i.e., not estimated. After checking the parameter specifications we confirm that they match the path diagram.

Next, it is helpful to assess the overall fit of the model by looking at the χ^2 and fit indices that appear in the section labeled `Goodness of Fit Statistics` (Table 14.13). The `Chi-Square for Independence Model with 55 Degrees of Freedom` is χ^2_{indep} (55, $N = 175$) = 516.237, $p < .01$. This χ^2 tests the hypotheses that the variables are unrelated; it should always be significant. If it is not, as is possible with very small samples, modeling should be reconsidered. The model chi square is significant, χ^2 (43, $N = 175$) = 70.24, $p = .005$. Ideally, a nonsignificant chi square is desired. The model χ^2 in this case is significant, but it is also less than two times the model degrees of freedom. This ratio gives a *very* rough indication that the model may fit the data. LISREL output includes many other fit indices, including the CFI = .94, GFI = .93, and the standardized RMSEA = .06. These indices all seem to indicate a good-fitting model.

TABLE 14.12   Syntax and Parameter Specifications for CFA Using LISREL

```
CONFIRMATORY FACTOR ANALYSIS OF THE WISC-R
DA NI = 13 NO = 175
RA FI = WISCSEM.DAT
LA
CLIENT, AGEMATE, INFO, COMP, ARITH, SIMIL, VOCAB,
DIGIT, PICTCOMP, PARANG, BLOCK, OBJECT, CODING
SE
INFO, COMP, ARITH, SIMIL, VOCAB,
DIGIT, PICTCOMP, PARANG, BLOCK, OBJECT CODING/
MO NX=11 NK=2
LK
VERBAL PERFORM
FR LX(1,1) LX(2,1) LX(3,1) LX(4,1) LX(5,1) LX(6,1)
FR LX(7,2) LX(8,2) LX(9,2) LX(10,2) LX(11,2)
VA 1 PH(1,1) PH(2,2)
OU SC SE TV RS SS MI ND=3

CONFIRMATORY FACTOR ANALYSIS OF THE WISC-R

                    Number of Input Variables 13
                    Number of Y - Variables    0
                    Number of X - Variables   11
                    Number of ETA - Variables  0
                    Number of KSI - Variables  2
                    Number of Observations   175

CONFIRMATORY FACTOR ANALYSIS OF THE WISC-R

        Covariance Matrix to be Analyzed

                INFO      COMP      ARITH     SIMIL     VOCAB     DIGIT
              --------  --------  --------  --------  --------  --------
      INFO     8.481
      COMP     4.034     8.793
     ARITH     3.322     2.684     5.322
     SIMIL     4.758     4.816     2.713    10.136
     VOCAB     5.338     4.621     2.621     5.022     8.601
     DIGIT     2.720     1.891     1.678     2.234     2.334     7.313
  PICTCOMP     1.965     3.540     1.052     3.450     2.456     0.597
    PARANG     1.561     1.471     1.391     2.524     1.031     1.066
     BLOCK     1.808     2.966     1.701     2.255     2.364     0.533
    OBJECT     1.531     2.718     0.282     2.433     1.546     0.267
    CODING     0.059     0.517     0.598    -0.372     0.842     1.344

        Covariance Matrix to be Analyzed

              PICTCOMP   PARANG     BLOCK    OBJECT    CODING
              --------  --------  --------  --------  --------
  PICTCOMP     8.610
    PARANG     1.941     7.074
     BLOCK     3.038     2.532     7.343
    OBJECT     3.032     1.916     3.077     8.088
    CODING    -0.605     0.289     0.832     0.433     8.249
```

(continued)

TABLE 14.12   Continued

CONFIRMATORY FACTOR ANALYSIS OF THE WISC-R

Parameter Specifications

LAMBDA-X

	VERBAL	PERFORM
INFO	1	0
COMP	2	0
ARITH	3	0
SIMIL	4	0
VOCAB	5	0
DIGIT	6	0
PICTCOMP	0	7
PARANG	0	8
BLOCK	0	9
OBJECT	0	10
CODING	0	11

PHI

	VERBAL	PERFORM
VERBAL	0	
PERFORM	12	0

THETA-DELTA

INFO	COMP	ARITH	SIMIL	VOCAB	DIGIT
13	14	15	16	17	18

THETA-DELTA

PICTCOMP	PARANG	BLOCK	OBJECT	CODING
19	20	21	22	23

Residuals are examined after evaluation of fit. Residual diagnostics are requested with RS on the OU (output line) of the syntax in Table 14.12. LISREL gives numerous residual diagnostics. Residuals in both the original scale of the variables, labeled FITTED RESIDUALS (not shown), and partially standardized residuals, labeled STANDARDIZED RESIDUALS, are included. For both types of residuals, the full residual covariance matrix, summary statistics, and a stem leaf plot are given. Partially standardized residual output appears in Table 14.14. Although the model fits the data well, there is a sizable residual (standardized residual 3.06) between picture arrangement (PARANG) and comprehension (COMP). This indicates that the model does not adequately estimate the relationship between these two variables. The large residual between PICTCOMP and COMP is also clearly indicated in the stem leaf plot. The stem leaf plot also shows that the residuals are centered around zero and symmetrically distributed. The median residual is zero. LISREL also provides

TABLE 14.13   Goodness-of-Fit Statistics for CFA Model Using LISREL (Syntax Appears in Table 14.12)

```
                    Goodness of Fit Statistics

                    Degrees of Freedom = 43
         Minimum Fit Function Chi-Square = 70.236 (P = 0.00545)
Normal Theory Weighted Least Squares Chi-Square = 71.045 (P = 0.00454)
            Estimated Non-centrality Parameter (NCP) = 28.045
         90 Percent Confidence Interval for NCP = (8.745; 55.235)


                 Minimum Fit Function Value = 0.404
        Population Discrepancy Function Value (F0) = 0.161
         90 Percent Confidence Interval for F0 = (0.0503; 0.317)
        Root Mean Square Error of Approximation (RMSEA) = 0.0612
     90 Percent Confidence Interval for RMSEA = (0.0342; 0.0859)
         P-Value for Test of Close Fit (RMSEA < 0.05) = 0.221


            Expected Cross-Validation Index (ECVI) = 0.673
         90 Percent Confidence Interval for ECVI = (0.562; 0.829)
                 ECVI for Saturated Model = 0.759
                 ECVI for Independence Model = 3.093


Chi-Square for Independence Model with 55 Degrees of Freedom = 516.237
                 Independence AIC = 538.237
                    Model AIC = 117.045
                  Saturated AIC = 132.000
               Independence CAIC = 584.050
                   Model CAIC = 212.835
                 Saturated CAIC = 406.876


            Root Mean Square Residual (RMR) = 0.468
                  Standardized RMR = 0.0585
            Goodness of Fit Index (GFI) = 0.931
        Adjusted Goodness of Fit Index (AGFI) = 0.894
       Parsimony Goodness of Fit Index (PGFI) = 0.606


               Normed Fit Index (NFI) = 0.864
             Non-Normed Fit Index (NNFI) = 0.924
        Parsimony Normed Fit Index (PNFI) = 0.675
            Comparative Fit Index (CFI) = 0.941
            Incremental Fit Index (IFI) = 0.942
             Relative Fit Index (RFI) = 0.826


                 Critical N (CN) = 168.123
```

**TABLE 14.14   Partially Standardized Residuals for CFA Using LISREL
(Syntax Appears in Table 14.12)**

STANDARDIZED RESIDUALS

	INFO	COMP	ARITH	SIMIL	VOCAB	DIGIT
INFO	- -					
COMP	-2.279	- -				
ARITH	2.027	0.054	- -			
SIMIL	-0.862	0.817	-0.748	- -		
VOCAB	2.092	-0.008	-1.527	-0.138	- -	
DIGIT	1.280	-0.753	0.897	-0.342	-0.168	- -
PICTCOMP	-0.734	3.065	-0.716	2.314	0.319	-0.942
PARANG	-0.176	-0.096	1.092	1.759	-1.488	0.579
BLOCK	-1.716	1.844	0.795	-0.441	-0.276	-1.341
OBJECT	-1.331	1.673	-2.394	0.633	-1.403	-1.443
CODING	-0.393	0.472	0.950	-1.069	1.051	2.146

STANDARDIZED RESIDUALS

	PICTCOMP	PARANG	BLOCK	OBJECT	CODING
PICTCOMP	- -				
PARANG	-0.779	- -			
BLOCK	-1.004	0.861	- -		
OBJECT	0.751	-0.319	0.473	- -	
CODING	-2.133	0.059	1.284	0.215	- -

Summary Statistics for Standardized Residuals

Smallest Standardized Residual =   -2.394
  Median Standardized Residual =   0.000
 Largest Standardized Residual =   3.065

Stemleaf Plot

```
 - 2|431
 - 1|755443310
 - 0|998877744333221000000000000
   0|1123556688899
   1|01133788
   2|0113
   3|1
```
Largest Positive Standardized Residuals
Residual for PICTCOMP and   COMP   3.065

a QPLOT of partially standardized residuals as shown in Figure 14.9. If the residuals are normally distributed the Xs hover around the diagonal. As in multiple regression, large deviations from the diagonal indicate nonnormality. Once again the large residual between PICTCOMP and COMP is clearly evident in the upper right hand corner of the plot.

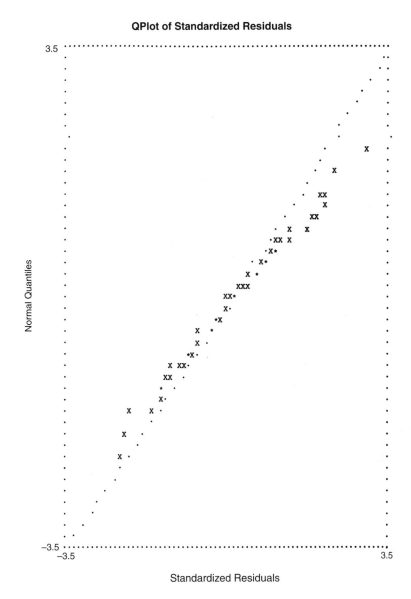

**FIGURE 14.9   Q Plot output of partially standardized residuals
for CFA model in LISREL. Syntax appears in Table 14.12.**

Finally, estimates of the parameters are examined (Table 14.15). In the section labeled `LIS-REL Estimates (Maximum Likelihood)` are, by row, the unstandardized regression coefficients, standard errors in parentheses, and z-scores (coefficient/standard error) for each indicator.[11]

---

[11]Pop quiz! With your knowledge of both SEM and EFA (Chapter 13), what are the regression coefficients in this CFA equivalent to in EFA? Answer: Elements in the patterns matrix.

TABLE 14.15   Output from LISREL of Parameter Estimates, Standard Errors, and _z_ Test and Partially Standardized Solution for CFA Model (Syntax Appears in Table 14.12)

```
LISREL Estimates (Maximum Likelihood)

        LAMBDA-X

               VERBAL      PERFORM
               --------    --------
     INFO        2.212       - -
               (0.201)
                10.997

     COMP        2.048       - -
               (0.212)
                 9.682

    ARITH        1.304       - -
               (0.173)
                 7.534

    SIMIL        2.238       - -
               (0.226)
                 9.911

    VOCAB        2.257       - -
               (0.202)
                11.193

    DIGIT        1.056       - -
               (0.213)
                 4.952

 PICTCOMP         - -        1.747
                           (0.244)
                             7.166

   PARANG         - -        1.257
                           (0.226)
                             5.566

    BLOCK         - -        1.851
                           (0.223)
                             8.287

   OBJECT         - -        1.609
                           (0.237)
                             6.781

   CODING         - -        0.208
                           (0.256)
                             0.811

          PHI

               VERBAL      PERFORM
               --------    --------
   VERBAL       1.000

  PERFORM       0.589       1.000
               (0.076)
                 7.792
```

740

TABLE 14.15   Continued

THETA-DELTA

	INFO	COMP	ARITH	SIMIL	VOCAB	DIGIT
	3.586	4.599	3.623	5.125	3.507	6.198
	(0.511)	(0.590)	(0.424)	(0.667)	(0.511)	(0.686)
	7.014	7.793	8.547	7.680	6.866	9.030

THETA-DELTA

	PICTCOMP	PARANG	BLOCK	OBJECT	CODING
	5.558	5.494	3.916	5.499	8.206
	(0.764)	(0.664)	(0.646)	(0.726)	(0.882)
	7.276	8.275	6.066	7.578	9.309

Completely Standardized Solution

LAMBDA-X

	VERBAL	PERFORM
INFO	0.760	- -
COMP	0.691	- -
ARITH	0.565	- -
SIMIL	0.703	- -
VOCAB	0.770	- -
DIGIT	0.390	- -
PICTCOMP	- -	0.595
PARANG	- -	0.473
BLOCK	- -	0.683
OBJECT	- -	0.566
CODING	- -	0.072

PHI

	VERBAL	PERFORM
VERBAL	1.000	
PERFOR	0.589	1.000

THETA-DELTA

	INFO	COMP	ARITH	SIMIL	VOCAB	DIGIT
	0.423	0.523	0.681	0.506	0.408	0.848

THETA-DELTA

	PICTCOMP	PARANG	BLOCK	OBJECT	CODING
	0.646	0.777	0.533	0.680	0.995

All of the indicators are significant ($p < .01$) with the exception of coding. Sometimes the different scales of the measured variables make the unstandardized coefficients difficult to interpret and often the scales of the measured variables lack inherent meaning. The standardized solution, labeled `Completely Standardized Solution` in Table 14.15, is often easier to interpret in such cases. Completely standardized output is requested with SC (on the OU − output line) in the syntax of Table 14.12. Note that this output is not completely standardized regarding error variances.

The first hypothesis, that the model fits the data, has been evaluated and supported, although there is a large residual between PICTCOMP and COMP. The final model with significant parameter estimates presented in standardized form appears in Figure 14.10. Other questions of interest are now examined. Is there a significant correlation between the Verbal and Performance factors? Looking at the completely standardized solution (or Figure 14.10), the Verbal and Performance factors are significantly correlated, $r = .589$, supporting the hypothesis of a relationship between the factors.

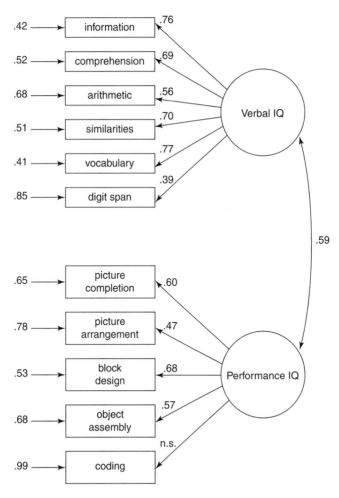

FIGURE 14.10   CFA model before modifications.

TABLE 14.16 Output of Squared Multiple Correlations for Indicators of Verbal and Performance Factor from LISREL (Syntax Appears in Table 14.12)

```
Squared Multiple Correlations for X - Variables

             INFO       COMP      ARITH      SIMIL      VOCAB      DIGIT
          --------   --------   --------   --------   --------   --------
             0.577      0.477      0.319      0.494      0.592      0.152

Squared Multiple Correlations for X - Variables

          PICTCOMP     PARANG      BLOCK     OBJECT     CODING
          --------   --------   --------   --------   --------
             0.354      0.223      0.467      0.320      0.005
```

LISREL provides estimates of the squared multiple correlations of the variables with the factors in the section labeled `Squared Multiple Correlations For X - Variables` in Table 14.16. It is also clear upon examining these SMCs that coding, with an SMC of .005, is not related to the performance factor.

14.6.1.4 Model Modification

At this point in the analysis there are several choices. The model fits the data, and we have confirmed that there is a significant correlation between the factors. Therefore, we could stop here and report the results. Generally, however, several additional models are examined that test further hypothesis (either *a priori* or *post hoc)* and/or attempt to improve the fit of the model. At least two *post hoc* hypotheses are of interest in this model: (1) Could the residual between the comprehension and picture completion be reduced by adding additional paths to the model? and (2) Could a good-fitting, more parsimonious model be estimated without data from the coding subtest?

Before demonstrating model modification, be warned that when adding and deleting parameters, there too often comes a point of almost total confusion: What have I added? What have I deleted? and What *am* I doing anyway? One hint for avoiding this sort of confusion is to diagram the estimated model, prior to any modifications, and make a few copies of it. Then, as parameters are added and deleted, draw the modifications on the copies. In this way, the model can be viewed at each stage without having to redraw the diagram each time. When one copy gets too messy, move to the next, and make more copies as necessary.

With copies of the diagram firmly in hand, modification indices are examined. Completely standardized modification indices are presented in Table 14.17. The largest univariate modification index is for the regression path predicting comprehension from the Performance factor, $\chi^2 = 9.767$, with an approximate completely standardized parameter value of .317. Because this path may also reduce the residual between the comprehension and picture completion subtests, a model is run with this path estimated, $\chi^2(42, N = 172) = 60.29$, $p = .03$, CFI = .96. The estimated (first) model and the newly modified model are nested within one another; the estimated model is a subset of this modified model. Therefore, a chi-square difference test was performed to see if the addition of this path significantly improves the model. The estimated (first) model has 43 dfs, and the modified model had 42 dfs; therefore, this is a 1 df test. The χ^2 for the first model was 70.236, for the second, 60.295. The

TABLE 14.17   Output from LISREL of Modification Indices for CFA Model (Syntax Appears in Table 14.12)

Modification Indices and Expected Change

Modification Indices for LAMBDA-X

	VERBAL	PERFORM
INFO	- -	4.451
COMP	- -	9.767
ARITH	- -	0.177
SIMIL	- -	2.556
VOCAB	- -	1.364
DIGIT	- -	1.852
PICTCOMP	1.763	- -
PARANG	0.020	- -
BLOCK	0.181	- -
OBJECT	1.174	- -
CODING	0.199	- -

Expected Change for LAMBDA-X

	VERBAL	PERFORM
INFO	- -	-0.597
COMP	- -	0.940
ARITH	- -	-0.106
SIMIL	- -	0.512
VOCAB	- -	-0.331
DIGIT	- -	-0.435
PICTCOMP	0.458	- -
PARANG	0.043	- -
BLOCK	-0.145	- -
OBJECT	-0.357	- -
CODING	0.148	- -

Standardized Expected Change for LAMBDA-X

	VERBAL	PERFORM
INFO	- -	-0.597
COMP	- -	0.940
ARITH	- -	-0.106
SIMIL	- -	0.512
VOCAB	- -	-0.331
DIGIT	- -	-0.435
PICTCOMP	0.458	- -
PARANG	0.043	- -
BLOCK	-0.145	- -
OBJECT	-0.357	- -
CODING	0.148	- -

TABLE 14.17 Continued

Completely Standardized Expected Change for LAMBDA-X

	VERBAL	PERFORM
INFO	- -	-0.205
COMP	- -	0.317
ARITH	- -	-0.046
SIMIL	- -	0.161
VOCAB	- -	-0.113
DIGIT	- -	-0.161
PICTCOMP	0.156	- -
PARANG	0.016	- -
BLOCK	-0.054	- -
OBJECT	-0.126	- -
CODING	0.052	- -

No Non-Zero Modification Indices for PHI

Modification Indices for THETA-DELTA

	INFO	COMP	ARITH	SIMIL	VOCAB	DIGIT
INFO	- -					
COMP	5.192	- -				
ARITH	4.110	0.003	- -			
SIMIL	0.744	0.668	0.559	- -		
VOCAB	4.378	0.000	2.332	0.019	- -	
DIGIT	1.637	0.567	0.804	0.117	0.028	- -
PICTCOMP	1.318	4.659	1.672	3.251	0.000	0.767
PARANG	0.087	1.543	2.081	3.354	4.122	1.208
BLOCK	1.415	1.205	2.561	2.145	0.325	0.923
OBJECT	0.101	2.798	6.326	0.803	0.686	0.873
CODING	0.762	0.035	0.832	3.252	1.509	4.899

Modification Indices for THETA-DELTA

	PICTCOMP	PARANG	BLOCK	OBJECT	CODING
PICTCOMP	- -				
PARANG	0.607	- -			
BLOCK	1.008	0.742	- -		
OBJECT	0.564	0.102	0.223	- -	
CODING	4.549	0.004	1.648	0.046	- -

(continued)

difference between the two chi-square values is a χ^2 equal to 9.941, $\chi^2_{diff}(1, N = 172) = 9.941$, $p < .01$, about the value of 9.767 anticipated from the modification index (LM test). We conclude that the addition of a path predicting comprehension from the Performance factor significantly improves the model. The largest standardized residual is now 2.614, and the plot of residuals is improved.

TABLE 14.17   Continued

Expected Change for THETA-DELTA

	INFO	COMP	ARITH	SIMIL	VOCAB	DIGIT
INFO	- -					
COMP	-0.990	- -				
ARITH	0.701	0.020	- -			
SIMIL	-0.403	0.391	-0.291	- -		
VOCAB	0.918	-0.003	-0.530	-0.065	- -	
DIGIT	0.544	-0.342	0.344	-0.165	-0.071	- -
PICTCOMP	-0.479	0.975	-0.498	0.865	0.003	-0.430
PARANG	0.117	-0.533	0.529	0.835	-0.802	0.514
BLOCK	-0.444	0.442	0.548	-0.627	0.213	-0.419
OBJECT	-0.130	0.741	-0.950	0.422	-0.338	-0.450
CODING	-0.404	0.093	0.392	-0.962	0.566	1.214

Expected Change for THETA-DELTA

	PICTCOMP	PARANG	BLOCK	OBJECT	CODING
PICTCOMP	- -				
PARANG	-0.419	- -			
BLOCK	-0.643	0.447	- -		
OBJECT	0.449	-0.165	0.283	- -	
CODING	-1.232	0.032	0.679	0.121	- -

Completely Standardized Expected Change for THETA-DELTA

	INFO	COMP	ARITH	SIMIL	VOCAB	DIGIT
INFO	- -					
COMP	-0.115	- -				
ARITH	0.104	0.003	- -			
SIMIL	-0.043	0.041	-0.040	- -		
VOCAB	0.107	0.000	-0.078	-0.007	- -	
DIGIT	0.069	-0.043	0.055	-0.019	-0.009	- -
PICTCOMP	-0.056	0.112	-0.074	0.093	0.000	-0.054
PARANG	0.015	-0.068	0.086	0.099	-0.103	0.072
BLOCK	-0.056	0.055	0.088	-0.073	0.027	-0.057
OBJECT	-0.016	0.088	-0.145	0.047	-0.041	-0.059
CODING	-0.048	0.011	0.059	-0.105	0.067	0.156

Completely Standardized Expected Change for THETA-DELTA

	PICTCOMP	PARANG	BLOCK	OBJECT	CODING
PICTCOMP	- -				
PARANG	-0.054	- -			
BLOCK	-0.081	0.062	- -		
OBJECT	0.054	-0.022	0.037	- -	
CODING	-0.146	0.004	0.087	0.015	- -

Maximum Modification Index is 9.77 for Element (2, 2) of LAMBDA-X

Additional paths could be added to the model but the decision is made to next test a third model with the coding subtest removed. The Wald test is unavailable in LISREL so we delete coding and estimate the third model, $\chi^2(33, N = 172) = 45.018$, $p = .08$, CFI = .974. By dropping the coding subtest completely we have changed *the data* and the parameters so the model is no longer nested and the chi-square difference test is no longer appropriate. Although a statistical test of an improvement is not available, other fit indices can be examined. The model AIC and CAIC can be compared between the models with small values indicating good-fitting, parsimonious models. The AIC for the model with the coding subtest is 108.30; without the coding subtest, the AIC drops to 89.018. The CAIC also drops after the coding subtest is deleted, CAIC with coding = 208.25 and without coding CAIC = 180.64. It is unclear if this drop is large enough as the AIC and CAIC are not normed; however, there does seem to be a sizable increase in fit and parsimony when the coding subtest is removed. The third model, with significant coefficients included in standardized form, is presented in Figure 14.11.

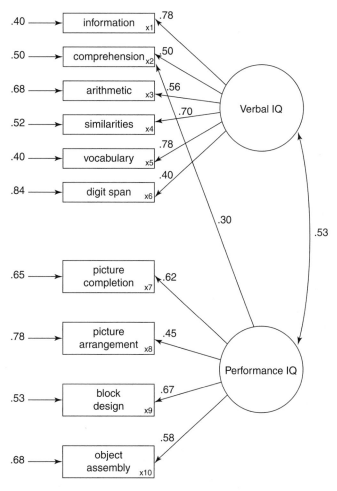

FIGURE 14.11   Final modified CFA model with significant coefficients presented in standardized form.

The model modifications in this example were post hoc and may have capitalized on chance. Ideally, these results would be cross-validated with a new sample. However, in the absence of a new sample, a helpful measure of the extent to which the parameters changed in the course of modifications is the bivariate correlation between the parameter estimates of the first and third models. This correlation, as calculated by SPSS CORRELATE, is $r(18) = .947$, $p < .01$, which indicates that, although model modifications were made, the relative size of the parameters hardly changed.

Table 14.28 (near the end of Section 14.6) contains a checklist for SEM. A Results section for the CFA analysis follows, in journal format.

Results

The Hypothesized Model

A confirmatory factor analysis, based on data from learning-disabled children, was performed through LISREL on the eleven subtests of the WISC-R. The hypothesized model is presented in Figure 14.8 where circles represent latent variables, and rectangles represent measured variables. Absence of a line connecting variables implies no hypothesized direct effect. A two factor model of IQ, Verbal and Performance, is hypothesized. The information, comprehension, arithmetic, similarities, vocabulary, and digit span subtests serve as indicators of the Verbal IQ factor. The picture comprehension, picture arrangement, block design, object assembly, and coding subtests serve as indicators of the Performance IQ factor. The two factors are hypothesized to covary with one another.

Assumptions

The assumptions of multivariate normality and linearity were evaluated through SPSS. One child had an extremely high score on the arithmetic subtest (19, $z = 4.11$, $p < .01$) and his data were deleted from the analysis. Using Mahalanobis distance, another child was a multivariate outlier, $p < .001$, and the data from this child were also deleted. This child had an extremely low comprehension subtest score and an extremely high arithmetic score. Structural equation modeling (SEM) analyses were performed using data from 175 children. There were no missing data.

Model Estimation

Maximum likelihood estimation was employed to estimate all models. The independence model that tests the hypothesis that all variables are uncorrelated was easily rejectable, $\chi^2(55, N = 175) = 516.24$, $p < .01$. The hypothesized model was tested next and support was found for the hypothesized model, $\chi^2(43, N = 175) = 70.24$, $p = .005$, comparative fix index (CFI) = .94. A chi-square difference test indicated a significant improvement in fit between the independence model and the hypothesized model.

Post hoc model modifications were performed in an attempt to develop a better fitting and possibly more parsimonious model. On the basis of the Lagrange multiplier test, a path predicting the comprehension subtest from the Performance factor was added, $\chi^2(42, N = 172) = 60.29$, $p = .03$, CFI = .96, CAIC = 108.25, AIC = 108.295. A chi square difference test indicated that the model was significantly improved by addition of this path, $\chi^2_{diff}(1, N = 172) = 9.941$, $p < .01$. Second, because the coefficient predicting the coding subscale from the Performance factor (.072) was not significant, SMC = .005, this variable was dropped and the model re-estimated, $\chi^2(33, N = 172) = 45.018$, $p = .08$, CFI = .974, CAIC = 180.643, AIC = 89.018. Both the CAIC and AIC indicated a better fitting, more parsimonious model after the coding subtest is dropped.

Because post hoc model modifications were performed, a correlation was calculated between the hypothesized model parameter estimates and the parameter estimates from the final model, $r(18) = .95$, $p < .01$; this indicates that parameter estimates were hardly changed despite modification of the model. The final model, including significant coefficients in standardized form, is illustrated in Figure 14.11.

14.6.2   SEM of Health Data

The second example demonstrates SEM of health and attitudinal variables.

14.6.2.1   SEM Model Specification

EQS 6.1 is used to assess the fit of the hypothesized model in Figure 14.12 using data in Appendix B.1. The model includes three hypothesized factors: Poor Sense of Self (with self-esteem–ESTEEM, satisfaction with marital status–ATTMAR, and locus of control–CONTROL, as indicators), Perceived Ill Health (with number of mental health problems–MENHEAL, and number of physical health problems–PHYHEAL as indicators), and Health Care Utilization (with number of visits to health professionals–TIMEDRS, and extent of drug use–DRUGUSE, as indicators). It is hypothesized that age and number of life stress units (STRESS), both measured variables, as well as poor sense of self (SELF), a latent variable, all predict perceived ill health (PERCHEAL) and health care

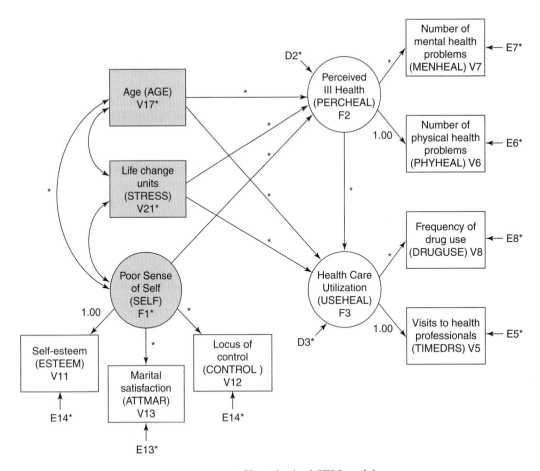

FIGURE 14.12   Hypothesized SEM model.

utilization (USEHEAL), both latent variables. Additionally perceived ill health (PERCHEAL) predicts health care utilization. As is typically done, all three independent variables (age, stress, and poor sense of self) are allowed, initially, to freely covary.

Several questions are of interest: (1) How well does this model estimate the population covariance matrix, i.e., reproduce the sample covariance matrix? (2) How well do the constructs predict the measured indicator variables, e.g., how strong is the measurement model? (3) Do age, stress, and poor sense of self directly predict perceived ill health and/or health care utilization? (4) Does perceived ill health directly predict health care utilization? (5) Does perceived ill health serve as in intervening variable between age, life stress units, poor sense of self, and health care utilization? Said another way, is there an indirect relationship between age, life stress units, poor sense of self, and health care utilization

As a preliminary check of the identifiability of the model, the number of data points and parameters to be estimated are counted. With 9 variables there are $9(9 + 1)/2 = 45$ data points. The hypothesized model contains 23 parameters to be estimated (10 regression coefficients, 3 covariance, 12 variances); therefore, the model is overidentified and is tested with 22 df. To set the scales of the factors, the path predicting number of physical health problems from Perceived Ill Health, the path predicting number of visits to health professionals from Health Care Utilization, and the path predicting self-esteem from poor sense of self are fixed to 1. EQS syntax and summary statistics appear in Table 14.18.

14.6.2.2   Evaluation of Assumptions for SEM

Output from the evaluation of assumptions is shown only when the procedure or output is different from that in other chapters or the CFA example.

14.6.2.2.1   Sample Size and Missing Data

The dataset contains responses from 459 participants. There are complete data for 443 participants on the nine variables of interest. Five participants (1.1%) are missing data on attitudes toward marriage (ATTMAR), 4 participants (.9%) are missing age (AGE), and 7 (1.5%) are missing the stress measure (STRESS). After examination of the pattern of missing data there is no evidence of a nonignorable missing data pattern (cf. Section 4.1.3). Although it would be reasonable to estimate the missing data, this analysis will use complete cases only. Given the number of measured variables and the hypothesized relationships the sample is adequate.

14.6.2.2.2   Normality and Linearity

Normality of the observed variables was assessed through examination of histograms using SPSS DESCRIPTIVES and EQS[12] and summary descriptive statistics in EQS. Eight of the ten observed variables were significantly skewed;

(1) TIMEDRS	$z = 24.84$	(2) PHYHEAL	$z = 9.28$
(3) MENHEAL	$z = 5.06$	(4) DRUGUSE	$z = 10.79$
(5) ESTEEM	$z = 4.22$	(6) CONTROL	$z = 4.46$
(7) STRESS	$z = 6.96$	(8) ATTMAR	$z = 8.75$

---

[12]Scatterplots also could have been examined through the EQS Windows program.

```
/TITLE
 Large sample example
/SPECIFICATIONS
 DATA = 'healthsem.ESS';
 VARIABLES=21; CASES=459; GROUPS=1;
 METHODS=ML,ROBUST;
 MATRIX=RAW
 ANALYSIS=COVARIANCE;
/LABELS
 V1=SUBNO; V2=EDCODE; V3=INCODE; V4=EMPLMNY; V5=TIMEDRS;
 V6=PHYHEAL; V7=MENHEAL; V8=DRUGUSE; V9=STRESS; V10=ATTMED;
 V11=ESTEEM; V12=CONTROL; V13=ATTMAR; V14=ATTROLE; V15=ATTHOUSE;
 V16=ATTWORK; V17=AGE; V18=SEL; V19=LTIMEDRS; V20=LPHYHEAL;
 V21=SCSTRESS; F1=SELF; F2=PERCHEAL; F3=USEHEAL;
/EQUATIONS
 !F1 Poor sense of self
  V11'=  1F1 + 1E11;
  V12 =  *F1 + 1E12;
  V13 = *F1 + E13;
  V14 = *F1 + E14;

 !F2 PERCHEAL
  V6 = 1F2 + 1E6;
  V7 = *F2 + 1E7;

 !F3 USEHEAL
  V5 = F3 + E5;
  V8 = *F3 + E8;

  F2 = *V21 + *V17 + *F1 + D2;
  F3 = *F2 + *V17 + D3;
  V21 = *V17 + E21;

 /VARIANCES
  F1 = *;
  D2,D3 = *;
  v17 = *;
  E21 = *;
  E6,E7 = *;
  E5,E8 = *;
  E11,E12,E13,E14 = *;

 /COVARIANCES
   F1,V17 = *;
 /PRINT
   FIT=ALL;
   TABLE=EQUATION;
   EFFECT = YES;
  /LMTEST
  /WTEST
   /END
```

TABLE 14.18   Continued

```
SAMPLE STATISTICS BASED ON COMPLETE CASES
                       UNIVARIATE STATISTICS
                       ----------------------
```

VARIABLE	TIMEDRS	PHYHEAL	MENHEAL	DRUGUSE	ESTEEM
MEAN	7.5730	4.9412	6.0871	8.5643	15.8301
SKEWNESS (G1)	2.9037	1.0593	.6175	1.2682	.4870
KURTOSIS (G2)	9.9968	1.2271	-.2605	1.0620	.2822
STANDARD DEV.	9.9821	2.3768	4.1858	9.0952	3.9513
VARIABLE	CONTROL	ATTMAR	ATTROLE	AGE	SCSTRESS
MEAN	6.7429	22.7298	35.1503	4.3638	2.0087
SKEWNESS (G1)	.4912	.7937	.0551	.0372	.7637
KURTOSTS (G2)	-.3978	.8669	-.4190	-1.1624	.2436
STANDARD DEV.	1.2657	8.8654	6.7708	2.2284	1.2967

```
                     MULTIVARIATE KURTOSIS
                     ---------------------

     MARDIA'S COEFFICIENT (G2,P) =        23.7537
     NORMALIZED ESTIMATE   =              16.4249


                 ELLIPTICAL THEORY KURTOSIS ESTIMATES
                 ------------------------------------

MARDIA-BASED KAPPA =   .1979 MEAN SCALED UNIVARIATE KURTOSIS =   .3813

MARDIA-BASED KAPPA IS USED IN  COMPUTATION.  KAPPA=    .1979

COVARIANCE  MATRIX TO BE ANALYZED: 10 VARIABLES
                          (SELECTED FROM 21 VARIABLES)
     BASED ON   459 CASES.
```

			TIMEDRS V 5	PHYHEAL V 6	MENHEAL V 7	DRUGUSE V 8	ESTEEM V 11
TIMEDRS	V	5	99.643				
PHYHEAL	V	6	10.912	5.649			
MENHEAL	V	7	10.705	4.957	17.521		
DRUGUSE	V	8	26.779	9.151	14.136	82.722	
ESTEEM	v	11	.726	.852	3.600	-1.541	15.613
CONTROL	V	12	.279	.328	1.490	.779	1.690
ATTMAR	V	13	4.332	1.683	9.026	7.262	10.251
ATTROLE	V	14	-5.460	-.814	-1.926	-5.897	4.947
AGE	V	17	-.336	.144	-.757	-.544	.031
SCSTRESS	V	21	3.315	.926	2.099	3.619	-.468

(continued)

TABLE 14.18 Continued

		CONTROL V 12	ATTMAR V 13	ATTROLE V 14	AGE V 17	SCSTRESS V 21
CONTROL	V 12	1.602				
ATTMAR	V 13	2.173	78.595			
ATTROLE	V 14	-.009	-3.804	45.844		
AGE	V 17	-.376	-1.762	3.423	4.966	
SCSTRESS	V 21	.099	1.251	-2.114	-.838	1.681

```
BENTLER-WEEKS STRUCTURAL REPRESENTATION:

    NUMBER OF DEPENDENT VARIABLES = 11
            DEPENDENT V'S :    5    6    7    8   11   12   13   14   21
            DEPENDENT F'S :    2    3

    NUMBER OF INDEPENDENT VARIABLES =  13
            INDEPENDENT V'S :   17
            INDEPENDENT F'S :    1
            INDEPENDENT E'S :    5    6    7    8   11   12   13   14   21
            INDEPENDENT D'S :    2    3

        NUMBER OF FREE PARAMETERS = 25
        NUMBER OF FIXED NONZERO PARAMETERS = 14

DETERMINANT OF INPUT MATRIX IS    .11552D+12

PARAMETER ESTIMATES APPEAR IN ORDER,
NO SPECIAL PROBLEMS WERE ENCOUNTERED DURING OPTIMIZATION.
```

EQS also provides information on multivariate normality (see Table 14.18). In the section labeled, MULTIVARIATE KURTOSIS, Mardia's coefficient and a normalized estimate of the coefficient are given; the normalized estimate can be interpreted as a z-score. In this example, after deletion of all outliers, NORMALIZED ESTIMATE = 16.42, suggesting that the measured variables are not distributed normally.

It is not feasible to examine all pairwise scatterplots to assess linearity; therefore, randomly selected pairs of scatterplots are examined using SPSS GRAPHS.[13] All observed pairs appear to be linearly related, if at all. Transformations are not made to these variables because it is reasonable to expect these variables to be skewed in the population (most women use few drugs, are not ill, and do not go to the doctor often). Instead, given the sample size ($N = 443$, a large sample) the decision is made to use provisions in the EQS program to take the nonnormality into account when assessing χ^2 statistics and standard errors by use of maximum likelihood estimation with the Satorra-Bentler scaled chi square and adjustment to the standard errors to the extent of the nonnormality. This analysis is requested from EQS by ME=ML,ROBUST (see Table 14.18).

---

[13]Scatterplots also could have been examined through the EQS Windows program.

14.6.2.2.3 Outliers

Using SPSS FREQUENCIES and GRAPHS there were no univariate outliers detected. Although there were *z*-scores on several variables greater than 3.3, these large scores were associated with naturally skewed distributions, i.e., most women don't go to the doctor often and a diminishing number go frequently. Using SPSS REGRESSION there were also no multivariate outliers.

14.6.2.2.4 Multicollinearity and Singularity

The determinant of the matrix, given in EQS (Table 14.18), as DETERMINANT OF INPUT MATRIX IS .1552D+12. This is much larger than 0, so there is no singularity.

14.6.2.2.5 Adequacy of Covariances

SEM programs have difficulty with computations if the scales of the variables, and therefore the covariances, are of vastly different sizes. In this example, the largest variance is 17,049.99 for STRESS, the smallest variance is 1.61 for CONTROL. This difference is large and a preliminary run of the model does not converge after 200 iterations. Therefore the STRESS variable was multiplied by .01. After this rescaling, the new variable SCSTRESS had a variance of 1.70 and there were no further convergence problems.

14.6.2.2.6 Residuals

Evaluation of residuals is performed as part of model evaluation.

14.6.2.3 SEM Model Estimation and Preliminary Evaluation

The syntax and edited output for SEM analysis are presented in Table 14.18. As a first step, look in the printout for the statement, PARAMETER ESTIMATES APPEAR IN ORDER, shown at the end of Table 14.18. If there are identification problems or other problems with estimation, this statement does not appear and, instead, a message about CONDITION CODES, as discussed in Section 14.5.1, is printed. This message is used to diagnose problems that come up during the analysis.

After checking the covariance matrix for reasonable relationships, examine the portion of the printout labeled BENTLER-WEEKS STRUCTURAL REPRESENTATION that lists the IVs and DVs as specified by the model. These should, and do in this case, match the path diagram.

Table 14.19 presents residuals and goodness-of-fit information for the estimated model in the section labeled GOODNESS OF FIT SUMMARY. A necessary condition for evaluating and interpreting a model is that the hypothesized model is a significant improvement over the independence model. The independence model tests the hypothesis that all the measured variables are independent of one another. Our proposed model hypothesizes that there are relationships among the measured variables therefore it is necessary that our hypothesized model is a significant improvement over the independence model. The test of improvement between independence and model chi squares is assessed with a chi-square difference test. Had the data been normal we simply could have subtracted the chi square test statistic values and evaluated the chi square with the dfs associated with the difference between the models. However, because the data were nonnormal and we used the Satorra-Bentler scaled chi square we need to make an adjustment as follows (Satorra & Bentler, 2002). First a scaling correction is calculated,

$$\text{Scaling correction} = \frac{(\text{df nested model})\left(\dfrac{\chi^2_{ML\,\text{nested model}}}{\chi^2_{S\text{-}B\,\text{nested model}}}\right) - (\text{df comparison model})\left(\dfrac{\chi^2_{ML\,\text{comparison model}}}{\chi^2_{S\text{-}B\,\text{comparison model}}}\right)}{(\text{df}_{\text{nested model}} - \text{df}_{\text{comparison model}})} \qquad \text{(eq. 14.38)}$$

$$\text{Scaling correction} = \frac{(36)\left(\dfrac{705.53}{613.17}\right) - (20)\left(\dfrac{99.94}{86.91}\right)}{(36 - 20)}$$

$$= 1.15$$

The scaling correction is then employed with the ML χ^2 values to calculate the S-B scaled χ^2 difference test statistic value,

$$\chi^2_{S\text{-}B\,\text{difference}} = \frac{\chi^2_{ML\,\text{nested model}} - \chi^2_{ML\,\text{comparison model}}}{\text{scaling correction}}$$

$$= \frac{705.53 - 99.942}{1.15}$$

$$= 525.91$$

This chi-square difference is evaluated with degrees of freedom equal to, $\text{df}_{\text{nested model}} - \text{df}_{\text{comparison model}} = 36 - 20 = 16$. The adjusted S-B $\chi^2(N = 443, 20) = 525.91, p < .01$. The chi-square difference test is significant, therefore, the model is a significant improvement over the independence model and model evaluation can continue. In practice, just about the only time this difference is not significant is when sample sizes are very small or there is a major problem with the hypothesized model.

TABLE 14.19   Standardized Residuals and Goodness-of-Fit Information from EQS Complete Example (Syntax Appears in Table 14.18)

```
STANDARDIZED RESIDUAL MATRIX:
                        TIMEDRS    PHYHEAL    MENHEAL    DRUGUSE    ESTEEM
                        V   5      V   6      V   7      V   8      V  11
   TIMEDRS   V   5       .000
   PHYHEAL   V   6       .094       .000
   MENHEAL   V   7      -.085      -.019       .000
   DRUGUSE   V   8       .000       .019      -.005       .000
   ESTEEM    V  11      -.079      -.065       .072      -.150       .000
   CONTROL   V  12      -.060      -.022       .159      -.023       .002
   ATTMAR    V  13      -.017      -.026       .145       .017       .021
      AGE    V  17       .005       .049      -.061      -.004       .073
   SCSTRESS  V  21      -.013      -.049       .061       .010      -.094

                        CONTROL    ATTMAR     AGE        SCSTRESS
                        V  12      V  13      V  17      V  21
   CONTROL   V  12       .000
   ATTMAR    V  13      -.034       .000
      AGE    V  17      -.075      -.042       .000
   SCSTRESS  V  21       .058       .107       .000       .000
```

TABLE 14.19   Continued

```
                     AVERAGE ABSOLUTE STANDARDIZED RESIDUALS   =   .0404
        AVERAGE OFF-DIAGONAL ABSOLUTE STANDARDIZED RESIDUALS   =   .0505
MAXIMUM LIKELIHOOD SOLUTION (NORMAL DISTRIBUTION THEORY)

LARGEST STANDARDIZED RESIDUALS:

     NO.    PARAMETER    ESTIMATE        NO.    PARAMETER    ESTIMATE
     ---    ---------    --------        ---    ---------    --------
      1     V12, V7         .159         11     V11, V7         .072
      2     V11, V7        -.150         12     V11, V6        -.065
      3     V13, V7         .145         13     V21, V7         .061
      4     V21, V13        .107         14     V17, V7        -.061
      5     V21, V11       -.094         15     V12, V5        -.060
      6     V6,  V5         .094         16     V21, V12        .058
      7     V7,  V5        -.085         17     V21, V6        -.049
      8     V11, V5        -.079         18     V17, V6         .049
      9     V17, V12       -.075         19     V17, V13       -.042
     10     V17, V11        .073         20     V13, V12       -.034
```

DISTRIBUTION OF STANDARDIZED RESIDUALS

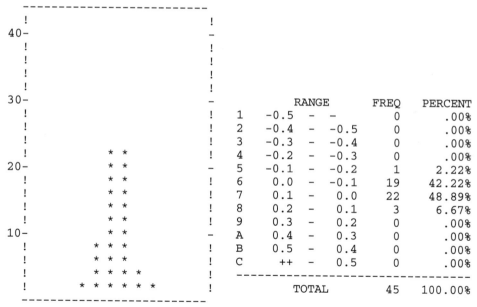

```
     !                           !
 40-                             -
     !                           !
     !                           !
     !                           !
     !                           !
 30-                             -       RANGE       FREQ    PERCENT
     !                           !   1  -0.5  -  -       0      .00%
     !                           !   2  -0.4  - -0.5     0      .00%
     !                           !   3  -0.3  - -0.4     0      .00%
     !             * *           !   4  -0.2  - -0.3     0      .00%
 20-               * *           -   5  -0.1  - -0.2     1     2.22%
     !             * *           !   6   0.0  - -0.1    19    42.22%
     !             * *           !   7   0.1  -  0.0    22    48.89%
     !             * *           !   8   0.2  -  0.1     3     6.67%
     !             * *           !   9   0.3  -  0.2     0      .00%
 10-               * *           -   A   0.4  -  0.3     0      .00%
     !           * * *           !   B   0.5  -  0.4     0      .00%
     !           * * *           !   C   ++   -  0.5     0      .00%
     !           * * * *         !     -------------------------------
     !         * * * * * *       !          TOTAL       45   100.00%

       1 2 3 4 5 6 7 8 9 A B C        EACH "*" REPRESENTS 2 RESIDUALS

MAXIMUM LIKELIHOOD SOLUTION (NORMAL DISTRIBUTION THEORY)

GOODNESS OF FIT SUMMARY FOR METHOD = ML

INDEPENDENCE MODEL CHI-SQUARE =        705.531 ON   36 DEGREES OF FREEDOM
```

(continued)

TABLE 14.19 Continued

```
INDEPENDENCE AIC =    633.53117   INDEPENDENCE CAIC =    448,88536
      MODEL AIC =     59.94157       MODEL CAIC =   -42.63943

CHI-SQUARE =       99.942 BASED ON   20 DEGREES OF FREEDOM
PROBABILITY VALUE FOR THE CHI-SQUARE STATISTIC IS      .00000

THE NORMAL THEORY RLS CHI-SQUARE FOR THIS ML SOLUTION IS     102.834.

FIT INDICES
-----------
BENTLER-BONETT     NORMED FIT INDEX =     .858
BENTLER-BONETT NON-NORMED FIT INDEX =     .785
COMPARATIVE FIT INDEX (CFI)         =     .881
BOLLEN   (IFI) FIT INDEX            =     .883
MCDONALD (MFI) FIT INDEX            =     .917
LISREL    GFI  FIT INDEX            =     .952
LISREL    AGFI FIT INDEX            =     .893
ROOT MEAN-SQUARE  RESIDUAL (RMR)    =    1.471
STANDARDIZED RMR                    =     .059
ROOT MEAN-SQUARE ERROR OF APPROXIMATION(RMSEA) = .093
90% CONFIDENCE INTERVAL OF RMSEA    (.075, .112)
GOODNESS OF FIT SUMMARY FOR METHOD = ROBUST

ROBUST INDEPENDENCE MODEL CHI-SQUARE = 613.174 ON  36 DEGREES OF FREEDOM
   INDEPENDENCE AIC = 541.17402   INDEPENDENCE  CAIC = 356.52821
         MODEL AIC = 46.90838        MODEL CAIC = -55.67262

   SATORRA-BENTLER SCALED CHI-SQUARE = 86.9084 ON   20 DEGREES OF FREEDOM
   PROBABILITY VALUE FOR THE CHI-SQUARE STATISTIC IS .00000

   RESIDUAL-BASED TEST STATISTIC                  = 70.815
   PROBABILITY VALUE FOR THE CHI-SQUARE STATISTIC IS .00000

YUAN-BENTLER RESIDUAL-BASED TEST STATISTIC        = 61.350
   PROBABILITY VALUE FOR THE CHI-SQUARE STATISTIC IS .00000

   YUAN-BENTLER RESIDUAL-BASED F-STATISTIC        = 3.394
   DEGREES OF FREEDOM                             = 20,439
   PROBABILITY VALUE FOR THE F-STATISTIC I   .00000

   FIT INDICES
   -----------
   BENTLER-BONETT     NORMED FIT INDEX =     .858
   BENTLER-BONETT NON-NORMED FIT INDEX =     .791
   COMPARATIVE FIT INDEX (CFI)         =     .884
   BOLLEN   (IFI) FIT INDEX            =     .887
   MCDONALD (MFI) FIT INDEX            =     .930
   ROOT-MEAN SQUARE  ERROR OF APPROXIMATION (RMSEA) = .085
   90% CONFIDENCE INTERVAL OF RMSEA      (.067, .104)
```

However, the Satorra-Bentler scaled chi-square test of the robust ML estimation is also significant, $\chi^2(20, N = 443) = 86.91, p < .001$, indicating a significant difference between the estimated and observed covariance matrices. In a large sample like this, trivial differences can produce a statistically significant χ^2 so fit indices often provide a better gauge of fit when sample size is large. However, none of the fit indices indicates a good fitting model. The residuals are symmetrically distributed around zero, but large. The largest standardized residual is .159, and a few are greater than .10. Because the hypothesized model does not fit the data, further inspection of parameters is deferred, and instead the Lagrange multiplier test is examined.

14.6.2.4 *Model Modification*

The hypothesized model does not fit. Models can be improved by adding paths so a first approach is to carefully examine the hypothesized model set-up to insure that important paths were not forgotten. No paths were obviously forgotten in our model so the next step is to examine the Lagrange multiplier test (LM Test). Doing model modifications to improve the fit of a model moves the analysis from a confirmatory analysis to an exploratory analysis and caution should be exercised in interpreting significance levels. Results of initial univariate and multivariate LM tests with default settings are presented in Table 14.20. Note that these tests are based on ML statistics because EQS does not yet print out Satorra-Bentler LM tests.

The multivariate LM test suggests that adding a path predicting V7 from F1 (predicting number of mental health problems from Poor Sense of Self) would significantly improve the model and lead to an approximate drop in model χ^2 of 32.199. This means that in addition to the relationship between poor sense of self and number of mental health problems through perceived ill health there is also a direct relationship between these two variables. This may be a reasonable parameter to add. It may be that women who have a poor sense of self also report more mental health problems over and above the relationship between poor sense of self and perceived ill health. The path is added and the model re-estimated (not shown).

Instead of the message, PARAMETER ESTIMATES APPEAR IN ORDER, however, the following message is found:

```
PARAMETER          CONDITION CODE
    D3,D3              CONSTRAINED AT LOWER BOUND
```

This indicates that during estimation, EQS held the disturbance (residual variance) of the third factor, the Health Care Utilization factor, at zero rather than permit it to become negative. This message may well indicate a potential problem with interpretation (negative error variance?). This path was not hypothesized and we add the path and re-estimate the model. Unfortunately we create a condition code (indicating a problem with the model). Therefore this path is not added, instead the correlations among the residuals are examined through the LM test.

Adding correlated residuals is conceptually and theoretically tricky. First and foremost this is dangerously close to data fishing! When we add correlated residuals we are correlating the parts of the dependent variable that are not predicted by the independent variable. So in essence we don't know exactly *what* we *are* correlating only what we are *not* correlating. Sometimes this makes good sense and other times it does not (Ullman, in press.)

TABLE 14.20   Edited EQS Output of Univariate and Multivariate Lagrange Multiplier Test (Syntax Appears in Table 14.18)

```
LAGRANGE MULTIPLIER TEST (FOR ADDING PARAMETERS)

          ORDERED UNIVARIATE TEST STATISTICS:
```

| | | | | | HANCOCK | | STANDARD- |
| | | | CHI- | | 20 DF | PARAMETER | IZED |
NO	CODE	PARAMETER	SQUARE	PROB.	PROB.	CHANGE	CHANGE
1	2 12	V7,F1	32.199	.000	.041	.573	.055
2	2 10	D3,D2	15.027	.000	.775	6.352	2.898
3	2 16	F3,F1	15.027	.000	.775	-.765	-.059
4	2 22	F2,F3	15.027	.000	.775	2.691	.295
5	2 11	V11,V21	14.052	.000	.828	-.596	-.116
6	2 11	V6,V21	9.991	.002	.968	-.364	-.118
7	2 11	V7,V21	9.991	.002	.968	.598	.110
8	2 20	V11,F3	9.956	.002	.969	-.168	-.008
9	2 20	V6,F3	9.212	.002	.960	1.385	.113
10	2 11	V11,V17	8.931	.003	.984	.283	.032

```
MAXIMUM LIKELIHOOD SOLUTION (NORMAL DISTRIBUTION THEORY)

MULTIVARIATE LAGRANGE MULTIPLIER TEST BY SIMULTANEOUS PROCESS IN STAGE  1

PARAMETER SETS (SUBMATRICES) ACTIVE AT THIS STAGE ARE:

  PVV PFV PFF PDD GVV GVF GFV GFF BVF BFF
```

		CUMULATIVE MULTIVARIATE STATISTICS				UNIVARIATE INCREMENT			
								HANCOCK'S	
								SEQUENTIAL	
STEP	PARAMETER	CHI-SQUARE	D.F.	PROB.	CHI-SQUARE	PROB.	D.F.	PROB.	
1	V7,F1	32.199	1	.000	32.199	.000	20	.041	
2	V7,F21	62.428	2	.000	30.229	.000	19	.049	
3	V11,V21	76.480	3	.000	14.052	.000	18	.726	
4	V11,V17	80.363	4	.000	3.883	.049	17	1.000	

The LM test is again employed to examine the usefulness of adding correlated errors to the model.[14] Correlated errors are requested in EQS by the inclusion of

```
/LMTEST
SET = PEE;
```

Table 14.21 shows that if the residuals between E6 and E5 are added the chi square will drop approximately 30.529 points. It may be that even after accounting for the common relationship

---

[14]Note: Adding post hoc paths is a little like eating salted peanuts—one is never enough. Extreme caution should be used when adding paths as they are generally post hoc and therefore potentially capitalizing on chance. Conservative p levels ($p <$ .001) may be used as a criterion for adding post hoc parameters to the model.

TABLE 14.21   Syntax Modifications and Edited EQS Multivariate LM Test for Adding Correlated Errors (Full Syntax Appears in Table 14.26)

```
/LMTEST
 set=pee;
  LAGRANGE MULTIPLIER TEST (FOR ADDING PARAMETERS)

            ORDERED UNIVARIATE TEST STATISTICS:

                                     HANCOCK              STANDARD-
                            CHI-       20 DF   PARAMETER     IZED
 NO    CODE   PARAMETER    SQUARE  PROB.  PROB.   CHANGE      CHANGE
 --    ----   ---------    ------  -----  -----  --------   ---------
  1    2  6     E6,E5      30.259  .000   .062     5.661       .418
  2    2  6     E7,E5      18.845  .000   .532    -7.581      -.295

MULTIVARIATE LAGRANGE MULTIPLIER TEST BY SIMULTANEOUS PROCESS IN STAGE  1

PARAMETER SETS (SUBMATRICES) ACTIVE AT THIS STAGE ARE:

  PEE

     CUMULATIVE MULTIVARIATE STATISTICS           UNIVARIATE INCREMENT
     ----------------------------------           --------------------

                                                             HANCOCK'S
                                                             SEQUENTIAL
 STEP  PARAMETER  CHI-SQUARE  D.F.  PROB.  CHI-SQUARE  PROB.  D.F.  PROB.
 ----  ---------  ----------  ----  -----  ----------  -----  ----  -----
   1    E6,E5       30.529     1   .000      30.529    .000    20   .062
   2    E11,E8      41.280     2   .000      10.750    .000    19   .932
   3    E12,E7      51.575     3   .000      10.295    .000    18   .922
   4    E13,E7      60.250     4   .000       8.675    .000    17   .950
   5    E11,E7      69.941     5   .000       9.690    .000    16   .882
```

between number of physical health problems and number of visits to health professionals through their respective factors there is still a unique significant relationship between these two measured variables. This seems reasonable therefore the path is added and the model re-estimated.

The model Satorra-Bentler χ^2 for the new model = 60.37, Robust CFI = .93, RMSEA = .07. The adjusted scaled χ^2 difference test is calculated using equation 14.38, Satorra-Bentler $\chi^2_{\text{difference}}$ (1, N = 459) = 17.54, $p < .05$. The model is significantly improved by adding this path (not shown). It would be feasible to stop adding paths at this point. However, the RMSEA is somewhat high and CFI is a little too low so the decision is made to examine the correlated residuals one more time with the goal of improving the model a little bit more if conceptually justifiable. The LM test is examined after the addition of the correlated residual between E6 and E5 and the test indicates that adding the covariance between E8 and E6 will be associated with an approximate drop in the model χ^2 of 23.63 (not shown). Again, it seems reasonable that there may be a unique relationship between frequency of drug use and number of physical health problems. A word of warning here about model modifications is necessary. These decisions to add paths are being made post hoc, *after* looking at the data. It is very easy to fool yourself into a convincing story about the theoretical importance of a path when you see that adding it would significantly improve the model. Exercise caution here!

The covariance between E8 and E6 is added and the model Satorra-Bentler χ^2 for the new model = 40.16, Robust CFI = .96, RMSEA = .05. The adjusted scaled χ^2 difference test is calculated using equation 14.38, Satorra-Bentler $\chi^2_{difference}(1, N = 459) = 20.60, p < .05$.

The final model goodness-of-fit information is presented in Table 14.22. The final model with significant parameter estimates presented in standardized form is diagrammed in Figure 14.13. Two paths were added that were not hypothesized therefore it is important to provide some evidence that the hypothesized model has not substantially changed. Ideally, the model should be tested on new data. No new data are available for analysis however so instead the bivariate correlation between the initial parameter estimates and the final parameter estimates is calculated. If this correlation is high (>.90) we can conclude that although paths were added, the model did not change substantially. The correlation between the final parameters and the hypothesized paths was calculated and exceeded .90 ($r = .97$) therefore although the model was changed it was not changed substantially.

Specific parameter estimates are now examined. The syntax for the final model, portions of the printout related to parameter estimates, and the standardized solution are shown in Table 14.23. The section labeled MEASUREMENT EQUATIONS WITH STANDARD ERRORS AND TEST STA-TISTICS contains the parameter estimates, standard errors, and, because robust estimation was employed, the robust statistics in parentheses. These are the standard errors and the z tests to interpret.

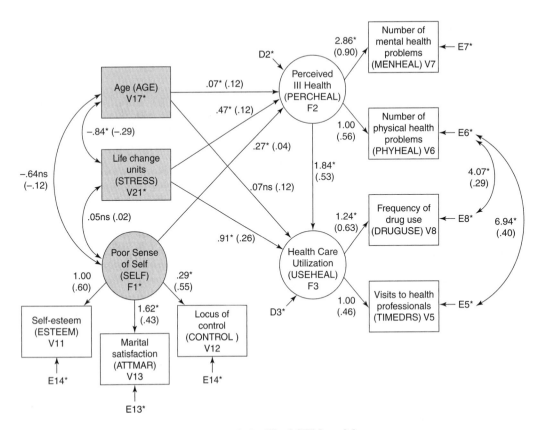

FIGURE 14.13   Final SEM model.

TABLE 14.22   Edited EQS Output for Final Model Goodness-of-Fit Summary

```
GOODNESS OF FIT SUMMARY FOR METHOD = ML

INDEPENDENCE MODEL CHI-SQUARE =          705.531 ON    36 DEGREES OF FREEDOM

INDEPENDENCE AIC =     633.53117    INDEPENDENCE CAIC =   448.88536
        MODEL AIC =       8.72072         MODEL CAIC =    -83.60219

CHI-SQUARE =       44.721 BASED ON     18 DEGREES OF FREEDOM
PROBABILITY VALUE FOR THE CHI-SQUARE STATISTIC IS        .00045

THE NORMAL THEORY RLS CHI-SQUARE FOR THIS ML SOLUTION IS         42.058.

FIT INDICES
-----------
BENTLER-BONETT     NORMED FIT INDEX =         .937
BENTLER-BONETT NON-NORMED FIT INDEX =         .920
COMPARATIVE FIT INDEX (CFI)        =          .960
BOLLEN (IFI)            FIT INDEX =          .961
MCDONALD (MFI)          FIT INDEX =          .971
LISREL GFI              FIT INDEX =          .980
LISREL AGFI             FIT INDEX =          .950
ROOT MEAN-SQUARE  RESIDUAL (RMR)   =          .992
STANDARDIZED RMR                   =          .044
ROOT MEAN-SQUARE ERROR OF APPROXIMATION(RMSEA) = .057
90% CONFIDENCE INTERVAL OF RMSEA (      .036,       .078)

GOODNESS OF FIT SUMMARY FOR METHOD = ROBUST

INDEPENDENCE MODEL CHI-SQUARE =          613.174 ON    36 DEGREES OF FREEDOM

INDEPENDENCE AIC =     541.17402    INDEPENDENCE CAIC =  356.52821
        MODEL AIC =      4.166650         MODEL CAIC =   -88.15640

SATORRA-BENTLER SCALED CHI-SQUARE =     40.1665 ON    18 DEGREES OF FREEDOM
   PROBABILITY VALUE FOR THE CHI-SQUARE STATISTIC IS        .00198

   RESIDUAL-BASED TEST STATISTIC                        =    51.064
   PROBABILITY VALUE FOR THE CHI-SQUARE STATISTIC IS        .00005

   YUAN-BENTLER RESIDUAL-BASED TEST STATISTIC           =    45.952
   PROBABILITY VALUE FOR THE CHI-SQUARE STATISTIC IS        .00030

   YUAN-BENTLER RESIDUAL-BASED F-STATISTIC     =     2.732
   DEGREES OF FREEDOM   =                         18,    441
   PROBABILITY VALUE FOR THE F-STATISTIC IS        .00018

FIT INDICES
-----------
BENTLER-BONETT     NORMED FIT INDEX =         .934
BENTLER-BONETT NON-NORMED FIT INDEX =         .923
COMPARATIVE FIT INDEX (CFI)        =          .962
BOLLEN (IFI)            FIT INDEX =          .963
MCDONALD (MFI)          FIT INDEX =          .976
ROOT MEAN-SQUARE ERROR OF APPROXIMATION(RMSEA) = .052
90% CONFIDENCE INTERVAL OF RMSEA (      .030,       .073)
```

```
/TITLE
  Large Sample Example Final Model
/SPECIFICATIONS
  DATA='healthsem 5th editon.ESS';
  VARIABLES=21; CASES=459; GROUPS=1;
  METHODS=ML,ROBUST;
  MATRIX=RAW;
  ANALYSIS=COVARIANCE;
/LABELS
   V1=SUBNO; V2=EDCODE; V3=INCODE; V4=EMPLMNY; V5=TIMEDRS;
   V6=PHYHEAL; V7=MENHEAL; V8=DRUGUSE; V9=STRESS; V10=ATTMED;
   V11=ESTEEM; V12=CONTROL; V13= ATTMAR; V14=ATTROLE; V15=ATTHOUSE;
   V16=ATTWORK; V17= AGE; V18=SEL; V19= LTIMEDRS; V20= LPHYHEAL;
   V21= SCSTRESS; F1=SELF; F2=PERCHEAL; F3=USEHEAL;
 /EQUATIONS
  !F1 Poor sense of self
  V11 = 1F1 + 1E11;
  V12 = *F1 + 1E12;
  V13 = *F1 + E13;

  !F2  PERCHEAL
   V6 = 1F2 + 1E6;
   V7 = *F2 + E7;

  !F3  USEHEAL
   V5 = F3 + E5;
   V8 = *F3 + E8;

   F2 = *V17 + *F1 + *V21 + D2;
   F3 = *F2 + *V17 + *V21 + D3;

  /VARIANCES
   F1 = *;
   D2,D3 = *;
   V17 = *;
   E6,E7 = *;
   E5,E8 = *;
   E11,E12,E13, = *;

  /COVARIANCES
   F1,V17 = *;
   F1,V21 = *;
   V17, V21 = *;
   E6,E5 = *;
   E8,E6 = *;
 /PRINT
  FIT=ALL;
  TABLE=EQUATION;
  EFFECT = YES;
/LMTEST
 SET = PEE;
/WTEST
/END
```

764

TABLE 14.23   Continued

MEASUREMENT EQUATIONS WITH STANDARD ERRORS AND TEST STATISTICS
STATISTICS SIGNIFICANT AT THE 5% LEVEL ARE MARKED WITH @.
(ROBUST STATISTICS IN PARENTHESES)

```
TIMEDRS =V5   =     1.000 F3      + 1.000 E5

PHYHEAL =V6   =     1.000*F2      + 1.000 E6

MENHEAL =V7   =     2.859*F2      + 1.000 E7
                     .364
                    7.846@
                 (   .368)
                 (  7.774@

DRUGUSE =V8   =     1.244*F3      + 1.000 E8
                     .207
                    5.998@
                 (   .212)
                 (  5.855@

ESTEEM  =V11  =     1.000 F1      + 1.000 E11

CONTROL =V12  =      .293*F1      + 1.000 E12
                     .046
                    6.355@
                 (   .052)
                 (  5.641@

ATTMAR  =V13  =     1.616*F1      + 1.000 E13
                     .278
                    5.808@
                 (   .268)
                 (  6.034@
```

CONSTRUCT EQUATIONS WITH STANDARD ERRORS AND TEST STATISTICS
STATISTICS SIGNIFICANT AT THE 5% LEVEL ARE MARKED WITH @.
(ROBUST STATISTICS IN PARENTHESES)

```
PERCHEAL=F2   =      .069*V17   +   .473*V21   +   .269*F1    + 1.000 D2
                     .030            .073           .054
                    2.297@          6.442@         4.962@
                 (   .030)       (   .081)      (   .059)
                 (  2.307@       (  5.870@      (  4.587@

USEHEAL =F3   =     1.837*F2    +   .070*V17   +   .905*V21   + 1.000 D3
                     .316            .119           .255
                    5.822@           .584          3.544@
                 (   .322)       (   .116)      (   .267)
                 (  5.697@       (   .602)      (  3.393@
```

(continued)

TABLE 14.23 Continued

```
COVARIANCES AMONG INDEPENDENT VARIABLES
----------------------------------------
STATISTICS SIGNIFICANT AT THE 5% LEVEL ARE MARKED WITH @.

                    V                           F
                   ---                         ---
F21 -  SCSTRESS       -.838*I                            I
V17 -  AGE             .141 I                            I
                     -5.958@I                            I
                   (  .136)I                             I
                   ( -6.143@I                            I
                          I                             I
F1  -  SELF           -.642*I                            I
V17 -  AGE             .338 I                            I
                     -1.899 I                            I
                   (  .343)I                             I
                   ( -1.873)I                            I
                          I                             I
F1  -  SELF            .049*I                            I
V21 -  SCSTRESS,       .194 I                            I
                       .254 I                            I
                   (  .199)I                             I
                   (  .248)I                             I
                          I                             I

                    E                           D
                   ---                         ---
E6  -PHYHEAL          6.941*I                            I
E5  -TIMEDRS          1.026 I                            I
                      6.673@I                            I
                   ( 1.363)I                             I
                   ( 5.094@I                             I
                          I                             I
E8  -DRUGUSE          4.074*I                            I
E6  -PHYHEAL           .910 I                            I
                      4.479@I                            I
                   (  .975)I                             I
                   ( 4.179@I                             I
                          I                             I

DECOMPOSITION OF EFFECTS WITH NONSTANDARDIZED VALUES
STATISTICS SIGNIFICANT AT THE 5% LEVEL ARE MARKED WITH @.

PARAMETER INDIRECT EFFECTS
--------------------------
TIMEDRS =V5 =    1.837 F2    +  .196 V17   - 1.775 V21   +  .494 F1
                  .316          .131          .309          .132
                 5.822@        1.495         5.744@        3.754@
               (  .322)      (  .137)      (  .345)      (  .144)
               ( 5.697@      ( 1.429)      ( 5.142@      ( 3.434@
```

TABLE 14.23   Continued

```
                    + 1.837 D2      + 1.000 D3
                       .316
                      5.822@
                    (  .322)
                    (  5.697@

PHYHEAL =V6  =       .069 V17    +   .473 V21   +   .269 F1    + 1.000 D2
                     .083            .073           .054
                    2.297@          6.442@         4.962@
                    (  .030)      (   .081)      (   .059)
                    (  2.307@     (   5.870@     (   4.587@

MENHEAL =V7  =       .197 V17    + 1.353 V21   +   .769 F1    + 2.859 D2
                     .083            .141           .128          .364
                    2.381@          9.592@         6.024@        7.846@
                    (  .097)      (   .381)      (   .240)     (   .368)
                    (  2.041@     (   3.551@     (   3.207@    (   7.774@

DRUGUSE =V8      = 2.284 F2     +   .244 V17   + 2.207 V21    +   .614 F1
                     .335            .161           .305           .147
                    6.823@          1.516          7.232@        4.167@
                    (  .377)      (   .170)      (   .594)     (   .171)
                    (  6.055@     (   1.434)     (   3.714@    (   3.591@

                    +2.284 D2     + 1.244 D2
                     .335            .207
                    6.823@          5.998@
                    (  .377)      (   .212)
                    (  6.055@     (   5.855@

USEHEAL =F3      =   .127*V17   +   .869*V21   +   .494 F1    + 1.837 D2
                     .059            .204           .132           .316
                    2.131@          4.257@         3.754@        5.822@
                    (  .060)      (   .227)      (   .144)     (   .322)
                    (  2.112@     (   3.826@     (   3.434@    (   5.697@
```

DECOMPOSITION OF EFFECTS WITH STANDARDIZED VALUES

PARAMETER INDIRECT EFFECTS

```
TIMEDRS =V5  =       .244 F2    +   .044 V17   +   .231 V21   +   .118 F1
                +    .185 D2    +   .340 D3

PHYHEAL =V6  =       .065 V17   +   .258 V21   +   .269 F1    +   .421 D2

MENHEAL =V7  =       .105 V17   +   .419 V21   +   .438 F1    +   .684 D2

DRUGUSE =V8  =       .333 F2    +   .060 V17   +   .315 V21   +   .161 F1
                +    .252 D2    +   .464 D3

USEHEAL =F3  =       .062*V17   +   .246*V21   +   .256 F1    +   .401 D2
```

(continued)

TABLE 14.23   Continued

```
STANDARDIZED SOLUTION:                                                    R-SQUARED

    TIMEDRS =V5     =   .461 F3   +   .887 E5                                   .213
    PHYHEAL =V6     =   .555*F2   +   .831 E6                                   .309
    MENHEAL =V7     =   .903 F2   +   .430 E7                                   .815
    DRUGUSE =V8     =   .629*F3   +   .777 E8                                   .396
    ESTEEM  =V11    =   .603*F1   +   .798 E11                                  .363
    CONTROL =V12    =   .552*F1   +   .834 E12                                  .305
    ATTMAR  =V13    =   .434*F1   +   .901 E13                                  .189
    PERCHEAL=F2     =   .116*V17  +   .464*V21  +  .485*F1  +  .758 D2          .426
    USEHEAL =F3     =   .529*F2   +   .034*V17  +  .256*V21  +  .737 D3         .457

    CORRELATIONS AMONG INDEPENDENT VARIABLES
    ----------------------------------------

                         V                            F
                        ---                          ---

    V21 -  SCSTRESS        -.290*I                      I
    V17 -  AGE                   I                      I
                                 I                      I
    F1  -  SELF            -.121*I                      I
    V17 -  AGE                   I                      I
                                 I                      I
    F1  -  SELF             .016*I                      I
    V21 -  SCSTRESS              I                      I
                                 I                      I

    CORRELATIONS AMONG INDEPENDENT VARIABLES
    ----------------------------------------

                         E                            D
                        ---                          ---

    E6   -PHYHEAL          .398*I                      I
    E5   -TIMEDRS               I                      I
                                I                      I
    E8   -DRUGUSE          .292*I                      I
    E6   -PHYHEAL               I                      I
                                I                      I
```

All of the path coefficients between measured variables and factors in the model are significant, $p < .05$. The section labeled CONSTRUCT EQUATIONS WITH STANDARD ERRORS contains the same information for the equations that relate one factor to another. The coefficients that are significant, 2 tailed, at $p < .05$ are marked with an @ sign. Increasing age, more life change units (stress), and poorer sense of self all significantly predict worse perceived ill health (unstandardized coefficient age [V17] = .069, stress [V21] = .473, poor sense of self [F1] = .269). Increased health care utilization is predicted by increased stress (unstandardized coefficient = .905) and perceived ill health (F2, unstandardized coefficient = 1.837). In this model, age does not significantly predict increased health care utilization. (Note: In the interest of developing a parsimonious model, it would have been reasonable to run a final model and drop all the nonsignificant paths.)

Evaluation of indirect effects is done from the section labeled DECOMPOSITION OF EFFECTS WITH NONSTANDARDIZED VALUES PARAMETER INDIRECT EFFECTS. Age, number of life change units, and Poor sense of Self all indirectly affect Health Care Utilization. Said another way, perceived ill health serves as an intervening variable between age, stress, Poor Sense of Self and Health Care Utilization. Increasing age, more stress, and Poor Sense of Self all predict greater Perceived Ill Health, which in turn predicts greater health care utilization over and above the direct effects of these variables on Health Care Utilization (age unstandardized indirect effect $= .127$, $z = 2.112$, stress unstandardized indirect effect $= .869$, $z = 3.83$, Poor Sense of Self unstandardized indirect effect $= .494$, $z = .49$). The standardized solution for the indirect effects appears in the section labeled DECOMPOSITION OF EFFECTS WITH STANDARDIZED VALUES. The standardized direct effects included in the model are shown in the section labeled STANDARDIZED SOLUTION.

The percent of variance in the dependent variables accounted for by the predictors is found in the R-SQUARED column in the STANDARDIZED SOLUTION section of Table 14.23: 42.6% of the variance in Perceived Ill Health is accounted for by age, stress, and Poor Sense of Self; 45.7% of the variance in Health Care Utilization is accounted for by Perceived Ill Health, age, and stress. The checklist for the SEM analysis is in Table 14.24. A Results section for the SEM analysis follows.

TABLE 14.24   Checklist for Structural Equations Modeling

1. Issues
   a. Sample size and missing data
   b. Normality of sampling distributions
   c. Outliers
   d. Linearity
   e. Adequacy of covariances
   f. Identification
   g. Path diagram–hypothesized model
   h. Estimation method
2. Major analyses
   a. Assessment of fit
      (1) Residuals
      (2) Model chi square
      (3) Fit indices
   b. Significance of specific parameters
   c. Variance in a variable accounted for by a factor
3. Additional analyses
   a. Lagrange Multiplier test
      (1) Tests of specific parameters
      (2) Addition of parameters to improve fit
   b. Wald test for dropping parameters
   c. Correlation between hypothesized and final model or cross-validate model
   d. Diagram–final model

Results

The Hypothesized Model

The hypothesized model is in Figure 14.12. Circles represent latent variables, and rectangles represent measured variables. Absence of a line connecting variables implies lack of a hypothesized direct effect.

The hypothesized model examined the predictors of health care utilization. Health care utilization was a latent variable with 2 indicators (number of visits to health professionals and frequency of drug use). It was hypothesized that perceived ill health (a latent variable with 2 indicators—number of mental health problems and number of physical health problems), age, and number of life stress units directly predicted increased health care utilization.

Additionally it was hypothesized that perceived ill health is directly predicted by poorer sense of self, greater number of life change units, and increasing age. Perceived ill health served as an intervening variable between age, life change units, poor sense of self and health care utilization.

Assumptions

The assumptions were evaluated through SPSS and EQS. The dataset contains responses from 459 women. There were complete data for 443 participants on the nine variables of interest. Five participants (1.1%) were missing data on attitudes toward marriage (ATTMAR), 4 participants (.9%) were missing age (AGE), and 7 (1.5%) are missing the stress measure (STRESS). This analysis used complete only cases (N = 443).

There were no univariate or multivariate outliers. There was evidence that both univariate and multivariate normality were violated. Eight of the measured variables (TIMEDRS, PHYHEAL, MENHEAL, DRUGUSE, ESTEEM, CONTROL, STRESS, and ATTMAR) were significantly univariately skewed, $p < .001$. Mardia's Normalized coefficient = 6.42, $p < .001$, indicating violation of multivariate normality. Therefore, the models were estimated with maximum likelihood estimation and tested with the Satorra-Bentler scaled chi square (Satorra & Bentler, 1988). The standard errors also were adjusted to extent of the nonnormality (Bentler & Dijkstra, 1985).

Model Estimation

Only marginal support was found for the hypothesized model Satorra-Bentler $\chi^2(20, N = 443) = 86.91$, $p < .05$, Robust CFI = .88, RMSEA = .08.

Post hoc model modifications were performed in an attempt to develop a better fitting model. On the basis of the Lagrange multiplier test, and theoretical relevance, two residual covariances were estimated (residual covariance between number of physical health problems and number of visits to health professionals and the residual covariance between frequency of drug use and number of visits to health professionals). The model was significantly improved with the addition of these paths, Satorra-Bentler $\chi^2_{\text{difference}}(2, N = 443) = 37.33$, $p < .05$.

The final model fit the data well, Satorra-Bentler $\chi^2(18, N = 443) = 40.17$, $p < .05$, Robust CFI = .96, RMSEA = .05. Because post hoc model modifications were performed, a correlation was calculated between the hypothesized model estimates and the estimates from the final model, $r(20) = .97$, $p < .01$. This high correlation indicates that the parameter estimates from the first and last models are highly related to each other. The final model with standardized and unstandardized coefficients is in Figure 14.13.

Direct Effects

Increased health care utilization was predicted by greater perceived ill health (unstandardized coefficient = 1.84, $p < .05$), and more stress (life change units, unstandardized coefficient = .90,

$p < .05$). Increasing age did not significantly predict increased health care utilization, (unstandardized coefficient = .07, $p > .05$).

Perceived ill health increased as stress increased (number of life change units, unstandardized coefficient = .47, $p < .05$), age increased (unstandardized coefficient = .07, $p < .05$), and women had a poorer sense of self (unstandardized coefficient = .27, $p < .05$).

Indirect Effects

The significance of the intervening variables was evaluated using tests of indirect effects through EQS (Sobel, 1988). This method of examining intervening variables has more power than the mediating variable approach (Baron & Kenny, 1986; MacKinnon, Lockwood, Hoffman, West, & Sheets, 2002).

Perceived ill health served as an intervening variable between age, life change units, and poor sense of self. Increased age predicted greater perceived ill health which predicted greater health care utilization (unstandardized indirect effect coefficient = .13, $p < .05$, standardized coefficient = .06). More life change units predicted worse perceived ill health, greater perceived ill health predicted greater health care utilization, (unstandardized indirect effect coefficient = .87, $p < .05$, standardized coefficient = .25). A poorer sense of self also predicted worse perceived ill health which was associated with greater health care utilization (unstandardized indirect effect coefficient = .49, $p < .05$, standardized coefficient = .26).

Almost half (45.7%) of the variance in health care utilization was accounted for by perceived ill health, age, stress, and poor sense of self. Poor sense of self, stress, and age accounted for 42.6% of the variance in perceived ill health.

14.7   Comparison of Programs

The four SEM programs discussed, EQS, LISREL, SAS CALIS, and AMOS, are full-service, multioption programs. A list of options included in each package is presented in Table 14.25.

14.7.1   EQS

EQS is the most user-friendly of the programs. The equation method of specifying the model is clear and easy to use and the output is well organized. Colors distinguish key words from user input in the equation method. In addition to the equation method of model specification there are also options to specify the model through a diagram or with a windows "point and click" method. EQS offers numerous diagnostics for evaluation of assumptions and handles deletion of cases very simply. Evaluation of multivariate outliers and normality can be performed within this program. Missing data can be imputed within EQS. EQS is the only program to impute nonnormal data. EQS reads data sets from a variety of other statistical and database programs. Several methods of estimation are offered. EQS is the only program that offers the correct adjusted standard errors and Satorra-Bentler scaled χ^2 for model evaluation and the Bentler-Yuan (1999) test statistic. This is the program of choice when data are nonnormal. A specific estimation technique is available if the measured variables are nonnormal with common kurtosis. A second method for treatment of nonnormal data, through estimation of polychoric or polyserial correlations, is also available. Additionally, EQS is able to analyze multilevel models.

EQS is also the program of choice if model modifications are to be performed. EQS is very flexible, offering both multivariate and univariate Lagrange multiplier tests as well as the multivariate Wald test. EQS offers several options for matrices to be considered for modification and allows specification of order of consideration of these matrices. Categorical variables can be included within the model without preprocessing. EQS allows multiple group models to be specified and tested easily. Diagrams are available in EQS. The entire EQS Manual is included within the program.

14.7.2   LISREL

LISREL is a set of three programs; PRELIS, SIMPLIS, and LISREL. PRELIS preprocesses data, e.g., categorical or nonnormal data, for SEM analyses through LISREL. SIMPLIS is a program that allows models to be specified with equations. SIMPLIS is very simple to use but is somewhat limited in options, and some output, e.g., a standardized solution, must be requested in LISREL output form. Models can be specified through diagrams or point and click methods with SIMPLIS. LISREL specifies SEM models with matrices and some models become quite complicated to specify when using this method. Missing data can be imputed with PRELIS. LISREL also is capable of estimating multilevel models.

LISREL offers residual diagnostics, several estimation methods, and many fit indices. LISREL includes two types of partially standardized solutions. The univariate Lagrange multiplier test is also available. Nonnormal and categorical data can be included in LISREL by first preprocessing the data in PRELIS to calculate polyseric and polychoric correlations. LISREL also calculates the SMC for each variable in the equations. Coefficients of determination are calculated for the latent DVs in the model. Diagrams are available in SIMPLIS.

TABLE 14.25   Comparison of Programs for Structural Equation Modeling

Feature	EQS	LISREL	AMOS	SAS CALIS
Input				
Covariance matrix				
Lower triangular	Yes	Yes	Yes	Yes
Full symmetric	Yes	Yes	Yes	Yes
Input stream	Yes	Yes	No	No
Asymptotic covariance matrix	Yes	Yes	No	No
Multiple covariance matrices	Yes	Yes	Yes	No
Correlation matrix				
Lower triangular	Yes	Yes	Yes	Yes
Full symmetric	Yes	Yes	Yes	Yes
Input stream	Yes	Yes	No	No
Matrix of polychoric, polyserial correlations	Yes	Yes	No	No
Correlation matrix based on optimal scores	No	Yes	No	No
Multiple correlation matrices	Yes	Yes	Yes	No
Moment matrices	Yes	Yes	Yes	Yes
Sum of squares and cross-products matrix	No	No	No	Yes
Raw data	Yes	Yes	Yes	Yes
User specified weight matrix	Yes	Yes	No	No
Categorical (ordinal) data	Yes	Yes[a]	No	No
Means and standard deviations	Yes	Yes	Yes	Yes
Delete cases	Yes	Yes[a]	No	Yes
Estimate model from diagram	Yes	Yes[b]	Yes	No
Windows "point and click" method	Yes	Yes[b]	Yes	No
Multilevel models	Yes	Yes	No	No
Estimation methods				
Maximum likelihood (ML)	Yes	Yes	Yes	Yes
Unweighted least squares (LS)	Yes	Yes	Yes	Yes
Generalized least squares (GLS)	Yes	Yes	Yes	Yes
Two-stage least squares	No	Yes	No	No
Diagonally weighted least squares	No	Yes	No	No
Elliptical least squares (ELS)	Yes	No	No	No
Elliptical generalized least squares (EGLS)	Yes	No	No	No
Elliptical reweighted least squares (ERLS)	Yes	No	No	No
Arbitrary distribution generalized least squares (AGLS)	Yes	Yes	Yes	No
Satorra-Bentler scaled chi square	Yes	Yes	No	No

TABLE 14.25 Continued

Feature	EQS	LISREL	AMOS	SAS CALIS
Estimation methods *(continued)*				
Bentler-Yuan (1999) F	Yes	No	No	No
Robust standard errors	Yes	No	No	No
Elliptical cor. chi square	No	No	No	Yes
Instrumental variables	No	Yes	No	No
Specify number of groups	Yes	Yes	Yes	No
Scale free least squares	No	No	Yes	No
Specify elliptical kurtosis parameter—kappa	Yes	No	No	No
Specify model with equations	Yes	Yes[b]	Yes	Yes
Specify model with matrix elements	No	Yes	No	Yes
Specify models with intercepts	Yes	Yes	Yes	Yes
Start values				
Automatic	Yes	Yes	Yes	Yes[c]
User specified	Yes	Yes	Yes	Yes
Specify confidence interval range	No	No	No	No
Automatically scale latent variables	No	Yes	No	No
Specify covariances	Yes	Yes	Yes	Yes
Specify general linear constraints	Yes	Yes	Yes	Yes
Nonlinear constraints	No	Yes	No	No
Specify cross-group constraints	Yes	Yes	Yes	No
Specify inequalities	Yes	Yes	No	Yes
Lagrange multiplier test—Univariate	Yes	Yes	Yes	Yes
Lagrange multiplier test—Multivariate	Yes	No	No	No
Lagrange multiplier options				
Indicate parameters to be considered first for addition	Yes	No	No	No
Indicate specific order for entry consideration	Yes	No	No	No
Specify LM testing process	Yes	No	No	No
Specify specific matrices only	Yes	Yes	No	No
Set probability value for criterion for inclusion	Yes	Yes	No	No
Specify parameters not to be included in LM test	Yes	Yes	No	No
Wald test—Univariate	Yes	No	No	Yes
Wald test—Multivariate	Yes	No	No	Yes
Wald test options				
Indicate parameters to be considered first for dropping	Yes	No	No	No
Indicate specific order for dropping consideration	Yes	No	No	No

(continued)

TABLE 14.25 Continued

Feature	EQS	LISREL	AMOS	SAS CALIS
Wald test options *(continued)*				
Set probability value	Yes	No	No	Yes
Specify parameters not to be included in Wald test	Yes	Yes	No	No
Specify number of iterations	Yes	Yes	Yes	Yes
Specify maximum CPU time used	No	Yes	No	No
Specify optimization method	No	No	No	Yes
Specific convergence criterion	Yes	Yes	Yes	Yes
Specify tolerance	Yes	No	No	No
Specify a ridge factor	No	Yes	No	Yes
Diagram	Yes	Yes	Yes	No
Effect decomposition	Yes	Yes	Yes	Yes
Simulation	Yes	Yes	No	Yes[d]
Bootstrapping	Yes	Yes	Yes	Yes
Missing data estimation	Yes	Yes	Yes	No
Output				
Means	Yes	No	No	Yes
Skewness and kurtosis	Yes	No	No	Yes
Mardia's coefficient	Yes	No	No	Yes
Normalized estimate	Yes	No	No	Yes
Mardia based kappa	Yes	No	No	Yes
Mean scaled univariate kurtosis	Yes[e]	No	No	Yes
Multivariate least squares kappa	Yes[e]	No	No	No
Multivariate mean kappa	Yes	No	No	No
Adjusted mean scaled univariate kurtosis	Yes	No	No	Yes
Relative multivariate kurtosis coefficient	No	No	No	Yes
Case numbers with largest contribution to normalized multivariate kurtosis	Yes	No	No	Yes
Sample covariance matrix	Yes	Yes	Yes	Yes
Sample correlation matrix	Yes	Yes	Yes	Yes
Estimated model covariance matrix	Yes	Yes	Yes	Yes
Correlations among parameter estimates	Yes	Yes	Yes	Yes
Asymptotic covariance matrix of parameters	No	Yes	No	No
Iteration summary	Yes	No	Yes	Yes
Determinant of input matrix	Yes	No	Yes	Yes
Residual covariance matrix	Yes	Yes	Yes	Yes
Largest raw residuals	Yes	Yes	Yes	Yes

TABLE 14.25   Continued

Feature	EQS	LISREL	AMOS	SAS CALIS
Output *(continued)*				
Completely standardized residual matrix	Yes	No	No	Yes
Largest completely standardized residuals	Yes	No	No	Yes
Frequency distribution of standardized residuals	Yes	No	No	Yes
Largest partially standardized residual	No	Yes	No	No
Partially standardized residual matrix	No	Yes	No	No
Q plot of partially standardized residuals	No	Yes	No	No
Frequency distribution of partially standardized residuals	No	Yes	No	No
Estimated covariance matrix	Yes	Yes	Yes	Yes
Estimated correlation matrix	Yes	Yes	Yes	Yes
Largest eigenvalue of $B*B'$	No	Yes	No	No
Goodness of fit indices				
Normed fit index (NFI) (Bentler & Bonett, 1980)	Yes	Yes	Yes	Yes
Non-normed fit index (NNFI) (Bentler & Bonett, 1980)	Yes	Yes	No	Yes
Comparative fit index (CFI) (Bentler, 1995)	Yes	Yes	Yes	Yes
Minimum of fit function	Yes	Yes	Yes	Yes
Non-centrality parameter (NCP)	No	Yes	Yes	No
Confidence interval of NCP	No	Yes	Yes	No
Goodness of fit index (GFI)	Yes	Yes	Yes	Yes
Adjusted goodness of fit index	Yes	Yes	Yes	Yes
Root mean square residual	Yes	Yes	Yes	Yes
Standardized root mean square residual	Yes	Yes	No	No
Population discrepancy function (PDF)	No	Yes	No	No
Confidence interval for PDF	No	Yes	No	No
Root mean square error of approximation (RMSEA)	Yes	Yes	Yes	No
Confidence interval for RMSEA	Yes	No	Yes	No
Akaikes information criterion model	Yes	Yes	Yes	Yes
Akaikes information criterion—independence model	Yes	Yes	Yes	No
Akaikes information criterion—saturated model	No	Yes	Yes	No
Consistent information criterion model	Yes	Yes	Yes	Yes
Consistent information criterion—independence model	Yes	Yes	Yes	No
Consistent information criterion—saturated model	No	Yes	Yes	No
Schwartz Bayesian criterion	No	No	No	Yes
McDonald's centrality (1989)	Yes	No	No	Yes
James, Mulaik, & Brett (1982) parsimony index	No	Yes	Yes	Yes
Z test (Wilson & Hilferty, 1931)	No	No	No	Yes

(continued)

TABLE 14.25 Continued

Feature	EQS	LISREL	AMOS	SAS CALIS
Goodness of fit indices *(continued)*				
Normed index rho 1 (Bollen, 1986)	No	Yes	No	Yes
Non-normed index delta 2 (Bollen, 1989a)	Yes	Yes	No	Yes
Expected cross validation index (ECVI)	No	Yes	Yes	No
Confidence interval for ECVI	No	Yes	Yes	No
ECVI for saturated model	No	Yes	Yes	No
ECVI for independence model	No	Yes	Yes	No
Hoelter's critical N	No	Yes	Yes	Yes
Brown-Cudeck criterion	No	No	Yes	No
Bayes information criterion	No	No	Yes	No
P for test of close fit	No	No	Yes	No
R^2 square for dependent variables	Yes	Yes	No	Yes
SMC for structural equations	No	Yes	Yes	No
Coefficient of determination for structural equations	No	Yes	No	No
Latent variable score regression coefficients	No	Yes	Yes	Yes
Unstandardized parameter estimates	Yes	Yes	Yes	Yes
Completely standardized parameter estimates	Yes	No	Yes	Yes
Partially standardized solution	No	Yes	No	No
Standard errors for parameter estimates	Yes	Yes	Yes	Yes
Variances of independent variables	Yes	Yes	Yes	Yes
Covariances of independent variables	Yes	Yes	Yes	Yes
Test statistics for parameter estimates	Yes	Yes	Yes	Yes
Save output to file				
Condition code flag	Yes	No	No	Yes
Convergence flag	Yes	No	No	No
Function minimum	No	No	No	Yes
Independence model χ^2	Yes	No	No	Yes
Model χ^2 value	Yes	No	No	Yes
Model degrees of freedom	Yes	No	No	Yes
Probability level	Yes	No	AMOS	Yes
Bentler-Bonett normed fit index	Yes	Yes	No	No
Bentler-Bonett non-normed fit index	Yes	Yes	No	No
Comparative fit index	Yes	Yes	No	Yes
GFI	No	Yes	No	Yes
AGFI	No	Yes	No	Yes
Root mean square residual	No	Yes	No	Yes

TABLE 14.25   Continued

Feature	EQS	LISREL	AMOS	SAS CALIS
Save output to file *(continued)*				
Number of parameters in model	No	No	No	Yes
AIC	No	Yes	No	Yes
CAIC	No	Yes	No	Yes
Schwartz's Bayesian criterion	No	No	No	Yes
James, Mulaik, & Hilferty parsimony index	No	Yes	No	Yes
z test of Wilson & Hilferty	No	No	No	Yes
Hoelter's critical N	No	Yes	No	Yes
Generated data	Yes	Yes[a]	No	Yes[d]
Derivatives	Yes	No	No	Yes
Gradients	Yes	No	No	Yes
Matrix analyzed	Yes	Yes	No	Yes
Means	No	No	No	Yes
Standard deviations	No	No	No	Yes
Sample size	No	No	No	Yes
Univariate skewness	No	No	No	Yes
Univariate kurtosis	No	No	No	Yes
Information matrix	No	No	No	Yes
Inverted information matrix	Yes	No	No	Yes
Weight matrix	Yes	No	No	No
Estimated population covariance matrix	No	Yes	No	Yes
Asymptotic covariance matrix	No	Yes	No	No
Asymptotic covariance matrix of parameter estimates	No	Yes	No	No
Parameter estimates	Yes	No	No	Yes
Residual matrix	Yes	No	No	No
Standard errors	Yes	No	No	Yes
LM test results	Yes	No	No	No
Wald test results	Yes	No	No	No
Updated start values	Yes	No	Yes	Yes
Automatic model modification	Yes	Yes	No	No

[a]In PRELIS.

[b]In SIMPLIS.

[c]Not available with COSAN model specification.

[d]With SAS IML.

[e]With AGLS estimation only.

[f]Kurtosis only.

14.7.3   AMOS

AMOS allows models to be specified through diagrams or equations. Different options are available for the equation method. AMOS makes clever use of colors in the equation specification method. Key words are presented in a different color than user-specified words. If the line is correct the color changes. If the line is incorrect the color remains. Several different estimation methods are available. Detailed goodness-of-fit information is given in output. Missing data can be estimated in AMOS. AMOS also has extensive bootstrapping capabilities. Multiple group models can be tested. Categorical data is not treated in AMOS. Table or text options are also presently available. AMOS has a clever output feature. If the cursor is placed over certain elements of the output within the AMOS program a little help screen pops up and explains that portion of the output. One limitation of AMOS is the inability to save output without transporting it to a word-processing program.

14.7.4   SAS System

SAS CALIS offers a choice of model specification methods: `lineqs` (Bentler-Weeks), `ram`, and `cosan` (a form of matrix specification). Diagnostics are available for evaluation of assumptions; for instance, evaluations of multivariate outliers and multivariate normality can be done within CALIS. If data are nonnormal, but with homogenous kurtosis, the chi-square test statistics can be adjusted within the program. Several different estimation techniques are available and lots of information about the estimation process is given. Categorical data are not treated in CALIS, nor can multiple group models be tested.

15 Multilevel Linear Modeling

15.1 General Purpose and Description

Multilevel (hierarchical) linear modeling[1] (MLM) is for research designs where the data for participants is organized at more than one level. For example, student achievement (the DV) is measured for pupils within classrooms, which are, in turn, organized within schools. Different variables may be available for each level of analysis; for example, there may be a measure of student motivation at the pupil level, a measure of teacher enthusiasm at the classroom level, and a measure of poverty at the school level. You may recognize this as the nested ANOVA design of Section 3.2.5.1. It also has many of the characteristics of the random effects ANOVA of Section 3.2.5.4, because most often the lower level units of analysis were not randomly assigned to higher levels of the hierarchy (pupils within classrooms or classrooms within schools).

MLM provides an alternative analysis to several different designs discussed elsewhere in this volume. Although the lowest level of data in MLM is usually an individual, it may instead be repeated measurements of individuals. For example, there may be measures of student achievement at the beginning, middle, and end of the school year, nested within pupils, nested within classrooms, nested within schools. Thus, MLM provides an alternative to univariate or multivariate analysis of repeated measures. Because there are separate analyses of each case over time, individual differences in *growth curves* may be evaluated. For example, Do students differ in their pattern of growth in achievement over the school year? If so, Are there variables, such as hours of homework, that predict these differences? MLM also has been developed within the framework of structural equation modeling to permit analysis of latent variables (where, for example, there might be several factors representing different aspects of teacher enthusiasm) and yet another approach to longitudinal data.

Another useful application of MLM is as an alternative to ANCOVA where DV scores are adjusted for covariates (individual differences) prior to testing treatment differences (Cohen, Cohen, West, & Aiken, 2003). MLM analyzes these experiments without the often-pesky assumption of homogeneity of regression, in which it is assumed that the relationship between the DV and CV(s) is the same for all treatment groups.

---

[1]We have chosen multilevel linear modeling (MLM) rather than hierarchical linear modeling (HLM) to avoid confusion with the HLM software package.

An advantage of MLM over alternative analyses is that independence of errors is not required. In fact, independence is often violated at each level of analysis. For example, students in a classroom influence each other, so are apt to be more alike than students in different classrooms; similarly, students within one school are apt to be more alike than students in different schools. In a repeated-measures[2] design, measurements made on occasions close in time are likely to be more highly correlated than measurements made on occasions farther apart in time (recall sphericity of Chapter 8). In addition, there may be interactions across levels of the hierarchy. For example, student motivation at the lowest level may well interact with teacher enthusiasm at the classroom level.

Analyzing data organized into hierarchies as if they are all on the same level leads to both interpretational and statistical errors. Suppose, for example, that data for student achievement are aggregated to the classroom level to see if groups[3] that differ in teacher enthusiasm have different mean scores. Interpretation is restricted to the classroom level, however, a common error of interpretation is to apply group level results to the individual level. This is called the *ecological fallacy*. Statistically, this type of analysis usually results in decreased power and loss of information because the unit of analysis for purposes of deriving the ANOVA error term is the group. That is, *n* is the number of groups, not the number of participants in each group.

A less common but equally misleading approach is to interpret individual-level analyses at the group level, leading to the *atomistic fallacy* (Hox, 2002). A multilevel model, on the other hand, permits prediction of individual scores adjusted for group differences as well as prediction of group scores adjusted for individual differences within groups. Statistically, if individual scores are used without taking into account the hierarchical structure, the Type I error rate is inflated because analyses are based on too many degrees of freedom that are not truly independent.

Multilevel linear modeling (MLM) addresses these issues by allowing intercepts (means) and slopes (IV–DV relationships) to *vary* between higher level units. For example, the relationship between student achievement (the DV) and student motivation (the IV) is allowed to vary between different classrooms. This *variability* is modeled by treating group intercepts and slopes as DVs in the next level of analysis. For the example, there is an attempt to predict differences in means and slopes within classrooms from differences in teacher enthusiasm between classrooms. These group differences, in turn, can vary across yet higher level units (e.g., schools), so that third-level equations can be built to model the variability between second-level units, and so on.

Multilevel models often are called *random coefficient* regression models. That is because the regression coefficients (the intercepts and predictor slopes) may vary across groups (higher-level units), which are considered to be randomly sampled from a population of groups. For the example, the regression coefficients for the relationship between student achievement and student motivation are considered to be randomly sampled from a population of classrooms. In garden-variety (OLS) regression, it is the individual participants who are considered to be a random sample from some population; in multilevel modeling the groups also are considered to be a random sample.

One advantage of the multilevel modeling approach over other ways of handling hierarchical data is the opportunity to include predictors at every level of analysis. For the example, a predictor of student achievement might include student motivation and/or study time and/or gender at the student level of analysis, teacher enthusiasm and/or teacher emphasis on homework at the classroom

---

[2]Repeated-measures terminology is customary in MLM rather than within-subjects terminology.

[3]Group, cluster, and context are terms are used synonymously in MLM to denote higher-level units of analysis. Thus, multilevel models are sometimes referred to as contextual or clustered models (not to be confused with cluster analysis).

level of analysis, and school poverty level and/or school type (public vs. private) and/or school size at the school level. Higher-level predictors may help explain lower-level differences in intercepts and slopes. For example, differences between average classroom achievement (the intercept) and the relationship between student achievement and student motivation (the slope) could be a function of school poverty level. Within-level interactions among predictors (e.g., between school poverty level and school size) can be modeled, as well as cross-level interactions (e.g., between teacher emphasis on homework and school type).

Rowan, Raudenbush, and Kang (1991) used MLM to study effects of organizational design on high school teachers. Teachers within schools were the lower-level unit of analysis, and provided ratings of structure or climate as the DVs and, as IVs, race, sex, years of education, years of experience, track (teacher-estimates of student achievement relative to school average), and dummy variables representing the teachers' most frequently taught course. School level variables were sector (public vs. Catholic), size, percentage minority enrollment, urbanicity, average student achievement, and average student SES. The researchers found that differences among high schools in perception of structure or climate are strongly related to whether they are public or private. However, there are also large differences between teachers within schools in perceptions of organizational design, and these are related to academic departments and curriculum tracks as well as demographic characteristics of teachers.

McLeod and Shanahan (1996) analyzed data from the National Longitudinal Survey of Youth in which children were assessed annually over several years on a variety of psychological, behavioral, and demographic measures. Data were analyzed using a two-level latent growth curve model (Section 15.5.1). The lowest level analysis was year of assessment, the second level was child. McLeod and Shanahan found higher levels of depression and antisocial behavior in children who were poor when first measured or who had prior histories of poverty. They also found that the number of years that the children are in poverty is correlated with the slope of their antisocial behavior over the assessment period, with rates of increase in antisocial behavior greater for children who had been in long-term poverty compared with those who were only transiently poor or who were not poor. The authors conclude that poverty experiences are related to the way that children develop and not just to scores measured during a single assessment period—the longer a child is poor, the greater the rate of behavioral disadvantages.

Barnett, Marshall, Raudenbush, and Brennan (1993) studied dual-earner couples in terms of the relationship between their psychological distress as DVs and features of their job experiences and demographic variables as IVs. Individuals within each of the 300 couples comprised the lower-level unit of analysis and provided scores on the DVs as well as lower-level predictors such as age, education, occupational prestige, quality of job role, job rewards, job concerns, quality of marital role, and gender. Couples were the higher-level unit of analysis and provided measures of income, number of years together, and a joint measure of parental status. They found that features of subjective experiences that individuals reported about the job were significantly associated with distress for both men and women, and the magnitude of effect depends little on gender.

Multilevel modeling is a highly complex set of techniques; we can only skim the surface of this fascinating topic. Several recent books address MLM in greater depth. One that is especially easy to follow and discusses software without being tied to any one package is Hox (2002). Snijders and Bosker (1999) offer an introduction to multilevel modeling as well as ample discussion of more advanced topics. The book is a rich source of examples in the social sciences, as well as discussion of more exotic transformations than we cover in this book to produce better-fitting models. The classic multilevel modeling text is by Raudenbush and Bryk (2001), who also are the authors of HLM, a stand-alone package. Kreft and DeLeeuw (1998) offer a helpful introductory guide to MLM that is tied to MlwiN, another stand-alone package. An introductory text by Heck and Thomas (2000) provides detailed information for multilevel factor and structural equation modeling as well as for the more common multilevel regression models emphasized in this chapter.

We (reluctantly) demonstrate only SPSS and SAS MIXED programs in this chapter. However, there are two excellent stand-alone packages, HLM and MLwiN, in addition to SYSTAT MIXED REGRESSION, that also handle MLM and have features absent in SPSS and SAS. Therefore, we discuss these programs throughout the text and compare their features in the final section of this chapter.

15.2   Kinds of Research Questions

The primary questions in MLM are like the questions in multiple regression: degree of relationship among the DV and various IVs (Is student achievement related to student motivation or teacher enthusiasm?); importance of IVs (How important is student motivation? Teacher enthusiasm?); adding and changing IVs (What happens when school poverty level is added to the equation?); contingencies among IVs (Once differences due to student motivation are factored into the equation, what happens to teacher enthusiasm?); parameter estimates (What is the slope of the equation that relates student motivation to student achievement?); and predicting DV scores for members of a new sample. However, additional questions can be answered when the hierarchical structure of the data is taken into account and random intercepts and slopes are permitted. Only these additional questions are discussed here.

15.2.1   Group Differences in Means

This question is answered as part of the first step in routine hierarchical analyses. Is there a significant difference in intercepts (means) for the various groups? For example, Is there a significant difference in mean student achievement in the different classrooms? As in ANOVA, this is a question about variability; is the variance between groups (between-subject variance in ANOVA) greater than would be expected by chance (within-subject variance in ANOVA)? Section 15.4.1.2 discusses analysis of first-level intercepts. These differences also are evaluated as precursors to MLM through calculation of intraclass correlations (Section 15.6.1).

15.2.2   Group Differences in Slopes

This question may also be answered as part of routine hierarchical analyses. Is there a significant difference in slopes for the various groups? For example, Is there a significant difference in the slope of the relationship between student achievement and student motivation among the different classrooms? Group differences in slope between a predictor and the DV are called a failure of homogeneity of regression in ANCOVA, but in MLM such differences are expected and included in the model. These differences are assessed separately for all first-level predictors if there is more than one. Section 15.4.2.2 discusses second-level analysis of first-level predictors.

15.2.3   Cross-Level Interactions

Does a variable at one level interact with a variable at another level in its effect on the DV? For example, does school-level poverty (a third-level variable) interact with student motivation (a first-level variable) to produce differences in student achievement? Or, does teacher level of enthusiasm (a second-level variable) interact with student motivation to produce differences in student achievement? Or, does school-level poverty interact with teacher enthusiasm to produce differences in student achievement? Addition of such cross-level interactions to the multilevel regression equation is discussed in Section 15.6.3. Cross-level interactions may be especially interesting in the context of experiments where the treated (as opposed to the control) group displays a different relationship between a predictor and the DV. Cohen et al. (2003) discuss an example in which the treatment moderates the relationship between weight loss (the DV) and motivation (a predictor)—treatment gives more highly motivated participants the means for effective dieting.

15.2.4   Meta-Analysis

MLM provides a useful strategy for meta-analyses in which the goal is to compare many studies from the literature that address the same outcome. For example, there may be hundreds of studies evaluating various aspects of student achievement. Original raw data usually are not available, but statistics for numerous studies are available in the form of effect sizes, p values, and often means and standard deviations. A common outcome measure is derived for the various studies, often a standardized effect size for the outcome measure (student achievement). When these problems are addressed through MLM, individual studies provide the lowest level of analysis. A simple analysis (Section 15.4.1) determines whether there are significant differences among studies in effect size. IVs (such as student motivation, teacher enthusiasm, or school poverty level) are then investigated to try to determine whether differences in the various studies are predicted by those IVs (cf. Hox, 2002, Chapter 6).

15.2.5   Relative Strength of Predictors at Various Levels

What is the relative size of the effect for individual-level variables versus group-level variables? Or, are interventions better aimed at the individual level or the group level? For example, if there is to be an intervention should it be directed at the motivation of individual students or the enthusiasm levels of teachers? Analytic techniques are available through SEM (cf. Chapter 14) to evaluate the relative strengths of individual versus group effects. Hox (2002), as well as Heck and Thomas (2000), demonstrates such multilevel factor and path analyses.

15.2.6   Individual and Group Structure

Is the factor structure of a model the same at the individual and group level? Do individual students and teachers have the same pattern of responses to a questionnaire? That is, do the same items regarding homework, extra curricular activities, and the like load on the same factors at the individual and group levels? These and similar questions can be answered through application of SEM techniques to analysis of covariance structures (variance-covariance matrices) aimed at data at the individual level and the group level. Section 15.5.3 discusses these models.

15.2.7   Path Analysis at Individual and Group Levels

What is the path model for prediction of the DV from level-one variables, level-two variables, and level-three variables? For example, what is the path model for predicting student achievement from student-level variables (e.g., student motivation, study time, and gender), teacher/classroom-level variables (teacher enthusiasm and teacher emphasis on homework), and school-level variables (poverty level, type of school, and school size)? Hox (2002) provides an example of this type of analysis in an educational setting. All of the power of path analysis and, indeed, latent factor analysis can be tapped by the application of SEM techniques to multilevel data.

15.2.8   Analysis of Longitudinal Data

What is the pattern of change over time on a measure? Do students show a linear trend of improvement over the school year or do improvements level off after a while? Do individuals differ in their trend of improvement (growth curves) over time? There are two MLM techniques that address this type of question without the restrictive assumptions of repeated-measures ANOVA: (1) direct application of MLM with occasions as the lowest level of analysis, and (2) latent growth modeling using the techniques of SEM (cf. Chapter 14). Section 15.5.1 demonstrates the first application and discusses the second. Section 15.7 provides a complete example of a three-level repeated-measures model through MLM techniques.

15.2.9   Multilevel Logistic Regression

What is the probability of a binary outcome when individuals are nested within several levels of a hierarchy? For example, what is the probability that a student will be retained when students are nested within classrooms and classrooms are nested within schools? Nonnormal, including binary, outcomes are discussed in Section 15.5.4.

15.2.10   Multiple Response Analysis

What are the effects of variables at different levels on multiple DVs at the individual level? For example, what are the effects of predictors at the student level, the teacher/classroom level, and the school level on several different types of student achievements (achievement in reading, achievement in math, achievement in problem solving, and so on—the DVs)? In these analyses, the multivariate DVs are presented as the lowest level of analysis. Section 15.5.5 discusses the multivariate form of MLM.

15.3   Limitations to Multilevel Linear Modeling

15.3.1   Theoretical Issues

Correlated predictors are even more problematic in MLM than in simple linear regression. In MLM, equations at multiple levels are solved and correlations among predictors at all levels are taken into account simultaneously. Because effects of correlated predictors are all adjusted for each other, it becomes increasingly likely that none of their regression coefficients will be statistically significant. The best advice, then, is to choose a very small number of relatively uncorrelated predictors. A strong theoretical framework helps limit the number of predictors and facilitates decisions about how to treat them.

If interactions are formed (Section 15.6.3), predictors in the interactions are bound to be correlated with their main effects. The problem of multicollinearity between interactions and their main effects can be solved by centering (see Section 15.6.2).

Raudenbush and Bryk (2001) recommend a build-up strategy for MLM analyses. First, a series of standard multiple regression analyses are run, starting with the most interesting (or theoretically important) predictor and adding predictors in order of importance. Then, predictors that do not enhance prediction are dropped unless they are components of cross-level interactions. This strategy is discussed more fully in Section 15.6.8.

Modeling high-level predictors at their own level is not always the best way to deal with them. If there are only a few of them, they are often best entered at the next lower level as categorical predictors. For example, if there are only a few schools, with several classrooms from each and many students in each class, then school can be considered a categorical predictor at the classroom (second) level of analysis, rather than a third level of analysis. The problem with considering schools at the third level is that there are too few of them to generalize to a population of schools (Rasbash et al., 2000).

15.3.2 Practical Issues

Multilevel linear modeling is an extension of multiple linear regression, so the limitations and assumptions of Section 5.2 apply to all levels of the analysis. Thus, conformity with distributional assumptions and outliers in the data and in the solution are considered using methods for multiple regression. The assumptions are evaluated for the set of predictors at each level and also for sets of predictors that are used within cross-level interactions. Raudenbush and Bryk (2001) recommend using exploratory multiple regression analyses to look for outliers among first-level predictors within second-level units. For the example, univariate and multivariate outliers are sought for student achievement, student motivation, study time, and student gender within each classroom. Ideally, all screening of first-level predictors should be within second-level units; however, this may be impractical when the number of second-level units is very large. In that case, they may be combined over the second-level units. Similarly, second-level predictors are examined within third-level units if possible, if not they are aggregated over the third-level units. Programs for MLM provide analyses of residuals or permit residuals to be saved to a file for analysis in other modules.

MLM uses maximum likelihood techniques, which pose some further problems, and the use of multiple levels creates additional complications, so issues of sample size and multicollinearity need to be addressed differently from garden-variety multiple regression. MLM also addresses independence of errors differently.

15.3.2.1 *Sample Size, Unequal-*n*, and Missing Data*

The price of large, complex models is instability and a requirement for a substantial sample size at each level. Even small models, with only a few predictors, grow rapidly as equations are added at higher levels of analysis. Therefore, large samples are necessary even if there are only a few predictors.

As a maximum likelihood technique, a sample size of at least sixty is required if only five or fewer parameters are estimated (Eliason, 1993). Parameters to be estimated include intercepts and slopes as well as effects of interest at each level. In practice, convergence often is difficult even with larger sample sizes. And you may be unpleasantly surprised at the lengthy processing time on even speedy computers when dealing with MLM.

Unequal sample sizes at each of the levels pose no problems and are, indeed, expected. Missing values can be tolerated in repeated-measures analyses (unlike the requirement for complete data at all levels of the repeated-measures IV in ANOVA [Rasbash et al., 2000, pp. 129–130]). In MLM, for example, some of the occasions for measurement may be missing for one or more cases. In other, nonrepeated measures designs, missing data are estimated using the techniques of Section 4.1.3 and inserted at the appropriate level of analysis. Group sizes may be as small as one, as long as other groups are larger (Snijders & Bosker, 1999). Group size itself predicts the DV, as in the example in which school size is one of the predictors.

As in most analyses, increasing sample sizes increases power while smaller effect sizes and larger standard errors decrease power. There are other issues that affect power in MLM, however, and their effects are not so easily predicted (Kreft & DeLeeuw, 1998). For example, power depends on compliance with assumptions of the analysis; with each type of assumption leading to a different relationship between the probability of rejecting the null hypothesis and the true effect size. Power issues also differ for first- versus higher-level effects, and whether effects are considered fixed or random (with tests of random effects usually less powerful because standard errors are larger). For example, Kreft and DeLeeuw (1998) conclude that power grows with the intraclass correlation (difference between groups relative to differences within groups, Section 15.6.1), especially for tests of second-level effects and cross-level interactions. Sufficient power for cross-level effects is obtained when sample sizes at the first level are not too small and the number of groups is twenty or larger. In general, simulation studies show that power is greater with more groups (second-level units) and fewer cases per group (first-level units) than the converse, although more of both leads to increased power. Hox (2002) devotes an entire chapter to power and sample size issues and provides guidelines for a simulation-based power analysis. Software for determining power and optimal sample size in MLM is available as a free download from Scientific Software International (Raudenbush, Liu, & Congdon, 2005).

15.3.2.2   *Independence of Errors*

MLM is designed to deal with the violation of the assumption of independence of errors expected when individuals within groups share experiences that may affect their responses. The problem is similar to that of heterogeneity of covariance (sphericity), in that events that are close in time are more alike than those farther apart. In MLM it usually is the individuals within groups who are closer to each other in space and experiences than to individuals in other groups. Indeed, when the lowest level of the hierarchy is repeated measures over time, multilevel modeling provides an alternative to the assumption of heterogeneity of covariance (sphericity) required in repeated-measures ANOVA.

The intraclass correlation (ρ, Section 15.6.1) is an explicit measure of the dependence of errors because it compares differences between groups to individual differences within groups. The larger the D, the greater the violation of independence of errors and the greater the inflation of Type I error rate if the dependence is ignored. If multilevel data are analyzed using non-MLM statistics and there is dependence of errors, Type I error can be dramatically increased. Barcikowski (1981) shows that Type I error rates at a nominal .05 level can be as high as .17 when group sample size is 100 and the intraclass correlation is as small as .01; the Type I error rate rises to .70 when the intraclass correlation is .20. Thus, significant effects of treatment cannot be trusted if independence of errors is assumed without justification. Instead, the hierarchical structure of the data must be considered when choosing the appropriate analysis.

15.3.2.3 *Absence of Multicollinearity and Singularity*

Collinearity among predictors is especially worrisome when cross-level interactions are formed, as is common in MLM, because these interactions are likely to be highly correlated with their component main effects. At the least, the problem often leads to a failure of significance for the main effect(s). At worst, multicollinearity can cause a failure of the model to converge on a solution. The solution is to center predictors, a practice with additional benefits, as discussed in Section 15.6.2.

15.4   Fundamental Equations

Small data sets are difficult to analyze through MLM. The maximum likelihood procedure fails to converge with multiple equations unless samples are large enough to support those equations. Nor is it convenient to apply MLM to a matrix (or a series of matrices). Therefore, we depart from our usual presentation of a small sample data set or a matrix of sample correlations and, instead, use a portion of a data set developed by others, with the names of variables changed to reflect our usual silly research applications.[4]

   Two ski resorts provide a total of 10 ski runs; one at Aspen Highlands (mountain = 1) and nine at Mammoth Mountain (mountain = 0). There are 260 skiers in all, with different skiers on each run at each mountain. Table 15.1 shows the data for a single run, labeled 7472, at Mammoth Mountain. The dependent variable is speed of skiing a run, with skill level of the skier as a level-one predictor.[5] The column labeled "skill deviation" is the skill score for each skier minus the average skill for skiers on that run (used in some later analyses).

   The hierarchy to be modeled is composed of skiers at the first level and runs (the grouping variable) at the second level. Skiers and runs are considered random effects. Skiers are nested within runs and runs are nested within mountains; however, with only two mountains, that variable is considered a fixed predictor at the second-level rather than specifying an additional third level of analysis. The major question is whether speed of skiing varies with mountain, after adjusting for average skier skill, differences in speed among runs, and differences in the relationship between skill and speed among runs.

   MLM often is conducted in a sequence of steps. Therefore, the equations and computer analyses are divided into three models of increasing complexity. The first is an intercepts-only ("null") model in which there are no predictors and the test is for mean differences between runs (groups— considered random) on the DV (skiing speed). The second is a model in which the first-level predictor, skill, is added to the intercepts-only model. The third is a model in which the second-level predictor, mountain, is added to the model with the first-level predictor. Table 15.2 identifies the 10

---

[4]The data set is a selection by Kreft and DeLeeuw (1998) of 260 cases from the NELS-88 data collected by the National Center for Educational Statistics of the US Department of Education. Actual variables are public vs. private sector schools for mountain, school for run, math achievement (rounded) for speed, and homework for skill level.

[5]The term covariate is used generically for predictors in MLM as in survival analysis. In this chapter we use the terms CVs for continuous predictors and IVs for categorical predictors; the term "predictors" here refers generically to a combination of CVs and IVs and/or random effects.

TABLE 15.1   Partial Listing of Sample Data

Run	Skier	Skill	Mountain	Speed	Skill Deviation
7472	3	1	0	5	−0.39
7472	8	0	0	5	−1.39
7472	13	0	0	6	−1.39
7472	17	1	0	5	−0.39
7472	27	2	0	5	0.61
7472	28	1	0	6	−0.39
7472	30	5	0	4	3.61
7472	36	1	0	7	−0.39
7472	37	1	0	4	−0.39
7472	42	2	0	6	0.61
7472	52	1	0	5	−0.39
7472	53	1	0	5	−0.39
7472	61	1	0	5	−0.39
7472	64	2	0	4	0.61
7472	72	1	0	6	−0.39
7472	83	4	0	4	2.61
7472	84	1	0	5	−0.39
7472	85	2	0	5	0.61
7472	88	1	0	5	−0.39
7472	93	1	0	5	−0.39
7472	94	1	0	4	−0.39
7472	96	1	0	5	−0.39
7472	99	1	0	5	−0.39

TABLE 15.2   Intercepts and Slopes for 10 Ski Runs in Sample Data

Run	Mountain	Intercept	Slope	Sample Size
7472	0	5.46835	−0.30538	23
7829	0	5.49309	−0.29493	20
7930	0	4.46875	0.81250	24
24725	0	3.92664	0.60039	22
25456	0	5.84116	−0.44765	22
25642	0	5.55777	−0.35458	20
62821	1	6.43780	0.11162	67
68448	0	4.09225	0.66052	21
68493	0	4.26857	0.62000	21
72292	0	4.25258	0.65464	20

runs (the grouping variable), the mountain for each run, and the intercept (mean speed) and slope of the relationship between skill and speed from separate bivariate regressions for each run, as well as sample size.

Table 15.3 summarizes the symbols that are typically used in MLM texts (e.g., Kreft & DeLeeuw, 1998) and software as a reference for demonstrating equations for the three models.

Prior to an MLM analysis, a number of choices have to be made. First is a decision about which predictors, if any, are to be included. Second is the choice between "fixing" value of a "parameter" to

TABLE 15.3   Equations and Symbols Typically Used in MLM

Symbol	Meaning
Level 1 Equation	$Y_{ij} = \beta_{0j} + \beta_{1j}(X_{ij}) + e_{ij}$
Y_{ij}	The DV score for a case at Level 1, i indexes the individual within a group, j indexes the group
X_{ij}	A Level-1 predictor
β_{0j}	The intercept for the DV in group j (Level 2)
β_{1j}	The slope for the relationship in group j (Level 2) between the DV and the Level-1 predictor
e_{ij}	The random errors of prediction for the Level-1 equation (sometimes called r_{ij})
	At Level 1, both the intercepts and the slopes in the j groups can be: 1. Fixed (all groups have the same values, but note that fixed intercepts are rare) 2. Nonrandomly varying (the intercepts and/or the slopes are predictable from an IV at Level 2) 3. Randomly varying (the intercept and/or the slopes are different in the different j groups, each with an overall mean and a variance)
Level 2 Equations	The DVs are the intercepts and slopes for the IV-DV Level-1 relationships in the j groups of Level 2. $$\beta_{0j} = \gamma_{00} + \gamma_{01}W_j + u_{0j}$$ $$\beta_{1j} = \gamma_{10} + u_{1j}$$
γ_{00}	The overall intercept; the grand mean of the DV scores across all groups when all predictors $= 0$
W_j	A Level-2 predictor
γ_{01}	The overall regression coefficient for the relationship (slope) between a Level-2 predictor and the DV
u_{0j}	Random error component for the deviation of the intercept of a group from the overall intercept; the unique effect of Group j on the intercept
γ_{10}	The overall regression coefficient for the relationship (slope) between a Level-1 predictor and the DV
u_{1j}	An error component for the slope; the deviation of the group slopes from the overall slope. Also the unique effect of Group j on slope when the value of the Level-2 predictor W is zero

(continued)

TABLE 15.3   Continued

Symbol	Meaning
Combined Equation with Cross-Level Interaction	$Y_{ij} = \gamma_{00} + \gamma_{01}W_j + \gamma_{10}X_{ij} + \gamma_{11}W_jX_{ij} + u_{0j} + u_{1j}X_{ij} + e_{ij}$
$\gamma_{01}W_j$	A Level-2 regression coefficient (γ_{01}) times a Level-2 predictor
$\gamma_{10}X_{ij}$	A Level-2 regression coefficient (γ_{10}) times a Level-1 predictor
$\gamma_{11}W_jX_{ij}$	A Level-2 regression coefficient (γ_{11}) times the cross-product of the Level-2 and Level-1 predictors; the cross-level interaction term
$u_{0j} + u_{1j}X_{ij}$ $+ e_{ij}$	The random error components for the combined equation
Variance Components	
τ (tau)	Variance-covariance matrix for the estimates of the values of random error components
τ_{00}	Variance among random intercepts (means)
τ_{11}	Variance among random slopes
τ_{10}	Covariance between slopes and intercepts

a constant over all groups or letting the value be a random effect (a different value for each group). For example, is the intercept to be considered a fixed effect over all groups or will it be allowed to vary over the groups. Typically, even random effects have their fixed component. That is, we are interested in looking at the overall mean of the DV (fixed effect, γ_{00}) as well as the difference in means over higher-level units (random effect, τ_{00}). Thus, we will have two parameters to estimate for the intercept: the fixed mean and the random variability in means. For predictors, we sometimes are interested in their variability in relationship with the DV over higher-level units (a random effect, e.g., τ_{11}) but are interested in their average relationship as well (a fixed effect, e.g., γ_{01}). This decision is made separately for each predictor at each level except that highest-level predictors may not be considered random because there is no higher level within which they can vary. A decision also is made as to whether to evaluate the covariance between slopes and intercepts (a random effect, e.g., τ_{10}) for each predictor that is considered random. Then, a decision is made to as to whether to evaluate covariance among slopes of different random-effect predictors if there is more than one (not shown in Table 15.3 because there is only one first-level predictor and the second-level predictor may not vary).

The parameters are the elements of Table 15.3 that are to be estimated (γs and τs as well as e_{ij}). Finally, there is the choice of the type of estimation to use (e.g., maximum likelihood or restricted maximum likelihood).

15.4.1   Intercepts-Only Model

The MLM is expressed as a set of regression equations. In the level-1 equation for the intercepts-only model (a model without predictors), the response (DV score) for an individual is predicted by an intercept that varies across groups. The intercepts-only model is of singular importance in MLM, because it provides information about the intraclass correlation (Section 15.6.1), a value helpful in determining whether a multilevel model is required.

15.4.1.1 The Intercepts-Only Model: Level-1 Equation

$$Y_{ij} = \beta_{0j} + e_{ij} \tag{15.1}$$

An individual score on the DV, Y_{ij}, is the sum of an intercept (mean), β_{0j}, that can vary over the j groups and individual error, e_{ij} (the deviation of an individual from her or his group mean).

The terms Y and e that in ordinary regression have a single subscript i, indicating case, now have two subscripts, ij, indicating case and group. The intercept, β_0 (often labeled A in a regression equation; cf. Sections 3.5.2 and 5.1) now also has a subscript, j, indicating that the coefficient varies over groups. That is, each group could have a separate Equation 15.1.

β_{0j} is not likely to have a single value because its value depends on the group. Instead, a parameter estimate (τ_{00}) and standard error are developed for the *variance* of a random effect. The parameter estimate reflects the degree of variance for a random effect; large parameter estimates reflect effects that are highly variable. For example, the parameter estimate for the groups represents how discrepant they are in their means. These parameter estimates and their significance tests are shown in computer runs to follow. The z test of the random component, τ_{00} divided by its standard error, evaluates whether the groups vary more than would be expected by chance.

15.4.1.2 The Intercepts-Only Model: Level-2 Equation

The second-level analysis (based on groups as research units) for the intercepts-only model uses the level-1 intercept (group mean) as the DV. To predict the intercept for group j:

$$\beta_{0j} = \gamma_{00} + u_{0j} \tag{15.2}$$

An intercept for a run β_{0j}, is predicted from the average intercept over groups when there are no predictors, γ_{00}, and group error, u_{0j} (deviation from average intercept for group j).

A separate Equation 15.2 could be written for each group. Substituting right-hand terms from Equation 15.2 into Equation 15.1:

$$Y_{ij} = \gamma_{00} + u_{0j} + e_{ij} \tag{15.3}$$

The average intercept (mean) is γ_{00} (a "fixed" component). The two random components are u_{0j} (the deviation in intercept for cases in group j) and e_{ij} (the deviation for case i from its group j).

The two-level solution to the intercepts-only model for the sample data is

$$Speed_{ij} = 5.4 + u_{0j} + e_{0j}$$

The unweighted mean speed (average of the means for each of the 10 groups) is 5.4. Thus, the skiing speed for an individual skier is the grand mean for all groups (5.4) plus the deviation of the skier's group from the grand mean plus the deviation of an individual's speed from his or her group. Those familiar with ANOVA recognize this way of thinking about an individual's score; it is conceived as the grand mean plus the deviation of the individual's group mean from the grand mean plus the deviation of the individual's score from her or his group mean.

In MLM, components that resolve to constants are called "fixed." In Equation 15.3, the grand mean is a fixed component and also an estimated parameter (5.4) with its own standard error. The ratio of the parameter estimate to its standard error is evaluated as a two-tailed z test at a predetermined α level. Here, the test is that the grand mean for speed differs from zero and is uninteresting; it is akin to the test of the intercept in standard bivariate or multiple regression. More interesting are the tests of the variances (u_{0j} and e_{0j}), shown in computer runs.

15.4.1.3 Computer Analyses of Intercepts-Only Model

Tables 15.4 and 15.5 show syntax and selected output for computer analysis of the data described in Section 15.4.1 through SAS MIXED and SPSS MIXED.

As seen in Table 15.4, SAS MIXED produces a "null" solution in a single level-2 run. The usual **model** instruction declares SPEED to be the DV; there is nothing after the "=" because there are no predictors in this model. The request for **solution** provides parameter estimates and significance tests for fixed effects. The **covtest** instruction provides hypothesis testing of the variance and covariance components of the random errors in the model. The maximum likelihood method (**method=ml**) has been chosen. Research units (**subject**) for the random part of the model are RUNs, which is identified as a **class** (categorical) variable. This indicates how the level-2 units (runs) are formed from the level-1 units (skiers). The **random** instruction sets the group **intercept** to be random. The fixed effect of intercept (the grand mean combined over groups) is included implicitly. The **type=un** instruction indicates that there are no assumptions made about the structure of the variance-covariance matrix (e.g., no assumption of sphericity).

The **Dimensions** section provides information useful for comparing models by showing the number of parameters in the model. **Covariance Parameters** refers to the two random effects in the model: group intercepts and residual. **Columns in X** refers to the single fixed effect in the model at this point: overall intercept. Thus, the total number of parameters for the model is 3, the two random effects and one fixed effect.

The **Covariance Parameter Estimates** section applies to the random components in the model. **UN(1,1)** is the variance in intercepts across runs (τ_{00} of Table 15.3), and with a one-tailed test (appropriate for testing whether intercepts vary *more* than expected by chance) at $\alpha = .05$ (critical value = 1.58) there is evidence that, indeed, the intercepts vary. This suggests the desirability of taking group differences into account when predicting speed.[6] The significant Residual indicates that there are individual differences among skiers within runs after accounting for differences between runs.

Fit statistics are useful for comparing models (recall Section 10.4.3). The **Null Model Likelihood Ratio Test** indicates that the data with groups identified differ significantly from a model with just a single fixed intercept.

---

[6]It is important also to evaluate the intraclass correlation, a measure of the strength of the between-group differences (Section 15.6.1), because inferential tests are highly influenced by sample size.

TABLE 15.4   Syntax and Selected Output for SAS MIXED Analysis of Intercepts-Only Model

```
proc mixed data=Sasuser.Ss_hlm covtest method=ml;
    class RUN;
    model SPEED= / solution;
    random intercept / type=un subject=RUN;
run;
```

Dimensions

Covariance Parameters	2
Columns in X	1
Columns in Z Per Subject	1
Subjects	10
Max Obs Per Subject	67

Covariance Parameter Estimates

Cov Parm	Subject	Estimate	Standard Error	Z Value	Pr Z
UN(1,1)	RUN	0.3116	0.1484	2.10	0.0178
Residual		0.7695	0.06872	11.20	<.0001

Fit Statistics

-2 Log Likelihood	693.5
AIC (smaller is better)	699.5
AICC (smaller is better)	699.6
BIC (smaller is better)	700.4

Null Model Likelihood Ratio Test

DF	Chi-Square	Pr > ChiSq
1	110.16	<.0001

Solution for Fixed Effects

Effect	Estimate	Standard Error	DF	t Value	Pr > |t|
Intercept	5.4108	0.1857	9	29.13	<.0001

The remaining output is for the fixed effects in the model. The $\mathtt{Intercept} = 5.4108$ is the unweighted mean of the 10 groups (mean of the 10 means, γ_{00} of Table 15.3). The fact that it differs significantly from zero is of no research interest.

Table 15.5 shows syntax and output for SPSS MIXED analysis. The DV is shown in the first line of syntax as speed; nothing more is specified when there are no predictors. The syntax for the FIXED equation also shows no predictors; the test for the fixed effect of the overall intercept is included by default. Method chosen is ML rather than the default REML. The PRINT instruction requests

TABLE 15.5   Syntax and Output for SPSS MIXED Analysis of Intercepts-Only Model

```
MIXED
  speed
  /CRITERIA = CIN(95) MXITER(100) MXSTEP(5) SCORING(1)
  SINGULAR(0.000000000001) HCONVERGE(0, ABSOLUTE) LCONVERGE(0, ABSOLUTE)
  PCONVERGE(0.000001, ABSOLUTE)
  /FIXED = | SSTYPE(3)
  /METHOD = ML
  /PRINT = SOLUTION TESTCOV
  /RANDOM INTERCEPT | SUBJECT(run) COVTYPE(VC) .
```

Mixed Model Analysis

Model Dimension[b]

		Number of Levels	Covariance Structure	Number of Parameters	Subject Variables
Fixed Effects	Intercept	1		1	
Random Effects	Intercept[a]	1	Variance Components	1	run
Residual				1	
Total		2		3	

[a]As of version 11.5, the syntax rules for the RANDOM subcommand have changed. Your command syntax may yield results that differ from those produced by prior versions. If you are using SPSS 11 syntax, please consult the current syntax reference guide for more information

[b]Dependent Variable: speed.

Information Criteria[a]

−2 Log Likelihood	693.468
Akraike's Information Criterion (AIC)	699.468
Hurvich and Tsai's Criterion (AICC)	699.562
Bozdogan's Criterion (CAIC)	713.150
Schwarz's Bayesian Criterion (BIC)	710.150

The information criteria are displayed in smaller-is-better forms.

[a]Dependent Variable: speed.

TABLE 15.5 Continued

Fixed Effects

Type III Tests of Fixed Effects[a]

Source	Numerator df	Denominator df	F	Sig.
Intercept	1	10.801	848.649	.000

[a]Dependent Variable: SPEED.

Estimates of Fixed Effects[a]

Parameter	Estimate	Std. Error	df	t	Sig.	95% Confidence Interval	
						Lower Bound	Upper Bound
Intercept	5.4108323	.1857377	10.797	29.132	.000	5.0010877	5.8205769

[a]Dependent Variable: SPEED.

Covariance Parameters

Estimates of Covariance Parameters[a]

Parameter	Estimate	Std. Error	Wald Z	Sig.	95% Confidence Interval	
					Lower Bound	Upper Bound
Residual	.7694610	.0687191	11.197	.000	.6459031	.9166548
Intercept [subject = run] Variance	.3115983	.1483532	2.100	.036	.1225560	.7922380

[a]Dependent Variable: speed.

parameter estimates and tests for the single fixed effect (**SOLUTION**) and the two random effects (**COVTEST**). The **RANDOM** equation explicitly lists the group intercept. Runs are declared to be "subjects" (i.e., units of analysis at 2nd level). Remaining syntax is produced by the menu system.[7]

The output begins with an indication of fixed and random effects, the type of covariance structure for the random effects, number of parameters (cf. Section 15.6.5.1), and the variable used to combine subjects (run). Helpfully, total number of parameters also is shown. Then information is provided that is useful to test differences between models, such as **−2 Log Likelihood**.

Fixed effects follow, with two tables: one for significance tests of combined effects and another for parameter estimates, single-df tests, and confidence intervals. Results match those of SAS, except for denominator df. The presence of random effects and varying levels can complicate the calculation of denominator df for fixed effects. SPSS applies corrections for random effects and multiple levels even in the simplest models, resulting in fractional values for denominator df. SAS does not apply these corrections unless requested (see Table 15.10).

The output concludes with a table for random effects, labeled **Estimates of Covariance Parameters**. The test for **Intercept** (whether runs differ in mean speed) is statistically significant at $z = 2.100$. Note that the **Sig.** value of .036 is for a two-tailed test rather than the more appropriate one-tailed test; we are concerned only with whether intercepts vary *more* than would be expected by chance. Therefore the appropriate p value for comparison is $.036/2 = .018$.

Table 15.6 summarizes the parameter estimates for both random and fixed effects and their interpretation.

TABLE 15.6   Summary of Symbols and Interpretations for Intercepts-Only Model

Parameter Estimate for Effect and Software Label	Symbol from Table 15.3	Sample-specific Interpretation	Generalized Interpretation
Random Effects (Covariance Parameter Estimates)			
Value = 0.3116 SPSS: Intercept [subject = run] Variance SAS:UN(1,1)	τ_{00}	The variance in the means of speed for the runs around the grand mean of speed	The variance in the group means on the DV around the grand mean on the DV (variance between groups)
Value = 0.7695 SPSS: Residual SAS: Residual	e_{ij}	Variance in speed for individual skiers within runs around the mean speed for the run	The variance among cases on the DV within groups around their own group means (variance within groups)
Fixed Effect (Parameter Estimate)			
Value = 5.4108 SPSS: Intercept SAS: Intercept	γ_{00}	The unweighted grand mean of speed for the runs	The overall intercept: unweighted mean of the means for the groups

[7]Note that there was a change in specification of COVTYP beginning with Version 11.5 of SPSS.

15.4.2   Model with a First-Level Predictor

The next model is one in which a predictor, skill of skier, is added to the equations to predict skiing speed.

15.4.2.1   Level-1 Equation for a Model with a Level-1 Predictor

The level-1 equation is now expanded so that the DV score for an individual is predicted by a random intercept that varies across groups (as in the previous section) and a random slope for the relationship between the DV and the level-1 predictor (that also varies across groups).

$$Y_{ij} = \beta_{0j} + \beta_{1j}X_{ij} + e_{ij} \tag{15.4}$$

An individual score on the DV, Y_{ij}, is a sum of an intercept, β_{0j}, that may vary over the j groups, a slope, β_{1j}, that may vary over groups times an individual's score on a predictor, X_{ij},[8] and error, e_{ij} (the deviation of an individual from his or her group mean).

Except for the subscripts, this is highly similar to the usual bivariate regression equation ($Y_i = \beta_0 + \beta_1 X_i + e_i$). All of the terms ($Y$, X, and e) that ordinarily would have a single subscript, i, indicating case, now have two subscripts, ij, indicating case and group. The regression coefficients, β_0 and β_1, now also have a subscript, j, indicating that each of these coefficients varies over groups.
  In terms of the example,

$$\text{Speed}_{ij} = \beta_{0j} + \beta_{1j}\text{Skill}_{ij} + e_{ij} \tag{15.5}$$

An individual's skiing speed is the sum of the intercept, β_{0j}, for that skier's run (group); a weighting, β_{1j}, for that skier's run (group) times the skier's skill level; and error, e_{ij} (deviation of individual score from its run).

Coefficients that vary across groups are treated as random. Thus, this is often called a random coefficients model, referring to the random coefficients for the intercept, β_{0j}, and for the slope, β_{1j}, that vary over groups (runs). Error always is considered random. These varying coefficients can be illustrated in scatterplots between skill and speed that differ between runs. For example, Figure 15.1 shows the skill-speed relationships for the run at Aspen Highlands and for the first run at Mammoth.
  There is a small positive relationship between skill and speed at the Aspen Highlands run, with an intercept at about 6.25, but a negative relationship for the first run at Mammoth Mountain, with an intercept at about 5.8. Thus, both the intercepts and slopes vary for these runs. A variety of other slopes and intercepts are noted for the remaining Mammoth runs, as seen in Figure 15.2, which shows all ten of the runs. You may recognize this as a failure of homogeneity of regression, as discussed in Chapter 6 (cf. Figure 6.2). That is, there is an interaction between skill (the predictor) and runs (the groups) on speed (the DV): slopes vary over runs.
  The varying intercepts and slopes are found through a separate regression analysis for each higher-level research unit, in this case each run. For the data in Table 15.1, a bivariate regression analysis is done, per Equations 3.30 through 3.32 with skill as X and speed as Y. The resulting slopes

---

[8]X is sometimes centered, for example, a deviation from the mean for the group may be used (cf. Section 15.6.2).

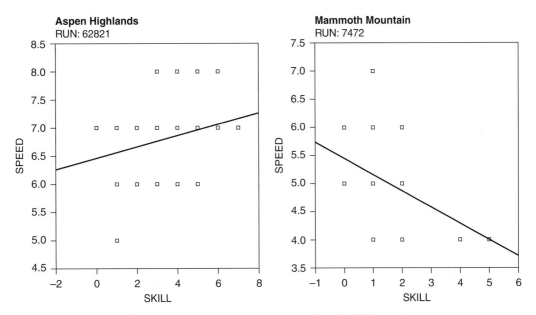

FIGURE 15.1 Relationships between speed and skill for two runs.
Generated in SYSTAT PLOT.

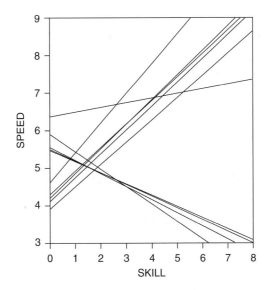

FIGURE 15.2 Relationship between speed and skill for all ten runs.
Generated in SYSTAT PLOT.

and intercepts are shown in Table 15.2. The procedures of Section 5.4 apply if there are several level-1 predictors.

15.4.2.2 Level-2 Equations for a Model with a Level-1 Predictor

Each random effect (other than individual error) requires a separate second-level equation. Therefore, the second-level analysis (where groups are research units) requires two equations with level-1 intercept and level-1 slope as DVs (see Table 15.2). To predict the random intercept:

$$\beta_{0j} = \gamma_{00} + u_{0j} \tag{15.6}$$

An intercept for a run β_{0j}, is predicted by the average intercept over groups when all predictors are zero, γ_{00} (a fixed effect), and error, u_{0j} (a random effect: deviation from average intercept for group j).

Note that this is the same as Equation 15.2 for the intercepts-only model.
  To predict the random slope:

$$\beta_{1j} = \gamma_{10} + u_{1j} \tag{15.7}$$

The slope for a run, β_{1j}, is predicted by a single intercept, γ_{10} (a fixed effect: the average IV-DV slope), and error, u_{1j} (a random effect: the deviation from average slope for group j).

Substituting right-hand terms from Equations 15.6 and 15.7 into Equation 15.4:

$$Y_{ij} = \gamma_{00} + u_{0j} + (\gamma_{10} + u_{1j})X_{ij} + e_{ij} \tag{15.8}$$

Thus the entire two-level equation, with terms rearranged, is:

$$Y_{ij} = \gamma_{00} + \gamma_{10}X_{ij} + u_{0j} + u_{1j}X_{ij} + e_{ij} \tag{15.9}$$

The two fixed components are γ_{00} (the average intercept) and γ_{10} (the average slope). The three random components are u_{0j} (the deviation in intercept for cases in group j), u_{1j} (the deviation in slope for cases in group j) times the first-level predictor score for case i in group j, and e_{ij} (the deviation for case i from its group j).

The solution to Equation 15.5 produced by SAS and SPSS software is

$$\beta_{0j} = 4.981 + u_{0j}$$

The intercept = 4.981 and is the unweighted mean for the 10 runs when predicted skill = 0. As usual, there is no research import to its test of significance.
  The solution to Equation 15.6 produced by software is

$$\beta_{1j} = 0.216 + u_{1j}$$

With a standard error = 0.151 for the average slope, $z = 0.216/0.151 = 1.43$. Thus, there is no evidence at $\alpha = .05$ of a relationship between the DV, speed, and the level-1 predictor, average skill. That is, skiing speed cannot be predicted from skill when it is averaged over all runs.

The entire two-level solution is:

$$Y_{ij} = 4.981 + 0.216(\text{Skill}) + (\beta_{0j} - 4.981) + (\beta_{1j} - 0.216)(\text{Skill}) + e_{ij}.$$

Recall that β_{0j} is the intercept for group j and β_{1j} is the slope for group j. As described for the intercepts-only model, parameter estimates for fixed effects are based on average values over runs, so that tests for them are tests of central tendency, using two-tailed z tests. Tests of some random effects are based on variances as parameter estimates and typically are tested using one-tailed z tests; is the variance greater than zero? That is, do intercepts and slopes vary more than expected over runs? Tests of whether intercepts and slopes differ among runs are shown in the following computer analyses, as is the test of whether slopes and intercepts are correlated (not included in these equations).

15.4.2.3   Computer Analysis of a Model with a Level-1 Predictor

The SAS MIXED **model** instruction in Table 15.7 declares SPEED to be the DV, with SKILL as an IV. Note that SKILL has been added to group intercept as a **random** effect and also has been included in the **model** instruction as a fixed effect. Remaining syntax is as for the intercepts-only model.

As before, the **Covariance Parameter Estimates** section applies to the random components in the model. **UN(1,1)** is the variance in group intercepts (means) across runs (τ_{00} of Table 15.3), and with a one-tailed test at $\alpha = .05$ (critical value = 1.58) there is still evidence that, indeed, the group intercepts vary across runs. **UN(2,1)** is the covariance between group intercepts and group slopes (τ_{10} of Table 15.3); there is no indication of a relationship between intercepts and slopes over runs (**Pr Z** = .0692). That is, there is no evidence that the effects of skill on speed differ depending on average speed on the run. Note that this is tested with a two-tailed probability value, because covariances can be either negative or positive. **UN(2,2)** is the variance in group slopes across runs (τ_{11} of Table 15.3); there is evidence that the relationship between skill and speed differs among runs, so a fixed effect of SKILL may not be interpretable. The significant **Residual** indicates that there are individual differences among skiers within runs after accounting for differences due to group membership and to skill. Note that the **−2 Log Likelihood** value is smaller than that of the intercepts-only model of Table 15.4. This difference can be evaluated through χ^2 to provide a test of model improvement by the addition of skill as a predictor, as seen in Section 15.6.5.1. The **Null Model Likelihood Ratio Test** indicates that the specified model differs significantly from a model with just a single fixed intercept.

The remaining output is for the fixed effects in the model. The **Intercept** = 4.9808 is the mean of the 10 groups when predicted skill = 0. The overall relationship between SPEED and SKILL is not statistically significant ($p = .1868$) but the result cannot be interpreted due to the significant random variance in slopes across runs, **UN(2,2)**.

Table 15.8 shows syntax and output for SPSS MIXED analysis. The continuous predictor, skill, is shown as a **WITH** variable (called a covariate on the menu). Skill appears as both a fixed and a random predictor, as in SAS. **COVTYP(UN)** specifies an unstructured covariance matrix and produces output that matches other software.[9] Remaining syntax is as for the intercepts-only model.

The output begins with an indication of fixed and random effects, the type of covariance structure for the random effects, and the variable used to combined subjects (RUN), as well as the number of parameters in the model. Then information is provided that is useful to test differences between models, such as −2 Log Likelihood. Fixed effects follow, with two tables: one for significance tests of

---

[9]Thanks to Jodie Ullman for uncovering this trick in Version 11.5.

TABLE 15.7   Syntax and Selected Output for SAS MIXED Analysis of Level-1 Predictor Model

```
proc mixed data=Sasuser.Ss_hlm covtest method=ml;
    class RUN;
    model SPEED= SKILL / solution;
    random intercept SKILL / type=un subject = RUN;
run;
```

                        The Mixed Procedure

                  Covariance Parameter Estimates

| | | | Standard | Z | |
Cov Parm	Subject	Estimate	Error	Value	Pr Z
UN(1,1)	RUN	0.6281	0.3050	2.06	0.0197
UN(2,1)	RUN	-0.2920	0.1607	-1.82	0.0692
UN(2,2)	RUN	0.2104	0.1036	2.03	0.0211
Residual		0.4585	0.04184	10.96	<.0001

                        Fit Statistics

            -2 Log Likelihood   587.9
            AIC (smaller is better)   599.9
            AICC (smaller is better)   600.2
            BIC (smaller is better)   601.7

                 Null Model Likelihood Ratio Test

              DF   Chi-Square   Pr > ChiSq

               3   139.50   <.0001

                  Solution for Fixed Effects

| | | Standard | | | |
Effect	Estimate	Error	DF	t Value	Pr > |t|
Intercept	4.9808	0.2630	9	18.94	<.0001
SKILL	0.2160	0.1512	9	1.43	0.1868

                Type 3 Tests of Fixed Effects

| | Num | Den | | |
Effect	DF	DF	F Value	Pr > F
SKILL	1	9	2.04	0.1868

TABLE 15.8   Syntax and Selected Output for SPSS MIXED Analysis of Model with Level-1 Predictor

```
MIXED
   speed WITH skill
   /CRITERIA = CIN(95) MXITER(100) MXSTEP(5) SCORING(1)
   SINGULAR(0.000000000001) HCONVERGE(0, ABSOLUTE) LCONVERGE(0, ABSOLUTE)
   PCONVERGE(0.000001, ABSOLUTE)
   /FIXED = skill | SSTYPE(3)
   /METHOD = ML
   /PRINT = SOLUTION TESTCOV
   /RANDOM INTERCEPT skill | SUBJECT(run) COVTYPE(UN) .
```

Model Dimension[b]

		Number of Levels	Covariance Structure	Number of Parameters	Subject Variables
Fixed Effects	Intercept	1		1	
	skill	1		1	
Random Effects	Intercept + skill[a]	2	Unstructured	3	run
Residual				1	
Total		4		6	

[a]As of version 11.5, the syntax rules for the RANDOM subcommand have changed. Your command syntax may yield results that differ from those produced by prior versions. If you are using SPSS 11 syntax, please consult the current syntax reference guide for more information.

[b]Dependent Variable: speed.

Information Criteria[a]

–2 Log Likelihood	587.865
Akraike's Information Criterion (AIC)	599.865
Hurvich and Tsai's Criterion (AICC)	600.197
Bozdogan's Criterion (CAIC)	627.229
Schwarz's Bayesian Criterion (BIC)	621.229

The information criteria are displayed in smaller-is-better forms.

[a]Dependent Variable: speed.

TABLE 15.8 Continued

Fixed Effects

Type III Tests of Fixed Effects[a]

Source	Numerator df	Denominator df	F	Sig.
Intercept	1	10.955	358.755	.000
skill	1	10.330	2.041	.183

a Dependent Variable: speed.

Estimates of Fixed Effects[a]

Parameter	Estimate	Std. Error	df	t	Sig.	95% Confidence Interval Lower Bound	Upper Bound
Intercept	4.9808208	.2629674	10.955	18.941	.000	4.4017434	5.5598981
skill	.2159816	.1511620	10.330	1.429	.183	-.1193769	.5513400

a Dependent Variable: speed.

Covariance Parameters

Estimates of Covariance Parameters[a]

Parameter		Estimate	Std. Error	Wald Z	Sig.	95% Confidence Interval Lower Bound	Upper Bound
Residual		.4585221	.0418398	10.959	.000	.3834324	.5483171
Intercept + skill [subject = run]	UN(1,1)	.6280971	.3050274	2.059	.039	.2424664	1.6270543
	UN(2,1)	-.2919760	.1606737	-1.817	.069	-.6068906	.0229386
	UN(2,2)	.2104463	.1036264	2.031	.042	.0801676	.5524383

a Dependent Variable: speed.

combined effects and another for parameter estimates, single-df tests, and confidence intervals. Results match those of SAS; SKILL differences are not statistically significant when averaged over runs.

Labeling and tests for random effects match those SAS. That is, UN(1,1) indicates statistically significant variance of the group intercept. Note that the $p = .039$ is incorrect for this one-tailed test, instead the obtained p value should be half that amount because the test is whether variance is *greater* than would be expected by chance. Similarly, the p value for UN(2,2), the test of the variance among group slopes, should be .021 rather than .042. The test for UN(1,2), the covariance between intercepts and slopes, is correctly interpreted as a two-tailed test because the relationship can be either negative or positive.

Table 15.9 summarizes the parameter estimates for both random and fixed effects of the model with a level-1 predictor and their interpretation.

TABLE 15.9 Summary of Symbols and Interpretations for Model with Level-1 Predictor

Parameter Estimate for Effect and Software Labels	Symbol from Table 15.3	Sample-specific Interpretation	Generalized Interpretation
Random Effects (Covariance Parameter Estimates)			
Value = 0.6281 SPSS: UN(1,1) SAS: UN(1,1)	τ_{00}	The variance in the means of speed for the runs around the grand mean of speed when skill is taken into account	The variance in the group means on the DV around the grand mean on the DV (variance between groups) when the predictor is taken into account
Value = −0.2920 SPSS: UN(2,1) SAS: UN(2,1)	τ_{10}	The covariance between means for runs and slopes (skill-speed association) for runs	The covariance between intercepts and slopes (predictor-DV association) for groups
Value = 0.2104 SPSS: UN(2,2) SAS: UN(2,2)	τ_{11}	The variance in the slopes for skill around the average slope for all runs	The variance in the slopes for a predictor around the average slope for all group
Value = 0.4585 SPSS: Residual SAS: Residual	e_{ij}	The variance in speed for individual skiers within runs around the mean speed for the run when skill is taken into account	The variance among cases on the DV within groups around their own group means (variance within groups) when the predictor is taken into account
Fixed Effect (Parameter Estimates)			
Value = 5.4108 SPSS: Intercept SAS: Intercept	γ_{00}	The unweighted grand mean of speed for the runs when skill level is zero	The overall intercept; unweighted mean of the means for the groups when the predictor level is zero
Value = 0.2160 SPSS: skill SAS: Skill	γ_{10}	The unweighted average of slopes for skill over all runs	The unweighted average of slopes for the predictor over all groups

15.4.3 Model with Predictors at First and Second Levels

The full model adds a second-level predictor, mountain, to the model with a first-level predictor, skill. Mountain is a categorical IV (although with only two levels it can be treated as either categorical or continuous).

15.4.3.1 Level-1 Equation for Model with Predictors at Both Levels

The level-1 equation for a model with predictors at both levels does not differ from that of the level-1 equations for a model with a predictor only at the first level. That is, the inclusion of the second-level IV does not affect the equations for the first level of analysis. However, it can affect the results of those equations, because all effects are adjusted for all other effects.

15.4.3.2 Level-2 Equations for Model with Predictors at Both Levels

The second-level analysis (based on groups as research units) uses all three variables of Table 15.2: level-1 intercept, level-1 slope, and the level-2 IV (mountain). For these analyses, level-1 slope and intercept again are considered DVs. To predict the random intercept:

$$\beta_{0j} = \gamma_{00} + \gamma_{01}W_j + u_{0j} \qquad (15.10)$$

An intercept for a run β_{0j}, is predicted by γ_{00} (the average intercept over groups when all predictors are zero), the slope γ_{01} (for the relationship between the intercepts of the level-1 analysis and the levels of the fixed level-2 IV) multiplied by W_j (the average value of the IV for the group), and error, u_{0j} (the deviation from average intercept for group j).

The coefficient, γ_{01}, is the relationship between the original DV, Y_{ij}, and the IV, W_j.
To predict the random slope for the level-1 predictor:

$$\beta_{1j} = \gamma_{10} + u_{1j} \qquad (15.11)$$

The slope for a run, β_{1j}, is predicted by a single intercept, γ_{10} (the average level-1 predictor-DV slope), and error u_{1j} (the deviation from average slope for group j).

Substituting right-hand terms from Equations 15.10 and 15.11 into Equation 15.4:

$$Y_{ij} = \gamma_{00} + \gamma_{01}W_j + u_{0j} + (\gamma_{10} + u_{1j})X_{ij} + e_{ij} \qquad (15.12)$$

Thus the entire two-level equation, with terms rearranged, is:

$$Y_{ij} = \gamma_{00} + \gamma_{01}W_j + \gamma_{10}X_{ij} + u_{0j} + u_{1j}X_{ij} + e_{ij} \qquad (15.13)$$

The three fixed components are γ_{00} (the average intercept-grand mean), $\gamma_{01}W_j$ (the overall slope for the relationship between a level-2 predictor and the DV times the second-level predictor score for group j), and $\gamma_{10}X_{ij}$ (the unique effect of group j on slope when the value of the level-2 predictor W is zero times the first-level predictor score for case i in group j). The three random components are u_{0j} (the deviation in intercept for cases in group j), $u_{1j} X_{ij}$ (the deviation in slope for cases in group j times the first-level predictor score for case i in group j), and e_{ij} (the deviation for case i from its group j).

The solution to Equation 15.9 produced by software is

$$\beta_{0j} + 4.837 + 1.472(\text{Mountain}) + u_{0j}$$

The intercept 4.837 is the unweighted mean for the Mammoth runs (mountain = 0) when skill is 0. The slope has a value of 1.472 and a standard error of 0.216; thus, $z = 1.472/0.197 = 7.467$, a significant result at $\alpha = .05$. That means that the intercepts (means) for Mammoth Mountain (coded 0) are lower than the mean for Aspen Highlands (coded 1). Skiing speed is significantly different for the two mountains.

The solution to Equation 15.10 produced by software is

$$\beta_{1j} = 0.209 + u_{1j}$$

The average slope is 0.209 with a standard error of 0.156; thus, $z = 0.209/0.156 = 1.343$. There is no evidence at $\alpha = .05$ of a relationship between the DV, speed, and the predictor, skill. That is, skiing speed cannot be predicted by skill when averaged over all runs.

The entire two-level (Equation 15.12) solution is

$$Y_{ij} = 4.837 + 1.472(\text{Mountain}) + 0.209(\text{Skill}) + (\beta_{0j} - 4.837)$$
$$+ (\beta_{1j} - 0.209)(\text{Skill}) + e_{ij}.$$

Recall that β_{0j} is the intercept for group j and β_{1j} is the slope for the relationship between the DV and level-1 predictor in group j.

Tests of whether intercepts and slopes differ among runs are shown in the following computer runs, as is the test of whether slopes and intercepts are correlated.

15.4.3.3   Computer Analyses of Model with Predictors at First and Second Levels

As seen in Table 15.10, SAS syntax adds MOUNTAIN to the `model` instruction. Because MOUNTAIN is a dichotomous variable, it may be treated as continuous, simplifying interpretation of output. Note that SKILL has been defined as a random effect, but MOUNTAIN remains only a fixed effect. There is no way to specify MOUNTAIN as a fixed level-2 (rather than level-1) variable (using runs rather than skiers as subjects), so that df for its tests need to be adjusted. The use of `ddfm = kenwardroger` approximates the appropriate df.

The total number of parameters in the model now is 7, as seen in the `Dimensions` section. The four `Covariance Parameters` are the variance in intercepts, variance in slopes, covariance between intercepts and slopes, and residual. The three fixed parameters (`Columns in X`) are the overall intercept, skill, and mountain.

TABLE 15.10 Syntax and Selected Output for SAS MIXED Analysis of Full Model

```
proc mixed data=Sasuser.Ss_hlm covtest method=ml;
    class RUN;
    model SPEED= SKILL MOUNTAIN / solution ddfm = kenwardroger;
    random intercept SKILL / type=un subject = RUN;
run;
```

Dimensions

Covariance Parameters	4
Columns in X	3
Columns in Z Per Subject	2
Subjects	10
Max Obs Per Subject	67

Number of Observations

Number of Observations Read	274
Number of Observations Used	260
Number of Observations Not Used	14

Covariance Parameter Estimates

Cov Parm	Subject	Estimate	Standard Error	Z Value	Pr Z
UN(1,1)	RUN	0.4024	0.2063	1.95	0.0256
UN(2,1)	RUN	-0.2940	0.1473	-2.00	0.0460
UN(2,2)	RUN	0.2250	0.1098	2.05	0.0202
Residual		0.4575	0.04167	10.98	<.0001

Fit Statistics

-2 Log Likelihood	570.3
AIC (smaller is better)	584.3
AICC (smaller is better)	584.8
BIC (smaller is better)	586.4

Null Model Likelihood Ratio Test

DF	Chi-Square	Pr > ChiSq
3	92.80	<.0001

Solution for Fixed Effects

Effect	Estimate	Standard Error	DF	t Value	Pr > |t|
Intercept	4.8367	0.2169	10.3	22.30	<.0001
SKILL	0.2090	0.1558	9.68	1.34	0.2105
MOUNTAIN	1.4719	0.2202	7.09	6.68	0.0003

(continued)

TABLE 15.10 Continued

Type 3 Tests of Fixed Effects

Effect	Num DF	Den DF	F Value	Pr > F
SKILL	1	9.68	1.80	0.2105
MOUNTAIN	1	7.09	44.68	0.0003

Again, UN(1,1) is the variance in intercepts (means) across runs; with a one-tailed test at $\alpha = .05$ (critical value $= 1.58$) there is evidence that the intercepts vary after adjusting for all other effects. UN(2,1) is the covariance between intercepts and slopes, and there now is indication at $\alpha = .05$ of a negative relationship between intercepts and slopes over runs. The negative parameter estimate of -0.2940 indicates that the higher the speed the lower the relationship between skill and speed, after adjusting for all other effects. This effect was not statistically significant before entry of MOUNTAIN into the model. UN(2,2) is the variance in slopes across runs, and there is evidence that the relationship between skill and speed differs among runs (making a fixed effect of SKILL difficult to interpret). The significant Residual indicates that there are individual differences among skiers within runs even after accounting for all other effects. Fit statistics are as described earlier.

The remaining output is for the fixed effects in the model. The statistically significant estimate of 1.4719 indicates greater speed for the mountain with the code of 1 (Aspen Highlands).

The Intercept $= 4.8367$ is the mean of the groups in the mountain coded 0 (Mammoth) when skill level is 0. The overall relationship between SPEED and SKILL is still not statistically significant ($p = .2120$) but is not interpretable, in any event, in the face of the significant random variance in slopes across runs, UN(2,2).

Table 15.11 shows syntax and output for SPSS MIXED analysis. Both skill and mountain are declared continuous (WITH) variables—as a dichotomous variable, MOUNTAIN may be treated as continuous. Subjects are nested within runs. Skill appears as both a FIXED and a RANDOM predictor; Mountain is only a fixed predictor. Remaining syntax is as previously described.

The output begins with specification of fixed and random effects (note that random effects are also listed in the fixed rows), the type of covariance structure for the random effects, and the variable used to combined subjects (RUN). Results for fixed effects are the same as those of SAS (except for df); MOUNTAIN differences are statistically significant but the average relationship between SKILL and SPEED is not.

The random effect test, UN(1,1), for Intercept (whether runs differ in mean speed) is statistically significant at $z = 1.95$ if a one-tailed criterion is used at $\alpha = .05$; the Sig. value of .051 is for a two-tailed test. The UN(2,2) test for skill (slope differences among runs) also shows significant differences, $z = 2.049$, as does the two-tailed test for the covariance between intercepts and slopes.

Table 15.12 summarizes the parameter estimates for both random and fixed effects of the model with a level-1 predictor and their interpretation.

TABLE 15.11   Syntax and Selected Output for SPSS MIXED Analysis of Full Model

```
MIXED
  speed WITH skill mountain
  /CRITERIA = CIN(95) MXITER(100) MXSTEP(5) SCORING(1)
  SINGULAR(0.000000000001) HCONVERGE(0, ABSOLUTE) LCONVERGE(0, ABSOLUTE)
  PCONVERGE(0.000001, ABSOLUTE)
  /FIXED = skill mountain | SSTYPE(3)
  /METHOD = ML
  /PRINT = SOLUTION TESTCOV
  /RANDOM INTERCEPT skill | SUBJECT(run) COVTYPE(UN) .
```

Model Dimension[b]

		Number of Levels	Covariance Structure	Number of Parameters	Subject Variables
Fixed Effects	Intercept	1		1	
	skill	1		1	
	mountain	1		1	
Random Effects	Intercept + skill[a]	2	Unstructured	3	run
Residual				1	
Total		5		7	

[a]As of version 11.5, the syntax rules for the RANDOM subcommand have changed. Your command syntax may yield results that differ from those produced by prior versions. If you are using SPSS 11 syntax, please consult the current syntax reference guide for more information.

[b]Dependent Variable: speed.

Information Criteria[a]

−2 Log Likelihood	570.318
Akraike's Information Criterion (AIC)	584.318
Hurvich and Tsai's Criterion (AICC)	584.762
Bozdogan's Criterion (CAIC)	616.242
Schwarz's Bayesian Criterion (BIC)	609.242

The information criteria are displayed in smaller-is-better forms.

[a]Dependent Variable: SPEED.

(continued)

TABLE 15.11   Continued

Fixed Effects

Type III Tests of Fixed Effects[a]

Source	Numerator df	Denominator df	F	Sig.
Intercept	1	12.206	500.345	.000
skill	1	11.826	1.805	.204
mountain	1	7.086	55.759	.000

[a]Dependent Variable: speed.

Estimates of Fixed Effects[a]

Parameter	Estimate	Std. Error	df	t	Sig.	95% Confidence Interval	
						Lower Bound	Upper Bound
Intercept	4.8366975	.2162292	12.206	22.368	.000	4.3664530	5.3069421
skill	.2089697	.1555458	11.826	1.343	.204	-.1304878	.5484272
mountain	1.4718779	.1971118	7.086	7.467	.000	1.0069237	1.9368322

[a]Dependent Variable: speed.

Covariance Parameters

Estimates of Covariance Parameters[a]

Parameter		Estimate	Std. Error	Wald Z	Sig.	95% Confidence Interval	
						Lower Bound	Upper Bound
Residual		.4575351	.0416711	10.980	.000	.3827358	.5469526
Intercept + skill [subject = run]	UN(1,1)	.4024587	.2063647	1.950	.051	.1473191	1.0994700
	UN(2,1)	-.2941195	.1474021	-1.995	.046	-.5830223	-.0052168
	UN(2,2)	.2250488	.1098392	2.049	.040	.0864633	.5857622

[a]Dependent Variable: speed.

TABLE 15.12 Summary of Symbols and Interpretations for Models with Predictors at Both Levels

Parameter Estimate for Effect and Software Labels	Symbol from Table 15.3	Sample-specific Interpretation	Generalized Interpretation
Random Effects (Covariance Parameter Estimates)			
Value = 0.6281 SPSS: UN(1,1) SAS: UN(1,1)	τ_{00}	The variance in the means of speed for the runs around the grand mean of speed when skill and mountain are taken into account	The variance in the group means on the DV around the grand mean on the DV (variance between groups) when predictors are taken into account
Value = –0.2920 SPSS: UN(2,1) SAS: UN(2,1)	τ_{01}	The covariance between means for runs and slopes (skill-speed association) for runs when mountain is taken into account	The covariance between means for groups and slopes (predictor-DV association) for runs when mountain is taken into account
Value = 0.2104 SPSS: UN(2,2) SAS: UN(2,2)	τ_{11}	The variance in the slopes for skill around the average slope for all runs when mountain is taken into account	The variance in the slopes for a predictor around the average slope for all groups when other predictors are taken into account
Value = 0.4585 SPSS: Residual SAS: Residual	n.a.	The variance in speed for individual skiers within runs around the mean speed for the run when skill and mountain are taken into account	The variance among cases on the DV within groups around their own group means (variance within groups) when predictors are taken into account
Fixed Effect (Parameter Estimates)			
Value = 5.4108 SPSS: Intercept SAS: Intercept	γ_{00}	The unweighted grand mean of speed for the runs when skill level and mountain are zero	The overall intercept; unweighted mean of the means for the groups when all predictor levels are zero
Value = 0.2160 SPSS: skill SAS: Skill	γ_{01}	The average slope for skill over all runs when mountain is taken into account	The average slope for the predictor over all groups when all other predictors are taken into account
Value = 1.472 SPSS: mountain SAS: MOUNTAIN	γ_{01}	The average slope for mountain over all runs when skill is taken into account	The average slope for the predictor over all groups when all other predictors are taken into account

15.5 Types of MLM

MLM is an extremely versatile technique that, like SEM, can be used for a variety of research designs. The ones discussed here are those that are implemented in one or more of the programs discussed. This chapter demonstrates repeated-measures designs and higher-order models. Other topics described in this section—latent variables, nonnormal outcome variables, multiple response models—are described only briefly, with references to other sources.

15.5.1 Repeated Measures

One of the more common uses of MLM is to analyze a repeated-measures design, which violates some of the requirements of repeated-measures ANOVA. Longitudinal designs (called growth curve data) are handled in MLM by setting measurement occasions as the lowest level of analysis with cases (e.g., students) the grouping variable. However, the repeated measures need not be limited to the first level of analysis (e.g., there could be repeated measurement of teachers and/or schools, as well). A big advantage of MLM over repeated-measures ANOVA is that there is no requirement for complete data over occasions (although it is assumed that data are missing at random), nor is there need for equal numbers of cases or equal intervals of measurements for each case. Another important advantage of MLM for repeated-measures data is the opportunity to test individual differences in growth curves (or other patterns of responses over the repeated measure). Are the regression coefficients the same for all cases? Because each case has its own regression equation when random slopes and intercepts are specified, it is possible to evaluate whether individuals do indeed differ in their mean response and/or in their pattern of responses over the repeated measure.

An additional advantage is that sphericity (uncorrelated errors over time) is not an issue because, as a linear regression technique, MLM tests trends for individuals over time (if individuals are the grouping variable). Finally, you may create explicit time-related level-1 predictors, other than the time period itself. Time-related predictors (a.k.a. time-varying covariates) come in a variety of forms: days in the study, age of participant, grade level of participant, and so on.

Unlike ANOVA, there is no overall test of the "repeated measures" factor unless one or more time-related predictors are explicitly entered. Once the time-related predictor is entered into the equation, it is evaluated as a single df test (e.g., linear relationship between time and the DV, a longitudinal growth curve), so that the assumption of sphericity is avoided. If other trends are of interest, they are coded and entered as separate predictors (e.g., time-squared for the quadratic trend). Thus, MLM can be used to provide all of the advantages of a trend analysis if relevant predictors are created and entered.

Table 15.13 shows a small, hypothetical data set prepared for SPSS MIXED with five cases in which number of books read per month serves as the DV and type of novel (science fiction, romance, mystery) serves as a fixed IV. The first column is the type of novel, the second is indicator of month, the third is the identification of the case, and the final column is the DV (number of books read that month). This corresponds to a two-way within-between-subjects design with month as the repeated-measures IV and novel as the between-subjects IV. The sample size is highly inadequate, especially for tests of random effects (although a solution miraculously emerged), but provides a convenient vehicle for demonstrating various facets of MLM repeated-measures analysis and has the further advantage of being sufficiently silly.

Figure 15.3 shows the layout of the data for this design.

**TABLE 15.13   Data Set for SPSS MIXED
Analysis of Repeated Measures Data**

Novel	Month	Case	Books
1	1	1	1
1	1	2	1
1	1	3	3
1	1	4	5
1	1	5	2
1	2	1	3
1	2	2	4
1	2	3	3
1	2	4	5
1	2	5	4
1	3	1	6
1	3	2	8
1	3	3	6
1	3	4	7
1	3	5	5
2	1	6	3
2	1	7	4
2	1	8	5
2	1	9	4
2	1	10	4
2	2	6	1
2	2	7	4
2	2	8	3
2	2	9	2
2	2	10	5
2	3	6	0
2	3	7	2
2	3	8	2
2	3	9	0
2	3	10	3
3	1	11	4
3	1	12	2
3	1	13	3
3	1	14	6
3	1	15	3
3	2	11	2
3	2	12	6
3	2	13	3
3	2	14	2
3	2	15	3
3	3	11	0
3	3	12	1
3	3	13	3
3	3	14	1
3	3	15	2

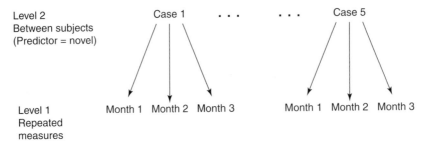

Level 2
Between subjects
(Predictor = novel)

Case 1   . . .   . . .   Case 5

Level 1
Repeated
measures

Month 1   Month 2   Month 3   Month 1   Month 2   Month 3

FIGURE 15.3   **Layout of Table 15.13 data.**

Table 15.14 shows syntax and partial output for analysis of effects of linear trend of month, novel type, and their interaction, as well as tests of mean differences among readers. The intercept is a RANDOM effect. The linear trend of month (1 to 2 to 3) is evaluated (rather than main effect) because MONTH is *not* declared to be a categorical variable (i.e., it is a WITH variable rather than a

TABLE 15.14   **Syntax and Selected Output from SPSS MIXED Analysis of Repeated-Measures Data**

```
MIXED
    books BY novel WITH month
    /CRITERIA = CIN(95) MXITER(100) MXSTEP(5) SCORING(1) SINGULAR(0.000000000001)
    HCONVERGE(0, ABSOLUTE) LCONVERGE(0, ABSOLUTE)
    PCONVERGE(0.000001, ABSOLUTE)
    /FIXED = novel month month*novel | SSTYPE(3)
    /METHOD = REML
    /PRINT = TESTCOV
    /RANDOM INTERCEPT month | SUBJECT(case) COVTYPE(UN).
```

Model Dimension[b]

		Number of Levels	Covariance Structure	Number of Parameters	Subject Variables
Fixed Effects	Intercept	1		1	
	novel	3		2	
	month	1		1	
	novel * month	3	Unstructured	2	
Random Effects	Intercept + month[a]	2		3	case
Residual				1	
Total		10		10	

[a]As of version 11.5, the syntax rules for the RANDOM subcommand have changed. Your command syntax may yield results that differ from those produced by prior versions. If you are using SPSS 11 syntax, please consult the current syntax reference guide for more information.

[b]Dependent Variable: books.

TABLE 15.14 Continued

Fixed Effects

Type III Tests of Fixed Effects[a]

Source	Numerator df	Denominator df	F	Sig.
Intercept	1	12.000	50.586	.000
novel	2	1.2000	11.321	.002
month	1	1.2000	.320	.582
novel * month	2	1.2000	20.540	.000

[a]Dependent Variable: books.

Covariance Parameters

Estimates of Covariance Parameters[a]

Parameter		Estimate	Std. Error	Wald Z	Sig.	95% Confidence Interval	
						Lower Bound	Upper Bound
Residual		1.355556	.494979	2.739	.006	.662678	2.772887
Intercept + month	UN(1,1)	.492593	1.887086	.261	.794	.000270	898.209913
[subject = case]	UN(2,1)	-.208333	.838257	-.249	.804	-1.851287	1.434620
	UN(2,2)	.155556	.420704	.370	.712	.000776	31.187562

[a]Dependent Variable: books.

817

BY variable). The linear trend of month by novel interaction as well as the main effect of novel are declared to be fixed by listing them in the FIXED instruction. Month is declared random by listing it in the RANDOM instruction. CASE is the group identifier.

No significant random effects are found. The test of UN(1,1), the variance of the intercept ($p = .794$), shows that there is no significant difference in the mean number of books read among the five cases. The random test of UN(2,2), $p = .804$, shows no significant difference among readers in the linear trend of month; this test is not available in ANOVA using either univariate or multivariate approaches. Finally, the test of UN(2,1) the random INTERCEPT by MONTH covariance ($p = .552$) shows no significant difference in the relationship between the mean number of books read and the linear trend of month across readers. If any of these were statistically significant, it might be worthwhile to explore some characteristics of individuals to "explain" those individual differences.

With respect to fixed effects, there is no significant fixed linear trend of month averaged over subjects ($p = .582$). The main effect of novel is statistically significant ($p = .002$) but this is interpreted with great caution in the presence of the significant month by novel interaction ($p < .001$). Thus, averaged over readers, the linear trend of month is different for the different types of novels. A plot of the interaction (as per Figure 8.1) would assist interpretation; cell means can be found by using the "split cases" instruction in SPSS, specifying NOVEL and MONTH as the grouping variables in a DESCRIPTIVES analysis.

This analysis shows that there is nothing special or different about repeated-measures vs. non-repeated-measures analysis in MLM. The repeated measures are simply treated as any other first-level unit of analysis, and participants become a second-level unit of analysis. If cases are nested within multiple units (e.g., students in classrooms), then classrooms become the third level unit of analysis, and so on. Thus, repeated measures add another, bottom-level, unit of analysis to any design. Because of this, models involving repeated measures often require more than two levels of analysis (see the complete example of Section 15.7.2).

Another issue that can arise in using MLM for repeated measures is the scale (coding) of the time variable. Section 15.6.2 addresses centering of predictors and how that affects interpretation. A problem in repeated measures is that the correlation between slopes and intercepts may be of particular interest if, for example, we want to know whether children who start high on the DV (e.g., reading achievement) have a steeper or shallower slope over time. The difficulty is that the correlation changes as a result of the coding of time. Therefore, the correlation can only be interpreted in light of the particular scaling of the occasions (Hox, 2002, pp. 84–86).

The example here has an IV for the level-2 analysis, novel, but no time-varying predictor for the level-1 analysis; that is, there is no variable that indicates case characteristics that change over time such as age, fatigue, or the like. A level-1, time-varying predictor can be especially useful in a longitudinal study because intervals between levels of the repeated measure can be unequal and different for each participant. For example, reading achievement can be evaluated on multiple occasions, with each child tested at different ages as well having as different numbers of tests. MLM permits evaluation of the relationship between age (the level-1 predictor) and reading (the DV) without requiring that each child be tested according to a standard schedule.

Additional level-2 predictors might be gender, or any other grouping variable that applies to cases and is stable over time, such as employment status, or some other stable case characteristic measured on a continuous scale, such as SES.

Another method for analyzing repeated-measures designs is through SEM as a latent growth (or curve) model. The time variable is defined in the measurement of the latent factors (Hox, 2002). There is one latent variable for the intercept and another for each slope (linear, quadratic, etc.). This also is a random coefficient model and Hox (2002) shows it to be equivalent to a two-level MLM as long as data are complete and time intervals are equally spaced.

The advantage of latent curve analysis is that it can be used for more complex two-level models, for example when slope is a predictor of some outcome. Although higher level models are possible, they require complicated program setups (Hox, 2002). Advantages of the MLM approach include automatic dealing with missing data and no requirement for equally spaced time intervals. Hox (2002) compares the two approaches in detail (pp. 273–274). Little, Schnabel, and Baumert (2000) also discuss analysis of growth models through MLM and SEM in detail. Singer and Willett (2003) concentrate on analyzing longitudinal data, including MLM strategies. Singer (1998) focuses on the use of SAS PROC MIXED to analyze individual growth models as well as other applications of MLM. The SAS and SPSS MIXED programs permit two approaches to repeated measures: the MLM approach illustrated here or the "repeated" approach in which the structure of the variance-covariance is specified.

15.5.2   Higher-Order MLM

The model described in Section 15.4 has two levels: skiers and runs. This is the most common type of model and, obviously, the easiest to analyze. Most MLM software is capable of analyzing three-level models; and some programs accommodate even more levels. An alternative strategy if the software is limited is to run a series of two-level models, using the slopes and intercepts from one level (as seen in Table 15.2) as DVs for the next higher level. A three-level example with repeated measures is demonstrated in Section 15.7.

15.5.3   Latent Variables

Latent variables are used in several ways in MLM—observed variables combined into factors (cf. Chapters 13 and 14), analysis of variables that are measured without error, analyses with data missing on one or more predictors, and models in which the latent factors are based on time.

The HLM manual (Raudenbush et al., 2004) shows examples of two applications. In the first, the latent variable regression option is chosen to analyze a latent variable measured without error, gender, which is entered on both levels of a two-level repeated-measures model (occasion and participant), with age as a level-1, occasion-level predictor. The DV is attitude toward deviant behaviors. Coefficients are available to test the linear growth rate in the DV (trend over occasions); the effect of gender on the growth rate; the effect of the initial value of the DV on the linear growth rate; and the total, direct, and indirect associations between gender and growth rate.

In the second example, the latent variable regression option in HLM is chosen to do garden-variety, single level, multiple regression with missing data. Data are reorganized so that a participant with complete data has as many rows as there are DVs; a participant with missing data has fewer rows. The value of the variable is entered in one column of the data set, and there are as many additional (indicator) columns as there are DVs. Each measure has a 1 in the column representing the variable that it measures and a zero in the remaining columns. Raudenbush et al. (2004) also show how to do HLM analyses with data that have multiply-imputed values (cf. Section 4.1.3.2).

SEM programs are designed to analyze multilevel models with latent variables (factors) composed of several measured variables (IVs). Variables associated with factors are specified as seen in Table 14.2 using EQS. Analysis of such models is based on a partition of variance-covariance matrices into within-group (first-level) and between-group (second-level) matrices. A single set of summary statistics provides information about fit of the combined multilevel model, and the usual information is provided to test the model against alternatives (cf. Chapter 14). Also available are parameter estimates for both levels of the model—the structures of the first-level (individual) and second-level (group) models.

Partitioning variance-covariance matrices into within-group and between-group matrices also permits MLM when groups are very small as, for example, when couples or twins form the level-2 grouping unit. With garden-variety MLM performed on such data, the separate regressions for each two-participant group may be problematic if there are level-2 predictors.

Another useful application of MLM with latent variables is confirmatory factor analysis which investigates the similarity of factor structures at different levels (e.g., at the individual and the group level). Do the loadings of variables on factors from individual-level data change substantively when group membership is taken into account or do we see the same factor structure when analyzing individual versus group covariance matrices? Put another way, does the same factor structure explain group differences as well as individual differences?

MLM confirmatory factor analysis is also useful in determining whether there are any group (higher level) differences worth taking into account before adding predictors to a model. By providing information to compute intraclass correlations (see Section 15.6.1) for factors, one can determine whether a multilevel model is necessary when adding predictors to a model.

Heck and Thomas (2000) devote fully half of their introductory MLM book to models with latent variables, including a demonstration of confirmatory factor analysis with a hierarchical data structure. Hox (2002) also devotes a chapter to multilevel factor models.

15.5.4 Nonnormal Outcome Variables

As a variant of the general linear model, MLM assumes multivariate normality so the usual diagnostic techniques and remedies of Section 4.1.5 can be applied. However some MLM programs also provide specialized techniques for dealing with nonnormal data (see Table 15.33). Models with nonnormal outcomes (DVs) are often referred to as multilevel generalized linear models.

MLwiN allows analysis of binomial and Poisson as well as normal error distributions for MLM. A binary response variable is analyzed using the binomial error distribution, as is a response variable that is expressed as a proportion. These are analogous to the two-outcome logistic regression models of Chapter 10. The Poisson distribution is used to model frequency count data. The MLwiN manual (Rasbash et al., 2000, Chapters 8 and 9) describes many choices of link functions and estimation techniques and demonstrates examples of these models. A special MLwiN manual (Yang et al., 1999) discusses categorical responses with ordered and unordered categories.

HLM also permits a variety of nonlinear options: Bernoulli (logistic regression for 0-1 outcomes), two types of Poisson distributions (constant or variable exposure), and binomial (number of trials), multinomial (logistic regression with more than two outcomes), and ordinal outcomes. The HLM6 manual (Raudenbush et al., 2004) demonstrates several of these models. Yet another approach offered through HLM is computation of standard errors that are robust to violation of normality, provided by default for all fixed effects, with a note describing their appropriateness. Hox (2002) devotes a chapter to the logistic model, in which generalized and multilevel generalized models are discussed thoroughly.

SAS has a separate nonlinear mixed models procedure (PROC NLMIXED) for dealing with such models. SPSS does not have nonnormal options for MLM at this time. However, as discussed in Chapter 10, SPSS COMPLEX SAMPLES LOGISTIC REGRESSION is available for two-level models with dichotomous outcomes by declaring groups (2nd-level units) to be clusters.

15.5.5   Multiple Response Models

The true multivariate analog of MLM is the analysis of multiple DVs as well as multiple predictors. These models are specified by providing an additional lowest level of analysis, defining the multivariate structure in a manner similar to that of repeated measures. That is, a case has as many rows as there are DVs, and some coding scheme is used to identify which DV is being recorded in that row. Snijders and Bosker (1994) discuss some of the advantages of multivariate multilevel models over MANOVA. First, missing data (assuming they are missing at random) pose no problem. This is a less restrictive assumption than required by MANOVA with imputed values for missing data—that data be missing completely at random (Hox, 2002). Second, tests are available to determine whether the effect of a predictor is greater on one DV than another. Third, if DVs are highly correlated, tests for specific effects on single DVs are more powerful because standard errors are smaller. Fourth, covariances among DVs can be partitioned into individual and group level, so that it is possible to compare size of correlations at the group vs. the individual level.

MLwiN and HLM have a special multivariate model technique for dealing with multiple DVs. The first level simply codes which response is being recorded. Chapter 11 of Rasbash et al. (2000) discusses multiple response models and demonstrates a model with two DVs (in which the level-1 predictor is a 0,1 dummy code indicating which response is recorded). That is, there is a separate row of data for each response in the multivariate set and a dummy variable to indicate whether it is one response or the other. This dummy variable is the level-1 predictor. Obviously, things get more complicated with more then two DVs. A separate dummy variable is required for every df—i.e., one fewer dummy variables than the number of DVs. The coding reflects the comparisons of interest. For example, if DVs were scores on arithmetic, reading, and spelling and the interest is in comparing reading with spelling, one of the dummy codes should be something like 0=reading, 1=spelling. The second comparison might be a contrast between arithmetic and the other two measures.

Chapter 9 of Hox (2002) discusses a meta-analysis with two responses, in which the lowest level is, again, a dummy code for response type, the second level is data collection condition (face-to-face, telephone, or mail), and the third level is the study. Each of the response types has a separate regression equation, and significance of differences between them is tested. Hox also discusses a model of a measurement instrument, in which item is the lowest level, student is second level, and school is the third level. This is similar to a repeated-measurement model as discussed in Section 15.5.2 although the emphases may differ.

15.6   Some Important Issues

15.6.1   Intraclass Correlation

The intraclass correlation (ρ)[10] is the ratio of variance *between* groups at the second level of the hierarchy (ski runs in the small-sample example) to variance *within* those groups. High values imply that the assumption of independence of errors is violated and that errors are correlated—that is, that the grouping level matters. An intraclass correlation is like η^2 in a one-way ANOVA, although in MLM the groups are not deliberately subjected to different treatments.

The need for a hierarchical analysis depends partially on the size of the intraclass correlation. If ρ is trivial, there is no meaningful average difference among groups on the DV, and data may be analyzed at the individual (first) level, unless there are predictors and groups differ in their relationships between predictors and the DV. In any event, Barcikowski (1981) shows that even small values of ρ can inflate Type I error rate with large groups. For example, with 100 cases per group, ρ of .01, and nominal α of .05, the actual α level is as high as .17; with 10 cases per group, ρ of .05, and nominal α of .05, the actual α level is .11. A practical strategy when the need for a hierarchical analysis is ambiguous is to do the analysis both ways to see whether the results differ substantively and then report the simpler analysis in detail if results are similar.

The intraclass correlation is calculated when there is a random intercept but no random slopes (because then there would be different correlations for cases with different values of a predictor). Therefore, ρ is calculated from the two-level intercept-only model (the "null" or unconditional model of Section 15.4.1). Such a model provides variances at each level; ρ is the level-2 variance (s_{bg}^2, between-group variability) divided by the sum of level-1 and level-2 variances ($s_{wg}^2 + s_{bg}^2$).

These components show up in the random effects portion of output. For example, in SAS (Table 15.4), the level-2 variance, 0.3116, is labeled `UN(1,1)` and the level-1 variance, 0.7695, is labeled `Residual`. Thus,

$$\rho = \frac{s_{bg}^2}{s_{bg}^2 + s_{wg}^2} = \frac{0.3116}{0.3116 + 0.7695} = .288 \tag{15.14}$$

That is, about 29% of the variability in the DV (skiing speed) is associated with differences between ski runs. Table 15.15 summarizes the labeling of information for intraclass correlations for several programs.

For a three-level design, there are two versions of the intraclass correlation, each with its own interpretation (Hox, 2002). For either version, a model is run with no predictors (a three-level intercepts-only model). The intraclass correlation at the second level is:

$$\rho_{\text{level 2}} = \frac{s_{bg2}^2}{s_{bg2}^2 + s_{bg3}^2 + s_{wg}^2} \tag{15.15}$$

---

[10]Note that intraclass correlation, the term conventionally used, is a misnomer; this really is a squared correlation or strength of association (effect size) measure.

TABLE 15.15   Labeling of Variances for Intraclass Correlations through Five Software Programs; Intercept-only Models for the Sample Data

Software (Table)	Within-Group Variance (s_{wg}^2)	Between-Group Variance (s_{bg}^2)
SAS (Table 15.4)	`Residual = 0.7695`	`UN(1,1) = 0.3116`
SPSS (Table 15.5)	Residual = .7694610	Intercept [subject = RUN] = .3115983
HLM	`level-1, R = 0.76946`	`INTRCPT1, U0 = 0.31160`
SYSTAT	`Residual variance = .769`	`Cluster variance = .312`
MLwiN	$e_{0skier,run} = 0.769$	$\mu_{0run} = 0.311$

where s_{bg2}^2 is the variance between the level-2 groups at level 2 and s_{bg3}^2 is the variance between the level-3 groups. The intraclass correlation at the third level is:

$$\rho_{\text{level 3}} = \frac{s_{bg3}^2}{s_{bg2}^2 + s_{bg3}^2 + s_{wg}^2} \tag{15.16}$$

Each is interpreted as a proportion of variance at the designated group level.

The second interpretation is as the expected shared variance between two randomly chosen elements in the same group (Hox, 2002). The equation for level 3 is the same for both interpretations (Equation 15.16). The level-2 equation for this interpretation is:

$$\rho_{\text{level 2}} = \frac{s_{bg2}^2 + s_{bg3}^2}{s_{bg2}^2 + s_{bg3}^2 + s_{wg}^2} \tag{15.17}$$

Intraclass correlations for a three-level model is demonstrated in the complete example of Section 15.7.

15.6.2   Centering Predictors and Changes in Their Interpretations

Subtracting a mean from each predictor score, "centering" it, changes a raw score to a deviation score. One major reason for doing this is to prevent multicollinearity when predictors are components of interactions or raised to powers, because predictors in their raw form are highly correlated with the interactions that contain them or with powers of themselves (cf. Section 5.6.6).

Centering is most commonly performed on level-1 predictors. Level-2 predictors are not usually centered (although it might be done if that enhanced interpretation). An exception would be when interactions are formed among two or more continuous level-2 predictors. Centering DVs is unusual because it is likely to make interpretation more difficult.

The meaning of the intercept changes when predictors are centered. For example, if all IVs are centered in multiple regression, the intercept is the mean of the DV. In multilevel models with no centering and viewed as a single equation (e.g., speed as a function of both skill and mountain), the intercept is the value of the DV (speed) when all IVs (skill and mountain) are zero. If, on the other hand, the level-1 IV (skill) is centered and 0 is the code for one of the mountains, then the intercept becomes the speed for skiers with an average skill level at that mountain.

Therefore, centering can facilitate interpretation when there is no meaning to a value of zero on a predictor. If, for example, IQ is a predictor, it does not make sense to interpret an intercept for a value of zero on IQ. On the other hand, if IQ is centered, the intercept becomes the value of the DV when IQ is equal to the mean of the sample. That is, with uncentered data, the intercept can be interpreted as the expected score of a student with an IQ of zero—an unrealistic value. On the other hand, centering changes the interpretation of the intercept as the expected score for a student with average IQ.

In multilevel models, there is a choice between centering a predictor around the grand mean (overall skill), centering around the means of the second-level units (mean skill at different ski runs) or even around some other value such as a known population mean. Centering around the grand mean reduces multicollinearity when interactions are introduced and produces models that are easily transformed back into models based on raw scores. Some values of parameters change, but the models have the same fit, the same predicted values, and the same residuals. Further, the parameter estimates are easily transformed into each other (Kreft & DeLeeuw, 1998). Thus, the goal of enhancing statistical stability by reducing multicollinearity is reached without changing the underlying model.

The more common practice of centering around group means has more serious consequences for interpretation unless group means are reintroduced as level-2 predictors. Differences in raw-score and group-centered models without reintroduction of group means as level-2 predictors can be large enough to change the direction of the findings, as illustrated by Kreft and DeLeeuw (1998, pp. 110–113), because important between-group information is lost. That is, mean differences between groups on an IV can be an important factor in prediction of the DV. Reintroducing the mean brings those between-group differences back into the model. In the small-sample example, this involves finding the mean skill for each group and adding it to the second level model, so that Equation 15.10 becomes

$$\beta_{0j} = \gamma_{00} + \gamma_{01}(\text{Mountain}) + \gamma_{02}(\text{MeanSkill}) + u_{0j}$$

The predictor at the first level here is group-centered at the second level—the "skill deviation" of Table 15.1. That is, DEV_SKL is formed by taking the raw score for each case and subtracting from it the group mean.

A model in which first-level predictor scores are centered around group means is shown through SAS MIXED in Table 15.16.

Conclusions have not changed substantively for this model. Notice that the new predictor **MEAN_SKL** is not a statistically significant addition to the model ($p = .4501$). Indeed, the larger value for **-2 Log Likelihood** suggests a poorer fit for the expanded and centered model (a direct comparison cannot be made because the models are not nested).

In this example, the level-1 intercept is the mean speed for a skier with a mean skill level (DEV_SKL coded 0) skiing on the mountain coded 0; the slope for the level-2 IV is the gain (or loss) in DV units for the other group (mountain coded 1). It is often worthwhile with a level-1 mean-centered IV to try a model with the group means introduced at level two to see if between-group (run) differences in the predictor are significant. And, of course, this is the method of choice if there is

TABLE 15.16   Syntax and Selected Output for Model with Mean Skill Added to Second-Level Equation for SAS MIXED Analysis

```
proc mixed data=Sasuser.Meanskl covtest method=ml;
    class RUN;
    model SPEED= DEV_SKL MOUNTAIN MEAN_SKL /
        solution ddfm = kenwardroger;
    random intercept DEV_SKL / type=un subject = RUN;
run;
```

The Mixed Procedure

Covariance Parameter Estimates

Cov Parm	Subject	Estimate	Standard Error	Z Value	Pr Z
UN(1,1)	RUN	0.08085	0.04655	1.74	0.0412
UN(2,1)	RUN	0.07569	0.05584	1.36	0.1752
UN(2,2)	RUN	0.2078	0.1025	2.03	0.0213
Residual		0.4594	0.04201	10.94	<.0001

Fit Statistics

-2 Log Likelihood	576.4
AIC (smaller is better)	592.4
AICC (smaller is better)	592.9
BIC (smaller is better)	594.8

Solution for Fixed Effects

Effect	Estimate	Standard Error	DF	t Value	Pr > |t|
Intercept	5.5328	0.3865	10.9	14.32	<.0001
DEV_SKL	0.2173	0.1505	9.69	1.44	0.1802
MOUNTAIN	1.9192	0.5026	8.85	3.82	0.0042
MEAN_SKL	-0.1842	0.2348	10.5	-0.78	0.4501

Type 3 Tests of Fixed Effects

Effect	Num DF	Den DF	F Value	Pr > F
DEV_SKL	1	9.69	2.09	0.1802
MOUNTAIN	1	8.85	14.58	0.0042
MEAN_SKL	1	10.5	0.62	0.4501

research interest in those between-group differences. Reintroduction of the group mean is unnecessary if centering is done around the grand mean.

The topic of centering in MLM is discussed and illustrated in much greater detail by Kreft and DeLeeuw (1998). Raudenbush and Bryk (2001) provide useful insights into interpretation of MLM parameters under various types of centering. Snijders and Bosker (1999) recommend that group-mean centering be used only when there is theory indicating that the DV is related not to the predictor but to the relative value of the predictor within a group. For example, if the DV is teacher's rating of student performance, the relative score within each group with the same teacher makes sense because teachers may use different rating criteria.

15.6.3 Interactions

Interactions of interest in MLM can be within a level or across levels. The small-sample example has only one predictor at each level. Had there been another level-1 predictor, such as skier age, the interaction between skill and age (if included in the model) would be a within-level interaction. However, if the interaction between skill and mountain is added to the small-sample example it is across levels, because skill is measured at level one and mountain at level two.

Inclusion of interactions is straightforward in MLM and follows the conventions of multiple regression: continuous predictors from which interactions are formed are centered and the interaction term is added. The interaction is formed in the small sample data set from the centered level-1 predictor, SKILL, and the level-2 predictor, MOUNTAIN (DEV_SKL*MOUNTAIN). This interaction tests whether the relationship between skill and speed (measured at the skier level) differs for the two mountains. Note that the interaction is added to the `model` equation after the main effects included in it. Changing the order of entry can affect parameter estimates for fixed effects even when Type III (default) sums of squares are used.

Table 15.17 shows SAS syntax and selected output for a model which includes the skill by mountain interaction. The table of fixed effects shows that there is no statistically significant difference in the relationship between skill and speed between the two mountains ($p = .8121$). (Note that this cross-level interaction is indeed predictive of the DV in the full NELS-88 data set from which this small sample was taken and relabeled.)

15.6.4 Random and Fixed Intercepts and Slopes

Multilevel modeling typically includes random intercepts because one of its goals is to deal with the increased Type I error rate that occurs in hierarchical data when groups differ in their average value of the DV. Random slopes, however, may or may not be appropriate in any given model. Inclusion of random slopes allows the relationships between the IV and DV to differ among groups.

Figure 15.4 illustrates some idealized combinations of random and fixed parameters with three groups (level-2 units). Figure 15.4(a) shows a need to include random intercepts in a model because the groups cross the Y axis at different places, but no need to include random slopes, because all of them are the same. That is, the rates of change in Y with change in X (the predictor) are the same for all groups. Figure 15.4(b) shows a need for both random intercepts and random slopes; groups cross the Y axis at different places, and the changes in Y from low to high values of the predictor (X) are different. Figure 15.4(c) shows a rare situation: a need for random slopes but a fixed intercept. That is, all of the groups have the same mean, but they differ in their change in Y with change in X from

TABLE 15.17   SAS MIXED Syntax and Output for Testing the Cross-Level Interaction for the Data of Table 15.1

```
proc mixed data=Sasuser.Meanskl covtest method=ml;
    class RUN ;
    model SPEED=DEV_SKL MOUNTAIN DEV_SKL*MOUNTAIN MEAN_SKL/
        solution ddfm=kenwardrogers;
    random intercept DEV_SKL / type=un subject = RUN;
run;
```

Covariance Parameter Estimates

Cov Parm	Subject	Estimate	Standard Error	Z Value	Pr Z
UN(1,1)	RUN	0.08068	0.04641	1.74	0.0411
UN(2,1)	RUN	0.07520	0.05546	1.36	0.1751
UN(2,2)	RUN	0.2061	0.1020	2.02	0.0216
Residual		0.4595	0.04202	10.93	<.0001

Fit Statistics

-2 Log Likelihood	576.3
AIC (smaller is better)	594.3
AICC (smaller is better)	595.0
BIC (smaller is better)	597.0

Null Model Likelihood Ratio Test

DF	Chi-Square	Pr > ChiSq
3	77.56	<.0001

Solution for Fixed Effects

Effect	Estimate	Standard Error	DF	t Value	Pr > |t|
Intercept	5.5366	0.3863	10.9	14.33	<.0001
DEV_SKL	0.2302	0.1587	9.83	1.45	0.1781
MOUNTAIN	1.8765	0.5111	10.3	3.67	0.0041
DEV_SKL*MOUNTAIN	-0.1186	0.4834	8.5	-0.25	0.8121
MEAN_SKL	-0.1839	0.2346	10.5	-0.78	0.4503

Type 3 Tests of Fixed Effects

Effect	Num DF	Den DF	F Value	Pr > F
DEV_SKL	1	9.83	2.10	0.1781
MOUNTAIN	1	10.3	13.48	0.0041
DEV_SKL*MOUNTAIN	1	8.5	0.06	0.8121
MEAN_SKL	1	10.5	0.61	0.4503

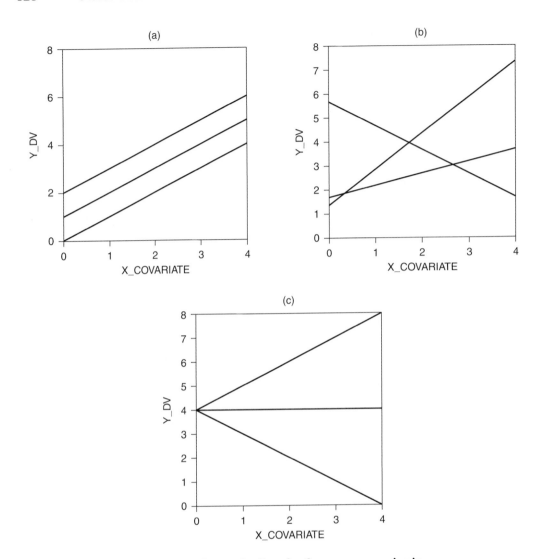

FIGURE 15.4 **Regression lines for three groups varying in**
(a) intercepts but not slopes, (b) both intercepts and slopes, and
(c) slopes but not intercepts. Generated in SYSTAT PLOT.

low to high values. If both intercepts and slopes can be fixed, there would be a single regression line, because regression lines for all groups would be superimposed. Garden-variety (single-level) regression is appropriate in such a case; there is no need for MLM.

The test of random intercepts in the small-sample example asks whether speed differs among runs. It may be the case, however, that those differences disappear with the addition of predictors. If, for example, different runs were chosen on the basis of skiing skill, then variance in intercepts in speed could go away once skill is taken into account. Remember that in any (standard) regression model, including a multilevel model, all effects are adjusted for each other. Note that in the three runs of Section 15.4 intercept variance doesn't disappear but it does become smaller as predictors are added.

In the small-sample example, is the relationship between skill (the IV) and speed (the DV) the same for all runs? If you fix the value for slopes in your model, you are assuming that the relationship between skill and speed is constant over runs. Unless you know that variability in slopes over runs is negligible, you should allow for random slope coefficients. In this example, the assumption that the relationship between skill and speed is constant (fixed) over runs is untenable because one would expect a much stronger relationship between skill and speed on more difficult ski runs. The price of inclusion of random slope coefficients is reduced power for the test for the DV–predictor association, because random effects usually have much larger standard errors than fixed effects.

Decisions about fixed versus random slopes apply separately to each predictor in a model. On the other hand, the decision about fixed versus random intercepts is made for the model as a whole, disregarding any predictors. Do the groups differ in average speed? That is, do runs have different speeds averaged over skiers? This is a question best answered through the intraclass correlation (Section 15.6.1).

The assumption that the slopes are constant is testable in a multilevel model in which random slope coefficients are specified as the test of the variance of slope. In the small-sample example with a level-1 predictor, slope variance $= 0.2104$ with a standard error of 0.1036 in the SAS and SPSS output of Tables 15.7 and 15.8, respectively. This produces a z value of $(0.2104/0.1036 =)$ 2.03, significant at the one-tailed $\alpha = .05$ level. Thus, random slope coefficients are appropriate in this model. The presence of significant heterogeneity of slopes also shows the difficulty in interpreting the nonsignificant DV–predictor association (here, between speed and skill) as unimportant. Recall from ANOVA that main effects (in this case the DV–IV association) cannot be unambiguously interpreted in the presence of interaction (violation of the assumption of homogeneity of slopes).

Failing to account for random slopes can have serious statistical and interpretational consequences. In the small-sample example, including skill as a fixed, but not random, effect (model not shown) produces a significant effect of skill. That is, one would conclude that speed is positively related to skill on the basis of such a model. However, as we have seen in Figure 15.2, that conclusion is incorrect for four of the runs.

Examination of the differences in slopes over the groups may be of substantive interest. For example, the slopes (and intercepts) for all of the runs in the small-sample example are illustrated in Figure 15.2 and suggest an interesting pattern. The slope for the Aspen run shows little relationship between skill and speed for that run. The remaining slopes, for the Mammoth runs, appear to group themselves into two patterns. Five of the runs have relatively low intercepts and positive relationships with skill. Runs are slow on average, but the greater the skill of the skier, the faster the speed. This suggests that the runs are difficult, but the difficulty can be overcome by skill. The remaining four runs have relatively high intercepts and negative relationships with skill. That is, these are fast runs but, for some reason, skilled skiers traverse them more slowly (perhaps to view the scenery). The pattern of negative relationships between intercepts and slopes is a common one, reflecting floor and ceiling effects. Those at the top have less room to grow than those at the bottom.

Higher-level variables also can have fixed or random slopes, but slopes are necessarily fixed for predictors in the highest level of analysis. Thus, mountain is considered a fixed level-2 IV in the small-sample example of Section 15.4.3. If it turns out that the intercept and slopes for *all* predictors at the first level can be treated as fixed, a single-level model may be a good choice.

15.6.5   Statistical Inference

Three issues arise regarding statistical inference in MLM. Is the model any good at all? Does making a model more complex make it any better? What is the contribution of individual predictors?

15.6.5.1   Assessing Models

Does a model predict the DV beyond what would be expected by chance, that is, does it do any better than an intercepts-only model? Is one model better than another? The same procedures are used to ask if the model is at all helpful and to ask whether adding predictors improves it. The familiar χ^2 likelihood-ratio test (e.g., Equation 10.7) of a difference between models is used as long as models are nested (all of the effects of the simpler model are in the more complex model) and full ML (not REML) methods are used to assess both models.

There are several ways of expressing the test, depending on the information available in the program used. Table 15.18 shows χ^2 equations using terminology from each of the software packages.

Degrees of freedom for the χ^2 equations of Table 15.18 are the difference in the number of parameters for the models being compared. Recall that SPSS provides the total number of parameters directly in the **Model Dimension** section of output (cf. Table 15.5). SAS presents the information in the **Dimensions** section in the form of **Covariance Parameters** plus **Columns in X** (cf., Table 15.4).

The test to answer the question, "Does the model predict better than chance?", pits the intercept-only model of Table 15.5 (with a **−2 Log Likelihood** value of 693.468 and 3 df) against the full model of Table 15.11 (with a **−2 Log Likelihood** value of 570.318 and 7 df). From Table 15.18:

$$\chi^2 = 693.468 - 570.318 = 123.15$$

This value is clearly statistically significant with $(7 - 3) = 4$ df, so the full model leads to prediction that is significantly better than chance. SAS MIXED provides the **Null Model Likelihood Ratio Test** routinely for all models.

To test for differences among nested models, chi-square tests are used to evaluate the consequences of making effects random, or to test a dummy-coded categorical variable as a single effect,

---

TABLE 15.18   Equations for Comparing Models Using Various Software Packages

Program	Equation
SAS MIXED	$\chi^2 = $ `(-2 Log Likelihood)`$_s - $ `(-2 Log Likelihood)`$_c$
SPSS MIXED	$\chi^2 = $ (−2 Log Likelihood)$_s - $ (−2 Log Likelihood)$_c$
MLwiN	$\chi^2 = (-2*loglikelihood)_s - (-2*loglikelihood)_c$
HLM	$\chi^2 = $ (Deviance)$_s - $ (Deviance)$_c$
SYSTAT MIXED REGRESSION	$\chi^2 = 2$ `(Log Likelihood)`$_c - 2$ `(Log Likelihood)`$_s$

Note: Subscript s = simpler model, c = more complex model.

or for model building in general (Section 15.6.8). Also, Wald tests of individual predictors can be verified, especially when samples are small, by testing the difference in models with and without them. Note, however, that if the test is for a random effect (variance component) other than covariance, the obtained p value associated with the chi-square difference test should be divided by two in order to create a one-tailed test of the null hypothesis that variance is no *greater* than expected by chance (Berkhof & Snijders, 2001).

Non-nested models can be compared using the AIC that is produced by SAS and SPSS (Hox, 2002). AIC can be calculated from the deviance (which is –2 times the log-likelihood) as:

$$AIC = d + 2p \qquad (15.18)$$

where d is deviance and p is the number of estimated parameters. Although no statistical test is available for differences in AIC between models, the model with a lower value of AIC is preferred.

15.6.5.2 Tests of Individual Effects

The programs reviewed provide standard errors for parameter estimates, whether random (variances and covariances[11]) or fixed. Some also provide z values (parameter divided by standard error—Wald tests) for those parameters, and some add p (probability) values as well. These test the contribution to the equation of the predictors represented by the parameters. However, there are some difficulties with these tests.

First, the standard errors are valid only for large samples, with no guidelines available as to how large is large enough. Therefore, it is worthwhile to verify the significance of a borderline predictor using the model-comparison procedure of Section 15.6.5.1. Second, Raudenbush and Bryk (2001) argue that, for fixed effects, the ratio should be interpreted as t (with df based on number of groups) rather than z. They also argue that the Wald test is not appropriate for variances produced by random effects (e.g., variability among groups) because the sampling distributions of variances are skewed. Therefore, their HLM program provides chi-square tests of random effects. That is, the tests as to whether intercepts differ among groups is a chi-square test.

When tests with and without individual predictors (Section 15.6.5.1) are used for random effects, recall that, except for covariances, one-tailed tests are appropriate. One wants to know if the predictor differs among higher level units (e.g., among level-2 groups) *more* than expected by chance? Therefore, the p value for the chi-square difference test should be divided by two (Hox, 2002).

An example of this test applied to a fixed predictor is available by comparing results of Sections 15.4.2 and 15.4.3, in which the level-2 IV, mountain, is added to a model that already has the level-1 predictor, skill:

$$\chi^2 = 587.865 - 570.318 = 17.55$$

With 1 df produced by adding the single level-2 predictor to the less complex model, this is clearly a statistically significant result. The model is improved by the addition of mountain as a predictor.

---

[11]Recall that variances represent differences among groups in slopes or intercepts. Covariances are relationships between slopes and intercepts or between slopes for two predictors if there is more than one predictor considered random.

An application of the test to addition of a random predictor is available by comparing the models of Sections 15.4.1 and 15.4.2, in which SKILL is added to the intercepts-only model. Using the most common form of the test:

$$\chi^2 = 693.468 - 587.865 = 105.6$$

This is clearly a significant effect with $6 - 3 = 3$ df. Remember to divide the probability level by two because the predictor is specified to be random.

15.6.6   Effect Size

Recall that effect size (strength of association) is the ratio of systematic variance associated with a predictor or set of predictors to total variance. How much of the total variance is attributable to the model?

Kreft and DeLeeuw (1998) point out several ambiguities in the ill-defined methods currently available for calculating effect size in MLM. Counterintuitively, the error variances on which these measures are based can increase when predictors are added to a model, so that there can be "negative effect sizes." Further, between-groups and within-groups estimates are confounded unless predictors are centered.

Kreft and DeLeeuw (1998) provide some guidelines for calculation of η^2, with the caution that these measures should only be applied to models with random intercepts and should not be applied to predictors with random slopes.[12] Further, separate calculations are done for the within-groups (level-1) and between-groups (level-2) portions of the MLM, because residual variances are defined differently for the two levels.

In general, for fixed predictors an estimate of effect size is found by subtracting the residual variance with the predictor (the larger model) from the residual variance of the intercepts-only model (the smaller model), and dividing by the residual variance without the predictor[13]:

$$\eta^2 = \frac{s_1^2 - s_2^2}{s_1^2} \tag{15.19}$$

where s_1^2 is the residual variance of the intercepts-only model and s_2^2 is the residual variance of the larger model (note that the larger model generally has the smaller residual variance). There is as yet no convenient method for finding confidence intervals around these effect sizes.

Refer to Kreft and DeLeeuw (1998, pp. 115–119) for a full discussion of the issues involved and definitions of these variances at the within-group and between-group levels of analysis for those relatively few instances when the measures can be applied and interpreted.

---

[12]Calculations for models with random slopes are much more difficult. Snijders and Bosker (1999, p. 105) point out that the required values are available in the HLM software. They also suggest that effect sizes calculated in Equation 15.19 do not differ much if the values are taken from a run in which only the fixed part of the slopes is included.

[13]This is different from ρ, the intraclass correlation. The intraclass correlation evaluates difference in variation between and within groups without consideration of predictors. The current effect size measure evaluates predicted variance–improvement in a model due to fixed predictors.

15.6.7   Estimation Techniques and Convergence Problems

As in all iterative procedures, a variety of estimation algorithms are available, with different programs offering a somewhat different choice among them. Table 15.19 shows the methods relevant to MLM in several programs.

The most common methods are maximum likelihood and restricted maximum likelihood. Maximum likelihood (ML) is a good choice when nested models are to be compared (e.g., when an effect has several parameter estimates to evaluate as when a categorical variable is represented as a series of dichotomous dummy variables, or when a comparison with an intercepts-only model is desired). In the case of categorical variables, models are compared with and without the set of dummy-coded predictors representing the categorical variable of interest. Maximum likelihood is the only method available in SYSTAT MIXED REGRESSION; the MLwiN form of ML is IGLS.

Restricted maximum likelihood (REML) estimates the random components averaged over all possible values of fixed effects, as opposed to ML, which estimates random components as well as fixed level-2 coefficients by maximizing their joint likelihood (Raudenbush et al., 2000). The advantage of REML is that the estimates of variances and covariances (random coefficients) depend on interval rather than fixed estimates of fixed effects. The method is more realistic and less biased because it adjusts for uncertainty about the fixed effects. The disadvantage is that the chi-square difference, likelihood-ratio test of Table 15.18, is available only for testing random coefficients. That is, REML cannot be used compare nested models, which differ in their fixed components. The two methods (ML and REML) produce very similar results when the number of level-2 units is large, but REML produces better estimates than ML when there are few level-2 units (Raudenbush & Bryk, 2001). The MLwiN form of REML is RIGLS. MLwiN also has some Bayesian modeling methods, fully discussed in the manual (Rasbash et al., 2000).

TABLE 15.19   Estimation Methods Available in Software Programs

Estimation Method	Programs Providing	Comments
Maximum Likelihood (ML)	SAS SPSS HLM SYSTAT (only)	Can be used for testing pairs of nested models.
Restricted Maximum Likelihood (REML)	SAS (default) SPSS (default) HLM (default)	Random components estimated averaging over all possible values of fixed effects. Recommended when not testing pairs of nested models.
Iterative Generalized Least Squares (IGLS)	MLwiN (default)	An iterative version of generalized least squares. Produces results congruent with ML.
Restrictive Generalized Least Squares (RIGLS)	MLwiN	Leads to unbiased estimates of random parameters. Equivalent to REML with normal random variables.
Minimum Variance Quadratic Unbiased Estimation (MIVQUE0)	SAS	Recommended only for large data sets and when ML and REML fail to converge

Convergence problems are common in MLM. All forms of maximum likelihood estimation require iterations, and either the number of iterations to convergence may be very large or convergence may never happen. What's more, different programs vary widely in the default number of iterations allowed and how that number is increased. Usually lack of convergence occurs simply because the model is bad. However, if samples are small even a good model may not produce convergence (or may require many, many iterations). Or there may be numerous random predictors with very small effects (e.g., the actual variance in slopes over groups may be negligible). A solution to this problem is to try changing random predictors to fixed.

Another solution in SAS is to use minimum variance quadratic unbiased estimation (`MIVQUE0`). The procedure does not require the normality assumption and does not involve iteration. Actually, SAS uses MIVQUE0 estimates as starting values for ML and REML. However, the procedure should be used with caution and only if there is difficulty in convergence with ML and REML (Searle, Casella, & McCulloch, 1992; Swallow & Monahan, 1984).

Hox (2002) points out that generalized least squares (GLS) estimates of coefficients are obtained from ML solutions when only one iteration is allowed. Thus, GLS estimates are the ML analog to MIVQUE0. These estimates are accurate when samples are very large. Although estimates produced through GLS are less efficient, and have inaccurate standard errors, they at least provide some information about the nature of the model, and may help diagnose failure to converge.

15.6.8   Exploratory Model Building

MLM often is conducted through a series of runs in which a model is built up. If there are numerous potential predictors, they are first screened through linear regression to eliminate some that are obviously not contributing to prediction. Hox (2002) provides a helpful step-by-step exploratory strategy to select an MLM model.

- Analyze the simplest intercept-only model (the Null model) and examine the intraclass correlation (Section 15.6.1).
- Analyze a model with all level-1 predictors (e.g., SKILL) fixed. Assess the contribution of each predictor and/or look at the differences in the models, using the techniques of Sections 15.6.5.1 and 15.6.5.2.
- Assess models in which the slope for each predictor is permitted to be random one at a time; include predictors that were nonsignificant in step two because they may vary among level-2 units, as was the case for SKILL in the small-sample example. If the intraclass correlation is sufficiently small and there are no significant random effects of predictors, a simple single-level regression analysis may be used.
- Test the difference between the model with all necessary random components and the model from the second step in which all predictors were fixed (see Section 15.6.5.1 for figuring the difference in the number of parameters to use as df).
- Add higher-level predictors and cross-level interactions. (Recall that ML rather than REML must be used to compare models unless comparisons are made only between random predictors.)

Alternatively, a top-down approach may be used for building a model. That is, you can start with the most complex model, which includes all possible random effects as well as higher-level predictors and cross-level interactions, and then systematically eliminate nonsignificant effects. Indeed, this is probably a more efficient strategy if it works, but may not work because the most complex model often is likely to result in a failure to converge on a solution.

If the overall sample is large enough to support it, exploratory model-building procedures are best tested with cross-validation. Half the sample is used to build the model, and the other half for cross-validation. Otherwise, the results of the exploratory technique may be unduly influenced by chance. Even with a model based on theory, the full model as hypothesized may well fail to provide a solution. These "exploratory" techniques may be used to tweak the model until an acceptable solution can be found. Such modification, of course, is reported in the results.

15.7   Complete Example of MLM

These data are from field studies of the effects of nighttime aircraft noise in the vicinity of two airports and one control site. Airports were Castle Air Force Base in Merced, CA (site 1), neighborhoods in the Los Angeles area that were not exposed to nighttime aircraft noise but were exposed to high levels of road traffic noise (site 2), and LAX (site 3). For the current analysis, 50 participants were selected in 24 households with at least two participants each providing data for at least 3 consecutive nights. Interior noise levels (NIGHTLEQ) were monitored between 10:00 PM and 8:00 AM. Each test participant used a palmtop computer at bedside to answer an evening and morning questionnaire, including items about time taken to fall asleep the previous night (LATENCY) and annoyance by aircraft noise the previous night (ANNOY). Latency was measured on a scale of 1 to 5 (1 = less than 10 minutes, 2 = 1 0–20 minutes, 3 = 20–30 minutes, 4 = 30–60 minutes, 5 = more than an hour).

Annoyance served as the DV for the MLM analysis, with nights as a repeated-measures first-level unit. ANNOY was measured on a scale of 0 to 5 (0 = not at all annoyed to 5 = extremely annoyed), First-level predictors were LATENCY and NIGHTLEQ. Participants served as the second-level unit, with AGE as the predictor. Households served as the third-level unit with SITE1 (Castle AFB vs. other sites) and SITE2 (control neighborhoods vs. other sites) as dummy-variable predictors. Thus, we have a complex three-level model with observed predictors at each level and no hypothesized interactions. Appendix B provides additional information about the Fidell, Pearsons, Tabachnick, Howe, Silvati, and Barber (1995) research. Data files are MLM.*.

Figure 15.5 shows the layout for the data to be analyzed in this example.

15.7.1   Evaluation of Assumptions

15.7.1.1   Sample Sizes, Missing Data, and Distributions

There were 747 nights of data collected from 50 participants (only those participants providing at least 3 nights of data were included in this analysis) residing in 24 households (only households with at least 2 participants were included). This is not a very large sample for MLM, so that convergence difficulties may be anticipated, particularly with this relatively large number of predictors. The existence of only 24 households (and few participants per household) is particularly problematic.

First-level variables were ANNOY (the DV), NIGHTLEQ, and LATENCY. Two participants each failed to provide latency values for one night; missing values were replaced with the average latency for that participant (1.22 for SUBNO = 219 and 2.5 for SUBNO = 323). SPSS FREQUENCIES provided descriptive statistics and histograms shown in Table 15.20.

All three variables have extreme positive skewness. LATENCY and NIGHTLEQ are considerably improved with logarithmic transformations, however, modeling with and without transformation of the variables produces results that do not differ substantively. Therefore, the decision was made to model untransformed predictors in the interest of interpretability. Various transformations of

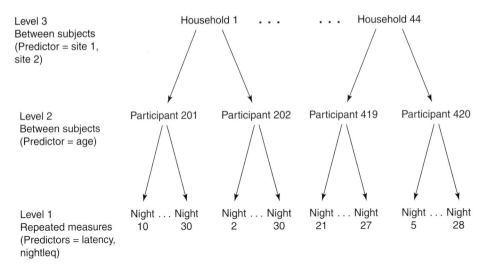

FIGURE 15.5 **Layout of data in complete example.**

TABLE 15.20 **Descriptive Statistics for First-level Variables Using SPSS Frequencies**

FREQUENCIES
  VARIABLES=nightleq latency annoy /FORMAT=NOTABLE
  /STATISTICS=STDDEV MINIMUM MAXIMUM MEAN SKEWNESS SESKEW KURTOSIS SEKURT
  /HISTOGRAM NORMAL
  /ORDER= ANALYSIS.

Frequencies

Histogram

Statistics

		NIGHTLEQ	LATENCY	ANNOY
N	Valid	747	745	747
	Missing	0	2	0
Mean		74.0855	1.7651	1.27
Std. Deviation		7.73557	1.00796	1.341
Skewness		1.090	1.447	.829
Std. Error of Skewness		.089	.090	.089
Kurtosis		1.572	1.705	-.218
Std. Error of Kurtosis		.179	.179	.179
Minimum		60.60	1.00	0
Maximum		111.50	5.00	5

TABLE 15.20 Continued

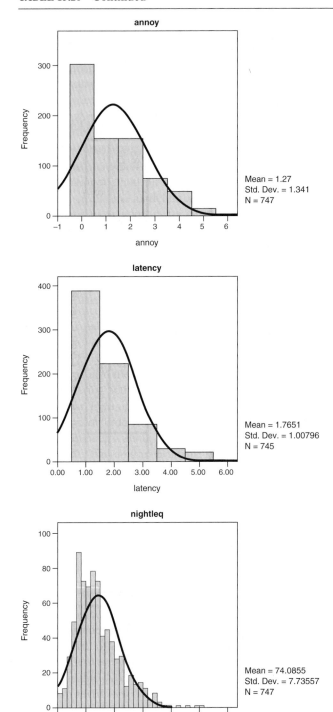

Mean = 1.27
Std. Dev. = 1.341
N = 747

Mean = 1.7651
Std. Dev. = 1.00796
N = 745

Mean = 74.0855
Std. Dev. = 7.73557
N = 747

ANNOY increase the negative kurtosis to unacceptable levels, so the decision is made to leave the DV untransformed as well.

AGE, the only second-level variable, is acceptably distributed, as seen in Table 15.21, using a reduced data set in which there is one record per participant (as differentiated from the major data set which has one record for each participant-night combination).

Frequency distributions are shown in Table 15.22 for the two dichotomous third-level predictors, SITE1 and SITE2, using a further reduced data set in which there is one record per household. Distributions are not optimal, but there are more than 10% of the households in the least frequent site (non-airport neighborhoods).

15.7.1.2   Outliers

At least one univariate outlier with extremely high noise level was noted in the transformed data (L_LEQ, $z = 4.19$, not shown). A check of multivariate outliers through SPSS REGRESSION (Table 15.23) shows three extreme cases (sequence numbers 73, 74, and 75 from participant #205) to be beyond the critical χ^2 of 13.815 for 2 df at $\alpha = .001$.

Examination of the original data revealed that several of the noise levels for participant #205 were highly discrepant from those of the housemate (participant #206) and were probably recorded erroneously. Thus, noise values for nights 17 through 23 for participant #205 were replaced with those recorded for participant #206. This produced acceptable Mahalanobis distance values for all cases.

TABLE 15.21   Descriptive Statistics for Second-level Predictor through SPSS Frequencies

FREQUENCIES
 VARIABLES=age /FORMAT=NOTABLE
 /STATISTICS=STDDEV MAXIMUM MEAN SKEWNESS SESKEW KURTOSIS SEKURT
 /HISTOGRAM NORMAL
 /ORDER= ANALYSIS.

Frequencies

Statistics

AGE		
N	Valid	50
	Missing	0
Mean		48.060
Std. Deviation		17.549
Skewness		.033
Std. Error of Skewness		.337
Kurtosis		−1.380
Std. Error of Kurtosis		.662
Minimum		19.0
Maximum		78.0

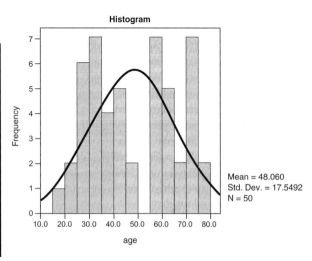

Histogram

Mean = 48.060
Std. Dev. = 17.5492
N = 50

TABLE 15.22   Frequency Distributions for the Third-level Predictors through SPSS Frequencies

```
FREQUENCIES
 VARIABLES=site1 site2
 /STATISTICS=STDDEV MEAN SKEWNESS SESKEW KURTOSIS SEKURT
 /ORDER= ANALYSIS .
```

Frequencies

Frequency Table

SITE1

		Frequency	Percent	Valid Percent	Cumulative Percent
Valid	0	16	66.7	66.7	66.7
	1				
	Total				

SITE2

		Frequency	Percent	Valid Percent	Cumulative Percent
Valid	0	21	87.5	87.5	87.5
	1	3	12.5	12.5	100.0
	Total	24	100.0	100.0	

Table 15.21 reveals no univariate outliers for the single second-level predictor, AGE. Splits were not too highly discrepant for the third-level predictors (SITE1 and SITE2, dichotomous variables) so that there were no outliers at that level.

15.7.1.3   *Multicollinearity and Singularity*

There are no interactions to be modeled, so no problems concerning collinearity are anticipated. A multiple regression run through SPSS REGRESSION that included the five predictors from all levels (Table 15.24) revealed no cause for concern about collinearity, despite the rather high condition index for the 6th dimension.

15.7.1.4   *Independence of Errors: Intraclass Correlations*

Intraclass correlation is evaluated by running a three-level (nights, subjects, and households) model through SPSS, with random intercepts but no predictors. Table 15.25 shows the syntax and relevant output. Note that SPSS changes the Covariance Structure from COVTYP(UN) to Identity whenever there is a random effect with only one level (i.e., `Intercept`).

The null model has four parameters, one each for the fixed intercept (grand mean), variability in participant intercepts, variability in household intercepts, and residual variance.

TABLE 15.23   Syntax and Selected SPSS REGRESSION Output for Multivariate Outliers for Level 1

```
REGRESSION
 /MISSING LISTWISE
 /STATISTICS COEFF OUTS R ANOVA COLLIN TOL
 /CRITERIA=PIN(.05) POUT(.10)
 /NOORIGIN
 /DEPENDENT annoy
 /METHOD=ENTER nightleq latency
 /RESIDUALS=OUTLIERS(MAHAL).
```

Outlier Statistics[a]

		Case Number	Statistic
Mahal. Distance	1	75	23.408
	2	74	21.026
	3	73	16.494
	4	72	13.491
	5	665	13.476
	6	212	12.354
	6	71	11.834
	8	554	11.768
	9	21	11.648
	10	645	11.557

[a]Dependent Variable: annoy

Applying Equation 15.15 for the second level:

$$\rho = \frac{s^2_{bg2}}{s^2_{bg2} + s^2_{bg3} + s^2_{wg}} = \frac{0.28607}{0.28607 + 0.41148 + 1.12096} = .16$$

With about 16% of the variability in annoyance associated with individual differences (differences among participants), an MLM is advisable.

Applying Equation 15.16 for the third level:

$$\rho = \frac{s^2_{bg2}}{s^2_{bg2} + s^2_{bg3} + s^2_{wg}} = \frac{0.41148}{0.28607 + 0.41148 + 1.12096} = .23$$

With about 23% of the variance in annoyance associated with the third level of the hierarchy (differences among households) a three-level MLM is advisable.

15.7.2   Multilevel Modeling

A three-level model is hypothesized with predictors at all three levels (noise, time to fall asleep, age, and site). Recall that no interactions are hypothesized, either within or between levels. Only nightleq

TABLE 15.24   Syntax and Selected SPSS REGRESSION Output for Multicollinearity

```
REGRESSION
 /MISSING LISTWISE
 /STATISTICS COEFF OUTS R ANOVA COLLIN TOL
 /CRITERIA=PIN(.05) POUT(.10)
 /NOORIGIN
 /DEPENDENT annoy
 /METHOD=ENTER age nightleq latency site1 site2 .
```

Regression

Coefficients[a]

Model		Unstandardized Coefficients B	Unstandardized Coefficients Std. Error	Standardized Coefficients Beta	t	Sig.	Collinearity Statistics Tolerance	Collinearity Statistics VIF
1	(Constant)	−2.110	.517		−4.079	.000		
	age	.012	.003	.151	3.903	.000	.802	1.247
	nightleq	.035	.007	.193	5.097	.000	.834	1.199
	latency	.192	.047	.144	4.104	.000	.967	1.034
	site1	−.183	.106	−.066	−1.733	.084	.824	1.214
	site2	−.431	.159	−.111	−2.713	.007	.707	1.413

[a]Dependent Variable: annoy

Collinearity Diagnostics[a]

Model	Dimension	Eigen-value	Condition Index	Variance Proportions (Constant)	age	night-leq	latency	site1	site2
1	1	4.281	1.000	.00	.00	.00	.01	.01	.00
	2	1.021	2.048	.00	.00	.00	.00	.11	.45
	3	.447	3.096	.00	.02	.00	.02	.77	.25
	4	.193	4.713	.00	.06	.00	.93	.00	.04
	5	.055	8.822	.03	.91	.03	.04	.04	.20
	6	.004	32.006	.97	.01	.96	.00	.06	.05

[a]Dependent Variable: annoy

is predicted to have a random slope; individual differences are expected in the relationship between noise and annoyance.

A model in which noise was treated as random failed to converge, even when number of iterations was increase to 500 and probability of convergence was relaxed to .001. Therefore, the decision was made to treat all predictors as fixed effects. Table 15.26 shows syntax and output for the full three-level model.

The full model has nine parameters, the four noted in Table 15.25 for the random effects of intercepts and residual as well as the fixed effect of the intercept, plus one parameter each for the five fixed predictors.

For random effects, there was significant variability in annoyance among participants within households ($p = .005/2 = .0025$) as well as significant variability in annoyance among households ($p = .048/2 = .024$). Unfortunately, there also was significant residual (unexplained) variance over nights after taking into account all effects in this hierarchical model ($p < .001$).

For fixed effects, nighttime noise level ($p < .001$) and time to fall asleep ($p = .010$) significantly predicted annoyance when averaged over participants and households. The parameter estimates show that annoyance is greater when nighttime noise was louder; for each leq unit increase in

TABLE 15.25   Three-level Intercepts-only Model through SPSS MIXED (Syntax and Selected Output)

```
MIXED
   annoy BY site1 site2 WITH nightleq latency
   /CRITERIA = CIN(95) MXITER(100) MXSTEP(5) SCORING(1)
   SINGULAR(0.000000000001) HCONVERGE(0, ABSOLUTE) LCONVERGE(0, ABSOLUTE)
   PCONVERGE(0.000001, ABSOLUTE)
   /FIXED = | SSTYPE(3)
   /METHOD = ML
   /PRINT = SOLUTION TESTCOV
   /RANDOM INTERCEPT | SUBJECT(subno) COVTYPE(UN)
   /RANDOM INTERCEPT | SUBJECT(house) COVTYPE(UN) .
```

Model Dimension[a]

		Number of Levels	Covariance Structure	Number of Parameters	Subject Variables
Fixed Effects	Intercept	1		1	
Random Effects	Intercept	1	Identity	1	subno
	Intercept	1	Identity	1	house
Residual				1	
Total		3		4	

[a]Dependent Variable: annoy.

Information Criteria[a]

−2 Log Likelihood	2307.764
Akraike's Information Criterion (AIC)	2315.764
Hurvich and Tsai's Criterion (AICC)	2315.818
Bozdogan's Criterion (CAIC)	2338.228
Schwarz's Bayesian Criterion (BIC)	2334.228

The information criteria are displayed in smaller-is-better forms.

[a]Dependent Variable: annoy.

TABLE 15.25   Continued

Covariance Parameters

Estimates of Covariance Parameters[a]

					95% Confidence Interval	
Parameter	Estimate	Std. Error	Wald Z	Sig.	Lower Bound	Upper Bound
Residual	1.120963	.0599997	18.683	.000	1.0093239	1.2449502
Intercept [subject Variance	.2860726	.1044476	2.739	.006	.1398605	.5851367
Intercept [subject Variance	.4114774	.1789685	2.299	.021	.1754378	.9650918

[a]Dependent Variable: annoy.

noise level, annoyance increased by about 0.04 on a scale of 0 to 5. Annoyance also increased on nights when the time to fall asleep was greater. Each one-unit increment in time to fall asleep increased annoyance by about 0.12. No statistically significant effects were found for age or either of the dummy variables for site.

Comparison between this model and the intercepts-only model of Table 15.25 shows that annoyance is predicted at better-than-chance level as a result of the set of two predictors. Following the equation in Table 15.18

$$\chi^2 = 2307.764 - 2260.286 = 47.48$$

A significant difference with $(9 - 4) = 5$ df at $\alpha = .05$.

A final, parsimonious, model eliminates the three nonsignificant effects (age and the two dummy-coded site variables). Table 15.27 shows the final model.

Comparison between this model and the full model of Table 15.26 shows that prediction of annoyance does not suffer when the three predictors are dropped. Following the equation in Table 15.18

$$\chi^2 = 2263.414 - 2260.286 = 3.13$$

a nonsignificant difference with $(9 - 6) = 3$ df at $\alpha = .05$. Table 15.28 compares the three models.

TABLE 15.26   Three-Level Model of Annoyance as Predicted by Noise Level, Time to Fall Asleep, Age, and Site (SPSS MIXED Syntax and Selected Output)

```
MIXED
 annoy BY site1 site2 WITH age nightleq latency
 /CRITERIA = CIN(95) MXITER(500) MXSTEP(10) SCORING(1)
 SINGULAR(0.000000000001) HCONVERGE(0, ABSOLUTE) LCONVERGE(0, ABSOLUTE)
 PCONVERGE(0.001, ABSOLUTE)
 /FIXED = age nightleq latency site1 site2 | SSTYPE(3)
 /METHOD = ML
 /PRINT = SOLUTION TESTCOV
 /RANDOM = INTERCEPT | SUBJECT(subno ) COVTYPE(UN)
 /RANDOM = INTERCEPT | SUBJECT(house ) COVTYPE(UN)
```

(continued)

TABLE 15.26 Continued

Mixed Model Analysis

Model Dimension[a]

		Number of Levels	Covariance Structure	Number of Parameters	Subject Variables
Fixed Effects	Intercept	1		1	
	age	1		1	
	nightleq	1		1	
	latency	1		1	
	site1	2		1	
	site2	2		1	
Random Effects	Intercept	1	Identity	1	subno
	Intercept	1	Identity	1	house
Residual				1	
Total		10		9	

[a]Dependent Variable: annoy.

Information Criteria[a]

–2 Log Likelihood	2260.286
Akraike's Information Criterion (AIC)	2278.286
Hurvich and Tsai's Criterion (AICC)	2278.530
Bozdogan's Criterion (CAIC)	2328.830
Schwarz's Bayesian Criterion (BIC)	2819.830

The information criteria are displayed in smaller-is-better forms.

[a]Dependent Variable: annoy

Fixed Effects

Type III Tests of Fixed Effects[a]

Source	Numerator df	Denominator df	F	Sig.
Intercept	1	163.137	16.794	.000
age	1	36.847	.671	.418
nightleq	1	718.546	37.464	.000
latency	1	744.681	6.627	.010
site1	1	22.899	.564	.460
site2	1	24.387	1.223	.280

[a]Dependent Variable: annoy.

TABLE 15.26 Continued

Estimates of Fixed Effects[a]

Parameter	Estimate	Std. Error	df	t	Sig.	95% Confidence Interval	
						Lower Bound	Upper Bound
Intercept	-3.195	.786	77.210	-4.063	.000	-4.761	-1.630
age	.007	.009	36.847	.819	.418	-.011	.025
nightleq	.043	.007	718.546	6.121	.000	.029	.057
latency	.122	.047	744.681	2.574	.010	.029	.214
[site1=0]	.249	.332	22.899	.751	.460	-.437	.935
[site1=1]	0[a]	0	.	.	.	.	.
[site2=0]	.569	.515	24.387	1.106	.280	-.492	1.630
[site2=1]	0[a]	0	.	.	.	.	.

[a]This parameter is set to zero because it is redundant.

[b]Dependent Variable: annoy.

Covariance Parameters

Estimates of Covariance Parameters[a]

Parameter		Estimate	Std. Error	Wald Z	Sig.	95% Confidence Interval	
						Lower Bound	Upper Bound
Residual		1.049	.056	18.661	.000	.944	1.165
Intercept [subject	Variance	.317	.114	2.780	.005	.157	.642
Intercept [subject	Variance	.332	.168	1.974	.048	.123	.895

[a]Dependent Variable: annoy.

845

Table 15.29 displays the results in a format for journal reporting.

Table 15.30 is a checklist of items to consider in MLM. An example of a Results section in journal format appears after Table 15.30.

TABLE 15.27 **Final Three-Level Model of Annoyance as Predicted by Noise Level and Time to Fall Asleep (SPSS MIXED Syntax and Selected Output)**

```
MIXED
  annoy BY site1 site2 WITH age nightleq latency
  /CRITERIA = CIN(95) MXITER(500) MXSTEP(10) SCORING(1)
  SINGULAR(0.000000000001) HCONVERGE(0, ABSOLUTE) LCONVERGE(0, ABSOLUTE)
  PCONVERGE(0.001, ABSOLUTE)
  /FIXED = nightleq latency | SSTYPE(3)
  /METHOD = ML
  /PRINT = SOLUTION TESTCOV
  /RANDOM = INTERCEPT | SUBJECT(subno ) COVTYPE(UN)
  /RANDOM = INTERCEPT | SUBJECT(house ) COVTYPE(UN).
```

Mixed Model Analysis

Model Dimension[a]

		Number of Levels	Covariance Structure	Number of Parameters	Subject Variables
Fixed Effects	Intercept	1		1	
Random Effects	nightleq	1		1	
	latency	1		1	
	Intercept	1	Identity	1	subno
	Intercept	1	Identity	1	house
Residual				1	
Total		5		6	

[a]Dependent Variable: annoy.

Information Criteria[a]

−2 Log Likelihood	2263.414
Akraike's Information Criterion (AIC)	2275.414
Hurvich and Tsai's Criterion (AICC)	2275.527
Bozdogan's Criterion (CAIC)	2309.110
Schwarz's Bayesian Criterion (BIC)	2303.110

The information criteria are displayed in smaller-is-better forms.

[a]Dependent Variable: annoy.

TABLE 15.27 Continued

Fixed Effects

Type III Tests of Fixed Effects[a]

Source	Numerator df	Denominator df	F	Sig.
Intercept	1	532.093	16.048	.000
nightleq	1	710.846	38.674	.000
latency	1	743.913	5.990	.015

[a]Dependent Variable: annoy.

Estimates of Fixed Effects[a]

Parameter	Estimate	Std. Error	df	t	Sig.	95% Confidence Interval	
						Lower Bound	Upper Bound
Intercept	-2.206	.551	532.093	-4.006	.000	-3.288	-1.124
nightleq	.044	.007	710.846	6.219	.000	.030	.058
latency	.116	.047	743.913	2.448	.015	.023	.208

[a]Dependent Variable: annoy.

Covariance Parameters

Estimates of Covariance Parameters[a]

Parameter		Estimate	Std. Error	Wald Z	Sig.	95% Confidence Interval	
						Lower Bound	Upper Bound
Residual		1.049	.056	18.665	.000	.944	1.165
Intercept [subject	Variance	.309	.110	2.797	.005	.153	.622
Intercept [subject	Variance	.427	.190	2.249	.025	.179	1.020

[a]Dependent Variable: annoy.

TABLE 15.28   Comparison of Multilevel Models for Annoyance Due to Noise

Model	−2 Log Likelihood	df	χ^2 Difference Test
Intercepts only	2307.764	4	
Full	2260.286	9	M1 − M2 = 47.48*
Final	2254.000	6	M3 − M2 = 3.13

TABLE 15.29   Results of Final Three-Level Model of Annoyance Due to Nighttime Noise Exposure (Excerpted from Table 15.27)

Random effect at Level 3 (Household Differences)

Effect	Parameter Estimate	Standard Error	Wald Z	p (1-sided)	95% Confidence Interval	
					Lower	Upper
Intercepts	0.427	0.190	2.25	.013	0.179	0.622

Random effect at Level 2 (Individual Differences)

Effect	Parameter Estimate	Standard Error	Wald Z	p (1-sided)	95% Confidence Interval	
					Lower	Upper
Intercepts	0.309	0.110	2.80	.003	0.153	0.622

Random effect at Level 1 (Nights)

Effect	Parameter Estimate	Standard Error	Wald Z	p (1-sided)	95% Confidence Interval	
					Lower	Upper
Residual	1.049	0.056	18.66	<.001	0.944	1.165

Fixed effects (Averaged over Participants and Households)

Effect	Parameter Estimate	Standard Error	t ratio	Approx df	p (2-sided)	95% Confidence Interval	
						Lower	Upper
Intercepts	−2.206	0.551	−4.01	532	<.001	−3.288	−1.124
Noise level	0.044	0.007	6.22	711	<.001	0.030	0.058
Latency	0.116	0.047	2.45	744	.008	0.023	0.208

TABLE 15.30   Checklist for Multilevel Modeling

1. Issues
   a. Adequacy of sample sizes and missing data
   b. Normality of distributions at each level
   c. Absence of outliers at each level
   d. Absence of multicollinearity and singularity
   e. Independence of errors (intraclass correlation)
2. Major analyses
   a. Analysis with first-level predictors
   b. Analysis with second-level predictors and significant first-level predictors, etc.
   c. Determination of final model
      (1) Parameter estimates for final model
      (2) Comparison of final with an intercepts-only model
3. Additional analyses
   a. Adding main effects and/or interactions
   b. Additional exploratory analyses

Results

Hypothesized Model

A three-level hierarchical model assessed the effects of nighttime noise exposure, latency to fall asleep, age, and location on annoyance due to nighttime aircraft noise. It was expected that annoyance would be positively related to noise exposure, latency, age, and proximity to an Air Force base.

First-level units were nights in which respondents participated in the study, with respondents limited to those who provided at least three nights of data, resulting in a total of 747 nights for analysis. Second-level units were the 50 participants residing in the 24 households, comprising the third-level units. Only households with at least two participants were selected for

analysis. Multilevel modeling was implemented through SPSS MIXED
MODELS, Version 13.

Hierarchical models are those in which data collected at different levels of analysis (e.g., people, households, sites) may be
studied without violating assumptions of independence in linear
multiple regression. For example, the fact that individuals respond
together and have the same exposure within a household means that
responses from individuals within each household are not independent of one another. Multilevel modeling takes account of these
dependencies by estimating variance associated with group (e.g.,
household) differences in average response (intercepts) and group
differences in associations (slopes) between predictors and DVs
(e.g., group differences in the relationship between annoyance and
noise). This is accomplished by declaring intercepts and/or slopes
to be random effects. Figure 15.5 shows the layout of the design.

In the hypothesized model, individuals and households are
declared random effects to assess variability among individuals
within households as well as variability among households. Also,
one of the predictors, noise level, was declared a random
effect, reflecting the hypothesis that there would be individual
differences in the association between noise level and
annoyance.

Assumptions

One missing latency value each for two participants was
replaced by the mean for that participant. Extreme positive skewness was noted for latency and noise level, but modeling with logarithmically transformed predictors did not substantively change
the results. Therefore the untransformed values were used. Transformation of annoyance produced unacceptable values of kurtosis,
so that DV also remained untransformed. Distributions for second-

and third-level predictors were acceptable. There were no outliers ($p < .001$) once presumably erroneous noise levels were
replaced by those recorded for the housemate. The intraclass correlations of .16 and .23 for second and third levels, respectively, indicate the value of including participants as a random
second-level unit and households as a random third-level unit.

Multilevel Modeling

One predictor, noise level, initially was entered as a random
effect, based on the hypothesis that there would be individual
differences in the relationship between noise and annoyance. That
model failed to converge, so that the full model considers all of
the predictors to be fixed effects.

The full model as a whole was significantly better than one
in which only the intercepts were included (i.e., differences
among individual and households), $\chi^2(6, N = 747) = 2307.764 -$
$2260.286 = 47.48$, $p < .001$. Thus, the predictors as a group
improved the model beyond that produced by considering variability in individuals and households.

Two of the five predictors were significantly associated with
annoyance, but age and the two indicators of site were not. Therefore, a final model was proposed in which only two fixed predictors were evaluated: nighttime noise level and time to fall
asleep. This model did not differ significantly from the full
model, $\chi^2(3, N = 747) = 2263.414 - 2260.286 = 3.13$, $p > .05$. Table
15.28 summarizes the three models evaluated.

Table 15.29 shows that there are individual and household
differences in intercepts (average annoyance varies for households and participants within households). Also noted is the statistically significant residual, indicating room to improve the
model.

On average, annoyance is positively related to nighttime noise level; for each leq unit increase in noise level, annoyance increased by about 0.04 on a scale of 0 to 5. Annoyance also increased on nights when the time to fall asleep was greater. Each 10 minute increment in time to fall asleep increased annoyance by about 0.12.

Thus, although annoyance differs among individuals and households, there is increased annoyance on average on nights when noise is louder and when it takes longer to fall asleep. No statistically significant associations were found between age and annoyance, nor between site and annoyance. That is, there is no evidence that annoyance due to nighttime noise increases with age, nor is there evidence that annoyance is greater in proximity to an Air Force base, after adjusting for the effect of the noise level itself.

15.8   Comparison of Programs

The programs discussed in this chapter vary widely in the kinds of models they analyze and even in the results of analyses of the same models. These programs have far less in common than those used for other statistical techniques. SAS MIXED is part of the SAS package and is used for many analyses other than MLM. Indeed, at the time of writing this chapter, neither the SAS manuals nor special SAS publications directly address the issue of MLM. SPSS MIXED MODEL is part of the SPSS package starting with Version 11 and has been substantially revised since Version 11.5. MLwiN and HLM are stand-alone programs for MLM. SYSTAT MIXED REGRESSION is part of the SYSTAT package starting with Version 10. Table 15.31 compares features of these programs.

15.8.1   SAS System

The program in the SAS system that handles multilevel modeling is PROC MIXED, although it is not specifically designed for that purpose. The program is so flexible, however, that judicious use of its RANDOM feature and nesting specifications can be applied to a wide variety of MLM models, including those with more than two levels (Suzuki & Sheu, 1999). However, it is probably a good idea to limit the use of PROC MIXED to relatively simple MLM models until there is more infor-

TABLE 15.31   Comparison of Programs for Prediction of Survival Time from Covariates

Feature	SAS MIXED	SPSS MIXED	HLM	MLwiN	SYSTAT MIXED REGRESSION
Input					
Multiple estimation techniques	Yes	Yes	Yes	Yes	No
Handles models with more than 2 levels	Yes	Yes	Yes	Yes	No
Specify random or fixed slopes	Yes	Yes	Yes	Yes	Yes
Accepts files from other software packages	Yes	Yes	Yes	No	Yes
Can be used for data simulation	No	No	No	Yes	No
Syntax mode available for input	Yes	Yes	Yes	No	Yes
Requires explicit column of data for constant	No	No	No	Yes	No
Specify categorical variables and interactions without recode	Yes	Yes	No	Yes	Yes
Specify nonnormal response variables and/or nonlinear model	No[a]	No[b]	Yes	Yes	No
Specify Bayesian modeling	Yes	No	No[c]	Yes	No[c]
Bootstrapping	No	No	No	Yes	No
Special specification for cross-classified models (overlapping groupings)	No	No	Yes	Yes	No
Specify multiple membership models (lower-level units belonging to more than one higher-level unit)	No	No	Yes	Yes	No
Specify latent variables	No	No	Yes	No	No
Specify known variance and covariance values	Yes	No	Yes	No	No
Specify structure of variance-covariance matrix	Yes	Yes	Yes	No	Yes
Specify design weights	Yes	Yes	Yes	No	No
Test specific hypotheses in a single run	Yes	Yes	Yes	No	No
Test and deal with failure of level-1 homogeneity of variance	Yes	No	Yes	No	No
Delete intercept from level-1 model	Yes	Yes	Yes	Yes	Yes
Constrain effects to be equal	Yes	No	Yes	Yes	No
Specify fixed intercept	Yes	Yes	No	Yes	Yes
Specify latent variables free of measurement error	No	No	Yes	No	No

(continued)

TABLE 15.31 Continued

Feature	SAS MIXED	SPSS MIXED	HLM	MLwiN	SYSTAT MIXED REGRESSION
Input *(continued)*					
Special procedure for single-level multiple regression analysis with missing data	No	No	Yes	No	No
Procedure to deal with multiply-imputed data	No	No	Yes	No	No
Control maximum number of iterations	Yes	Yes	Yes	Yes	Yes
Additional controls on iterations	Yes	Yes	Yes	Yes	No
Control number of units for OLS equations	NA	NA	Yes	NA	NA
Determine convergence criterion and/or tolerance	Yes	Yes	Yes	Yes	Yes
Alternatives for correcting unacceptable start values	No	No	Yes	No	No
Alternative approaches for repeated-measures analysis	Yes	Yes	No	No	Yes
Data restructuring for repeated measures	No	No	No	Yes	Yes
Output					
Parameter estimates and standard errors	Yes	Yes	Yes	Yes	Yes
Parameter estimates with robust standard errors	No	No	Yes	No	No
Log-likelihood, −2*log-likelihood value, and/or deviance	Yes	Yes	Yes	Yes	Yes
Other fit statistics	Yes	Yes	No	No	No
t or z ratio or χ^2 for effects	Yes	Yes	Yes	No	Yes
df for effects	Yes[d]	Yes	Yes	No	No
Probability value for tests of effects	Yes	Yes	Yes	No	Yes
Printed confidence limits for fixed and random effects	No	Yes	No	No	No
Number of estimated parameters	No	Yes	Yes	No	No
Summary of model in equation format	No	No	Yes	Yes	No
Reliability for least-squares estimates of level-1 coefficients across set of level-2 units	No	No	Yes	No	No

TABLE 15.31   Continued

Feature	SAS MIXED	SPSS MIXED	HLM	MLwiN	SYSTAT MIXED REGRESSION
Output *(continued)*					
Null/independence model test	Yes	No	No	No	No
Level-1 intercepts and slopes and summary over level-1 units	No	No	Yes	No	No
Results for an OLS analysis and with robust SEs	No	No	Yes	No	No
Prints start values and/or progress of iterations	Yes	No	Yes	No[e]	Yes
Prints variance-covariance (and/or correlation) matrix of parameters	Yes	Yes	Yes	No	Yes
Residual variance-covariance matrices	No	Yes	No	No	No
Intraclass (intracluster) correlation for fixed-coefficient models	No	No	No	No	Yes
Summary of hierarchical structure with sample sizes	No	No	No	Yes	Yes
Univariate statistics	No[f]	Yes	Yes	Yes	Yes
Analysis of residuals	No	No	Yes	Yes	No
Plots					
Scattergrams of variables	No[f]	No	No	Yes	No
Plots of residuals	No[f]	No	Yes	Yes	No
Graph of predicted values	No[f]	No	Yes	Yes	No
Trellis plot	No	No	No	Yes	No
Additional diagnostic plots	No[f]	No	Yes	Yes	Yes
Saved on request					
Residuals and predicted scores	Yes[g]	Yes	Yes	No	Yes
Variance-covariance matrices of estimates and parameters	Yes	No	Yes	No	No
Confidence limits for fixed and random effects	Yes	No	No	Yes	No

[a]PROC NLMIXED handles nonlinear MLM.
[b]Two-level models with dichotomous DVs may be analyzed through SPSS COMPLEX SAMPLES LOGISTIC REGRESSION.
[c]Empirical Bayes estimates are routinely provided of all randomly varying level-1 coefficients.
[d]Appropriate df for fixed effects provided by ddfm=kenwardroger.
[e]Can be seen interactively on screen as analysis progresses.
[f]Available in other programs in package.
[g]Any table from SAS MIXED can be converted to an SAS data set.

mation about its applicability. Only the features of the program that apply to MLM are reviewed here.

SAS MIXED has many options for specifying the structure of a variance-covariance matrix (unstructured is the most common choice for MLM) and permits input of known values. Bayesian modeling can be specified through the `PRIOR` instruction. A number of fit statistics are provided, including deviance (–2 log-likelihood), AIC, AICC, and BIC. SAS also shows the null/independence goodness-of-fit test.

Categorical predictors may be specified directly as `CLASS` variables; however, the default coding is unusual and produces results that may be more difficult to interpret than those of other programs unless adjustments are made, such as recoding the categorical variable in the data set. Degrees of freedom for higher-level fixed effects require adjustment (cf. Table 15.10).

15.8.2   SPSS Package

The MIXED MODEL module of SPSS is a full-featured MLM program. Options are available for specifying fixed and random effects as well as alternative methods for dealing with repeated measures (including specification of the variance-covariance matrix, but not known values within it). As usual, output is well formatted and easy to follow; however, the menu system is somewhat confusing, with rather subtle ways to specify multiple levels of the hierarchy. This is the only program that lets you specify size of confidence intervals for effects and routinely prints them out.

A variety of fit indices (information criteria) are shown. A handy feature is a listing of the number of parameters in a model, and whether each is fixed or random. This is the only program reviewed that makes available the residual variance-covariance matrix.

15.8.3   HLM Program

The HLM program reviewed here is Version 6 (Raudenbush et al., 2004) and is designed to handle both 2-level and 3-level data. Indeed, there are separate modules for the two models (labeled HLM2 and HLM3). The manual makes extensive reference to the textbook on hierarchical linear models written by two of its authors: Raudenbush and Bryk (2001).

The program permits input of SAS, SPSS, STATA, and SYSTAT as well as ASCII data and may use the same file for all levels or separate files for each level. In any event, variables have to be defined for each level, a sometimes confusing process. Analyses with and without robust standard errors are routinely provided in output.

HLM has procedures for dealing with variables measured without error (e.g., gender) and for single-level garden-variety multiple regression analysis with missing data; both are considered latent variable techniques in HLM. The manual also shows how to do an MLM analysis using multiply-imputed data. Also available is a procedure for 2-level MLM when variances and covariances are known rather than to be estimated. HLM provides for analysis of a large variety of nonnormal and nonlinear models.

Parameter estimates and variance-covariances matrices as well as residuals can be printed and saved to file. The residuals file contains Mahalanobis' distances for the level-2 groups.

Level-1 parameters such as intercept and slope(s) are printed as part of output. Also printed are OLS results for fixed effects; comparing these with the final results shows the distortion that would have resulted from the use of simple multiple regression instead of MLM. Extensive facilities are available for Bayesian modeling.

15.8.4 MLwiN Program

This program was developed as part of a project in the Institute of Education at the University of London (Rasbash et al., 2000). It is a comprehensive program that permits a large variety of models to be fit, with up to 5 levels permissible for a model. There is extensive capability for producing graphs to explore, diagnose, and interpret models.

The manual is extremely helpful in setting up both simple and complex models, with numerous examples and special handling of such models as multivariate, repeated-measures, and nonnormal (binary and count) data. There is an extensive simulation facility, including Bayesian modeling and bootstrapping. The manual also shows how to deal with cross-classified data, in which cases are partly but not fully nested, for example when children belong to neighborhoods and schools, but there is overlap between the neighborhoods and schools. Multiple membership models are also possible, in which lower level units can belong to more than one higher level unit, as for example, when students in a longitudinal study change schools.

15.8.5 SYSTAT System

The MIXED REGRESSION module of SYSTAT, as are most modules in SYSTAT, is simple to use and produces output that is easy to interpret. There are even a few special features not widely available in other programs; conversion of repeated measures data to that required for MLM is simple and the intraclass (intracluster) correlation is provided by default when a model is specified without predictors. Although not rich in special features, all the basics are provided as well as two approaches to repeated measures. The major limitation is that only 2-level models can be analyzed.

16 Multiway Frequency Analysis

16.1 General Purpose and Description

Relationships among three or more discrete (categorical, qualitative) variables are studied through multiway frequency analysis or an extension of it called log-linear analysis. Relationships between two discrete variables, say, area of psychology (clinical, general experimental, developmental) and average number of publications a year (0, 1, 2, 3, and 4 or more), are studied through the two-way χ^2 test of association. If a third variable is added, such as number of statistics courses taken (two or fewer vs. more than two), two- and three-way associations are sought through multiway frequency analysis. Is number of publications related to area of psychology and/or to number of statistics courses taken? Is number of statistics courses taken related to area of psychology? Is there a three-way relationship among number of publications, area of psychology, and number of statistics courses taken?

To do a multiway frequency analysis, tables are formed that contain the one-way, two-way, three-way, and higher-order associations. A linear model of (the logarithm of) expected cell frequencies is developed. The log-linear model starts with all of the one-, two-, three-, and higher way associations and then eliminates as many of them as possible while still maintaining an adequate fit between expected and observed cell frequencies. In the preceding example, the three-way association among number of publications, area of psychology, and number of statistics courses is tested first and then eliminated if not statistically significant. Then the two-way associations (number of publications and area of psychology, number of publications and number of statistics courses, area of psychology and number of statistics courses) are tested and, if not significant, eliminated. Finally, there is a one-way test for each of the variables against the hypothesis that frequencies are equal in each cell (e.g., that there are equal numbers of psychologists in each area—a test analogous to equal frequency goodness-of-fit tests in χ^2 analysis).

The researcher may consider one of the variables a DV whereas the others are considered IVs. For example, a psychologist's success as a professional (successful vs. unsuccessful) is studied as a function of number of publications, area of psychology, number of statistics courses taken, and their interactions. Used this way, multiway frequency analysis is like a nonparametric analysis of variance with a discrete DV as well as discrete IVs. However, the method of choice with a discrete DV usually is logistic regression (Chapter 10).

Uses of multiway frequency analysis (MFA) include stability in types of errors made on math tests by boys and girls (Marshall, 1983). Analyzing only incorrect responses, a three-way contingency table was formed for each item: sex by year (1976 through 1979) by distractor (three incorrect alternatives plus no answer). The clear winner among models was limited to second-order (two-way) effects. The distractor by sex association was significant as hypothesized, as was the distractor by year association. This was the best-fit model for 128 of the 160 items.

Pope and Tabachnick (1995) used $2 \times 2 \times 2$ MFAs to investigate each of 10 types of responses (e.g., a client that claimed recovered memories of childhood sex abuse and found external validation that it had occurred?) among 382 licensed psychologists as a function of (a) having at least one male client with that characteristic or not, (b) having at least one female client with that characteristic or not, and (c) therapist orientation—psychodynamic or not. No statistically significant three-way associations were found. Indeed, the only significant relationships found were between male and female clients reporting the experience. For example, therapists who believed the abuse had not really occurred in one or more male clients was more common among therapists who believed the abuse had not occurred in one or more female clients.

16.2   Kinds of Research Questions

The purpose of multiway frequency analysis is to discover associations among discrete variables. Once a preliminary search for associations is complete, a model is fit that includes only the associations necessary to reproduce the observed frequencies. Each cell has its own combination of parameter estimates for the associations retained in the model. The parameter estimates are used to predict cell frequency, and they also reflect the importance of each effect to the frequency in that cell. If one of the variables is a DV, the odds that a case falls into one of its categories can be predicted from the cell's combination of parameter estimates. The following questions, then, are addressed by multiway frequency analysis.

16.2.1   Associations among Variables

Which variables are associated with one another? By knowing which category a case falls into on one variable, can you predict the category it falls into on another? The procedures of Section 16.4 show, for a simple data set, how to determine statistically which variables are associated and how to decide on the level of complexity of associations necessary to describe the relationships.

As the number of variables increases so do the number of potential associations and their complexity. With three variables there are seven potential associations: one three-way association, three two-way associations, and three one-way associations. With four variables there is a potential four-way association, four three-way associations, and so on. With more variables, then, the highest-level associations are tested and eliminated if nonsignificant until a preliminary model is found with the fewest required associations.

In the previous example, the three-way association between number of publications, number of statistics courses, and area in psychology might be ruled out in preliminary analysis. The set of two-way associations is then tested to see which of these might be ruled out. Number of statistics courses and number of publications might be associated, as well as area of psychology and number of statistics courses, but not area of psychology and number of publications. Finally, one-way "associations" are tested. For example, there is a one-way association for area of psychology if numbers of psychologists differ significantly among areas.

16.2.2   Effect on a Dependent Variable

In the usual multiway frequency table, cell frequency is the DV that is influenced by one or more discrete variables and their associations. Sometimes, however, one of the variables is considered a DV. In this case, questions about association are translated into tests of main effects (associations between the DV and each IV) and interactions (association between the DV and the joint effects of two or more IVs).

Under most circumstances this type of data is more efficiently analyzed through logistic regression, the subject of Chapter 10. Logistic regression uses a discrete DV but has the flexibility to include both discrete and continuous IVs. Thus, this chapter is limited to analyses in which none of the variables is considered a DV.

16.2.3   Parameter Estimates

What is the expected frequency for a particular combination of categories of variables? First, statistically significant effects are identified, and then coefficients, called *parameter estimates,* are found for each level of all the statistically significant effects. Section 16.4.3.2 shows how to calculate parameter estimates and use them to find expected frequencies.

16.2.4   Importance of Effects

Because parameter estimates are developed for each level (or combinations of levels) of each significant effect, the relative importance of each effect to the frequency in each cell can be evaluated. Effects with larger standardized parameter estimates are more important in predicting that cell's frequency than effects with smaller standardized parameter estimates. If, for instance, number of statistics courses has a higher standardized parameter estimate than number of publications in the cell for successful psychologists, it is the more important effect.

16.2.5   Effect Size

How well does a model fit the observed frequencies? Effect size measures typically are not available in statistical packages used for log-linear analysis. The χ^2 value that is a measure of the fit between the model and observed frequencies can be considered a measure of effect size, considering that the expected value of χ^2/df is 1 when there is no association among variables. Section 16.6.2.3 demonstrates software that calculates the confidence interval around a χ^2 value.

Bonett and Bentler (1983) describe the use of a normed fit index (NFI). Although influenced by sample size, NFI may give a better notion of how well a model fits observed frequencies than is available from formal goodness-of-fit tests such as chi square. See Section 14.5.3.1 for further discussion of NFI and other indices of model fit.

16.2.6   Specific Comparisons and Trend Analysis

If a significant association is found, it may be of interest to decompose the association to find its significant components. For example, if area of psychology and number of publications, both with several levels, are associated, which areas differ in number of publications? These questions are analogous to those of analysis of variance where a many-celled interaction is investigated in terms of simpler interactions or in terms of simple effects (cf. Section 8.5.2). Similarly, if the categories of one of the variables differ in quantity (e.g., number of publications), a trend analysis often helps one understand the nature of its relationship with other variables. SAS CATMOD and SPSS LOGLINEAR (available only in syntax) provide procedures for specifying contrasts.

16.3   Limitations to Multiway Frequency Analysis

16.3.1   Theoretical Issues

As a nonparametric statistical technique with no assumptions about population distributions, multiway frequency analysis is remarkably free of limitations. The technique can be applied almost universally, even to continuous variables that fail to meet distributional assumptions of parametric statistics if the variables are cut into discrete categories.

   With the enormous flexibility of current programs for log-linear analysis, many of the questions posed by highly complex data sets can be answered. However, the greatest danger in the use of this analysis is inclusion of so many variables that interpretation boggles the mind—a danger frequently noted in multifactorial analysis of variance, as well.

16.3.2   Practical Issues

The only limitations to using multiway frequency analysis are the requirement for independence, adequate sample size, and the size of the expected frequency in each cell. During interpretation, however, certain cells may turn out to be poorly predicted by the solution.

16.3.2.1   Independence

Only between-subjects designs may be analyzed in most circumstances, so that the frequency in each cell is independent of the frequencies in all other cells. If the same case contributes values to more than one cell, those cells are not independent. *Verify that the total N is equal to the number of cases.*

   Sometimes the restriction to between-subjects designs is circumvented by inclusion of a time variable, as in McNemar's test for two-way χ^2. A case is in a particular combination of cells over the time periods. Similarly, "yes–no" variables may be developed. For example, in a 2×2 design, a person attends karate classes but does not take piano lessons (yes on karate, no on piano), or does neither (no on both), or does both (yes on both), or takes piano lessons but not karate (no on karate, yes on piano). Each case is in only one of four cells, despite having "scores" on both karate and piano. SAS CATMOD has a procedure for analyzing designs in which a discrete DV is measured repeatedly. SPSS COMPLEX SAMPLES also may be used for repeated-measures designs with a dichotomous DV when cases are defined as clusters.

16.3.2.2   Ratio of Cases to Variables

A number of problems may occur when there are too few cases relative to the number of variables. Log-linear analysis may fail to converge when combinations of variables result in too many cells with no cases. *You should have at least five times the number of cases as cells in your design.* In the example, area of psychology has three levels and number of publications five levels, so $3 \times 5 \times 5$ or 75 cases are needed. Software programs are available to aid in estimating required sample sizes for two-way but not multiway frequency analysis.

16.3.2.3 *Adequacy of Expected Frequencies*

The fit between observed and expected frequencies is an empirical question in tests of association among discrete variables. Sample cell sizes are observed frequencies; statistical tests compare them with expected frequencies derived from some hypothesis, such as independence between variables. The requirement in multiway frequency analysis is that expected frequencies are large enough. Two conditions produce expected frequencies that are too small: a small sample in conjunction with too many variables with too many levels (as discussed in Section 16.3.2.2) and rare events.

　　When events are rare, the marginal frequencies are not evenly distributed among the various levels of the variables. For example, there are likely to be few psychologists who average four or more publications a year. A cell from a low-probability row and/or a low-probability column will have a very low expected frequency. The best way to avoid low expected frequencies is to attempt to determine in advance of data collection which cells will be rare, and then sample until those cells are adequately filled.

　　In any event, *examine expected cell frequencies for all two-way associations to assure that all are greater than one, and that no more than 20% are less than five.* Inadequate expected frequencies generally do not lead to increased Type I error (except in some cases with use of the Pearson χ^2 statistic; cf. Section 16.5.2). But power can be so drastically reduced with inadequate expected frequencies that the analysis is worthless. Reduction of power becomes notable as expected frequencies for two-way associations drop below five in some cells (Milligan, 1980).

　　If low expected frequencies are encountered despite care in obtaining your sample, several choices are available. First, you can simply choose to accept reduced power for testing effects associated with low expected frequencies. Second, you can collapse categories for variables with more than two levels. For example, you could collapse the "three" and "four or more" categories for number of publications into one category of "three or more." The categories you collapse depend on theoretical considerations as well as practical ones because it is quite possible that associations will disappear as a result. Because this is equivalent to a complete reduction in power for testing those associations, nothing has been gained.

　　Finally, you can delete variables to reduce the number of cells. Care is taken to delete only variables that are not associated with the remaining variables. For example, in a three-way table, you might consider deleting a variable if there is no three-way association and if at least one of the two-way associations with the variable is nonsignificant (Milligan, 1980). The common practice of adding a constant to each cell is not recommended because it has the effect of further reducing power. Its purpose is to stabilize Type I error rate, but as noted before, that is generally not the problem and when it is, other remedies are available (Section 16.5.2). Some of the programs, such as SPSS LOG-LINEAR and HILOGLINEAR, add the constant by default anyway under circumstances that do not affect the outcome of the analysis.

　　Section 16.6.1. demonstrates procedures for screening a multidimensional frequency table for expected cell frequencies.

16.3.2.4 *Absence of Outliers in the Solution*

Sometimes there are substantial differences between observed and expected frequencies derived from the best-fitting model for some cells. If the differences are large enough, there may be no model that adequately fits the data. Levels of variables may have to be deleted or collapsed or new variables added before a model is fit. But whether or not a model is fit, examination of residuals in search of discrepant cells leads to better interpretation of the data set. Analysis of residuals is discussed in Sections 16.4.3.1 and 16.6.2.3.

16.4 Fundamental Equations for Multiway Frequency Analysis

Analysis of multiway frequency tables typically requires three steps: (1) screening, (2) choosing and testing appropriate models, and (3) evaluating and interpreting the selected model. A small-sample example of hypothetical data with three discrete variables is illustrated in Table 16.1. The first variable is type of preferred reading material, READTYP, with two levels: science fiction (SCIFI) and spy novels (SPY). The second variable is SEX; the third variable is three levels of profession, PROFESS: politicians (POLITIC), administrators (ADMIN), and belly dancers (BELLY).

     In this section, the simpler calculations are illustrated in detail, and the more complex arithmetic is covered only enough to provide some idea of the methods used to model multidimensional data sets. The computer packages used in this section are also the most straightforward. With real data sets, the various computer packages allow choice of strategy on the basis of utility rather than simplicity. Computer analyses of this data set through SAS CATMOD, SPSS GENLOG, and HILOGLINEAR are in Section 16.4.4.

TABLE 16.1 Small Sample of Hypothetical Data for Illustration of Multiway Frequency Analysis

Profession	Sex		Reading Type		Total
			SCIFI	*SPY*	**Total**
Politicians	Male		15	15	30
	Female		10	15	25
		Total	25	30	55
Administrators	Male		10	30	40
	Female		5	10	15
		Total	15	40	55
Belly dancers	Male		5	5	10
	Female		10	25	35
		Total	15	30	45

If only a single association is of interest, as is usually the case in the analysis of a two-way table, the familiar χ^2 statistic is used:

$$\sum_{ij} (f_o - F_e)^2 / F_e \tag{16.1}$$

where f_o represents observed frequencies in each cell of the table and F_e represents the expected frequencies in each cell under the null hypothesis of independence (no association) between the two variables. Summation is over all cells in the two-way table.

If the goodness-of-fit tests for the two marginal effects are also computed, the usual χ^2 tests for the two one-way and one two-way effects do not sum to total χ^2. This situation is similar to that of unequal-n ANOVA, where F tests of main effects and interactions are not independent (cf. Chapter 6). Because overlapping variance cannot be unambiguously assigned to effects, and because overlapping variance is repeatedly analyzed, interpretation of results is not clear-cut. In multiway frequency tables, as in ANOVA, nonadditivity of χ^2 becomes more serious as additional variables produce higher-order (e.g., three-way and four-way) associations.

An alternative strategy is to use the likelihood ratio statistic, G^2. The likelihood ratio statistic is distributed as χ^2, so the χ^2 tables can be used to evaluate significance. However, under conditions to be described in Section 16.4.2, G^2 has the property of additivity of effects. For example, in a two-way analysis,

$$G_T^2 = G_A^2 + G_B^2 + G_{AB}^2 \tag{16.2}$$

The test of overall association within a two-way table, G_T^2, is the sum of the first-order goodness-of-fit tests, G_A^2 and G_B^2, and the test of association, G_{AB}^2.

G^2, like χ^2, has a single equation for its various manifestations that differ among themselves only in how the expected frequencies are found.

$$G^2 = 2 \sum (f_o) \ln (f_o / F_e) \tag{16.3}$$

For each cell, the natural logarithm of the ratio of obtained to expected frequency is multiplied by the obtained frequency. These values are summed over cells, and the sum is doubled to produce the likelihood ratio statistics.

16.4.1   Screening for Effects

Screening is done if the researcher is data snooping and wishes simply to identify statistically significant effects. Screening is also done if the researcher hypothesizes a full model, a model with all possible effects included. Screening is not done if the researcher has hypothesized an incomplete model, a model with some effects included and others eliminated; in this case, the hypothesized model is tested and evaluated (see Section 16.4.2) followed by, perhaps, post hoc analysis.

The first step in screening is to determine if there are any effects to investigate. If there are, then screening progresses to a computation of F_e for each effect, a test of the reliability (significance) of each effect (finding G^2 for the first-order effects, the second-order or two-way associations, the third-order or three-way associations, and so on), and an estimation of the size of the statistically significant effects. Because Equation 16.3 is used for all tests of the observed frequencies (f_o), the trick is to find the F_e necessary to test the various hypotheses, as illustrated in what follows using the data of Table 16.1.

16.4.1.1 Total Effect

If done by hand, the process starts by calculation of overall G_T^2, which is used to test the hypothesis of no effects in the table (the hypothesis that all cells have equal frequencies). If this hypothesis cannot be rejected, there is no point to proceeding further. (Note that when all effects are tested simultaneously, as in computer programs, one can test either G_T^2 or G^2 for each of the effects, but not both, because degrees of freedom limit the number of hypotheses to be tested.)

For the test of total effect,

$$F_e = N/rsp \qquad (16.4)$$

Expected frequencies, F_e, for testing the hypothesis of no effects are the same for each cell in the table and are found by dividing the total frequency (N) by the number of cells in the table, i.e., the number of levels of READTYP (represented by r) times the number of levels of SEX (s) times the number of levels of PROFESS (p).

For these data, then,

$$F_e = 155/(2)(2)(3) = 12.9167$$

Applying Equation 16.3 for the test of overall effect,

$$G_T^2 = 2 \sum_{ijk} (f_o) \ln (f_o/F_e) \qquad df = rsp - 1$$

where $I = 1, 2,\ldots, r; j = 1, 2,\ldots, s$; and $k = 1, 2,\ldots, p$.

Filling in frequencies for each of the cells in Table 16.1, then,

$$
\begin{aligned}
G^2 &= 2[15 \ln (15/12.9167) + 15 \ln (15/12.9167) + 10 \ln (10/12.9167) \\
&\quad + 15 \ln (15/12.9167) + 10 \ln (10/12.9167) + 30 \ln (30/12.9167) + 5 \ln (5/12.9167) \\
&\quad + \cdots + 25 \ln (25/12.9167)] \\
&= 2[2.243 + 2.243 + (-2.559) + 2.243 + (-2.559) + 25.280 + (-4.745) \\
&\quad + (-2.559) + (-4.745) + (-4.745) + (-2.559) + 16.509] \\
&= 48.09
\end{aligned}
$$

With df $= 12 - 1 = 11$ and critical χ^2 at $\alpha = .05$ equal to 19.68 (cf. Table C.4 in Appendix C), there is a statistically significant departure from equal frequencies among the 12 cells.[1] Further

---

[1] Throughout this section, calculations may differ slightly from those produced by computer programs due to rounding error.

analysis is now required to screen the table for sources of this departure. In the normal course of data analysis the highest-order association is tested first, and so on. Because, however, of the greater complexity for finding expected frequencies with higher-order associations, the presentation here is in the reverse direction, from the first-order to highest-order associations.

16.4.1.2   First-Order Effects

There are three first-order effects to test, one for each of the discrete variables. Starting with READ-TYP, a goodness-of-fit test evaluates the equality of preference for science fiction and spy novels. Only the marginal sums for the two types of reading material are relevant, producing the following observed frequencies:

f_o	
SCIFI	SPY
55	100

Expected frequencies are found by dividing the total frequency by the number of relevant "cells," that is, $r = 2$, yielding $F_e = 155/2 = 77.5$. The expected frequencies, then, are

F_e	
SCIFI	SPY
77.5	77.5

and the test for goodness of fit is

$$G_T^2 = 2 \sum_i (f_o) \ln (f_o/F_e) \qquad df = r - 1$$

$$= 2 \left[55 \ln \left(\frac{55}{77.5} \right) + 100 \ln \left(\frac{100}{77.5} \right) \right] = 13.25 \qquad df = 1$$

Because critical χ^2 with $df = 1$ at $\alpha = .05$ is 3.84, a significant preference for spy novels is suggested. As in ANOVA, however, significant lower-order (main) effects cannot be interpreted unambiguously if there are higher-order (interaction) effects involving the same variable.

Similar tests for main effects of SEX and PROFESS produce $G_S^2 = 0.16$ with 1 df and $G_P^2 = 1.32$ with 2 df, suggesting no statistically significant difference in the number of men (80) and women (75), nor a significant difference in the numbers of politicians (55), administrators (55), and belly dancers (45), and an interesting sampling strategy.

16.4.1.3 Second-Order Effects

Tests of partial associations use an iterative procedure to develop a full set of expected frequencies in which all marginal sums (except the one to be tested) match the observed marginal frequencies.[2] First, the three-way table is collapsed into three two-way tables, one for each two-way interaction. For the $R \times S$ association, for instance, the cells for each combination of reading type and sex are summed over the three levels of profession (P), forming as the observed frequencies:

	f_o		
	SCIFI	**SPY**	
MEN	30	50	80
WOMEN	25	50	75
	55	100	155

The expected frequencies are found as in the usual way for a two-way χ^2 test of association:

$$\text{Cell } F_e = (\text{row sum})(\text{column sum})/N \tag{16.5}$$

for the appropriate row and column for each cell; that is, for the first cell, men preferring science fiction,

$$F_e = (80)(55)/155 = 28.3871$$

After the computations are completed for the remaining cells, the following table of expected frequencies is found:

	F_e		
	SCIFI	**SPY**	
MEN	28.3871	51.6129	80
WOMEN	26.6129	48.3871	75
	55	100	155

Once found, the expected frequencies are duplicated at each level of the other variable. The results of this iteration for the partial test of the $R \times S$ association appear in Table 16.2. Notice that computation of the expected frequencies is repeated for politicians, administrators, and belly dancers.

All the entries are too large because the two-way table has simply been duplicated three times. That is, $N = 465$ instead of 155, there are 80 male politicians instead of 30, and so on. A second iteration is performed to adjust the values in Table 16.2 for another two-way association, in this case the

---

[2]Other methods for finding partial associations are based on differences in G^2 between hierarchical models.

TABLE 16.2   First Iteration Estimates of Expected Frequencies for the Partial Test of the READTYP × SEX Association

| Profession | Sex | | Reading Type | | Total |
			SCIFI	SPY	
Politicians	Male		28.3871	51.6129	80
	Female		26.6129	48.3871	75
		Total	55	100	155
Administrators	Male		28.3871	51.6129	80
	Female		26.6129	48.3871	75
		Total	55	100	155
Belly dancers	Male		28.3871	51.6129	80
	Female		26.6129	48.3871	75
		Total	55	100	155

$R \times P$ association. This iteration begins with the $R \times P$ table of observed frequencies and relevant marginal sums:

| | f_o | |
	SCIFI	SPY
POLITIC	25	30
ADMIN	15	40
BELLY	15	30
	55	100

Note that the actual number of politicians preferring science fiction is 25, whereas after the first iteration (Table 16.2), the number is $(28.3871 + 26.6129) = 55$. The goal is to compute a proportion that, when applied to the relevant numbers in Table 16.2 (in this case, both male and female politicians who prefer science fiction), eliminates the effects of any $R \times P$ interaction:

$$f_o/F_e^{\#1} = 25/55 = 0.45455$$

producing

$$F_e^{\#2} = F_e^{\#1}(0.45455) = (28.3871)(0.45455) = 12.9032$$

and

$$F_e^{\#2} = F_e^{\#1}(0.45455) = (26.6129)(0.45455) = 12.0968$$

for male and female politicians preferring science fiction, respectively.

To find second iteration expected frequency for female belly dancers preferring spy stories, the last cell in the table,

$$f_o/F_e^{\#1} = 30/100 = 0.3$$

$$F_e^{\#2} = (48.3871)(0.3) = 14.5161$$

Table 16.3 shows the results of applying this procedure to all cells of the data matrix.

Notice that correct totals have been produced for overall N, for R, P, and S, and for $R \times P$, but that the $S \times P$ values are incorrect. The third and final iteration, then, adjusts the $S \times P$ expected values from the second iteration for the $S \times P$ matrix of observed values. These $S \times P$ matrices are:

	f_o			F_e	
	Men	**Women**		**Men**	**Women**
POLITIC	30	25	POLITIC	28.3871	26.6129
ADMIN	40	15	ADMIN	28.3871	26.6129
BELLY	10	35	BELLY	23.22258	21.7742

For the first cell, male politicians preferring to read science fiction, the proportional adjustment (rounded off) is

$$f_o/F_e^{\#2} = 30/28.3871 = 1.0568$$

TABLE 16.3   Second Iteration Estimates of Expected Frequencies for the Partial Test of the READTYP × SEX Association

Profession	Sex		Reading Type		
			SCIFI	*SPY*	**Total**
Politicians	Male		12.9032	15.4839	28.3871
	Female		12.0968	14.5161	26.6129
		Total	25	30	55
Administrators	Male		7.7419	20.6452	28.3871
	Female		7.2581	19.3548	26.6129
		Total	15	40	55
Belly dancers	Male		7.7419	15.4839	23.2258
	Female		7.2581	14.5161	21.7742
		Total	15	30	45

to produce

$$F_e^{\#3} = F_e^{\#2}(1.0568) = (12.9032)(1.0568) = 13.6363$$

And for the last cell, female belly dancers who prefer spy stories,

$$f_o/F_e^{\#2} = 35/21.7742 = 1.6074$$

$$F_e^{\#3} = (14.5161)(1.6074) = 23.3333$$

Following this procedure for the remaining 10 cells of the matrix produces the third iteration estimates, as shown in Table 16.4. These values fulfill the requirement that all expected marginal frequencies are equal to observed marginal frequencies except for $R \times S$, the association to be tested. At this point, we have the F_e necessary to calculate G_{RS}^2

$$G_{RS}^2 = 2 \sum_{ij} (f_o) \ln (f_o/F_e)$$

$$= 2[(f_o) \ln (f_o/F_e)]$$

$$= 2[15 \ln (15/13.6363) + \cdots + 25 \ln (25/23.3333)]$$

$$= 2.47$$

However, a final adjustment is made for the three-way association, G_{RSP}^2 (as computed in what follows). The partial likelihood ratio statistic for the association between READTYP and SEX, then, is

$$G_{RS(\text{part})}^2 = G_{RS}^2 - G_{RSP}^2 \qquad df = (r - 1)(s - 1)$$

$$= 2.47 - 1.85 = 0.62 \qquad df = 1$$

This partial test shows a lack of association.

TABLE 16.4 Third Iteration Estimates of Expected Frequencies for the Partial Test of the READTYP × SEX Association

Profession	Sex		Reading Type		Total
			SCIFI	*SPY*	
Politicians	Male		13.6363	16.3637	30
	Female		11.3637	13.6363	25
		Total	25	30	55
Administrators	Male		10.9090	29.0910	40
	Female		4.0909	10.9091	15
		Total	15	40	55
Belly dancers	Male		3.3333	6.6666	10
	Female		11.6667	23.3333	35
		Total	15	30	45

The same process is (tediously) followed for the partial tests of the $R \times P$ and the $S \times P$ associations. The resultant partial likelihood ratio statistic for the $R \times P$ association is

$$G^2_{RP(part)} = 4.42 \qquad df = 2$$

showing lack of association between reading preferences and profession. For the $S \times P$ association, the partial likelihood ratio result is

$$G_{SP(part)} = 27.12 \qquad df = 2$$

a statistically significant association.

Corresponding partial tests of intermediate associations in this example produce the same conclusions and interpretation is clear-cut: There is a statistically significant association between sex and profession and no evidence of association between sex and reading preferences or between reading preferences and profession. In some situations, however, interpretation is more problematic because the results of marginal and partial tests differ. Procedures for dealing with such situations are discussed in Section 16.5.3.

16.4.1.4   Third-Order Effect

The test for the three-way $R \times S \times P$ association requires a much longer iterative process because all marginal expected frequencies must match observed frequencies ($R, S, P, R \times S, R \times P$, and $S \times P$). Ten iterations are required to compute the appropriate F_e for the 12 cells (not shown in the interests of brevity and avoidance of terminal boredom), producing

$$G^2_{RSP} = 2 \sum_{ijk} (f_o) \ln (f_o/F_e) \qquad df = (r-1)(s-1)(p-1)$$

$$= 1.85 \qquad\qquad\qquad df = 2$$

The three-way association, then, shows no statistical significance.

A summary of the results of the calculations for all effects appears in Table 16.5. At the bottom of the table is the sum of all one-, two-, and three-way effects using partial methods for calculating G^2. As can be seen, this fails to match G^2_T; the sum is too large. Further, depending on the data, either over- or underadjustment of each effect may occur. Therefore, additional modeling may be required (see Section 16.5.3).

16.4.2   Modeling

In some applications of multiway frequency analysis, results of screening provide sufficient information for the researcher. In the current example, for instance, the results are clear-cut. One first-order effect, preference for reading type, is statistically significant, as is the sex-by-profession association. Often, however, the results are not so evident and consistent, and/or the goal is to find the best model for predicting frequencies in each cell of the design.

TABLE 16.5 Summary of Screening Tests for Small-Sample Example of Multiway Frequency Analysis

Effect	df	G^2	Prob
All (total)	11	48.09	<.05
READTYP	1	13.25	<.05
SEX	1	0.16	>.05
PROFESS	2	1.32	>.05
$R \times S$	1	0.62	>.05
$R \times P$	2	4.42	>.05
$S \times P$	2	27.12	<.05
$R \times S \times P$	2	1.85	>.05
Sums	11	48.74	

A log-linear model is developed where an additive regression-type equation is written for (the log of) expected frequency as a function of the effects in the design. The procedure is similar to multiple regression where a predicted DV is obtained by combining the effects of several IVs.

A full[3] model includes all possible effects in a multiway frequency analysis. The full model for the three-way design of the example is

$$\ln F_{e_{ijk}} = \theta + \lambda_{A_i} + \lambda_{B_j} + \lambda_{C_k} + \lambda_{AB_{ij}} + \lambda_{AC_{ik}} + \lambda_{BC_{jk}} + \lambda_{ABC_{ijk}} \qquad (16.6)$$

For each cell (the natural logarithm of) the expected frequency, $\ln F_e$, is an additive sum of the effect parameters, λs, and a constant, θ.

For each effect in the design, there are as many values of λ—called effect parameters—as there are levels in the effect, and these values sum to zero. In the example, there are two levels of READTYP, so there is a value of λ_R for SCIFI and for SPY, and the sum of these two values is zero. For most cells, then, the expected frequency is derived from a different combination of effect parameters.

The full (saturated) model always provides a perfect fit to data so that expected frequencies exactly equal observed frequencies. The purpose of modeling is to find the *incomplete* model with the fewest effects that still closely mimics the observed frequencies. Screening is done to avoid the necessity of exploring all possible incomplete models, an inhumane effort with large designs, even with computers. Effects that are found to be nonsignificant during the screening process are often omitted during modeling.

Model fitting is accomplished by finding G^2 for a particular incomplete model and evaluating its significance. Because G^2 is a test of fit between observed and expected frequencies, a good model

---

[3]Full models are also called saturated models.

is one with a *nonsignificant* G^2. Because there are often many "good" models, however, there is a problem in choosing among them. The task is to compare nonsignificant models with one another.

Models come in two flavors, hierarchical and nonhierarchical. Hierarchical (nested) models include the highest-order statistically significant association and all its component parts; nonhierarchical models do not necessarily include all the components (see Section 16.5.1). For hierarchical models, the optimal model is one that is not significantly worse than the next most complex one. Therefore, the choice among hierarchical models is made with reference to statistical criteria. There are no statistical criteria for choosing among nonhierarchical models and they are not recommended.

Several methods for comparing models are available, as discussed in Section 16.5.3. In the simplest method, illustrated here, a few hierarchical models are selected on the basis of screening results and compared using the significance of the difference in G^2 between them. When the models are hierarchical, *the difference between the two G^2s is itself a G^2*. That is,

$$G_1^2 - G_2^2 = G^2 \qquad (16.7)$$

if Model 1 is a subset of Model 2 in which all the effects in Model 1 are included in Model 2. For the example, a Model 1 with $R \times P$, R, and P effects is nested within a Model 2 with $R \times S$, $R \times P$, R, S, and P effects.

To simplify description of models, the preceding Model 1 is designated (RP) and Model 2 (RS,RP). This is a fairly standard notation for hierarchical models. Each association term (e.g., RS) implies that all lower-order effects (R and S) are included in the model. In the example, the most obvious model to choose is (SP,R), which includes the $S \times P$ association and all three first-order effects.

In practice, the first step is to evaluate the highest-order effect before sequentially testing lower-order effects. During screening on the example, the three-way association is ruled out but at least one of the two-way associations is statistically significant. Because there are only three effects in the design, it would not be difficult by computer to try out a model with all three two-way associations (RS,RP,SP) and compare that with models with all pairwise combinations of two-way associations. If there are ambiguities in the partial tests of effects, models with and without the ambiguous effects are compared.

In the example, lack of significance for partial tests of the RP and RS effects would ordinarily preclude their consideration in the set of models to be tested. The RP effect is included in a model to be tested here for illustrative purposes only.

For each model to be tested, expected frequencies and G^2 are found. To obtain G^2 for a model, the G^2 for each of the effects is subtracted from total G^2 to yield a test of residual frequency that is not accounted for by effects in the model. If the residual frequencies are not significant, there is a good fit between obtained and expected frequencies from the reduced model.

For the example, G^2 values for the (SP,R) model are available from the screening tests shown in Table 16.5. For the two-way effects, the G^2 values from the partial tests are used. G^2 for the (SP,R) model is, then

$$G_{(SP,R)}^2 = G_T^2 - G_{SP}^2 - G_S^2 - G_P^2 - G_R^2$$
$$= 48.09 - 27.12 - 0.16 - 1.32 - 13.25$$
$$= 6.24$$

Degrees of freedom are those associated with each of the effects as in Section 16.4.1, so that df $= 11 - 2 - 1 - 2 - 1 = 5$. Because residuals from this model are not statistically significant, the model is adequate.

For the example, a more complex model includes the $R \times P$ association. Following the earlier procedures, the (SP, RP) model produces $G^2 = 2.48$ with 3 df. The test of the difference between (SP, R) and (SP, RP) is simply the difference between G^2s (Equation 16.7) for the two models, using the difference between degrees of freedom to test for significance:

$$G^2_{(diff)} = G^2_{(SP,R)} - G^2_{(SP,RP)}$$
$$= 6.24 - 2.48 = 3.76 \quad \text{with} \quad df = 5 - 3 = 2$$

a nonsignificant result. Because the difference between models is not statistically significant, the more parsimonious (SP, R) model is preferred over the more complex (SP, RP) model. The model of choice, then, is

$$\ln F_e = 0 + \lambda_R + \lambda_S + \lambda_P + \lambda_{SP}$$

16.4.3   Evaluation and Interpretation

The optimal model, once chosen, is evaluated in terms of both the degree of fit to the overall data matrix (as discussed in the previous section) and the amount of deviation from fit in each cell.

16.4.3.1   Residuals

Once a model is chosen, expected frequencies are computed for each cell and the deviation between the expected and observed frequencies in each cell (the residual) is used to assess the adequacy of the model for fitting the observed frequency in that cell. In some cases, a model predicts the frequencies in some cells well, and in others very poorly, to give an indication of the combination of levels of variables for which the model is and is not adequate.

For the example, the observed frequencies are in Table 16.1. Expected frequencies under the (SP, R) model, derived through an iterative procedure as demonstrated in Section 16.4.1.3, are shown in Table 16.6. Residuals are computed as the cell-by-cell differences between the values in the two tables.

Rather than trying to interpret raw differences, residuals usually are standardized by dividing the difference between observed and expected frequencies by the square root of the expected frequency to produce a z value. Both raw differences and standardized residuals for the example are in Table 16.7. The most deviant cell is for male politicians preferring science fiction, with 4.4 fewer cases expected than observed and a standardized residual of $z = 1.3$. Although the discrepancies for men are larger than those for women, none of the cells is terribly discrepant; so this seems to be an acceptable model.

16.4.3.2   Parameter Estimates

There is a different linear combination of parameters for most cells, and the sizes of the parameters in a cell reflect the contribution of each of the effects in the model to the frequency found in that cell.

TABLE 16.6   Expected Frequencies under the Model

| Profession | Sex | Reading Type | | Total |
		SCIFI	*SPY*	
Politicians	Male	10.6	19.4	30.0
	Female	8.9	16.1	25.0
	Total	19.5	35.5	55.0
Administrators	Male	14.2	25.8	40.0
	Female	5.3	9.7	15.0
	Total	19.5	35.5	55.0
Belly dancers	Male	3.5	6.5	10.0
	Female	12.4	22.6	35.0
	Total	16.0	29.0	45.0

One can evaluate, for example, how important READTYP is to the number of cases found in the cell for female politicians who read science fiction.

Parameters are estimated for the model from the F_e in Table 16.6 in a manner that closely follows ANOVA. In ANOVA, the size of an effect for a cell is expressed as a deviation from the grand mean. Each cell has a different combination of deviations that correspond to the particular combination of levels of the statistically significant effects for that cell.

In MFA, deviations are derived from natural logarithms of proportions: $\ln(P_{ijk})$. Expected frequencies for the model (Table 16.6) are converted to proportions by dividing F_e for each cell by $N = 155$, and then the proportions are changed to natural logarithms. For example, for the first cell, male politicians who prefer science fiction:

$$\ln(P_{ijk}) = \ln(F_{e_{ijk}}/155)$$

$$= \ln(10.6/155)$$

$$= -2.6825711$$

Table 16.8 gives all the resulting values.

The values in Table 16.8 are then used in a three-step process that culminates in parameter estimates, expressed in standard deviation units, for each effect for each cell. The first step is to find both the overall mean and the mean (in natural logarithm units) for each level of each of the effects in the model. The second step is to express each level of each effect as a deviation from the overall mean. The third step is to convert the deviations to standard scores to compare the relative contributions of various parameters to the frequency in a cell.

In the first step, various means are found by summing $\ln(P_{ijk})$ across appropriate cells and dividing each sum by the number of cells involved. For example, to find the overall mean,

$$\bar{x}... = (1/rsp) \sum_{ijk} \ln(P_{ijk})$$

$$= (1/12)[-2.6825711 + (-2.0781521) + (-2.8573738) + \cdots + (-1.9254752)]$$

$$= -2.6355346$$

TABLE 16.7 Raw and Standardized Residuals for Hypothetical Data Set under Model (SP,R)

Profession	Sex	Reading Type	
		SCIFI	*SPY*
Raw residuals $(f_o - F_e)$:			
Politicians	Male	4.4	−4.4
	Female	1.1	−1.1
Administrators	Male	−4.2	4.2
	Female	−0.3	0.3
Belly dancers	Male	1.5	−1.5
	Female	−2.4	2.4
Standardized residuals $(f_o - F_e)/F_e^{1/2}$:			
Politicians	Male	1.3	−1.0
	Female	0.4	−0.3
Administrators	Male	−1.1	0.8
	Female	−0.1	0.1
Belly dancers	Male	0.8	−0.6
	Female	−0.7	0.5

TABLE 16.8 Expected ln P_{ijk} for Model (SP,R)

Profession	Sex	Reading Type	
		SCIFI	*SPY*
Politicians	Male	−2.6825711	−2.0781521
	Female	−2.8573738	−2.2646058
Administrators	Male	−2.3901832	−1.7930506
	Female	−3.3757183	−2.7712992
Belly dancers	Male	−3.7906621	−3.1716229
	Female	−2.5257286	−1.9254752

Note: $\ln(P_{ijk}) = (F_{ijk}/155) = \ln(F_{ijk}) - \ln(155)$.

To find the mean for SCIFI, the first level of READTYP:

$$\bar{x}_{1..} = (1/sp) \sum_{jk} \ln(P_{ijk})$$

$$= (1/6)[-2.6825711 + (-2.8573738)$$
$$+ (-2.3901832) + (-3.3757183) + (-3.7906621) + (-2.5257286)]$$
$$= -2.9370395$$

The mean for belly dancers is

$$\bar{x}_{..3} = (1/rs) \sum_{ij} \ln(P_{ijk})$$

$$= (1/4)[-3.7906621 + (-3.1716229) + (-2.5257286) + (-1.9254752)]$$
$$= -2.8533722$$

and so on for the first-order effects.

The means for second-order effects are found in a similar manner. For instance, for the $S \times P$ association, the mean for male politicians is

$$\bar{x}_{.11} = (1/r) \sum_{i} \ln(P_{ijk})$$

$$= (1/2)[-2.6825711 + (-2.0781521)]$$
$$= -2.3803616$$

In the second step, parameter estimates are found by subtraction. For first-order effects, the overall mean is subtracted from the mean for each level. For example, λ_{R_1}, the parameter for SCIFI, the first level of READTYP is

$$\lambda_{R_1} = \bar{x}_{1..} - \bar{x}_{...}$$

$$= -2.9370395 - (-2.6355346)$$
$$= -.302$$

For belly dancers, the third level of PROFESS

$$\lambda_{P_3} = \bar{x}_{3..} - \bar{x}_{...}$$

$$= -2.8533722 - (-2.63555346)$$
$$= -.218$$

and so on.

To find λ for a cell in two-way effect, the two appropriate main effect means are subtracted from the two-way mean, and the overall mean is added (in a pattern that is also familiar from ANOVA). For example, $\lambda_{SP_{23}}$, the parameter for female belly dancers (second level of sex, third level of profession), is found by subtracting from the female belly dancer mean (averaged over the two types of reading material) the mean for women and the mean for belly dancers, and then adding the overall mean.

$$\lambda_{SP_{23}} = \bar{x}_{23\cdot} - \bar{x}_{\cdot\cdot2} - \bar{x}_{\cdot\cdot3} + \bar{x}_{\cdot\cdot\cdot}$$
$$= -2.2256019 - (-2.6200335) - (-2.8533722) + (-2.6355346)$$
$$= .612$$

All the λ values, as shown in Table 16.9, are found in a similar, if tedious, fashion. In the table, θ is the conversion of the overall mean from proportion to frequency units by addition of $\ln(N)$:

$$\theta = \bar{x}_{\cdot\cdot\cdot} + \ln(155)$$
$$= 2.4079$$

The expected frequency generated by the model for each cell is then expressed as a function of the appropriate parameters. For example, the expected frequency (19.40) for male politicians who read spy novels is

$$\ln F_e = \theta + \lambda_{R_2} + \lambda_{S_1} + \lambda_{P_1} + \lambda_{SP_{11}}$$
$$= 2.4079 + .302 + (-.015) + .165 + .106$$
$$= 2.9659 \approx \ln(19.40)$$

within rounding error.

TABLE 16.9   Parameter Estimates for Model (SP,R). θ (MEAN) = 2.4079

Effect	Level	λ	λ/SE
READTYP	SCIFI	−.302	−3.598
	SPY	.302	3.598
SEX	MALE	−.015	−0.186
	FEMALE	.015	0.186
PROFESSION	POLITICIAN	.165	2.045
	ADMINISTRATOR	.053	0.657
	BELLY DANCER	−.218	−2.702
SEX BY PROFESS	MALE POLITICIAN	.106	1.154
	FEMALE POLITICIAN	−.106	−1.154
	MALE ADMINISTRATOR	.506	5.510
	FEMALE ADMINISTRATOR	−.506	−5.510
	MALE BELLY DANCER	.612	7.200
	FEMALE BELLY DANCER	−.612	−7.200

These parameters are used to find expected frequencies for each cell but are not interpreted in terms of magnitude until step 3 is taken. During step 3, parameters are divided by their respective standard errors to form standard normal deviates that are interpreted according to their relative magnitudes. Therefore, the parameter values in Table 16.9 are given both in their λ form and after division by their standard errors.

Standard errors of parameters, SE, are found by squaring the reciprocal of the number of levels for the set of parameters, dividing by the observed frequencies, and summing over the levels. For example, for READTYP:

$$SE^2 = \sum (1/r_i)^2/f_o$$
$$= (1/2)^2/55 + (1/2)^2/100$$
$$= (.25)/55 + (.25)/100$$
$$= .0070455$$

and

$$SE = .0839372$$

Note that this is the simplest method for finding SE (Goodman, 1978) and does not weight the number of levels by unequal marginal frequencies, as do some other methods.

To find the standard normal deviate for SCIFI (the first level of READTYP), λ for SCIFI is divided by its standard error

$$\lambda_{R_1}/SE = -.302/.0839372$$
$$= -3.598$$

This ratio is interpreted as a standard normal deviate (z) and compared with critical z to assess the contribution of an effect to a cell. The relative importance of the various effects to a cell is also derived from these values. For female belly dancers preferring spy novels, for example, the standard normal deviates for the parameters are 3.598 (SPY), 0.186 (FEMALE), -2.702 (BELLY), and -7.200 (FEMALE BELLY). The most important influences on cell frequency are, in order, the sex by profession association, preferred type of reading material, and profession—all statistically significant at $p < .01$ because they exceed 2.58. Sex contributes little to the expected frequency in this cell and is not statistically significant.

Because of the large number of effects produced in typical loglinear models, a conservative criterion should be used if statistical significance is evaluated. A criterion z of 4.00 often is considered reasonable.

Further insights into interpretation are provided in Section 16.6.2.4. Conversion of parameters to odds when one variable is a DV is discussed in Section 10.6.3.

16.4.4   Computer Analyses of Small-Sample Example

Syntax and selected output for computer analyses of the data in Table 16.1 appear in Tables 16.10 through 16.12. SPSS HILOGLINEAR and GENLOG[4] are in Tables 16.10 and 16.11, respectively, and SAS CATMOD is in Table 16.12.

The syntax of SPSS HILOGLINEAR (Model Selection on Loglinear menu) in Table 16.10 produces output appropriate for screening a hierarchical multiway frequency analysis. Additional instructions are necessary to test models. The instruction PRINT=FREQ produces the table of

TABLE 16.10   Multiway Frequency Analysis of Small-Sample Example through SPSS HILOGLINEAR (Syntax and Selected Output)

```
WEIGHT by freq.
HILOGLINEAR
    profess(1 3) sex(1 2) readtyp(1 2) /METHOD=BACKWARD
    /CRITERIA MAXSTEPS (10) p(.05) ITERATION(20) DELTA(.5)
    /PRINT=FREQ ASSOCIATION ESTIM
    /DESIGN.
```

HiLog

Observed, Expected Frequencies and Residuals.

Factor	Code	OBS count	EXP count
profess	1		
sex	1		
readtyp	1	15.5	15.5
readtyp	2	15.5	15.5
sex	2		
readtyp	1	10.5	10.5
readtyp	2	15.5	15.5
profess	2		
sex	1		
readtyp	1	10.5	10.5
readtyp	2	30.5	30.5
sex	2		
readtyp	1	5.5	5.5
readtyp	2	10.5	10.5
profess	3		
sex	1		
readtyp	1	5.5	5.5
readtyp	2	5.5	5.5
sex	2		
readtyp	1	10.5	10.5
readtyp	2	25.5	25.5

[4]Another program, SPSS LOGLINEAR, is available only through syntax and is demonstrated in Section 16.6.2.

TABLE 16.10 Continued

Goodness-of-fit test statistics

Likelihood ratio chi square =	.00000	DF = 0	P =	.
Pearson chi square =	.00000	DF = 0	P =	.

Tests that K-way and higher order effects are zero.

K	DF	L.R. Chisq	Prob	Pearson Chisq	Prob	Iteration
3	2	1.848	.3969	1.920	.3828	3
2	7	33.353	.0000	32.994	.0000	2
1	11	48.089	.0000	52.097	.0000	0

Tests that K-way effects are zero.

K	DF	L.R. Chisq	Prob	Pearson Chisq	Prob	Iteration
1	4	14.737	.0053	19.103	.0008	0
2	5	31.505	.0000	31.073	.0000	0
3	2	1.848	.3969	1.920	.3828	0

Tests of PARTIAL associations.

Effect Name	DF	Partial Chisq	Prob	Iter
profess*sex	2	27.122	.0000	2
profess*readtyp	2	4.416	.1099	2
sex*readtyp	1	.621	.4308	2
profess	2	1.321	.5166	2
sex	1	.161	.6879	2
readtyp	1	13.255	.0003	2

Note: For saturated models .500 has been added to all observed cells.
This value may be changed by using the CRITERIA = DELTA subcommand.

(continued)

TABLE 16.10 Continued

```
Estimates for Parameters.
profess*sex*readtyp
```

Parameter	Coeff.	Std. Err.	Z-Value	Lower 95 CI	Upper 95 CI
1	.0259458833	.11956	.21702	-.20839	.26028
2	-.1763513737	.12929	-1.36397	-.42977	.07706

```
profess*sex
```

Parameter	Coeff.	Std. Err.	Z-Value	Lower 95 CI	Upper 95 CI
1	.1038757056	.11956	.86884	-.13046	.33821
2	.4347541617	.12929	3.36255	.18134	.68817

```
profess*readtyp
```

Parameter	Coeff.	Std. Err.	Z-Value	Lower 95 CI	Upper 95 CI
1	.1517793544	.11956	1.26951	-.08255	.38611
2	-.1790991017	.12929	-1.38522	-.43251	.07432

```
sex*readtyp
```

Parameter	Coeff.	Std. Err.	Z-Value	Lower 95 CI	Upper 95 CI
1	.0714203084	.09098	.78501	-.10690	.24974

```
profess
```

Parameter	Coeff.	Std. Err.	Z-Value	Lower 95 CI	Upper 95 CI
1	.1935846228	.11956	1.61918	-.04075	.42792
2	.0064171131	.12929	.04963	-.24700	.25983

```
sex
```

Parameter	Coeff.	Std. Err.	Z-Value	Lower 95 CI	Upper 95 CI
1	-.0065095139	.09098	-.07155	-.18483	.17181

```
readtyp
```

Parameter	Coeff.	Std. Err.	Z-Value	Lower 95 CI	Upper 95 CI
1	-.2491455461	.09098	-2.73844	-.42747	-.07082

Observed, Expected Frequencies and Residuals. Because no model is specified in the syntax, a full model (all effects included in the model) is produced in which expected and observed frequencies are identical. SPSS adds 0.5 to each observed frequency for a full model; however, this has no effect on subsequent values. The table of Goodness-of-fit test statistics also reflects a perfectly fitting model.

The next three tables are produced by the **ASSOCIATION** instruction and consist of tests of all effects individually, effects combined at each order, and effects combined at each order and higher orders. The table labeled `Tests of PARTIAL associations` shows tests of each two-way and one-way effect. These values are the same as those of Table 16.5, produced by hand calculation. Tests of the combined associations at each order are presented in the table labeled `Tests that K-way effects are zero`. In the row labeled 2 is the test of the three two-way associations combined which, in this case, shows statistical significance using both the likelihood ratio (`L.R.`) and `Pearson chisq` criteria. This output suggests that at least one of the two-way associations is significant by both criteria. The test of the single three-way association is also provided in this table when $k = 3$; it is not significant. In the table labeled `Tests that K-way and higher order effects are zero`, the row labeled 1 contains the test of the combination of all one-way, two-way, and three-way associations, significant in this case by both likelihood ratio and Pearson chi-square criteria. The row labeled 2 contains the test of the combination of all two- and three-way associations, and so on.

The final section of Table 16.10 contains parameter estimates, an alternative way of testing effects. Instead of a partial test for each effect, parameter estimates for the effect are tested by dividing each `Coeff.` by its standard error (`Std. Err.`) to produce a `Z-Value` and a 95% confidence interval (`Lower 95 CI` and `Upper 95 CI`).[5] These parameter estimates are available only for saturated models—models that include all possible effects. Note that if an effect has more than 1 df, a single test for the effect is not provided because the parameter estimate for each df is tested separately.

Table 16.11 shows the results of an unspecified (full, saturated) model run through SPSS GENLOG. Note that specification of cell weight occurs outside the GENLOG procedure. Output begins with a description of observed and expected cell frequencies and percentages for the specified model, spelled out in footnote b of the first output table shown. All cell counts are automatically incremented by 0.5. Although the title of the table includes residuals, they do not appear because this is a saturated model; the observed and expected counts are equal. The final table shows parameter estimates, each shown with its standard error (**Std. Error**), as well as z value: the **Estimate** divided by Std. Error. The final two columns show the 95% confidence interval for each parameter estimate. Note that the difference between the parameters and their standard errors for SPSS HILOGLINEAR and GENLOG are due to the different ways that the models are parameterized in the two programs.

SAS CATMOD syntax and output for MFA appear in Table 16.12. The full model is specified by listing the three-way association, `PROFESS*SEX*READTYP` equal to `_response_`, a keyword that induces a log-linear model. Unneeded output is suppressed with `noiter`. The `loglin` instruction is used; the instructions in this syntax specify that all variables—PROFESS, SEX, READTYP—are to be treated the same, that none is the DV.

After information on description of the design, CATMOD provides details about the response profiles. The `Maximum Likelihood Analysis of Variance` table contains likelihood

---

[5]These parameter estimates differ somewhat from those produced by hand calculation (Table 16.9) because of the different algorithm used by this program.

ratio Chi-Square tests of each effect individually. Note that due to differences in the algorithms used, these estimates differ a bit from those of SPSS HILOGLINEAR and a great deal from those of SPSS GENLOG.

There are also tests of individual parameter estimates in the following section (Analysis of Maximum Likelihood Estimates), although some of these differ from both the ones shown for hand calculation (Table 16.9) and those produced by SPSS HILOGLINEAR and GENLOG (Tables 16.10 and 16.11). Chi-Square tests (rather than z) are given for each of the parameter estimates.

TABLE 16.11   Multiway Frequency Analysis of Small-Sample Example through SPSS GENLOG (Syntax and Selected Output)

```
GENLOG
    profess sex readtyp
    /MODEL = POISSON
    /PRINT = FREQ ESTIM
    /CRITERIA = CIN(95) ITERATE(20) CONVERGE(.001) DELTA(.5)
    /DESIGN
```

Cell Counts and Residuals[a,b]

profess	sex	readtyp	Observed Count	Observed %	Expected Count	Expected %
1.00	1.00	1.00	15.500	9.6%	15.500	9.6%
		2.00	15.500	9.6%	15.500	9.6%
	2.00	1.00	10.500	6.5%	10.500	6.5%
		2.00	15.500	9.6%	15.500	9.6%
2.00	1.00	1.00	10.500	6.5%	10.500	6.5%
		2.00	30.500	18.9%	30.500	18.9%
	2.00	1.00	5.500	3.4%	5.500	3.4%
		2.00	10.500	6.5%	10.500	6.5%
3.00	1.00	1.00	5.500	3.4%	5.500	3.4%
		2.00	5.500	3.4%	5.500	3.4%
	2.00	1.00	10.500	6.5%	10.500	6.5%
		2.00	25.500	15.8%	25.500	15.8%

[a]Model: Poisson

[b]Design: Constant + profess + sex + readtyp + profess * sex + profess* readtyp + sex * readtyp + profess * sex * readtyp

TABLE 16.11   Continued

Parameter Estimates[b,c]

Parameter	Estimate	Std. Error	Z	Sig.	95% Confidence Interval Lower Bound	95% Confidence Interval Upper Bound
Constant	3.239	.198	16.355	.000	2.851	3.627
[profess = 1.00]	−.498	.322	−1.546	.122	−1.129	.133
[profess = 2.00]	.887	.367	−2.420	.016	−1.605	−.169
[profess = 3.00]	0[a]					
[sex = 1.00]	−1.534	.470	−3.262	.001	−2.455	−.612
[sex = 2.00]	0[a]					
[readtyp = 1.00]	−.887	.367	−2.420	.016	−1.606	−.169
[readtyp = 2.00]	0[a]					
[profess = 1.00]*[sex = 1.00]	1.534	.592	2.593	.010	.374	2.694
[profess = 1.00]*[sex = 2.00]	0[a]					
[profess = 2.00]*[sex = 1.00]	2.600	.591	4.401	.000	1.442	3.758
[profess = 2.00]*[sex = 2.00]	0[a]					
[profess = 3.00]*[sex = 1.00]	0[a]					
[profess = 3.00]*[sex = 2.00]	0[a]					
[profess = 1.00]*[readtyp = 1.00]	.498	.542	.918	.359	−.565	1.561
[profess = 1.00]*[readtyp = 2.00]	0[a]					
[profess = 2.00]*[readtyp = 1.00]	.241	.641	.375	.708	−1.017	1.498
[profess = 2.00]*[readtyp = 2.00]	0[a]					
[profess = 3.00]*[readtyp = 1.00]	0[a]					
[profess = 3.00]*[readtyp = 2.00]	0[a]					
[sex = 1.00]*[readtyp = 1.00]	.887	.706	1.257	.209	−.496	2.271
[sex = 1.00]*[readtyp = 2.00]	0[a]					
[sex = 2.00]*[readtyp = 1.00]	0[a]					
[sex = 2.00]*[readtyp = 2.00]	0[a]					
[profess = 1.00]*[sex = 1.00]* [readtyp = 1.00]	−.498	.887	-.561	.575	−2.236	1.241
[profess = 1.00]*[sex = 1.00]* [readtyp = 2.00]	0[a]					
[profess = 1.00]*[sex = 1.00]* [readtyp = 2.00]	0[a]					
[profess = 1.00]*[sex = 2.00]* [readtyp = 1.00]	0[a]					
[profess = 1.00]*[sex = 2.00]* [readtyp = 2.00]	−1.307	.950	−1.375	.169	−3.170	.556
[profess = 2.00]*[sex = 1.00]* [readtyp = 2.00]	0[a]					
[profess = 2.00]*[sex = 2.00]* [readtyp = 1.00]	0[a]					
[profess = 2.00]*[sex = 2.00]* [readtyp = 2.00]	0[a]					
[profess = 3.00]*[sex = 1.00]* [readtyp = 1.00]	0[a]					
[profess = 3.00]*[sex = 1.00]* [readtyp = 2.00]	0[a]					
[profess = 3.00]*[sex = 2.00]* [readtyp = 1.00]	0[a]					
[profess = 3.00]*[sex = 2.00]* [readtyp = 2.00]	0[a]					

[a]This parameter is set to zero because it is redundant.   [b]Model: Poisson
[c]Design: Constant + profess + sex + readtyp + profess * sex + profess * readtyp + sex * readtyp + profess * sex *readtyp

TABLE 16.12 Multiway Frequency Analysis of Small-Sample Example through SAS CATMOD (Syntax and Selected Output)

```
proc  catmod data=SASUSER.SSMFA;
      weight  freq;
      model  PROFESS*SEX*READTYP=_response_/
             noiter;
             loglin  PROFESS|SEX|READTYP;
run;
```

The CATMOD Procedure
Data Summary

Response	PROFESS*SEX*READTYP	Response Levels	12
Weight Variable	FREQ	Populations	1
Data Set	SSMFA	Total Frequency	155
Frequency Missing	0	Observations	12

Population Profiles

Sample	Sample Size
1	155

Response Profiles

Response	PROFESS	SEX	READTYP
1	1	1	1
2	1	1	2
3	1	2	1
4	1	2	2
5	2	1	1
6	2	1	2
7	2	2	1
8	2	2	2
9	3	1	1
10	3	1	2
11	3	2	1
12	3	2	2

Maximum Likelihood Analysis

Maximum likelihood computations converged.

TABLE 16.12   **Continued**

Maximum Likelihood Analysis of Variance

Source	DF	Chi-Square	Pr > ChiSq
PROFESS	2	3.46	0.1777
SEX	1	0.01	0.9256
PROFESS*SEX	2	17.58	0.0002
READTYP	1	7.61	0.0058
PROFESS*READTYP	2	2.62	0.2691
SEX*READTYP	1	0.66	0.4168
PROFESS*SEX*READTYP	2	1.89	0.3894
Likelihood Ratio	0	.	.

Analysis of Maximum Likelihood Estimates

Parameter		Estimate	Standard Error	Chi-Square	Pr > ChiSq
PROFESS	1	0.2081	0.1229	2.87	0.0903
	2	0.00538	0.1337	0.00	0.9679
SEX	1	−0.00878	0.0940	0.01	0.9256
PROFESS*SEX	1 1	0.1101	0.1229	0.80	0.3700
	2 1	0.4567	0.1337	11.67	0.0006
READTYP	1	−0.2595	0.0940	7.61	0.0058
PROFESS*READTYP	1 1	0.1581	0.1229	1.66	0.1982
	2 1	−0.1885	0.1337	1.99	0.1586
SEX*READTYP	1 1	0.0764	0.0940	0.66	0.4168
PROFESS*SEX*READTYP	1 1 1	0.0250	0.1229	0.04	0.8387
	2 1 1	−0.1777	0.1337	1.77	0.1837

16.5   Some Important Issues

16.5.1   Hierarchical and Nonhierarchical Models

A model is hierarchical, or nested, if it includes all the lower effects contained in the highest-order association that is retained in the model. A hierarchical model for a four-way design, $ABCD$, with a significant three-way association, ABC, is $A \times B \times C$, $A \times B$, $A \times C$, $B \times C$, and A, B, and C. The hierarchical model might or might not also include some of the other two-way associations and the D first-order effect. A nonhierarchical model derived from the same four-way design includes only the significant two-way associations and first-order effects along with the significant three-way association; that is, a nonsignificant $B \times C$ association is included in a hierarchical model that retains the ABC effect but is not included automatically in a nonhierarchical model.

In log-linear analysis of multiway frequency tables, hierarchical models are the norm (e.g., Goodman, 1978; Knoke & Burke, 1980). Nonhierarchical models are suspect because higher order effects are confounded with lower order components. Therefore, it is best to explicitly include component lower order associations when specifying models in general log-linear programs.

One major advantage of hierarchical models is the availability of a significance test for the difference between models, so that the most parsimonious adequately fitting model can be identified using inferential procedures. With nonhierarchical models, a statistical test for the difference between models is not available unless one of the candidate models happens to be nested in the other.

SPSS LOGLINEAR and GENLOG, and SAS CATMOD have the Newton-Raphson algorithm for assessing models and are considered general log-linear programs because they do not automatically impose hierarchical modeling. SPSS HILOGLINEAR is restricted to hierarchical models.

16.5.2   Statistical Criteria

A potential source of confusion is that tests of models look for statistical *non*significance while tests of effects look for statistical significance. Both kinds of tests commonly use the same statistics—forms of χ^2. This is usual practice in model-fitting techniques, as in Chapter 10 (Logistic Regression) and Chapter 14 (Structural Equation Modeling).

16.5.2.1   *Tests of Models*

Both Pearson χ^2 and the likelihood ratio statistic G^2 are often available for screening for the complexity of model necessary to fit data and for testing overall fit of models. Between the two, consistency favors use of G^2 because it is available for testing overall fit, screening, and testing for differences among hierarchical models. Also, under some conditions, inadequate expected frequencies can inflate Type I error rate when Pearson χ^2 is used (Milligan, 1980).

In assessing goodness-of-fit for a model, you look for a nonsignificant G^2 where the frequencies estimated from the model are similar to the observed frequencies. Thus, retention of the null hypothesis is the desired outcome—an unhappy state of affairs for choosing an appropriate alpha level. In order to avoid finding too many "good" models, you need a less strict criterion for α, say .10 or .25.

Further, with very large samples, small discrepancies between expected and observed frequencies often result in statistical significance. A significant model, even at $\alpha = .05$, may actually have adequate fit. With very small samples, on the other hand, large discrepancies often fail to reach statistical significance so that a nonsignificant model, even at $\alpha = .25$, actually has a poor fit. Choice of a significance level, then, is a matter of considering both sample size and the nature of the test. With larger samples, smaller tail probability values are chosen.

16.5.2.2   *Tests of Individual Effects*

Two types of tests typically are available for testing individual effects in multiway frequency tables: chi-square tests of partial effects and z tests for single df parameter estimates.

SPSS HILOGLINEAR and SAS CATMOD provide partial G^2 tests of all effects in a full model. In addition, all programs print parameter estimates and their standard errors, which are converted to z tests of parameters or, in the case of SAS, χ^2 tests. However, SPSS HILOGLINEAR prints these only for saturated models.

SPSS LOGLINEAR and GENLOG provide parameter estimates and their associated z tests, but no omnibus test for any effect that has more than one degree of freedom. If an effect has more than two levels, there is no single inferential test of that effect. Although one can attribute statistical significance to an effect if any of its single df tests is significant, no overall tail probability level is available. Also, an effect may be statistically significant even though none of its single df parameters reaches significance. With only the single df z tests of parameters, such an effect is not identified.

16.5.3   Strategies for Choosing a Model

If you have one or more models hypothesized a priori, then there is no need for the strategies discussed in this section. The techniques in this section are used if you are building a model, or trying to find the most parsimonious incomplete model. As in all exploratory modeling, care should be taken in overgeneralizing results which may be subject to overfitting and inflated Type I error.

Strategies for choosing a model differ depending on whether you are using SPSS or SAS. Options and features differ among programs. You may find it handy to use one program to screen and another to evaluate models. Recall that hierarchical programs automatically include lower order components of higher order associations; general log-linear programs require that you explicitly include lower order components when specifying candidate hierarchical models.

16.5.3.1   *SPSS HILOGLINEAR (Hierarchical)*

This program provides a test of each individual effect (with partial χ^2 reported where appropriate), simultaneous tests of all k-way effects (all one-way effects combined, all two-way effects combined, and so on), and simultaneous tests of all k- and higher-way effects (with a four-way model, all three- and four-way effects combined, all two-, three-, and four-way effects combined, and so on). Both Pearson and likelihood ratio χ^2 (G^2) are reported. A strategy that follows the recommendations of Benedetti and Brown (1978) proceeds as follows.

Consider the *ABC* effect in a four-way design with *ABCD*. First, look at the tests of all three-way effects combined and three-way and four-way effects combined because combined results take precedence over tests of individual effects. If both combined tests are nonsignificant, the *ABC* association is deleted regardless of its partial test unless this specific three-way interaction has been hypothesized beforehand. If the combined test is significant, and the *ABC* effect is significant, the *ABC* effect is retained in the final model. If some of the tests are significant while others are not, further screening is recommended. This process is demonstrated in Section 16.6.2.1. (Recall that the cutoff p values for assessing significance depend on sample size. Larger samples are tested with smaller p values to avoid including statistically significant but trivial effects.)

Further screening of effects with ambiguous results (disagreement, say, between G^2 and Pearson χ^2, or a result between $\alpha = .01$ and $\alpha = .05$) proceeds stepwise. SPSS HILOGLINEAR provides only backward stepping, in which one starts with all the unambiguously significant effects plus all ambiguous effects from the initial screening of the full model. The term that is least helpful to the model is deleted first, followed by assessment of the remaining terms of the same order. χ^2 for the difference between simpler and more complex models is reported. Terms that do not significantly degrade the model when deleted are excluded.

Note that this stepwise procedure, like others, violates rules of hypothesis testing. Therefore, don't take the χ^2 and probability values produced by the stepping procedure too seriously. View this as a search for the most reasonable model, with χ^2 providing guidelines for choosing among models, as opposed to a stricter view that some models are truly significantly better or worse than others.

16.5.3.2   *SPSS GENLOG (General Log-Linear)*

This program provides neither simultaneous tests for associations nor a stepping algorithm. Therefore, the procedure for choosing an appropriate model is simpler but less flexible.

A preliminary run with a full model is used to identify effects whose parameters differ significantly from zero. Recall that each cell of a design has a parameter for each effect and that, if the effect has more than two levels, the size of the parameter for the same effect may be different in the different cells. If an effect has a parameter that is highly significant for any cell, the effect is retained. If all the parameters for an effect are clearly nonsignificant, the effect is deleted.

Ambiguous cases occur when some parameters are marginally significant. Subsequent runs are made with and without ambiguous effects. In these runs, the significance of parameters is assessed along with the fit of the overall model. The strategy of backward elimination of simple effects, as described above, is followed for the safest route to the most reasonable model.

16.5.3.3 SAS CATMOD and SPSS LOGLINEAR (General Log-Linear)

Although these programs have no provision for stepwise model building and no simultaneous tests of association for each order, they do provide separate tests for each effect in a model, including effects with more than one df. A preliminary run with a full model, then, is used to identify candidates for model testing through the maximum likelihood chi-square test of association. Evaluation of models follows the spirit of backward elimination of simple effects as described in Section 16.5.3.1.

16.6 Complete Example of Multiway Frequency Analysis

Data to illustrate multiway frequency analysis were taken from the survey of clinical psychologists described in Appendix B, Section B.3. The example is a hierarchical analysis of five dichotomous variables: whether the therapists thought (1) that their clients were aware of the therapist's attraction to them (AWARE), (2) the attraction was beneficial to the therapy (BENEFIT), and (3) the attraction was harmful to the therapy (HARM), as well as whether the therapists had (4) sought consultation when attracted to a client (CONSULT), or (5) felt uncomfortable as a result of the attraction (DISCOMF). This is an exploratory analysis, attempting to fit a model as opposed to a model in which hypothesized effects are specified. Concerns regarding overfitting apply as in all atheoretical models. Files are MFA.*.

16.6.1 Evaluation of Assumptions: Adequacy of Expected Frequencies

There are 585 psychologists in the sample. Of these, 151 are excluded from the analysis because of missing data and because only therapists who had felt attraction to at least one client answered the questions used for the analysis. The usable sample, then, consists of 434 psychologists for the hierarchical analysis, as seen in the SPSS CROSSTABS run of Table 16.13. The first part of syntax COMPUTEs a FILTER to assure that cases missing data on any of the variables are omitted from the analysis. Then the CROSSTABS instructions request observed frequency COUNTs and EXPECTED frequencies for all combinations of 2×2 tables. Only a few tables are shown.

Sample sizes are adequate for the analysis. The $2 \times 2 \times 2 \times 2 \times 2$ data table contains 32 cells, for which a sample of 434 should be sufficient; more than five cases are expected per cell if the dichotomous splits are not too bad. All the two-way contingency tables of Table 16.13, are examined to determine the adequacy of expected frequencies. The smallest expected frequency, 41.3 for the cell in which clients probably were aware of the attraction and the attraction was beneficial, is well in excess of the required minimum of 5 cases. Discussion of outliers in the solution appears in the section on adequacy of fit of the selected model that follows the section on selection of a model.

16.6.2   Hierarchical Log-Linear Analysis

16.6.2.1   Preliminary Model Screening

The full model is proposed because there are no a priori reasons to eliminate any associations. Therefore, screening and model building are used to eliminate associations that do not contribute to observed cell frequencies. Table 16.14 contains the information needed to start the model-building procedure; the simultaneous tests for effects of each order, each order and higher, and the tests of individual association, all requested through the ASSOCIATION instruction.

Both likelihood ratio and Pearson criteria are used to evaluate the k-way and higher order effects and the k-way effects. Note that the probability levels for more than two-way associations are greater than 0.05 for the simultaneous tests of both k-way effects and k-way and higher order effects. The two sets of simultaneous tests agree that variables are independent in three-way and higher-order effects. Thus the model need contain no associations greater than two-way.[6]

TABLE 16.13   Syntax and Partial Output from Preliminary SPSS CROSSTABS RUN for Hierarchical Loglinear Analysis

```
USE ALL.
COMPUTE filter_$=(aware < 3 and benefit < 3 and harm < consult < 3 and
  discomf < 3).
VARIABLE LABEL filter_$ 'aware < 3 and benefit < 3 and harm < consult < 3 and'+
  ' discomf < 3 (FILTER)'.
VALUE LABELS filter_$  0 'Not Selected' 1 'Selected'.
FORMAT filter_$ (f1.0).
FILTER BY filter_$.
EXECUTE .

CROSSTABS
 /TABLES=aware benefit harm  BY consult discomf
 /FORMAT= AVALUE TABLES
 /CELLS= COUNT EXPECTED ROW COLUMN .
CROSSTABS
 /TABLES=aware BY benefit harm
 /FORMAT= AVALUE TABLES
 /CELLS= COUNT EXPECTED ROW COLUMN .
CROSSTABS
 /TABLES=benefit BY harm
 /FORMAT= AVALUE TABLES
 /CELLS= COUNT EXPECTED ROW COLUMN .
CROSSTABS
 /TABLES=consult BY discomf
 /FORMAT= AVALUE TABLES
 /CELLS= COUNT EXPECTED ROW COLUMN .
```

(continued)

---

[6]Although one three-way effect, BENEFIT by HARM by AWARE, approaches the $p < .01$ criterion, the three-way associations are not considered for inclusion because the simultaneous tests take precedence over the component associations.

TABLE 16.13   Continued

Crosstabs

Was client aware of attraction? * Was there consultation about attraction? Crosstabulation

			Was there consultation about attraction?		
			NEVER	YES	Total
Was client aware of attraction?	PROB_NOT	Count	155	151	306
		Expected Count	126.9	179.1	306.0
		% within Was client aware of attraction?	50.7%	49.3%	100.0%
		% within Was there consultation about attraction?	86.1%	59.4%	70.5%
	YES	Count	25	103	128
		Expected Count	53.1	74.9	128.0
		% within Was client aware of attraction?	19.5%	80.5%	100.0%
		% within Was there consultation about attraction?	13.9%	40.6%	29.5%
Total		Count	180	254	434
		Expected Count	180.0	254.0	434.0
		% within Was client aware of attraction?	41.5%	58.5%	100.0%
		% within Was there consultation about attraction?	100.0%	100.0%	100.0%

The final portion of the table provides the basis of a search for the best model of one- and two-way effects. Among the two-way effects, several associations are clearly significant ($p < .01$): AWARE by BENEFIT, AWARE by CONSULT, AWARE by HARM, BENEFIT by CONSULT, HARM by DISCOMF, and CONSULT by DISCOMF. Two of the two-way associations are clearly nonsignificant: AWARE by DISCOMF and BENEFIT by DISCOMF. The remaining two-way effects—BENEFIT by HARM and HARM by CONSULT—are ambiguous ($.01 < p < .05$) and are tested through a stepwise analysis.

All first-order effects need to be included in the final hierarchical model, most because they are highly significant, and HARM because it is part of a significant two-way association. Recall that in a hierarchical model a term automatically is included if it is a part of an included higher-order association.

TABLE 16.13   Continued

Was client aware of attraction? * Was there discomfort due to attraction? Crosstabulation

			Was there discomfort due to attraction?		
			NEVER	YES	Total
Was client aware of attraction?	PROB_NOT	Count	119	187	306
		Expected Count	107.9	198.1	306.0
		% within Was client aware of attraction?	38.9%	61.1%	100.0%
		% within Was there discomfort due to attraction?	77.8%	66.5%	70.5%
	YES	Count	34	94	128
		Expected Count	45.1	82.9	128.0
		% within Was client aware of attraction?	26.6%	73.4%	100.0%
		% within Was there discomfort due to attraction?	22.2%	33.5%	29.5%
Total		Count	153	281	434
		Expected Count	153.0	281.0	434.0
		% within Was client aware of attraction?	35.3%	64.7%	100.0%
		% within Was there discomfort due to attraction?	100.0%	100.0%	100.0%

(continued)

16.6.2.2   *Stepwise Model Selection*

Stepwise selection by simple deletion from the model with 8 of the 10 two-way terms is illustrated in the SPSS HILOGLINEAR run of Table 16.15. Although 10 steps are permitted by the instruction MAXSTEPS(10), the selection process stops after the second step because the criterion probability (.01) is reached.

Recall that each potential model generates a set of expected frequencies. The goal of model selection is to find the model with the smallest number of effects that still provides a fit between expected frequencies and observed frequencies. First, the optimal model must have a nonsignificant Likelihood ratio chi square value (cf. Section 16.5.2.1, for choice between Pearson and likelihood ratio values). Second, the selected model should not be significantly worse than the next more complicated model. That is, if an effect is deleted from a model, that model should not be significantly worse than the model with the term still in it.

TABLE 16.13 Continued

Was client aware of attraction? * Was attraction beneficial to therapy? Crosstabulation

			Was attraction beneficial to therapy?		
			NEVER	YES	Total
Was client aware of attraction?	PROB_NOT	Count	129	177	306
		Expected Count	98.7	207.3	306.0
		% within Was client aware of attraction?	42.2%	57.8%	100.0%
		% within Was attraction beneficial to therapy?	92.1%	60.2%	70.5%
	YES	Count	11	117	128
		Expected Count	41.3	86.7	128.0
		% within Was client aware of attraction?	8.6%	91.4%	100.0%
		% within Was attraction beneficial to therapy?	7.9%	39.8%	29.5%
Total		Count	140	294	434
		Expected Count	140.0	294.0	434.0
		% within Was client aware of attraction?	32.3%	67.7%	100.0%
		% within Was attraction beneficial to therapy?	100.0%	100.0%	100.0%

Notice first in Table 16.15 that the first model (before Step 1) includes 8 effects, certain and ambiguous, that might be included. This model is not significant, meaning that it provides an acceptable fit between expected and observed frequencies, $\chi^2(18) = 24.549$, $p = .138$. At Step 1, effects are deleted one at a time. CONSULT by HARM is deleted at Step 1 because eliminating it produces the least Chisq Change with $p = .0366$. This model also is nonsignificant, $\chi^2(19) = 28.917, p = .067$.

Any further deletion of effects violates the criterion $p = .01$; deletion of BENEFIT by HARM has $p = .0053$ for Chisq Change. Therefore, the model at the end of Step 1 is retained.

However, the second criterion is that the model should not be significantly different from the next more complicated model. The next more complicated model is the initial model that contains CONSULT by HARM. Deletion of CONSULT by HARM at Step 1 results in a significant difference between the models, $\chi^2(1) = (28.917 - 24.492) = 4.37$, $p < .05$. Therefore, the model at Step 1 is unsatisfactory because it is significantly worse than the next more complicated model. (Use of a more conservative alpha, for example $p < .01$, would lead to a decision in favor of the best model at Step 1 with seven effects.)

The best model (8 two-way effects) is satisfactory by all criteria. Observed and expected frequencies based on this model do not differ significantly. Remember that this model includes all one-way effects because all variables are represented in one or more associations.

The model of choice for explaining the observed frequencies, then, includes all first-order effects and the two-way associations between benefit and harm, benefit and awareness, benefit and consultation, harm and awareness, harm and discomfort, harm and consultation, awareness and consultation, and discomfort and consultation. Not required in the model are the two-way associations between benefit and discomfort or discomfort and harm.

16.6.2.3   Adequacy of Fit

Overall evaluation of the model is made on the basis of the likelihood ratio χ^2, which, as seen in Table 16.16, indicates a good fit between observed and expected frequencies. For the model of choice, the likelihood ratio value is 24.55 with 18 df and $p = .138$. Confidence limits around χ^2 (recall that likelihood ratio is a form of χ^2) are found by entering χ^2 and df for the selected model and the percentage for the desired confidence interval into Smithson's (2003) NoncChi.sav and running it through NoncChi.sps. Results are added to NoncF.sav, as seen in Table 16.16. Confidence limits are 0 to 27.49. Even the upper value is less than the critical value of 28.87 with 18 df at $\alpha = .05$. This again shows inability to reject the null hypothesis of a good fit between observed and expected frequencies.

Assessment of fit of the model in individual cells proceeds through inspection of the standardized residuals for each cell (cf. Section 16.4.3.1). These residuals, as produced by SPSS HILOG-LINEAR, are shown in Table 16.17. The table displays the observed frequencies for each cell, the expected frequencies for each cell (**EXP count**), the differences between observed and expected

TABLE 16.14   Syntax and Edited Output for SPSS HILOGLINEAR Preliminary Run of Simultaneous and Component Associations

```
HILOGLINEAR
  aware(1 2) benefit(1 2) harm(1 2) consult(1 2) discomf(1 2)
  /CRITERIA ITERATION(20) DELTA(0)
  /PRINT=ASSOCIATION
  /DESIGN.
```

Tests that K-way and higher order effects are zero.

K	DF	L.R. Chisq	Prob	Pearson Chisq	Prob	Iteration
5	1	.295	.5869	.164	.6859	3
4	6	10.162	.1180	11.141	.0841	4
3	16	24.086	.0876	21.854	.1480	6
2	26	253.506	.0000	364.212	.0000	2
1	31	436.151	.0000	491.346	.0000	0

Tests that K-way effects are zero.

K	DF	L.R. Chisq	Prob	Pearson Chisq	Prob	Iteration
1	5	182.645	.0000	127.133	.0000	0
2	10	229.420	.0000	342.358	.0000	0
3	10	13.924	.1765	10.713	.3803	0
4	5	9.867	.0791	10.978	.0518	0
5	1	.295	.5869	.164	.6859	0

(continued)

TABLE 16.14 Continued

* * * * * * * H I E R A R C H I C A L L O G L I N E A R * * * * * * * *

Tests of PARTIAL associations.

Effect Name	DF	Partial Chisq	Prob	Iter
aware*benefit*harm*consult	1	3.068	.0798	3
aware*benefit*harm*discomf	1	3.594	.0580	3
aware*benefit*consult*discomf	1	1.200	.2733	3
aware*harm*consult*discomf	1	2.059	.1513	4
benefit*harm*consult*discomf	1	.430	.5120	3
aware*benefit*harm	1	6.089	.0136	4
aware*benefit*consult	1	.660	.4165	4
aware*harm*consult	1	.613	.4338	4
benefit*harm*consult	1	.412	.5210	4
aware*benefit*discomf	1	.157	.6920	3
aware*harm*discomf	1	.745	.3879	4
benefit*harm*discomf	1	1.065	.3022	4
aware*consult*discomf	1	2.202	.1379	4
benefit*consult*discomf	1	.423	.5153	4
harm*consult*discomf	1	.055	.8139	4
aware*benefit	1	31.954	.0000	6
aware*harm	1	11.708	.0006	6
benefit*harm	1	4.688	.0304	5
aware*consult	1	15.947	.0001	6
benefit*consult	1	9.769	.0018	5
harm*consult	1	4.283	.0385	5
aware*discomf	1	.263	.6081	5
benefit*discomf	1	.313	.5760	5
harm*discomf	1	28.987	.0000	5
consult*discomf	1	21.474	.0000	5
aware	1	75.203	.0000	2
benefit	1	55.854	.0000	2
harm	1	.590	.4425	2
consult	1	12.679	.0004	2
discomf	1	38.318	.0000	2

TABLE 16.15   Syntax and Partial Output from SPSS HILOGLINEAR Model Selection RUN for Hierarchical Log-Linear Analysis

```
HILOGLINEAR
 aware(1 2) benefit(1 2) harm(1 2) consult(1 2) discomf(1 2) /METHOD=BACKWARD
 /CRITERIA MAXSTEPS(10) P(.01) ITERATION(20) DELTA(0)
 /PRINT=ASSOCIATION
 /DESIGN aware*benefit aware*consult aware*harm benefit*consult benefit*harm
 consult*discomf consult*harm discomf*harm.
```

```
 * * * * * * H I E R A R C H I C A L   L O G   L I N E A R * * * * * * 
```

Backward Elimination (p = .010) for DESIGN 1 with generating class

    aware*benefit
    aware*consult
    aware*harm
    benefit*consult
    benefit*harm
    consult*discomf
    consult*harm
    discomf*harm

Likelihood ratio chi square =   24.54925   DF = 18 P = .138

If Deleted Simple Effect is	DF	L.R. Chisq Change	Prob	Iter
aware*benefit	1	31.828	.0000	5
aware*consult	1	15.824	.0001	4
aware*harm	1	11.589	.0007	5
benefit*consult	1	11.267	.0008	4
benefit*harm	1	5.762	.0164	4
consult*discomf	1	23.481	.0000	5
consult*harm	1	4.368	.0366	4
discomf*harm	1	30.463	.0000	5

Step 1

   The best model has generating class

       aware*benefit
       aware*consult
       aware*harm
       benefit*consult
       benefit*harm
       consult*discomf
       discomf*harm

Likelihood ratio chi square =   28.91701   DF = 19 P = .067

If Deleted Simple Effect is	DF	L.R. Chisq Change	Prob	Iter
aware*benefit	1	30.474	.0000	4
aware*consult	1	19.227	.0000	4
aware*harm	1	14.992	.0001	4
benefit*consult	1	13.289	.0003	4
benefit*harm	1	7.784	.0053	4

(continued)

TABLE 16.15   **Continued**

```
* * * * * * H I E R A R C H I C A L    L O G    L I N E A R * * * * * *
If Deleted Simple Effect is            DF   L.R. Chisq Change   Prob    Iter

 consult*discomf                        1              31.545   .0000      4
 discomf*harm                           1              38.527   .0050      4

Step 2

  The best model has generating class

        aware*benefit
        aware*consult
        aware*harm
        benefit*consult
        benefit*harm
        consult*discomf
        discomf*harm

  Likelihood ratio chi square =     28.91701    DF = 19  P = .067
-------------------------------------------------------------------

   * * * * * * H I E R A R C H I C A L    L O G    L I N E A R * * * * * *
The final model has generating class

        aware*benefit
        aware*consult
        aware*harm
        benefit*consult
        benefit*harm
        consult*discomf
        discomf*harm


The Iterative Proportional Fit algorithm converged at iteration 0.
The maximum difference between observed and fitted marginal totals is  .071
and the convergence criterion is  .250
-------------------------------------------------------------------

Goodness-of-fit test statistics

     Likelihood ratio chi square =   28.91701    DF = 19   P = .067
                Pearson chi square =   28.04255    DF = 19   P = .083
-------------------------------------------------------------------
```

TABLE 16.16   **Data Set Output from NoncChi.sps for Likelihood Ratio (Chi Square) with 95%
Confidence Limits (lc2 and uc2).**

	chival	df	conf	lc2	ucdf	uc2	lcdf	power
1	24.5500	18	.950	.0000	.8622	27.4869	.0250	.8292

TABLE 16.17   Syntax and Partial Output of SPSS HILOGLINEAR RUN to Evaluate Residuals

```
HILOGLINEAR
 aware(1 2) benefit(1 2) harm(1 2) consult(1 2) discomf(1 2)
 /CRITERIA ITERATION(20) DELTA(0)
 /PRINT=FREQ RESID
 /PLOT=RESID NORMPROB
 /DESIGN aware*benefit aware*consult aware*harm benefit*consult benefit*harm
 consult*discomf consult*harm discomf*harm.
```

Observed, Expected Frequencies and Residuals.

Factor	Code	OBS count	EXP count	Residual	Std Resid
aware	PROB NOT				
benefit	NEVER				
harm	NEVER				
consult	NEVER				
discomf	NEVER	43.0	37.1	5.92	.97
discomf	YES	20.0	21.8	-1.84	-.39
consult	YES				
discomf	NEVER	10.0	9.8	.17	.06
discomf	YES	16.0	16.7	- .65	-.16
harm	YES				
consult	NEVER				
discomf	NEVER	4.0	7.5	-3.50	-1.28
discomf	YES	14.0	14.8	-.81	-.21
consult	YES				
discomf	NEVER	4.0	3.2	.81	.45
discomf	YES	18.0	18.1	-.14	-.03
benefit	YES				
harm	NEVER				
consult	NEVER				
discomf	NEVER	27.0	28.0	-1.04	-.20
discomf	YES	14.0	16.5	-2.51	-.62
consult	YES				
discomf	NEVER	13.0	16.0	-2.97	-.74
discomf	YES	30.0	27.1	2.93	.56
harm	YES				
consult	NEVER				
discomf	NEVER	11.0	9.8	1.17	.37
discomf	YES	22.0	19.4	2.59	.59
consult	YES				
discomf	NEVER	7.0	9.0	-2.00	-.67
discomf	YES	53.0	51.1	1.89	.26

(continued)

TABLE 16.17   **Continued**

Observed, Expected Frequencies and Residuals (continued)

Factor	Code	OBS count	EXP count	Residual	Std Resid
aware	YES				
benefit	NEVER				
harm	NEVER				
consult	NEVER				
discomf	NEVER	.0	1.3	-1.27	-1.13
discomf	YES	.0	.7	-.75	-.87
consult	YES				
discomf	NEVER	1.0	.9	.06	.06
discomf	YES	.0	1.6	-1.60	-1.26
harm	YES				
consult	NEVER				
discomf	NEVER	1.0	.6	.41	.54
discomf	YES	3.0	1.2	1.84	1.71
consult	YES				
discomf	NEVER	.0	.7	-.70	-.83
discomf	YES	6.0	4.0	2.05	1.03
benefit	YES				
harm	NEVER				
consult	NEVER				
discomf	NEVER	3.0	5.4	-2.36	-1.02
discomf	YES	7.0	3.2	3.85	2.17
consult	YES				
discomf	NEVER	10.0	8.5	1.49	.51
discomf	YES	15.0	14.4	.57	.15
harm	YES				
consult	NEVER				
discomf	NEVER	5.0	4.3	.73	.35
discomf	YES	6.0	8.4	-2.44	-.84
consult	YES				
discomf	NEVER	14.0	10.9	3.08	.93
discomf	YES	57.0	62.0	-4.99	-.63

Goodness-of-fit test statistics

```
    Likelihood ratio chi square = 24.54925   DF = 18   P = .138
               Pearson chi square = 22.01415   DF = 18   P = .231
```

frequencies (`Residual`), and the standardized deviates (`Std Resid`, the standardized residual values from which discrepancies are evaluated).

Most of the standardized residual values are quite small; only one cell has a value that exceeds the critical z value of 1.96. Since the classification table has 32 cells, a standardized residual value of 2.17 (the largest of the standardized residuals) for one of them is not unexpected; this cell is not deviant enough to be considered an outlier. However, the fit of the model is least effective for this cell, which contains therapists who felt their attraction to clients was beneficial to the therapy, who thought their clients were aware of the attraction, and who felt uncomfortable about it, but who never felt it harmful to the therapy or sought consultation about it. As seen from the observed frequency table, seven of the 434 therapists responded in this way. The expected frequency table shows that, according to the model, only about three were predicted to provide this pattern of responses.

The syntax of Table 16.17 also requests a normalized probability plot of residuals (/PLOT=RESID NORMPROB). Figure 16.1 shows the output produced by this request, in which observed standardized residuals are seen to be acceptably close to those that are expected (the diagonal line).

16.6.2.4   *Interpretation of the Selected Model*

Two types of information are useful in interpreting the selected model: parameter estimates for the model and marginal observed frequency tables for all included effects.

The log-linear parameter estimate, lambda (`Coeff.`), and the `Z-Value`—ratio of the `Coeff.`/`Std. Err.` (cf. Section 16.4.3.2)—from SPSS LOGLINEAR (available only in syntax) are shown in Table 16.18 for each effect included in the model—recall that these are not available through HILOGLINEAR for an unsaturated model. Because there are only two levels of each

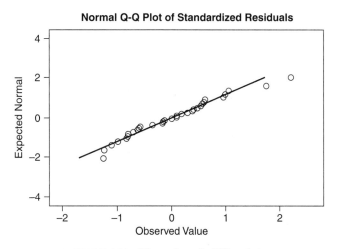

**FIGURE 16.1   Normal probability plot
for selected model.**

variable, each effect is summarized by a single parameter value where one level of the effect has the positive value of the parameter and the other the negative value of the parameter.

Especially useful for interpretation are the standardized parameter estimates (Z-Values). Effects with the largest standardized parameter estimates are the most important in influencing the frequency in a cell. If the effects are rank ordered by the sizes of their standardized parameter estimates, the relative importance of the various effects becomes apparent. With a standardized parameter estimate of 8.815, the strongest predictor of cell size is whether or not the therapist thought the client was aware of the therapist's attraction. The least predictive of all the effects in the model, with a standardized parameter estimate of 0.508, is whether the therapist's attraction to the client was believed to be harmful to the therapy. (Recall from Table 16.17 that this one-way effect is included in the hierarchical model only because it is a component of at least one two-way association; it was not statistically significant by itself.)

Parameter estimates are useful in determining the relative strength of effects and in creating a prediction equation, but they do not provide a simple view of the direction of effects. For interpretation of direction, the marginal tables of observed frequencies for each effect in the model are useful, as illustrated in the CROSSTABS output of Table 16.13.

The results as displayed in Table 16.13 are best interpreted as proportions of therapists responding in a particular way. For example, the BENEFIT marginal subtable (see third subtable) shows that 32% $(140/434)$ of the therapists believe that there was never any benefit to be gained from the therapist being attracted to a client. The BENEFIT by HARM marginal subtable (next to last subtable) shows that, among those who believe that there was no benefit, 64% $(90/140)$ also believe there was no harm. Of those who believe there was at least some benefit, 59% $(175/294)$ also believe there was at least some harm.

Table 16.19 summarizes significance tests and their confidence intervals, as found per Smithson (2003). Recall that the expected value of chi square when the null hypothesis is true is equal to the df. Table 16.20 summarizes parameter estimates.

TABLE 16.18   Syntax and Partial Output for SPSS LOGLINEAR Run on Parameter Estimates

```
LOGLINEAR
   aware(1 2) benefit(1 2) harm(1 2) consult(1 2) discomf(1 2)
/PRINT=ESTIM
/DESIGN aware*benefit aware*consult aware*harm benefit*consult benefit*harm
   consult*discomf consult*harm discomf*harm aware benefit consult harm discomf.


*  *  *  *  *  *  *  *  *  L O G    L I N E A R    A N A L Y S I S * * * * * * * * *

  Estimates for Parameters

  aware * benefit

   Parameter       Coeff.        Std. Err.      Z-Value Lower 95 CI Upper 95 CI

        1      .4275530136         .08643        4.94693      .25815        .59695

  aware * consult

   Parameter       Coeff.        Std. Err.      Z-Value Lower 95 CI Upper 95 CI

        2      .2562875811         .06667        3.84412      .12561        .38696
```

TABLE 16.18   Continued

aware * harm

Parameter	Coeff.	Std. Err.	Z-Value	Lower 95 CI	Upper 95 CI
3	.2055847749	.06127	3.35516	.08549	.32568

benefit * consult

Parameter	Coeff.	Std. Err.	Z-Value	Lower 95 CI	Upper 95 CI
4	.1915142665	.05724	3.34580	.07932	.30370

benefit * harm

Parameter	Coeff.	Std. Err.	Z-Value	Lower 95 CI	Upper 95 CI
5	.1381059298	.05766	2.39509	.02509	.25112

consult * discomf

Parameter	Coeff.	Std. Err.	Z-Value	Lower 95 CI	Upper 95 CI
6	.2649530804	.05507	4.81159	.15702	.37288

consult * harm

Parameter	Coeff.	Std. Err.	Z-Value	Lower 95 CI	Upper 95 CI
7	.1182768367	.05644	2.09545	.00765	.22891

discomf * harm

Parameter	Coeff.	Std. Err.	Z-Value	Lower 95 CI	Upper 95 CI
8	.3022378758	.05582	5.41421	.19282	.41165

aware

Parameter	Coeff.	Std. Err.	Z-Value	Lower 95 CI	Upper 95 CI
9	.7935949466	.09003	8.81471	.61713	.97006

benefit

Parameter	Coeff.	Std. Err.	Z-Value	Lower 95 CI	Upper 95 CI
10	-.617075264	.08537	-7.22819	-.78440	-.44975

consult

Parameter	Coeff.	Std. Err.	Z-Value	Lower 95 CI	Upper 95 CI
11	-.166275582	.07171	-2.31865	-.30683	-.02572

harm

Parameter	Coeff.	Std. Err.	Z-Value	Lower 95 CI	Upper 95 CI
12	.0351320444	.06924	.50742	-.10057	.17084

discomf

Parameter	Coeff.	Std. Err.	Z-Value	Lower 95 CI	Upper 95 CI
13	-.301825163	.05544	-5.44463	-.41048	-.19317

TABLE 16.19 Significance Tests for Hierarchical Model of Therapists' Attraction to Clients, $N = 434$

Effect	Partial Association Chi Square df = 1	95% Confidence Interval for Chi Square	
		Lower	*Upper*
First-order effects:			
Aware	72.20**	42.73	109.35
Benefit	55.85**	30.39	88.98
Discomfort	38.32**	17.91	66.43
Consult	12.68**	2.56	30.48
Harm	0.59	0	7.42
Second-order effects:			
Benefit by aware	31.95**	13.64	57.95
Harm by discomfort	28.99**	11.72	53.94
Discomfort by consult	21.47**	7.15	43.47
Aware by consult	15.95**	4.14	35.45
Harm by aware	11.71**	2.14	28.97
Benefit by consult	9.77**	1.36	25.86
Benefit by harm	4.69*	0	17.02
Harm by consult	4.28*	0	16.23

*$p < .05$.

**$p < .01$.

TABLE 16.20 Parameter Estimates for Hierarchical Model of Therapists' Attraction to Clients; $N = 434$, Constant = 1.966

Effect	Log-linear Parameter Estimate (Lambda)		Lambda/*SE*	
First-order effects:				
	Prob. not	Yes	Prob. not	Yes
Aware	0.794	−0.794	8.815	−8.815
	Never	Yes	Never	Yes
Benefit	−0.617	0.617	−7.228	7.228
	Never	Yes	Never	Yes
Discomfort	−0.302	0.302	−5.445	5.445
	Never	Yes	Never	Yes
Consult	−0.166	0.166	−2.319	2.319
	Never	Yes	Never	Yes
Harm	0.035	−0.035	0.508	−0.508

TABLE 16.20 Continued

Effect		Log-linear Parameter Estimate (Lambda)		Lambda/*SE*	
Second-order effects:					
		Prob. not	Yes	Prob. not	Yes
Benefit by aware	Never	0.428	−0.428	4.947	−4.947
	Yes	−0.428	0.428	−4.947	4.947
		Never	Yes	Never	Yes
Harm by discomfort	Never	0.302	−0.302	5.414	−5.414
	Yes	−0.302	0.302	−5.414	5.414
		Never	Yes	Never	Yes
Discomfort by consult	Never	0.265	−0.265	4.812	−4.812
	Yes	−0.265	0.265	−4.812	4.812
		Never	Yes	Never	Yes
Aware by consult	Prob. not	0.256	−0.256	3.844	−3.844
	Yes	−0.256	0.256	−3.844	3.844
		Prob. not	Yes	Prob. not	Yes
Harm by aware	Never	0.206	−0.206	3.555	−3.555
	Yes	−0.206	0.206	−3.555	3.555
		Never	Yes	Never	Yes
Benefit by consult	Never	0.191	−0.191	3.345	−3.345
	Yes	−0.191	0.191	−3.345	3.345
		Never	Yes	Never	Yes
Benefit by harm	Never	0.138	−0.138	2.397	−2.397
	Yes	−0.138	0.138	−2.397	2.397
		Never	Yes	Never	Yes
Harm by consult	Never	0.118	−0.138	2.095	−2.095
	Yes	−0.118	0.118	−2.095	2.095

**TABLE 16.21   Checklist for Hierarchical
Multiway Frequency Analysis**

1. Issues
    a. Adequacy of expected frequencies
    b. Outliers in the solution
2. Major analysis
    a. Model screening
    b. Model selection
    c. Evaluation of overall fit. If adequate:
        (1) Significance tests for each model effect
            and their confidence intervals
        (2) Parameter estimates
3. Additional analyses
    a. Interpretation via proportions
    b. Identifying extreme cells (if fit inadequate)

A checklist for hierarchical multiway frequency analysis appears in Table 16.21. A Results section, in journal format, follows for the analysis described.

Results

A five-way exploratory frequency analysis was performed to develop a hierarchical log-linear model of attraction of therapists to clients. Dichotomous variables analyzed were whether the therapist (1) believed the attraction to be beneficial to the client, (2) believed the attraction to be harmful to the client, (3) thought the client was aware of the attraction, (4) felt discomfort, and (5) sought consultation as a result of the attraction.

Four hundred thirty-four therapists provided usable data for this analysis. All two-way contingency tables provided expected frequencies in excess of five. After the model was selected, none of the 32 cells was an outlier.

Stepwise selection by simple deletion of effects using SPSS HILOGLINEAR produced a model that included all first-order effects and eight of the ten possible two-way associations. The model had a likelihood ratio χ^2 (18) = 24.55 with 95% confidence limits from 0 to 27.49, p = .14, indicating a good fit between observed frequencies and expected frequencies generated by the model. A summary of the model with results of tests of significance (partial likelihood ratio χ^2) and their 95% confidence limits is in Table 16.19. A summary of log-linear parameter estimates in raw and standardized form appears in Table 16.20.

Most of the therapists (68%) reported that the attraction they felt for clients was at least occasionally beneficial to therapy, while a slight majority (52%) also reported that it was at least occasionally harmful. Seventy-one percent of the therapists thought that clients were probably aware of the attraction. Most therapists (65%) felt at least some discomfort about the attraction, and more than half (58%) sought consultation as a result of the attraction.

Of those therapists who thought the attraction beneficial to the therapy, 60% also thought it harmful. Of those who thought the attraction never beneficial, 36% thought it harmful. Perception of benefit was also related to client's awareness. Of those who thought their clients were aware of the attraction, 91% thought it beneficial. Among those who thought clients unaware, only 58% thought it beneficial.

Those who sought consultation were also more likely to see the attraction as beneficial. Of those seeking consultation, 78% judged the attraction beneficial. Of those not seeking consultation, 53% judged it beneficial.

Lack of harm was associated with lack of awareness. Fifty-seven percent of therapists who thought their clients unaware felt the attraction was never harmful. Only 28% of those who thought their clients aware considered it never harmful. Discomfort was more likely to be felt by those therapists who considered the attraction harmful to therapy (80%) than by those therapists who thought it was not harmful to therapy (49%). Similarly, consultation was more likely to be sought by those who felt the attraction harmful (71%) than by those who did not feel it harmful (45%).

Seeking consultation was also related to client awareness and therapist discomfort. Therapists who thought clients were aware of the attraction were more likely to seek consultation (80%) than those who thought the client unaware (43%). Those who felt discomfort were more likely to seek consultation (69%) than those who felt no such discomfort (39%).

No statistically significant two-way associations were found between benefit and discomfort or between awareness and discomfort. None of the higher-order associations reached statistical significance.

16.7   Comparison of Programs

Five programs are available in SAS, SPSS, and SYSTAT for analysis of multiway frequency tables. There are two types of programs for log-linear analysis, those that deal exclusively with hierarchical models and general log-linear programs that can handle nonhierarchical models as well (cf. Section 16.5.1). SPSS GENLOG and LOGLINEAR, SYSTAT LOGLIN and SAS CATMOD are general programs for nonhierarchical as well as hierarchical models (cf. Section 16.5.1). SPSS HILOGLINEAR deals only with hierarchical models, but includes features for stepwise model building (cf. Section 16.5.3). All five programs provide observed and expected cell frequencies, tests of fit of incomplete models, and parameter estimates accompanied by their standard errors. Beyond that, the programs differ widely. Features of the five programs appear in Table 16.22.

TABLE 16.22   Comparisons of Programs for Multiway Frequency Analysis

Feature	SPSS GENLOG	SPSS HILOG-LINEAR	SPSS LOG-LINEAR	SAS CATMOD	SYSTAT LOGLIN
Input					
Individual case data	Yes	Yes	Yes	Yes	Yes
Cell frequencies and indices	No[c]	WEIGHT	WEIGHT	WEIGHT	FREQ
Cell weights (structural zeros)	CSTRUCTURE	CWEIGHT	CWEIGHT	Yes	ZERO CELL
Convergence criteria	CONVERGE	CONVERGE	CONVERGE	EPSILON	CONV, LCONV
Tolerance	No	No	No	No	TOL
Level of confidence interval	CIN	No	No	No	No
Epsilon value for redundancy checking	EPS	No	No	No	No
Specify maximum number of iterations	ITERATE	ITERATE	ITERATION	MAXITER	ITER
Maximum number of halvings	No	No	No	No	HALF
Stepping options	N.A.	Yes	N.A.	N.A.	N.A.
Specify maximum no. of steps	N.A.	Yes	N.A.	N.A.	N.A.
Specify significance level for adequate fit	N.A.	P	N.A.	N.A.	N.A.
Specify maximum order of terms	N.A.	MAXORDER	N.A.	N.A.	N.A.
Force terms into stepping model	N.A.	No	N.A.	N.A.	N.A
Covariates (continuous)	Yes	No	Yes	No	No
Logit model specification	Yes	No	Yes	Yes	No
Single df partitions & contrasts	No	No	Yes	Yes	No
Specify delta for each cell	DELTA	DELTA	DELTA	ADDCELL	DELTA
Include cases with user-missing values	INCLUDE	INCLUDE	INCLUDE	No	No
Specify a repeated measures factor (DV only)	No[d]	No[d]	No[d]	Yes	No
Specify ordered factor(s)	No[a]	No	Yes	Yes	No
Poisson model	Default	No	No	No	No
Multinomial logit model	Yes	Yes	Yes	Yes	Yes
Specify weighted least-squares method	No	No	No	Yes	No

(continued)

TABLE 16.22 Continued

Feature	SPSS GENLOG	SPSS HILOG-LINEAR	SPSS LOG-LINEAR	SAS CATMOD	SYSTAT LOGLIN
Output					
Nonhierarchical models	Yes	No	Yes	Yes	Yes
Tests of partial association	No	Yes	No	Yes	No
Tests of models with and without each item	No	No	No	No	Yes
Maximum likelihood (χ^2) tests of association (ANOVA)	No	No	No	Yes	No
Tests of k-way effects	No	Yes	No	No	No
Tests of k-way & higher effects	No	Yes	No	No	No
Pearson model tests	Yes	Yes	Yes	No	Yes
Likelihood ratio model tests	Yes	Yes	Yes	Yes	Yes
Observed & expected (predicted) frequencies	Yes	Yes	Yes	Yes	Yes
Observed & expected probabilities or percentages	Yes	No	Yes	Yes	Yes
Raw residuals	Yes	Yes	Yes	Yes	Yes
Standardized residuals	Yes	Yes	Yes	No	Yes
Deviation residuals	Yes	No	No	No	No
Generalized residuals	Yes	No	Yes	No	No
Adjusted residuals	Yes	No	Yes	No	No
Freeman-Tukey residuals	No	No	No	No	Yes
Pearson χ^2 residuals	No	No	No	No	Yes
Likelihood ratio components	No	No	No	No	Yes
Contribution to log likelihood for each cell	No	No	No	No	Yes
Log-linear parameter estimates	Parameter	Coeff[a]	Coeff	Estimate	Param
Standard error of parameter estimate	SE	Std. Err[a]	Std. Err.	Standard error	SE(Param)
Ratio of parameter estimate to standard error (z or t)	Z-value	Z-value[a]	Z-value	No	Param/SE
Confidence limits for parameter estimates	Yes	Yes[a]	Yes	No	No
Chi-square tests for parameter estimate	No	No	No	Yes	No

TABLE 16.22   Continued

Feature	SPSS GENLOG	SPSS HILOG-LINEAR	SPSS LOG-LINEAR	SAS CATMOD	SYSTAT LOGLIN
Output *(continued)*					
Multiplicative parameter estimates	No	No	No	No	Yes
Index of dissimilarity	No	No	No	No	Yes
Correlation matrix for parameter estimates	Yes	No	Yes	Yes	Yes
Covariance matrix for parameter estimates	Yes	No	No	Yes	Yes
Design matrix	Yes	No	Yes	Yes	No
Plots of standardized or adjusted residuals vs. observed and expected frequencies	Yes	Yes	Yes	No	No
Normal plots of adjusted residuals	Yes	Yes	Yes	No	No
Detrended normal plots of adjusted and deviance residuals	Yes	Yes	Yes	No	No[b]
Raftery's BIC	No	No	No	No	Yes
Dissimilarity	No	No	No	No	Yes

[a]Saturated model only.

[b]Available through PPLOT.

[c]Done outside the program (see Table 16.11).

[d]Done through SPSS COMPLEX SAMPLES LOGISTIC REGRESSION.

16.7.1   SPSS Package

Currently there are two programs for handling multiway frequency tables in the SPSS package: HILOGLINEAR, which deals with only hierarchical models and GENLOG, which deals with hierarchical and nonhierarchical models.

SPSS HILOGLINEAR, labeled Model Selection in the Loglinear menu, is well suited to choosing among hierarchical models, with several options for controlling stepwise selection of effects. Simultaneous tests of all k-way effects and of all k-way and higher effects are available for a quick screening of the complexity of the model from which to start stepwise selection. Parameter estimates and partial tests of association are available, but only for full models.

SPSS GENLOG does not provide stepwise selection of hierarchical models, although it can be used to compare user-specified models of any sort. The program permits specification of continuous covariates. Also available is a simple specification of a logit model (in which one factor is a DV). Specification of a cell weighting variable must occur outside the GENLOG program.

No inferential tests of model components are provided in the program. Parameter estimates and their z tests are available for any specified model, along with their 95% confidence intervals. However, the parameter estimates are reported by single degrees of freedom, so that a factor with more than two categories has no omnibus significance test reported for either its main effect or its association with other effects (cf. Section 16.5.2.2). No quick screening for k-way tests is available. Screening information can be gleaned from a full model run, but identifying an appropriate model may be tedious with a large number of factors. Both SPSS programs offer residuals plots. SPSS GENLOG is the only program offering specification of Poisson models, which do not require that the analysis be conditional on total sample size.

SPSS LOGLINEAR, available only through syntax, fills in where SPSS HILOGLINEAR leaves off when developing a model, providing parameter estimates for models that are not saturated. SPSS LOGLINEAR also may be used for nonhierarchical models and permits specification of continuous covariates and contrasts. SPSS COMPLEX SAMPLES LOGISTIC REGRESSION may be used when a dichotomous DV is repeatedly measured on the same cases by defining cases as clusters.

16.7.2 SAS System

SAS CATMOD is a general program for modeling discrete data, of which log-linear modeling is only one type. The program is primarily set up for logit analyses where one variable is the DV but provision is made for log-linear models where no such distinction is made. The program offers simple designation of logit models, contrasts, and single df tests of parameters as well as maximum likelihood tests of more complex components. The program lacks provision for continuous covariates and stepwise model building procedures.

SAS CATMOD uses different algorithms from the other three programs both for parameter estimation and model testing. The output in Table 16.12 compared with that of Tables 16.10 and 16.11 demonstrates some of the inconsistencies.

This is the only program that allows specification of factors that are ordered. Also, this is the only program that permits multiple DVs that are defined as repeated measurements of the same variable.

16.7.3 SYSTAT System

SYSTAT LOGLIN is a general program for log-linear analysis of categorical data. The program uses its typical MODEL statement to set up the full, saturated, model (i.e., observed frequencies) on the left-hand side of the equation, and the desired model to be tested on the right. Structural zeros can be specified, and several options are available for controlling the iterative processing of model estimation. All of the usual descriptive and parameter estimate statistics are available, as well as multiple tests of effects in the model, both hierarchical and nonhierarchical. The program also prints outlying cells, designated "outlandish." Estimated frequencies and parameter estimates can be saved to a file.

17 An Overview of the General Linear Model

17.1 Linearity and the General Linear Model

To facilitate choice of the most useful technique to answer your research question, the emphasis has been on differences among statistical methods. We have repeatedly hinted, however, that most of these techniques are special applications of the general linear model (GLM). The goal of this chapter is to introduce the GLM and to fit the various techniques into the model. In addition to the aesthetic pleasure provided by insight into the GLM, an understanding of it provides a great deal of flexibility in data analysis by promoting use of more sophisticated statistical techniques and computer programs. Most data sets are fruitfully analyzed by one or more of several techniques. Section 17.3 presents an example of the use of alternative research strategies.

Linearity and additivity are important to the GLM. Pairs of variables are assumed to have a linear relationship with each other; that is, it is assumed that relationships between pairs of variables are adequately represented by a straight line. Additivity is also relevant, because if one set of variables is to be predicted by a set of other variables, the effects of the variables within the set are additive in the prediction equation. The second variable in the set adds predictability to the first one, the third adds to the first two, and so on. In all multivariate solutions, the equation relating sets of variables is composed of a series of weighted terms added together.

These assumptions, however, do not prevent inclusion of variables with curvilinear or multiplicative relationships. As is discussed throughout this book, variables can be multiplied together, raised to powers, dichotomized, transformed, or recoded so that even complex relationships are evaluated within the GLM.

17.2 Bivariate to Multivariate Statistics and Overview of Techniques

17.2.1 Bivariate Form

The GLM is based on prediction or, in jargon, regression. A regression equation represents the value of a DV, Y, as a combination of one or more IVs, Xs, plus error. The simplest case of the GLM, then, is the familiar bivariate regression:

$$A + BX + e = Y \tag{17.1}$$

where B is the change in Y associated with a one-unit change in X; A is a constant representing the value of Y when X is 0; and e is a random variable representing error of prediction.

If X and Y are converted to standard z-scores, z_x and z_y, they are now measured on the same scale and cross at the point where both z-scores equal 0. The constant A automatically becomes 0 because z_y is 0 when z_x is 0. Further, after standardization of variances to 1, slope is measured in equal units (rather than the possibly unequal units of X and Y raw scores) and now represents strength of the relationship between X and Y; in bivariate regression with standardized variables, β is equal to the Pearson product-moment correlation coefficient. The closer β is to 1.00 or -1.00, the better the prediction of Y from X (or X from Y). Equation 17.1 then simplifies to

$$\beta z_x + e = z_y \tag{17.2}$$

As discussed in Chapters 1 and 2, one distinction that is sometimes important in statistics is whether data are continuous or discrete.[1] There are, then, three forms of bivariate regression for situations where X and Y are (1) both continuous, analyzed by Pearson product-moment correlation, (2) mixed, with X dichotomous and Y continuous, analyzed by point biserial correlation, and (3) both dichotomous, analyzed by phi coefficient. In fact, these three forms of correlation are identical. If the dichotomous variable is coded 0–1, all the correlations can be calculated using the equation for Pearson product-moment correlation. Table 17.1 compares the three bivariate forms of the GLM.

17.2.2   Simple Multivariate Form

The first generalization of the simple bivariate form of the GLM is to increase the number of IVs, Xs, used to predict Y. It is here that the additivity of the model first becomes apparent. In standardized form:

$$\sum_{i=1}^{k} \beta_i z_{x_i} + e = z_y \tag{17.3}$$

That is, Y is predicted by a weighted sum of Xs. The weights, β_i, no longer reflect the correlation between Y and each X because they are also affected by correlations among the Xs. Here, again, as seen in Table 17.1, there are special statistical techniques associated with whether all Xs are continuous; here also, with appropriate coding, the most general form of the equation can be used to solve all the special cases.

If Y and all Xs are continuous, the special statistical technique is multiple regression. Indeed, as seen in Chapter 5, Equation 17.3 is used to describe the multiple regression problem. But if Y is continuous and all Xs are discrete, we have the special case of regression known as analysis of variance. The values of X represent "groups" and the emphasis is on finding mean differences in Y

---

[1]When discrete variables have more than two levels, they are dummy variable coded into $k - 1$ (df) dichotomous variables to eliminate the possibility of nonlinear relationships. In this section, when we speak of statistical techniques using discrete variables, we imply that recoding is unnecessary or is handled internally in computer programs designed for the particular analysis.

TABLE 17.1   Overview of Techniques in the General Linear Model

A. Bivariate form (Eq. 17.2)

   1. Pearson product-moment correlation: X continuous, Y continuous

   2. Point biserial correlation: X dichotomous, Y continuous

   3. Phi coefficient: X dichotomous, Y dichotomous

B. Simple multivariate form (Eq. 17.3)

   1. Multiple regression: all Xs continuous, Y continuous

   2. ANOVA: all Xs discrete, Y continuous

   3. ANCOVA: some Xs continuous and some discrete, Y continuous

   4. Two-group discriminant analysis: all Xs continuous, Y dichotomous

   5. Multiway frequency analysis: all Xs discrete, Y is category frequency (or dichotomous in logit analysis)

   6. Two-group logistic regression analysis: Xs continuous and/or discrete, Y dichotomous

   7. Multilevel modeling: Xs at each level may be continuous or discrete. Ys at each level are continuous

   8. Survival analysis: Xs continuous and/or dichotomous, Y continuous (time)

   9. Time series analysis: Xs continuous (time) and dichotomous, Y continuous

C. Full multivariate form (Eq. 17.4)

   1. Canonical correlation: all Xs continuous, all Ys continuous

   2. MANOVA: all Xs discrete, all Ys continuous

   3. MANCOVA: some Xs continuous and some discrete, all Ys continuous

   4. Profile analysis: all Xs discrete, all Ys continuous and commensurate

   5. Discriminant analysis: all Xs continuous, all Ys discrete

   6. Factor analysis (FA)/principal component analysis (PCA): all Ys continuous, all Xs latent

   7. Structural equations modeling: Xs continuous and/or latent, Ys continuous and/or latent

   8. Multiway frequency analysis: all Xs discrete, Y is category frequency

   9. Polychotomous logistic regression analysis: Xs continuous and/or discrete, Y discrete

---

among groups rather than on predicting Y, but the basic equation is the same. A significant difference among groups implies that knowledge of X can be used to predict performance on Y.

    Analysis of variance problems can be solved through multiple regression computer programs. There are as many Xs as there are degrees of freedom for the effects. For example, in a one-way design, three groups are recoded into two dichotomous Xs, one representing the first group versus the other two and the second representing the second group versus the other two. The third group is those who are not in either of the other two groups. Inclusion of a third X would produce singularity because it is perfectly predictable from the combination of the other two.

If IVs are factorially combined, main effects and interactions are still coded into a series of dichotomous X variables. Consider an example of one IV, anxiety level, divided into three groups and a second IV, task difficulty, divided into two groups. There are two X components for the 2 df associated with anxiety level and one X component for the 1 df associated with task difficulty. An additional two X components are needed for the 2 df associated with the interaction of anxiety level and task difficulty. The five X components are combined to test each of the two main effects and the interaction or are tested individually if the comparisons coded into each component are of interest. Detailed description of analysis of variance through multiple regression is available in Tabachnick and Fidell (2007) as well as in such books as Cohen et al. (2003) and Keppel and Zedeck (1989).

If some Xs are continuous and others are discrete, with Y continuous, we have analysis of covariance. The continuous Xs are the covariates and the discrete ones are the IVs. The effects of IVs on Y are assessed after adjustments are made for the effects of the covariates on Y. Actually, the GLM can deal with combinations of continuous and discrete Ys in much more general ways than traditional analysis covariance, as alluded to in Chapters 5 and 6.

If Y is dichotomous (two groups), with Xs continuous, we have the simple multivariate form of discriminant analysis. The aim is to predict group membership on the basis of the Xs. There is a reversal in terminology between ANOVA and discriminant analysis; in ANOVA the groups are represented by X; in discriminant analysis the groups are represented by Y. The distinction, although confusing, is trivial within the GLM. As seen in forthcoming sections, all the special techniques are simply special cases of the full GLM.

If Y and all Xs are discrete we have multiway frequency analysis. The log-linear, rather than simple linear, model is required to evaluate relationships among variables. Logarithmic transforms are applied to cell frequencies and the weighted sum of these cell frequencies is used to predict group membership. Because the equation eventually boils down to a weighted sum of terms, it is considered here to be part of the GLM.

If Y is dichotomous and Xs are continuous and/or discrete, we have logistic regression analysis. Again a nonlinear model, in this case the logistic model, is required to evaluate relationships among variables. Y is expressed in terms of the probability of being in one or the other level. The linear regression equation is the (natural log of the) probability of being in one group divided by the probability of being in the other group. Because the linear regression equation does appear in the model, it can be considered part of the GLM.

Multilevel modeling deals with a hierarchy of Ys and Xs and equations to relate them. At the first level, Ys may be individual scores for each case on a single DV, individual scores for each case at a particular time (repeated-measures application of MLM) on a single DV, or scores for each case on multiple DVs. Ys at subsequent levels are intercepts and/or slopes over units at lower levels. Xs at each level are predictors of scores *at that level*. Although there are multiple Ys at each level except the first one (and even at the first one if there is more than one DV) and may be multiple Xs, they are never formed into combinations. Therefore, this is not a true multivariate strategy.

If Y is continuous and is the time it takes for something to happen, we have survival analysis. Xs can be continuous covariates and/or treatment(s), dichotomously coded. Here the equation is based on a log-linear rather than a linear model, but like logistic regression may be considered part of the GLM. The difference between logistic regression and survival analysis is that the Y in logistic regression is the probability of something happening and in survival analysis the Y is how long it takes to happen.

In time-series analysis, Y is continuous, and one X is always time. Intervention studies also require at least one dichotomous X, usually a treatment but rarely experimentally manipulated.

17.2.3   Full Multivariate Form

The GLM takes a major leap when the Y side of the equation is expanded because more than one equation may be required to relate the Xs to the Ys:

Root

$$1: \qquad \sum_{i=1}^{k} \beta_{i1} z_{xi1} = \sum_{i=1}^{p} \gamma j 1_{yj1}^{z}$$

$$2: \qquad \sum_{i=1}^{k} \beta_{i2} z_{xi2} = \sum_{i=1}^{p} \gamma j_{yj2}^{2z} \qquad\qquad (17.4)$$

$$\vdots$$

$$m: \qquad \sum_{i=1}^{k} \beta_{im} z_{xim} + e = \sum_{i=1}^{p} \gamma j m_{yjm}^{z}$$

where m equals k or p, whichever is smaller, and γ are regression weights for the standardized Y variables.

In general, there are as many equations as the number of X or Y variables, whichever is smaller. When there is only one Y, Xs are combined to produce one straight-line relationship with Y. Once there is more than one Y, however, combined Ys and combined Xs may fit together in several different ways. Section 9.1 and Figure 9.1 show how the combination of two or three groups (Ys) and three predictors (Xs) might fit together.

Each combination of Ys and Xs is a root. Roots are called by other names in the special statistical technique in which they are developed: discriminant functions, principal components, canonical variates, and so forth. Full multivariate techniques need multidimensional space to describe relationships among variables. With 2 df, two dimensions might be needed. With 3 df, up to three dimensions might be needed, and so on.

The number of roots necessary to describe the relationship between two sets of variables may be smaller than the number of roots maximally available. For this reason, the error term for Equation 17.4 is not necessarily associated with the mth root. It is associated with the last necessary root, with "necessary" statistically or psychometrically defined.

As with simpler forms of the GLM, specialized statistical techniques are associated with whether variables are continuous, as summarized in Table 17.1. Canonical correlation is the most general form and the noble ancestor of the GLM where all Xs and Ys are continuous. With appropriate recoding, all bivariate and multivariate problems (with the exceptions of PCA, FA, MFA, logistic regression, survival, and time series) could be solved through canonical correlation. Practically, however, the programs for canonical correlation tend not to give the kinds of information usually desired when one or more of the X or Y variables is discrete. Programs for the "multivariate general linear model" tend to be rich, but much more difficult to use.

With all Xs discrete and all Ys continuous, we have multivariate analysis of variance. The discrete X variables represent groups, and combinations of Y variables are examined to see how their centroids differ as a function of group membership. If some Xs are continuous, they can be analyzed as covariates, just as in ANCOVA; MANCOVA is used to discover how groups differ on Ys after adjustment for the effects of covariates.

If the Ys are all measured on the same scale and/or represent levels of a within-subjects IV, profile analysis is available—a form of MANOVA that is especially informative for these kinds of data. And if there are multiple DVs at each level of a within-subjects IV, doubly-multivariate analysis of variance is used to discover the effects of the IVs on the Ys.

When Y is discrete (more than two groups) and Xs are continuous, the full multivariate form of discriminant analysis is used to predict membership in Y.

There is a family of procedures—FA and PCA—in which the continuous Ys are measured empirically but the Xs are latent. It is assumed that a set of roots underlies the Ys; the purpose of analysis is to uncover the set of roots, or factors, or Xs.

In structural equations modeling, continuous and latent variables are acceptable on both sides of the equations, the X side as well as the Y side. For each Y, whether continuous (an observed indicator variable) or latent (a factor composed of multiple observed indicator variables) there is an equation involving continuous and/or latent Xs. Ys for some equations may serve as Xs for other equations, and vice versa. It is these equations that render structural equations modeling part of the GLM.

Finally, if Y is discrete and Xs are continuous and/or discrete, we have logistic regression analysis. As for MFA, a nonlinear model, the logistic model, is required to evaluate relationships among variables. Y is expressed in terms of the probability of being in one versus any of the other levels, with a separate equation for each level of Y but one. For each equation, the linear regression equation is the (natural log of the) probability of being in one group divided by the probability of being in any of the other groups. Because the model includes the linear regression equation, it can be considered part of the GLM.

Tables 17.2 and 17.3 show how each technique could be set up in SPSS and SAS GLM, along with the interpretation of the B weights produced by the program. In some cases, such as logistic regression or survival analysis, the variables require transformation to counteract the nonlinear nature of the relationships within that technique. And, of course, GLM programs do not necessarily present the information of greatest interest in the technique. For example, GLM programs do not show correlations between Y variables and roots for canonical correlation or factor analyses.

17.3   Alternative Research Strategies

For most data sets, there is more than one appropriate analytical strategy, and choice among them depends on considerations such as how the variables are interrelated, your preference for interpreting statistics associated with certain techniques, and the audience you intend to address.

A data set for which alternative strategies are appropriate has groups of people who receive one of three types of treatments, behavior modification, short-term psychotherapy, or a waiting-list control group. Suppose a great many variables are measured: self-reports of symptoms and moods,

TABLE 17.2   Syntax and Interpretation of Coefficients for Techniques in GLM: Bivariate and Simple Multivariate Forms

Technique	SPSS GLM Syntax	SAS GLM Syntax	Interpreting Weights

Pearson product-moment correlation, point-biserial correlation, phi coefficient	GLM
	Y WITH X
	/METHOD = SSTYPE(3)
	/PRINT = PARAMETER
	/DESIGN = X.
		`proc glm;`
		  `model Y = X;`
		`run;`
			B for *X* is the increase in *Y* for every one-unit increase in *X*.

Multiple regression	GLM
	Y WITH X1 X2
	/METHOD = SSTYPE(3)
	/PRINT = PARAMETER
	/DESIGN = X1 X2.
		`proc glm;`
		  `model Y = X1 X2;`
		`run;`
			B for each *X* is the increase in *Y* for every one-unit increase in that *X*, holding all other *X*s constant.

ANOVA	GLM
	Y BY X
	/METHOD = SSTYPE(3)
	/PRINT = PARAMETER
	/DESIGN = X.
		`proc glm;`
		  `class = X;`
		  `model Y = X;`
		`run;`
			B for each df of *X* is the increase in *Y* for every one-unit increase in that df of *X*, holding all other dfs of *X* constant.

ANCOVA	GLM
	Y BY X2 WITH X1
	/METHOD = SSTYPE(3)
	/PRINT = PARAMETER
	/DESIGN = X1 X2.
		`proc glm;`
		  `class = X2;`
		  `model Y = X1 X2;`
		`run;`
			B for each df of *X2* is the increase in *Y* for every one-unit increase in that df of *X2*, holding all other dfs of *X2* and *X1* constant. *B* for *X1* is the increase in *Y* for each one-unit increase in *X1*, holding *X2* constant.

Two-group discriminant analysis: *Y* represents groups, coded 0,1	GLM
	Y WITH X1 X2
	/METHOD = SSTYPE(3)
	/PRINT = PARAMETER
	/DESIGN = X1 X2.
		`proc glm;`
		  `model Y = X1 X2;`
		`run;`
			B for each *X* is the increase in *Y* for every one-unit increase in that *X*, holding all other *X*s constant. If *Y* is greater than 0.5, then case is predicted to be in group coded 1, otherwise in group coded 0.

Multiway frequency analysis, *Y* represents natural logarithm of frequency, *X* represents cell combination in contingency table	GLM
	Y WITH X1 X2
	/METHOD = SSTYPE(3)
	/PRINT = PARAMETER
	/DESIGN = X1 X2
	
	GLM not feasible for saturated model (no error term available)
		`proc glm;`
		  `model Y = X1 X2;`
		`run;`
			B for each *X* is the increase in expected frequency for a cell by considering the effect of that cell combination.

(continued)

TABLE 17.2 Continued

Technique	SPSS GLM Syntax	SAS GLM Syntax	Interpreting Weights
Two-group logistic regression analysis, Y is the natural logarithm of the odds of being in a group (see Equation 10.3)	GLM Y WITH X1 X2 /METHOD = SSTYPE(3) /PRINT = PARAMETER /DESIGN = X1 X2.	proc glm; model Y = X1 X2; run;	e^B for each X is the increase in the odds of being in one of the groups for every one-unit increase in that X, holding all other Xs constant.
Multilevel modeling: after first level, Ys are intercepts and slopes from lower levels; there are separate equations for each Y	GLM Y WITH X1 X2 /METHOD = SSTYPE(3) /PRINT = PARAMETER /DESIGN = X1 X2.	proc glm; model Y = X1 X2; run;	Separately for each equation: B for each X is the increase in Y for every one-unit increase in that X, holding all other Xs constant over all equations.
Survival analysis: Y is the probability of survival	GLM Y WITH X1 X2 /METHOD = SSTYPE(3) /PRINT = PARAMETER /DESIGN = X1 X2.	proc glm; model Y = X1 X2; run;	e^B for each X is the increase in the probability of survival for each one-unit increase in that X, holding all other Xs constant.
Time-series analysis: Xs are ARIMA parameters (see Section 18.4.1), time, and, if present, intervention	GLM Y WITH X1 X2 /METHOD = SSTYPE(3) /PRINT = PARAMETER /DESIGN = X1 X2.	proc glm; model Y = X1 X2; run;	B for each X is the increase in Y for every one-unit increase in that X, holding all other Xs constant.

reports of family members, therapist reports, and a host of personality and attitudinal tests. The major goal of analysis is probably to find out if, and on which variable(s), the groups differ after treatment.

The obvious strategy is MANOVA, but a likely problem is that the number of variables exceeds the number of clients in some group, leading to singularity. Further, with so many variables, some are likely to be highly related to combinations of others. You could choose among them or combine them on some rational basis, or you might choose first to look at empirical relationships among them.

A first step in reducing the number of variables might be examination of squared multiple correlations of each variable with all the others through regression or factor analysis programs. But the

TABLE 17.3 Syntax and Interpretation of Coefficients for Techniques in GLM: Full Multivariate Forms

Technique	SPSS GLM Syntax	SAS GLM Syntax	Interpreting Weights

Canonical correlation: Separate analyses for each set of variables considered Ys. Canonical variate analysis requires specialized software.	GLM
	  Y1 Y2 WITH X1 X2
	  /METHOD = SSTYPE(3)
	  /PRINT = PARAMETER
	  /DESIGN = X1 X2.
		`proc glm;`
		`  model Y1 Y2`
		`   = X1 X2;`
		`run;`
			Separately for each Y: B for each X is the increase in Y for every one-unit increase in that X, holding all other Xs and Ys constant.

MANOVA	GLM
	  Y1 Y2 BY X
	  /METHOD = SSTYPE(3)
	  /PRINT = PARAMETER
	  /DESIGN = X.
		`proc glm;`
		`  class = X;`
		`  model Y1 Y2 = X;`
		`run;`
			Separately for each Y: B for each df of X is the increase in Y for every one-unit increase in that df of X, holding all other Ys and all other dfs of X constant.

MANCOVA	GLM
	  Y1 Y2 BY X2 WITH X1
	  /METHOD = SSTYPE(3)
	  /PRINT = PARAMETER
	  /DESIGN = X1 X2.
		`proc glm;`
		`  class = X2;`
		`  model Y1 Y2 = X1 X2;`
		`run;`
			Separately for each Y: B for each df of $X2$ is the increase in Y for every one-unit increase in that df of $X2$, holding all other Ys and all other dfs of $X2$ and $X1$ constant. B for $X1$ is the increase in Y for each one-unit increase in $X1$, holding $X2$ constant.

Profile analysis:	GLM
	  Y1 Y2 BY X
	  /WSFACTOR = factor1 2
	  Polynomial
	  /METHOD = SSTYPE(3)
	  /PRINT = PARAMETER
	  /WSDESIGN = factor1
	  /DESIGN = X.
		`proc glm;`
		`  class = X;`
		`  model Y1 Y2 = X;`
		`  repeated factor1 2`
		`   profile;`
		`run;`
			Separately for each Y: B for each df of X is the increase in Y for every one-unit increase in that df of X, holding all other Ys and all other dfs of X constant.

(continued)

TABLE 17.3 Continued

Technique	SPSS GLM Syntax	SAS GLM Syntax	Interpreting Weights
Discriminant analysis: *Y*s are dummy-coded groups	GLM Y1 Y2 WITH X1 X2 /METHOD = SSTYPE(3) /PRINT = PARAMETER /DESIGN = X1 X2.	proc glm; model Y1 Y2 = X1 X2;run;	Separately for each contrast on *Y* groups (e.g., group 1 vs. groups 2 and 3): *B* for each *X* is the increase in *Y* for every one-unit increase in that *X*, holding all other *Y*s and all other *X*s constant. If *Y* is greater than 0.5, then case is predicted to be in group coded 1, otherwise in one of groups coded 0.
Factor and principal components analysis: Each *X* is a factor score based on a linear combination of *Y*s.	GLM Y1 Y2 WITH X1 X2 /METHOD = SSTYPE(3) /PRINT = PARAMETER /DESIGN = X1 X2.	proc glm; model Y1 Y2 = X1 X2; run;	Separately for each *Y*: *B* for each *X* is the increase in *Y* for every one-unit increase in that *X*, holding all other *Y*s and all other *X*s constant.
Structural equations modeling: some *X* and/or some *Y*s are factor scores based on a linear combination of the other.	GLM Y1 Y2 WITH X1 X2 /METHOD = SSTYPE(3) /PRINT = PARAMETER /DESIGN = X1 X2	proc glm; model Y1 Y2 = X1 X2; run;	Separately for each *Y*: *B* for each *X* is the increase in *Y* for every one-unit increase in that *X*, holding all other *Y*s and all other *X*s constant.
Multiway frequency analysis: *Y* is the natural logarithm of frequency, *X* represents a cell combination in contingency table	GLM Y WITH X1 X2 /METHOD = SSTYPE(3) /PRINT = PARAMETER /DESIGN = X1 X2 GLM not feasible for saturated model (no error term available)	proc glm; model Y = X1 X2; run;	*B* for each *X* is the increase in expected frequency for a cell by considering the effect of that cell combination.
Polychotomous logistic regression analysis, each *Y* is the natural logarithm of the odds of being in a group (see Equation 10.3), groups are dummy-coded.	GLM Y1 Y2 WITH X1 X2 /METHOD = SSTYPE(3) /PRINT = PARAMETER /DESIGN = X1 X2.	proc glm; model Y1 Y2 = X1 X2; run;	Separately for each *Y*, e^B for each *X* is the increase in the odds of being in one of the groups for every one-unit increase in that *X*, holding all other *Y*s and all other *X*s constant.

SMCs might or might not provide sufficient information for a judicious decision about which variables to delete and/or combine. If not, the next likely step is a principal component analysis on the pooled within-cells correlation matrix.

The usual procedures for deciding the number of components and type of rotation are followed. Out of this analysis come scores for each client on each component and some idea of the meaning of each component. Depending on the outcome, subsequent strategies might differ. If the principal components are orthogonal, the component scores can serve as DVs in a series of univariate ANOVAs, with adjustment for experimentwise Type I error. If the components are correlated, then MANOVA is used with component scores as DVs. The stepdown sequence might well correspond to the order of components (the scores on the first component enter first, and so on).

Or you might want to analyze the component scores through a discriminant analysis to learn, for instance, that differences between behavior modification and short-term psychotherapy are most notable on components loaded heavily with attitudes and self-reports, but differences between the treated groups and the control group are associated with components loaded with therapist reports and personality measures.

You could, in fact, solve the entire problem through discriminant analysis or logistic regression. Both types of analyses protect against multicollinearity and singularity by setting a tolerance level so that variables that are highly predicted by the other variables do not participate in the solution. Logistic regression is especially handy when the predictors are a mix of many different types of variables.

These strategies are all "legitimate" and simply represent different ways of getting to the same goal. In the immortal words spoken one Tuesday night in the Jacuzzi by Sanford A. Fidell, "You mean you only know one thing, but you have a dozen different names for it?"

A Skimpy Introduction to Matrix Algebra

The purpose of this appendix is to provide readers with sufficient background to follow, and duplicate as desired, calculations illustrated in the fourth sections of Chapters 5 through 14. The purpose is not to provide a thorough review of matrix algebra or even to facilitate an in-depth understanding of it. The reader who is interested in more than calculational rules has several excellent discussions available, particularly those in Tatsuoka (1971), Carroll, Green, and Chaturvedi (1997), and Rummel (1970).

Most of the algebraic manipulations with which the reader is familiar—addition, subtraction, multiplication, and division—have counterparts in matrix algebra. In fact, the algebra that most of us learned is a special case of matrix algebra involving only a single number, a scalar, instead of an ordered array of numbers, a matrix. Some generalizations from scalar algebra to matrix algebra seem "natural" (i.e., matrix addition and subtraction) while others (multiplication and division) are convoluted. Nonetheless, matrix algebra provides an extremely powerful and compact method for manipulating sets of numbers to arrive at desirable statistical products.

The matrix calculations illustrated here are calculations performed on square matrices. Square matrices have the same number of rows as columns. Sums-of-squares and cross-products matrices, variance-covariance matrices, and correlation matrices are all square. In addition, these three very commonly encountered matrices are symmetrical, having the same value in row 1, column 2, as in column 1, row 2, and so forth. Symmetrical matrices are mirror images of themselves about the main diagonal (the diagonal going from top left to bottom right in the matrix).

There is a more complete matrix algebra that includes nonsquare matrices as well. However, once one proceeds from the data matrix, which has as many rows as research units (subjects) and as many columns as variables, to the sum-of-squares and cross-products matrix, as illustrated in Section 1.6, most calculations illustrated in this book involve square, symmetrical matrices. A further restriction on this appendix is to limit the discussion to only those manipulations used in the fourth sections of Chapters 5 through 14. For purposes of numerical illustration, two very simple matrices, square, but not symmetrical (to eliminate any uncertainty regarding which elements are involved in calculations), will be defined as follows:

$$\mathbf{A} = \begin{bmatrix} a & b & c \\ d & e & f \\ g & h & i \end{bmatrix} = \begin{bmatrix} 3 & 2 & 4 \\ 7 & 5 & 0 \\ 1 & 0 & 8 \end{bmatrix}$$

$$\mathbf{B} = \begin{bmatrix} r & s & t \\ u & v & w \\ x & y & z \end{bmatrix} = \begin{bmatrix} 6 & 1 & 0 \\ 2 & 8 & 7 \\ 3 & 4 & 5 \end{bmatrix}$$

A.1 The Trace of a Matrix

The trace of a matrix is the sum of the numbers on the diagonal that runs from the upper left to lower right. For matrix $\mathbf{A}$, the trace is 16 ($3 + 5 + 8$); for matrix $\mathbf{B}$ it is 19. If the matrix is a sum-of-squares and cross-products matrix, then the trace is the sum of squares. If it is a variance-covariance matrix, the trace is the sum of variances. If it is a correlation matrix, the trace is the number of variables (each having contributed a value of 1 to the trace).

A.2 Addition or Subtraction of a Constant to a Matrix

If one has a matrix, $\mathbf{A}$, and wants to add or subtract a constant, k, to the elements of the matrix, one simply adds (or subtracts) the constant to every element in the matrix.

$$\mathbf{A} + k = \begin{bmatrix} a & b & c \\ d & e & f \\ g & h & i \end{bmatrix} + k = \begin{bmatrix} a+k & b+k & c+k \\ d+k & e+k & f+k \\ g+k & h+k & i+k \end{bmatrix} \tag{A.1}$$

If $k = -3$, then

$$\mathbf{A} + k = \begin{bmatrix} 0 & -1 & 1 \\ 4 & 2 & -3 \\ -2 & -3 & 5 \end{bmatrix}$$

A.3 Multiplication or Division of a Matrix by a Constant

Multiplication or division of a matrix by a constant is a straightforward process.

$$kA = k\begin{bmatrix} a & b & c \\ d & e & f \\ g & h & i \end{bmatrix}$$

$$kA = \begin{bmatrix} ka & kb & kc \\ kd & ke & kf \\ kg & kh & ki \end{bmatrix} \tag{A.2}$$

and

$$
\frac{1}{k}\mathbf{A} = \begin{bmatrix} \dfrac{a}{k} & \dfrac{b}{k} & \dfrac{c}{k} \\[2mm] \dfrac{d}{k} & \dfrac{e}{k} & \dfrac{f}{k} \\[2mm] \dfrac{g}{k} & \dfrac{h}{k} & \dfrac{i}{k} \end{bmatrix}
$$

(A.3)

Numerically, if $k = 2$, then

$$
k\mathbf{A} = \begin{bmatrix} 6 & 4 & 8 \\ 14 & 10 & 0 \\ 2 & 0 & 16 \end{bmatrix}
$$

A.4   Addition and Subtraction of Two Matrices

These procedures are straightforward, as well as useful. If matrices **A** and **B** are as defined at the beginning of this appendix, one simply performs the addition or subtraction of corresponding elements.

$$
\mathbf{A} + \mathbf{B} = \begin{bmatrix} a & b & c \\ d & e & f \\ g & h & i \end{bmatrix} = \begin{bmatrix} r & s & t \\ u & v & w \\ x & y & z \end{bmatrix} = \begin{bmatrix} a+r & b+s & c+t \\ d+u & e+v & f+w \\ g+x & h+y & i+z \end{bmatrix}
$$

(A.4)

and

$$
= \begin{bmatrix} a-r & b-s & c-t \\ d-u & e-v & f-w \\ g-x & h-y & i-z \end{bmatrix}
$$

(A.5)

For the numerical example:

$$
\mathbf{A} + \mathbf{B} = \begin{bmatrix} 3 & 2 & 4 \\ 7 & 5 & 0 \\ 1 & 0 & 8 \end{bmatrix} + \begin{bmatrix} 6 & 1 & 0 \\ 2 & 8 & 7 \\ 3 & 4 & 5 \end{bmatrix}
$$

$$
\mathbf{A} + \mathbf{B} = \begin{bmatrix} 9 & 3 & 4 \\ 9 & 13 & 7 \\ 4 & 4 & 13 \end{bmatrix}
$$

Calculation of a difference between two matrices is required when, for instance, one desires a residuals matrix, the matrix obtained by subtracting a reproduced matrix from an obtained matrix (as in factor analysis, Chapter 13). Or, if the matrix that is subtracted happens to consist of columns with appropriate means of variables inserted in every slot, then the difference between it and a matrix of raw scores produces a deviation matrix.

A.5 Multiplication, Transposes, and Square Roots of Matrices

Matrix multiplication is both unreasonably complicated and undeniably useful. Note that the ijth element of the resulting matrix is a function of row i of the first matrix and column j of the second.

$$\mathbf{AB} = \begin{bmatrix} a & b & c \\ d & e & f \\ g & h & i \end{bmatrix} \begin{bmatrix} r & s & t \\ u & v & w \\ x & y & z \end{bmatrix} = \begin{bmatrix} ar + bu + cx & as + bv + cy & at + bw + cz \\ rd + eu + fx & ds + ev + fy & dt + ew + fz \\ gr + hu + ix & gs + hv + iy & gt + hw + iz \end{bmatrix} \quad (A.6)$$

Numerically,

$$\mathbf{AB} = \begin{bmatrix} 3 & 2 & 4 \\ 7 & 5 & 0 \\ 1 & 0 & 8 \end{bmatrix} \begin{bmatrix} 6 & 1 & 0 \\ 2 & 8 & 7 \\ 3 & 4 & 5 \end{bmatrix}$$

$$= \begin{bmatrix} 3 \cdot 6 + 2 \cdot 2 + 4 \cdot 3 & 3 \cdot 1 + 2 \cdot 8 + 4 \cdot 4 & 3 \cdot 0 + 2 \cdot 7 + 4 \cdot 5 \\ 7 \cdot 6 + 5 \cdot 2 + 0 \cdot 3 & 7 \cdot 1 + 5 \cdot 8 + 0 \cdot 4 & 7 \cdot 0 + 5 \cdot 7 + 0 \cdot 5 \\ 1 \cdot 6 + 0 \cdot 2 + 8 \cdot 3 & 1 \cdot 1 + 0 \cdot 8 + 8 \cdot 4 & 1 \cdot 0 + 0 \cdot 7 + 8 \cdot 5 \end{bmatrix}$$

$$= \begin{bmatrix} 34 & 35 & 34 \\ 52 & 47 & 35 \\ 30 & 33 & 40 \end{bmatrix}$$

Regrettably, $\mathbf{AB} \neq \mathbf{BA}$ in matrix algebra. Thus

$$\mathbf{BA} = \begin{bmatrix} 6 & 1 & 0 \\ 2 & 8 & 7 \\ 3 & 4 & 5 \end{bmatrix} \begin{bmatrix} 3 & 2 & 4 \\ 7 & 5 & 0 \\ 1 & 0 & 8 \end{bmatrix} = \begin{bmatrix} 25 & 17 & 24 \\ 69 & 44 & 64 \\ 42 & 26 & 52 \end{bmatrix}$$

If another concept of matrix algebra is introduced, some useful statistical properties of matrix algebra can be shown. The transpose of a matrix is indicated by a prime ($'$) and stands for a rearrangement of the elements of the matrix such that the first row becomes the first column, the second row the second column, and so forth. Thus

$$\mathbf{A}' = \begin{bmatrix} a & d & g \\ b & e & h \\ c & f & i \end{bmatrix} = \begin{bmatrix} 3 & 7 & 1 \\ 2 & 5 & 0 \\ 4 & 0 & 8 \end{bmatrix} \qquad (A.7)$$

When transposition is used in conjunction with multiplication, then some advantages of matrix multiplication become clear, namely,

$$\mathbf{AA}' = \begin{bmatrix} a & b & c \\ d & e & f \\ g & h & i \end{bmatrix} \begin{bmatrix} a & d & g \\ b & e & h \\ c & f & i \end{bmatrix}$$

$$= \begin{bmatrix} a^2 + b^2 + c^2 & ad + be + cf & ag + bh + ci \\ ad + be + cf & d^2 + e^2 + f^2 & dg + eh + fi \\ ag + bh + ci & dg + eh + fi & g^2 + h^2 + i^2 \end{bmatrix} \qquad (A.8)$$

The elements in the main diagonal are the sums of squares and those off the diagonal are cross products.

Had $\mathbf{A}$ been multiplied by itself, rather than by a transpose of itself, a different result would have been achieved.

$$\mathbf{AA} = \begin{bmatrix} a^2 + bd + cg & ab + be + ch & ac + bf + ci \\ da + ed + fg & db + e^2 + fh & dc + ef + fi \\ ga + hd + ig & gb + he + ih & gc + hf + i^2 \end{bmatrix}$$

If $\mathbf{AA} = \mathbf{C}$, then $\mathbf{C}^{1/2} = \mathbf{A}$. That is, there is a parallel in matrix algebra to squaring and taking the square root of a scalar, but it is a complicated business because of the complexity of matrix multiplication. If, however, one has a matrix $\mathbf{C}$ from which a square root is desired (as in canonical correlation, Chapter 12), one searches for a matrix, $\mathbf{A}$, which, when multiplied by itself, produces $\mathbf{C}$. If, for example,

$$\mathbf{C} = \begin{bmatrix} 27 & 16 & 44 \\ 56 & 39 & 28 \\ 11 & 2 & 68 \end{bmatrix}$$

then

$$\mathbf{C}^{1/2} = \begin{bmatrix} 3 & 2 & 4 \\ 7 & 5 & 0 \\ 1 & 0 & 8 \end{bmatrix}$$

A.6 Matrix "Division" (Inverses and Determinants)

If you liked matrix multiplication, you'll love matrix inversion. Logically, the process is analogous to performing division for single numbers by finding the reciprocal of the number and multiplying by the reciprocal: if $a^{-1} = 1/a$, then $(a)(a^{-1}) = a/a = 1$. That is, the reciprocal of a scalar is a number that, when multiplied by the number itself, equals 1. Both the concepts and the notation are similar in matrix algebra, but they are complicated by the fact that a matrix is an array of numbers.

To determine if the reciprocal of a matrix has been found, one needs the matrix equivalent of the 1 as employed in the preceding paragraph. The identity matrix, $\mathbf{I}$, a matrix with 1s in the main diagonal and zeros elsewhere, is such a matrix. Thus

$$\mathbf{I} = \begin{bmatrix} 1 & 0 & 0 \\ 0 & 1 & 0 \\ 0 & 0 & 1 \end{bmatrix} \tag{A.9}$$

Matrix division, then, becomes a process of finding $\mathbf{A}^{-1}$ such that

$$\mathbf{A}^{-1}\mathbf{A} = \mathbf{A}\mathbf{A}^{-1} = \mathbf{I} \tag{A.10}$$

One way of finding $\mathbf{A}^{-1}$ requires a two-stage process, the first of which consists of finding the determinant of $\mathbf{A}$, noted $|\mathbf{A}|$. The determinant of a matrix is sometimes said to represent the generalized variance of the matrix, as most readily seen in a 2×2 matrix. Thus we define a new matrix as follows:

$$\mathbf{D} = \begin{bmatrix} a & b \\ c & d \end{bmatrix}$$

where

$$|\mathbf{D}| = ad - bc \tag{A.11}$$

If $\mathbf{D}$ is a variance-covariance matrix where a and d are variances while b and c are covariances, then $ad - bc$ represents variance minus covariance. It is this property of determinants that makes them useful for hypothesis testing (see, for example, Chapter 7, Section 7.4, where Wilks' lambda is used in MANOVA).

Calculation of determinants becomes rapidly more complicated as the matrix gets larger. For example, in our 3 by 3 matrix,

$$|\mathbf{A}| = a(ei - fh) + b(fg - di) + c(dh - eg) \tag{A.12}$$

Should the determinant of $\mathbf{A}$ equal 0, then the matrix cannot be inverted because the next operation in inversion would involve division by zero. Multicollinear or singular matrices (those with variables that are linear combinations of one another, as discussed in Chapter 4) have zero determinants that prohibit inversion.

A full inversion of **A** is

$$\mathbf{A}^{-1} = \begin{bmatrix} a & b & c \\ d & e & f \\ g & h & i \end{bmatrix}^{-1}$$

(A.13)

$$= \frac{1}{|\mathbf{A}|} \begin{bmatrix} ei - fh & ch - bi & bf - ce \\ fg - di & ai - cg & cd - af \\ dh - eg & bg - ah & ae - bd \end{bmatrix}$$

Please recall that because **A** is not a variance-covariance matrix, a negative determinant is possible, even somewhat likely. Thus, in the numerical example,

$$|\mathbf{A}| = 3(5 \cdot 8 - 0 \cdot 0) + 2(0 \cdot 1 - 7 \cdot 8) + 4(7 \cdot 0 - 5 \cdot 1) = -12$$

and

$$= \frac{1}{-12} \begin{bmatrix} 5 \cdot 8 - 0 \cdot 0 & 4 \cdot 0 - 2 \cdot 8 & 2 \cdot 0 - 4 \cdot 5 \\ 0 \cdot 1 - 7 \cdot 8 & 3 \cdot 8 - 4 \cdot 1 & 4 \cdot 7 - 3 \cdot 0 \\ 7 \cdot 0 - 5 \cdot 1 & 2 \cdot 1 - 3 \cdot 0 & 3 \cdot 5 - 2 \cdot 7 \end{bmatrix}$$

$$= \begin{bmatrix} \dfrac{40}{-12} & \dfrac{-16}{-12} & \dfrac{-20}{-12} \\ \dfrac{-56}{-12} & \dfrac{20}{-12} & \dfrac{28}{-12} \\ \dfrac{-5}{-12} & \dfrac{2}{-12} & \dfrac{1}{-12} \end{bmatrix}$$

$$= \begin{bmatrix} -3.33 & -1.33 & 1.67 \\ 4.67 & -1.67 & -2.33 \\ 0.42 & -0.17 & -0.08 \end{bmatrix}$$

Confirm that, within rounding error, Equation A.10 is true. Once the inverse of **A** is found, "division" by it is accomplished whenever required by using the inverse and performing matrix multiplication.

A.7   Eigenvalues and Eigenvectors: Procedures for Consolidating Variance from a Matrix

We promised you a demonstration of computation of eigenvalues and eigenvectors for a matrix, so here it is. However, you may well find that this discussion satisfies your appetite for only a couple of hours. During that time, round up Tatsuoka (1971), get the cat off your favorite chair, and prepare for an intelligible, if somewhat lengthy, description of the same subject.

Most of the multivariate procedures rely on eigenvalues and their corresponding eigenvectors (also called characteristic roots and vectors) in one way or another because they consolidate the variance in a matrix (the *eigenvalue*) while providing the linear combination of variables (the *eigenvector*) to do it. The coefficients applied to variables to form linear combinations of variables in all the multivariate procedures are rescaled elements from eigenvectors. The variance that the solution "accounts for" is associated with the eigenvalue and is sometimes called so directly.

Calculation of eigenvalues and eigenvectors is best left up to a computer with any realistically sized matrix. For illustrative purposes, a 2×2 matrix will be used here. The logic of the process is also somewhat difficult, involving several of the more abstract notions and relations in matrix algebra, including the equivalence between matrices, systems of linear equations with several unknowns, and roots of polynomial equations.

Solution of an eigenproblem involves solution of the following equation:

$$(\mathbf{D} - \lambda\mathbf{I})V = 0 \tag{A.14}$$

where λ is the eigenvalue and V the eigenvector to be sought. Expanded, this equation becomes

$$\left[\begin{bmatrix} a & b \\ c & d \end{bmatrix} - \lambda \begin{bmatrix} 1 & 0 \\ 0 & 1 \end{bmatrix} \right] \begin{bmatrix} v_1 \\ v_2 \end{bmatrix} = 0$$

or

$$\left[\begin{bmatrix} a & b \\ c & d \end{bmatrix} - \begin{bmatrix} \lambda & 0 \\ 0 & \lambda \end{bmatrix} \right] \begin{bmatrix} v_1 \\ v_2 \end{bmatrix} = 0$$

or, by applying Equation A.5,

$$\begin{bmatrix} a - \lambda & b \\ c & d - \lambda \end{bmatrix} \begin{bmatrix} v_1 \\ v_2 \end{bmatrix} = 0 \tag{A.15}$$

If one considers the matrix $\mathbf{D}$, whose eigenvalues are sought, a variance-covariance matrix, one can see that a solution is desired to "capture" the variance in $\mathbf{D}$ while rescaling the elements in $\mathbf{D}$ by v_1 and v_2 to do so.

It is obvious from Equation A.15 that a solution is always available when v_1 and v_2 are 0. A nontrivial solution may also be available when the determinant of the leftmost matrix in Equation A.15 is 0.[1] That is, if (following Equation A.11)

$$(a - \lambda)(d - \lambda) - bc = 0 \tag{A.16}$$

---

[1]Read Tatsuoka (1971); a matrix is said to be positive definite when all $\lambda_i > 0$, positive semidefinite when all $\lambda_1 \geq 0$, and ill-conditioned when some $\lambda_i < 0$.

then there may exist values of λ and values of v_1 and v_2 that satisfy the equation and are not 0. However, expansion of Equation A.16 gives a polynomial equation, in λ, of degree 2:

$$\lambda^2 - (a + d)\lambda + ad - bc = 0 \tag{A.17}$$

Solving for the eigenvalues, λ, requires solving for the roots of this polynomial. If the matrix has certain properties (see footnote 1), there will be as many positive roots to the equation as there are rows (or columns) in the matrix.

If Equation A.17 is rewritten as $x\lambda^2 + y\lambda + z = 0$, the roots may be found by applying the following equation:

$$\lambda = \frac{-y \pm \sqrt{y^2 - 4xz}}{2x} \tag{A.18}$$

For a numerical example, consider the following matrix.

$$\mathbf{D} = \begin{bmatrix} 5 & 1 \\ 4 & 2 \end{bmatrix}$$

Applying Equation A.17, we obtain

$$\lambda^2 - (5 + 2)\lambda + 5 \cdot 2 - 1 \cdot 4 = 0$$

or

$$\lambda^2 - 7\lambda + 6 = 0$$

The roots to this polynomial may be found by Equation A.18 as follows:

$$\lambda = \frac{-(-7) + \sqrt{(-7)^2 - 4 \cdot 1 \cdot 6}}{2 \cdot 1} = 6$$

and

$$\lambda = \frac{-(-7) - \sqrt{(-7)^2 - 4 \cdot 1 \cdot 6}}{2 \cdot 1} = 1$$

(The roots could also be found by factoring to get $[\lambda - 6][\lambda - 1]$.)

Once the roots are found, they may be used in Equation A.15 to find v_1 and v_2, the eigenvector. There will be one set of eigenvectors for the first root and a second set for the second root. Both solutions require solving sets of two simultaneous equations in two unknowns, to wit, for the first root, 6, and applying Equation A.15.

$$\begin{bmatrix} 5 - 6 & 1 \\ 4 & 2 - 6 \end{bmatrix} \begin{bmatrix} v_1 \\ v_2 \end{bmatrix} = 0$$

or

$$\begin{bmatrix} -1 & 1 \\ 4 & -4 \end{bmatrix} \begin{bmatrix} v_1 \\ v_2 \end{bmatrix} = 0$$

so that

$$-1v_1 + 1v_2 = 0$$

and

$$4v_1 - 4v_2 = 0$$

When $v_1 = 1$ and $v_2 = 1$, a solution is found.

For the second root, 1, the equations become

$$\begin{bmatrix} 5-1 & 1 \\ 4 & 2-1 \end{bmatrix} \begin{bmatrix} v_1 \\ v_2 \end{bmatrix} = 0$$

or

$$\begin{bmatrix} 4 & 1 \\ 4 & 1 \end{bmatrix} \begin{bmatrix} v_1 \\ v_2 \end{bmatrix} = 0$$

so that

$$4v_1 + 1v_2 = 0$$

and

$$4v_1 + 1v_2 = 0$$

When $v_1 = -1$ and $v_2 = 4$, a solution is found. Thus the first eigenvalue is 6, with $[1, 1]$ as a corresponding eigenvector, while the second eigenvalue is 1, with $[-1, 4]$ as a corresponding eigenvector.

Because the matrix was 2×2, the polynomial for eigenvalues was quadratic and there were two equations in two unknowns to solve for eigenvectors. Imagine the joys of a matrix 15×15, a polynomial with terms to the 15th power for the first half of the solution and 15 equations in 15 unknowns for the second half. A little more appreciation for your computer, please, next time you use it!

B

Research Designs for Complete Examples

B.1   Women's Health and Drug Study

Data used in most of the large-sample examples were collected with the aid of a grant from the National Institute on Drug Abuse (#DA 00847) to L. S. Fidell and J. E. Prather in 1974–1976. Methods of collecting the data and references to the measures included in the study are described here approximately as they have been previously reported (Hoffman & Fidell, 1979).

Method

A structured interview, containing a variety of health, demographic, and attitudinal measures, was given to a randomly selected group of 465 female, 20- to 59-year-old, English-speaking residents of the San Fernando Valley, a suburb of Los Angeles, in February 1975. A second interview, focusing primarily on health variables but also containing the Bem Sex Role Inventory (BSRI; Bem, 1974) and the Eysenck Personality Inventory (EPI; Eysenck & Eysenck, 1963), was conducted with 369 (79.4%) of the original respondents in February 1976.

The 1975 target sample of 703 names was approximately a .003 probability sample of appropriately aged female residents of the San Fernando Valley and was randomly drawn from lists prepared by listers during the weeks immediately preceding the sample selection. Lists were prepared for census blocks that had been randomly drawn (proportional to population) from 217 census tracks, which were themselves randomly drawn after they were stratified by income and assigned probabilities proportional to their populations. Respondents were contacted after first receiving a letter soliciting their cooperation. Substitutions were not allowed. A minimum of four callbacks was required before the attempt to obtain an interview was terminated. The completion rate for the target sample was 66.1%, with a 26% refusal rate and a 7.9% "unobtainable" rate.

The demographic characteristics of the 465 respondents who cooperated in 1975 confirmed the essentially white, middle- and working-class composition of the San Fernando Valley and agreed, for the most part, with the profile of characteristics of women in the valley that was calculated from 1970 Census Bureau data. The final sample was 91.2% white, with a median family income (before taxes) of $17,000 per year and an average Duncan scale (Featherman, 1973) socioeconomic level (SEL) rating of 51. Respondents were also well educated (13.2 years of school completed, on average) and predominantly Protestant (38%), with 26% Catholic, 20% Jewish, and the remainder "none" or "other." A total of 52.9% worked (either full-time—33.5%—or part-time—19.4%). Seventy-eight percent were living with husbands at the time of the first interview, with 9% divorced, 6% single, 3% separated, 3% widowed, and fewer than 1% "living together." Altogether, 82.4% of the women had children; the average number of children was 2.7, with 2.1 children, on the average, still living in the same house as the respondent.

Of the original 465 respondents, 369 (79.4%) were re-interviewed a year later. Of the 96 respondents who were not re-interviewed, 51 refused, 36 had moved and could not be relocated, 8 were known to be in the Los Angeles area but were not contacted after a minimum of 5 attempts, and 1 was deceased. Those who were and were not re-interviewed were similar (by analyses of variance) on health and attitudinal variables. They differed, however, on some demographic measures. Those who were re-interviewed tended to be higher-SEL, higher-income white women who were better-educated, were older, and had experienced significantly fewer life change units (Rahe, 1974) in 1975.

The 1975 interview schedule was composed of items assessing a number of demographic, health, and attitudinal characteristics (see Table B.1). Insofar as possible, previously tested and validated items and measures were used, although time constraints prohibited including all items from some measures. Coding on most items was prearranged so that responses given large numbers reflected increasingly unfavorable attitudes, dissatisfaction, poorer health, lower income, increasing stress, increasing use of drugs, and so forth.

The 1976 interview schedule repeated many of the health items, with a shorter set of items assessing changes in marital status and satisfaction, changes in work status and satisfaction, and so forth. The BSRI and EPI were also included, as previously mentioned. The interview schedules for both 1975 and 1976 took 75 minutes on average to administer and were conducted in respondents homes by experienced and trained interviewers.

To obtain median values for the masculine and feminine scores of the BSRI for a comparable sample of men, the BSRI was mailed to the 369 respondents who cooperated in 1976, with instructions to ask a man near to them (husband, friend, brother, etc.) to fill out and return it. The completed BSRI was received from 162 (46%) men, of whom 82% were husbands, 8.6% friends, 3.7% fiances, 1.9% brothers, 1.2% sons, 1.2% ex-husbands, 0.6% brothers-in-law, and 0.6% fathers. Analyses of variance were used to compare the demographic characteristics of the men who returned the BSRI with those who did not (insofar as such characteristics could be determined by responses of the women to questions in the 1975 interview). The two groups differed in that, as with the reinterviewed women, the men who responded presented an advantaged socioeconomic picture relative to those who did not. Respondents had higher SEL2 ratings, were better educated, and enjoyed higher income. The unweighted averages of the men's and women's median masculine scores and median feminine scores were used to split the sample of women into those who were feminine, masculine, androgynous, and undifferentiated.

B.2 Sexual Attraction Study

Data used in the large-sample multiway frequency analysis example (Section 7.6) were collected in 1984 as part of a survey assessing issues surrounding the nature of sexual attraction to clients among

TABLE B.1 Description of Some of the Variables Available from 1975–1976 Interviews

Variable	Abbreviation	Brief Description	Source
Demographic variables			
Socioeconomic level	SEL, SEL2	Measure of deference accorded employment categories (SEL2 from second interview)	Featherman (1973), update of Duncan Scale
Education	EDUC	Number of years completed	
	EDCODE	Categorical variable assessing whether or not education proceeded beyond high school	
Income	INCOME	Total family income before taxes	
	INCODE	Categorical variable assessing family income	
Age	AGE	Chronological age in 5-year categories	
Marital status	MARITAL	A categorical variable assessing current marital status	
	MSTATUS	A dichotomous variable assessing whether or not currently married	
Parenthood	CHILDREN	A categorical variable assessing whether or not one has children	
Ethnic group membership	RACE	A categorical variable assessing ethnic affiliation	
Employment status	EMPLMNT	A categorical variable assessing whether or not one is currently employed	
	WORKSTAT	A categorical variable assessing current employment status and, if not, attitude toward unemployed status	
Religious affiliation	RELIGION	A categorical variable assessing religious affiliation	
Attitudinal variables			
Attitudes toward housework	ATTHOUSE	Frequency of experiencing various favorable and unfavorable attitudes toward homemaking	Derived from Johnson (1955)
Attitudes toward paid work	ATTWORK	Frequency of experiencing various favorable and unfavorable attitudes toward paid work	Johnson (1955)
Attitudes toward role of women	ATTROLE	Measure of conservative or liberal attitudes toward role of women	Spence and Helmreich (1972)
Locus of control	CONTROL	Measure of control ideology; internal or external	Rotter (1966)

TABLE B.1   Continued

Variable	Abbreviation	Brief Description	Source
Attitudinal variables *(cont.)*			
Attitudes toward marital status	ATTMAR	Satisfaction with current marital status	From Burgess & Locke (1960); Locke & Wallace (1959); and Rollins & Feldman (1970)
Personality variables			
Self-esteem	ESTEEM	Measures of self-esteem and confidence in various situations	Rosenberg (1965)
Neuroticism-stability index	NEUROTIC	A scale derived from factor analysis to measure neuroticism vs. stability	Eysenck & Eysenck (1963)
Introversion-extraversion index	INTEXT	A scale derived from factor analysis to measure introversion vs. extraversion	Eysenck & Eysenck (1963)
Androgyny measure	ANDRM	A categorical variable based on femininity and masculinity	Derived from Bem (1974)
Health variables			
Mental health	MENHEAL	Frequency count of mental health problems (feeling somewhat apart, can't get along, etc.)	Langer (1962)
Physical health	PHYHEAL	Frequency count of problems with various body systems (circulation, digestion, etc.), general description of health	
Number of visits	TIMEDRS	Frequency count of visits to physical and mental health professionals	
Use of psycho-tropic drugs	DRUGUSE	A frequency, recency measure of involvement with prescription and nonprescription major and minor tranquilizers, sedatives-hypnotics, antidepressants, and stimulants	Balter & Levine (1971)
Use of psycho-tropic and over-the-counter drugs	PSYDRUG	DRUGUSE plus a frequency, recency measure of over-the-counter mood-modifying drugs	
Attitudes toward medication	ATTDRUG	Items concerning attitudes toward use of medication	
Life change units	STRESS	Weighted items reflecting number and importance of change in life situation	Rahe (1974)

clinical psychologists. Data-collection methods and demographic characteristics that follow are approximately as they appear in an *American Psychologist* paper (Pope, Keith-Spiegel, & Tabachnick, 1986).

Method

A cover letter, a brief 17-item questionnaire (15 structured questions and 2 open-ended questions), and a return envelope were sent to 1,000 psychologists (500 men and 500 women) randomly selected from the 4,356 members of Division 42 (Psychologists in Private Practice) as listed in the 1983 Membership Register of the American Psychological Association.

The questionnaire requested respondents to provide information about their gender, age group, and years of experience in the field. Information was elicited about the respondents' incidence of sexual attraction to male and female clients; clients' reactions to this experience of attraction; beliefs about the clients' awareness of and reciprocation of the attraction; the impact of the attraction on the therapy process; how such feelings were managed; the incidence of sexual fantasies about clients; why, if relevant, respondents chose to refrain from acting out their attraction through actual sexual intimacies with clients; what features determined which clients would be perceived as sexually attractive; incidence of actual sexual activity with clients; and the extent to which the respondents' graduate training and internship experiences had dealt with issues related to sexual attraction to clients.

Questionnaires were returned by 585 respondents. Of these 59.7% were men. Sixty-eight percent of the male respondents returned their questionnaires as compared with 49% of the female respondents. Approximately half of the respondents were 45 years of age and under. The sample's median age was approximately 46 years as compared with the median age of 40 years reported in a 1983 survey of mental health service providers (VandenBos & Stapp, 1983).

Respondents averaged 16.99 (SD = 8.43) years of professional experience with no significant differences between male and female psychologists. Younger therapists averaged 11.36 (SD = 8.43) years of experience and older therapists averaged 21.79 (SD = 8.13) years of experience. Only 77 of the 585 therapists reported never being attracted to any client.

B.3   Learning Disabilities Data Bank

Data for the large-sample example in Section 8.6.1 (profile analysis of subscales of the WISC) were taken from a data bank developed at the California Center for Educational Therapy (CCET) in the San Fernando Valley.

All children who were referred to the CCET were given an extensive battery of psychodiagnostic tests to measure current intellectual functioning, perceptual development, psycholinguistic abilities, visual and auditory functioning, and achievement in a number of academic subjects. In addition, an extensive Parent Information Outline queried parents about demographic variables, family health history, as well as the child's developmental history, strengths, weaknesses, preferences, and the like. The entire data bank consisted of 112 variables from the testing battery plus 155 variables from the Parent Information Outline.

Data collection began in July 1972 and continued to 1993. The Chapter 8 sample includes children tested before February 1984 who were administered the Wechsler Intelligence Scale for Children, who were diagnosed as learning-disabled, whose parents agreed to be included in the data bank, and whose parents answered a question about the child's preference for playmates' age. Available answers to this question were: (1) older, (2) younger, (3) same age, and (4) no preference. The latter two categories were combined into a single one for the Chapter 8 analysis because either category by itself would have been too small. Of the 261 children tested between 1972 and 1984, 177 were eligible for inclusion in the Chapter 8 sample.

For the entire sample of 261 cases, average age is 10.58 years with a range of 5 to 61 years. (The Chapter 8 sample consists of school-age children only.) About 75% of the entire sample is male. At the time of testing, 63% of the entire sample attended public school; 33% were enrolled in various types of private schools. Of the 94% of parents who revealed their educational level, mothers had completed an average of 13.6 years of schooling and fathers had completed an average of 14.9 years.

B.4   Reaction Time to Identify Figures

Data for this study were collected by Damos (1989). Twenty right-handed males were required to respond "standard" or "mirror" on a keypad as quickly as possible to a stimulus (the letter G or a symbol) and its mirror image. Each trial consisted of 30 presentations of the stimulus and 30 presentations of its mirror image, five presentations in each of six orientations: upright (0), 60, 120, 180, 240, and 300 degrees of rotation. In all, the experiment required ten sessions over two consecutive weeks. Each session had four blocks of nine trials distributed over morning and afternoon periods. Thus, each subject made 21,600 responses during the study. Half the subjects were given stimulus G in the first week and a symbol in the second week; the others participated only for one week and were given the symbol stimulus. Order of presentation of all stimuli was random on all trials.

The two DVs were average correct reaction time and error rate. Responses to 60° of absolute rotation were averaged with those of 300°, and responses to 120° were averaged with those of 240°. Averages were calculated separately for standard and mirror-image trials. Linear regression on the averages for each subject in each session yielded a slope and intercept.

Data selected for analysis in Section 8.6.2 were slopes and intercepts from the first four sessions: morning and afternoon sessions for days 1 and 2 of the first week, so that practice effects could be observed. Only the trials on which the subject responded "standard" were used. Thus, each of the 20 subjects provided eight scores, and a slope and an intercept for each of the four sessions.

B.5   Field Studies of Noise-Induced
Sleep Disturbance

These data are from field studies of the effects of nighttime noise on sleep disturbance reported by Fidell et al. (1995). Measurements were taken in the vicinity of two airports and one control site. Airports were Castle Air Force Base in Merced, CA, neighborhoods in the Los Angeles area that were not exposed to nighttime aircraft noise but were exposed to high levels of road traffic noise, and LAX. Indoor and outdoor noise exposure was measured for periods of approximately one month in 45 homes of 82 test participants for a total of 1,887 subject nights. Responses included awakenings, time to fall asleep, number of recalled awakenings, behavioral awakenings, sleep quality, sleep time, annoyance, tiredness in the morning, reported time awake, and awakenings by aircraft noise.

A number of personal characteristics also were recorded, including age, gender, spontaneous awakening rate, duration of residence, use of alcohol and medications, and tiredness upon retiring. Event-based awakening was predicted in a logistic regression analysis by event noise level, ambient noise level, rate of spontaneous awakening, age, time since retiring, duration of residence, and tiredness on retiring. However, the slope of the relationship between awakening and noise exposure was rather shallow. Results were found to be in good agreement with those of other field studies, but quite different from laboratory studies.

B.6   Clinical Trial for Primary Biliary Cirrhosis

Data in this example were collected during a double-blind randomized clinical trial conducted at the Mayo Clinic January 1974 and May 1984. Of the 424 patients who met eligibility requirements for the study, 312 (with their physicians) agreed to participate in the randomized trial. A number of clinical, biochemical, serologic, and histologic measurements were made for the 312 patients, in addition to the date of random assignment to either a placebo group or to a group treated with the drug D-penicillamine. Data from the clinical trial were analyzed in 1986 for presentation in the clinical literature.

Data from the 312 cases used in the survival analysis in Chapter 11 included 125 who had died (the 11 patients whose deaths were not attributable to PBC were not distinguished) and 19 who had undergone liver transplant. The data and some analyses are described by Fleming and Harrington (1991, Chapters 1 and 4) and by Markus et al. (1989). The latter paper focuses on the efficacy of liver transplantation, but also discusses the Mayo model which determined the variables used in the survival analysis of Section 11.7.

Fleming and Harrington describe PBC as ". . . a rare but fatal chronic liver disease of unknown cause, with a prevalence of about 50 cases per million population. The primary pathologic event appears to be the destruction of interlobular bile ducts, which may be mediated by immunologic mechanisms" (p. 2).

B.7   Impact of Seat Belt Law

Rock (1992) applied an ARIMA approach to evaluating the impact of the Illinois seat belt use law on accidents, deaths, and various degrees of injury. Data were collected monthly starting January 1980 and continued through December 1990, with the seat belt law taking effect in 1985. Accident statistics are available from the Illinois Department of Transportation (IDOT). Rock considered other statistical techniques in addition to ARIMA, which he concluded was preferable (less biased).

Dependent variables, evaluated separately, were monthly fatalities, accidents, and A-, B-, and C-level injuries. A-level injuries, the ones chosen for demonstration in Section 18.7 (Chapter 18 is available online at ablongman.com/tabachnick5e) are the most serious, defined as incapacitating; C-level injuries are the least severe, possibly not visible. Different ARIMA models were applied to the five DVs, with a statistically significant impact of the law found only for A-level injuries. Rock concluded that overall he net impact of the law was a reduction in injury severity.

C Statistical Tables

TABLE C.1   Normal Curve Areas

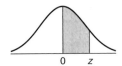

z	.00	.01	.02	.03	.04	.05	.06	.07	.08	.09
0.0	.0000	.0040	.0080	.0120	.0160	.0199	.0239	.0279	.0319	.0359
0.1	.0398	.0438	.0478	.0517	.0557	.0596	.0636	.0675	.0714	.0753
0.2	.0793	.0832	.0871	.0910	.0948	.0987	.1026	.1064	.1103	.1141
0.3	.1179	.1217	.1255	.1293	.1331	.1368	.1406	.1443	.1480	.1517
0.4	.1554	.1591	.1628	.1664	.1700	.1736	.1772	.1808	.1844	.1879
0.5	.1915	.1950	.1985	.2019	.2054	.2088	.2123	.2157	.2190	.2224
0.6	.2257	.2291	.2324	.2357	.2389	.2422	.2454	.2486	.2517	.2549
0.7	.2580	.2611	.2642	.2673	.2704	.2734	.2764	.2794	.2823	.2852
0.8	.2881	.2910	.2939	.2967	.2995	.3023	.3051	.3078	.3106	.3133
0.9	.3159	.3186	.3212	.3238	.3264	.3289	.3315	.3340	.3365	.3389
1.0	.3413	.3438	.3461	.3485	.3508	.3531	.3554	.3577	.3599	.3621
1.1	.3643	.3665	.3686	.3708	.3729	.3749	.3770	.3790	.3810	.3830
1.2	.3849	.3869	.3888	.3907	.3925	.3944	.3962	.3980	.3997	.4015
1.3	.4032	.4049	.4066	.4082	.4099	.4115	.4131	.4147	.4162	.4177
1.4	.4192	.4207	.4222	.4236	.4251	.4265	.4279	.4292	.4306	.4319
1.5	.4332	.4345	.4357	.4370	.4382	.4394	.4406	.4418	.4429	.4441
1.6	.4452	.4463	.4474	.4484	.4495	.4505	.4515	.4525	.4535	.4545
1.7	.4554	.4564	.4573	.4582	.4591	.4599	.4608	.4616	.4625	.4633
1.8	.4641	.4649	.4656	.4664	.4671	.4678	.4686	.4693	.4699	.4706
1.9	.4713	.4719	.4726	.4732	.4738	.4744	.4750	.4756	.4761	.4767
2.0	.4772	.4778	.4783	.4788	.4793	.4798	.4803	.4808	.4812	.4817
2.1	.4821	.4826	.4830	.4834	.4838	.4842	.4846	.4850	.4854	.4857
2.2	.4861	.4864	.4868	.4871	.4875	.4878	.4881	.4884	.4887	.4890
2.3	.4893	.4896	.4898	.4901	.4904	.4906	.4909	.4911	.4913	.4916
2.4	.4918	.4920	.4922	.4925	.4927	.4929	.4931	.4932	.4934	.4936
2.5	.4938	.4940	.4941	.4943	.4945	.4946	.4948	.4949	.4951	.4952
2.6	.4953	.4955	.4956	.4957	.4959	.4960	.4961	.4962	.4963	.4964
2.7	.4965	.4966	.4967	.4968	.4969	.4970	.4971	.4972	.4973	.4974
2.8	.4974	.4975	.4976	.4977	.4977	.4978	.4979	.4979	.4980	.4981
2.9	.4981	.4982	.4982	.4983	.4984	.4984	.4985	.4985	.4986	.4986
3.0	.4987	.4987	.4987	.4988	.4988	.4989	.4989	.4989	.4990	.4990

Source: Abridged from Table 1 of *Statistical Tables and Formulas,* by A. Hald. Copyright © 1952, John Wiley & Sons, Inc. Reprinted by permission of A. Hald.

TABLE C.2   Critical Values of the t Distribution for $\alpha = .05$ and $.01$, Two-Tailed Test

Degrees of Freedom	.05	.01
1	12.706	63.657
2	4.303	9.925
3	3.182	5.841
4	2.776	4.604
5	2.571	4.032
6	2.447	3.707
7	2.365	3.499
8	2.306	3.355
9	2.262	3.250
10	2.228	3.169
11	2.201	3.106
12	2.179	3.055
13	2.160	3.012
14	2.145	2.977
15	2.131	2.947
16	2.120	2.921
17	2.110	2.898
18	2.101	2.878
19	2.093	2.861
20	2.086	2.845
21	2.080	2.831
22	2.074	2.819
23	2.069	2.807
24	2.064	2.797
25	2.060	2.787
26	2.056	2.779
27	2.052	2.771
28	2.048	2.763
29	2.045	2.756
30	2.042	2.750
40	2.021	2.704
60	2.000	2.660
120	1.980	2.617
∞	1.960	2.576

TABLE C.3   **Critical Values of the *F* Distribution**

df_1 df_2		1	2	3	4	5	6	8	12	24	∞
1	0.1%	405284	500000	540379	562500	576405	585937	598144	610667	623497	636619
	0.5%	16211	20000	21615	22500	23056	23437	23925	24426	24940	25465
	1%	4052	4999	5403	5625	5764	5859	5981	6106	6234	6366
	2.5%	647.79	799.50	864.16	899.58	921.85	937.11	956.66	976.71	997.25	1018.30
	5%	161.45	199.50	215.71	224.58	230.16	233.99	238.88	243.91	249.05	254.32
	10%	39.86	49.50	53.59	55.83	57.24	58.20	59.44	60.70	62.00	63.33
2	0.1	998.5	999.0	999.2	999.2	999.3	999.3	999.4	999.4	999.5	999.5
	0.5	198.50	199.00	199.17	199.25	199.30	199.33	199.37	199.42	199.46	199.51
	1	98.49	99.00	99.17	99.25	99.30	99.33	99.36	99.42	99.46	99.50
	2.5	38.51	39.00	39.17	39.25	39.30	39.33	39.37	39.42	39.46	39.50
	5	18.51	19.00	19.16	19.25	19.30	19.33	19.37	19.41	19.45	19.50
	10	8.53	9.00	9.16	9.24	9.29	9.33	9.37	9.41	9.45	9.49
3	0.1	167.5	148.5	141.1	137.1	134.6	132.8	130.6	128.3	125.9	123.5
	0.5	55.55	49.80	47.47	46.20	45.39	44.84	44.13	43.39	42.62	41.83
	1	34.12	30.81	29.46	28.71	28.24	27.91	27.49	27.05	26.60	26.12
	2.5	17.44	16.04	15.44	15.10	14.89	14.74	14.54	14.34	14.12	13.90
	5	10.13	9.55	9.28	9.12	9.01	8.94	8.84	8.74	8.64	8.53
	10	5.54	5.46	5.39	5.34	5.31	5.28	5.25	5.22	5.18	5.13
4	0.1	74.14	61.25	56.18	53.44	51.71	50.53	49.00	47.41	45.77	44.05
	0.5	31.33	26.28	24.26	23.16	22.46	21.98	21.35	20.71	20.03	19.33
	1	21.20	18.00	16.69	15.98	15.52	15.21	14.80	14.37	13.93	13.46
	2.5	12.22	10.65	9.98	9.60	9.36	9.20	8.98	8.75	8.51	8.26
	5	7.71	6.94	6.59	6.39	6.26	6.16	6.04	5.91	5.77	5.63
	10	4.54	4.32	4.19	4.11	4.05	4.01	3.95	3.90	3.83	3.76
5	0.1	47.04	36.61	33.20	31.09	29.75	28.84	27.64	26.42	25.14	23.78
	0.5	22.79	18.31	16.53	15.56	14.94	14.51	13.96	13.38	12.78	12.14
	1	16.26	13.27	12.06	11.39	10.97	10.67	10.29	9.89	9.47	9.02
	2.5	10.01	8.43	7.76	7.39	7.15	6.98	6.76	6.52	6.28	6.02
	5	6.61	5.79	5.41	5.19	5.05	4.95	4.82	4.68	4.53	4.36
	10	4.06	3.78	6.62	3.52	3.45	3.40	3.34	3.27	3.19	3.10
6	0.1	35.51	27.00	23.70	21.90	20.81	20.03	19.03	17.99	16.89	15.75
	0.5	18.64	14.54	12.92	12.03	11.46	11.07	10.57	10.03	9.47	8.88
	1	13.74	10.92	9.78	9.15	8.75	8.47	8.10	7.72	7.31	6.88
	2.5	8.81	7.26	6.60	6.23	5.99	5.82	5.60	5.37	5.12	4.85
	5	5.99	5.14	4.76	4.53	4.39	4.28	4.15	4.00	3.84	3.67
	10	3.78	3.46	3.29	3.18	3.11	3.05	2.98	2.90	2.82	2.72
7	0.1	29.22	21.69	18.77	17.19	16.21	15.52	14.63	13.71	12.73	11.69
	0.5	16.24	12.40	10.88	10.05	9.52	9.16	8.68	8.18	7.65	7.08
	1	12.25	9.55	8.45	7.85	7.46	7.19	6.84	6.47	6.07	5.65
	2.5	8.07	6.54	5.89	5.52	5.29	5.12	4.90	4.67	4.42	4.14
	5	5.59	4.74	4.35	4.12	3.97	3.87	3.73	3.57	3.41	3.23
	10	3.59	3.26	3.07	2.96	2.88	2.83	2.75	2.67	2.58	2.47
8	0.1	25.42	18.49	15.83	14.39	13.49	12.86	12.04	11.19	10.30	9.34
	0.5	14.69	11.04	9.60	8.81	8.30	7.95	7.50	7.01	6.50	5.95

TABLE C.3 Continued

df$_1$ df$_2$		1	2	3	4	5	6	8	12	24	∞
	1%	11.26	8.65	7.59	7.01	6.63	6.37	6.03	5.67	5.28	4.86
	2.5%	7.57	6.06	5.42	5.05	4.82	4.65	4.43	4.20	3.95	3.67
	5%	5.32	4.46	4.07	3.84	3.69	3.58	3.44	3.28	3.12	2.93
	10%	3.46	3.11	2.92	2.81	2.73	2.67	2.59	2.50	2.40	2.29
9	0.1	22.86	16.39	13.90	12.56	11.71	11.13	10.37	9.57	8.72	7.81
	0.5	13.61	10.11	8.72	7.96	7.47	7.13	6.69	6.23	5.73	5.19
	1	10.56	8.02	6.99	6.42	6.06	5.80	5.47	5.11	4.73	4.31
	2.5	7.21	5.71	5.08	4.72	4.48	4.32	4.10	3.87	3.61	3.33
	5	5.12	4.26	3.86	3.63	3.48	3.37	3.23	3.07	2.90	2.71
	10	3.36	3.01	2.81	2.69	2.61	2.55	2.47	2.38	2.28	2.16
10	0.1	21.04	14.91	12.55	11.28	10.48	9.92	9.20	8.45	7.64	6.76
	0.5	12.83	8.08	7.34	6.87	6.54	6.12	5.66	5.17	4.64	
	1	10.04	7.56	6.55	5.99	5.64	5.39	5.06	4.71	4.33	3.91
	2.5	6.94	5.46	4.83	4.47	4.24	4.07	3.85	3.62	3.37	3.08
	5	4.96	4.10	3.71	3.48	3.33	3.22	3.07	2.91	2.74	2.54
	10	3.28	2.92	2.73	2.61	2.52	2.46	2.38	2.28	2.18	2.06
11	0.1	19.69	13.81	11.56	10.35	9.58	9.05	8.35	7.63	6.85	6.00
	0.5	12.23	8.91	7.60	6.88	6.42	6.10	5.68	5.24	4.76	4.23
	1	9.65	7.20	6.22	5.67	5.32	5.07	4.74	4.40	4.02	3.60
	2.5	6.72	5.26	4.63	4.28	4.04	4.88	3.66	3.43	3.17	2.88
	5	4.84	3.98	3.59	3.36	3.20	3.09	2.95	2.79	2.61	2.40
	10	3.23	2.86	2.66	2.54	2.45	2.39	2.30	2.21	2.10	1.97
12	0.1	18.64	12.97	10.80	9.63	8.89	8.38	7.71	7.00	6.25	5.42
	0.5	11.75	8.51	7.23	6.52	6.07	5.76	5.35	4.91	4.43	3.90
	1	9.33	6.93	5.95	5.41	5.06	4.82	4.50	4.16	3.78	3.36
	2.5	6.55	5.10	4.47	4.12	3.89	3.73	3.51	3.28	3.02	2.72
	5	4.75	3.88	3.49	3.26	3.11	3.00	2.85	2.69	2.50	2.30
	10	3.18	2.81	2.61	2.48	2.39	2.33	2.24	2.15	2.04	1.90
13	0.1	17.81	12.31	10.21	9.07	8.35	7.86	7.21	6.52	5.78	4.97
	0.5	11.37	8.19	6.93	6.23	5.79	5.48	5.08	4.64	4.17	3.65
	1	9.07	6.70	5.74	5.20	4.86	4.62	4.30	3.96	3.59	3.16
	2.5	6.41	4.97	4.35	4.00	3.77	3.60	3.39	3.15	2.89	2.60
	5	4.67	3.80	3.41	3.18	3.02	2.92	2.77	2.60	2.42	2.21
	10	3.14	2.76	2.56	2.43	2.35	2.28	2.20	2.10	1.98	1.85
14	0.1	17.14	11.78	9.73	8.62	7.92	7.43	6.80	6.13	5.41	4.60
	0.5	11.06	7.92	6.68	6.00	5.56	5.26	4.86	4.43	3.96	3.44
	1	8.86	6.51	5.56	5.03	4.69	4.46	4.14	3.80	3.43	3.00
	2.5	6.30	4.86	4.24	3.89	3.66	3.50	3.27	3.05	2.79	2.49
	5	4.60	3.74	3.34	3.11	2.96	2.85	2.70	2.53	2.35	2.13
	10	3.10	2.73	2.52	2.39	2.31	2.24	2.15	2.05	1.94	1.80
15	0.1	16.59	11.34	9.30	8.25	7.57	7.09	6.47	5.81	5.10	4.31
	0.5	10.80	7.70	6.48	5.80	5.37	5.07	4.67	4.25	3.79	3.26
	1	8.68	6.36	5.42	4.89	4.56	4.32	4.00	3.67	3.29	2.87
	2.5	8.20	4.77	4.15	3.80	3.58	3.41	3.20	2.96	2.70	2.40

(continued)

TABLE C.3 Continued

df₁ df₂		1	2	3	4	5	6	8	12	24	∞
	5%	4.54	3.80	3.29	3.06	2.90	2.79	2.64	2.48	2.29	2.07
	10%	3.07	2.70	2.49	2.36	2.27	2.21	2.12	2.02	1.90	1.76
16	0.1	16.12	10.97	9.00	7.94	7.27	6.81	6.19	5.55	4.85	4.06
	0.5	10.58	7.51	6.30	5.64	5.21	4.91	4.52	4.10	3.64	3.11
	1	8.53	6.23	5.29	4.77	4.44	4.20	3.89	3.55	3.18	2.75
	2.5	6.12	4.69	4.08	3.73	3.50	3.34	3.12	2.89	2.63	2.32
	5	4.49	3.63	3.24	3.01	2.85	2.74	2.59	2.42	2.24	2.01
	10	3.05	2.67	2.46	2.33	2.24	2.18	2.09	1.99	1.87	1.72
17	0.1	15.72	10.66	8.73	7.68	7.02	6.56	5.96	5.32	4.63	3.85
	0.5	10.38	7.35	6.16	5.50	5.07	4.78	4.39	3.97	3.51	2.98
	1	8.40	6.11	5.18	4.67	4.34	4.10	3.79	3.45	3.08	2.65
	2.5	6.04	4.62	4.01	3.66	3.44	3.28	3.06	2.82	2.56	2.25
	5	4.45	3.59	3.20	2.96	2.81	2.70	2.55	2.38	2.19	1.96
	10	3.03	2.64	2.44	2.31	2.22	2.15	2.06	1.96	1.84	1.69
18	0.1	15.38	10.39	8.49	7.46	6.81	6.35	5.76	5.13	4.45	3.67
	0.5	10.22	7.21	6.03	5.37	4.96	4.66	4.28	3.86	3.40	2.87
	1	8.28	6.01	5.09	4.58	4.25	4.01	3.71	3.37	3.00	2.57
	2.5	5.98	4.56	3.95	3.61	3.38	3.22	3.01	2.77	2.50	2.19
	5	4.41	3.55	3.16	2.93	2.77	2.66	2.51	2.34	2.15	1.92
	10	3.01	2.62	2.42	2.29	2.20	2.13	2.04	1.93	1.81	1.66
19	0.1	15.08	10.16	8.28	7.26	6.61	6.18	5.59	4.97	4.29	3.52
	0.5	10.07	7.09	5.92	5.27	4.85	4.56	4.18	3.76	3.31	2.78
	1	8.18	5.93	5.01	4.50	4.17	3.94	3.63	3.30	2.92	2.49
	2.5	5.92	4.51	3.90	3.56	3.33	3.17	2.96	2.72	2.45	2.13
	5	4.38	3.52	3.13	2.90	2.74	2.63	2.48	2.31	2.11	1.88
	10	2.99	2.61	2.40	2.27	2.18	2.11	2.02	1.91	1.79	1.63
20	0.1	14.82	9.95	8.10	7.10	6.46	6.02	5.44	4.82	4.15	3.38
	0.5	9.94	6.99	5.82	5.17	4.76	4.47	4.09	3.68	3.22	2.69
	1	8.10	5.85	4.94	4.43	4.10	3.87	3.56	3.23	2.86	2.42
	2.5	5.87	4.46	3.86	3.51	3.29	3.13	2.91	2.68	2.41	2.09
	5	4.35	3.49	3.10	2.87	2.71	2.60	2.45	2.28	2.08	1.84
	10	2.97	2.59	2.38	2.25	2.16	2.09	2.00	1.89	1.77	1.61
21	0.1	14.59	9.77	7.94	6.95	6.32	5.88	5.31	4.70	4.03	3.26
	0.5	9.83	6.89	5.73	5.09	4.68	4.39	4.01	3.60	3.15	2.61
	1	8.02	5.78	4.87	4.37	4.04	3.81	3.51	3.17	2.80	2.36
	2.5	5.83	4.42	3.82	3.48	3.25	3.09	2.87	2.64	2.37	2.04
	5	4.32	3.47	3.07	2.84	2.68	2.57	2.42	2.25	2.05	1.81
	10	2.96	2.57	2.36	2.23	2.14	2.08	1.98	1.88	1.75	1.59
22	0.1	14.38	9.61	7.80	6.81	6.19	5.76	5.19	4.58	3.92	3.15
	0.5	9.73	6.81	5.65	5.02	4.61	4.32	3.94	3.54	3.08	2.55
	1	7.94	5.72	4.82	4.31	3.99	3.76	3.45	3.12	2.75	2.31
	2.5	5.79	4.38	3.78	3.44	3.22	3.05	2.84	2.60	2.33	2.00
	5	4.30	3.44	3.05	2.82	2.66	2.55	2.40	2.23	2.03	1.78
	10	2.95	2.56	2.35	2.22	2.13	2.06	1.97	1.86	1.73	1.57

TABLE C.3   Continued

df₁ / df₂		1	2	3	4	5	6	8	12	24	∞
23	0.1%	14.19	9.47	7.67	6.69	6.08	5.65	5.09	4.48	3.82	3.05
	0.5%	9.63	6.73	5.58	4.95	4.54	4.26	3.88	3.47	3.02	2.48
	1%	7.88	5.66	4.76	4.26	3.94	3.71	3.41	3.07	2.70	2.26
	2.5%	5.75	4.35	3.75	3.41	3.18	3.02	2.81	2.57	2.30	1.97
	5%	4.28	3.42	3.03	2.80	2.64	2.53	2.38	2.20	2.00	1.76
	10%	2.94	2.55	2.34	2.21	2.11	2.05	1.95	1.84	1.72	1.55
24	0.1	14.03	9.34	7.55	6.59	5.98	5.55	4.00	4.00	3.74	2.97
	0.5	9.55	6.66	5.52	4.89	4.49	4.20	3.83	3.42	2.97	2.43
	1	7.82	5.61	4.72	4.22	3.90	3.67	3.36	3.03	2.66	2.21
	2.5	5.72	4.32	3.72	3.38	3.15	2.99	2.78	2.54	2.27	1.94
	5	4.26	3.40	3.01	2.78	2.62	2.51	2.36	2.18	1.98	1.73
	1.0	2.93	2.54	2.33	2.19	2.10	2.04	1.94	1.83	1.70	1.53
25	0.1	13.88	9.22	7.45	6.49	5.88	5.46	4.91	4.31	3.66	2.89
	0.5	9.48	6.60	5.46	4.84	4.43	4.15	3.78	3.37	2.92	2.38
	1	7.77	5.57	4.68	4.18	3.86	3.63	3.32	2.99	2.62	2.17
	2.5	5.69	4.29	3.69	3.35	3.13	2.97	2.75	2.51	2.24	1.91
	5	4.24	3.38	2.99	2.76	2.60	2.49	2.34	2.16	1.96	1.71
	1.0	2.92	2.53	2.32	2.18	2.09	2.02	1.93	1.82	1.69	1.52
26	0.1	13.74	9.12	7.36	6.41	5.80	5.38	4.83	4.24	3.59	2.82
	0.5	9.41	6.54	5.41	4.79	4.38	4.10	3.73	3.33	2.87	2.33
	1	7.72	5.53	4.64	4.14	3.82	3.59	3.29	2.96	2.58	2.13
	2.5	5.66	4.27	3.67	3.33	3.10	2.94	2.73	2.49	2.22	1.88
	5	4.22	3.37	2.98	2.74	2.59	2.47	2.32	2.15	1.95	1.69
	10	2.91	2.52	2.31	2.17	2.08	2.01	1.92	1.81	1.68	1.50
27	0.1	13.61	9.02	7.27	6.33	5.73	5.31	4.76	4.17	3.52	2.75
	0.5	9.34	6.49	5.36	4.74	4.34	4.06	3.69	3.28	2.83	2.29
	1	7.68	5.49	4.60	4.11	3.78	3.56	3.26	2.93	2.55	2.10
	2.5	5.63	4.24	3.65	3.31	3.08	2.92	2.71	2.47	2.19	1.85
	5	4.21	3.35	2.96	2.73	2.57	2.46	2.30	2.13	1.93	1.67
	10	2.90	2.51	2.30	2.17	2.07	2.00	1.91	1.80	1.67	1.49
28	0.1	13.50	8.93	7.19	6.25	5.66	5.24	4.69	4.11	3.46	2.70
	0.5	9.28	6.44	5.32	4.70	4.30	4.02	3.65	3.25	2.79	2.25
	1	7.64	5.45	4.57	4.07	3.75	3.53	3.23	2.90	2.52	2.06
	2.5	5.61	4.22	3.63	3.29	2.06	2.90	2.69	2.45	2.17	1.83
	5	4.20	3.34	2.95	2.71	2.56	2.44	2.29	2.12	1.91	1.65
	10	2.89	2.50	2.29	2.16	2.06	2.00	1.90	1.79	1.66	1.48
29	0.1	13.39	8.85	7.12	6.19	5.59	5.18	4.64	4.05	3.41	2.64
	0.5	9.23	6.40	5.28	4.66	4.26	3.98	3.61	3.21	2.76	2.21
	1	7.60	5.42	4.54	4.04	3.73	3.50	3.20	2.87	2.49	2.03
	2.5	5.59	4.20	3.61	3.27	3.04	2.88	2.67	2.43	2.15	1.81
	5	4.18	3.33	2.93	2.70	2.54	2.43	2.28	2.10	1.90	1.64
	10	2.89	2.50	2.28	2.15	2.06	1.99	1.89	1.78	1.65	1.47
30	0.1	13.29	8.77	7.05	6.12	5.53	5.12	4.58	4.00	3.36	2.59
	0.5	9.18	6.35	5.24	4.62	4.23	3.95	3.58	3.18	2.73	2.18

(continued)

TABLE C.3 Continued

df₁ df₂		1	2	3	4	5	6	8	12	24	∞
	1%	7.56	5.39	4.51	4.02	3.70	3.47	3.17	2.84	2.47	2.01
	2.5%	5.57	4.18	3.59	3.25	3.03	2.87	2.65	2.41	2.14	1.79
	5%	4.17	3.32	2.92	2.69	2.53	2.42	2.27	2.09	1.89	1.62
	10%	2.88	2.49	2.28	2.14	2.05	1.98	1.88	1.77	1.64	1.46
40	0.1	12.61	8.25	6.60	5.70	5.13	4.73	4.21	3.64	3.01	2.23
	0.5	8.83	6.07	4.98	4.37	3.99	3.71	3.35	2.95	2.50	1.93
	1	7.31	5.18	4.31	3.83	3.51	3.29	2.99	2.66	2.29	1.80
	2.5	5.42	4.05	3.46	3.13	2.90	2.74	2.53	2.29	2.01	1.64
	5	4.08	3.23	2.84	2.61	2.45	2.34	2.18	2.00	1.79	1.51
	10	2.84	2.44	2.23	2.09	2.00	1.93	1.83	1.71	1.57	1.38
60	0.1	11.97	7.76	6.17	5.31	4.76	4.37	3.87	3.31	2.69	1.90
	0.5	8.49	5.80	4.73	4.14	3.76	3.49	3.13	2.74	2.29	1.69
	1	7.08	4.98	4.13	3.65	3.34	3.12	2.82	2.50	2.12	1.60
	2.5	5.29	3.93	3.34	3.01	2.79	2.63	2.41	2.17	1.88	1.48
	5	4.00	3.15	2.76	2.52	2.37	2.25	2.10	1.92	1.70	1.39
	10	2.70	2.30	2.10	2.04	1.05	1.37	1.77	1.66	1.51	1.20
120	0.1	11.38	7.31	5.79	4.95	4.42	4.04	3.55	3.02	2.40	1.56
	0.5	8.18	5.54	4.50	3.92	3.55	3.28	2.93	2.54	2.09	1.43
	1	6.85	4.79	3.95	3.48	3.17	2.96	2.66	2.34	1.95	1.38
	2.5	5.15	3.80	3.23	2.89	2.67	2.52	2.30	2.05	1.76	1.31
	5	3.92	3.07	2.68	2.45	2.29	2.17	2.02	1.83	1.61	1.25
	10	2.75	2.35	2.13	1.99	1.90	1.82	1.72	1.60	1.45	1.19
∞	0.1	10.83	6.91	5.42	4.62	4.10	3.74	3.27	2.74	2.13	1.00
	0.5	7.88	5.30	4.28	3.72	3.35	3.09	2.74	2.36	1.90	1.00
	1	6.64	4.60	3.78	3.32	3.02	2.80	2.51	2.18	1.79	1.00
	2.5	5.02	3.69	3.12	2.79	2.57	2.41	2.19	1.94	1.64	1.00
	5	3.84	2.99	2.60	2.37	2.21	2.09	1.94	1.75	1.52	1.00
	10	2.71	2.30	2.08	1.94	1.85	1.77	1.67	1.55	1.35	1.00

TABLE C.4 Critical Values of Chi Square (χ^2)

df	0.250	0.100	0.050	0.025	0.010	0.005	0.001
1	1.32330	2.70554	3.84146	5.02389	6.63490	7.87944	10.828
2	2.77259	4.60517	5.99147	7.37776	9.21034	10.5966	13.816
3	4.10835	6.25139	7.81473	9.34840	11.3449	12.8381	16.266
4	5.38527	7.77944	9.48773	1.1433	13.2767	14.8602	18.467
5	6.62568	9.23635	11.0705	12.8325	15.0863	16.7496	20.515
6	7.84080	10.6446	12.5916	14.4494	16.8119	18.5476	22.458
7	9.03715	12.0170	14.0671	16.0128	18.4753	20.2777	24.322
8	10.2188	13.3616	15.5073	17.5346	20.0902	21.9550	26.125
9	11.3887	14.6837	16.9190	19.0228	21.6660	23.5893	27.877
10	12.5489	15.9871	18.3070	20.4831	23.2093	25.1882	29.588
11	13.7007	17.2750	19.6751	21.9200	24.7250	26.7569	31.264
12	14.8454	18.5494	21.0261	23.3367	26.2170	28.2995	32.909
13	15.9839	19.8119	22.3621	24.7356	27.6883	29.8194	34.528
14	17.1770	21.0642	23.6848	26.1190	29.1413	31.3193	36.123
15	18.2451	22.3072	24.9958	27.4884	30.5779	32.8013	37.697
16	19.3688	23.5418	26.2962	28.8454	31.9999	34.2672	39.252
17	20.4887	24.7690	27.5871	30.1910	33.4087	35.7185	40.790
18	21.6049	25.9894	28.8693	31.5264	34.8053	37.1564	42.312
19	22.7178	27.2036	30.1435	32.8523	36.1908	38.5822	43.820
20	23.8277	28.4120	31.4104	34.1696	37.5662	39.9968	45.315
21	24.9348	29.6151	32.6705	35.4789	38.9321	41.4010	46.797
22	26.0393	30.8133	33.9244	36.7807	40.2894	42.7956	48.268
23	27.1413	32.0069	35.1725	38.0757	41.6384	44.1813	49.728
24	28.2412	33.1963	36.4151	39.3641	42.9798	45.5585	51.179
25	29.3389	34.3816	37.6525	40.6465	44.3141	46.9278	52.620
26	30.4345	35.5631	38.8852	41.9232	45.6417	48.2899	54.052
27	31.5284	36.7412	40.1133	43.1944	46.9630	49.6449	55.476
28	32.6205	37.9159	41.3372	44.4607	48.2782	50.9933	56.892
29	33.7109	39.0875	42.5569	45.7222	49.5879	52.3356	58.302
30	34.7998	40.2560	43.7729	46.9792	50.8922	53.6720	59.703
40	45.6160	51.8050	65.7585	59.3417	63.6907	66.7659	73.402
50	56.3336	63.1671	67.5048	71.4202	76.1539	79.4900	86.661
60	66.9814	74.3970	79.0819	83.2976	88.3794	91.9517	99.607
70	77.5766	85.5271	90.5312	95.0231	100.425	104.215	112.317
80	88.1303	96.5782	101.879	106.629	112.329	116.321	124.839
90	98.6499	107.565	113.145	118.136	124.116	128.299	137.208
100	109.141	118.498	124.342	129.561	135.807	140.169	149.449

TABLE C.5   Critical Values for Squared Multiple Correlation (R^2) in Forward Stepwise Selection: $a = .05$

k	F	10	12	14	16	18	20	25	30	35	40	50	60	80	100	150	200
									$N-k-1$								
2	2	43	38	33	30	27	24	20	16	14	13	10	8	6	5	3	2
2	3	40	36	31	27	24	22	18	15	13	11	9	7	5	4	2	2
2	4	38	33	29	26	23	21	17	14	12	10	8	7	5	4	3	2
3	2	49	43	39	35	32	29	24	21	18	16	12	10	8	7	4	2
3	3	45	40	36	32	29	26	22	19	17	15	11	9	7	6	4	3
3	4	42	36	33	29	27	25	20	17	15	13	11	9	7	5	4	3
4	2	54	48	44	39	35	33	27	23	20	18	15	12	10	8	5	4
4	3	49	43	39	36	33	30	25	22	19	17	14	11	8	7	5	4
4	4	45	39	35	32	29	27	22	19	17	15	12	10	8	6	5	3
5	2	58	52	47	43	39	36	31	26	23	21	17	14	11	9	6	5
5	3	52	46	42	38	35	32	27	24	21	19	16	13	9	8	5	4
5	4	46	41	38	35	52	29	24	21	18	16	13	11	9	7	5	4
6	2	60	54	50	46	41	39	33	29	25	23	19	16	12	10	7	5
6	3	54	48	44	40	37	34	29	25	22	20	17	14	10	8	6	5
6	4	48	43	39	36	33	30	26	23	20	17	14	12	9	7	5	4
7	2	61	56	51	48	44	41	35	30	27	24	20	17	13	11	7	5
7	3	59	50	46	42	39	36	31	26	23	21	18	15	11	9	7	5
7	4	50	45	41	38	35	32	27	24	21	18	15	13	10	8	6	4
8	2	62	58	53	49	46	43	37	31	28	26	21	18	14	11	8	6
8	3	57	52	47	43	40	37	32	28	24	22	19	16	12	10	7	5
8	4	51	46	42	39	36	33	28	25	22	19	16	14	11	9	7	5
9	2	63	59	54	51	47	44	38	33	30	27	22	19	15	12	9	6
9	3	58	53	49	44	41	38	33	29	25	23	20	16	12	10	7	6
9	4	52	46	43	40	37	34	29	25	23	20	17	14	11	10	7	6
10	2	64	60	55	52	49	46	39	34	31	28	23	20	16	13	10	7
10	3	59	54	50	45	42	39	34	30	26	24	20	17	13	11	8	6
10	4	52	47	44	41	38	35	30	26	24	21	18	15	12	10	8	6
12	2	66	62	57	54	51	48	42	37	33	30	25	22	17	14	10	8
12	3	60	55	52	47	44	41	36	31	28	25	22	19	14	12	9	7
12	4	53	48	45	41	39	36	31	27	25	22	19	16	13	11	9	7
14	2	68	64	60	56	53	50	44	39	35	32	27	24	18	15	11	8
14	3	61	57	53	49	46	43	37	32	29	27	23	20	15	13	10	8
14	4	43	49	46	42	40	37	32	29	26	23	20	17	13	11	9	7
16	2	69	66	61	58	55	53	46	41	37	34	29	25	20	17	12	9
16	3	61	58	54	50	47	44	38	34	31	28	24	21	17	14	11	8
16	4	53	50	46	43	40	38	33	30	27	24	21	18	14	12	10	8
18	2	70	67	65	60	57	55	49	44	40	36	31	27	21	18	13	9
18	3	62	59	55	51	49	46	40	35	32	30	26	23	18	15	12	9
18	4	54	50	46	44	41	38	34	31	28	25	22	19	15	13	11	8
20	2	72	68	64	62	59	56	50	46	42	38	33	28	22	19	14	10
20	3	62	60	56	52	50	47	42	37	34	31	27	24	19	16	12	9
20	4	54	50	46	44	41	37	35	32	29	26	23	20	16	14	11	8

TABLE C.5 Continued $\alpha = .01$

								$N-k-1$									
k	F	10	12	14	16	18	20	25	30	35	40	50	60	80	100	150	200
2	2	59	53	48	43	40	36	30	26	23	20	17	14	11	9	7	5
2	3	58	52	46	42	38	35	30	25	22	19	16	13	10	8	6	4
2	4	57	49	44	39	36	32	26	22	19	16	13	11	8	7	5	4
3	2	67	60	55	50	46	42	35	30	27	24	20	17	13	11	7	5
3	3	63	58	52	47	43	40	34	29	25	22	19	16	12	10	7	5
3	4	61	54	48	44	40	37	31	26	23	20	16	14	11	9	6	5
4	2	70	64	58	53	49	46	39	34	30	27	23	19	15	12	8	6
4	3	67	62	56	51	47	44	37	32	28	25	21	18	14	11	8	6
4	4	64	58	52	47	43	40	34	29	26	23	19	16	13	11	7	6
5	2	73	67	61	57	52	49	42	37	32	29	25	21	16	13	9	7
5	3	70	65	59	54	50	46	39	34	30	27	23	19	15	12	9	7
5	4	65	60	55	50	46	43	36	31	28	25	20	17	14	12	8	6
6	2	74	69	63	59	55	51	44	39	34	31	26	23	18	14	10	8
6	3	72	67	61	56	51	48	41	36	32	28	24	20	16	13	10	7
6	4	66	61	56	52	48	45	38	33	29	26	22	19	15	13	9	7
7	2	76	70	65	60	56	53	46	40	36	33	28	25	19	15	11	9
7	3	73	68	62	57	53	50	42	37	33	30	25	21	17	14	10	8
7	4	67	62	58	54	49	46	40	35	31	28	23	20	16	14	10	8
8	2	77	72	66	62	58	55	48	42	38	34	29	26	20	16	12	9
8	3	74	69	63	58	54	51	44	39	34	31	26	22	18	15	11	9
8	4	67	63	59	55	50	47	41	36	32	29	24	21	17	15	11	9
9	2	78	73	67	63	60	56	49	43	39	36	31	27	21	17	12	10
9	3	74	69	64	59	56	52	45	40	35	32	27	23	19	16	12	9
9	4	68	63	60	56	51	48	42	37	33	30	25	22	18	16	12	9
10	2	79	74	68	65	61	58	51	45	40	37	32	28	22	18	13	10
10	3	74	69	65	50	57	53	47	41	37	33	28	24	20	17	13	10
10	4	68	64	61	56	52	49	43	38	34	31	26	23	19	17	13	9
12	2	80	75	70	66	63	60	53	48	43	39	34	30	24	20	14	11
12	3	74	70	66	62	58	55	48	43	39	35	30	26	21	18	14	10
12	4	69	65	61	57	53	50	44	40	35	32	27	24	20	18	13	10
14	2	81	76	71	68	65	62	55	50	45	41	36	32	25	21	15	11
14	3	74	70	67	63	60	56	50	45	41	37	31	27	22	19	15	11
14	4	69	65	61	57	54	52	45	41	36	33	28	25	21	19	14	10
16	2	82	77	72	69	66	63	57	52	47	43	38	34	27	22	16	12
16	3	74	70	67	64	61	58	52	47	42	39	33	29	23	20	15	11
16	4	70	66	62	58	55	52	46	42	37	34	29	26	22	20	14	11
18	2	82	78	73	70	67	65	59	54	49	45	39	35	28	23	17	12
18	3	74	70	67	65	62	59	53	48	44	41	35	30	24	21	16	12
18	4	70	65	62	58	55	53	47	43	38	35	30	27	23	20	15	11
20	2	82	78	74	71	68	66	60	55	50	46	41	36	29	24	18	13
20	3	74	70	67	65	62	60	55	60	46	42	36	32	26	22	17	12
20	4	70	66	62	58	55	53	47	43	39	36	31	28	24	21	16	11

Note: Decimals are omitted; k = number of candidate predictors; N = sample size; F = criterion F-to-enter.

Source: Adapted from Tables 1 and 2 in "Tests of significance in forward selection regression," by L. Wilkinson and G. E. Dallal, *Technometrics,* 1981, *23*(4), 377–380. Reprinted with permission from *Technometrics.* Copyright 1981 by the American Statistical Association and the American Society for Quality Control. All rights reserved.

TABLE C.6   Critical Values for F_{MAX} (S^2_{MAX}/S^2_{MIN}) Distribution for $a = .05$ and $.01$

$a = .05$

k df	2	3	4	5	6	7	8	9	10	11	12
4	9.60	15.5	20.6	25.2	29.5	33.6	37.5	41.1	44.6	48.0	51.4
5	7.15	10.8	13.7	16.3	18.7	20.8	22.9	24.7	26.5	28.2	29.9
6	5.82	8.38	10.4	12.1	13.7	15.0	16.3	17.5	18.6	19.7	20.7
7	4.99	6.94	8.44	9.70	10.8	11.8	12.7	13.5	14.3	15.1	15.8
8	4.43	6.00	7.18	8.12	9.03	9.78	10.5	11.1	11.7	12.2	12.7
9	4.03	5.34	6.31	7.11	7.80	8.41	8.95	9.45	9.91	10.3	10.7
10	3.72	4.85	5.67	6.34	6.92	7.42	7.87	8.28	8.66	9.01	9.34
12	3.28	4.16	4.79	5.30	5.72	6.09	6.42	6.72	7.00	7.25	7.48
15	2.86	3.54	4.01	4.37	4.68	4.95	5.19	5.40	5.59	5.77	5.93
20	2.46	2.95	3.29	3.54	3.76	3.94	4.10	4.24	4.37	4.49	4.59
30	2.07	2.40	2.61	2.78	2.91	3.02	3.12	3.21	3.29	3.36	3.39
60	1.67	1.85	1.96	2.04	2.11	2.17	2.22	2.26	2.30	2.33	2.36
∞	1.00	1.00	1.00	1.00	1.00	1.00	1.00	1.00	1.00	1.00	1.00

$a = .01$

k df	2	3	4	5	6	7	8	9	10	11	12
4	23.2	37	49	59	69	79	89	97	106	113	120
5	14.9	22	28	33	38	42	46	50	54	57	60
6	11.1	15.5	19.1	22	25	27	30	32	34	36	37
7	8.89	12.1	14.5	16.5	18.4	20	22	23	24	26	27
8	7.50	9.9	11.7	13.2	14.5	15.8	16.9	17.9	18.9	19.8	21
9	6.54	8.5	9.9	11.1	12.1	13.1	13.9	14.7	15.3	16.0	16.6
10	5.85	7.4	8.6	9.6	10.4	11.1	11.8	12.4	12.9	13.4	13.9
12	4.91	6.1	6.9	7.6	8.2	8.7	9.1	9.5	9.9	10.2	10.6
15	4.07	4.9	5.5	6.0	6.4	6.7	7.1	7.3	7.5	7.8	8.0
20	3.32	3.8	4.3	4.6	4.9	5.1	5.3	5.5	5.6	5.8	5.9
30	2.63	3.0	3.3	3.4	3.6	3.7	3.8	3.9	4.0	4.1	4.2
60	1.96	2.2	2.3	2.4	2.4	2.5	2.5	2.6	2.6	2.7	2.7
∞	1.00	1.0	1.0	1.0	1.0	1.0	1.0	1.0	1.0	1.0	1.0

Note: S^2_{MAX} is the largest and S^2_{MIN} the smallest in a set of k independent mean squares, each based on degrees of freedom (df).

Source: Adapted from Table 31 in *Biometrika Tables for Statisticians,* vol. 1, 3d ed., edited by E. S. Pearson and H. O. Hartley (New York: Cambridge University Press, 1958). Reproduced with the permission of the *Biometrika* trustees.

REFERENCES

Aiken, L. S., Stein, J. A., and Bentler, P. M. (1994). Structural equation analysis of clinical subpopulation differences and comparative treatment outcomes: Characterizing the daily lives of drug addicts. *Journal of Consulting and Clinical Psychology, 62*(3), 488–499.

Aiken, L. S. and West, S. G. (1991). *Multiple Regression: Testing and Interpreting Interactions.* Newbury Park, CA: Sage Publications.

Akaike, H. (1987). Factor analysis and AIC. *Psychometrika, 52,* 317–332.

Algina, J., and Swaminathan, H. (1979). Alternatives to Simonton's analyses of the interrupted and multiple-group time-series design. *Psychological Bulletin, 86,* 919–926.

Allison, P. D. (1987). Estimation of linear models with incomplete data. In C. Cogg (Ed.), *Sociological Methodology 1987* (pp. 71–103). San Francisco: Jossey-Bass.

Allison, P. D. (1995). *Survival Analysis Using the SAS System: A Practical Guide.* Cary, NC: SAS Institute Inc.

American Psychological Association. (2001). *Publication Manual of the American Psychological Association* (5th ed.). Washington, DC: Author.

Anderson, J. C., and Gerbing, D. W. (1984). The effect of sampling error on convergence, improper solutions, and goodness-of-fit indices for maximum likelihood confirmatory factor analysis. *Psychometrika, 49,* 155–173.

Asher, H. B. (1976). *Causal Modeling.* Beverly Hills, CA: Sage Publications.

Baldry, A. C. (2003). Bullying in schools and exposure to domestic violence. *Child Abuse and Neglect, 27*(7), 713–732.

Balter, M. D., and Levine, J. (1971). Character and extent of psychotropic drug usage in the United States. Paper presented at the Fifth World Congress on Psychiatry, Mexico City.

Barcikowski, R. S. (1981). Statistical power with group mean as the unit of analysis. *Journal of Educational Statistics, 6,* 267–285.

Baron, R. M., & Kenny, D. A. (1986). The moderator-mediator variable distinction in social psychological research: Conceptual, strategic, and statistical considerations. *Journal of Personality and Social Psychology, 51,* 1173–1182.

Barnett, R. C., Marshall, N. L., Raudenbush, S., and Brennan, R. (1993). Gender and the relationship between job experiences and psychological distress: A study of dual-earner couples. *Journal of Personality and Social Psychology, 65*(5), 794–806.

Bartlett, M. S. (1941). The statistical significance of canonical correlations. *Biometrika, 32,* 29–38.

Bartlett, M. S. (1954). A note on the multiplying factors for various chi square approximations. *Journal of the Royal Statistical Society, 16* (Series B), 296–298.

Bearden, W. O., Sharma, S., and Teel, J. E. (1982). Sample size effects on chi-square and other statistics used in evaluating causal models. *Journal of Marketing Research, 19,* 425–430.

Belsley, D. A., Kuh, E., and Welsch, R. E. (1980). *Regression Diagnostics: Identifying Influential Data and Sources of Collinearity.* New York: John Wiley & Sons.

Bem, S. L. (1974). The measurement of psychological androgyny. *Journal of Consulting and Clinical Psychology, 42,* 155–162.

Bendel, R. B., and Afifi, A. A. (1977). Comparison of stopping rules in forward regression. *Journal of the American Statistical Association, 72,* 46–53.

Benedetti, J. K., and Brown, M. B. (1978). Strategies for the selection of log-linear models. *Biometrics, 34,* 680–686.

Bentler, P. M. (1980). Multivariate analysis with latent variables: Causal modeling. *Annual Review of Psychology, 31,* 419–456.

Bentler, P. M. (1983). Some contributions to efficient statistics in structural models: Specifications and estimation of moment structures. *Psychometrika, 48,* 493–517.

Bentler, P. M. (1988). Comparative fit indexes in structural models. *Psychological Bulletin, 107,* 238–246.

Bentler, P. M. (1995). *EQS: Structural Equations Program Manual.* Encino, CA: Multivariate Software, Inc.

Bentler, P. M., and Bonett, D. G. (1980). Significance tests and goodness of fit in the analysis of covariance structures. *Psychological Bulletin, 88,* 588–606.

Bentler, P. M., and Dijkstra, T. (1985). Efficient estimation via linearization in structural models. In P. R. Krishnaiah (Ed.), *Multivaraite analysis VI* (pp. 9–42). Amsterdam: North-Holland.

Bentler, P. M., and Weeks, D. G. (1980). Linear structural equation with latent variables. *Psychometrika, 45,* 289–308.

Bentler, P. M., and Yuan, K.-H. (1999). Structural equation modeling with small samples: Test statistics. *Multivariate Behavioral Research, 34*(2), 181–197.

Berkhof, J., and Snijders, T. A. B. (2001). Variance component testing in multilevel models. *Journal of Educational and Behavioral Statistics, 26,* 133–152.

Berry, W. D. (1993). *Understanding Regression Assumptions.* Newbury Park, CA: Sage Publications.

Blalock, H. M. (Ed.) (1985). *Causal Models in Panel and Experimental Designs.* New York: Aldine Publishing.

Bock, R. D. (1966). Contributions of multivariate experimental designs to educational research. In Cattell, R. B. (Ed.), *Handbook of Multivariate Experimental Psychology.* Chicago: Rand McNally.

Bock, R. D. (1975). *Multivariate Statistical Methods in Behavioral Research.* New York: McGraw-Hill.

Bock, R. D., and Haggard, E. A. (1968). The use of multivariate analysis of variance in behavioral research. In D. K. Whitla (Ed.), *Handbook of Measurement and Assessment in Behavioral Sciences.* Reading, MA: Addison-Wesley.

Bollen, K. A. (1986). Sample size and Bentler and Bonnett's nonnormed fit index. *Psychometrika, 51,* 375–377.

Bollen, K. A. (1989a). A new incremental fit index for general structural equation models. *Sociological Methods & Research, 17,* 303–316.

Bollen, K. A. (1989b). *Structural Equations with Latent Variables.* New York: John Wiley & Sons.

Bonett, D. G., and Bentler, P. M. (1983). Goodness-of-fit procedures for evaluation and selection of log-linear models, *Psychological Bulletin, 93*(1), 149–166.

Boomsma, A. (1983). On the robustness of LISREL (maximum likelihood estimation) against small sample size and nonnormality. Ph.D. Thesis, University of Groningen, The Netherlands.

Box, G. E. P., and Cox, D. R. (1964). An analysis of transformations. *Journal of the Royal Statistical Society, 26*(Series B), 211–243.

Box, G. E. P., and Jenkins, G. M. (1976). *Time Series Analysis: Forecasting and Control.* Rev. ed. Oakland, CA: Holden-Day.

Box, G. E. P., Jenkins, G. M., and Reinsel, G. (1994). *Time Series Analysis: Forecasting and Control* (3rd ed.). Englewood Cliffs, NJ: Prentice-Hall.

Bozdogan, H. (1987). Model selection and Akaike's information criteria (AIC): The general theory and its analytical extensions. *Psychometrika, 52,* 345–370.

Bradley, J. V. (1982). The insidious L-shaped distribution. *Bulletin of the Psychonomic Society, 20*(2), 85–88.

Bradley, J. V. (1984). The complexity of nonrobustness effects. *Bulletin of the Psychonomic Society, 22*(3), 250–253.

Bradley, R. H., and Gaa, J. P. (1977). Domain specific aspects of locus of control: Implications for modifying locus of control orientation. *Journal of School Psychology, 15*(1), 18–24.

Brambilla, P., Harenski, K., Nicoletti, M., Sassi, R. B., Mallinger, A. G., Frank, E., and Kupfer, D. J. (2003). MRI investigation of temporal lobe structures in bipolar patients. *Journal of Psychiatric Research, 37*(4), 287–295.

Brown, C. C. (1982). On a goodness-of-fit test for the logistic model based on score statistics. *Communications in Statistics, 11,* 1087–1105.

Brown, D. R., Michels, K. M. and Winer, B. J. (1991). *Statistical Principles in Experimental Design* (3rd ed.). New York: McGraw-Hill.

Brown, M. B. (1976). Screening effects in multidimensional contingency tables. *Applied Statistics, 25,* 37–46.

Browne, M. W. (1975). Predictive validity of a linear regression equation. *British Journal of Mathematical and Statistical Psychology, 28,* 79–87.

Browne, M. W., and Cudeck, R. (1993). Alternative ways of assessing model fit. In K. A. Bollen and J. S. Long (Eds.), *Testing Structural Models.* Newbury Park, CA: Sage Publications.

Burgess, E., and Locke, H. (1960). *The Family* (2nd ed.). New York: American Book.

Byrne, B. M., Shavelson, R. J., and Muthen, B. (1989). Testing for the equivalence of factor covariance and mean structures: The issue of partial measurement invariance. *Psychological Bulletin, 105,* 456–466.

Campbell, D. R., and Stanley, J. C. (1966). *Experimental and Quasi-experimental Designs for Research.* New York: Rand McNally.

Cantor, A. (1997). *Extending SAS® Survival Analysis Techniques for Medical Research.* Cary, NC: SAS Institute Inc.

Carroll, J. D., Green, P. E., and Chaturvedi, A. (1997). *Mathematical tools for applied analysis,* rev. ed., San Diego, CA: Academic Press.

Carver, R. P. (1978). The case against statistical significance testing. *Harvard Educational Review, 48,* 378–399.

Carver, R. P. (1993). The case against statistical significance testing, revisited. *The Journal of Experimental Education, 61*(4), 287–292.

Cattell, R. B. (1957). *Personality and Motivation Structures and Measurement.* Yonkers-on-Hudson, NY: World Book.

Cattell, R. B. (1966). The scree test for the number of factors. *Multivariate Behavioral Research, 1,* 245–276.

Cattell, R. B., and Baggaley, A. R. (1960). The salient variable similarity index for factor matching. *British Journal of Statistical Psychology, 13,* 33–46.

Cattell, R. B., Balcar, K. R., Horn, J. L., and Nesselroade, J. R. (1969). Factor matching procedures: An improvement of the *s* index; with tables. *Educational and Psychological Measurement, 29,* 781–792.

Cattin, P. (1980). Note on the estimation of the squared cross-validated multiple correlation of a regression model. *Psychological Bulletin, 87*(1), 63–65.

Chinn, S. (2000). A simple method for converting an odds ratio to effect size for use in meta-analysis. *Statistics in Medicine, 19*(22), 3127–3131.

Chou, C.-P., and Bentler, P. M. (1993). Invariant standardized estimated parameter change for model modification in covariance structure analysis. *Multivariate Behavioral Research, 28,* 97–110.

Cohen, J. (1965). Some statistical issues in psychological research. In B. B. Wolman (Ed.), *Handbook of Clinical Psychology* (pp. 95–121). New York: McGraw-Hill.

Cohen, J. (1988). *Statistical Power Analysis for the Behavioral Sciences* (2nd ed.). Mahwah, NJ: Lawrence Erlbaum Associates.

Cohen, J. (1994). The earth is round (p < .05). *American Psychologist, 49,* 997–1003.

Cohen, J., and Cohen, P. (1975). *Applied Multiple Regression/Correlation Analysis for the Behavioral Sciences.* New York: Lawrence Erlbaum Associates.

Cohen, J., and Cohen, P. (1983). *Applied Multiple Regression/Correlation Analysis for the Behavioral Sciences* (2nd ed.). New York: Lawrence Erlbaum Associates.

Cohen, J., Cohen, P., West, S.G., and Aiken, L. S. (2003). *Applied Multiple Regression/Correlation Analysis for the Behavioral Sciences* (3rd ed.). Mahwah, NJ: Lawrence Erlbaum Associates.

Collins, R. P., Litman, J.A., and Spielberger, C.D. (2004). The measurement of perceptual curiosity. *Personality and Individual Differences, 36*(5), 1127–1141.

Comrey, A. L. (1962). The minimum residual method of factor analysis. *Psychological Reports, 11,* 15–18.

Comrey, A. L., and Lee, H. B. (1992). *A First Course in Factor Analysis* (2nd ed.). Hillsdale, NJ: Lawrence Erlbaum Associates.

Conger, A. J. (1974). A revised definition for suppressor variables: A guide to their identification and interpretation. *Educational and Psychological Measurement, 34,* 35–46.

Cook, T. D., and Campbell, D. T. (1979). *Quasi-experimentation: Design and Analysis Issues for Field Settings.* Chicago: Rand McNally College Publishing Co.; Boston: Houghton Mifflin.

Cooley, W. W., and Lohnes, P. R. (1971). *Multivariate Data Analysis.* New York: Wiley.

Copland, L. A., Blow, F. C., and Barry, K. L. (2003). Health care utilization by older alcohol-using veterans: Effects of a brief intervention to reduce at-risk drinking. *Health Education & Behavior, 30*(3), 305–321.

Cornbleth, T. (1977). Effects of a protected hospital ward area on wandering and nonwandering geriatric patients. *Journal of Gerontology, 32*(5), 573–577.

Costanza, M. C., and Afifi, A. A. (1979). Comparison of stopping rules in forward stepwise discriminant analysis. *Journal of the American Statistical Association, 74,* 777–785.

Cox, D. R., and Snell, D. J. (1989). *The Analysis of Binary Data* (2nd ed.). London: Chapman and Hall.

Cromwell, J., Hannan, M., Labys, W. C., and Terraza, M. (1994). *Multivariate Tests for Time Series Models.* Thousand Oaks, CA: Sage Publications.

Cropsey, K. L., and Kristeller, J. L. (2003). Motivational factors related to quitting smoking among prisoners during a smoking ban. *Addictive Behaviors, 28,* 1081–1093.

Cryer, J. D. (1986). *Time Series Analysis.* Boston: Duxbury Press.

Cumming, G., and Finch, S. (2005). Inference by eye: Confidence intervals and how to read pictures of data. *American Psychologist, 60*(2), 170–180.

Damos, D. L. (1989). *Transfer of mental rotation skills.* Los Angeles: University of Southern California, Department of Human Factors.

D'Amico, E. J., Neilands, T. B., and Zambarano, R. (2001). Power analysis for multivariate and repeated measures designs: A flexible approach using the SPSS MANOVA procedure. *Behavior Research Methods, Instruments, & Computers, 33,* 479–484.

Dempster, A. P., Laird, N. M., and Rubin, D. B. (1977). Maximum likelihood from incomplete data via the EM algorithm. *Journal of the Royal Statistical Society,* Serial B, *39,* 1–38.

De Vaus, D. (Ed.) (2002). *Social Surveys* (4 vol. set). London: Sage Publications.

Diefenbach, G. J., Hopko, D. R., Feigon, S., Stanley, M. A., Novy, D. M., Beck, J. G., and Averill, P. M. (2003). "Minor GAD": Characteristics of subsyndromal GAD in older adults. *Behavior Research and Therapy, 41,* 481–487.

Dillon, W. R., and Goldstein, M. (1984). *Multivariate Analysis: Methods and Applications.* New York: Wiley.

Dixon, W. J. (Ed.) (1992). *BMDP Statistical Software: Version 7.0.* Berkeley, CA: University of California Press.

Duffy, D. L., Martin, N. G., Battistutta, D., Hopper, J. L., and Mathews, J. D. (1990). Genetics of asthma and hay fever in Australian twins. *American Review of Respiratory Disease, 142,* 1351–1358.

Duncan, T. E., Alpert, A., and Duncan, S. C. (1998). Multilevel covariance structure analysis of sibling antisocial behavior. *Structural Equation Modeling, 5*(3), 211–228.

Edwards, A. L. (1976). *An Introduction to Linear Regression and Correlation.* San Francisco: Freeman.

Egan, W. J., and Morgan, S. L. (1998). Outlier detection in multivariate analytical chemical data. *Analytical Chemistry, 70,* 2372–2379.

Elashoff, J. D. (2000). *nQuery Advisor® Version 4.0 User's Guide.* Cork, Ireland: Statistical Solutions, Ltd.

Eliason, S. R. (1993). *Maximum Likelihood Estimation: Logic and Practice.* Newbury Park, CA: Sage Publications.

Era, P., Jokela, J., Qvarnverg, Y., and Heikkinen, E. (1986). Pure-tone threshold, speech understanding, and their correlates in samples of men of different ages. *Audiology, 25,* 338–352.

Eysenck, H. J., and Eysenck, S. B. G. (1963). *The Eysenck Personality Inventory.* San Diego, CA: Educational and Industrial Testing Service; London: University of London Press.

Fava, J. L., and Velicer, W. F. (1992). The effects of overextraction on factor and component analysis. *Multivariate Behavioral Research, 27,* 387–415.

Featherman, D. (1973). Metrics of occupational status reconciled to the 1970 Bureau of Census Classification of Detailed Occupational Titles (based on Census Technical Paper No. 26, "1970 Occupation and Industry Classification Systems in Terms of their 1960 Occupational and Industry Elements"). Washington, DC: Government Printing Office. (Update of Duncan's socioeconomic status metric described in Reiss, J., et al., *Occupational and Social Status.* New York: Free Press of Glencoe.)

Feldman, P., Ullman, J. B., and Dunkel-Schetter, C. (1998). Women's reactions to rape victims: Motivational processes associated with blame and social support. *Journal of Applied Social Psychology, 6,* 469–503.

Fidell, S., Pearsons, K., Tabachnick, B., Howe, R., Silvati, L., and Barber, D. (1995). Field study of noise-induced sleep disturbance. *Journal of the Acoustical Society of America, 98,* 1025–1033.

Fleming, T. R., and Harrington, D. P. (1991). *Counting processes and survival analysis.* New York: Wiley.

Fox, J. (1991). *Regression Diagnostics.* Newbury Park, CA: Sage Publications.

Frane, J. W. (August 3, 1977). Personal communication. Health Sciences Computing Facility, University of California, Los Angeles.

Frane, J. W. (1980). The univariate approach to repeated measures—foundation, advantages, and caveats.

BMD Technical Report No. 69. Health Sciences Computing Facility, University of California, Los Angeles.

Friedrich, R. G. (1982). In defense of multiplicative terms in multiple regression equations. *American Journal of Political Science, 26,* 797–833.

Fullagar, C., McCoy, D., and Shull, C. (1992). The socialization of union loyalty. *Journal of Organization Behavior, 13,* 13–26.

Gebers, M. A., and Peck, R. C. (2003). Using traffic conviction correlates to identify high accident-risk drivers. *Accident Analysis & Prevention, 35*(6), 903–912.

Gini, C. (1912). *Variabiliteé Mutabilitá: Contributo allo Studio delle Distribuzioni e delle Relazioni Statistiche.* Bologna: Cuppini.

Glass, G. V., Wilson, V. L., and Gottman, J. M. (1975). *Design and Analysis of Time-Series Experiments.* Boulder, CO: Colorado Associated University Press.

Glock, C., Ringer, B., and Babbie, E. (1967). *To Comfort and to Challenge.* Berkeley: University of California Press.

Goodman, L. A. (1978). *Analyzing Qualitative/Categorical Data.* Cambridge, MA: Abt Books.

Gorsuch, R. L. (1983). *Factor Analysis.* Hillsdale, NJ: Lawrence Erlbaum.

Graham, J. W., Cumsille, P. E., and Elek-Fisk, E. (2003). Methods for handling missing data. Chapter 4 in J. A. Schinka and W. F. Velicer (Vol. Eds.), *Comprehensive Handbook of Psychology: Volume 2. Research Methods in Psychology.* New York: Wiley.

Gray-Toft, P. (1980). Effectiveness of a counseling support program for hospice nurses. *Journal of Counseling Psychology, 27,* 346–354.

Green, S. B. (1991). How many subjects does it take to do a regression analysis? *Multivariate Behavioral Research, 26,* 449–510.

Greenhouse, S. W., and Geisser, S. (1959). On the methods in the analysis of profile data. *Psychometrika, 24,* 95–112.

Guadagnoli, E., and Velicer, W. F. (1988). Relation of sample size to the stability of component patterns. *Psychological Bulletin, 103,* 265–275.

Haberman, S. J. (1982). Analysis of dispersion of multinomial responses. *Journal of the American Statistical Association, 77,* 568–580.

Hadi, A. S., and Simonoff, J. W. (1993). Procedures for the identification of multiple outliers in linear models. *Journal of the American Statistical Association, 88,* 1264–1272.

Hakstian, A. R., Roed, J. C., and Lind, J. C. (1979). Two-sample T^2 procedure and the assumption of homogeneous covariance matrices. *Psychological Bulletin, 86,* 1255–1263.

Hall, S. M., Hall, R. G., DeBoer, G., and O'Kulitch, P. (1977). Self and external management compared with psychotherapy in the control of obesity. *Behavior Research and Therapy, 15,* 89–95.

Harlow, L. L. (2002). Using multivariate statistics (4th ed.). [Review of the book]. *Structural Equation Modeling, 9*(4), 621–656.

Harlow, L. L., and Newcomb, M. D. (1990). Towards a general hierarchical model of meaning and satisfaction in life. *Multivariate Behavioral Research, 25,* 387–405.

Harman, H. H. (1967). *Modern Factor Analysis* (2nd ed.). Chicago: University of Chicago Press.

Harman, H. H. (1976). *Modern Factor Analysis* (3rd ed.). Chicago: University of Chicago Press.

Harman, H. H., and Jones, W. H. (1966). Factor analysis by minimizing residuals (Minres). *Psychometrika, 31,* 351–368.

Harris, R. J. (2001). *Primer of Multivariate Statistics* (3rd ed.). New York: Academic Press.

Hay, P. (2003). Quality of life and bulimic eating disorder behaviors: Findings from a community-based sample. *International Journal of Eating Disorders, 33*(4), 434–442.

Heck, R. H., and Thomas, S. L. (2000). *An Introduction to Multilevel Modeling Techniques.* Mahwah, NJ: Lawrence Erlbaum Associates.

Heise, D. R. (1975). *Causal Analysis.* New York: Wiley.

Hensher, D., and Johnson, L. W. (1981). *Applied discrete choice modelling.* London: Croom Helm.

Hershberger, S. L., Molenaar, P. C. M., and Corneal, S. E. (1996). A hierarchy of univariate and multivariate structural time series models. In G. A. Marcoulides and R. E. Schumacker (Eds.), *Advanced structural equation modeling: Issues and techniques* (pp. 159–194). Mahwah, NJ: Lawrence Erlbaum Associates.

Hintze, J. L. (2002). *PASS User's Guide—II: PASS 2000 Power Analysis and Sample Size for Windows.* Kaysville, UT: NCSS [documentation on software CD].

Hoffman, D., and Fidell, L. S. (1979). Characteristics of androgynous, undifferentiated, masculine and feminine middle class women. *Sex Roles, 5*(6), 765–781.

Holmbeck, G. N. (1997). Toward terminological, conceptual, and statistical clarity in the study of mediators and moderators: Examples from the child-clinical and pediatric psychology literatures. *Journal of Consulting and Clinical Psychology, 65,* 599–661.

Honigsfeld, A., and Dunn, R. (2003). High school male and female learning-style similarities and differences in diverse nations. *Journal of Educational Research, 96*(4), 195–205.

Horn, J. L. (1965). A rationale and test for the number of factors in factor analysis. *Psychometrika, 30,* 179–185.

Hosmer, D. W., and Lemeshow, S. (2000). *Applied Logistic Regression* (2nd ed.). New York: Wiley.

Hox, J. J. (2002). *Multilevel Analysis.* Mahwah, NJ: Lawrence Erlbaum Associates.

Hu, L., and Bentler, P. M. (1999). Cutoff criteria for fit indexes in covariance structure analysis: Conventional criteria versus new alternatives. *Structural Equation Modeling, 6,* 1–55.

Hu, L.-T., Bentler, P. M., and Kano Y. (1992). Can test statistics in covariance structure analysis be trusted? *Psychological Bulletin, 112,* 351–362.

Hull, C. H., and Nie, N. H. (1981). SPSS Update 7–9. New York: McGraw-Hill.

Huynh, H., and Feldt, L. S. (1976). Estimation of the Box correction for degrees of freedom from sample data in the randomized block and split-plot designs. *Journal of Educational Statistics, 1,* 69–82.

James, L. R., Mulaik, S. A., and Brett, J. M. (1982). *Causal Analysis: Assumptions, Models, and Data.* Beverly Hills, CA: Sage Publications, Inc.

Jamieson, J. (1999). Dealing with baseline differences: Two principles and two dilemmas. *International Journal of Psychophysiology, 31,* 155–161.

Jamieson, J. & Howk, S. (1992). The law of initial values: A four factor theory. *International Journal of Psychophysiology, 12,* 53–61.

Johnson, G. (1955). An instrument for the assessment of job satisfaction. *Personnel Psychology, 8,* 27–37.

Jöreskog, K. G., and Sörbom, D. (1988). *LISREL 7: A Guide to the Program and Applications.* Chicago: SPSS Inc.

Kaiser, H. F. (1970). A second-generation Little Jiffy. *Psychometrika, 35,* 401–415.

Kaiser, H. F. (1974). An index of factorial simplicity. *Psychometrika, 39,* 31–36.

Keppel, G., and Wickens, T. D. (2004). *Design and Analysis: A Researcher's Handbook* (4th ed.). Upper Saddle River, NJ: Prentice-Hall.

Keppel, G., and Zedeck, S. (1989). *Data Analysis for Research Designs: Analysis of Variance and Multiple Regression/Correlation Approaches.* New York: W. H. Freeman.

Kirk, R. E. (1995). *Experimental Design* (3rd ed.). Pacific Grove, CA: Brooks/Cole.

Kirkpatrick, B., and Messias, E. M. (2003). Substance abuse and the heterogeneity of schizophrenia in a population-based study. *Schizophrenia Research, 62,* 293–294.

Kish, L. (1965). *Survey Sampling.* New York: Wiley.

Knecht, W. (2003). Personal communication.

Knoke, D., and Burke, P. J. (1980). *Log-linear Models.* Beverly Hills, CA: Sage Publications.

Kreft, I., and DeLeeuw, J. (1998). *Introducing Multilevel Modeling.* Thousand Oaks, CA: Sage Publications.

Lachin, J. M., and Foulkes, M. A. (1986). Evaluation of sample size and power for analyses of survival with allowance for nonuniform patient entry, losses to follow-up, noncompliance, and stratification. *Biometrics, 42,* 507–516.

Langer, T. (1962). A 22-item screening score of psychiatric symptoms indicating impairment. *Journal of Health and Human Behavior, 3,* 269–276.

Larzelere, R. E., and Mulaik, S. A. (1977). Single-sample test for many correlations. *Psychological Bulletin, 84*(3), 557–569.

Lawley, D. N., and Maxwell, A. E. (1963). *Factor Analysis as a Statistical Method.* London: Butterworth.

Lee, S. Y., Poon, W. Y., and Bentler, P. M. (1994). Covariance and correlation structure analyses with continuous and polytomous variables. *Multivariate Analysis and Its Applications, 24,* 347–356.

Lee, W. (1975). *Experimental Design and Analysis.* San Francisco: Freeman.

Levine, M. S. (1977). *Canonical Analysis and Factor Comparison.* Beverly Hills: Sage Publications.

Leviton, L. C., and Whitely, S. E. (1981). Job seeking patterns of female and male Ph.D. recipients. *Psychology of Women Quarterly, 5,* 690–701.

Levy, P. S., and Lemeshow, S. (1999). *Sampling Populations* (3rd ed.). New York: Wiley Interscience.

Lin, P. M., and Crawford, M. H. (1983). A comparison of mortality patterns in human populations residing under diverse ecological conditions: A time series analysis. *Human Biology, 55*(1), 35–62.

Little, R. J. A., and Rubin, D. B. (1987). *Statistical Analysis with Missing Data.* New York: John Wiley and Sons.

Little, T. D., Schnabel, K. U., and Baumert, J. (Eds.) (2000). *Modeling Longitudinal and Multilevel Data.* Mahwah, NJ: Lawrence Erlbaum Associates.

Locke, H., and Wallace, K. (1959). Short marital-adjustment and prediction tests: Their reliability and validity. *Marriage and Family Living, 21,* 251–255.

Lunneborg, C. E. (1994). *Modeling Experimental and Observational Data.* Belmont, CA: Duxbury Press.

MacCallum, R. (1986). Specification searches in covariance structure modeling. *Psychological Bulletin, 100,* 107–120.

MacCallum, R. C., Brown, M. W., Sugawara, H. M. (1996). Power analysis and determination of sample size for covariance structure modeling. *Psychological Methods, 1,* 130–149.

MacKinnon, D. P., Lockwood, C. M., Hoffman, J. M., West, S. G., and Sheets, V. (2002). A comparison of methods to test mediation and other intervening variable effects. *Psychological Methods, 7,* 83–104.

Maddala, G. S. (1983). *Limited-Dependent and Qualitative Variables in Econometrics.* Cambridge, U.K.: Cambridge University Press.

Mann, M. P. (2004). The adverse influence of narcissistic injury and perfectionism on college students' institutional attachment. *Personality and Individual Differences, 36*(8), 1797–1806.

Marascuilo, L. A., and Levin, J. R. (1983). *Multivariate Statistics in the Social Sciences: A Researcher's Guide.* Monterey, CA: Brooks/Cole.

Mardia, K. V. (1971). The effect of nonnormality on some multivariate tests and robustness to nonnormality in the linear model. *Biometrika, 58*(1), 105–121.

Markus, B. H., Dickson, E. R., Grambsch, P. M., Fleming, T. R., Mazzaferro, V., Klintmalm, G. B. G., Wiesner, R. H., Van Thiel, D. H., and Starzl, T. E. (1989). Efficacy of liver transplantation in patients with primary biliary cirrhosis. *New England Journal of Medicine, 320*(26), 1709–1713.

Marshall, S. P. (1983). Sex differences in mathematical errors: An analysis of distractor choices. *Journal for Research in Mathematics Education, 14,* 325–336.

Martis, B., Alam, D. Dowd, S. M., Hill, S. K., Sharma, R. P., Rosen, C., Pliskin, N., Martin, E., Carson, V., and Janicak, P. G. (2003). Neurocognitive effects of repetitive transcranial magnetic stimulation in severe major depression. *Clinical Neurophysiology, 114,* 1125–1132.

Mason, L. (2003). High school students' beliefs about maths, mathematical problem solving, and their achievement in maths: A cross-sectional study. *Educational Psychology, 23*(1), 73–85.

Mayo, N. E., Korner-Bitensky, N. A., and Becker, R. (1991). Recovery time of independent function post-stroke. *American Journal of Physical Medicine and Rehabilitation, 70*(1), 5–12.

McArdle, J. J., and McDonald, R. P. (1984). Some algebraic properties of the Reticular Action Model for moment structures. *British Journal of Mathematical and Statistical Psychology, 37,* 234–251.

McCain, L. J., and McCleary, R. (1979). The Statistical Analysis of the Simple Interrupted Time-Series Quasi-Experiment. Chapter 6 in Cook, T. D., and Campbell, D. T. *Quasi-Experimentation: Design & Analysis Issues for Field Settings.* Chicago: Rand-McNally.

McCleary, R., and Hay, R. A. Jr. (1980). *Applied Time Series Analysis for the Social Sciences.* Beverly Hills, CA: Sage.

McDonald, R. P. (1989). An index of goodness-of-fit based on noncentrality. *Journal of Classification, 6,* 97–103.

McDonald, R. P., and Marsh, H. W. (1990). Choosing a multivariate model: Noncentrality and goodness of fit. *Psychological Bulletin, 107,* 247–255.

McDowall, D., McCleary, R., Meidinger, E. E., and Hay, R. A. Jr. (1980). *Interrupted Time Series Analysis.* Thousand Oaks, CA: Sage.

McLean, J. E. and Ernest, J. M. (1998). The role of statistical significance testing in educational research. *Research in the Schools, 5*(2), 15–22.

McLeod, J. D., & Shanahan, M. J. (1996). Trajectories of poverty and children's mental health. *Journal of Health and Social Behavior, 37,* 207–220.

Menard, S. (2001). *Applied Logistic Regression Analysis* (2nd ed.). Thousand Oaks, CA: Sage Publications.

Miller, J. K., and Farr, S. D. (1971). Bimultivariate redundancy: A comprehensive measure of interbattery relationship. *Multivariate Behavioral Research, 6,* 313–324.

Milligan, G. W. (1980). Factors that affect Type I and Type II error rates in the analysis of multidimensional contingency tables. *Psychological Bulletin, 87,* 238–244.

Milligan, G. W., Wong, D. S., and Thompson, P. A. (1987). Robustness properties of nonorthogonal analysis of variance. *Psychological Bulletin, 101*(3), 464–470.

Mitchell, L. K., and Krumboltz, J. D. (1984, April). The effect of training in cognitive restructuring on the inability to make career decisions. Paper presented at the meeting of the Western Psychological Association, Los Angeles, CA.

Moser, C. A., and Kalton, G. (1972). *Survey Methods in Social Investigation.* New York: Basic Books.

Mosteller, F., and Tukey, J. W. (1977). *Data analysis and Regression.* Reading, MA: Addison-Wesley.

Mudrack, P. E. (2004). An outcomes-based approach to just world beliefs. *Personality and Individual Differences, 38*(7), 380–384.

Mulaik, S. A. (1972). *The Foundation of Factor Analysis.* New York: McGraw-Hill.

Mulaik, S. A., James, L. R., Van Alstine, J., Bennett, N., Lind, S., and Stillwell, C. D. (1989). An evaluation of goodness of fit indices for structural equation models. *Psychological Bulletin, 105,* 430–445.

Muthén, B., Kaplan, D., and Hollis, M. (1987). On structural equation modeling with data that are not missing completely at random. *Psychometrika, 52,* 431–462.

Muthén, L. K., and Muthén, B. O. (2001). *Mplus User's Guide* (2nd ed.). Los Angeles, CA: Muthén & Muthén.

Myers, J. L., and Well, A. D. (2002). *Research Design and Statistical Analysis* (2nd ed.). New York: HarperCollins.

Nagelkerke, N. J. D. (1991). A note on a general definition of the coefficient of determination. *Biometrika, 78,* 691–692.

Nie, N. H., Hull, C. H., Jenkins, J. G., Steinbrenner, K., and Bent, D. H. (1975). *Statistical Package for the Social Sciences* (2nd ed.). New York: McGraw-Hill.

Nolan, T., Debelle, G., Oberklaid, F., and Coffey, C. (1991). Randomised trial of laxatives in treatment of childhood encopresis. *Lancet 338*(8766), 523–527.

Norušis, M. J. (1990). *SPSS Advanced Statistics User's Guide.* Chicago: SPSS.

Nunn, S. (1993). Computers in the cop car: Impact of the Mobile Digital Terminal technology on motor vehicle theft clearance and recovery rates in a Texas city. *Evaluation Review, 17*(2), 182–203.

O'Connor, B. P. (2000). Using Parallel Analysis and Velicer's MAP Test. *Behavior Research Methods, Instruments, & Computers, 32,* 396–402.

Olson, C. L. (1976). On choosing a test statistic in multivariate analysis of variance. *Psychological Bulletin, 83*(4), 579–586.

Olson, C. L. (1979). Practical considerations in choosing a MANOVA test statistic: A rejoinder to Stevens. *Psychological Bulletin, 86,* 1350–1352.

Overall, J. E., and Spiegel, D. K. (1969). Concerning least squares analysis of experimental data. *Psychological Bulletin, 72*(5), 311–322.

Overall, J. E., and Woodward, J. A. (1977). Nonrandom assignment and the analysis of covariance. *Psychological Bulletin, 84*(3), 588–594.

Pankrantz, A. (1983). *Forecasting with Univariate Box-Jenkins Models: Concepts and Cases.* New York: Wiley.

Parinet, B., Lhote, A., and Legube, B. (2004). Principal component analysis: an appropriate tool for water quality evaluation and management-application to a tropical lake system. *Ecological Modeling, 178*(3–4), 295–311.

Pisula, W. (2003). The Roman high- and low-avoidance rats respond differently to novelty in a familiarized environment. *Behavioural Processes, 63,* 63–72.

Pope, K. S., Keith-Spiegel, P., and Tabachnick, B. (1986). Sexual attraction to clients: The human therapist and the (sometimes) inhuman training system. *American Psychologist, 41,* 147–158.

Pope, K. S., and Tabachnick, B. G. (1995). Recovered memories of abuse among therapy patients: A national survey. *Ethics and Behavior, 5,* 237–248.

Preacher, K. J., & Hayes, A. F. (2004). SPSS and SAS procedures for estimating indirect effects in simple mediation models. *Behavior Research Methods, Instruments, & Computers, 36*(4), 717–731.

Prentice, R. L. (1976). A generalization of the probit and logit methods for dose response curves. *Biometrika, 32,* 761–768.

Price, B. (1977). Ridge regression: Application to nonexperimental data. *Psychological Bulletin, 84*(4), 759–766.

Rahe, R. H. (1974). The pathway between subjects' recent life changes and their near-future illness reports: Representative results and methodological issues. In B. S. Dohrenwend and B. P. Dohrenwend (eds.), *Stressful Life Events: Their Nature and Effects.* New York: Wiley.

Rangaswamy, M., Porjesz, B., Chorlian, D. B., Want, K., Jones, K. A., Bauer, L. O., Rohrbaugh, J., O'Connor, S. J., Kuperman, S., Reich, T., and Begleiter, H. (2002). Beta power in the EEG of alcoholics. *Biological Psychiatry, 51,* 831–842.

Rao, C. R. (1952). *Advanced Statistical Methods in Biometric Research.* New York: Wiley.

Rasbash, J., Browne, W., Goldstein, H., Yang, M., Plewis, I., Healy, M., Woodhouse, G., Draper, D., Langford, I., and Lewis, T. (2000). *A User's Guide to MLwiN.* Multilevel Models Project, Institute of Education, University of London.

Raudenbush, S. W., and Bryk, A. S. (2001). *Hierarchical Linear Models* (2nd ed.). Newbury Park, CA: Sage Publications.

Raudenbush, S., Bryk, A., Cheong, Y. F., and Congdon, R. (2004). *HLM 6: Hierarchical and Nonlinear Modeling.* Lincolnwood, IL: Scientific Software International.

Raudenbush, S. W., Spybrook, J., Liu, X.-F., and Congdon, R. (2005). Optimal design for longitudinal and multilevel research, Version 1.55 [Computer software and manual]. Retrieved October 20, 2005 from http://sitemaker.umich.edu/group-based/optimal_design_software.

Rea, L. M., and Parker, R. A. (1997). *Designing and Conducting Survey Research: A comprehensive guide.* San Francisco: Jossey-Bass.

Rock, S. M. (1992). Impact of the Illinois seat belt use law on accidents, deaths, and injuries. *Evaluation Review, 16*(5), 491–507.

Rollins, B., and Feldman, H. (1970). Marital satisfaction over the family life cycle. *Journal of Marriage and Family, 32,* 29–38.

Rosenthal, R. (2001). *Meta-analytic Procedures for Social Research.* Belmont, CA: Sage.

Rosenberg, M. (1965). *Society and the Adolescent Self-Image.* Princeton, NY: Princeton University Press.

Rossi, J. S. (1990). Statistical power of psychological research: What have we gained in 20 years? *Journal of Consulting and Clinical Psychology, 58,* 646–656.

Rotter, J. B. (1966). Generalized expectancies for internal versus external control of reinforcement. *Psychological Monographs, 80*(1, Whole No. 609).

Rousseeuw, P. J., and van Zomeren, B. C. (1990). Unmasking multivariate outliers and leverage points. *Journal of the American Statistical Association, 85,* 633–639.

Rowen, B., Raudenbush, S., and Kang, S. J. (1991). Organizational design in high schools: A multilevel analysis. *American Journal of Education, 99,* 238–266.

Rozeboom, W. W. (1979). Ridge regression: Bonanza or beguilement? *Psychological Bulletin, 82*(6), 242–249.

Rubin, D. (1987). *Multiple Imputation for Nonresponse in Surveys.* New York: Wiley.

Rubin, D. (1996). Multiple imputation after 18+ years. *Journal of the American Statistical Association, 91,* 473–489.

Rubin, Z., and Peplau, L.A. (1975). Who believes in a just world? *Journal of Social Issues, 31,* 65–90.

Rummel, R. J. (1970). *Applied Factor Analysis.* Evanston, IL: Northwestern University Press.

St. Pierre, R. G. (1978). Correcting covariables for unreliability. *Evaluation Quarterly, 2*(3), 401–420.

Sapnas, K. G., and Zeller, R. A. (2002). Minimizing sample size when using exploratory factor analysis for measurement. *Journal of Nursing Measurement, 10*(2), 135–153.

Satorra, A., and Bentler, P. M. (1988). Scaling corrections for chi-square statistics in covariance structure analysis. *Proceedings of the American Statistical Association,* 308–313.

Satorra, A., and Bentler, P. M. (2001). A scaled difference chi-square test statistic for moment structure analysis. *Psychometrika, 66,* 507–514.

Schafer, J. L. (1999). *NORM: Multiple Imputation of Incomplete Multivariate Data under a Normal Model, Version 2.* Software for Windows 95/98/NT, available from www.stat.psu.edu/~jls/misoftwa.html.

Schafer, J. L., and Graham, J. W. (2002). Missing Data: Our view of the state of the art. *Psychological Methods, 7*(2), 147–177.

Schall, J. J., and Pianka, E. R. (1978). Geographical trends in numbers of species. *Science, 201,* 679–686.

Scheffé, H. A. (1953). A method of judging all contrasts in the analysis of variance. *Biometrika, 40,* 87–104.

Searle, S.R., Casella, G., and McCulloch, C. E. (1992). *Variance Components.* New York: Wiley.

Sedlmeier, P., and Gigerenzer, G. (1989). Do studies of statistical power have an effect on the power of studies? *Psychological Bulletin, 105,* 309–316.

Seo, T., Kanda, T., and Fujikoshi, Y. (1995). The effects of nonnormality on tests for dimensionality in canonical correlation and MANOVA models, *Journal of Multivariate Analysis, 52,* 325–337.

Shannon, C. E. (1948). A mathematical theory of communication. *Bell System Technical Journal, 50,* 379–423 and 623–656.

Singer, J. (1998). Using SAS PROC MIXED to fit multilevel models, hierarchical models, and individual growth models. *Journal of Educational and Behavioral Statistics, 24*(4), 323–355.

Singer, J. D., and Willett, J. B. (2003). *Applied Longitudinal Data Analysis.* New York: Oxford University Press.

Smith, R. L., Ager, J. W. Jr., and Williams, D. L. (1992). Suppressor variables in multiple regression/correlation. *Educational and Psychological Measurement, 52,* 17–29.

Smithson, M. J. (2003). *Confidence Intervals.* Belmont, CA: Sage.

Snijders, T. A. B., and Bosker, R. (1994). Modeled variance in two-level models. *Sociological Methods & Research, 22,* 342–363.

Snijders, T. A. B., and Bosker, R. (1999). *Multilevel Analysis: An Introduction to Basic and Advanced Multilevel Modeling.* London: Sage Publications, Ltd.

Sobel, M. E. (1982). Asymptotic intervals for indirect effects in structural equations models. In S. Leinhart (Ed.), *Sociological methodology 1982.* San Francisco: Jossey-Bass (pp. 290–312).

Sörbom, D. (1974). A general method for studying differences in factor means and factor structures between groups. *British Journal of Mathematical and Statistical Psychology, 27,* 229–239.

Sörbom, D. (1982). Structural equation models with structured means. In K. G. Jöreskog & H. Wold (Eds.), *Systems under Indirect Observation: Causality, Structure, Prediction I* (pp. 183–195). Amsterdam: North-Holland.

Sörbom, D., and Jöreskog, K. G. (1982). The use of structural equation models in evaluation research. In C. Fornell (Ed.), *A second generation of multivariate anaysis: Vol 2. Measurement and Evaluation* (pp. 381–418). New York: Praeger.

Spence, J., and Helmreich, R. (1972). The attitudes toward women scale: An objective instrument to measure attitude towards rights and roles of women in contemporary society. *Journal Supplementary Abstract Service (Catalogue of selected Documents in Psychology), 2,* 66.

Statistical Solutions, Ltd. (1997). *SOLAS for Missing Data Analysis 1.0.* Cork, Ireland: Statistical Solutions, Ltd.

Stefl-Mabry, J. (2003). A social judgment analysis of information source preference profiles: An exploratory study to empirically represent media selection patterns. *Journal of the American Society for Information Science and Technology, 54*(9), 879–904.

Steiger, J. H. (1980). Tests for comparing elements of a correlation matrix. *Psychological Bulletin, 87*(2), 245–251.

Steiger, J. H., and Browne, M. S. (1984). The comparison of interdependent correlations between optimal linear composites. *Psychometrika, 49,* 11–24.

Steiger, J. H., and Fouladi, R. T. (1992). R2: A Computer Program for Interval Estimation, Power Calculation, and Hypothesis Testing for the Squared Multiple Correlation. *Behavior Research Methods, Instruments, and Computers, 4,* 581–582.

Stein, J. A., Newcomb, M. D., and Bentler, P. M. (1993). Differential effects of parent and grandparent drug use on behavior problems of male and female children. *Developmental Psychology, 29,* 31–43.

Steinberg, D., and Colla, P. (1991). *LOGIT: A Supplementary Module for SYSTAT.* Evanston, IL: SYSTAT Inc.

Stewart, D., and Love, W. (1968). A general canonical index. *Psychological Bulletin, 70,* 160–163.

Strober, M. H., and Weinberg, C. B. (1977). Working wives and major family expenditures. *Journal of Consumer Research, 4*(3), 141–147.

Suzuki, S., and Sheu, C.-F. (1999, November). *Fitting Multilevel Linear Models using SAS.* Paper presented at the meeting of the Society for Computers in Psychology, Los Angeles, CA.

Swallow, W. H., and Monahan, J. F. (1984). Monte Carlo comparison of ANOVA, MIVQUE, REML and ML estimators of variance components. *Technometrics, 26*(1), 47–57.

SYSTAT Software Inc. (2004). *SYSTAT 11: Statistics II manual.* Richmond, CA: SYSTAT Software Inc.

Tabachnick, B. G., and Fidell, L. S. (2007). *Experimental Designs Using ANOVA.* Belmont, CA: Duxbury Press.

Tabachnick, B. G., and Fidell, L. S. (1989). *Using Multivariate Statistics* (2nd ed.). New York: Harper and Row.

Tabachnick, B. G., and Fidell, L. S. (1996). *Using Multivariate Statistics* (3rd ed.). New York: HarperCollins.

Tabachnick, B. G., and Fidell, L. S. (2001). *Using Multivariate Statistics* (4th ed.). New York: Allyn and Bacon.

Tanaka J. S. (1993). Multifaceted conceptions of fit. In K. A. Bollen and J. S. Long (eds.), *Testing Structural Models.* Newbury Park: Sage Publications.

Tanaka J. S., and Huba G. J. (1989). A general coefficient of determination for covariance structure models under arbitrary GLS estimation. *British Journal of Mathematical and Statistical Psychology, 42,* 233–239.

Tatsuoka, M. M. (1971). *Multivariate Analysis: Techniques for Educational and Psychological Research.* New York: Wiley.

Tatsuoka, M. M. (1975). Classification procedures. In D. J. Amick and H. J. Walberg (Eds.), *Introductory Multivariate Analysis.* Berkeley: McCutchan.

Theil, H. (1970). On the estimation of relationships involving qualitative variables. *American Journal of Sociology, 76,* 103–154.

Thurstone, L. L. (1947). *Multiple Factor Analysis.* Chicago: University of Chicago Press.

Tzelgov, J., and Henik, A. (1991). Suppression situations in psychological research: Definitions, implications, and applications. *Psychological Bulletin, 109,* 524–536.

Ullman, J. B. (2001). Structural equation modeling. In B. G. Tabachnick and L. S. Fidell, *Using Multivariate Statistics* (4th ed., pp. 653–771). New York: Allyn & Bacon.

Ullman, J. B. (in press). Structural equation modeling: Demystified and applied. *Journal of Personality Assessment.*

VandenBos, G. R., and Stapp, J. (1983). Service providers in psychology: Results of the 1982 APA human resources survey. *American Psychologist, 38,* 1330–1352.

Van der Pol, J. A., Ooms, E. R., van't Hof, T., and Kuper, F. G. (1998). Impact of screening of latent defects at electrical test on the yield-reliability relation and application to burn-in elimination. *1998 IEEE International Reliability Physical Symposium Proceedings: 36th Annual* (pp. 370–377). Piscataway, NJ: Institute of Electrical and Electronic Engineers, Inc.

Vaughn, G. M., and Corballis, M. C. (1969). Beyond tests of significance: Estimating strength of effects in selected ANOVA designs. *Psychological Bulletin, 72*(3), 204–213.

Velicer, W. F. (1976). Determining the number of components from the matrix of partial correlations. *Psychometrika, 41,* 321–327.

Velicer, W. F., and Fava, J. L. (1998). Affects of variable and subject sampling on factor pattern recovery. *Psychological Methods, 3,* 231–251.

Velicer, W. F., and Jackson, D. N. (1990). Component analysis versus common factor analysis: Some issues in selecting an appropriate procedure. *Multivariate Behavioral Research, 25,* 1–28.

Vevera, J., Zukov, I., Morcinek, T., and Papezová, H. (2003). Cholesterol concentrations in violent and non-violent women suicide attempters. *European Psychiatry, 18,* 23–27.

Wade, T. C., and Baker, T. B. (1977). Opinions and use of psychological tests: A survey of clinical psychologists. *American Psychologist, 32*(10), 874–882.

Waternaux, C. M. (1976). Asymptotic distribution of the sample roots for a nonnormal population. *Biometrika, 63*(3), 639–645.

Wesolowsky, G. O. (1976). *Multiple Regression and Analysis of Variance.* New York: Wiley-Interscience.

Wherry, R. J. Sr. (1931). A new formula for predicting the shrinkage of the coefficient of multiple correlation. *Annals of Mathematical Statistics, 2,* 440–457.

Wiener, Y., and Vaitenas, R. (1977). Personality correlates of voluntary midcareer change in enterprising occupation. *Journal of Applied Psychology, 62*(6), 706–712.

Wilkinson, L. (1979). Tests of significance in stepwise regression. *Psychological Bulletin, 86*(1), 168–174.

Wilkinson, L. (1990). *SYSTAT: The System for Statistics.* Evanston, IL: SYSTAT, Inc.

Wilkinson, L., and Dallal, G. E. (1981). Tests of significance in forward selection regression with an *F*-to-enter stopping rule. *Technometrics, 23*(4), 377–380.

Wilkinson, L., and Task Force on Statistical Inference (1999). Statistical methods in psychology journals: Guidelines and explanations. *American Psychologist, 54*(8), 594–604.

Williams, L. J., and Holahan, P. J. (1994). Parsimony-based fit indices for multiple-indicator models: Do they work? *Structural Equation Modeling, 1,* 161–189.

Willis, J. W., and Wortman, C. B. (1976). Some determinants of public acceptance of randomized control group experimental designs. *Sociometry, 39*(2), 91–96.

Wilson, E. B., and Hilferty, M. M. (1931). The distribution of chi-square. *Proceeding of the National Academy of Science, 17,* 694.

Woodward, J. A., Bonett, D. G., and Brecht, M.-L. (1990). *Introduction to Linear Models and Experimental Design.* San Diego: Harcourt Brace Jovanovich.

Woodward, J. A., and Overall, J. E. (1975). Multivariate analysis of variance by multiple regression methods. *Psychological Bulletin, 82*(1), 21–32.

Yang, M., Rasbash, J., Goldstein, H., and Barbosa, M. (1999). *MLwiN Macros for Advanced Multilevel Modeling, Version 2.0.* London, Multilevel Models Project, Institute of Education.

Young, R. K., and Veldman D. J. (1981). *Introductory Statistics for the Behavioral Sciences* (4th ed.). New York: Holt, Rinehart and Winston.

Zeller, R. A. (2005). *How Few Cases Is Enough to Do a Credible Factor Analysis? A Monte Carlo Simulation.* Manuscript submitted for publication.

Zoski, K. W., and Jurs, S. (1996). An objective counterpart to the visual scree test for factor analysis: The standard error scree. *Educational and Psychological Measurement, 56,* 443–451.

Zwick, W. R., and Velicer, W. F. (1986). Comparison of five rules for determining the number of components to retain. *Psychological Bulletin, 99,* 432–442.

INDEX